Practical Process Engineering

Other McGraw-Hill Books in Process Engineering

Chopey and Hicks · HANDBOOK OF CHEMICAL ENGINEERING CALCULATIONS (1984)
Considine · PROCESS INSTRUMENTS AND CONTROLS HANDBOOK, 3D ED. (1985)
Davidson · HANDBOOK OF WATER SOLUBLE GUMS AND RESINS (1980)
Dean · LANGE'S HANDBOOK OF CHEMISTRY, 13TH ED. (1985)
Grant · GRANT & HACKH'S CHEMICAL DICTIONARY, 5TH ED. (1987)
Harper · HANDBOOK OF PLASTICS AND ELASTOMERS (1975)
Hicks · STANDARD HANDBOOK OF ENGINEERING CALCULATIONS, 2D ED. (1985)
Hopp and Hennig · HANDBOOK OF APPLIED CHEMISTRY (1983)
Juran · QUALITY CONTROL HANDBOOK, 4TH ED. (1987)
Lyman · HANDBOOK OF CHEMICAL PROPERTY ESTIMATION METHODS (1982)
Meyers · HANDBOOK OF CHEMICALS PRODUCTION PROCESSES (1986)
Meyers · HANDBOOK OF PETROLEUM REFINING PROCESSES (1986)
Meyers · HANDBOOK OF SYNFUELS TECHNOLOGY (1984)
Perry · ENGINEERING MANUAL (1976)
Perry and Green · PERRY'S CHEMICAL ENGINEERS' HANDBOOK, 6TH ED. (1984)
Rohsenow et al., · HANDBOOK OF HEAT TRANSFER APPLICATIONS (1985)
Rohsenow et al., · HANDBOOK OF HEAT TRANSFER FUNDAMENTALS (1985)
Rosaler and Rice · STANDARD HANDBOOK OF PLANT ENGINEERING (1983)
Schweitzer · HANDBOOK OF SEPARATION TECHNIQUES FOR CHEMICAL ENGINEERS (1979)
Shinskey · DISTILLATION CONTROL, 2D ED. (1983)
Shinskey · PROCESS CONTROL SYSTEMS, 2D ED. (1979)

Practical Process Engineering
A Working Approach to Plant Design

Henry J. Sandler
Senior Process Engineer, United Engineers & Constructors Inc.

Edward T. Luckiewicz
Adjunct Professor, Drexel University

McGraw-Hill Book Company
New York St. Louis San Francisco Auckland Bogotá
Hamburg Johannesburg London Madrid Milan
Mexico Montreal New Delhi Panama
Paris São Paulo Singapore
Sydney Tokyo Toronto

Library of Congress Cataloging-in-Publication Data
Sandler, Henry J.
  Practical process engineering.

  Includes bibliographies and index.
  1. Chemical processes.   I. Luckiewicz, Edward T.
II. Title.
TP155.7.S26 1987   660.2   86-10378

Copyright © 1987 by McGraw-Hill, Inc. All rights reserved.
Printed in the United States of America. Except as permitted
under the United States Copyright Act of 1976, no part of this
publication may be reproduced or distributed in any form or by
any means, or stored in a data base or retrieval system, without
the prior written permission of the publisher.

1234567890   DOC/DOC   8932109876

ISBN 0-07-078595-3

The editors for this book were Betty Sun and Beatrice E. Eckes,
the designer was Naomi Auerbach, and the production
supervisor was Thomas G. Kowalczyk. It was set in Plantin
by University Graphics, Inc.

Printed and bound by R. R. Donnelley & Sons Company.

*This book is dedicated to our children
for their encouragement and help and to Rhoda and Kaye
for guarding our health and sanity
during its preparation.*

*In Memoriam:
Lewis C. Knox (1921–1986), a true engineer and mentor*

Contents

Preface xiii

Chapter 1 Introduction 1
  The Role of the Process Engineer 2
  The Process Engineer as a Problem Solver 3
  Economic Analysis 4
    Preliminary Estimates 4
    Detailed Estimates 8
    Cost Indexing 10
    Economic Decisions 10
  Nomenclature and Dimensional Units 18
  References 20

Section 1 Process Documentation 21

Chapter 2 Block Diagrams and Process Flow Sheets 23
  Block Diagrams 23
    Other Modes of Representation 26
  Process Flow Sheets 27
    Data Sources for Process Flow Sheets 27
    Process Information Required 29
    Presentation 29
    Flow Lines 30
    Material-Balance Table 32
    Notes 34

Chapter 3 Engineering Flow Diagrams 35
  Types of Engineering Flow Diagrams 36
    The Lead Sheet, or Piping and Instrumentation Symbols Diagram 36
    Processing Sections 37
    Utility Sections 39
    Specialty Flow Diagrams 40
  Relationship to Disciplines 40
    Chemical Engineers and Engineering Flow Diagrams 40
    Mechanical Engineers and Engineering Flow Diagrams 41
    Instrument Engineers and Engineering Flow Diagrams 41
    Electrical Engineers and Engineering Flow Diagrams 43
    Structural Engineers and Engineering Flow Diagrams 43
  Preparation of Engineering Flow Diagrams 43
    Informational Basis of Engineering Flow Diagrams 44
    Formats of Engineering Flow Diagrams 44
    Development of an Engineering Flow Diagram 45
    Equipment Identification 50
    Piping Identification 52
    Notes 62
    Special Legends and Tables 63

Line Tabulations	63
Checklists	68

Chapter 4 Logistics of Flow Sheets and Flow Diagrams — 109

Types of Drawing Issues	109
Preliminary Issues	110
The Zero Issue	112
Revision Issues	112
Interim Issues	115
Internal Communication	116
Master Sets	116
Provisional Dating of Tracings	117
Interim-Issue Designation	118
Archives	119
Calculation Sheets	119
Notes of Conference	119
Drawings	120

Section 2 Materials Selection and Piping Calculations — 121

Chapter 5 Materials Selection — 123

Mechanical Properties	123
Materials	129
Metals	129
Polymeric Materials	149
Ceramic Materials	158
Graphite	159
Glasses	160
Corrosion	162
Factors Affecting Corrosion	163
Types of Corrosion, Causes, and Cures	163
Material Selection for Corrosion Resistance	174
Bibliography	177
Appendix 5.1	179

Chapter 6 Piping Calculations, Part 1: Limitations and Theoretical Considerations — 186

Limitations on Piping	187
Available Piping	187
Tubing and Other Flow Conduits	190
Economical Sizing of Pipe	192
Velocity Constraints	195
Friction Losses	198
Types of Fluids	199
Single-Phase Flows	199
Two-Phase Flows	200
Theoretical Considerations	201
Viscosity	201

Types of Flow Patterns	211
Reynolds Number	214
Velocity Head	218

Chapter 7 Piping Calculations, Part 2: Friction Losses for Incompressible Fluids — 223

Friction Factors	223
Laminar-Flow Friction	225
Transition- and Turbulent-Flow Friction Factors	228
Aids to Determine Head Losses	230
Simplified Moody Charts	232
Nomographs and Alignment Charts	232
Slide Rules	234
Friction-Loss Tables	234
Pipe Flow Diagrams	251
Mnemonic Devices	251
Friction Factors for Other Than Clean Steel Pipe	254
Effect of Roughness Parameter	254
Roughness Friction Factor	255
Friction Loss in Fittings and Piping Components	257
Resistance Coefficient K	257
Equivalent Lengths of Fittings	257
L/D Values in Laminar Region	259
Effect of Surface Roughness on L/D Values	260
Entrance, Discharge, and Pipe-Size-Change Losses	260
Valve Coefficient C_v	261
Cumulative Head Losses in Piping Systems	266
Typical Evaluations	266
Siphon Effects	266
References	273

Chapter 8 Piping Calculations, Part 3: Compressible Fluids — 274

Properties of Compressible Fluids	277
Nonideal Behavior	278
Thermodynamic Effects	278
Typical Evaluation of Compressible-Fluid Friction Loss	284
Compressible-Fluid Flow: Shortcut Methods	285
Review of Loss Due to Kinetic-Energy Change	286
Friction Loss: The Incompressible Analogy	290
Maximum Equivalent Length and Flow Rates	297
Compressible-Fluid Flow: A Rigorous Approach	299
Compressible-Fluid-Flow Parameters	299
Frictional Flow in Pipes and Ducts	314
Analyzing Nozzles and Orifices	318
Shock-Wave Formation	330
References	333

Chapter 9 Piping Calculations, Part 4: Complex Fluids — 335

Gas-Liquid Systems	335
Gas-Liquid Flow Regimes	335

Pressure-Drop Calculations for Gas-Liquid Systems	338
Design Criteria for Two-Phase (Gas-Liquid) Flow Systems	343
Fluid-Solid Systems	348
Terminal Velocity	348
Fluid-Solid-System Flow Regimes	352
Pressure Gradients for Fluid-Solid Systems	356
Design Criteria for Fluid-Solid Systems	361
References	361

Section 3 Equipment Selection, Sizing, and Related Subjects — 363

Chapter 10 Vessels — 365

Vessel-Design Criteria	365
Vessel Stresses	365
Welded Joints	376
Vessel Inspection	376
Vessel-Design Codes	378
Vessel Selection	379
Storage Vessels	379
Process Vessels	381
References	383
Bibliography	383

Chapter 11 Pumps — 384

The Pump	384
Pump Classification and Types	384
Reciprocating Pumps	384
Rotary Pumps	387
Centrifugal Pumps	390
Turbine Pumps	395
Pump-Performance Characteristics	396
General Pump Parameters	396
General Performance Characteristics of Pumps	402
Packing and Mechanical Seals	422
Packing	422
Mechanical Seals	425
Pumping-System Design	429
Pump Priming	429
Minimum Flow through Pumps	431
Sump Design	432
References	435
Bibliography	435

Chapter 12 Fans, Blowers, and Compressors — 436

Fans	436
Propeller Fans	437
Centrifugal Fans	437
Axial Fans	440
Fan Controls	441

Fan-System Arrangements	443
Blowers and Compressors	444
Compressor Mechanical Design	445
Performance Characteristics	450
Surge Control	458
References	460
Bibliography	460

Chapter 13 Vacuum Equipment — 461

Equipment for Producing Vacuums	464
High-Vacuum Equipment	464
Low-Vacuum Equipment	465
Medium- or Industrial-Vacuum Equipment	466
Vacuum-Equipment Sizing	483
Inlet Pressure to Vacuum Equipment	483
Flow Rates of Vapors and Gases	484

Chapter 14 Heat Exchangers — 491

Heat-Exchange Fundamentals	491
Conduction	491
Convection	495
Radiation	517
Heat-Exchanger Design	524
Mechanical Design	525
Specialty Heat-Exchange Equipment	540
References	556
Bibliography	559

Chapter 15 Thermal Insulation and Tracing — 560

Thermal Insulation	560
Usages for Thermal Insulation	564
Types of Insulation	566
Recommended Thickness of Insulation	567
Tracing	573
Steam Tracing	574
Electric Tracing	575
Jacketing and Bundling	579
References	579

Chapter 16 Mixers and Agitators — 580

Agitators: Impeller-Type Mixers	580
Fluid Mechanics of Agitators	581
Agitator-System Design	582
Mechanical Design of Agitators	593
Motionless Mixers	593
References	597
Bibliography	597

Chapter 17 Electric Power and Motors — 598

- Power Distribution — 598
  - Single-Phase and Multiphase Alternating-Current Systems — 601
  - Direct-Current Power Systems — 605
- Electrical Codes and Standards — 605
- Electric Motors — 608
  - DC Motors — 610
  - AC Motors — 610
- References — 621
- Bibliography — 621

Index 622

Preface

The modern university engineering curricula must treat a wide variety of subjects in order to prepare the undergraduates to take their place in industry. The courses must cover those common to all the engineering disciplines as well as an appreciation of the humanities. During the remaining period, the chemical and mechanical engineering courses range from fundamentals such as heat and mass transfer, thermodynamics, and fluid flow to the more specialized aspects of the respective disciplines and computer applications. In addition, both the mechanical and the chemical engineering students must take an introduction to all their sister engineering disciplines.

So much time must be allotted to the theory underlying the many principles of the engineering disciplines that it is the rare program that can cover the practical aspects of the professions. While many chemical and mechanical engineering graduates enter the field of process engineering for chemical facilities, refineries, and power plants, the broad range of the practical aspects of modern process engineering is barely covered by the academic curriculum. The burden of providing the young engineer with hands-on experience typically rests on industry. Starting engineers, in their initial positions, are usually "apprenticed" to more experienced engineers who induct them in the practical arts of process engineering. It is taken for granted that the graduates have a thorough theoretical background.

Too often, however, this apprenticeship is slipshod and proceeds without a coherent plan. Training is usually given only in those areas immediately relevant to the specific job at hand. The introduction of young engineers to the many facets of practical process engineering proceeds by fits and starts over a long period. Although the new engineers make definite contributions to their employers during this period, their efficiency is somewhat limited until the full apprenticeship has been completed.

Some companies have already perceived the need for a basic course in practical process engineering to supplement the theoretical work covered in the formal undergraduate course. Some curricula offer a portion of the material in design courses. However, there is no one reference book available at this time to present the information which covers the range of practical process engineering. This is the purpose of the present volume.

Various aspects of the role of a process engineer in chemical facilities, refineries, and power plants are examined. Examples of how the roles can be organized logically and performed in an expeditious manner are presented. Illustrations of how the various tools available to the process engineer can be simplified and applied within the limitations of good engineering practice are given. The various classes of processing equipment are reviewed, and emphasis is placed on the salient features of those of most interest to the process engineer.

It is obvious that *Perry's Chemical Engineers' Handbook, Marks' Standard Handbook for Mechanical Engineers,* and other standard general reference books cover much of the material being presented here. However, the chapters in those books all too frequently intersperse practical information with the theoretical development needed to appreciate the application. In addition, practical information is often presented in the form of exact equations. These usually require considerable effort to solve for a particular application when reference to a graph or a table gives the desired result with a sufficient degree of accuracy for practical applications.

Much of the information regarding practical process engineering that is presented in this book is already available in the journals servicing the field. However, by the very nature of the publications, the information and topics have appeared sporadically over a considerable period of time. It is highly unlikely that a new graduate or a young engineer has had subscriptions to the various journals for a sufficient period of time or has been knowledgeable or conscientious enough to collect all the relevant articles.

A *Manual for Process Engineering Calculations* by Clarke and Davidson presents a collection of equations, nomographs, tables, and graphs for rigorous and shortcut approaches for many calculations that are encountered by process engineers. The *Standard Handbook of Engineering Calculations* by Tyler Hicks, and Chopey and Hicks's companion volume entitled *Handbook of Chemical Engineering Calculations* illustrate many typical problems with examples and their solutions. However, only summary background information is given on the equipment or unit operations, and little is presented on the organization of a project.

The authors, with combined process engineering experience of over 50 years in large engineering and construction firms, frequently have served as mentors to younger engineers entering the process design field. In addition, one of them has taught chemical-plant design courses at a large university. The courses include practical information on equipment and its application. The materials thus developed to introduce undergraduates and new engineers to the art and science of process engineering have been enlarged upon in this book. The volume should prove useful as a text for a one-semester course for university seniors to balance theoretical considerations with practical applications. It should be helpful as a reference tool to graduating engineers who plan to enter design and construction companies. The volume could also be the basis upon which a design and construction company could build its own process engineering training program.

A modern trend is for process engineers to lean more and more on computer software programs to obtain answers to their fluid-flow and equipment-sizing problems. The practical approaches taken in this work for these areas not only enable process engineers to appreciate software programs and check solutions thus produced but enable the engineers to obtain acceptable results when a computer is not available or its use is not required.

The authors have enjoyed satisfactory, productive, and fulfilling careers as process engineers. This has been due, in part, to the many fine engineers who have aided us on our way and imparted their secrets and shortcuts. We wish this volume to do the same for its readers and in that way serve as thanks to our mentors.

The authors wish to express their appreciation to Paul Dan, John DeLeone, Cheryl Robertson, Alexandra Sandler, and Dana Schrader for reviewing large segments of the rough drafts and for their valued comments and suggestions. We also thank Donald Dallal, Oleg Dudkin, Norman Greenberg, Henry Kaminsky, Jeff Kerner, Anderson McCabe, Robert Moore, William Pallaver, Charles Rehrig, Milton Schwartz, Jack Trechock, Nelson Whitney, and Malcolm Woodman for lending their expertise in selected areas. We are grateful to Dr. Michel A. Saad for his scrutiny of our work on compressible-fluid flow. We wish to acknowledge United Engineers & Constructors Inc. for use of materials from its files and particularly thank its Computer-Assisted Design and Drafting Department and the department's staff for adapting the flow sheets and diagrams presented in Chaps. 2 and 3 and for preparing many of the figures used throughout the text. We would like to thank all those who did so many favors, large and small, for us; your kindnesses are well remembered. Our sincere gratitude and admiration go to Sally Pekora and Mary Lou Tarallo, whose excellent secretarial skills put the whole work together. The assistance of Dena Sher with the index was a valued contribution.

Process engineering for chemical facilities, refineries, and power plants concerns itself with many different aspects, unit operations, and types of equipment. It has not been possible to cover the full range of the subjects involved in their design in the limited size of this book. The authors have therefore confined themselves to those topics which are common to 90 to 95 percent of the process engineering efforts for chemical, refinery, and power-plant facilities. We would be pleased to learn from readers of any omissions from this book that should be included in future editions. We, of course, would appreciate being notified of any inaccuracies in the text.

H.J.S.
E.T.L.
Philadelphia, 1986

Chapter 1

Introduction

Modern engineering sciences have their roots, according to J. Bronowski,[1] in events in the evolutionary process which occurred about 10,000 years ago. During that period, humans began the gradual change from a nomadic hunter-forager existence to a settled agriculture-based life. The agrarian framework led to the emergence of technologies from which the present-day sciences and engineering disciplines are derived. Rudimentary civil engineering arose as permanent shelters were required and irrigation systems developed. The roots of mechanical engineering lay in the application of the lever in the fashioning of plows and in the invention of the wheel and the pulley to facilitate the movement of burdens.

During the millennia that followed, other engineering technologies along with mathematics, the sciences, the arts, and written means for communication of thought and knowledge gradually developed to meet the increasing needs of evolving civilizations. Included in the process was the development of formal educational institutions which served to define and delineate the various branches of the evolving sciences and engineering fields.

As emerging technologies grew, their application to modern industry became more and more complex. At one time, it was relatively easy for a person to conceive and bring into existence a modest manufacturing operation individually or with the assistance of only a few others. However, as cottage industry evolved into the complex plants of modern technology, individuals with more and more specialized knowledge and skills were required in all phases of the planning and operation of the resulting facilities. This trend is reflected in the various curricula that are available in modern universities for the engineering disciplines.

While the general emphases of the various disciplines differ greatly from one another, there are often overlapping areas of study between the various engineering disciplines so that graduates of several branches can fulfill the needs of certain functions in some specializations. A good example is that of process engineers for the design or operation of chemical and related types of plants, refineries, powerhouses, and similar installations. Large segments of the training received by undergraduates in mechanical and chemical engineering are similar. In addition to the first year's courses in mathematics, physics, chemistry, and basic engineering, which they share with all other engineering disciplines, the mechanical and chemical engineering curricula emphasize fluid flow, heat transfer, and equipment used in unit operations for associated industrial applications. These areas of studies are basic for a process engineer. Thus, it is common to find a person with either a mechanical or a chemical engineering background functioning as the process engineer on a project. In projects such as boiler-house or power stations, where conventional mechanical equipment is involved, a mechanical engineer is more likely to be the process engineer. If, however, the process involves chemical-equilibrium evaluations, the application of reaction kinetics, or the application of diffusional operations, a chemical engineer usually serves as the process engineer.

The Role of the Process Engineer

Process engineers have responsibilities in six major areas of a project. In the order in which they occur during the design of a facility, these are:

1. Economic-feasibility studies and estimates of equipment costs.
2. Preparation of a series of documents to define the overall view of the project, present the material and energy balances, and outline the tasks required of the other engineers and the piping designers. The latter documents also define the initiating work to be performed by the instrument, electrical, and structural disciplines.
3. Selection of materials of construction for equipment and piping.
4. Process calculations for flows, pressure drops, and heat transfer in piping and for the sizing of equipment.
5. Selection of equipment based on suitability, availability, and cost effectiveness.
6. Purchase of equipment and engineering follow-up.

This book concerns itself with the practical aspects of these major responsibilities of process engineering. It is divided into three sections to coincide with these practical aspects. Chapters 2, 3, and 4 deal with process flow sheets, engineering flow diagrams, and the logistics of flow diagrams, respectively, and together constitute Sec. 1. Chapter 3 also contains a checklist of major items that can be reviewed for inclusion in the preparation of flow diagrams and equipment specifications.

The selection of materials of construction for equipment and piping is reviewed in Chap. 5, which along with various aspects of piping calculations in Chaps. 6, 7, 8, and 9 makes up Sec. 2. Piping calculations have been singled out for special and detailed consideration since the evaluation of pressure loss and other flow parameters is involved in the sizing of almost all process equipment. Even if flow through the unit is not effected in pipes, the theoretical considerations regarding fluid flow reviewed in Chap. 6 can be adapted for analogous situations.

Chapter 7 develops the basic equations necessary for evaluating friction loss in pipes and ducts and their application to incompressible fluids. These equations may be used with compressible and complex fluids when conditions are such that these fluids may be treated as incompressible. Included in this chapter are shortcut methods and a review of various aids for obtaining friction losses with explanations of how they may be extrapolated beyond their stated range. Simplified equations for the quick calculation of flow velocity and Reynolds number are given for application to both liquid and gaseous fluids. Simple mnemonic devices are presented to permit the rapid estimation of flow rates and pressure drops.

The piping calculations for compressible fluids are reviewed in Chap. 8. In addition to reviewing the rigorous methods for determining pressure drops in adiabatic, isothermal, and polytropic flow, the chapter derives a set of simplified relationships that are analogous to incompressible-fluid flow and are suitable for use in most practical applications. Included are means of determining whether or not the approach to sonic velocity and the associated phenomenon of choking is such that use of the rigorous method is required.

Chapter 9 concludes the review of piping calculations by summarizing practical methods for determining flow characteristics of various two-phase-fluid systems. This includes pressure-drop calculations, determination of flow regimes, and minimum-velocity criteria.

Section 3, consisting of Chaps. 10 through 17, reviews the most frequently used major categories of equipment common to most chemical plants, refineries, and power facilities. These include vessels, pumps, blowers and compressors, vacuum equipment, heat exchangers, and mixers and agitators. While the treatment afforded the equipment is not comprehensive, sufficient descriptions of the items and their applications are given to cover their use in the large majority of the cases in which they are required. Equipment related to unit operations such as distillation, extraction, filtration, centrifugation, crystallization, drying, and similar fields requires special and detailed treatment. While these operations are important and are used extensively in some facilities, their nature is such that their individual use represents only a small portion of the operations in the overall spectrum of the processing field. Treatment of these specialized subjects had to be omitted from this general volume, and it is recommended that the reader refer to the specific references books available for the respective topics.

Included in Sec. 3 are brief expositions regarding heat tracing, thermal insulation, and electric power and motors. While responsibility for the subject matter in these two chapters lies primarily with other disciplines, process engineers should be sufficiently conversant with these subjects to make logical and economic choices in the application of the respective items.

Once the process engineers have designed equipment, they have a role in completing specification sheets by which the equipment is presented to potential suppliers. Responses to equipment inquiries must be analyzed by process engineers not only for compliance with the process specifications but for determining which of the proposed equipment is the most cost-effective. The basis for making economic analyses and decisions is reviewed in a subsequent subsection of this chapter.

After the equipment has been ordered, the process engineers examine the vendor-supplied drawings and literature pertaining to the purchased items to ensure that they meet the specifications in the purchase contract. The process engineers mark necessary corrections on the circulating copies of the drawings and consult with the designers with respect to the required nozzle orientation and location so that the vendor can eventually issue certified drawings for the project. The certified drawings are used by the process engineers to update the engineering flow diagrams.

The Process Engineer as a Problem Solver

The process engineering tasks just outlined are common to all projects which require a process engineer. However, this by no means implies that process engineering work is routine and can be done by rote. On the contrary, it is rare that any two projects are alike. Even the duplication of a facility to prepare a given product at the same production rate may involve adapting to different utility conditions. Process engineers must be able to formulate and define the problems to be solved, evaluate the alternatives, and make sustainable decisions. They call on their theoretical background and practical knowledge and combine these with imagination and ingenuity to resolve problems and effect decisions to meet the challenge of satisfactorily completing a project on schedule and within its budget. A logical, planned approach is required to resolve both objective and subjective issues.

Process engineers face problems which require either qualitative or quantitative solutions. Questions relating to the selection of processes, process components, or materials are best answered by past experience, a literature search, or consultation with knowledgeable persons. Mathematical models, on the other hand, are used to obtain quantitative solutions. Depending on the complexity of the problem and the degree of accuracy demanded for the solution, one can resort to a rule of thumb, a shortcut method, a nomograph, a graph, a table, an easily solvable equation, or a set of more complex relationships which requires a programmable calculator or a computer to solve. A judgment as to practicality and cost must be made in choosing the method. In addition, process engineers must be thoroughly familiar with any analytical method used, especially with computer software not written by themselves, to ensure that its constraints or limitations are compatible with the problem being solved.

Economic Analyses

Process engineers are called upon to make economic analyses and decisions in all phases of a project. These range from preliminary cost estimates for "scoping" a project to detailed estimates which can be used for judging whether or not to undertake the project and then be used for its cost control, to economic studies for guiding choices among processing alternatives as the project progresses, to economic analyses for selecting the most cost-effective equipment offering in response to equipment specifications.

Preliminary estimates

Preliminary estimates serve several functions and, depending upon their use and how they are prepared, have various degrees of accuracy. Table 1.1 lists several types of preliminary estimates and some of the more popular ways of preparing them. These

TABLE 1.1 Estimating Techniques

Type	Basis	Accuracy range, %
A. Preliminary		
1. Rough capacity	($/ton) (tonnage)	-30 to $+50$
2. Scaled rough capacity	($/ton) (tonnage)^n	-25 to $+40$
3. Major-equipment factored installed costs		
a. Equipment costs: gross installation factors	$(1 + F_G)(\Sigma$ equipment costs)	-20 to $+30$
b. Equipment costs: grouped installation factors	$\Sigma[(1 + F_i)$(equipment costs)]	-15 to $+25$
B. Detailed estimates		
1. Factored equipment: estimated installation	(Equipment costs)^n	-10 to $+20$
2. Quoted equipment: estimated installation	....	-5 to $+15$

types of estimates can be made rather quickly. The easiest estimates to make are listed at the top of the table, with those requiring more research and calculation given further down. As shown in Table 1.1, those preliminary estimates for which more effort is expended are likely to be more nearly accurate in their predictions.

Product-based estimates. The estimating techniques which require the least effort and which, in turn, yield estimates with the least degree of accuracy involve the simple expedient of multiplying the installed cost per annual unit of production by the expected annual production rate for the proposed facility. These are referred to as the "rough" estimating Methods A-1 and A-2 in Table 1.1. They are usually required by top management when it wishes a quick approximation of costs for several projects that it might be scoping for investment possibilities.

Method A-2 is the more refined of the two, in that it corrects the estimate for the anticipated scale of the proposed facility. This is done by taking the results of Method A-1 and multiplying them by a factor obtained by raising the ratio of the annual capacities of the proposed plant and that of the plant used in setting the installed cost per annual unit of production to a power n which is peculiar to the product under investigation. Equation (1.1) summarizes the procedure.

$$\text{New installed costs} = \left(\frac{\$}{\text{annual unit}}\right)_{\text{base}} \times \text{number of new units} \times \left(\frac{\text{new capacity}}{\text{base capacity}}\right)^n \quad (1.1)$$

The value of the cost per annual unit of production for various products can be obtained from several sources. The best and probably the most reliable would be data from company files for a similar plant constructed within the past decade. However, most process engineers rarely have access to this type of information since it is unlikely that their company has the necessary data. Therefore, most engineers must rely on published statistics. These can be found in texts on chemical- and power-plant economics* and in articles that appear in trade journals.† Table 1.2 lists typical installed costs per annual production tonnage for several selected basic chemicals. Included are recommended exponents to be used in scaling for various annual capacities.

The accuracy of these forecasting methods is decreased in that the techniques usually have to rely on installed-cost data that are from 5 to 20 years old. There is no difficulty in taking into account the escalating effects of inflation since costs can be indexed, a subject to be reviewed in the subsection "Cost Indexing." However, indexing doesn't take into account improvements in process technology for better yields, development of more efficient equipment, less expensive materials of construction, or labor-reduction techniques. Therefore, it is important to use the most recent data available.

If a capacity exponent is not available for a particular product being considered, the figure 0.70 may be used as its approximate value. This is the approximate

---

*Typical is K. M. Guthrie, *Process Plant Estimating, Evaluation and Control*, Craftsman Book Co. of America, Solana Beach, Calif., 1974.

†K. M. Guthrie, "Capital and Operating Costs for 54 Chemical Processes," *Chemical Engineering*, June 15, 1970, pp. 140–156, is an example.

TABLE 1.2 Selected Chemical-Plant Costs and Capacity Exponents*

Product	Plant capacity, tons/year	Installed costs, $/(ton·year)	Capacity exponent n
Acetic acid	10	540,000	0.68
Acetone	100	200,000	0.45
Acetylene	10	2,500,000	0.65
Ammonia	100	190,000	0.58
Butanol	100	460,000	0.40
Chlorine	100	280,000	0.45
Ethanol	10	3,100,000	0.83
Ethylene	100	200,000	0.73
Ethylene oxide	100	740,000	0.78
Glycol	10	1,900,000	0.75
Hydrochloric acid	10	510,000	0.68
Methanol	100	150,000	0.60
Nitric acid	100	55,000	0.60
Phenol	100	460,000	0.75
Polyethylene	10	2,100,000	0.65
PVC	10	1,900,000	0.60
Styrene	100	330,000	0.60
Sulfuric acid	100	250,000	0.65
Urea	100	100,000	0.70
Vinyl chloride	100	230,000	0.80

*Based on K. M. Guthrie, "Capital and Operating Costs for 54 Chemical Processes," *Chemical Engineering,* June 15, 1970, pp. 140–156. Installed costs are adjusted from *Chemical Engineering* plant index = 126 in spring, 1970, to cost equivalent to *Chemical Engineering* plant index = 326 in the first quarter of 1986.

weighted exponential average of the n factors in Table 1.2. Fortunately, the calculated results are not sensitive to small inaccuracies in the value of n, and in most cases the use of 0.70 yields results within the accuracy range of the predicting method.

Major-equipment factored installation-cost estimates. The accuracy of a preliminary estimate can be improved by regarding the proposed facility as a distinct operation composed of individual components rather than scaling a consolidated value. Such a method entails determining costs for the major equipment that will be required for the project and using suitable factors to account for subsidiary equipment, for installation, and for other necessary expenditures.

Two ways to prepare major-equipment factored estimates are frequently used. They require the preparation of a rough process-flow schematic which shows the sequence of the main components such as tanks, columns, pumps, compressors, vacuum equipment, heat exchangers, and any expensive specialty equipment, e.g., ion exchangers, incinerators, etc. Noted on the process schematic should be sufficient material- and energy-balance information to allow first approximations of the sizings for the various major pieces of equipment and indications of their materials of construction.

Shortcut calculation methods may be used to determine the sizes of the various pieces of equipment,* which are then tabulated in groups according to type. The next

---

*Sizings by shortcut methods for many types of equipment are given by Ulrich.[2]

step is to assign a realistic uninstalled cost to each item. Company files can be very helpful in finding the price of similar equipment if the cost control department has prepared the necessary documentation.

If these data are available, they will, in all probability, need to be adjusted for price escalation and have a scaling exponent applied for size. The current prices of many items of equipment are listed in commercial services such as Richardson's *Process Plant Construction Estimating Standards.** Informal contact may be made with suppliers of the equipment at this time to obtain approximate current prices for the determined sizes. However, for most preliminary estimates of this nature, it usually is sufficient to use price data from surveys that are published frequently by the various periodicals. Typical is an article in the April 5, 1982, issue of *Chemical Engineering* which lists 18 different types of equipment commonly used in chemical facilities and gives prices as of January 1982 for various size ranges for different materials of construction and temperature and pressure ranges. An extensive listing of equipment prices is presented in Chap. 25 of the sixth edition of *Perry's Chemical Engineers' Handbook*. In this case a price predicated on a Marshall and Swift index of 1000† is given for each piece of a certain size, and a scaling exponent is shown for use in obtaining the price of another piece in a designated size range.

TABLE 1.3 Installed-Costs Gross Factors F_G*

Type of installation	F_G
Grass-roots plants†	
Processing solids materials	0.8
Processing solids and fluids streams	1.2
Processing fluids	1.5
Battery-limits installation and expansions‡	
Processing solids materials	1.6
Processing solids and fluid streams	1.9
Processing fluids	2.5

*Based on data from *Perry's Chemical Engineers' Handbook*, 6th ed., McGraw-Hill Book Company, New York, 1984, Table 25-50.
†Entirely new plant on unimproved land.
‡Adjacent to existing plant which furnishes main utilities.

A simplified method of estimating the installed cost of a facility C_I is to multiply the total cost for the major uninstalled equipment C_E by a factor $(1 + F_G)$ in accordance with the following equation:

$$C_I = C_E(1 + F_G) \tag{1.2}$$

It is best to use a value for the gross factor F_G that is peculiar to a similar category of facility. Typical values for F_G are given in Table 1.3. The gross factors in the table are composed of the installation costs for the components in the group as a whole. Thus, the accuracy of a factored estimate can be improved by treating the components classifications separately and using the installation factor F_I for each type of equipment. Equation (1.2) then becomes

$$C_I = \Sigma[(C_{E,i})(1 + F_{I,i})] \tag{1.3}$$

---

*Richardson Engineering Services, Inc., San Marcos, Calif. 92069.
†Refer to subsection "Cost Indexing" and Fig. 1.1.

TABLE 1.4 Typical Installed-Equipment-Cost Factors $F_{I,i}$*

	$F_{I,i}$		
Type of equipment	All carbon steel	All 304 stainless steel	All 316 stainless steel
Vertical vessels	2.0	1.2	1.1
Horizontal vessels	1.1	0.8	0.7
Storage tanks	0.3	0.2	0.2
Exchangers piped on both sides	1.3	0.8	0.7
Air-cooler exchangers	0.8	0.4	0.4
Pumps and drivers	1.4	0.9	0.9
Compressors and drivers	1.2	0.8	0.7

*Based on data from K. M. Guthrie, *Chemical Engineering*, Mar. 24, 1969, pp. 114–142.

Table 1.4 gives values for the various factors $F_{I,i}$ for typical component groups in common materials of construction.

The values represented by Eqs. (1.2) and (1.3) are the costs of the equipment and the expense of installing it along with the necessary foundations and the electrical and instrumentation requirements. The installed costs thus obtained must be further multiplied by a factor of 1.8 for grass-roots facilities or by 1.4 for battery-limits installations or plant expansions. These values account for site preparation, excavations, and necessary buildings and services as well as engineering and field costs, taxes, insurance, contractors' fees, and contingencies.

Detailed estimates

A detailed estimate differs from a preliminary estimate primarily in that it is a joint effort of all disciplines. After equipment has been realistically sized, the costs of as many items as possible are obtained from vendors. In addition, the costs of installation of the various items of equipment plus the necessary piping and auxiliaries such as tracing and insulation are actually calculated instead of being included as factors. It is the responsibility of the process engineers to develop rudimentary flow schematics showing all the major equipment and auxiliaries with their approximate sizes and materials of construction. They and the other discipline engineers and designers lay out a preliminary equipment arrangement in the space allotted for the project. An approximate takeoff of the piping and fittings requirements in each category is made from the initial pipe sizings on the flow schematics and from the placement of equipment on the initial arrangements. The approximate number of valves according to type, size, and material is obtained from the rudimentary flow diagrams, while the number of the various fittings is judged from an approximate routing on the equipment arrangement.

The process engineers work with the other discipline engineers to assist them in developing their portions of the estimate. A list of the major equipment and auxiliaries is prepared by the process engineers. Together with the mechanical engineers, they contact manufacturers to obtain the current costs for the equipment corresponding to their preliminary descriptions. Staff mechanical engineers can also use the flow diagrams, equipment arrangements, and equipment lists to develop the number of piping drawings, isometrics, and specifications required for the project.

A tabulation of all motors in accordance with the categories required and preliminary estimates of their horsepowers is given to the electrical engineers so that they can plan their motor-control centers and substation needs. The rudimentary flow schematics should indicate any electrical tracing of lines and any special electrical-classification requirements. The equipment arrangement can be used by the electrical engineers to prepare the preliminary routing of electrical conduit and cable trays. The electrical engineers can also use the information in the flow schematics and equipment layouts to project the amount of design work, drawings, and specifications that the project will entail.

Likewise, the instrument engineers can use the rudimentary flow schematics to make a takeoff of the sensing, measuring, and analysis points as well as the number and size of control-valve stations. The elementary equipment arrangements permit them to plan the placement of the local panelboards, the control rooms, and the runs of instrumentation pneumatic tubing or electric-signal wires connecting the transmitting and control points to one another and to the control panels or control rooms. This, in turn, permits the instrument engineers to estimate the number of instrumentation flow diagrams, drawings, logic diagrams, and specifications required for the project.

The equipment layout together with a listing of all equipment with its size and approximate operating weights allows the structural engineers to estimate foundations and supporting steel and concrete work as well as accompanying underground and grading tasks. From these, the structural engineers can predict the number of drawings and specifications that they must prepare for the project.

If the project is large enough, a cost estimator may be assigned to translate the information regarding equipment and auxiliaries costs developed by each discipline into the total installed cost for the facility. If cost estimators are not available, each discipline must estimate the installed cost for the portion of work for which it is responsible. The process engineer usually cooperates with the mechanical engineers in determining these values for the equipment, piping, fittings, valves, and insulation.

It was noted earlier that current uninstalled costs for equipment can be obtained directly from manufacturers. If necessary, the techniques for scaling equipment costs described in the subsection "Preliminary Estimates" may be employed if contacts with vendors can't be made. The structural group can detail the cubic yards of excavation and concrete work required for the various foundations. These values can then be expressed in terms of hours of labor required through factors given in Richardson's *Process Plant Construction Estimating Standards,* which also list typical hourly wage rates for various regions of the United States. These estimating standards likewise give costs for installing a large variety of equipment, piping, and fittings.

Once the information from each discipline has been transposed into installed costs for the facility, this information is given to the project manager. Each discipline also presents to the project manager an estimate of the number of drawings required for its portion of the project. The disciplines also estimate the number of workhours needed to prepare the drawings, to perform the needed calculations, and to produce the specifications required to define the project. The project manager or a designated engineer reviews and assembles the installed costs from the various disciplines and adds markup, contingency, and escalation figures. A companion piece to the detailed estimate prepared by the process engineers is the annual requirements for raw materials, intermediates, and utilities based on hourly requirements and hours-per-year usage. Also listed are the yields of products and by-products together with an estimate of labor-force requirements.

Cost indexing

Several well-known systems for keeping track of costs with time are in current use and can be employed by process engineers to estimate current costs based on available past charges. The indices are:

1. *Engineering News-Record* (ENR) index, which is reported weekly by the *Engineering News-Record*, published by McGraw-Hill, Inc., New York, New York 10020. This index reflects costs in the construction industry since it is based upon a set amount of building materials and common labor. The base for the ENR index is 100 in 1913.
2. The Marshall and Swift (M&S) Publishing Co., Los Angeles, California 90026, compiles quarterly indices for equipment costs in 47 different industries. *Chemical Engineering*, also published by McGraw-Hill, Inc., reports the M&S indices for equipment costs in eight different process industries plus several related industries, respectively, in addition to the average M&S value for all the industries they survey. The M&S index is based on 100 for the year 1926.
3. *Chemical Engineering* also reports its own CE plant-cost index monthly plus indices for the various components which compose installed-equipment costs for the elements required in a preliminary or detailed estimate. A weighted-average value of the components is the CE plant index. The base for this index is an average value of 100 for the years 1957–1959.

Figure 1.1 shows the change in the various indices over the past two decades. These indices have different absolute values since their base of 100 reflects different initial reference years. However, their slopes on a semilog curve are relatively the same, although the ENR index has a slightly higher average escalation rate. The indices reflect the general change of costs in the United States during the time period in Fig. 1.1. Any of the indices is suitable for use with preliminary estimates. The M&S index is best suited to scaling equipment costs in process industries. It would be even more suitable if reference were made to the individual industry similar to the one under study. The breakdown of the various items in the CE plant-cost index is useful in transposing equivalent costs from a prior year to those required for a current detailed estimate.

Economic decisions

During the stages when a project is being contemplated and when it is being undertaken, the process engineers must make economic decisions at all levels of their activities. These range from alternatives to carrying out portions of the process to choosing between several offered pieces of equipment for a particular task. These decisions require both subjective and objective analysis. The subjective considerations can vary from compliance with environmental and safety regulations or local ordinances to space limitations or aesthetic values. The objective consideration is usually economic. The annual cost of one process or item is measured against the annual cost of another to determine which is the more cost-effective.

Annual fixed costs. The annual cost of a process or an item consists of the "annual fixed costs" plus the "annual operating costs." The annual fixed costs are a portion of the initial costs that can be assigned to a particular item or portion of the process

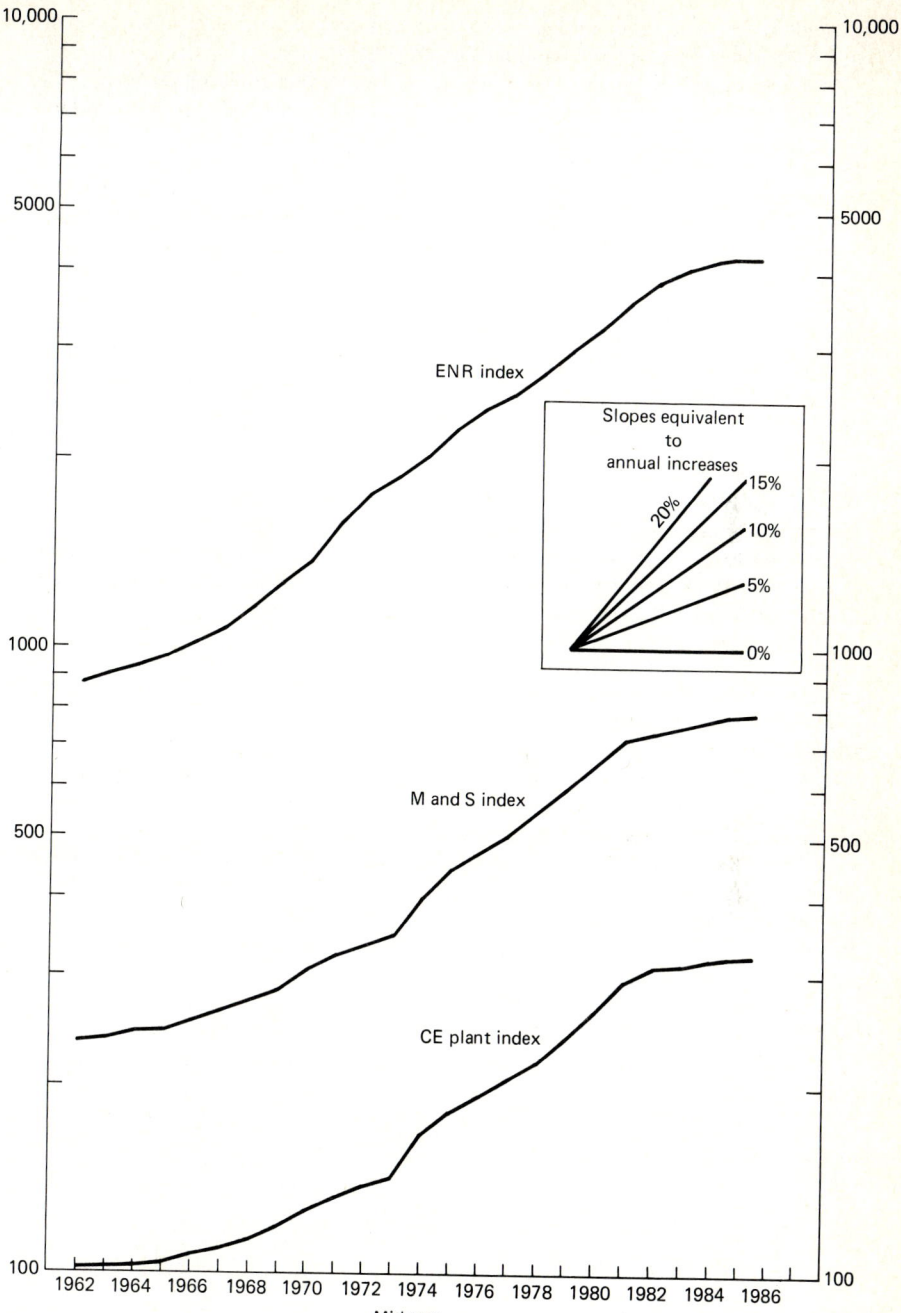

Figure 1.1 Histories of construction, process, and equipment cost indices.

regardless of the amount of its use during a year. The fixed costs, in turn, are composed of two major factors, the "annual carrying charge" and a "burden."

The annual carrying charge reflects the costs of capital used to purchase equipment and build the project. Its evaluation by financial analyzers takes into account parameters such as current interest rates, yields on stocks or bonds used to finance the project, depreciation allowances, anticipated salvage values, and the tax structure of the sponsoring group. It is outside the limits of this book to define completely the relationships which are used to calculate annual carrying charges. However, they can run from 0.10 to 0.25 of installed costs, depending mainly on interest rates and depreciation life. As a rule of thumb, the annual carrying charge may be estimated by adding one-half of the current prime rate, as a fraction, to the inverse of the depreciation life. Thus, if a plant is to be depreciated in 10 years and the prime rate is 12 percent, the annual carrying charge is about 0.16.

Other items such as local taxes, insurance, and general maintenance must be paid whether or not the plant operates, and these also are fixed costs, known as burden. The burden varies with the type of plant and the location. Its rate also is determined by the financial offices of the sponsoring group and usually falls between 0.06 and 0.09 of the installed costs.

The sum of the annual carrying charge and the burden may be taken as the general annual-fixed-costs factor; it can be applied with acceptable accuracy to large segments of the facility. The accuracy is less when dealing with any one piece of equipment. However, it is reasonably sufficient to allow the factor to be applied when making economic comparisons.

Payback time. From the above considerations, it can be seen that the combination of annual carrying charge and burden generally ranges from 0.16 to 0.34. More commonly, the fixed-charge rate is about 0.20 for conditions in the spring of 1986. An interesting concept is the reciprocal of the fixed charge which is known as "payback time." For the fixed-charge rates just cited, it would be $6\frac{1}{4}$ years, about 3 years, and 5 years, respectively. The particular payback time for a profit-making project allows management to decide whether or not a project is viable. In this case, total operating costs forecast for the project during the period defined by the payback time must be less than total installed costs for the project by an amount sufficient to allow acceptable earnings on investments. If a project is sponsored by a nonprofit organization or is adjunct to another process and is needed to comply with environmental or regulatory requirements, the fixed-cost concept and payback time are used to make economic decisions for selecting processes or equipment. Needless to say, payback time is also used for similar decisions for profit-making projects.

It should be noted that annual fixed charges remain relatively constant for long periods of time and change only if there is a large deviation in one of the constituents. Payback time, on the other hand, can fluctuate greatly over short spans of time owing to variations in the cash-flow position for the project. For example, the fixed-charge rate on a project may be 0.20, indicating a payback time of 5 years. Yet it may be decided by those in control of the finances of a project that most of the funds for the project have already been allocated, so that changes to the process for improvements or other benefits cannot be incorporated unless payback time is 1 year or even less. At that point, the process engineer should note those changes which meet the original criterion but not the revised payback time and try to ensure that space is allocated in the layout to accommodate the necessary equipment so that the changes can be incorporated when the cash-flow situation improves and payback time increases.

Annual operating costs. The other main piece of information required to make an economic decision regarding a process or a piece of equipment is the annual operating costs ascribed to it. The annual operating costs that must be taken into account by most process engineers in making localized economic decisions are:

1. Raw-material usages and costs
2. Product and by-product rates and values
3. Labor needs and rates
4. Utility requirements and costs
5. Extraordinary maintenance costs

When alternative processes are compared, it may be necessary to take into account all these items plus special considerations for the particular instance. However, when comparing pieces of equipment for a given use, it is usually sufficient to evaluate the utility costs attributed to the various units since there will be but minor variations in the other factors in the annual operating costs.

Economic selection. Economic decisions in the selection of processes or equipment are based, on the one hand, on a comparison of total annual fixed costs plus annual operating costs for the several units and selecting the one which costs the least. If benefits for products or by-products are derived, they must be taken into account, and selection is then based on the greatest net benefit or the least cost if there are no net benefits. On the other hand, the decision can also be made by comparing two units to determine whether or not the net difference in annual operating costs divided into the net difference in installed costs between the units is less than the payback time specified for the project.

The following examples illustrate the application of these methods.

Example 1.1 In an MgO regeneration plant outlined in Chap. 2, Sec. 400 in Fig. 2.2 deals with waste-heat recovery. Three methods were proposed to reduce the temperature of a major process-gas stream from 1750 to 750°F:

1. Quench the process gas by injection and vaporization of process water.
2. Pass the hot gas stream through a waste-heat boiler to make 450-psig, 650°F superheat steam for export and credit.
3. Interchange a portion of the heat from the hot process gas to preheat another airstream in the process and use the remainder of the heat to prepare export steam.

Which of these methods is the economical choice if the annual carrying charge rate is 15.5 percent and the burden is taken as 7.5 percent?

solution   The capital costs and operating requirements for the various methods as prepared by the process engineers for the project are shown in Table 1.5, and Table 1.6 gives the various fixed costs and annual operating costs for the methods. An inspection of Table 1.6 shows that the use of a waste-heat boiler plus a fixed preheater for the process air offered the best net annual credit, and this method was incorporated in the project.

In Example 1.1 it should be noted that the combination of preheating the air with the production of some export steam appeared to have good economic potential. However, preliminary testing had shown that there could be blockage of the exchanger tubes, and the sum of $30,000 per year for a maintenance differential reduced the economic feasibility of that method.

TABLE 1.5 Waste-Heat Recovery Methods for Example 1.1: Installed Costs and Operating Requirements*

Method	No heat recovery		Waste-heat boiler		Shell-and-tube interchanger	
	Conditions	Cost	Conditions	Cost	Conditions	Cost
1. Waste-heat boiler						
Duty, Btu/h	......		11.6×10^6		6.1×10^6	
Off gas, in/out, °F	......		1750/700		1750/1250	
Steam, lb/h	......		9200		4800	
Steam, °F/psig	......		650/450		650/450	
Installed costs		......		$160,000		$115,000
2. Boiler-feed conditioning						
Duty, gal/min	......		23		15	
Dissolved solids, in/out, ppm	......		170/3		170/3	
Temperature, in/out, °F	......		Ambient/ambient		Ambient/ambient	
Chemicals, HCl/NaOH, lb/day	......		300/115		170/70	
Installed costs		......		$100,000		$75,000
3. Fired preheater						
Air duty, Btu/h	4.8×10^6		4.8×10^6			
Air temperature, in/out, °F	150/1000		150/1000			
No. 6 fuel oil, gal/min	0.8		0.8			
Installed costs		$110,000		$110,000		
4. Shell-and-tube interchangers						
Duty, Btu/h	......		......		5.5×10^6	
Off-gas temperature, in/out, °F	......		......		1250/700	
Air temperature, in/out, °F	......		......		150/1000	
Installed costs		......		......		$200,000

*Basis: autumn, 1980. Adapted by permission of United Engineers & Constructors Inc., Philadelphia.

TABLE 1.6 Waste-Heat Recovery Methods for Example 1.1: Installed and Annual Costs*

Costs	No heat recovery, $	Waste-heat boiler, $	Shell-and-tube interchanger, $
1. Total installed costs	110,000	370,000	390,000
2. Annual costs			
Capital carrying charges	17,100	57,400	60,500
Burden	8,300	27,800	29,300
No. 6 fuel oil	148,000	148,000	
Municipal water	1,800	2,000	1,000
Chemicals	......	5,500	3,000
Labor differential	......	8,000	8,000
Maintenance differential	......	......	30,000
3. Total annual costs	175,200	248,000	131,800
4. Steam credit	......	298,000	155,000
5. Net annual cost	175,200		
6. Net annual credit	......	50,000	24,000

*Basis: autumn, 1980. Adapted by permission of United Engineers & Constructors Inc., Philadelphia.

Consideration is given in Example 1.2 to choosing between offered pieces of equipment on the basis of payback time.

Example 1.2 A 316 stainless-steel pump and spare were required in another section of the MgO regeneration project of Example 1.1 to circulate 3000 gal/min of scrubbing liquor with a total dynamic head (TDH) of 75 ft. Of the several pumps bid, one type was offered for $14,275 each with a 100-hp motor, while another pump cost $18,900 but had a 75-hp motor. The second pump had a higher operating efficiency so that it required 68.2 brake horsepower (bhp), while the first used 82.8 bhp. Which set of pumps should have been purchased if the annual carrying charge was 15.5 percent and the burden was 7.5 percent, with electricity costing 7 cents per kilowatthour?

solution The difference in initial costs for a pair of pumps was $9350. Since the costs of foundations and installation of piping would be about equal and the costs for the electrical portion of the installation would also be about the same, the $9350 figure may be taken as the difference in installed costs between the second and the first set of pumps. However, the second set of pumps used 14.6 less horsepower. Since it was intended to operate a pump for 6609 h/year, the second set of pumps would effect a saving of $5032 per year.* Thus, the payback time for the increased costs of the second set of pumps was $9350/$5032, or 1.86 years. This was considerably less than the payback time of $4\frac{1}{3}$ years, equivalent to a total annual-fixed-cost rate of 0.23, and the second set of pumps was purchased for the project.

Had the difference in initial costs in Example 1.2 been greater, i.e., above $21,800, or the annual savings in electricity been less than $2150 per year, the payback time for the second set of pumps would have been more than $4\frac{1}{3}$ years, and the first set of pumps would have been purchased, all other considerations being equal. Of course,

---

*14.6 hp $\times$ 0.745 kW/hp $\times$ 6609 h/year $\times$ 0.07 cents per kilowatthour.

had there been extraordinary circumstances so that the payback time under the original conditions had been reduced from 4½ years to less than 22 months, then the first set of pumps would have been purchased.

Useful economic parameters. Sometimes economic decisions concern the spending of additional capital to effect a higher production yield or a greater recovery of product. Most economic evaluations, however, involve additional expenditures in operating costs to justify a saving in capital expenditure, or vice versa. The price of most raw materials and intermediate chemicals can be obtained from a current issue of *Chemical Marketing Reporter*.* A list of many chemicals and their manufacturers is given in *Chemical Engineering Catalog*,† and up-to-date prices and conditions can be obtained directly from the producers.

Utility costs are often peculiar to a particular plant and depend greatly upon its location, the amount required, and the contract price for which outside power, fuel, or municipal water may be purchased. The current price for the utility must be obtained from the local authorities or suppliers. Other commonly used services are produced by the facility itself from basic purchased utilities through the use of auxiliary equipment such as boilers, pumps, compressors, and specialty equipment, e.g., cooling towers, water softeners, or demineralizers. The cost of the various prepared services is a function of the annual fixed costs of the equipment required to generate or circulate the required utility, the annual operating costs for the basic fuel, power, source material, or replacement material, and the number of units required per year.

Nomenclature and Dimensional Units

The nomenclature and symbols given to various parameters throughout the book are those in general use in the United States of America. The symbols are defined and their dimensions given in each chapter as they are introduced in that chapter. They may be repeated in that chapter if required for clarity.

It has been almost a decade since the engineering societies in the United States recommended that the country gradually convert from the traditional engineering units based on the English, or U.S. customary, system to the International System of Units (SI). Equipment catalogs which show dimensions and parameters in both systems of units are now beginning to appear. However, it is still the general practice for most engineers in the United States to perform their hand calculations and prepare equipment specifications in English units.

The use of SI units as the generally accepted system in the United States still appears to be quite a while away. Therefore, as a practical matter, English units of measurement are generally used throughout the book. This should be less confusing than using the two systems concurrently or alternately. In addition, many of the shortcut relationships and mnemonic devices would be rendered useless by translation from English to SI units. (It should be noted that metric units are used for viscosity since data are more readily available in those units.) Table 1.7 is included to facilitate

---

*Published weekly by Schnell Publishing Co., Inc., New York, N.Y. 10007.

†Published annually by Reinhold Publishing, Division Penton/IPC, Stamford, Conn. 06904.

TABLE 1.7 Useful Relationships among Common English Units (*Continued*)

Multiply	By	To obtain
acres	43,500	square feet (ft^2)
atmospheres (atm)	33.9	feet (ft) of water (at 4°C)
atmospheres (atm)	29.92	inches of mercury (inHg) (at 0°C)
atmospheres (atm)	14.7	pounds per square inch (lbf/in^2; psi)
bars	0.987	atmospheres (atm)
barrels (bbl) (oil)	42.0	gallons (gal) (oil)
British thermal units (Btu)	778.2	foot pounds (ft·lbf)
British thermal units (Btu)	0.000393	horsepower hours (hp·h)
British thermal units (Btu)	0.000293	kilowatthours (kWh)
British thermal units per hour (Btu/h)	0.000393	horsepower (hp)
British thermal units per hour (Btu/h)	0.393	watts (W)
centipoises (cP)	2.42	pounds per foot hour [lbf/(ft·h)]
centipoises (cP)	0.000672	pounds per foot second [lbf/(ft·s)]
circumference	6.283	radians (rad)
cubic feet (ft^3)	1,728	cubic inches (in^3)
cubic feet (ft^3)	7.481	gallons (gal)
cubic feet (ft^3)	62.43	pounds (lb) of water
cubic inches (in^3)	0.0005787	cubic feet (ft^3)
cubic yards (yd^3)	27.0	cubic feet (ft^3)
days	1,440	minutes (min)
degrees (°) (angle)	0.01745	radians (rad)
fathoms	6.0	feet (ft)
feet (ft) of water	0.0295	atmospheres (atm)
feet (ft) of water	0.8826	inches of mercury (inHg)
feet (ft) of water	62.43	pounds per square foot (lbf/ft^2)
feet (ft) of water	0.4335	pounds per square inch (lbf/in^2)
foot pounds (ft·lbf)	0.001286	British thermal units (Btu)
foot pounds (ft·lbf)	5.050×10^{-7}	horsepower hours (hp·h)
foot pounds per minute [(ft·lbf/min]	3.030×10^{-5}	horsepower (hp)
foot pounds per minute [(ft·lbf)/min]	2.260×10^{-5}	kilowatts (kW)
gallons (gal)	0.1337	cubic feet (ft^3)
gallons (gal)	231	cubic inches (in^3)
gallons (gal) (British imperial liquid)	1.20095	gallons (gal) (U.S. liquid)
gallons (gal) (U. S. liquid)	0.83267	gallons (gal) (British imperial liquid)
gallons (gal) of water	8.337	pounds (lb) of water
grains (gr)	0.0001429	pounds (lb)
horsepower (hp)	2,547	British thermal units per hour (Btu/ h)
horsepower (hp)	33,000	foot pounds per minute [(ft·lbf)/min]
horsepower (hp)	0.7457	kilowatts (kW)
horsepower (hp) (boiler)	33,542	British thermal units per hour (Btu/h)
horsepower (hp) (boiler)	9.803	kilowatts (kW)
horsepower hours (hp·h)	2,547	British thermal units (Btu)
horsepower hours (hp·h)	0.7457	kilowatthours (kWh)
inches of mercury (inHg)	0.0342	atmospheres (atm)
inches of mercury (inHg)	1.133	feet (ft) of water
inches of mercury (inHg)	0.4912	pounds per square inch (lbf/in^2; psi)
inches (in) of water (4°C)	0.03613	pounds per square inch (lbf/in^2; psi)

TABLE 1.7 Useful Relationships among Common English Units (Continued)

Multiply	By	To obtain
kilowatts (kW)	3,413	British thermal units per hour (Btu/h)
kilowatts (kW)	44,260	foot pounds per minute [(ft·lbf)/min]
kilowatts (kW)	1.341	horsepower (hp)
ln n	0.4343	$\log_{10} n$
$\log_{10} n$	2.303	ln n
miles (mi) (statute)	5,280	feet (ft)
miles per hour (mi/h)	88	feet per second (ft/s)
mils	0.001	inches (in)
pints (pt) (liquid)	0.125	gallons (gal)
pints (pt) (liquid)	0.5	quarts (qt) (liquid)
poises (P)	1.0	grams per centimeter second [g/(cm·s)]
poundals	0.03108	pounds (lb)
pounds (lb)	7,000	grains (gr)
pounds (lb)	16	ounces (oz)
pounds (lb)	32.17	poundals
pounds (lb) of water	0.01602	cubic feet (ft^3)
pounds (lb) of water	27.68	cubic inches (in^3)
pounds (lb) of water	0.1198	gallons (gal)
pounds per square inch (lbf/in^2; psi)	0.6804	atmospheres (atm)
pounds per square inch (lbf/in^2; psi)	2.307	feet (ft) of water
pounds per square inch (lbf/in^2; psi)	2.036	inches of mercury (inHg)
pounds per square inch (lbf/in^2; psi)	144	pounds per square foot (lbf/ft^2)
quarts (qt) (liquid)	0.03342	cubic feet (ft^3)
quarts (qt) (liquid)	57.75	cubic inches (in^3)
quarts (qt) (liquid)	0.25	gallons (gal)
radians (rad)	57.296	degrees (°)
revolutions (r)	6.283	radians (rad)
slugs	32.17	pounds (lb)
square feet (ft^2)	144	square inches (in^2)
square feet (ft^2)	0.1111	square yards (yd^2)
square inches (in^2)	0.006944	square feet (ft^2)
square miles (mi^2)	640	acres
square miles (mi^2)	2.788 × 10^7	square feet (ft^2)
square yards (yd^2)	9	square feet (ft^2)
temperature (°F + 460)	1.0	absolute temperature (°R)
temperature (°F − 32)	5/9	temperature (°C)
tons (long)	2,240	pounds (lb)
tons (short)	2,000	pounds (lb)
tons (refrigeration)	12,000	British thermal units per hour (Btu/h)
watts (W)	3.413	British thermal units per hour (Btu/h)
watts (W)	44.27	foot pounds per minute [(ft·lbf)/min]
watts (W)	0.001341	horsepower (hp)
watts (W)	0.001	kilowatts (kW)
watts (W) (absolute)	1.0	joules per second (J/s)
watthours (Wh)	3.413	British thermal units (Btu)
watthours (Wh)	2,646	foot pounds (ft·lbf)
watthours (Wh)	0.001341	horsepower hours (hp·h)
weeks	168	hours (h)
years	365.256	days (mean solar)
years	8,766.1	hours (h) (mean solar)

TABLE 1.8 Relationship of Common English and Metric Terms with the International System of Units (SI)

Multiply	By	To obtain
atmospheres (atm) (mean sea level)	101,325.0	pascals (Pa)
barrels (bbl) (42 gal)	0.1599	cubic meters (m^3)
British thermal units (Btu)	1,055.1	joules (J)
British thermal units per hour (Btu/h)	0.293	watts (W)
British thermal units per hour per square foot [Btu/(h·ft^2)]	3.155	joules per second per square meter [J/(s·m^2)]
British thermal units per hour per square foot per degree Fahrenheit [Btu/(h·ft^2·°F)]	5.678	joules per second per square meter per kelvin [J/(s·m^2·K)]
British thermal units per hour per square foot per degree Fahrenheit per foot {Btu/[h·ft^2·(°F/ft)]}	1.731	joules per second per meter per kelvin [J/(s·m·K)]
British thermal units per pound per degree Fahrenheit [Btu/(lb·°F)]	4,187	joules per kilogram per kelvin [J/(kg·K)]
British thermal units per second (Btu/s)	1,055.1	watts (W)
calories (cal)	4.187	joules (J)
calories per gram per degree Celsius [cal/(g·°C)]	4,187	joules per kilogram per kelvin [J/(kg·K)]
centimeters (cm) of mercury (0°C)	1,333.2	pascals (Pa)
centimeters (cm) of water (4°C)	98.06	pascals (Pa)
centipoises (cP)	0.001	pascal seconds (Pa·s)
centistokes (cSt)	1.0×10^{-6}	square meters per second (m^2/s)
cubic feet (ft^3)	0.028317	cubic meters (m^3)
cubic inches (in^3)	1.63871×10^{-5}	cubic meters (m^3)
degrees Fahrenheit (°F) + 460	5/9	kelvins (K)
degrees Rankine (°R)	5/9	kelvins (K)
dynes (dyn)	1.0×10^{-5}	newtons (N)
ergs	1.0×10^{-7}	joules (J)
feet (ft)	0.3048	meters (m)
feet of water (4°C)	2,988.98	pascals (Pa)
feet per second per second [ft/(s·s)]	0.3048	meters per square seconds (m/s^2)
foot pound-force (ft·lbf)	1.3558	joules (J)
gallon (gal) (U.S. liquid)	0.003785	cubic meters (m^3)
horsepower (hp)	745.7	watts (W)
inches (in)	0.0254	meters (m)
inches of mercury (inHg) (60°F)	3,376.9	pascals (Pa)
inches of water (60°F)	248.84	pascals (Pa)
kilocalories (kcal)	4,186.8	joules (J)
kilograms-force (kgf)	9.80665	newtons (N)
mils	2.54×10^{-5}	meters (m)
miles (mi) (U.S. statute)	1,609.34	meters (m)
miles per hour (mi/h)	0.44704	meters per second (m/s)
millimeters of mercury (mmHg) (0°C)	133.322	pascals (Pa)
ounces (oz) (mass)	0.0283495	kilograms (kg)
pints (pt) (U.S. liquid)	4.7318×10^{-4}	cubic meters (m^3)
poises (P)	0.1	pascal seconds (Pa·s)
poundals	0.13825	newtons (N)
pounds-force (lbf)	4.4482	newtons (N)

TABLE 1.8 Relationship of Common English and Metric Terms with the International System of Units (SI) *(Continued)*

Multiply	By	To obtain
pounds-force per second per square foot [lbf/(s·ft^2)]	47.8803	pascal seconds (Pa·s)
pounds-force per square inch (lbf/in^2)	6,894.76	pascals (Pa)
pounds-mass (lb or lbm)	0.453592	kilograms (kg)
pounds-mass per cubic foot (lb/ft^3)	16.0185	kilograms per cubic meter (kg/m^3)
pounds-mass per foot per second [lb/(ft·s)]	1.48816	pascal seconds (Pa·s)
slugs	14.5939	kilograms (kg)
square feet (ft^2)	0.092903	square meters (m^2)
square feet per hour (ft^2/h)	2.5806 × 10^{-5}	square meters per second (m^2/s)
square inches (in^2)	6.4516 × 10^{-4}	square meters (m^2)
stokes (St)	0.0001	square meters per second (m^2/s)
tons (short)	907.18	kilograms (kg)
torr (mmHg, 0°C)	133.322	pascals (Pa)
watthours (Wh)	3,600	joules (J)

conversion from one English unit to another, while Table 1.8 presents English and SI units to allow conversion from one system to the other. Reverse conversions in Tables 1.7 and 1.8 can be obtained by applying the following relationship: If A multiplied by B gives C, then C divided by B gives A, or C multiplied by the inverse of B gives A.

REFERENCES

1. J. Bronowski, *The Ascent of Man*, Little, Brown and Company, Boston, 1973, pp. 59ff.
2. G. D. Ulrich, *A Guide to Chemical Engineering Process Design and Economics*, John Wiley & Sons, Inc., New York, 1984, pp. 61ff.

Section

1

Process Documentation

Chapter 2

Block Diagrams and Process Flow Sheets

The single most important task that process engineers must perform is preparing the different types of diagrams required to depict the project to the various levels of responsibility of the project. An overall view of the process is presented as distinct sections in block diagrams. These diagrams are usually employed for presentation to upper management in order to provide quick explanations of the process and are generally included in portions of a process description. Block diagrams can be used for the entire process or for any subsection of the process.

Process flow sheets are somewhat simplified schematics which delineate the material and energy balances as well as the temperature and pressure profiles for the project's major equipment and its process and utility flows. These sheets are designed to provide the discipline engineers (mechanical, electrical, and instrumentation as well as process) with sufficient information to enable them to begin their own work on the project. As such, data on various parameters such as density, viscosity, etc., are frequently included in the material-balance tabulations. The complete process diagram, once it has been approved by the client or the authorizing section, becomes the basis of the scope of the project.

The importance of process flow sheets cannot be overemphasized. These flow sheets are expanded by the process engineers into the engineering flow diagrams, which, in turn, are used by the piping designers to prepare the piping drawings for the construction of the plant. Depending upon the type being prepared, the engineering flow diagrams contain varying amounts of instrumentation information.

Block Diagrams

There are essentially no hard-and-fast rules for preparing block diagrams. There can be almost as many forms of presentation as there are engineers preparing them. The format varies from simple to elaborate and is limited only by the imagination of the process engineer. Regardless of format, however, the function of the diagram is to depict each of the main processing steps involved in a given process.

The simplest format, and the one most commonly used, is a series of rectangles or blocks each of which is labeled to represent a step in the process. This is illustrated in Fig. 2.1 for a flue-gas-desulfurization (FGD) process for a coal-burning electricity-generating station. In actuality, several processing plants are involved. The sulfur dioxide is removed by reaction with a magnesium oxide solution and slurry. The resultant magnesium sulfite is concentrated and dried and is then transported to a regeneration plant located in or adjacent to a sulfuric acid production facility for regeneration of magnesium oxide and recovery of sulfur dioxide for processing to acid.

The process operations for the MgO regeneration plant are delineated further by additional block diagrams as shown in Fig. 2.2. The operating steps are given here as separate boxes. For convenience, each operating step may be assigned a block or section number which can later be cross-referenced with flow-stream numbers in conjunction with the process flow sheets or for equipment numbering when the engineering piping and instrumentation drawings are prepared.

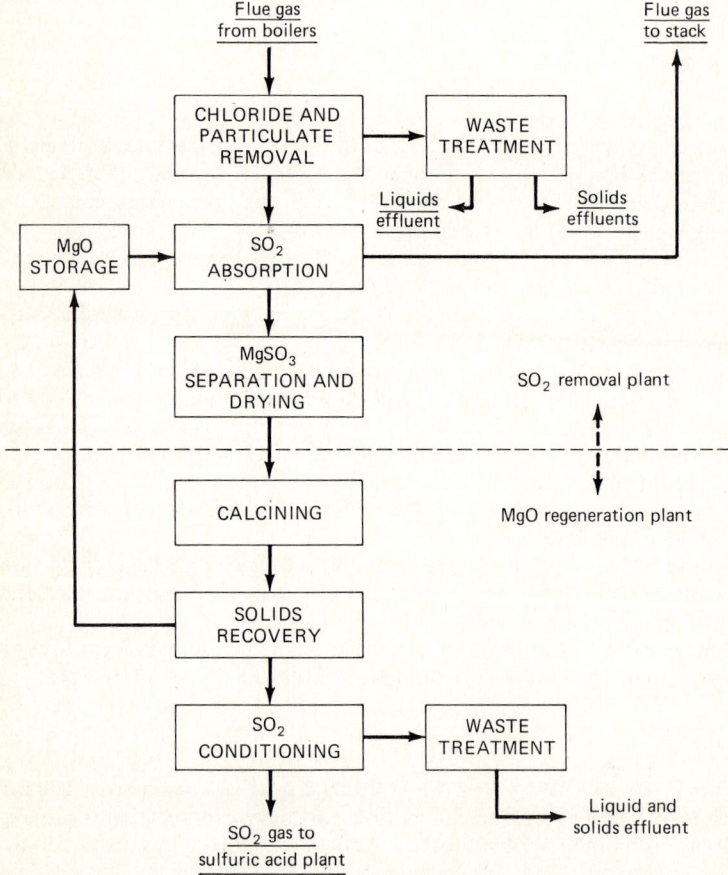

Figure 2.1 Block diagram for an MgO flue-gas-desulfurization plant.

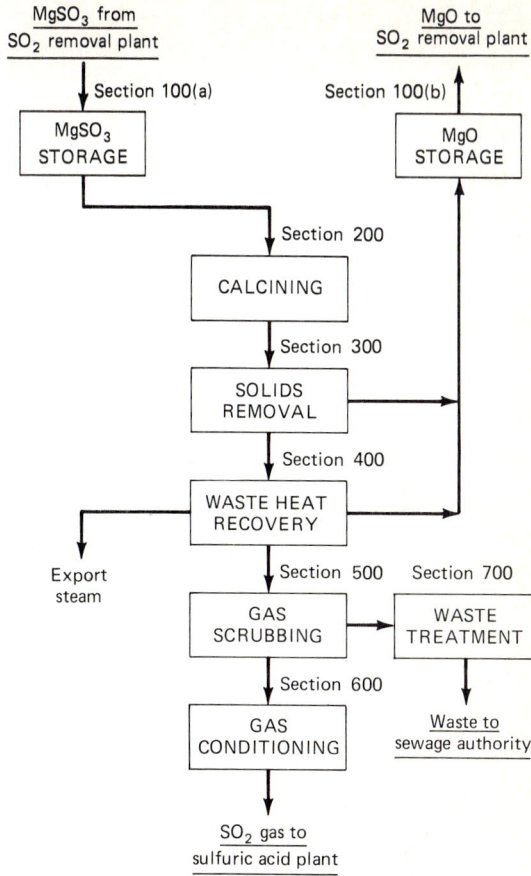

Figure 2.2 Block diagram for an MgO regeneration plant.

Included with the regeneration plant is a waste-treatment plant to manage the continuous and intermittent waste streams that are or could be generated in the plant. The exact treatment that must be accomplished depends on the requirements of the agency or authority that has jurisdiction over the receiving body to which the waste will go.

The various utilities such as fuels, cooling water, refrigeration, compressed air or nitrogen, steam, etc., are not usually represented as separate elements in a block diagram. There may be exceptions, of course, if the production or handling of the particular utility represents a significant operation in the process. In Fig. 2.2, the production of steam, a large percentage of which is used in the process itself, is given a block instead of that step's being assigned to gas cooling. The storage and pumping of the fired oils to the calciners, on the other hand, have been omitted from the block diagram since they are normal, straightforward operations. Should it have been necessary to provide unusual treatment or conditioning of the fuel oil or other utility, a block would have been assigned in the diagram to that step.

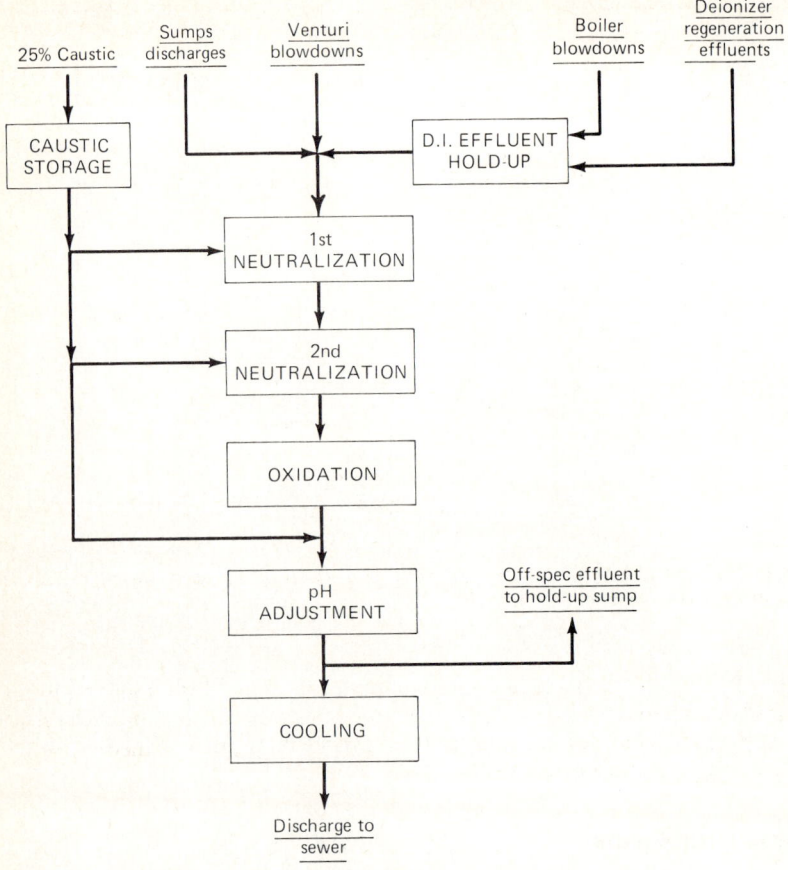

Figure 2.3 Block diagram of Sec. 700, waste treatment.

Any one of the steps in a block diagram can be further represented by its own block diagram. For example, Sec. 700 in the regeneration plant shown in Fig. 2.2 is termed "waste treatment." The steps within that section are depicted in a block diagram as Fig. 2.3. This level of diagram requires little descriptive explanation since it reduces this section of the project to its most nearly rudimentary elements.

Other modes of representation

A variation of the block diagram to present the elementary steps of the plant is the use of semigraphical symbols instead of boxes. An example is given in Fig. 2.4 for the regeneration portion of the plant. The semigraphical format makes a clearer presentation and gives the uninitiated observer a better appreciation of the equipment involved.

The most descriptive presentation, however, is that of an isometric representation of the plant as it will be upon completion. The regeneration plant is shown in this manner in Fig. 2.5. This type of presentation is the most vivid and is useful to those

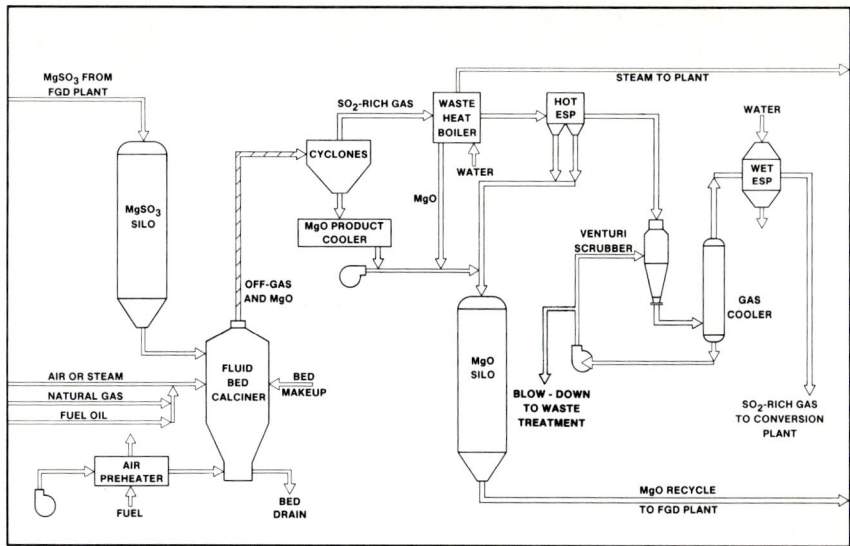

Figure 2.4 Schematic diagram of an MgO regeneration plant. *(Adapted with permission of United Engineers & Constructors Inc., Philadelphia.)*

concerned with detail. As such, it may be a valuable tool in training sessions for future operators and supervisors. The isometric representations, on the other hand, are more costly to prepare, and they usually cannot be initiated until the engineering of the plant has progressed to a considerable degree.

Process Flow Sheets

The preparation of a process flow sheet involves several stages: collection of data, calculating material and energy balances, and laying them out so that the information regarding the process is presented in a logical fashion and in sufficient detail to permit a full understanding of the elementary steps of the process.

Data sources for process flow sheets

The material- and energy-balance information which comes to the process engineers may derive from one or more sources. Frequently, the development section of the engineers' company or that of a client will already have prepared the process flow sheet on the basis of pilot-plant or bench-scale studies. In that case, the process engineers need only review the client's basic process flow sheets to provide internal consistency and to assure themselves that the required information is present. Any missing information may be found in the compilations of relevant data or process data books which usually accompany the process flow sheets. Should any information be lacking, it can be requested from those responsible or be compiled by the process engineers from standard reference sources. Process data books are needed not only to prepare the process flow sheets but to provide necessary data for completion of the equipment specifications.

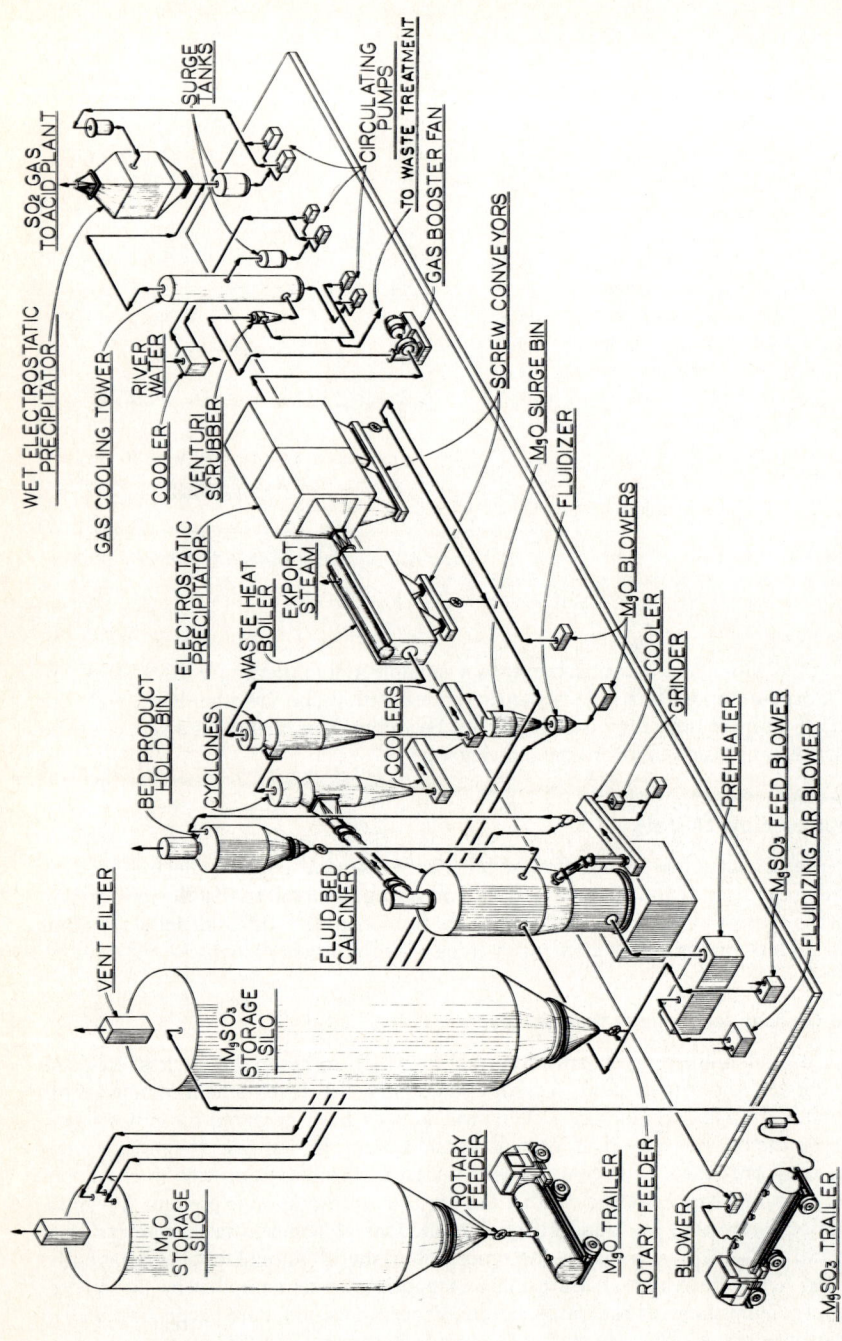

Figure 2.5 Isometric diagram of an MgO regeneration plant. (Adapted with permission of United Engineers & Constructors Inc., Philadelphia.)

Often the process engineers become involved with the project as consultants to the pilot-plant or bench-scale experiments as the process is being developed. In this case, the process engineers must become thoroughly familiar with the pilot or test work being undertaken. At the same time, they must visualize the basic schematic that will eventually evolve for the process. With the latter in mind, the process engineers can influence the direction of the bench-scale or pilot operations to yield the necessary data for inclusion in the process flow sheets and, if necessary, for the process data book.

Infrequently, development of a new process is accomplished without resort to bench-scale testing but is based instead on literature data only. In that instance, a team of technical personnel (chemists, physical chemists, and process engineers) surveys the published articles and reference works for the data necessary for the process flow sheets and process data book. It is the responsibility of the process engineers to ensure that the group preparing the process evolves the required data.

It becomes a matter of judgment whether or not the reaction kinetics obtained from the literature is sufficiently trustworthy to serve as the design basis for a large-scale plant. Similarly, consideration should be given to the effect of recycle streams and the buildup of impurities. If necessary, pilot-plant studies should be designed to clarify these and similar problems.

Process information required

The process flow sheets present a rudimentary schematic of the process and basic information regarding the process. The schematic should show all the major equipment for the unit operations in the process. The equipment is connected by lines with flow arrows to designate the paths of major streams as they pass through the process or are recycled. The material balance for the process must record the flow rate of each component on a weight and/or molar or volumetric basis for each of the flow paths or streams. Important physical parameters such as temperature, pressure, specific gravity, viscosity, etc., should be included in the recorded data where required.

The energy balances are defined by entering exchanger duties near such units and by showing heats of vaporization or fusion if a physical change of state takes place. Also required are net endothermic or exothermic heats of reaction at any unit where a chemical reaction occurs. Approximate brake horsepowers should be shown for important pumps, compressors, or blowers.

All the material-balance information should be scaled or calculated from the pilot-plant data or from the literature studies upon which the process is based. The various physical parameters can be obtained from the pilot-plant work or from the process data book. Energy-balance values are calculated from the material balances, the physical parameters, and relevant information in the process data book.

Presentation

The schematic presentation of the process and essential information regarding it can be accomplished in any of several manners. In some instances, the entire process flow sheet is shown on one sheet which is of a convenient width and is as long as required. This is satisfactory if the number of elements in the process is not too great. However, for a process with a considerable number of steps, a single process flow sheet could become inordinately long and inconvenient to handle, requiring manipulation of the entire flow sheet if information were wanted on only one of the subsections of the process.

A more acceptable presentation is the use of a series of process flow sheets in order to avoid a lengthy sheet. In this case, the size of the tracing (linen, Mylar, or other standard material) should be the same as that to be used for the engineering flow diagram and piping drawings. Most operating and design companies have their own preferred sizes. These tracings are usually about 23 by 35 in (known as D size) or 27 by 40 in (known as F size).*

A logical breakdown of the process for division among the several tracings is that of the block diagram for the overall process. One or several of the blocks or major steps can become a separate flow sheet. Thus, the waste-treatment block in the regeneration block diagram (Fig. 2.2) is shown in Fig. 2.6 as a process flow sheet. The waste-treatment facility receives all the liquid effluents originating in the regeneration plant and processes them so that they are suitable with respect to pH, chemical oxygen demand, and temperature for discharge to the local sewage authority.

Prior to preparation of the process flow sheet, the process must be analyzed and careful consideration given to all the essential equipment that will be required to convert the reactants to the products or, in this case, the discharge effluent. These major pieces of equipment are taken to be those apparatuses in which unit operations such as reaction, distillation, evaporation, heat exchange, adsorption, grinding, drying, dehumidification, and similar steps occur. They are usually those steps in which a chemical or important physical change takes place. Items such as storage tanks, intermediate tanks, mixers, agitators, and strainers are generally omitted from the process flow sheet. They may, however, be included should they be crucial to the process itself. For example, a tank which provides residence time to complete a reaction after flows leave a reactor–mixing chamber or which is used for storage when the flow is being converted from continuous to batch or vice versa should be shown if needed for clarity.

A piece of equipment that is to be included in the flow sheet can be shown as a graphic representation of the unit with its identification placed near it. However, the use of a simple rectangular box to identify the piece allows a clear and more easily readable presentation and is used in Fig. 2.6. On the other hand, on this type of flow sheet transport equipment such as pumps, compressors, blowers, etc., is usually depicted graphically to facilitate understanding of the movement of the various streams through the flow sheet.

Flow lines

All pieces of equipment are connected with lines to designate major flows and recycles between the pieces of equipment or to and from the boundaries of the plant. In the interest of clarity and simplicity, minor flows such as pump recycles and side streams which will flow through instruments or which may be used as seal fluid can be omitted from the process flow sheet. These are recycled among themselves and do not affect the material and energy balances.

Only major utilities such as cooling water are shown on the process flow sheet. Others that do not appreciably affect the material and energy balances are omitted

---

*The letter designations for tracing sizes have never been standardized in the industry. Thus, the sizes cited could also be 27 by 36 in and 30 by 40 in, respectively, at some companies.

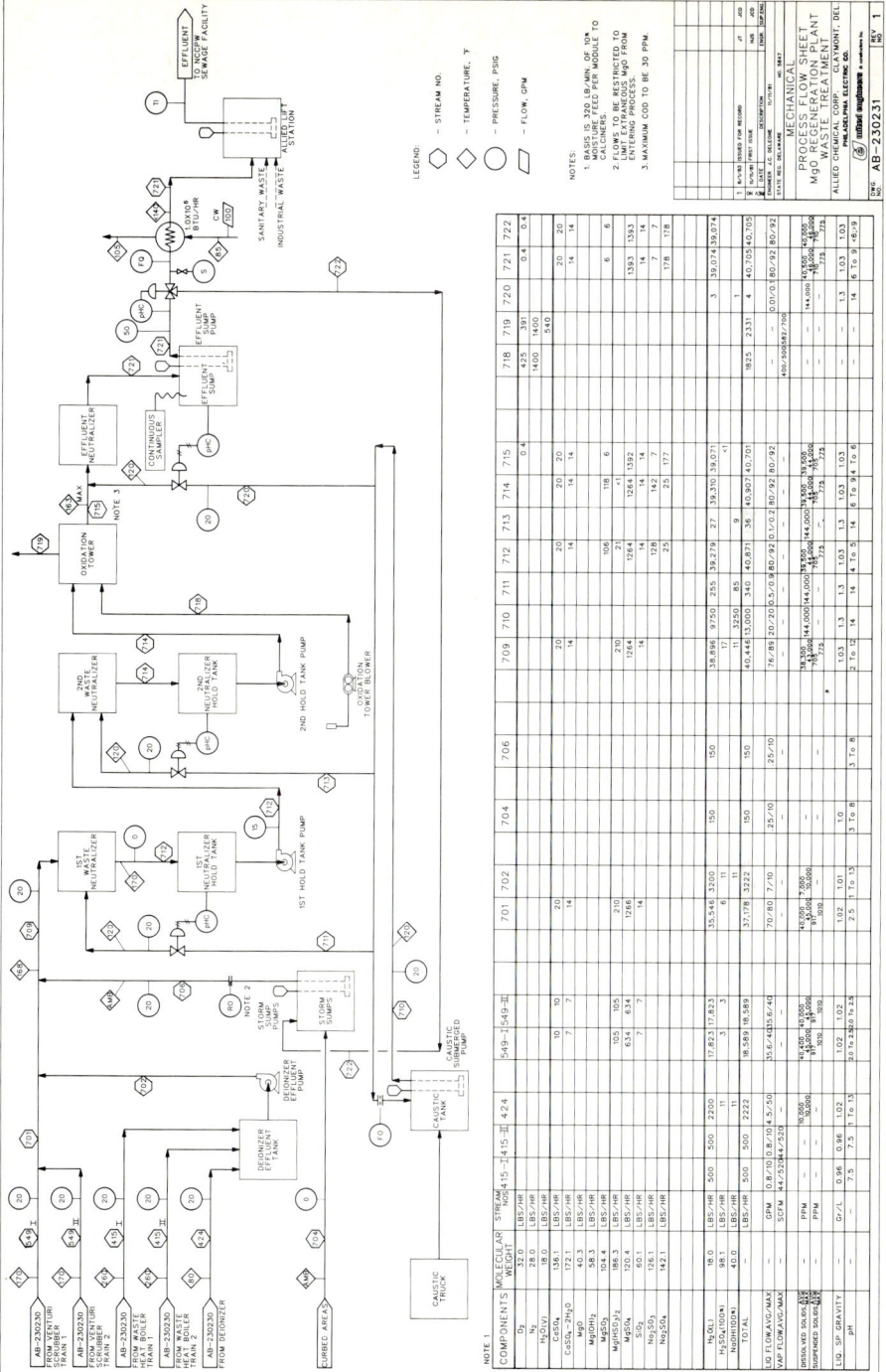

Figure 2.6 Typical process flow sheet. (Adapted with permission of United Engineers & Constructors Inc, Philadelphia.)

from the flow sheet. However, if it is considered sufficiently important, a utility-use diagram similar to a process flow sheet may be devoted to a particular utility or to a set of utilities. Utility-use diagrams are included in documents for regulatory agencies that wish, for example, to know the source of a given material such as water and its eventual disposition in the proposed plant.

To avoid unwarranted complexity, most instrumentation is omitted from process flow sheets. However, the more important control functions must be included on the flow sheets. They should indicate what parameter is being sensed and which stream is being controlled.

Stream numbers. Each flow stream with a distinct and different composition should be assigned a separate stream number. These numbers are keyed to the material balance, which appears on the bottom half of the drawing. It is convenient to adopt a convention whereby the initial digit or set of digits for a stream number identifies the block-diagram box in which the stream originates.* The waste-treatment step in Fig. 2.2 for the regeneration plant is termed Sec. 700. Thus, all streams on the waste-treatment process flow sheet (Fig. 2.6) are in the 700 number series, i.e., 700 to 799. A flow line which originates outside waste treatment carries the number of the stream on its particular flow sheet. Thus, it can be deduced that stream 549 originates in Sec. 500, the gas-scrubbing step. An exception to this procedure is utilities, which should be given stream numbers corresponding to the sheet on which they are used. If a utility-use diagram is included, it can carry these utility stream numbers, with those streams internal to the utility-use diagram itself being assigned a separate, distinct number series.

Keyed parameters. Important temperatures and pressures are keyed as separate symbols to the flow lines. They are usually repeated from line to line only if there is a significant change. Although pressure increases across a pump, the downstream pressure usually is not noted unless it is of importance to the process.

Additional physical parameters may be keyed to the flow lines at the discretion of the process engineer. This can be done if it is thought that recording them on the flow lines instead of in the material-balance grid adds necessary importance to them or clarity to the process.

Each parameter keyed to a flow line is assigned an individual and distinctly recognizable symbol. In addition to hexagons to represent stream numbers, circles for pressure, and diamonds for temperature, parallelograms are used to indicate flow rates of certain utilities. These symbols are usually affixed to lines or may be shown within or attached to the boxes representing equipment. Special important parameters for individual streams may be specified directly on the flow line or as a note.

Material-balance table

The first column in the material-balance table lists all components in the stream composition plus the several important physical parameters for any or all of the streams. It is useful to group the components together in the gaseous or vaporous, liquid, or solids states according to their phases. Rows should be provided to indicate the total

---

*Should there be more than nine boxes in the block diagram, the section numbers should be continued into the thousands. In that event the first two digits identify the section.

flow rate and the average molecular weight (MW), if needed or useful, of each of the states as well as the total flow rate and the average molecular weight for the entire stream.

The total flow rate in pounds per hour or moles per hour, if required, reflects the average flows for the basis of design of the equipment in the process. These average weight flow rates are translated, by means of appropriate conversion factors, into average volumetric flow rates in terms of gallons per minute (gal/min)* for the liquid components and standard cubic feet per minute (scfm)† for components which are gaseous or in the vapor state. However, average flow rates are essentially fictitious values, set by a desired annual production rate and an assumed number of on-stream hours per year. Thus, they should be used only as guides, and the compositions are most useful in prorating the components for a similar stream at a different flow rate.

More important than the average flow rate is the range of flow rates for which the piping and equipment must be designed. This is reflected by showing a maximum flow rate for the stream and a minimum flow rate if that is important. The maximum flow rate for a continuous step in the process is defined as the average design rate for the basis of design multiplied by a design factor. This latter value, which usually ranges from 1.10 to 1.25 for the continuous portion of the process, is set by project management after consultation with the staff functions and disciplines involved in the process. The design factors are given in the notes. Factors influencing their values will be discussed in the subsection "Notes."

It is important to show maximum flow rates for a batch step or process. These rates must take into account not only the design factor but also the minimum time allotted to transfer a maximum batch from one piece of equipment to another. The maximum flow rate shown on the process flow sheet for transfer equipment may not necessarily represent the actual design rate for this equipment and the piping associated with it. Additional flows may be required for recycle streams to maintain constant flow through a pump or a compressor or to a reactor or to provide a test stream for an instrument sensor.

Only rarely is a minimum flow rate needed on a process flow sheet. When a minimum flow rate is required through a reactor, an exchanger, a pump, or other piece of equipment to prevent damage or other undesired consequences, this need should be emphasized by showing an appropriate minimum value in the flow-rate row.

The physical parameters are listed below the flow-rate values. Additional rows may be allowed for a parameter if it must be defined for more than one state. In that event, the letter g or v, l, or s for gas or vapor, liquid, or solid, respectively, is sometimes used after the parameter to ensure the identity of the particular state.

The molecular weight of each component is given in the next column. One decimal figure is usually considered sufficient for molecular-weight values.

$$*\text{Gallons per minute} = \frac{\text{pounds per hour liquid}}{(60)(8.33)(\text{sp gr})} = \frac{\text{pounds per hour liquid}}{(500)(\text{sp gr})}$$

$$= \frac{2}{1000} \times \frac{\text{pounds per hour liquid}}{\text{sp gr}}$$

$$\dagger\text{Standard cubic feet per minute} = \frac{(379)(\text{pounds per hour gas})}{(60)(\text{MW})}$$

$$= \frac{(6.32)(\text{pounds per hour gas})}{\text{MW}}$$

A column is used to delineate the units for the row associated with it. The material-balance portion of the column must normally be given in a weight-per-unit-of-time dimension such as pounds per hour, tons per day, kilograms per minute, etc. In rare instances, a volume-per-unit-of-time dimension may be employed. In any event, units for all the component rows as well as the total rows should be the same. It may be that the common dimension for the material balance is meaningless or is not applicable for one or more components. For example, while a component may be present only in parts per million (ppm), it may be significant that it be recorded as such for several flow streams in the balance. The flow dimension for the component would then be omitted from the column and ppm would be entered under the value in the appropriate flow-stream column.

The ppm's usually do not affect the value in the rows designating total material flows. However, were the ppm large enough or the total flow great enough, the flow dimension should be shown for the component and its flow value entered in the appropriate flow-stream column. If it is important to know the ppm of the component for certain flow streams, that value can be added directly on the flow line or as a note to the flow line.

Notes

The notes section of the process flow sheet is used to record important general or supplementary information about the process that cannot readily be expressed by means of the diagram or the material-balance table. Although the notes for the flow sheet can be placed in any convenient location, a preferred position is in the lower left-hand corner or above the title block. Each separate note is numbered. Those of a general nature are not referenced to the flow sheet, but all notes which pertain to supplementary information about a specific point in the diagram or the material-balance table must be so referenced by indicating the number of the note at the appropriate point.

Chapter 3

Engineering Flow Diagrams

The greatest responsibility assigned to process engineers is that of initiating, completing, and maintaining the engineering flow diagrams for a project. It is only through engineering flow diagrams that all the discipline engineers involved in the project are provided with the information needed to initiate and continue their work together as a team. The various disciplines supplement the engineering flow diagrams with their own drawings, but all require these diagrams as their starting documents.

Engineering flow diagrams are analogous to detailed road maps. They show all the salient features of the "landscape," which in the case of a chemical plant, power station, or refinery consist of all the equipment involved in the process. They indicate the equivalent of the main arteries and secondary connections, i.e., the piping which exists between the salient features and between the various routes. Included on the paths are symbols which indicate regulation of the flow of traffic between the salient features and along the paths. Sufficient notation is provided to clarify any ambiguities that might arise.

The expression "engineering flow diagram" may cause confusion, as it often is synonymous with another frequently used term, "piping and instrumentation diagram (P&ID)," which, as its title implies, carries full information regarding instrumentation as well as mechanical requirements. At other times, the information normally included on the P&ID is divided between two drawings.* The detailed instrumentation work is placed on one sheet, usually called an "instrument application diagram" or "instrumentation flow diagram," while the mechanical information with only primary instrumentation is shown on the other sheet. The latter is also called an engineering flow diagram. Thus, regardless of whether or not the drawing contains detailed instrumentation, the term engineering flow diagram can be used in reference to the portion of work for which process engineers are responsible. The term engineering flow diagram will be used throughout the following discussions regardless of whether it is synonymous with a P&ID or is the equivalent portion of

---

*The subsection "Use of Separate Engineering and Instrumentation Flow Diagrams" explains the circumstances for the use of two drawings instead of one P&ID.

an instrument application diagram. The term piping and instrumentation diagram will be used only when it is needed for clarity.

This chapter reviews the several types of engineering flow diagrams that are normally encountered and their relationship to the various disciplines that are associated with the project. A summary will be given of information that must be included on engineering flow diagrams with respect to the various disciplines. Possible sources of information and data are reviewed. This is followed by a discussion of the logical preparation of flow diagrams with systems for formats, equipment identification, line numbering, and tabulation. Included is a checklist of items to assist the process engineer in preparing engineering flow diagrams.

Types of Engineering Flow Diagrams

The process engineer is responsible for preparing the various types of engineering flow diagrams that are required for the project. The most obvious kind is flow diagrams which deal with the process itself, i.e., those required to indicate the routing of raw materials or reactants, the processing steps themselves, the storage and transfer of products, and the treatment of effluent streams. Similar flow diagrams are required for the preparation or conditioning and distribution of the various utilities required in the project. Specialized flow diagrams are provided for specific networks that service large portions of the plant.

A different type of flow diagram is known as the "lead sheet." It contains the legends by which the other engineering flow diagrams are constructed or understood, and thus it must be prepared at an early stage or concurrently with the evolution of the other engineering flow diagrams. A discussion of the engineering flow diagrams therefore begins with consideration of the lead sheet or, more technically speaking, with the piping and instrumentation symbols diagram.

The lead sheet, or piping and instrumentation symbols diagram

A complete road map contains a key to all the symbols and acronyms employed on it to ensure that the user can interpret correctly the objects or conditions that may be encountered. A similar key is required to interpret all the symbols and acronyms that are used for the various flow diagrams. Because many symbols and acronyms must be interpreted for a series of flow diagrams, a separate drawing is usually set aside as the lead sheet. It contains essentially all the symbols and notations which are used on the other engineering flow diagrams. This does not preclude specialized symbols or notes which may be required on any individual engineering flow diagram from being entered on the particular drawing.

There are no set rules for the preparation of the lead sheet, i.e., its layout, format, or general presentation. The only requirement is that it contain all the information needed to understand the engineering flow diagrams or piping and instrumentation drawings which follow. Information is grouped by various logical categories representing symbols for valves and their actuators, for piping fittings and specialties, and for instrumentation or equipment used in connection with piping. Included should be a legend distinguishing the various kinds of piping lines along with an explanation of a set of typical line-identification symbols and a list of piping-system designations used in the line-identification system. All abbreviations used on the engineering flow diagrams should be defined on the lead sheet.

Major areas of the lead sheet are devoted to instrumentation and controls. All the instrumentation symbols must be presented, accompanied by a list of instrument designations defining the usage of the first letter and succeeding letters of the nomenclature entered as part of the instrument symbols. The instrument symbols and definitions of the identifying letters are usually consistent with those specified by the Instrument Society of America (ISA).* Portions of these sections may be omitted if reference is made to the nomenclature system developed by the ISA. However, their inclusion is frequently helpful for clarity.

General notations are also included on the lead sheet.

Figure 3.1 presents the piping and instrumentation symbols drawing or lead sheet that was part of the engineering-flow-diagram package for the MgO regeneration facility referred to in the subsection "Block Diagrams" of Chap. 2. The drawing is typical of the information presented and is an example of a format that may be employed. The arrangement of the data is usually a matter of choice by the process engineer in conformity with the arrangement normally utilized by or acceptable to the group with which the engineer is associated.

Most organizations have a prototype of a lead sheet that includes nearly all the symbols and information required for a project. Process engineers should review the prototype and anticipate the inclusion of additional material needed for the task at hand. Normally the greater part of the information and symbols on the prototype applies to all projects, and additional portions are usually required only for those sections dealing with piping-system designations, abbreviations, special equipment, or notes.

It should be emphasized that a complete lead sheet is not necessary at the initiation of the project. Frequent reference should be made to the lead sheet during the project to ensure that all symbols, conventions, or abbreviations used on the engineering flow diagrams do in fact appear on the lead sheet. It is the duty of the process engineer to maintain the lead sheet as nearly correct as possible.

If no model exists, the lead sheet given in Fig. 3.1 could serve as a prototype, with the process engineer making changes as required. Some groups initiate their lead sheet and add to it continuously for new projects. Thus, the lead sheet may contain superfluous symbols or information. This is of little consequence, as too much is usually better than too little. If the lead sheet is built up in this manner, however, it is only a matter of time before many of the abbreviations and piping-system descriptions are not in alphabetical order. This fault can be alleviated by allowing gaps in the original listings so that data originating with the same letter can be grouped together. Of course, the entire sheet could be revised at frequent intervals to rearrange the information as needed. Computer-assisted drafting facilitates movement of information on a drawing.

For convenience, it is best to use the same size of tracing for the lead sheet as will be used for the engineering flow diagrams describing the process.

Processing sections

The processing sections of the project are covered by a series of engineering flow diagrams. While it is conceivable that a process could be entered on one long drawing,

---

Instrumentation Symbols and Identification, ISA-55.1, Instrument Society of America, P.O. Box 12277, Research Triangle Park, N.C. 27701.

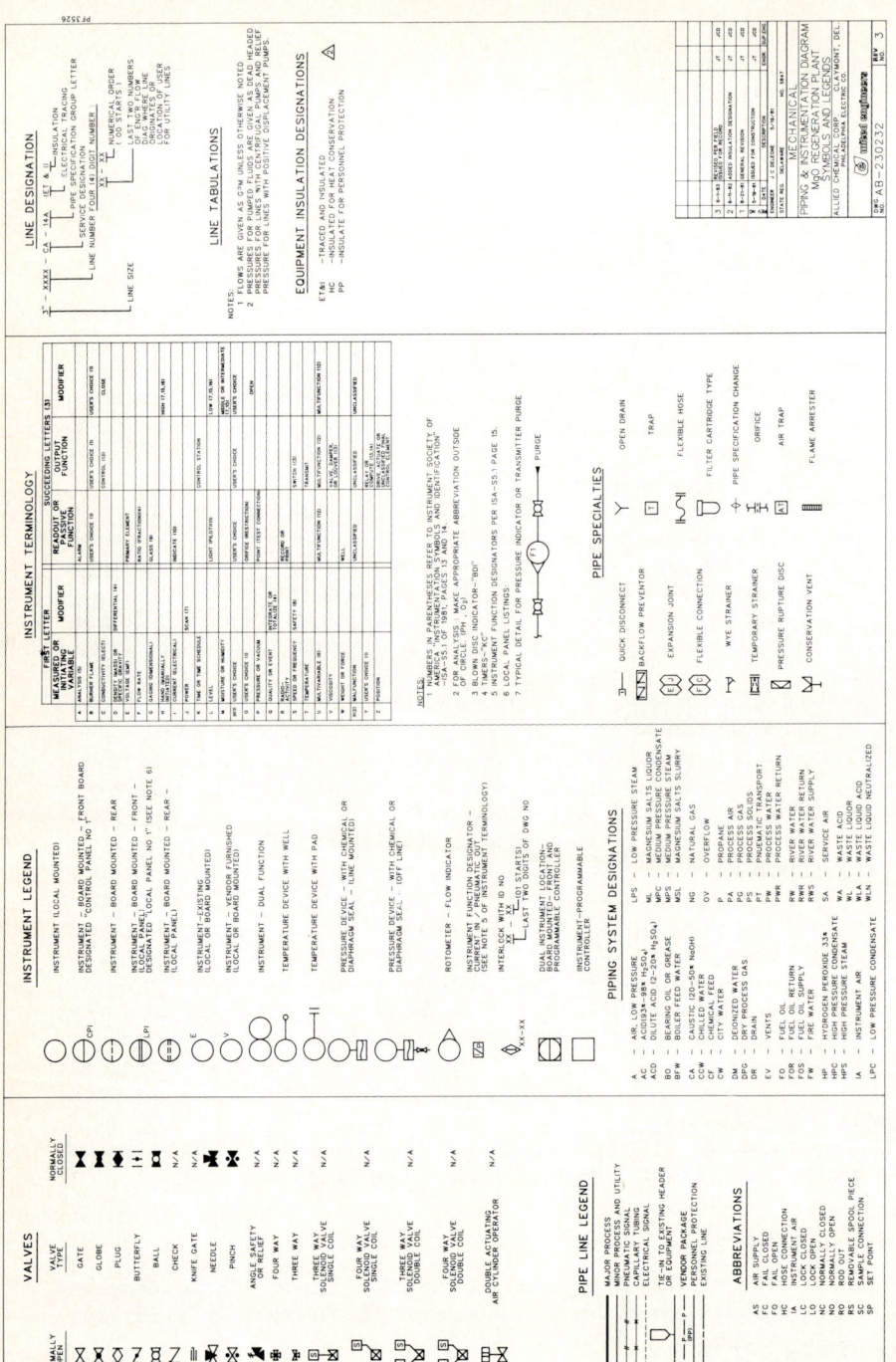

Figure 3.1 Typical lead sheet. *(Adapted with permission of United Engineers & Constructors Inc., Philadelphia.)*

the diagram would probably be unwieldy and difficult to follow. Instead, the process is broken into series of logical processing steps. It is convenient to have these areas coincide with the sections outlined in a block diagram of the process.* Even in this case it may not be possible to represent the process for a particular section on only one sheet, so that two or more drawings are required.

The flow diagrams for the processing sections of the plant must contain all the piping and instrumentation information pertaining to the various pieces of equipment that will be installed. These drawings start with the reception, storage, and conditioning of all raw materials and reactants. They then progress through the complete processing operations, showing all utilities, and conclude with the conditioning, storage, and disposition of all products and waste streams.

Utility sections

Essentially all chemical plants, refineries, or powerhouses require utilities in addition to electricity for the processing steps needed to convert raw materials into products. The MgO regeneration facility already cited under "Block Diagrams" in Chap. 2 typifies the various utility requirements for many plants. A review of the utilities for this case with some comments regarding them follows.

River water which has passed through the adjacent acid plant is collected and then used in the several heat exchangers where heat is extracted from the regeneration process. A closed-loop cooling system is included in the facility so that municipal water can be used to extract heat from certain conveyors handling hot solids and then reject it by exchange with river water. This cascaded heat-transfer scheme is required since the river water, being high in chlorides, is not acceptable as a cooling medium in the hot-solids conveyors. An economic evaluation has indicated that annual fixed and operating charges for the exchangers are less than the cost of the use of once-through municipal water. River water is also required as hose-down water.

The water utilities also include a municipal-water–process-water system. In this case, municipal water is purchased from a water authority whose water main runs close to the plant. This water is metered for accounting purposes and distributed throughout the plant for various uses such as sanitary facilities (potable, lavatory, and laboratory), safety showers, and process water.† Process water in the MgO regeneration plant is required for both continuous and intermittent users. In this case, the continuous uses include flows to a wet electrostatic precipitator to humidify the main gas stream and to limit solids concentration in a waste blowdown stream, feedwater to the deionizer system for boiler-water supply, and water to the pump seals and bearings. Intermittent users are those items of equipment which must be filled when the process is brought on stream initially or after a shutdown. Distributors for the process water are segregated from sanitary users to ensure that no process fluids enter the potable-water or safety-shower supply.

A portion of an engineering flow diagram is used to show the fire-water pumping station and the distribution of fire water to the various hydrants throughout the plant and to the standpipes in the control and maintenance buildings.

Another utility flow diagram is devoted to the storage, pumping, and heating of the

---

*See Fig. 2.2 for an example.

†Fire-protection water, if taken from a municipal water supply, usually employs a separate meter.

high- and low-sulfur fuel oils that are required in the process. Included on this drawing are the recirculating-oil systems with methods to hold a given pressure to the oil users and another pressure-control system to prevent the reciprocating oil pumps from overpressurizing the circulating system. A portion of a drawing is devoted to the reception, boosting, and distribution of natural gas as an alternative fuel.

The remaining utility flow diagrams for the MgO regeneration project deal with the plant and instrument air systems. Two drawings are devoted to the compression and drying of atmospheric air and several others to the distribution of instrument and plant air. Other utilities such as steam, refrigeration, etc., would be assigned separate utility flow diagrams or portions of flow diagrams.

Specialty flow diagrams

In addition to the drawings for process and utility piping and instrumentation, it may be necessary to provide flow diagrams of a special nature in a given project. For example, there may be an elaborate pneumatic conveying network including manifolded pickup and discharge points with several switching stations to service large numbers of pieces of equipment. This network would warrant its own engineering flow diagram.

An underground drainage system for the disposal of cooling water and nonprocess drains can normally be referenced on the regular processing and utility engineering flow diagrams. The structural engineers would then be responsible for showing the collection and disposal of these drains in their drawings. However, should a situation arise where, for example, spills or other conditions could cause a portion of the drainage piping to become contaminated, these liquids may not be allowed to become part of the normal effluent until they have been tested and treated as required. In this case, the isolation, diversion and storage, and treatment facilities must be represented on an engineering flow diagram.

Relationship to Disciplines

All engineering disciplines use flow diagrams and are involved in their preparation. A review of the relationship between the disciplines and flow diagrams and how the various disciplines commence their work on the project from them are discussed in the following subsections.

Chemical engineers and engineering flow diagrams

Should a project require chemical reactions, diffusional operations such as distillations, or similar separation techniques or have extensive material and energy balances, chemical engineers are usually assigned to the project to prepare the process flow sheets. They would probably continue on the project as process engineers with development of the piping and instrument drawing or engineering flow diagrams as their prime responsibility.

Once the basic engineering flow diagrams have been initiated, the process engineers participate with the mechanical engineering group and the structural engineers in setting plant layouts and plot plans. The process engineers' responsibility in this task is to ensure that there is an orderly, logical flow of materials through the processing steps. This includes, for example, provision for the gravity flow of liquids or solids where needed and for sufficient barometric legs in vacuum steps.

The engineering flow diagrams show equipment that will be purchased for the project by the project team, the owner, or the mechanical contractor. The process engineers size all equipment and, in conjunction with the mechanical engineers, determine necessary dimensions and calculate all pump heads and blower or compressor compression ratios after sizing all lines on the flow diagrams.

As the mechanical-piping group prepares piping and equipment-arrangement drawings, using the engineering flow diagrams as the basis for its work, the process engineers interpret and clarify the flow diagrams when needed. The mechanical-piping drawings, as well as those from the other disciplines, should be reviewed frequently to assure adherence to the engineering flow diagrams.

The equipment specifications are based to a large extent upon the information on the engineering flow diagrams as well as on the process flow sheets. The process engineers prepare the process portions of the specifications in conjunction with the mechanical engineers responsible for them.

Mechanical engineers and engineering flow diagrams

Many projects are of such a nature that a chemical engineer is not assigned as part of the engineering group. This occurs when the process flow sheet as well as the material and energy balances is simple. An example would be a boiler house where only minimal treatment is required for the feedwater and little or no waste treatment of side streams is needed. In this case, a mechanical engineer could fulfill the duties of the process engineer. The functions to be performed with respect to the engineering flow diagrams would then be the same as those outlined in the preceding subsection when a chemical engineer is assigned to the project. Regardless of who is responsible for preparation of the engineering flow diagrams, mechanical engineers use them in their other duties on the project. Engineering flow diagrams are the basis for the plot plans, equipment arrangements, and piping drawings being prepared and detail the heat tracing and insulation to be specified. Information on engineering flow diagrams is the basis for preparation of specifications for the purchase of equipment.

Instrument engineers and engineering flow diagrams

As their title implies, piping and instrumentation drawings show a large percentage of the instrumentation that is required for the project. By means of instrumentation symbols, they indicate all the primary sensors and the controllers plus the transmitters required between the sensors and the controllers. The P&IDs show which valves have limit switches to indicate whether or not a particular valve (or door or other opening) is open or closed or is permissive to some other function. It should be noted that the instrumentation information shown on P&IDs represents only a part of the work required from the instrumentation group. The greater portion and all the details appear on the individual instrumentation drawings, specifications, and logic diagrams prepared under the direction of instrument engineers.

It is the responsibility of the process engineers to determine which variables are to be measured, indicated, or controlled and at which points in the process. Process engineers must also determine whether or not a measurement is to be indicated locally at the sensing point, on an area control board, or on a central control panel. They also indicate placement points for flow-restriction orifices. The instrument engineers complete work on the piping and instrumentation drawings by determining the type of instrumentation to be used for the sensors and controllers, by elaborating on the transmitters required between the sensors and controllers, and by determining whether the instrumentation loops will be pneumatic or electrical. The instrument engineers are also responsible for providing instrument numbers on the P&IDs and for instrument tabulations.

Use of separate engineering and instrumentation flow diagrams. It can be appreciated that the instrumentation portion of a P&ID is a coordinated effort and the responsibility of both the process and the instrument engineering disciplines. However, some organizations divide responsibilities between the two disciplines by having two completely separate documents. One document is under the direction of the process engineers and is known as an engineering flow diagram* or piping diagram. The other is known as an instrumentation flow (or application) diagram, and the instrument engineer is responsible for it.

All the normal mechanical information required for a P&ID as described earlier is given on the engineering flow diagram. However, the only instrumentation to be shown on the diagram is the location and identification of the instrumentation devices mounted directly in the piping or requiring a process connection in the piping or on a piece of equipment. On the other hand, the instrumentation flow diagram is in reality a rudimentary engineering flow diagram with most of the information extraneous to the instrumentation discipline omitted from it or presented in a muted fashion. It does contain the instrument and control-valve information on its companion engineering flow diagram plus the remainder of the instrumentation that would normally be presented on a piping and instrumentation drawing.

The use of two sets of drawings can be an asset to a project since it eliminates or reduces the prominence of unneeded or confusing information from each set while leaving intact that information required by each of the respective disciplines. However, for this to be effective there must be rigorous control of the respective tracings so that changes or revisions to one tracing appear simultaneously on the other if required. This becomes a difficult task if responsibility for the basic documents, i.e., the engineerng flow diagrams and the instrumentation-application diagrams, is divided between two discipline engineers and their respective designers and interaction between them depends on the respective resulting documents.

If, on the other hand, a computer graphics system (CGS) or a computer-assisted drafting and design (CADD) system is available, it can be used to facilitate preparation of separate engineering and instrumentation flow diagrams. In this case, all information stored in the computer for replication on both the engineering and the instrumentation flow diagrams is internally coded to show up in full tone, muted or dashed, or not at all depending on the flow diagram that is being drawn. This technique, known as "layering," permits the preparation and simultaneous updating of both the engineering and the instrumentation flow diagrams whenever a change is made in one that affects the other.

---

*It was noted earlier that some groups refer to piping and instrumentation drawings as engineering flow diagrams. In any event, process engineers' responsibilities are the same for both documents.

Electrical engineers and engineering flow diagrams

The electrical engineering group needs the piping and instrumentation drawings to begin its work. Probably the most useful information for the electrical group is the identification of all motors on the project. Their enumeration on the engineering flow diagrams permits the electrical engineers to plan the motor-control centers and required electrical substations as well as to begin the distribution and wiring drawings.

Other typical information of interest to the electrical engineering group shown on the engineering flow diagrams is electrical tracing of piping or equipment, electrical instrumentation, location and numbering of running lights, limit switches, hand switches, recorders, alarms, interlocks, and specialized equipment which requires electricity for its operation. The P&ID also indicates whether a microprocessor or a programmable controller is to be included in the project and which units are to be interlocked.

The electrical engineers must use the information contained on the initial preliminary engineering flow diagrams with caution, particularly with respect to the motor-control centers. The motor horsepowers on the initial flow diagrams are only first approximations by the process engineers which will be refined as the piping is sized and vendors respond to the specifications developed for the equipment.

Structural engineers and engineering flow diagrams

The structural group probably has the least relationship of all the engineering disciplines on the project to the engineering flow diagrams. However, there is considerable interaction between process and structural engineers in planning equipment layouts within the limitations imposed by structural constraints. Process engineers must ensure that as much working and equipment area is allocated at the start of the project as is feasible. If not, the result might be an excessively cramped plant which could be difficult to operate and maintain.

The engineering flow diagrams themselves provide the structural group with several important sources of information for their work, e.g., for drains, sumps, dikes, berms, or curbing as well as cranes and hoists when pertinent to the process. The types of supports for major pieces of equipment are sometimes indicated, and special clearances between pieces of equipment as well as equipment liftout or lay-down areas should be noted on engineering flow diagrams as guides for the structural group.

Preparation of Engineering Flow Diagrams

As mentioned previously, primary responsibility for preparation of engineering flow diagrams rests with the process engineer. If the project is large enough, it may be advisable to assign two or more process engineers, each responsible for the engineering flow diagrams relevant to different sections of the plant. It is important to note that "large enough" is not necessarily measured in terms of dollars or even of real estate but in numbers of major pieces of equipment. In the MgO regeneration facility referred to in the subsection "Block Diagrams" of Chap. 2, there are more than 150 pieces of major equipment such as tanks, columns, reactors, pumps, blowers, compressors, and exchangers. The process engineering work was staffed by two senior engineers, who were each responsible for about 10 engineering flow diagrams. They were assisted from time to time by junior engineers.

Informational basis of engineering flow diagrams

For most projects, the process flow sheet is the main source of information for the preparation of engineering flow diagrams. Although it gives only the basic elements of the process, the function of process engineers is to flesh out this skeleton and develop the engineering flow diagrams.

Process engineers can always call upon their own experience or that of their colleagues with respect to similar processing equipment and situations to complete the engineering flow diagrams. If a company or a group has an archive of engineering flow diagrams from past projects, a review of the pertinent diagrams can refresh the memory or provide necessary guidance for portions of the project at hand. In that event, it is important that the process engineers be aware of any proprietary information on the older diagrams and not violate previous confidential agreements.

In other instances, the project may be a near duplication of an existing process at the particular site or at another plant. The engineering flow diagrams for the original plant can be used as the starting point for flow diagrams for the processing sections of the new plant. However, caution must be exercised in using this approach since newer materials of construction may have been developed for use in equipment or piping, more efficient types of equipment may be available, more sophisticated instrumentation may be attainable, and the utilities for the current project may not be the same in temperature, pressure, or quality. It is always the wisest course of action to use older flow diagrams only as guides to the required piping. The process engineers for the current project should always size the equipment and piping anew to guard against errors that may have been committed in previous designs.

For those areas of flow diagrams for which there is no precedent or previous experience, process engineers can obtain guidance from text or reference books, equipment literature, or vendor representatives.

In the remaining portions of this subsection the formats of engineering flow diagrams will be discussed, and important engineering considerations for various types of equipment as they affect flow diagrams will be pointed out.

Formats of engineering flow diagrams

It cannot be stressed too often that engineering flow diagrams must show diagrammatically every piece of equipment and every pipeline and duct as well as every instrument* required for the project. Several formats have been evolved to represent this information. These will be presented and evaluated.

In each of the formats, the equipment, instrumentation, and process piping are shown in the center portion of the drawing. In all cases, process streams leaving the sheet to be continued onto another drawing are brought to the edge of the one sheet and started on the subsequent sheet at approximately the same level. Information

---

*This is so when a P&ID is also termed an engineering flow diagram. See subsection "Use of Separate Engineering and Instrumentation Flow Diagrams" for instrumentation to be included when separate flow diagrams are used.

should be given at the edge of the originating sheet regarding the drawing where the line is continued and its destination, and vice versa.

The major difference between particular conventional engineering-flow-diagram representations involves the presentation of utilities and the location of line tabulations. Sometimes all the utilities are carried across the entire width of the processing sections of the engineering flow diagrams at the top or bottom of the drawings. This method of displaying utility distributions has the advantage of showing graphically which utilities are required on each drawing and is useful when the number of utilities or their users is limited. However, this method can require excessive space at the top or bottom of engineering flow diagrams to show the runs of utility lines regardless of whether or not they are required on the diagrams. Locating the equipment on the diagrams may be a problem if distribution of the various utilities represented is to be realistic.

An alternative method of showing utility distributions is to divorce them from the processing sections completely and indicate on the individual processing engineering flow diagrams only those utilities that are respectively used. This is accomplished by having a utility shown only at its individual users, with the distribution network being given on a utility flow diagram. Typical of this method is the engineering flow diagram shown as Fig. 3.2 in the subsection "Development of an Engineering Flow Diagram." Several utilities, namely, process air, city water, steam supply, river-water supply and return, and bearing and seal water, interface with this flow diagram. In most instances, utility lines originate from a convenient level at the near right or left margin, whichever is closer. An exception is bearing- and seal-water supplies, which are shown as a box next to the equipment to which they apply. A note refers to a utility diagram showing the details of the bearing- and seal-water piping and instrumentation. Another acceptable method of preparing engineering flow diagrams is to treat each utility as a box next to its user rather than bringing it from a margin.

The second major difference among formats for engineering flow diagrams is that of placement of line tabulations. The tabulation is an integral part of the information contained on an engineering flow diagram, and its composition is discussed in the subsection "Line Tabulations." Some formats carry line tabulations as separate sheets or smaller drawings which are auxiliary to the flow diagram itself. Other types of engineering flow diagrams contain line tables as an integral part of the drawing. Figure 3.2 is an example of this format. Although the table occupies a large portion of the available space on the drawing, it is the consensus among process engineers that this type of presentation has advantages that make it preferable. Its greatest advantage is that the table is always available with the drawing for reference when required. When line tabulations are separate sheets (an average project could have 50 to 75 individual tables), they are usually stapled together and carried separately as a packet. Having the tables as an integral part of the drawings facilitates making changes in the line tabulations as revisions are made to the drawings themselves. The greatest advantage, however, is the certainty that a tabulation, when part of a flow diagram, is current with the revisions to the drawing. This format avoids the possibility of mixing the line table of an earlier revision with that of a current flow diagram.

Development of an engineering flow diagram

The layout of an engineering flow diagram should be planned so that it is not crowded or could become crowded owing to later changes or additions. It is preferable to have too much open space on a flow diagram rather than to have a crowded drawing which is confusing and can be misinterpreted. A rough rule of thumb is to limit the number of main pieces of equipment such as tanks, vessels, reactors, or columns to about three or four for the processing section of the project on any engineering flow diagram that measures about 30 by 40 in. These pieces with their accompanying pumps and exchangers, if of a reasonable number, make a balanced and uncrowded presentation. For smaller-sized sheets (24 by 36 in), the number of main pieces of equipment may have to be limited to one or two.

An engineering flow diagram should show the relative elevation of process equipment even if it can't give plan relationships. A "grade line" should be established on the drawing to provide a reference for the approximate elevations. It should be drawn 3 to 5 in above the line tabulations if they are on the diagram or from the bottom of the drawing if the tabulations are separate. This leaves space for sumps or underground lines to be shown and still allows room for equipment-identification names and numbers as well as basic technical descriptive information.

The main pieces of equipment such as vessels, tanks, columns, or other large items should be spaced equally across the diagram. Other pieces of equipment associated with them, i.e., pumps, exchangers, filters, conveyors, in-line mixers, etc., are then placed in approximate proximity to the larger equipment as would be expected in the plant. The grade line is used as the base for such items as pumps or exchangers where applicable.* Connecting piping and instrumentation are added as required.

Figure 3.2 is offered as an example of the method by which engineering flow sheets are developed. It was prepared for the waste-treatment section of the MgO regeneration plant and illustrates some of the points regarding the development of flow sheets. The sheet represents a portion of Sec. 700 in the block diagram in Fig. 2.2 and some of the main processing steps outlined in a sub-block diagram, Fig. 2.3. The original flow diagram has been altered to some extent to provide additional examples of points to be discussed regarding engineering flow diagrams.

All equipment should be represented on engineering flow diagrams by configurations, shapes, or symbols which suggest the individual equipment itself. Journal articles, texts, and references present various symbols for equipment, but no engineering society or authoritative body has yet originated a group of symbols to be adhered to by industry. Thus, each design organization develops its own symbols and adds to or changes them for new projects.

Figure 3.2 contains symbols typical of those used on the MgO regeneration project. Abstracted on Fig. 3.3 is a series of symbols, by no means complete, that is being used by a large design and construction firm in its computer-assisted design and drafting operations.

Equipment should be entered on engineering flow diagrams according to relative size where feasible. A linear scale is not practical. If an exchanger 1 ft in diameter is represented by a 1-in symbol, it is unrealistic to show a 20-ft-high storage tank as 20 in tall on the drawing. Any arbitrary scale may be used as long as it conveys a sense of proportion. A convenient one is based on the square root of the dimensions. Thus, if a 1-ft-diameter exchanger is shown by a 1-in symbol, a 20-ft tank is about $4\frac{1}{2}$ in high (the approximabe square root of 20).

Regardless of the scaling factor, any piece of equipment may be represented out of scale to provide sufficient space in and around it to accommodate the necessary piping

---

*It may be necessary or convenient to place a piece of equipment elsewhere than on the main grade line. In this event a baseline should be shown for that piece.

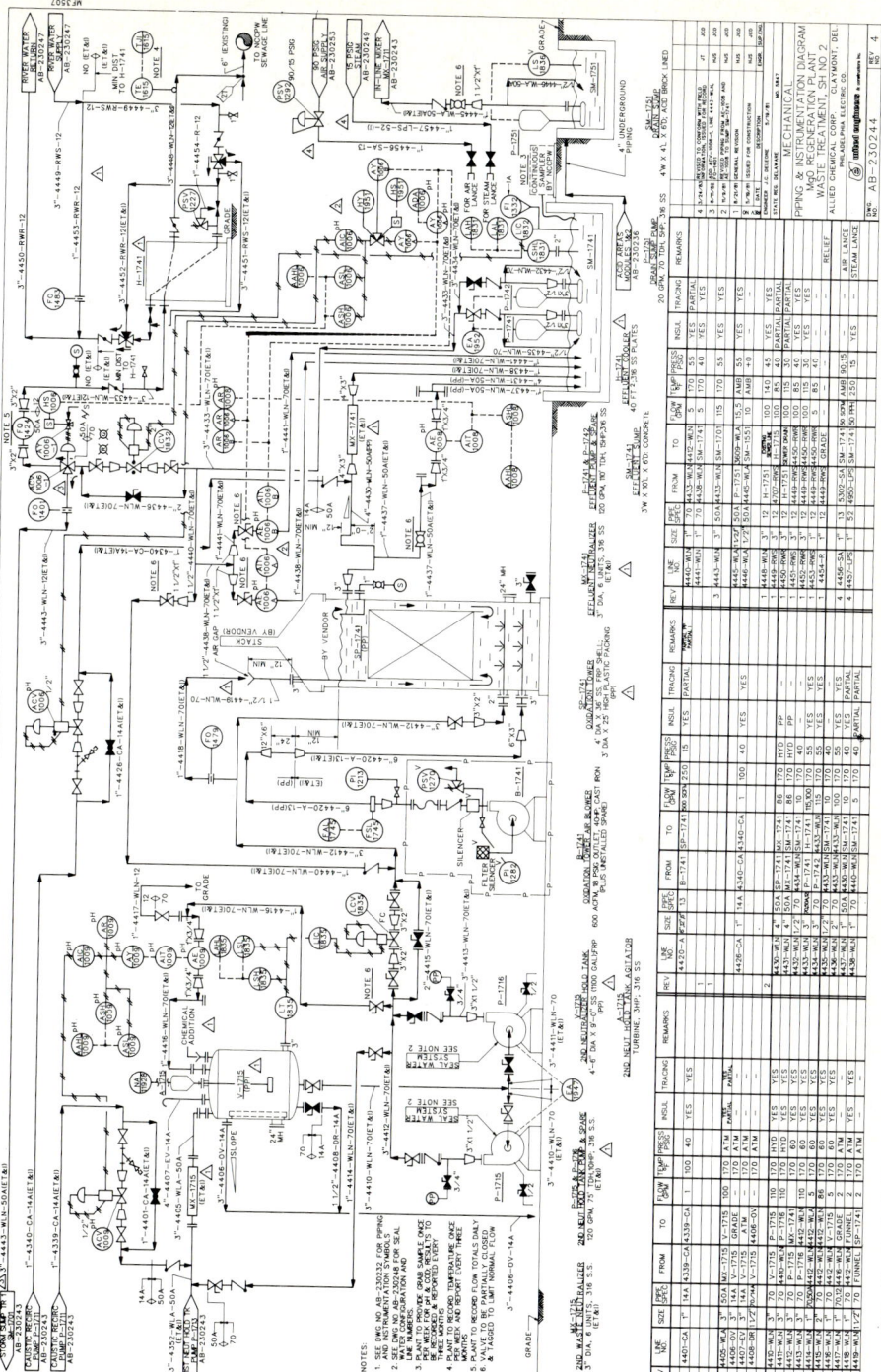

Figure 3.2 Typical piping and instrumentation drawing. Equipment shown in the foreground of the cover photograph is depicted in this flow diagram. (*Adapted with permission of United Engineers & Constructors Inc., Philadelphia.*)

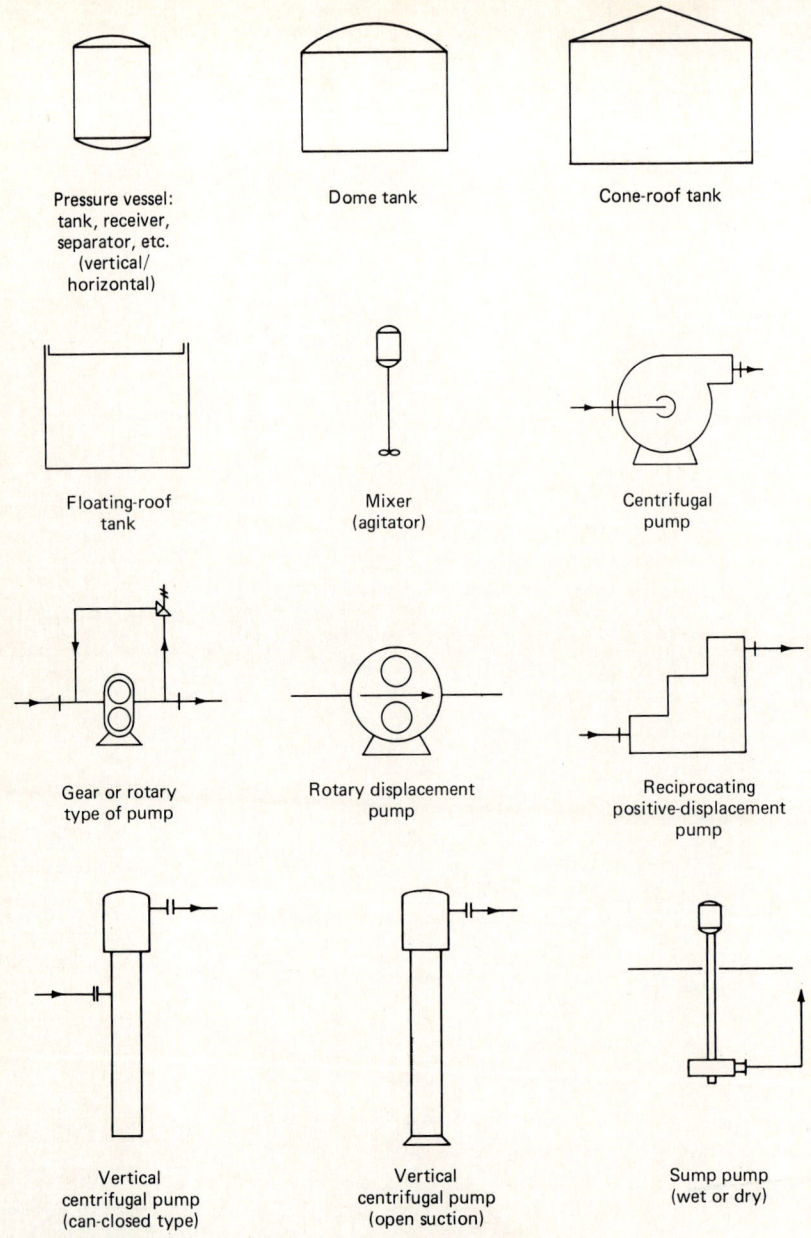

Figure 3.3 Typical equipment symbols.

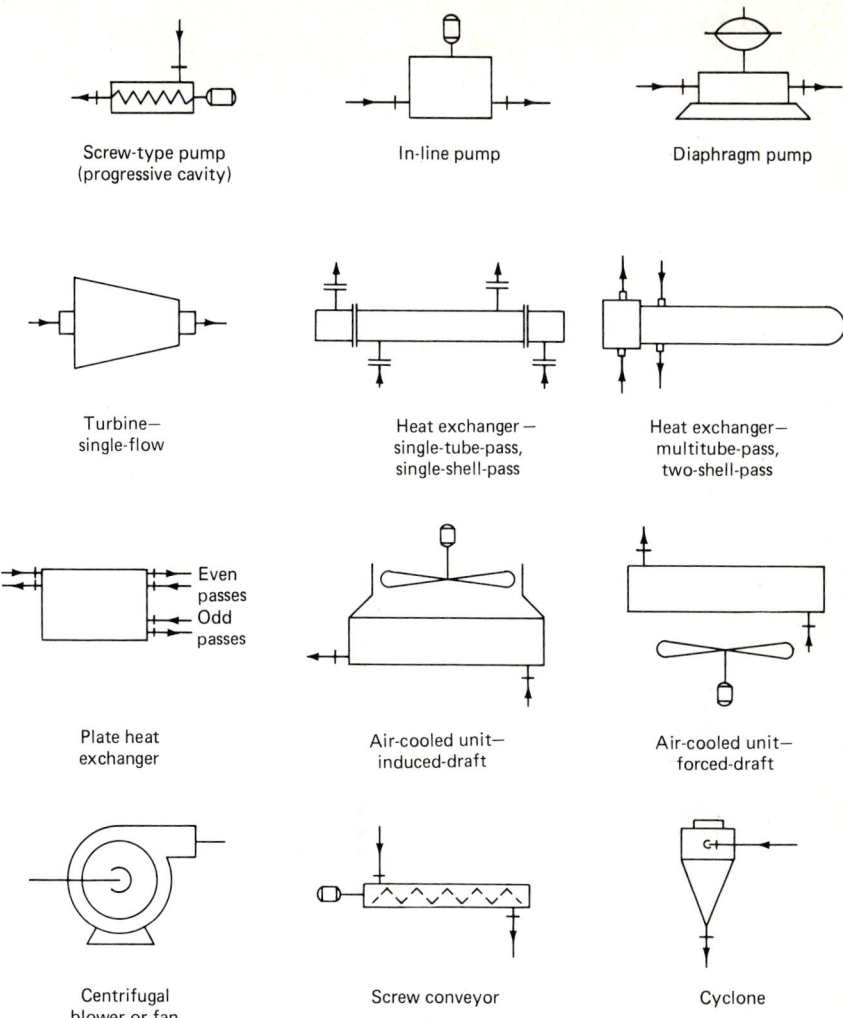

and instrumentation connections as well as other appurtenances plus identification information and avoid crowding or confusion. There may be only a limited area in which to position additional required equipment due to a revision without excessive rearrangement of large portions of the drawing. In this case, it is acceptable to insert the symbol in the available space regardless of scale as long as the accompanying information is clear. It will be understood that this is an exception to the scaling process, particularly if the equipment identification number and name are accompanied by dimensional information in the short technical description.

Equipment identification

Each piece of equipment for a project is usually assigned an item number which is a combination of alpha and numeric terms plus a name to give it a unique identification code.

Item numbers. The alpha portion of the identification code is usually descriptive of the nature of the piece of equipment itself. Typical letter combinations for the MgO regeneration project are given in Table 3.1. Thus, the types of equipment presented on Fig. 3.2 can easily be understood from the pictorial representation plus the alpha portion of the respective codes.

TABLE 3.1 Typical Item-Number Descriptions

A	Agitator	P	Pump
B	Blower	RV	Rotary valve
C	Compressor	SC	Conveyor-cooler
H	Heater, boiler	SM	Sump
LB	Live bottom	SP	Specialty product
MX	In-line mixer	T	Tank, column, vessel

Various sources have suggested sets of alpha identifications, but none have been ever established as a standard. Thus, it is rare to find that two operating or design companies use identical codes in their work.

The numeric portion of equipment identification is generally a three- or four-digit number. Three digits are normally sufficient since there are usually fewer than 99 pieces of equipment in a section of the block diagram of the process.* The fourth digit of the code, generally a prefix but possibly a suffix, usually has a special meaning or purpose of its own. For example, the initial digit of the four numbers for all the equipment on Fig. 3.2 is a 1. This indicates that the equipment is part of module 1 of three nearly identical modules. The equipment in the other two modules was identified by prefixes 2 and 3 respectively.

The second digit for this project or the first in a three-digit series is equivalent to the process block-diagram section. The 7 as the second digit in the code for equipment on Fig. 3.2 indicates that the particular piece is a part of the waste-treatment Sec. 700 on Fig. 2.2. The last two digits are numbers between 01 and 99 which give each piece of equipment its own unique identification. Some design groups do not repeat these last two digits in a given block-diagram section. However, it is generally accepted that two-digit numbers can be duplicated within a section as long as the alpha portion of the code is different.

Some projects elect to use a numerical code only in accordance with an index set up by the accounting department prior to the start of engineering. Two or three initial numbers can be used, for example, to identify a portion of the project equivalent to a section in a process block diagram, while the last three digits are unique to the particular piece of equipment in that section.

An alpha or a numeric digit is sometimes used as a suffix to an item number to indicate a spared unit or to break down individual units in a set of equipment purchased as a package.

---

*The number X00 is rarely used as an equipment-identification code.

The item number for each piece of equipment should be entered within the equipment symbol or immediately adjacent to it and underlined. It is then repeated at the bottom of the diagram in the area left vacant below the grade line. The combination may be placed in line with the piece of equipment or offset somewhat to accommodate all the codes within the available space. The item numbers and the names in this area are underlined.

Some design groups choose to place the item-number–name combinations across the top of the diagram, while others locate them throughout the sheet near the equipment. If these latter conventions are used, the grade line can be lowered. Showing the item-number–name combinations across the bottom or top of the diagram facilitates locating equipment on a drawing.

Item-number assignments. Item numbers for equipment in a given processing section or in the utilities preparation and conditioning portions of the process should be assigned in a logical manner to assist in locating the equipment quickly on the flow diagrams. One of the duties of the process engineer is that of initiating the "Equipment List," an index of all equipment on the project by item numbers and name, and keeping it up to date.

The item numbers should be assigned sequentially in the order of flow progress through the process. When use is made of the alphabetic portion, it is an accepted procedure to assign a set of numeric digits to a major piece of equipment such as a tank, a column, or an exchanger and then repeat the same digits as part of the item numbers of all the peripheral equipment as long as none of the alpha portions is duplicated. In the event of a set of spared equipment, the same item number may be used for both but with a suffix such as A added for one and B for the other. Often the suffix is omitted, and the next item number in the series is used for the second piece.

To provide item numbers in proper order for spares and to accommodate equipment which might later become part of the project, it is best to assign item numbers in clusters. It is good practice to set aside 10 item numbers for use about each major piece of equipment when the flow diagrams are initiated. If, however, a separate set of numeric digits must be assigned to each piece of equipment regardless of type or designation, the process engineer must limit unused numbers between clusters.

Similar care should be exercised when assigning item numbers to equipment associated with the preparation or conditioning of utilities. Since there will usually be several or more types of utilities associated with the project, it is best to treat each type as if it were a piece of equipment and assign 10 or 20 item numbers to a utility according to its requirements.

Once an item number has been assigned to a piece of equipment, it is imperative that it remain with that piece as long as the paperwork associated with it remains intact. The integrity of the item number for a piece of equipment should be maintained even if it becomes necessary to relocate the unit on another flow diagram. A piece of equipment may, of course, be deleted from the project owing to revisions in the process. Its item number then is deleted from the records and paperwork. Item numbers should never be reassigned to another piece of equipment until all other available numbers have been used.

Names. Accompanying the item number in the bottom section is the name by which the equipment is referred to in all matters dealing with the project. The name should be relatively short but descriptive of the type of equipment and its function in the project. Every attempt should be made to see that each piece has its own unique name in order to avoid ambiguity. The equipment names shown in Fig. 3.2 are typical. Vessel V-1715 is termed a "Neutralizer Hold Tank" since it provides retention time for the neutralization step which takes place in it. However, since there is a prior neutralization step, differentiation is achieved by adding the term "1st" or "2nd" to the respective vessel. It should also be noted that P-1715 and P-1716 are both termed "2nd Neut. Hold Tank Pump," but one is a spare for the other.*

Short descriptions. Many design groups accompany the identification code and name with a succinct description of pertinent physical or design information regarding the equipment. Typical information presented for several common types of equipment is:

Vessels: materials of construction, volume, dimensions

Pumps: materials of construction, flow rate, pump head, brake horsepower

Compressors: materials of construction, actual volumetric rate, inlet or discharge pressure, compression ratio, brake horsepower

Exchangers or heaters: materials of construction (both sides), dimensions, duty, transfer area

Columns or towers: materials of construction, dimensions, type of packing or distillation trays (Note: Simulated packing or trays are usually shown on equipment symbols, with end, feed, and draw-off trays being numbered or heights of packed sections indicated.)

In-line mixers: materials of construction, diameter

Instead of positioning the descriptive information directly below the item number and name, the descriptive material is sometimes placed in tables on the drawing, usually at the bottom of the tracing or above the revision boxes if they are located at the right or left margin. In this case, the equipment is usually listed by item number in a table for the respective categories of equipment on the diagram. The use of tables often permits the inclusion of additional data regarding the equipment such as design pressure and temperature, operating temperature and pressures, or number of compressor stages.

Information regarding required equipment heat tracing and insulation is indicated below the item-number–name combination. Examples of this can be seen for the MgO regeneration project in Fig. 3.2 with the explanation for the abbreviations used being found on the lead sheet (see Fig. 3.1).

When all the descriptive information is not available initially, space should be allocated for it. Likewise, process engineers may enter approximate values for dimensions and correct them at a later date.

Piping identification

The analogy between engineering flow diagrams and road maps can be continued by considering the piping or tubing to represent the various routes interconnecting the equipment or instruments as salient features on a map. A map uses different types of lines for various kinds of highways and roads. Similarly, flow diagrams employ different-value lines to express process piping, pneumatic lines, pneumatic tubing, etc.

---

*By a different convention they could have been designated P-1715-A and P-1715-B.

(see Fig. 3.1). Major highways on a map are usually emphasized with somewhat wider lines than those used for secondary roads. Similarly, it is customary to use a broader line to represent the main process-flow piping than that used for subsidiary lines. To avoid confusion process engineers must exercise care that not too many lines are emphasized as process lines. For example, the main function of the MgO regeneration process is to prepare solid MgO and gaseous SO_2 from the incoming $MgSO_3$. The waste-treatment section depicted in Fig. 3.2 does not show any bold process lines since neither recoverable MgO nor SO_2 is conveyed in any of its lines.

Piping formats. The analogy between flow diagrams and maps ends with piping routing and line numbering. Distance on a road map can be obtained from the length of the route and the scale of the map, but flow diagrams have no scale. Thus, the relative length of lines on the diagrams is not indicative of their actual lengths in the completed project. For example, a line between a water-metering station and a tank is only several inches long on a utilities flow diagram, yet it can represent several hundred or thousand feet of piping between a city-water takeoff at a highway and the process-water break tank. Conversely, a crossover between a point in the center of one flow diagram and another on a second sheet could be almost 2 ft in length on the diagrams. Yet, in reality, the line may not exist at all since the two process lines may be joined by mating the flanges of tees in the respective lines for the crossover. However, it is necessary to show a separate line on the flow diagrams to lead the designers from one flow diagram to the other. In addition, showing a separate line number ensures the proper connection since process engineers have no means of knowing the proximity of the various lines when the flow diagrams are in preparation.

The piping portions of flow diagrams are initiated by connecting the various pieces of equipment with the necessary piping in order to provide the desired flow patterns to accomplish the objectives of the process. Other lines are then added to interconnect the processing lines as required. Additional lines are placed on the diagrams to provide the necessary recycles, bypasses, vents, reliefs, drains, and other secondary lines required to finish the piping requirements.

Line-description procedure. Each line on engineering flow diagrams must be identified to show its size, have a unique line number assigned to it, indicate the fluid contained in the pipe, register the pipe's construction-material specification, and denote information regarding any heat tracing and insulation. These elements are needed to ensure clear definitions of the various lines for use by the designers in the various engineering disciplines as well as for preparation of the specifications for the project. A typical identification set, as used for the MgO regeneration project, is given in Fig. 3.1. Its application may be seen in Fig. 3.2. In this case, the order of the information is that listed above. Other design groups may choose to present the same information in a different order.

Line sizes. The various pipes are sized by the process engineers in accordance with the principles and procedures outlined in Chaps. 6 through 9. The nominal line size given to a pipe at its point of origin is shown in the line-designation set and is carried to its end unless there is an indication that a change in pipe size is required for the process or for connection to a piece of equipment. The symbol most frequently used to indicate a change in line size is the "reducer." A reducer is usually represented by the configuration in Fig. 3.4a if the line size is being reduced or by that in Fig. 3.4b for instances when the line size is being increased. The application of these configurations can be seen on several lines in Fig. 3.2.

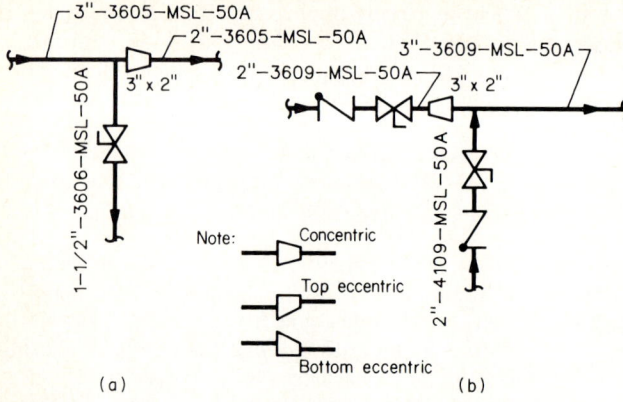

Figure 3.4 Line-size reductions.

The full line-identification code should be entered on the line on both sides of the reducer symbol to ensure clarity. This duplication is not required when the reduction or increase takes place at a nozzle or connection to a piece of equipment. In this event, the reducer symbol is placed immediately adjacent to the equipment connection and the size of the nozzle or connection placed next to it, with the line-identification set being located on the line close to the connection. The 3-in line 4412-WLN connecting to a $1\frac{1}{2}$-in nozzle on P-1715 in Fig. 3.2 is an example.

If a line of a given size starts or terminates at another line of a larger size, it is normally not necessary to use the reducer symbol on the smaller line near the interception point on the flow diagram. The identification sets for each of the respective lines should be so located that there is no confusion regarding the sizes of the two lines or which line is the continuous one. An example of this procedure can be seen in Fig. 3.2, in which line 1″-4416-WLN branches from line 3″-4412-WLN just after pump P-1716. However, reducer symbols should be included if they are required for clarity.

A size change may take place owing to a flow split or an additional flow entering a line. Great care should be exercised in showing these intersections on a flow diagram to ensure that it always is clear which of the lines is the continuous one. It is best to indicate the continuous line as being the 180° portion of the intersection, with the other line being perpendicular to it. This avoids misinterpretation owing to the eye following the wrong line if the identifications cannot be placed close to the intersection. An illustration of these points is given in Fig. 3.5.

Line number. All lines representing piping in engineering flow diagrams, with the exception of nipples for instrument root valves, are assigned distinct and unique line numbers. Separate line numbers are given in the following general order to lines which originate and end in the following typical situations:

1. From equipment to equipment
2. From equipment to another numbered line
3. From a numbered line to a piece of equipment
4. From a numbered line to another numbered line
5. From equipment or a numbered line to or from the atmosphere

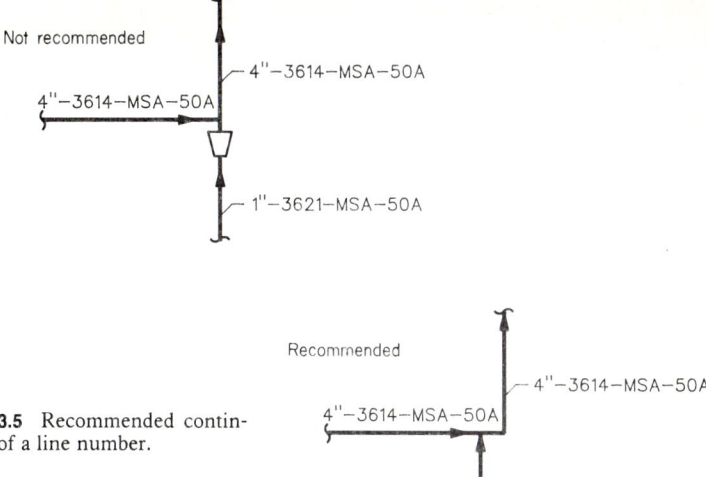

Figure 3.5 Recommended continuation of a line number.

6. From or to equipment or lines to or from the boundary limits
7. From equipment or lines as drains or vents unless a drain or vent valve is directly connected by means of a flange or a pipe nipple to the point to be drained or vented

Identification portions of a line number. A practical method of composing a line number is to have it contain two sets of digits. The first set serves to reference lines originating on an engineering flow diagram to that drawing. The second set makes the line number unique and identifies it as a particular line on the flow diagram to which the first set is referenced.

Initial digits. A convention in widespread use is to relate the initial digits in a line number to the engineering-flow-diagram drawing number. This is generally taken as the last two digits of the diagram's drawing number. Two digits are usually sufficient since there are rarely more than 90 engineering flow diagrams on a project. It is recommended that the digit sets 00 through 09 not be used as the initial digits for a line-numbering system since line numbers starting with the digit 0 are awkward to refer to and may also be confusing.

Since the line-numbering system depends upon the drawing number, care must be exercised in assigning drawing numbers to the drawings on the project. It is recommended that a block of drawing numbers be set aside for the engineering flow diagrams so that the numbers are sufficient to allow for the required initial flow diagrams plus several spares in sequential order with none ending in 00 through 09.

Figure 3.2 is typical of this type of relationship between drawing numbers and initial digits for line numbers. The drawing number for the diagram in the figure is AB-230244. Thus, all lines originating on the diagram are termed 44--. This procedure aids in determining the drawing on which a line originates.

An exception to referencing the initial digits of line numbers to the drawing of origin is that of utility or auxiliary chemical feed lines. In these cases it is good practice to assign line numbers according to the engineering flow diagram where the lines terminate since data regarding utility or chemical usage is more frequently required at the user than at the distribution header. All other utility or chemical lines involved in their preparation, conditioning, or storage as well as their distribution headers should be assigned a line number in accordance with the drawing number of the respective utility engineering flow diagram where they originate.

If the drawing numbers assigned to the flow diagrams are not suitable, in that the last two digits of the drawing could not serve as the first two digits of the line numbers, a substitute artificial numbering system should be used for the engineering flow diagrams.

Last two digits. The use of two digits at the end of a line number enables 100 lines, --00 through --99, to be given separate and unique identification numbers on any given engineering flow diagram. In most cases this is sufficient to cover the requirements for line numbers on any sheet, particularly if the number of pieces of equipment placed on the diagram is limited. Usually considerably fewer than 100 lines are on any one engineering flow diagram. Thus, gaps between sets of numbers can be left for clarity so that additional line numbers related to new lines can be entered within the numbering section.

If, however, an engineering flow diagram has more than 100 lines, it is necessary to devise a means of assigning numbers to all of them and, at the same time, to ensure that each has its unique identification. This situation often arises when multiple similar units are placed on one flow diagram. In this case, a large number of lines have the same respective size and service, differing only with reference to the individual item with which they are associated. Thus, the lines about the first unit could be given individual line numbers but have the suffix -a. The corresponding lines about the second unit would carry the same line numbers as their counterparts about the first one but have the suffix -b instead. The procedure is repeated for other similar units. The listings in the line tabulation reflect all the suffixes for the one numerical line number.

At other times, there could be more than 100 essentially unrelated lines on a flow diagram. In this case, an expedient of converting line-identification numbers from four digits to five digits by using a prefix such as 1 for the excess lines can be utilized.

The use of more than 100 line numbers on any engineering flow diagram causes no problems with respect to line tabulation if separate tabulation sheets are used. In this event, extra sheets are inserted behind the original ones associated with the given engineering flow diagram. When line tabulations are entered across the bottom or along the side of the diagram,* additional tables are entered in the body of the drawing where convenient. It may be necessary to relocate lines on the flow diagram to make space for the additional tables.

One and only one line number is usually assigned to a given line from the point at which it starts to the point at which it ends. Thus, in Fig. 3.2, line 4352-WLA originates on Drawing 230243 and continues onto Drawing 230244, retaining its 4352-WLA number until it ends at MX-1715. The particular stream continues beyond MX-1715; however, the line number changes at a piece of equipment, and it emerges from MX-1715 as 4405-WLA since that line starts on FS-230244.

There are occasions when the number of a line may be changed during its run. This is done for clarity. For example, in the hot-oil system for the MgO regeneration

---

*Usually, there are spaces for 60 lines in the normal tabulations on a C- or D-size tracing and 100 on an F-size tracing.

project oil was taken from an oil storage tank by a rotary pump and passed through a heat exchanger to a header from which oil was withdrawn to the calciners and fired heaters before the unused portion of the pumped-oil stream was recirculated back to the storage tank. To maintain the oil to the users at a required pressure, a restriction orifice was placed in the circulating line just after the last user takeoff. To ensure an absence of confusion between the high- and low-pressure portions of the line, the line number was changed at the restrictive orifice and so noted with a change mark on the pertinent utility flow diagram as shown in Fig. 3.6. The portion of the line up to the restrictive orifice is called a hot-oil supply line and is known as 4638-HOS. After the restrictive orifice it is termed a return line and is labeled 4639-HOR.

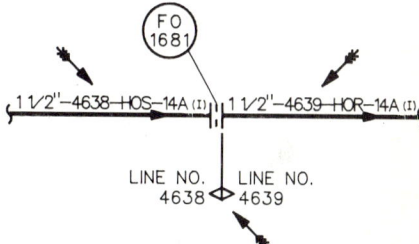

Figure 3.6 Changing line numbers in the same pipe run.

Piping-system designation. In the line-designation system the service of the particular line should be indicated. This service usually is expressed as several characters, generally alphabetical, which are an abbreviation of or stand for the fluid being transported or contained in the line. The piping-system designations (also known as piping-service designations) used for the MgO regeneration project, which are listed in Fig. 3.1, are typical of the type of contractions that are employed. For services such as air or steam, the nominal pressure level, usually expressed as pounds per square inch gauge (psig), is often given as a number following the alphabetical designation.

The piping-system designation is usually retained for the entire length of the line to which a given line number is assigned. This is particularly true of designations expressing process services. However, a utility line such as air or steam changes pressure level as it passes through a pressure-control valve and thus changes service designation also. This is illustrated in Fig. 3.7.

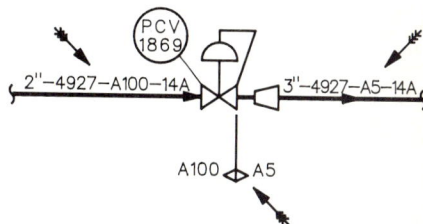

Figure 3.7 Changing the pipe-service designation for a pipe run; 100-psig plant air is reduced to 5 psig with the line number remaining the same. Line size is increased to accommodate the greater specific volume of 5-psig air.

TABLE 3.2 Typical Piping Specifications for Figs. 3.8 and 3.14

System designation	Service	Material of construction	Temperature, °F	Pressure, psig	Specification class
WLN	Waste liquor neutralized	Carbon steel	225	150	12
CA	Caustic (20–50% NaOH)	Carbon steel (all welded)	300	150	14A
DR	Drains	Carbon steel (all welded)	300	150	14A
EV	Vents	Carbon steel (all welded)	300	150	14A
DV	Overflows	Carbon steel (all welded)	300	150	14A
WA	Waste acid	Stainless Type 316L	250	150	50A
WLA	Waste-liquor acid	Stainless Type 316L	250	150	50A
LPS	Low-pressure steam	Carbon steel	350	150	52
ACD	Sulfuric acid (2–20%)	Carbon steel, TFE-lined	200	150	70

Piping-specification class. Of particular interest to mechanical designers who translate engineering flow diagrams into the piping and isometric drawings which show the actual routing of the pipes is the piping-specification-class information that is contained in the line-identification sets. The primary function of the piping-class designation, usually several alphanumeric characters, is to indicate the specification for the pipe and fittings. A piping specification itself is a listing which details the materials of construction for the pipe and fittings as well as the maximum temperature-pressure relationship for which the particular piping material may be employed.

Each piping specification carries its own set of piping thicknesses or schedule as a function of pipe size and the types of fittings, valves, and flanges that are to be used for the various sizes of pipe as well as the requirements for bolts, nuts, and gaskets for the given set of temperature-pressure* conditions.

Preparation of piping specifications is a joint effort of the process engineers and the mechanical engineers in charge of piping on the project. The process engineers determine the material of constuction and advise the mechanical engineers regarding the several temperature and pressure levels at which the fluids in their various states will be encountered in the project as well as the types of valves that might be required for each set of conditions.

The piping-specification designations are usually omitted from the lead sheet since a set of piping specifications is circulated to every engineer and designer on the project when they are prepared and they become part of the mechanical specifications for bidding and constructing the project. Table 3.2 presents a summary of relevant information for several typical specifications which were used for the MgO regeneration project and may be employed to interpret the line-identification sets in Fig. 3.2.

A line or a portion of a line is assigned a specification class in accordance with the most stringent materials of construction for the fluid and temperature-pressure con-

---

*The normal interpretation of the combination is that the stated pressure is the maximum allowed at the given temperature. The piping could be used at a higher pressure if the maximum process temperature were lower or at a higher temperature if the maximum process pressure were lower. However, the combination should be checked with the piping mechanical engineer before applying the specification for a condition that exceeds the listed temperature or pressure.

ditions to which it may be subjected. Thus, different portions of a line may carry different specifications if there is a change in fluid within the pipe or a change in the temperature or pressure values to warrant it.

The change in materials of construction for a line takes place when a second fluid with more stringent specification requirements enters the original line or because the original line ties into another line or piece of equipment carrying fluid which requires a higher specification so that the original line must be protected for a portion of its length. The change in line specification due to an actual or potential fluid change usually takes place at a control valve, a shutoff valve, or a check valve. The valve and the downstream portion of the line are included in the higher-specification section. In the event of gravity flow from a line with a low specification through a dip leg into a vessel with a fluid carrying a higher material specification, a specification change is made where the dip leg begins.

A convenient method of showing a change in line specification consists of using a suitable marker at the point where the change takes place and repeating the line identification on both sides of the marker. A typical marker, as used on the MgO regeneration project, is defined in the pipe-specialties section of its lead sheet (see Fig. 3.1). Examples of the various types of specification changes making use of this marker are found in Fig. 3.2. Several examples from the diagrams on the project are simplified and repeated for illustrative purposes in Fig. 3.8.

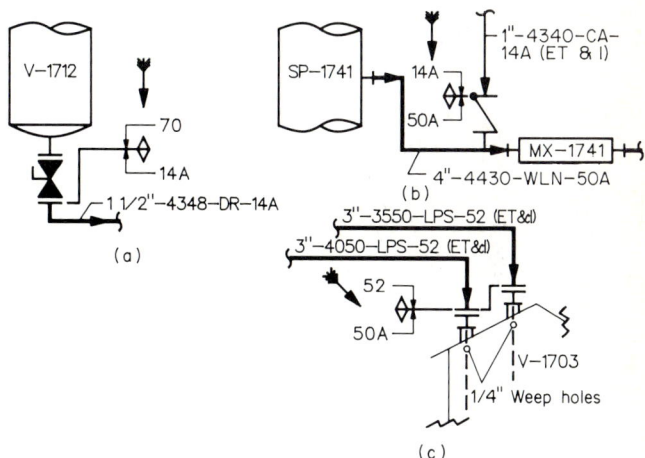

Figure 3.8 Specification change in pipe runs.

In Fig. 3.8a, the drain valve on tank V-1712 must be Type 316 stainless steel in accordance with Specification 70 since the fluid in the tank is mildly acidic. However, the drain line itself after the valve may be carbon steel, Specification 14A, since the tank is rarely drained and carbon steel is satisfactory for intermittent contact with the mildly acidic fluid. The saving over a stainless-steel line justifies the small risk of a leak occurring in the carbon steel line. The specification-change marker is placed just downstream of the valve. Since the valve is immediately adjacent to the tank, there is no need to place a line-identification set between the marker and the tank. One is obviously placed on the line downstream of the marker.

The carbon steel line 4340-CA (Specification 14A) in Fig. 3.8*b* brings caustic from the control valve ACV-1006 to line 4430-WLN to neutralize any acid in the overflow line from the Oxidation Tower SP-1741. Because of the pressure drop across the in-line mixer MX-1741, the acid medium could be forced into line 4340-CA as far as the check valve. Therefore, the check valve and that portion of line 4340-CA downstream of it must be Type 316 stainless steel, Specification 50A, as shown by the specification change.

Lines 3550-LPS and 4050-LPS carry boiler blowdown to the Deionizer Effluent Tank V-1703, as shown in Fig. 3.8*c*, and may be carbon steel, Specification 52. However, V-1703 also receives intermittent regenerants from a deionizer. Since these are acidic at times, V-1703 and its accompanying equipment and piping must be stainless steel. A specification change from 52 to 50A is therefore shown for the dip-leg sections of the boiler blowdown lines where they enter V-1703. The weep holes in each dip leg prevent a vacuum from forming in the blowdown lines, so that the acidic water in the tank cannot be drawn up beyond the tank. Without the weep holes a considerable portion of the blowdown lines themselves would have to be stainless steel.

Heat-tracing and insulation information. Each line should carry information relating to any tracing and insulation required by its various segments. Chapter 15 presents a review of the requirements for heat tracing, the various types that are used, and their applications. It also explains the various uses for insulation, the types available, and criteria for sizing insulation thicknesses.

Notations for tracing. Notification that a line on a flow diagram must be heat-traced can be effected in several ways. One method is to enter an abbreviation such as ST for steam or ET for electrical tracing in parentheses at the end of the line-identification code. With few exceptions, lines that are traced are also insulated. In this case, the designation at the end of the identification code is (ST&I) or (ET&I) to signify both tracing and insulation. This method is shown in Fig. 3.9*a*.

In some cases, insulation is not put over electrical tracing. This occurs, for example, when a short length of flexible metallic hose or tubing is traced when conducting a fluid but the hose or tubing is removed frequently. In this situation, insulation is omitted and a guard is provided for personnel protection.

Often only a portion of a line requires tracing. An example is a 25 percent caustic line that runs outdoors where the ambient temperature could be lower than its freezing point of 6°F and the line then enters a building which is maintained at a minimum of 50°F. The outdoors portion of the line requires tracing and insulation, while sections of the line inside the building do not. A demarcation line similar to the specification-class-change symbol is placed on the line and the notation "Outdoors (ST&I) [or (ET&I)]" located on one side and "Indoors (no ST&I) [or (no ET&I)]" on the other side. An example is shown as Fig. 3.9*b*.

Another popular method of denoting the use of heat tracing on engineering flow diagrams is to place a dashed line parallel to the pipeline and insert a notation such as S for steam or E for electrical tracing at frequent intervals, as shown in Fig. 3.9*c*. The dashed-line method, however, can cause confusion with electrical-instrumentation connectors if many lines are traced on a crowded portion of the diagram.

Notations for insulation. One commonly used design system indicates that insulation is required by simulating it on a portion of each of the lines to be insulated. An example of this method is given in Fig. 3.10*a*. This representation can be confusing when there are partially insulated lines. Further confusion could result if a

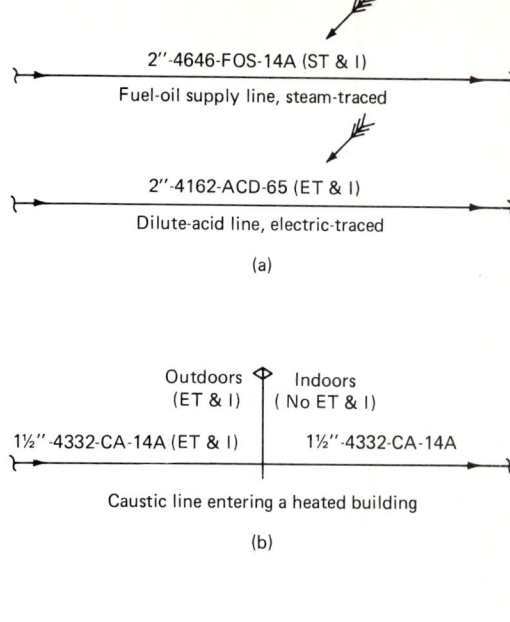

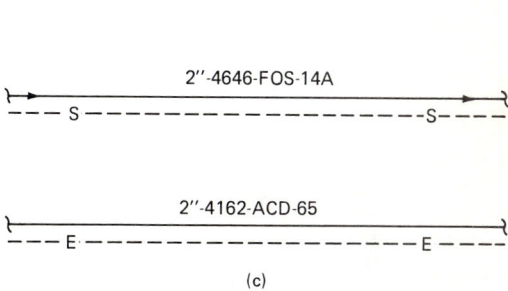

Figure 3.9 Identification of tracing. (*a*) Use of parentheses. (*b*) Use of parentheses with change of requirements. (*c*) Use of dashed lines for tracing. Insulation is given in line tabulation only.

large number of adjacent lines in a crowded section on the flow diagram were insulated.

Another design system that is sometimes used omits all reference to insulation in the body of the engineering flow diagram and includes it only in the line tabulation for that flow diagram instead.

While the insulation requirements are given in the line tables in flow diagrams similar to Fig. 3.2, it is good engineering practice to include a notation for the insulation in the line-identification code as required. This convention facilitates identification of lines requiring insulation. It also aids piping designers in planning the allocation of space for insulation in pipe runs.

A popular method to indicate insulation in the body of the flow diagram is the use of a symbol in parentheses after the line identification. A typical symbol is (I), as

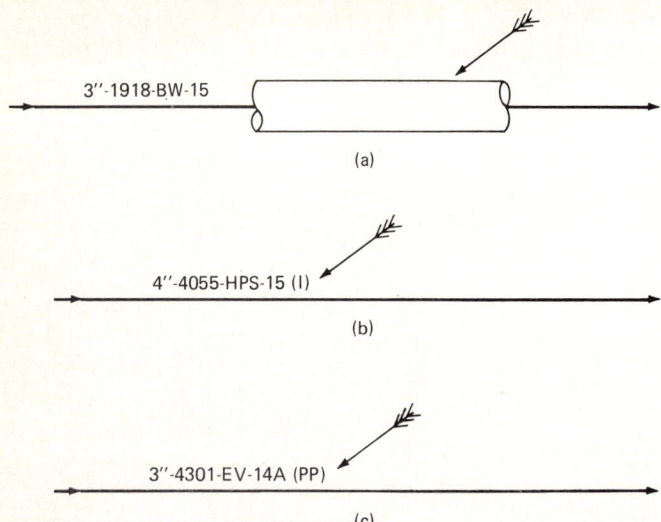

Figure 3.10 Insulation notation. (*a*) Simulated insulation. (*b*) Use of parentheses for heat or energy conservation. (*c*) Use of parentheses for personnel protection.

shown in Fig. 3.10*b*. When insulation is used in conjunction with electrical or steam tracing, the symbol becomes (ET&I) or (ST&I), respectively.

A pipe run may require insulation for only a portion of its length; e.g., a line may require tracing outdoors but not inside, tracing a line to its high point but not thereafter, or antisweat insulation which is needed indoors but not outside. In these cases, the insulated and noninsulated portions are delineated by use of the notation (I) and "No (I)" on the respective sides of a demarcation line.

Insulation for personnel protection is symbolized by (PP). This indicates that only portions of the line normally accessible to workers, maintenance people, or visitors must be insulated so that the outer-surface temperature is 140°F or less. Examples of these notations can be seen in Fig. 3.10*c*.

Notes

Sometimes standardized symbols are not sufficient to convey important information regarding the design or operation of the process, and notes must be added to the diagram. An area, usually at the bottom or along one margin of the flow diagram, is set aside to record the necessary notations. Several illustrations of notations are given on Fig. 3.2.

The first note usually references the lead sheet, which explains the symbols, abbreviations, and conventions used on the diagrams. Other notes may be general in nature or specific to individual situations. General notes need not be referred to in the body of the flow diagram. However, specific notes which are applicable to one or several situations in the diagram have a phrase such as "See Note 2" placed adjacent to the spot in the body of the diagram where the circumstance applies. Figure 3.2 shows several examples of these references. Other specific notes can be used to modify piping specifications for special purposes, indicate particular dimensions or instructions, or clarify further conditions for any segment of the diagram.

Special legends and tables

The lead sheet contains sufficient information to define normal symbology of engineering flow diagrams. However, there may be special situations which require particular symbols or additional abbreviations not included on the lead sheet. This information may be important enough to be referenced on the respective diagram as a special legend. This special legend emphasizes its importance and assures that the information will not be overlooked. Special tables are sometimes included on engineering flow diagrams to outline critical start-up and operating procedures or present listings of pertinent information.

Line tabulations

Line tabulations, also called line schedules or line tables, collate and summarize key pieces of information for a line segment. These include size, specification, tracing, and insulation requirements. Supplemental data regarding pertinent operating factors for each line are also included. The tables tally line numbers current on an engineering flow diagram and expedite locating lines in the body of the diagram.

Line-tabulation methods vary considerably from company to company. A typical line-tabulation format is given on Fig. 3.2. If the tables are not placed on the body of the diagram, they are presented on separate $8\frac{1}{2}$- by 11-in supplements or grouped on a separate drawing for each diagram. In these cases they are usually prepared with 25 lines per table instead of the 20 that are used in Fig. 3.2, and the separate sheets or groups are referenced to the original diagram. It is advisable to emphasize each fifth line in the tabulations as a visual aid in locating line numbers quickly and in following lines across the table.

A review of the main elements of the line tabulations, as shown in Fig. 3.2, follows.

Line number. The first column should be reserved for the line number since the lines are indexed by number. The piping-system designation usually accompanies the line number in the first column. The system designation is entered as a prefix or as a suffix depending on the order in which it appears in the identification sets in the body of the diagram. For lines which have more than one system designation, all designations should be included with the line number, but the number should be listed only once.

During preliminary stages in the development of engineering flow diagrams and in subsequent revisions, there are occasions when lines may be eliminated from the body of a flow diagram. If this occurs, the respective line number and its row of information are deleted from the line tabulation. The row should remain vacant until a subsequent issue of the drawing. This precaution reduces the chance of confusion that might occur if the line number were to be reused immediately.

Size. All nominal pipe sizes of a line, unless specified otherwise, are entered in the size column for each line. It is not customary to note reducer-fitting sizes if a line originates or ends at another line of differing size or as size reductions and enlarge-

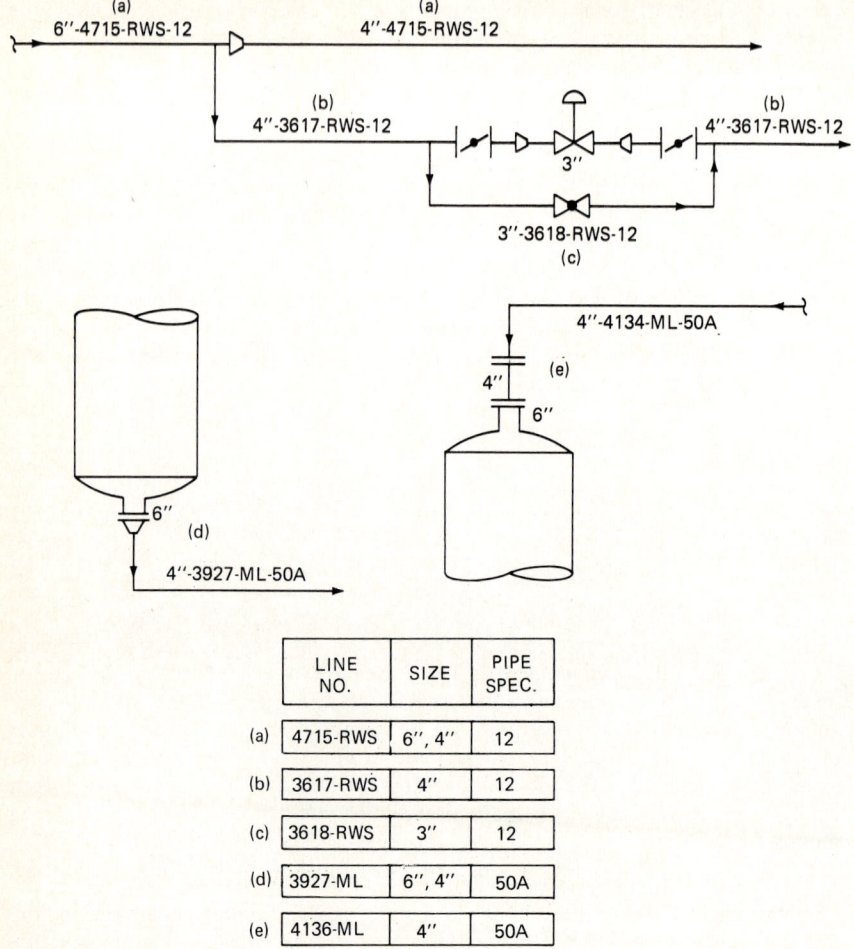

Figure 3.11 Size representation of line tabulations.

ments back to the original size that take place at control valves.* Examples of reporting line sizes for various typical cases appear in Fig. 3.11.

In the design of electric-generating stations, it is common practice that different-sized sequential segments of a steam-pipe run are assigned different line numbers by the addition of a numerical suffix to a baseline number. Any change in piping specification for a given run is accompanied by a similar change in line number. This process is followed to ensure proper identification of the pipe-run segments for stress analysis and quality control. In contrast, the convention followed for designing chemical plants, refineries, and small boiler houses is to retain a line number throughout

---

*Reducers are shown, however, in the body of the drawing for size changes at control valves and the sizes enumerated.

an entire pipe run regardless of changes in size or piping specification. In these cases, it is good practice to enter the pipe sizes in the size column in the order in which they occur from the point of origin to the end of the particular line number. It is rare that more than three or four sizes are applicable to most process or utility lines. On the other hand, utility-distribution headers, particular cooling-water or steam and condensate systems, may progress through a considerable number of size changes from their origin to their termination at the last user. In this event, it is satisfactory to enter the initial and final sizes and separate them with dashes, e.g., 10 in,---, 2 in.

Pipe specification. All piping-specification classes pertaining to a particular pipeline and its valves and fittings are entered in this column. These include classes listed in the piping-identification codes on the line plus any that appear on either side of the mark denoting a specification change.

From and to. The two columns labeled "From" and "To" indicate the respective points of origin and termination of a pipe run with a given line number. Generally, these points refer to pieces of equipment or to other pipelines. For equipment, only the item number usually is entered. However, the word "jacket," "seal," or that for any other special part of the equipment is added if it will help to identify any special interface point on the equipment. If another line is the point of origin or termination, then the term "Line XXXX" or "XXXX" and the line-system designation may be entered in the respective column. Other points at which lines may begin or end are, typically, the atmosphere for intakes or discharges, bodies of water, tank trucks or railcars for raw materials or products, grade or trenches for drains, hose stations, and eyewashes and safety showers. In these cases, an appropriate phrase is added to the respective column.

The From and To columns are helpful for locating lines in the body of an engineering flow diagram. The lines easiest to locate are those which originate or terminate at a piece of equipment. If a line originates and terminates only at other lines, it often is necessary to find the antecedent lines to locate the original line.

Utility-distribution-header systems are developed after the engineering flow diagrams have been prepared. Thus, the word "Header" is placed in the From or To column for the utility lines from or to their users. Once the distribution headers have been laid out and line numbers assigned to them, these identification numbers replace "Header" in the appropriate column for lines bringing utilities from or to a user.

Flow. The value to be reported in the "Flow" column is the design flow rate at which the pressure drop for the line is calculated. For most processing lines, this flow rate is listed as maximum on the accompanying process flow sheet. Flow-rate values are usually rounded off to one or two significant figures unless it is necessary to carry more figures for clarity.

Units of various dimensions are used to report flow rates. However, the greater number of lines in the average chemical plant or refinery transport liquids. Therefore, it is customary to report flows in terms of gallons per minute (gal/min), "GPM," unless otherwise noted. This is so stated on the lead sheet with "GPM" accompanying "Flow" at the head of the column. Flows in the other terms must be accompanied by a relevant dimension. Gas or vapor flows are normally reported as standard cubic feet per minute (scfm), "SCFM." Actual cubic feet per minute (acfm), "ACFM," is sometimes used as long as the temperature and pressure of the system are clearly understood; acfm is frequently used with such items as suction lines to compressors.

The dimension for steam flow is usually pounds per hour (lb/h), "PPH." This unit is also used for solids flow and for any other stream for which it would be more appropriate to have the flow designated in this term.

Flow rates are not reported for lines such as vents, drains, equalizing lines, etc., for which there normally is no flow. A dash is entered in the column instead to indicate that the rate has been considered. Even though no flow rate is stated, these lines are sized by the process engineer in accordance with accepted engineering practice for the individual situation. The function of the line is sometimes entered in the "Remarks" column.

Temperature. The temperature entered in this column is the value at normal flowing conditions. For most process lines, the value is obtained from the process flow sheet. However, for process lines such as recycle or bypass lines, equalizing lines, and vent, drains, or relief lines, the assigned temperature values are equivalent to those of the lines from which they originate. Lines which undergo a significant temperature change due to the confluence of two fluids with different temperatures carry both the upstream and the downstream temperatures in order.

A line that is normally stagnant can be subject to a range of temperatures along its length according to the originating point, the point where it ends, ambient conditions, and whether or not it is traced or jacketed. An accepted convention is to assign a temperature value according to the temperature of the fluid in the equipment or in the line from which flow to the normally stagnant line would originate.

The temperature of many utilities such as plant air, instrument air, nitrogen, and carbon dioxide varies with the season and may be listed as ambient.* Municipal water also has a seasonal variation, and it is listed as ambient if it is used as process or seal water. However, if the water serves as a cooling medium, its highest summer temperature is recorded for the lines bringing it to the exchangers or jackets. River water and cooling-tower water are treated similarly. Well water usually has the same temperature at its source throughout the year. The temperatures for chilled water, brines, and glycols are given as the design temperature at their source.

The temperatures for lines transporting steam are taken as the saturation temperature of the steam. However, if the steam is deliberately superheated, the superheated temperature is listed. If there is a pressure reduction in the line, the saturation temperature of the steam downstream of the reduction valve should be included along with upstream temperature. Both upstream and downstream temperatures for condensate lines including condensate traps are noted.

Small temperature changes due to line friction or to energy input across pumps are normally disregarded. However, if the net flow across a pump is small compared with horsepower input, there may be a significant temperature rise, which is reflected in the tabulation.

The units for temperature are usually given as degrees Fahrenheit. However, if it is the convention of the client to use degrees Celsius for temperature values, this should be noted in the column heading.

Pressure (psig). The pressure value is difficult to define since it could be the process design pressure, the maximum process pressure, the pipe design pressure, or even the test pressure of the pipe. A recommended practice is to record the pressure value

---

*Most distribution systems for these gases are long enough that the gases in uninsulated lines soon attain ambient temperatures regardless of their initial temperatures.

at the source of the fluid normally flowing through the line. Thus, for gases the value should be the pressure in the header or equipment from which the stream originates. The pressure for steam is reported as that at the boiler where it is generated or from the header in the distribution system.

Unless otherwise stated, pressure is presented as psig, as noted at the head of the column, for all positive pressures and no dimensions are shown.* The term "Atm." is used for gases when the point of origin is essentially atmospheric pressure. For lines that are slightly below or above atmospheric pressure, entries in terms of inches of water column (WC) are used. If the pressure is below atmospheric, the expression should carry a minus sign and be enclosed in parentheses, e.g., $(-5'' WC)$.

The pressures for liquids in drains or in lines to pump suctions are termed "Hyd." for "hydrostatic," to signify that the originating vessel or line is essentially at atmospheric pressure and the maximum pressure in the line is the head of fluid. However, if the equipment is pressurized, the pressure of the equipment is reported for liquid lines originating from them. The pressure for discharge lines of a pump is shown as the suction-line pressure plus the design total dynamic head (TDH) of the pump. The pressure thus obtained is used for all lines associated with the discharge section of a pump until they terminate at a piece of equipment or pass through a pressure-control valve.

It should be noted that the discharge pressure of a centrifugal pump increases somewhat while the discharge pressure of a positive-displacement pump rises considerably when the pumps are deadheaded or throttled to shutoff. In some instances the deadheaded pressure is reported in the line tabulation. If it is not, it is important that the process engineer recognize this situation and examine the requirements for relief valves to protect piping or equipment.

Despite the friction losses that may occur in a line, it is generally not the practice to take them into account when listing the pressure for the line. Similarly, the changes in pressure due to various static heads of liquid are not recorded unless they make a considerable difference in the pressure exerted by the fluid. In this event, the initial pressure and the highest pressure in the line are given. On the other hand, pressures before and after a pressure-control valve should be shown. If a line has a flow-restriction orifice with a significant drop, it is a good practice to include both the upstream and the downstream pressures.

The pressure values shown in the line tabulations are useful only as guides to the pressures in the various lines rather than as absolute values. Other engineers should always consult the process engineer responsible for a given engineering flow diagram to establish the exact pressure at any point on a given line when such information is required in their work.

It is good practice to report vacuum conditions in terms of millimeters of mercury absolute, and it is understood that the use of the dimension mmHg always implies absolute pressure. Sometimes the term "inches of mercury," which can mean either absolute pressure or the degree of pressure below atmospheric, is used. Because of this ambiguity, it is best to only use mmHg with its connotation of absolute pressure.

Insulation. The insulation column in the line tabulation is used to indicate the presence of insulation or to record the type of insulation to be used and its thickness. Several conventions may be used in completing the insulation column. The examples

---

*The basic unit for countries customarily using the metric system would be kilograms per cubic centimeter (kg/cm^2).

shown in Fig. 3.2 merely place the word "Yes" in the space for lines that are insulated for heat or energy conservation or the indication PP for personnel protection. Had insulation been required for antisweat purposes, the abbreviation AS would have been entered. The presence of these indications alerts designers and engineers to consult the insulation specifications for the type of insulation and its thickness. Other groups prefer to indicate the type of insulation by an appropriate abbreviation and possibly the required thickness directly in the insulation column.

If no insulation is needed for a line, a dash is placed in the column. Should a line be only partially insulated, as denoted by an insulation-break indicator on the line in the body of the flow diagram, the word "Partial" or "Part." should be entered in the insulation column or in the remarks column.

Tracing. The tracing column is used to indicate which lines are traced and, if necessary, the type of tracing to be employed. A line not requiring tracing is denoted by a dash in the tracing column. "St." or "Steam" is placed in the column for a line that is steam-traced, while "Elec." is generally used for a line that is electric-traced. As with insulation, if only a portion of the line is traced, the word "Partial" or "Part." is entered in the appropriate space and the break point shown on the lines in the diagram. Tracing, along with insulation, is discussed in Chap. 15.

Remarks. The remarks section of the line tabulation is used to call attention to special conditions not readily discernible from the other columns in the tabulation. Reference can be made to the notes section of the diagram if there is insufficient space in the remarks column to describe the condition. In this case "See Note X" is entered in the column.

Checklists

The placement of the required equipment and necessary connecting piping with identifications on an engineering flow diagram is only the initial phase of the work required to make the diagram complete. A great many details must be added so that the process being described on the diagram fulfills its function in an economical and safe manner and is in compliance with environmental requirements. While the details entered vary considerably from project to project, many items are common to most projects. The following checklists regarding major components found on many engineering flow diagrams should be kept in mind while preparing the diagrams. The lists concern the appurtenances on equipment and piping and also review primary instrumentation as well as important safety considerations.

The items reviewed in the checklists are given in Table 3.3. Although the lists are not fully comprehensive, the items in them relate to situations most likely to be encountered and alert the process engineer to the type of details to be considered for inclusion on engineering flow diagrams.

Short elaborations of the various items in the checklists follow to give guidance for their representation on an engineering flow diagram.

Vessels

Connections. (See Fig. 3.12.)

1. Couplings, generally for connections of $1\frac{1}{2}$-in diameter and smaller.
2. Flanged nozzles, for any size of connection. Nozzles with a diameter smaller than

TABLE 3.3 Checklists for Engineering Flow Diagrams

Vessels	*Fans, blowers, and compressors*	*Instrumentation and safety*
Nozzle types	Types of gas as movers	Relief valves
Vents	Types of drivers	Rupture disks
Drains	Coolers	Vacuum breakers
Vortex breakers	Knockout pots	Conservation vents
Overflow pipes	Valved drains	Flame arresters
Sample connections	Accumulators	Pressure sensors
Dip legs	Screens, filters, and silencers	Temperature sensors
Gooseneck inlets	Controls and interlocks	Sight glasses and lights
Baffles	*Heat exchangers*	Level indicators
Manholes	Types and configurations	Flexible connections
Handholes	Sloping exchangers	Expansion joints
Air locks	Position of nozzles	Fire valves
Rotary feeders	Backflushing exchangers	Spring-closure valves
Live bottoms	Impingement protection	Locked valves
Vessel jackets	Vents and drains	Limit switches
Heat-transfer panels	Tracing and insulation	Automatic switch-over of pumps
Internal coils	*Vacuum equipment*	Failure alarms
Vessel tracing	Types of vacuum equipment	Interlocks
Insulation	Drivers and motive forces	*Miscellaneous*
Removable spool pieces	Cooling vacuum equipment	Expansion tanks
Position	Knockout pots and traps	Backflow preventers
Vessel supports	*Filters and centrifuges*	Air gaps
Agitators or mixers	Types of filters and centrifuges	Safety showers, eyewashes
Location of agitators	Types of drivers	Utility service stations
Types of agitators	Additives and precoating	Fire protection
Steady bearings	Recycle and sampling	Seal fluid systems
Seals	Auxiliary lines and equipment	
Pumps	*Piping*	
Types of pumps	Valves	
Types of drivers	Bypasses for control valves	
Valved vents and drains	Vent and drain sizes	
Quench system	Condensate traps	
Flushing and seal fluid systems	Protective screens	
Jacketed pumps	Jacketing and bundling	
Tracing and insulation	Boundary definitions	
Pump recycle lines		

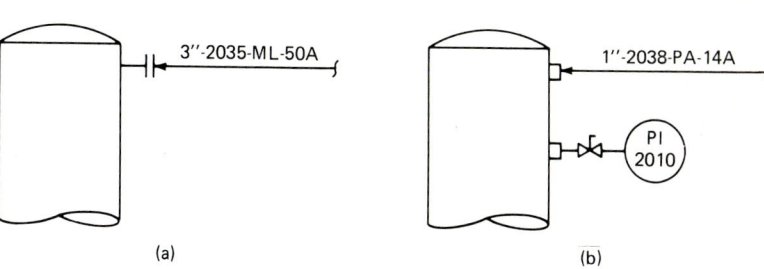

Figure 3.12 Vessel connections. (*a*) Nozzle. (*b*) Couplings.

2 in on low-pressure vessels are fragile, and connections should be made by means of reducing fittings or flanges.
3. Nonmetallic connections are often of limited size. Manufacturer catalogs should be consulted.

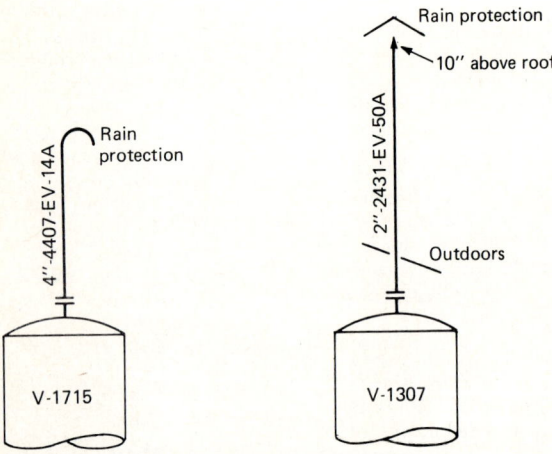

Figure 3.13 Vents.

Vents. (See Fig. 3.13.)

1. Closed vessels are generally vented.
2. Vents with only air or low amounts of water vapor terminate above the top of the vessel.
3. Vents with vapors that may be harmful but are not toxic or lethal, i.e., hot gases or hot vapors, may terminate outdoors above adjacent structures in accordance with pertinent regulations.
4. Dangerous vapors or gases pass to a flare system or to a collection system for further treatment.

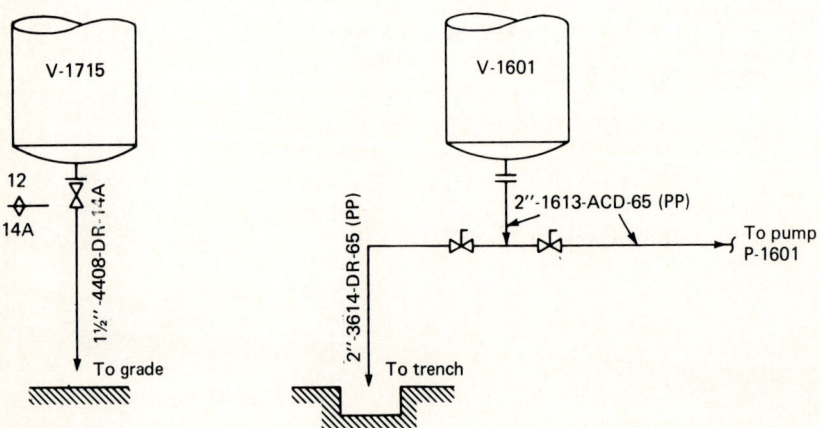

Figure 3.14 Drains.

5. An invert or "rain hat" protects vessel contents from precipitation.
6. The diameter of the vent line is generally made equal at least to that of the largest liquid line entering the vessel.

Drains. (See Fig. 3.14.)

1. Most vessels are provided with drains.
2. Drain connections can originate from a discharge line or from the vessel itself.
3. The destination of a drain should be indicated.

Vortex breakers. (See Fig. 3.15.)

1. A vessel with a low liquid level may require a vortex breaker to prevent gases from entering a pump suction.

Overflow pipes. (See Fig. 3.16.)

1. Open or low-pressure vessels handling liquids require an overflow connection.
2. The top invert of the overflow nozzle must be sufficiently below the top tangent line to accommodate the constriction head at the maximum flow rate through the nozzle.
3. Overflow lines for closed, blanketed vessels or those under a slight negative pressure must be sealed in a suitable liquid loop seal or by means of a mechanical sealing trap.

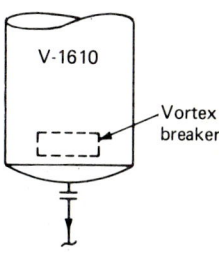

Figure 3.15 Vortex breaker.

4. Seal-leg heights should be included on the engineering flow diagram if required to prevent overpressuring or collapse of the vessel.
5. A high-level switch can be used to close valves in inlet lines or to stop supply pumps.

Sample connections

1. Sample connections may be placed on a vessel, on a discharge line at the bottom of the vessel, or on a continuously flowing line from a pump taking suction from the vessel.
2. Samples may be taken from the top of a vessel through a suitable sampling mechanism.
3. A series of connections is required for vessels which may have stratified layers.

Dip legs. (See Fig. 3.17.)

1. Dip legs are required to diminish the generation of static-electricity effects with flammable liquids, reduce foaming, introduce reactants into an agitated vessel, return heated, viscous fluids to the suction nozzle of a circulating pump, create a barometric leg in a vacuum system, or prevent corrosive effects due to aeration.
2. A weep hole or a siphon break is frequently required in dip legs other than those acting as barometric legs.
3. Weep holes in lines with vapor-liquid flow should be of minimum size and located in the neck of the nozzle.

Gooseneck inlets. (See Fig. 3.18.)

1. A gooseneck inlet is an alternative to a dip leg in a metallic vessel to reduce static-electricity effects.
2. A weep hole is not required with a gooseneck inlet.

72 Process Documentation

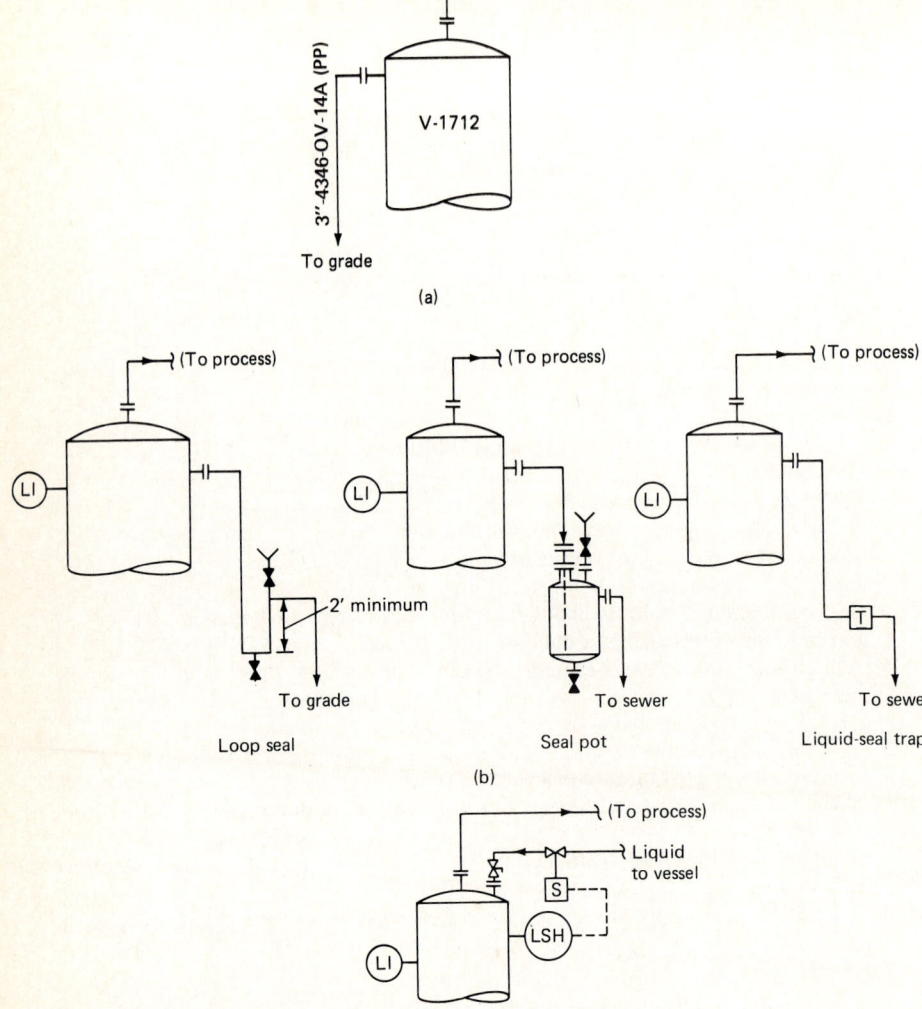

Figure 3.16 Overflow connections. (*a*) Overflow from atmospheric tank. (*b*) Sealed overflows. (*c*) Automatic shutoff to prevent overflow.

Baffles

1. Baffles are frequently required to improve agitation.
2. The vendor of agitation equipment usually recommends the baffle sizing and configuration best suited for the respective vessel dimensions.
3. V-1715 in Fig. 3.2 shows a typical representation.

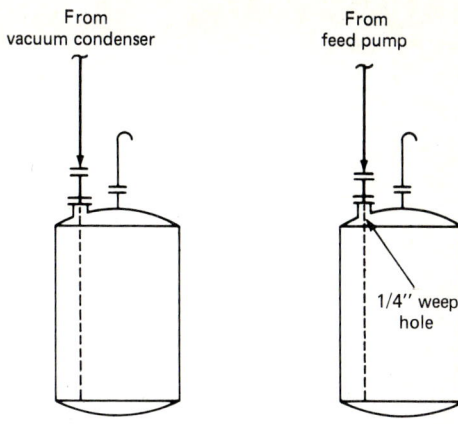

Figure 3.17 Dip legs.

Manholes

1. Manholes are required for inspections, cleanout, and maintenance.
2. For ventilation and as an escape or rescue port, one manhole should be in the head or top of a vessel with another on the side or bottom.
3. Manholes are normally of 24-in diameter but never smaller than a 16- by 24-in ellipsoid for use in special instances.
4. Unwanted dead space in a manhole can be avoided by use of an internal "hat" or a curved flush cover.
5. Number, sizes, and special configurations are to be shown on engineering flow diagrams.

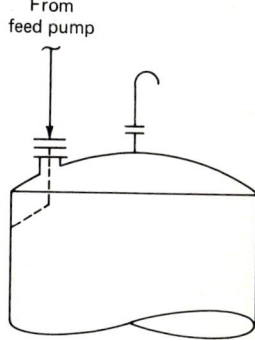

Figure 3.18 Gooseneck inlet.

Handholes

1. Handholes are used on small vessels for inspecting limited areas, feeling for the integrity of a vessel, and charging or removing catalysts, desiccants, or packing from columns or vessels.
2. Number, sizes, and special configurations are to be called out on flow diagrams.

Air locks. (See Fig. 3.19.)

1. An air lock is a small vessel mounted at the top or directly beneath a bin or reactor to introduce or remove solids or liquid intermittently.
2. Suitable valves are placed upstream and downstream on the air lock. One or both valves are always closed to isolate the bin or reactor from its surroundings.
3. An inert purge may be used to preclude air or water vapor from a bin or reactor or to prevent the atmosphere in the vessel from escaping to the surroundings.

Rotary feeders. (See Fig. 3.19.)

1. A rotary or star feeder permits a continuous, controlled, adjustable flow of solids into or from a vessel while isolating it from its surroundings.

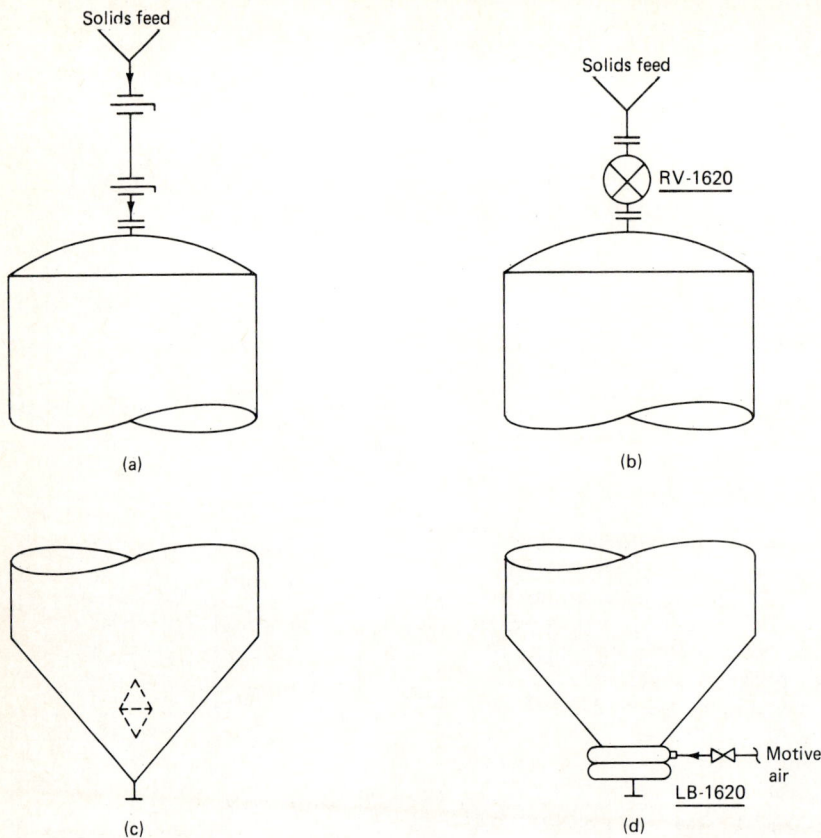

Figure 3.19 Solids handling for vessels. (*a*) Airlock with slide gates. (*b*) Rotary valve. (*c*) Internal inverted cones. (*d*) Live bottom.

2. An inert purge may be used to preclude air or water vapor from a vessel or to prevent the atmosphere in the vessel from escaping to the surroundings.

Live bottoms and other solids-flow devices. (See Fig. 3.19.)

1. The containing vessel should have a coned bottom whose angle is greater than the angle of repose of the solids.
2. An internal pair of small cones, one inverted on the other, facilitates the outflow of solids.
3. An air lance is frequently used to prevent or disrupt bridging.
4. One or more mechanical vibrators attached to the outside of the cone bottom helps to induce the flow of solids.
5. A "live bottom," or pulsating section, is often inserted between the cone section and the discharge valve for difficult solids-flow problems.

Vessel jackets. (See Fig. 3.20.)

1. Carbon steel vessel jackets offer a low-cost means of transferring heat to glass-lined or high-alloy vessels.

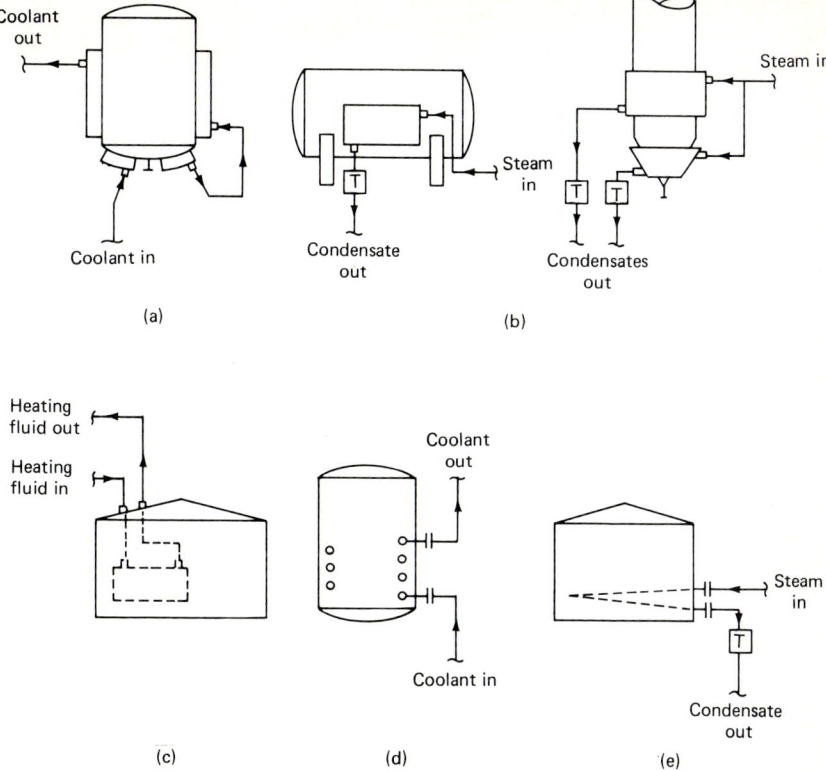

Figure 3.20 Heat-transfer methods for vessels. (*a*) Jacketed vessel. (*b*) Vessels with external heating panels. (*c*) Tank with internal heating panel. (*d*) Vessel with coils. (*e*) Tank with bottom coil.

2. Liquid media enter the bottom of a jacket and follow a tortuous path through the baffled or dimpled annulus to exit at the top.
3. Condensing media enter at the top, and the condensate leaves at the bottom; baffling is not needed. A vent should be provided.
4. Steam and water may be used alternately, but vents and drains must be provided in utility piping so that the two media do not contact one another. Controls should be provided in utility piping to glass-lined tanks to prevent high temperature differentials between the vessel and the jacket contents.

Heat-transfer panels. (See Fig. 3.20.)

1. The panels are used to transfer modest amounts of heat or in retrofitting a vessel. They are frequently used to maintain the temperature of vessel contents against heat loss or gain.
2. External panels may be constructed of carbon steel or galvanized sheet regardless of the vessel material.
3. They are available in a variety of shapes to fit various contours. They may be supplied with different internal paths, serpentine configurations, and positions for inlet and discharge connections.
4. Usually they are given item numbers separate from that of the vessel.

Internal coils. (See Fig. 3.20.)

1. These coils are placed inside agitated vessels for improved heat transfer. They may be used by themselves or in addition to a jacket.
2. They may consist of more than one concentric set of coils provided there is adequate space for circulation between coils and rows.
3. The materials of construction must be compatible with the contents of the vessel at the extremes of temperature expected within the coil.

Vessel tracing

1. Heat needed to hold a storage vessel at a required temperature or to provide freeze protection is sometimes supplied by tracing the vessel with electrical tape. Heat transfer is enhanced by the use of a special grout or mastic.
2. Sheets of metallic foil can be used to cover vessels constructed of plastic, resinous, or other nonmetallic materials before the application of electric-tracing tape to prevent the generation of localized hot spots.
3. The presence of electric tracing on a vessel is normally not indicated in the body of an engineering flow diagram but is noted with the item number, name, and description.

Insulation

1. Some design groups indicate insulation by portraying a section of it on a vessel in the body of the flow diagram; others include a notation "Insulate" or the purpose of the insulation in the section with the item description.

Removable spool pieces. (See Fig. 3.21.)

1. Vessels which personnel can enter should be provided with removable spool pieces at liquid inlet nozzles servicing the vessel.
2. Spool pieces are removed and piping or valves blanked off whenever a person is in the vessel.

Vessel position

1. Engineering flow diagrams should show whether vessels are in a vertical, horizontal, or inclined position.

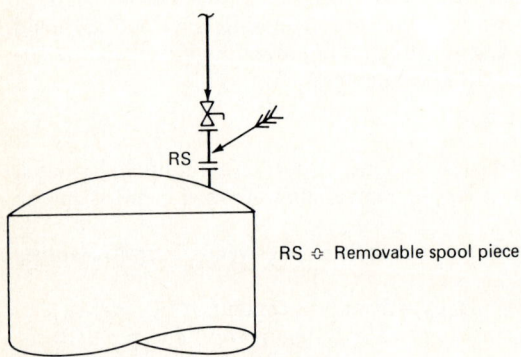

Figure 3.21 Removable spool pieces.

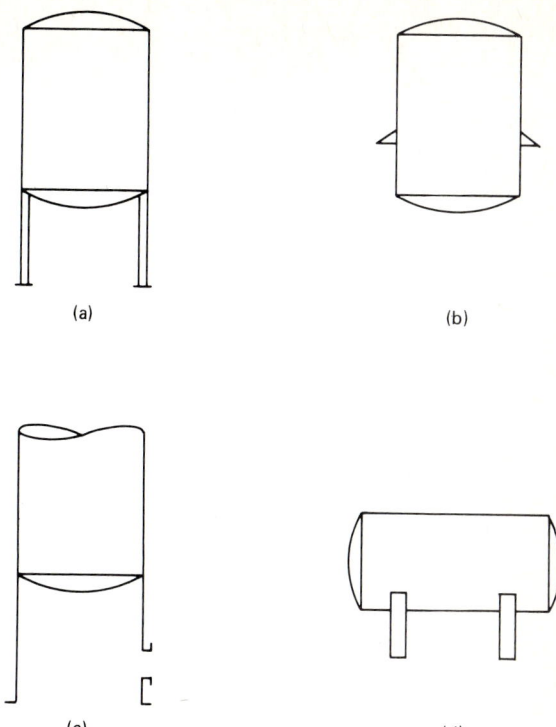

Figure 3.22 Vessel supports. (*a*) Legs. (*b*) Lugs. (*c*) Skirt. (*d*) Saddles.

Vessel supports. (See Fig. 3.22.)

1. Representations of vessel supports should be included in flow diagrams.
2. Chapter 10 discusses criteria for choosing the appropriate type of support.

Agitators or mixers

Location of agitators

1. Most agitators enter at the top, and the shaft extends straight down or at an angle.
2. A side-entering agitator is mounted through a nozzle below the liquid level.
3. In some instances, the shaft may enter through the bottom head.

Types of agitators. (See Fig. 3.23.)

1. The type of agitator (propeller, turbine, paddle, anchor, etc.) should be depicted on the flow diagram.
2. Multiple sets of blades are sometimes required and should be shown.
3. Auxiliary internals such as vessel baffles or draft tubes are often associated with agitators.
4. In the absence of prior similar or pilot-plant experience, a reputable agitator or mixer manufacturer should be consulted to determine the type of agitator or auxiliaries.

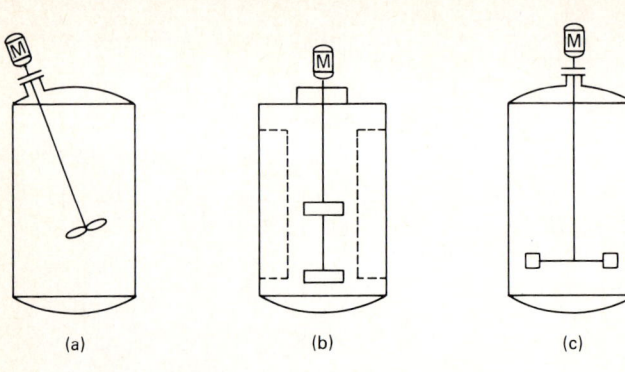
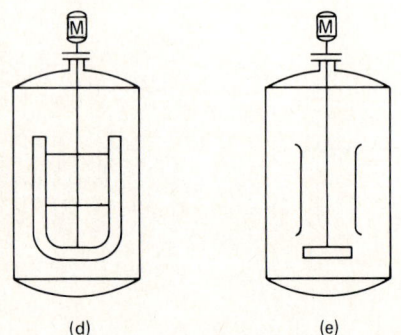

Figure 3.23 Typical agitator types. (*a*) Propeller-type agitator. (*b*) Turbine-type agitator with baffles. (*c*) Paddle-type agitator. (*d*) Anchor-type agitator. (*e*) Turbine agitator with draft tube.

Steady bearings

1. A steady bearing should be denoted on a flow diagram if one is included to reduce the size of the shaft diameter.

Seals. (See Fig. 3.24.)

1. Packing or a mechanical seal is used to prevent leakage of air into or vapor out of a closed vessel.
2. Shaft packing or a mechanical seal is required for a side- or bottom-entering agitator.
3. While shaft packing or mechanical seals generally are not noted on engineering flow diagrams, special lubrication or cooling systems are depicted.

Pumps

Types of pumps

1. A large variety of pumps is available for use in a processing plant. Representations of several of the more common types are shown in Fig. 3.25.

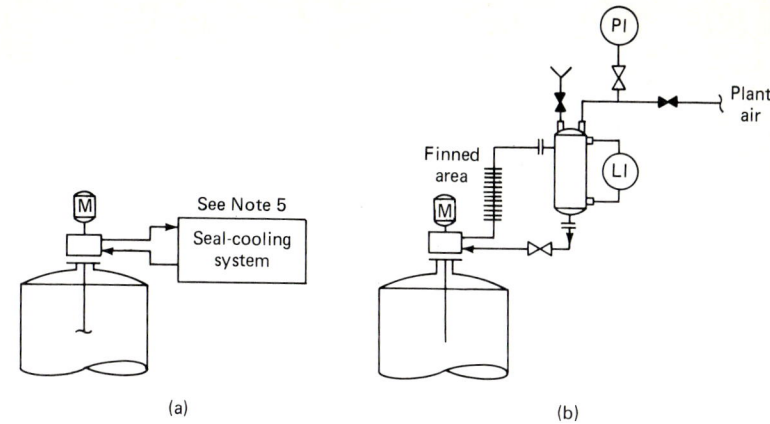

Figure 3.24 Typical seal-cooling systems. (*a*) External cooling system. (*b*) Thermosiphon cooling system.

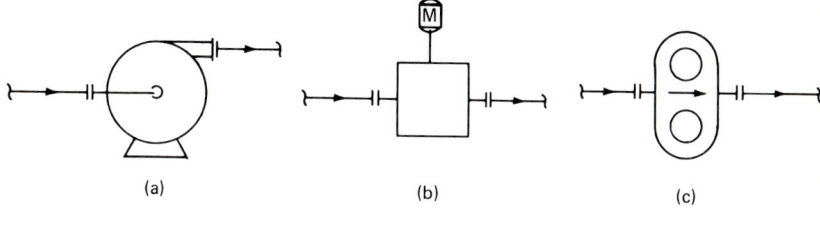

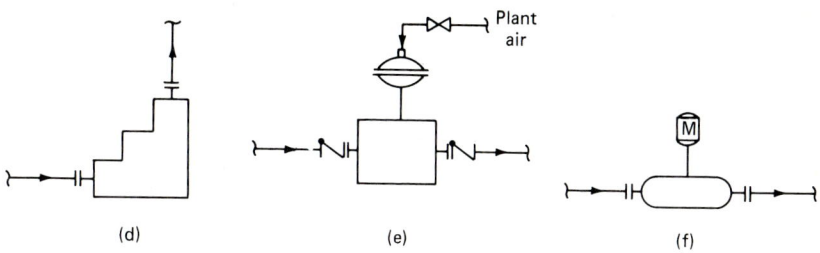

Figure 3.25 Several pump types. (*a*) Centrifugal pump. (*b*) Reciprocal pump. (*c*) Gear pump. (*d*) Metering pump. (*e*) Diaphragm pump (air-operated). (*f*) In-line pump.

Types of drivers. (See Fig. 3.26.)

1. Since most pumps are driven by electric motors, many design groups usually omit a driver symbol if the driver is an electric motor.
2. An air motor may be employed if the power requirement is low or for safety considerations in electrically hazardous areas.
3. Hydraulic motors are used for high-torque requirements in special applications.
4. Steam turbines are used as an economy measure when exhaust steam is available. They are also used for critical pumps or compressors when flow is to be maintained during a power failure.

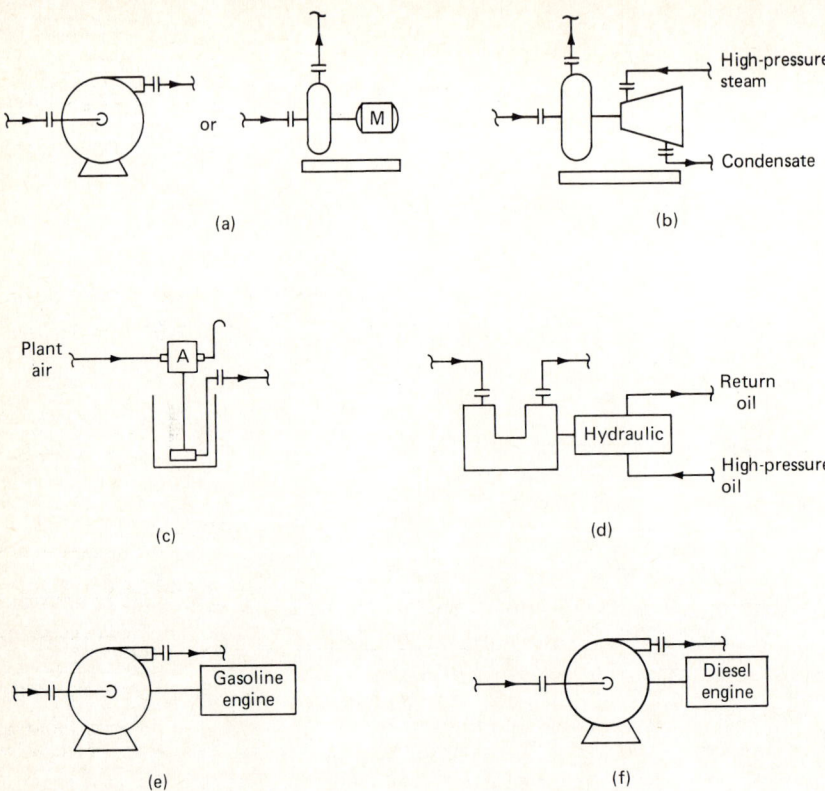

Figure 3.26 Several drive types. (*a*) Electric-motor drive. (*b*) Steam-turbine drive. (*c*) Air-motor drive. (*d*) Hydraulic drive. (*e*) Gasoline-engine drive. (*f*) Diesel-engine drive.

5. Diesel and gasoline engines are used as alternatives to steam for critical services. They are also used as prime movers in remote locations.

Valved vents and drains

1. Vents and drains can be provided on a pump casing rather than on the discharge or suction piping.
2. The type and size of valve are to be shown at the appropriate location on the pump symbol (see P-1715 and P-1716 on Fig. 3.2).
3. T-bar handles can be welded to casing plugs. They should be indicated on the flow diagram.

Quench system

1. Pump packings or seals must be protected from high-temperature fluids by a suitable cooling-water quench.
2. Some types of seals require special jacketing.
3. Water piping to and from pump quench or seal jackets is shown on the flow diagram.

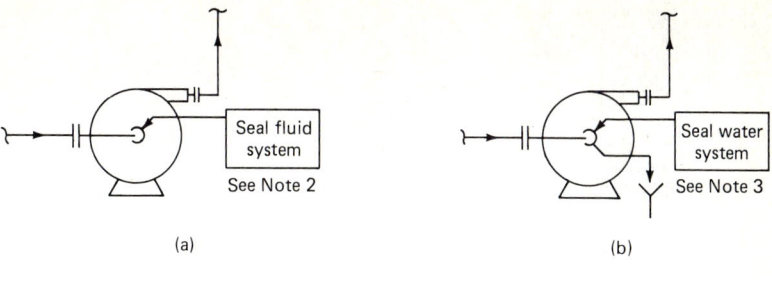
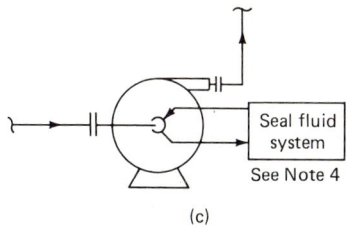

Figure 3.27 Typical seal fluid systems. (*a*) Labyrinth ring with packing or single mechanical seal. (*b*) Double mechanical seal with water as seal fluid. (*c*) Double mechanical seal with recirculating fluid. NOTE: A notation at each box refers to a note on the flow diagram that references the drawing giving details of the seal system.

Flushing and seal fluid systems. (See Fig. 3.27.)

1. Water or another compatible fluid is sometimes required to flush slurries, crystallizing solutions, or corrosive liquids from shaft packings.
2. Liquids may be needed to flush a mechanical seal to remove frictional heat or to prevent a slurry from reaching a seal face.

Jacketed pumps. (See Fig. 3.28.)

1. Jacketed casings are used to maintain the fluid in the pump at an elevated or a chilled temperature.
2. Jacketing is also used to protect the pump casing from excessive temperatures.

Tracing and insulation

1. Heat tracing may be required to prevent residual fluid in a pump from freezing when the pump is not operating.
2. Insulation may be required over tracing or jackets to conserve heat or energy. It may also be needed for personnel protection or antisweat purposes.
3. Tracing and insulation requirements are included with the equipment description (see P-1715 and P-1716 on Fig. 3.2).

Pump recycle line

1. A continuous recycle line is sometimes installed to maintain minimum flow through a pump.
2. A restriction orifice is normally placed in the recycle line.
3. Lines 4414-WLN and 4416-WLN on Fig. 3.2, which are installed for other purposes, also serve as protective recycle lines for pumps P-1715 and P-1716.

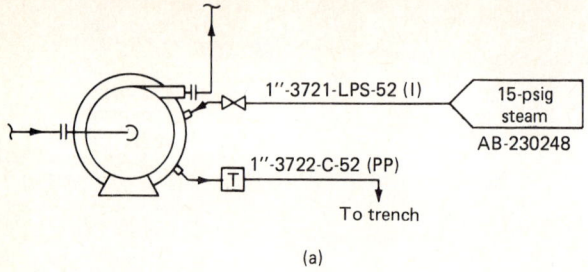

(a)

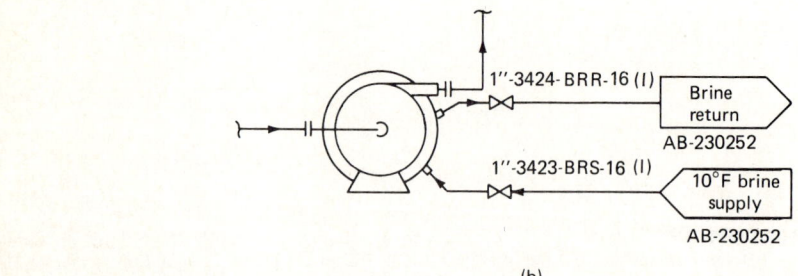

(b)

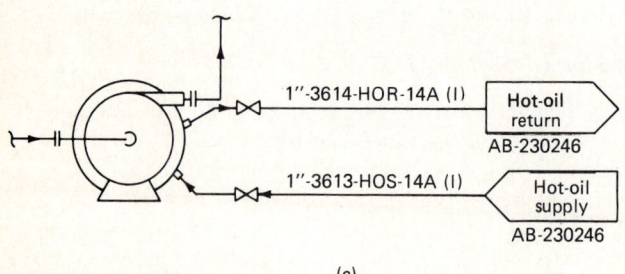

(c)

Figure 3.28 Typical jacketed pumps. (*a*) Steam-jacketed pump. (*b*) Brine-jacketed pump. (*c*) Oil-jacketed pump.

Heat exchangers

Types and configurations. (See Fig. 3.29.)

1. The representation of the exchanger on the flow diagram should reflect as much as possible its particular type and configuration.
2. The location of the nozzles should approximate those on the exchanger as it is to be placed in service.

Sloped exchangers. (See Fig. 3.30.)

1. Single-tube-pass horizontal exchangers in condensing service are frequently sloped in the direction of the condensate outlet.

Position of nozzles. (See Fig. 3.31.)

1. Liquids being heated should enter exchangers at the low point and leave at a high point to prevent buildup of air or other gases which may come out of solution.

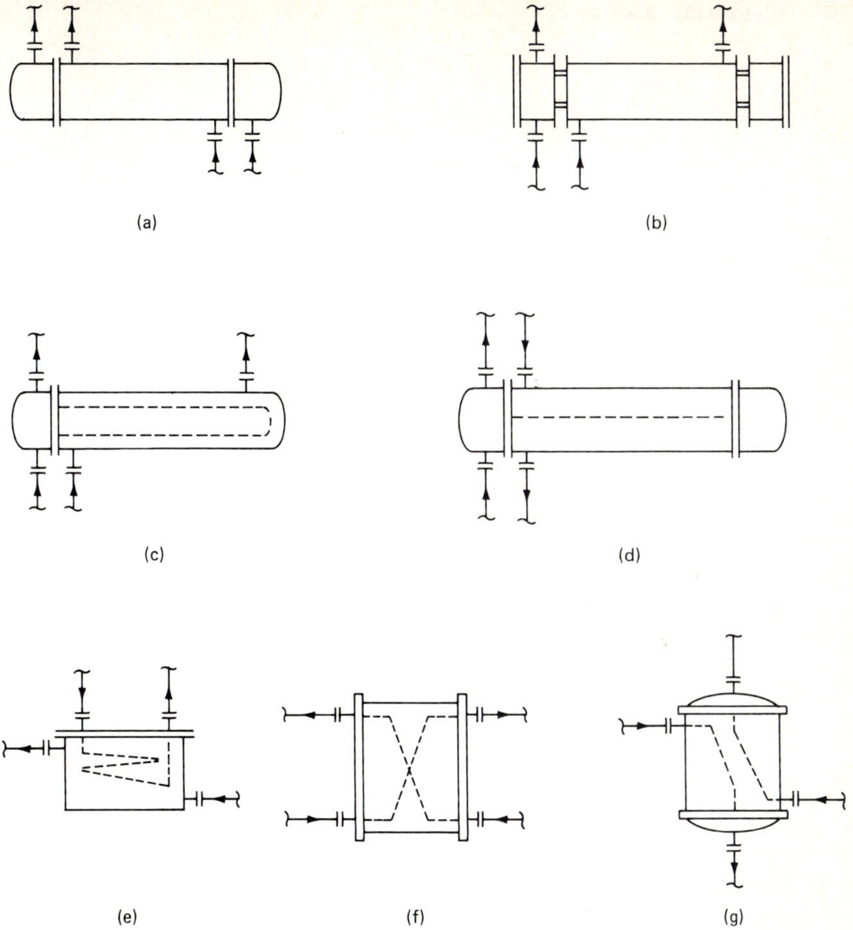

Figure 3.29 Heat-exchanger types. (*a*) Single shell-and-tube passes. Fixed tube sheets; bonnet ends. Shell-and-tube exchanger (cocurrent flow). (*b*) Fixed double-tube sheets. Multipass tube side; single-pass shell side; channel ends. (*c*) U tube. Shell-and-tube exchanger. (*d*) Two-pass shell side. (*e*) Helical exchanger. (*f*) Plate exchanger. (*g*) Spiral exchanger.

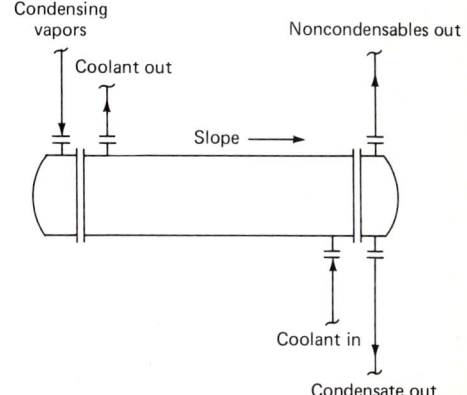

Figure 3.30 Sloped exchanger.

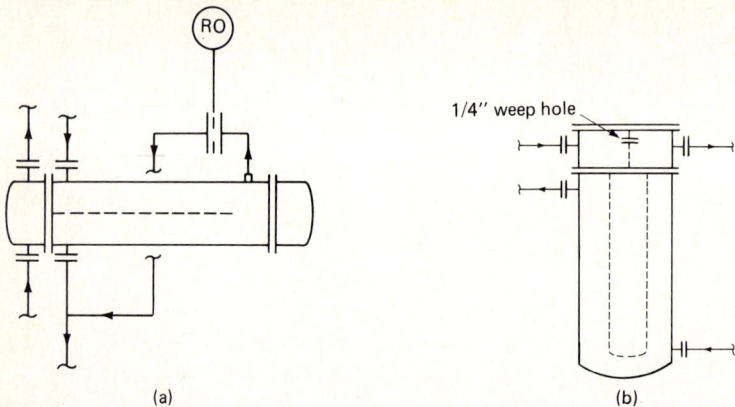

Figure 3.31 Venting accumulated gases. (a) Two-shell-pass exchanger with restricted shell vent. (b) Internal venting of U-tube exchanger.

2. If suspended solids are present, the flow should be from the top down to prevent a buildup of solids.
3. Gases may be removed from a two-shell-pass exchanger by an external restricted vent or from the channel of a vertical U-tube exchanger by an internal weep hole. The orifices are sized to preclude excessive bypassing of fluid.

Backflushing exchangers. (See Fig. 3.32.)

1. Periodic backflushing is needed if a fluid contains suspended matter and there is a wide variation in its velocity as it passes through the exchanger.
2. Cross piping and valving are provided to direct normal inlet fluid to the discharge nozzle while effluent leaves the inlet nozzle carrying dislodged suspended material to the discharge piping.

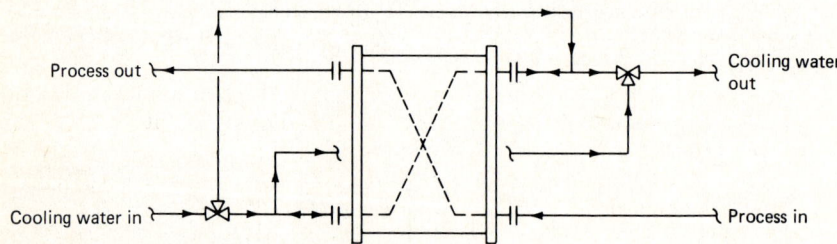

Figure 3.32 Backflushing an exchanger.

Impingement protection. (See Fig. 3.33.)

1. In many exchangers, it is necessary to protect the tubes or internals from erosion by high-velocity fluids entering the shell. The exchanger manufacturer should be consulted to determine whether or not such protection is required and what form it should have.
2. Protection is afforded by an impingement baffle beyond the fluid inlet nozzle, a dome at the inlet nozzle, or an oversized nozzle.

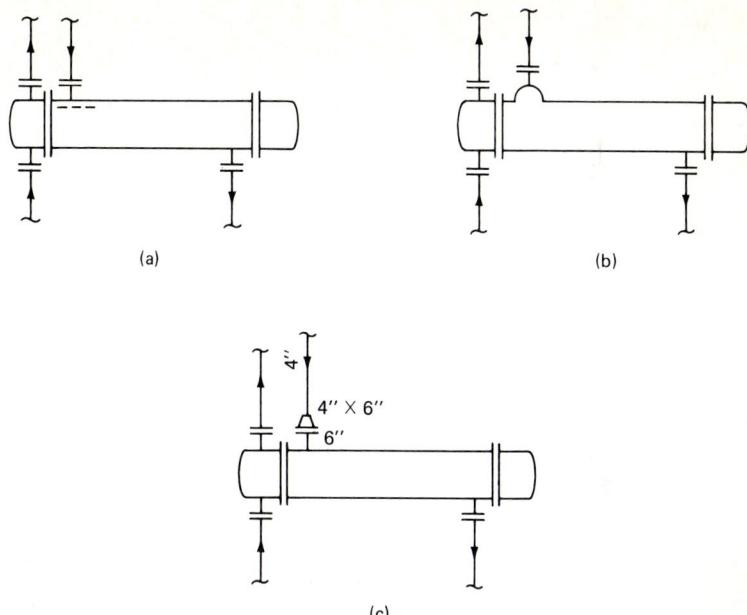

Figure 3.33 Examples of exchanger impingement protection. (*a*) Exchanger with impingement baffle. (*b*) Exchanger with inlet dome. (*c*) Exchanger with oversized inlet nozzle.

Vents and drains

1. Vents and drains, when required, are usually placed in the inlet or discharge piping to an exchanger; however, they can be made to connections on the inlet and outlet discharge nozzles or to a boring through the tubesheet.

Tracing and insulation

1. Tracing and insulation are required to prevent one or both sides from freezing unless a small continuous flow is sufficient to prevent freezing or it is permissible to drain the affected side of the exchanger when it is not in use.
2. Insulation may be required for heat or energy conservation, personnel protection, or antisweat reasons.
3. Requirements for tracing and insulation are indicated on a flow diagram as part of the equipment description.
4. It is not feasible to trace or insulate some exchangers such as plate-type units. Shrouds must be indicated for them for personnel protection, and they must be drained when not in use to prevent freezing.

Fans, blowers, and compressors

Types of fans, blowers, and compressors. (See Fig. 3.34.)

1. Chapter 12 discusses various general types of gas movers; an appropriate symbol should be entered on the flow diagram.

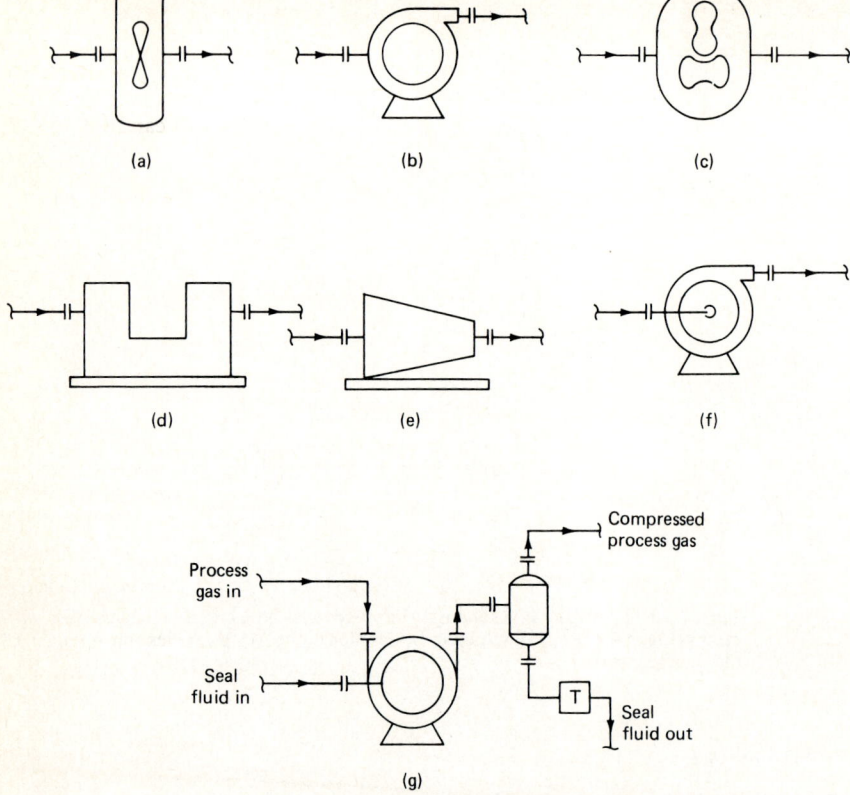

Figure 3.34 Typical fans, blowers, and compressors. (*a*) Fan. (*b*) Blower. (*c*) Rotary blower. (*d*) Two-stage reciprocating compressor. (*e*) Screw compressor. (*f*) Centrifugal compressor. (*g*) Liquid-ring compressor.

Types of drivers. (Refer to Fig. 3.26.)

1. Gas movers are usually driven by electric motors.
2. Compressors sometimes are operated by steam turbines if excess steam is available or for emergency use during an electrical interruption.
3. Gasoline or diesel engines are employed for emergency conditions or at remote locations.

Coolers

1. The heat developed in a compressor must be removed if the discharge temperature exceeds the unit's maximum allowable value, usually about 350°F.
2. Some compressors are designed to circulate cooling water in jackets or internal passages. Filtration is often required to prevent clogging if municipal water is not used as the cooling medium.
3. Intercoolers are frequently used between compressor stages with an aftercooler following the unit; provision is made to remove condensate.

4. Large compressors usually have separate lubrication systems including external lubricating-oil coolers which require municipal or filtered cooling water.

Knockout pots

1. These are required in the suction line to a blower or compressor if there is liquid carryover from the preceding step of the process or if condensate can form in the suction piping.

Valved drains

1. Despite the use of knockout pots, moisture may enter a blower or compressor and coalesce.
2. Drains also are needed to remove wash fluid that may be injected intermittently to clean the blower or compressor blades.

Accumulators

1. Accumulators are needed to smooth suction gas flow if the upstream volume is relatively small or if there are upstream pressure variations.
2. They are needed to smooth pulsations from a reciprocating compressor and reduce downstream pressure variations.

Screens, filters, and silencers

1. A screen may be needed in the inlet line of an air blower or compressor to prevent foreign objects from entering the unit.
2. A filter may be required if suction gas contains small solid particles which would cause excessive wear or a buildup of deposits. Nonlubricating reciprocating compressors are particularly subject to the abrasion of piston rings.
3. Filters are used after oil knockout pots on lubricated compressors to remove fine oil mists.
4. Silencers are sometimes needed on the suction or discharge sides of blowers or compressors to reduce noise to an acceptable level. Some high-speed centrifugals must be provided with a noise-abatement envelope.

Controls and interlocks

1. Primary sensors and control valves are shown on engineering flow diagrams to ensure economical and satisfactory operation of blowers and compressors and proper functioning of auxiliary cooling and lubrication systems.

Vacuum equipment

Types of vacuum equipment. (See Fig. 3.35.)

1. The particular type of vacuum equipment used should be represented by an appropriate symbol.

Drivers and motive forces

1. Mechanical types of vacuum equipment are usually operated by electric motors.
2. Vacuum equipment with nonmoving parts is powered by a high-pressure fluid, usually steam, air, or water. Piping for the motivating fluid is shown on the flow diagram.

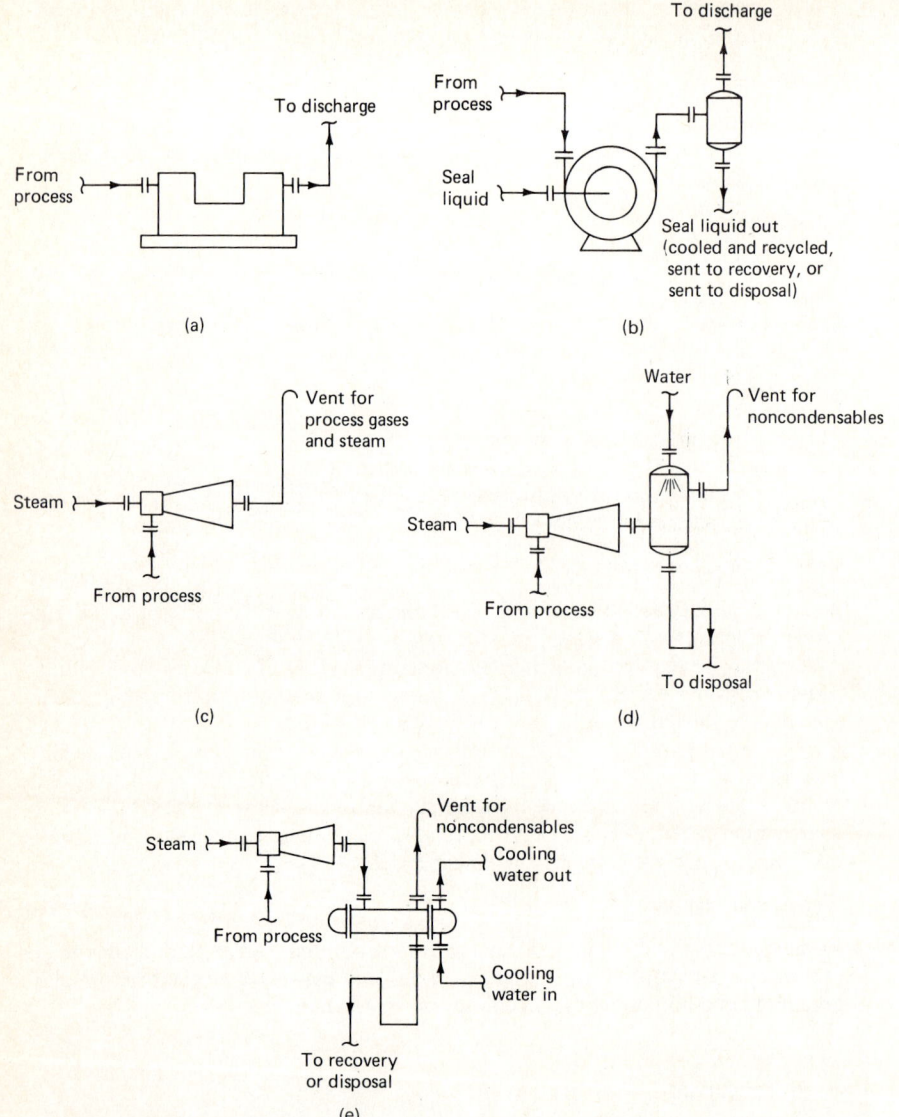

Figure 3.35 Vacuum equipment. (*a*) Two-stage reciprocating vacuum pump. (*b*) Liquid-ring vacuum pump. (*c*) Single-stage ejector. (*d*) Single-stage ejector with barometric condenser. (*e*) Single-stage ejector with aftercondenser.

Cooling vacuum equipment

1. The heat of compression is removed from nonwetted mechanical equipment by circulating water through jackets or internal passages. Interstage cooling is required if there is more than one pumping stage.
2. Liquid-ring vacuum pumps are cooled by a modest flow of fresh water or a cooled

recycle and compatible sealant fluid. There is usually a sufficient pressure differential between the discharge and suction sides to accomplish the recycle through an exhanger.
3. Steam-jet-ejector systems are supplied with interstage condensers once the interstage dew-point temperature is above that of the cooling medium. Condensation may take place either by direct contact with the cooling medium or indirectly in a heat exchanger.
4. Condensate from an interstage condenser is subatmospheric and is collected through a closed dip leg in a hot well or is pumped to disposal. The flow diagram should include a note regarding the minimum height of the interstage condenser above the hot well, taking into account the specific gravity of the continuous phase of the condensate.

Knockout pots and traps

1. Inlet knockout pots are provided to prevent entrained liquids from entering rotary, screw, or reciprocating-type vacuum pumps.
2. Oil-lubricated units are followed by a knockout pot to recover and recycle oil. Knockout pots follow liquid-ring vacuum pumps to separate the sealant from the discharge gas.
3. Motive-steam lines to jet ejectors should contain a condensate separator and be well trapped to prevent intermittent condensate particles from damaging the orifices in the jets.

Filters and centrifuges

Types of filters and centrifuges. (See Fig. 3.36.)

1. Representations of filters and centrifuges in an engineering flow diagram should be typical of their configurations.

Types of drivers

1. Centrifuges and rotating filters usually are powered by electric motors; some centrifuges use hydraulic motors to drive solids-removing plows.
2. Compressed air or nitrogen is frequently used to dislodge solids from filter socks in a baghouse or to inflate flexible members in some plate-and-frame filters for the compression of the cake or to assist in its discharge.

Additives and precoating

1. Some liquid-solids separations require the addition of surfactants or coagulants before filtration.
2. It is often necessary to precoat a filter or to mix filter aid with the filter feed.
3. All mix tanks, agitators, pumps, and controls are to be shown on the flow diagrams as auxiliaries to the filtration.

Recycle and sampling

1. A recycle line is included in batch filtration operations to ensure sufficient buildup of cake or filter medium to give the desired clarity.
2. A sampling point should be provided in a filtrate discharge line if sampling ports are not provided on the filter.

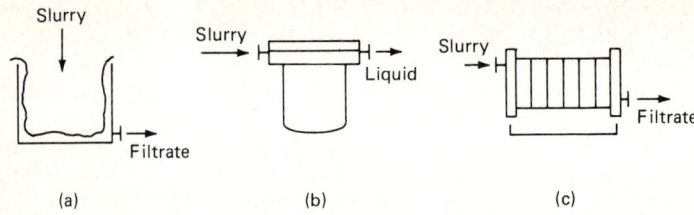
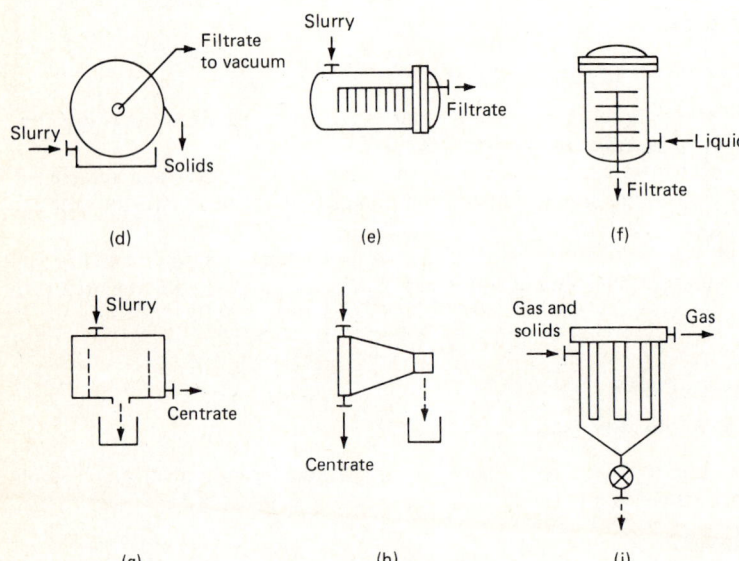

Figure 3.36 Filtration equipment. (*a*) Nutsche (batch) filter. (*b*) Cartridge filter. (*c*) Plate-and-frame filter, (*d*) Drum filter. (*e*) Leaf filter. (*f*) Sparkler filter. (*g*) Batch centrifuge. (*h*) Solid-bowl centrifuge. (*i*) Baghouse.

Auxiliary lines and equipment

1. A wash system is frequently required for filters and centrifuges.
2. A means to contain and transfer segregated solids is needed.
3. Provision must be made to collect and transfer mother liquor from a centrifuge.
4. Drum filters require vacuum equipment as an auxiliary.
5. Piping and equipment for all auxiliaries are shown on the flow diagrams.

Piping. A considerable amount of the work represented on engineering flow diagrams is concerned with piping. The following are several important aspects which should be kept in mind when preparing the diagrams.

Valves. Valves are used to isolate piping or equipment from other portions of the process and to throttle or divert flows; thus, they are an essential part of any piping system. Valves should be designated on engineering flow diagrams by a symbol which not only shows where a valve is required but indicates the type of valve. The process engineer selects the type of valve for each use from among a large variety in accordance with its (1) suitability for the service, (2) availability in the required size, pressure and temperature rating, and material of construction, and (3) installed cost.

Figure 3.37 lists the more common valves used in various processes together with suggested symbology for representing them on flow diagrams. Figure 3.38 illustrates the size ranges for some of the valves, and Table 3.4 lists their available materials of construction. However, manufacturers' engineering data should be consulted for size, temperature and pressure ranges, and materials of construction for a given valve.

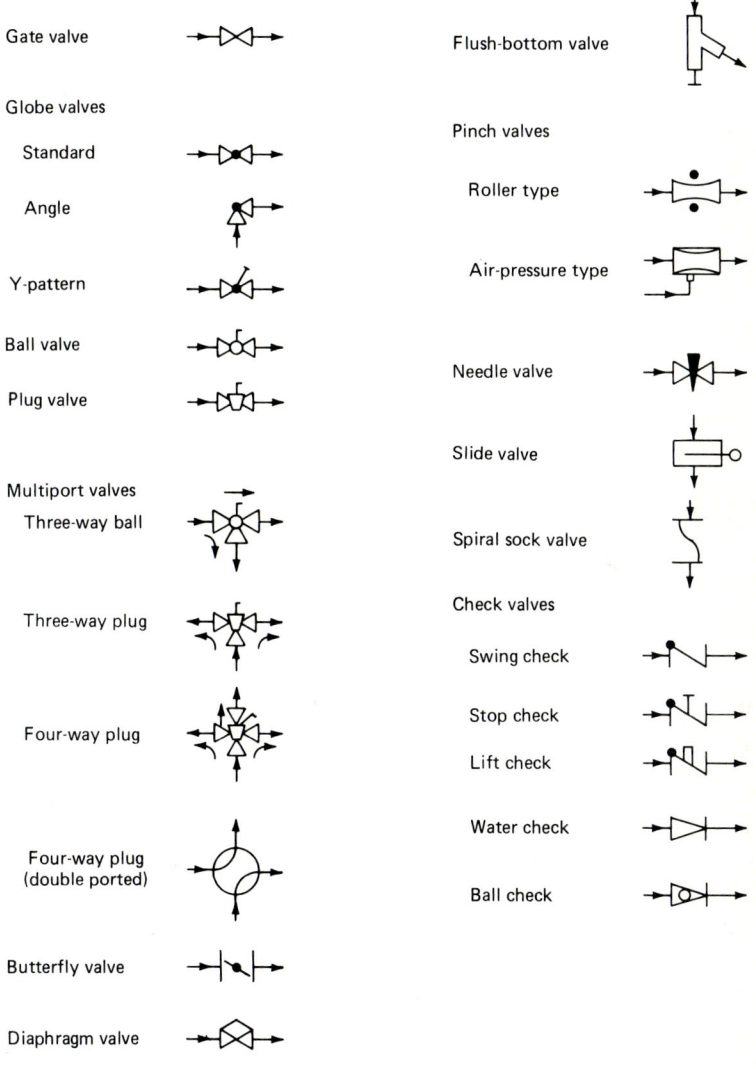

Figure 3.37 Types of common valves.

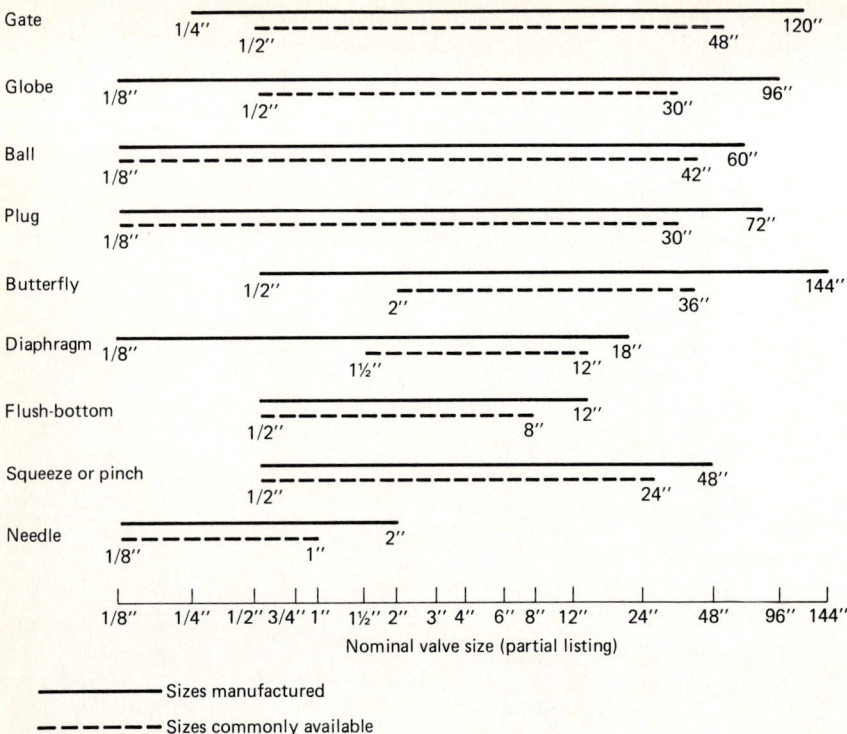

Figure 3.38 Sizes of common valves. All sizes may not be available in all ranges of materials and temperature and pressure ratings.

TABLE 3.4 Common Materials of Construction for Valves*

Type of valve	Cast iron	Ductile iron	Cast steel	Ferritic steel	Brass	Stainless steel	High alloy	Lined	Plastic	Glass ceramic
Gate	X	X	X	X	X	X	X		X	
Globe	X	X	X	X	X	X	X	X	X	
Ball	X	X	X	X	X	X	X	X	X	
Plug	X		X		X	X		X		X
Butterfly	X	X	X			X	X	X	X	
Diaphragm	X	X	X		X		X	X	X	
Flush-bottom		X	X			X	X	X		
Pinch or squeeze	X	X	X			X		X		
Needle			X		X	X	X		X	
Check	X	X	X	X	X	X	X	X		

*Based on *Chemical Processing*, June 1983, pp. 47–62.

Table 3.5 presents guidelines for selecting valves for a particular usage. The usage is first classified as being a liquid, gas, or solid flow and then categorized by the nature of the fluid, i.e., neutral, corrosive, hygienic, slurry, etc. The function of the valve, i.e., on-off or regulating, must then be evaluated. Other practical considerations take into account safety, the type of actuators available, or special installation problems

TABLE 3.5 Primary Usages for Various Common Valve Types*

Type	Liquid flows											Gaseous flows						Solid flows	
	Neutral (water, oil, etc.)		Corrosive (alkaline, acid, etc.)		Hygienic (beverages, foods, drugs)		Slurry		Fibrous suspensions	Neutral (air, steam, N₂, etc.)		Corrosive (acid vapors, chloride, etc.)		Vacuum		Abrasive powder (silica, etc.)	Lubricating powder (graphite, talc, etc.)
	On-off	Reg.	On-off	Reg.	On-off	Reg.	On-off	Reg.	On-off and reg.	On-off	Reg.	On-off	Reg.	On-off	Reg.	On-off and reg.	On-off and reg.
Gate	X		X							X	X			X			
Globe		X		X				X	X	X	X		X	X			
Ball	X		X				X			X		X					
Plug	X		X				X			X		X					
Butterfly	X	X	X	X	X	X	X	X		X	X	X	X	X	X		
Diaphragm	X	X	X	X	X	X	X	X	X	X		X					
Flush-bottom							X	X	X								
Squeeze						X	X	X	X							X	X
Pinch						X	X	X	X							X	X
Needle		X									X		X				
Slide gate																X	
Spiral sock																	X

*Based on Chart 3.2, "Valve Selection Guide," D. A. Sherwood and D. J. Whistance, *The Piping Guide*, Syentek Books, San Francisco, 1973.

that could make a certain valve incompatible with the piping requirements, e.g., a need for all-welded connections.

A review of the more common valves follows. For specific descriptions and illustrations of the variations within each category as well as details of construction, available packing and seal materials, and types of standard connections and flanges, the reader should consult standard references on the subject,* special reports in professional journals,† or manufacturers' catalogs and engineering data.

Gate valves

1. These valves afford straight-through flow passage which can be blocked by a disk perpendicular to the line of flow.
2. The disk is lifted through the line of flow by a stem (rising or nonrising).
3. The stem passes through a bonnet fitted with packing to prevent leakage of fluid along the stem.
4. The disk may be a one-piece construction which doesn't afford tight shutoff, or it may have two pieces, enabling complete closure except for flows with particulate matter. A gate valve with a knife disk can cut through solids to effect a tight closure.
5. Gate valves are used for infrequent on-off service; they are rarely used for throttling except for some slurries or fluids with suspended fibers.
6. They are large in size and weight but are relatively inexpensive and have a low pressure drop when fully open.

Globe valves

1. These valves are available in several designs: standard, angle, and Y-pattern. In all cases, the flow passes through a seat into which a tapered plug is fitted.
2. A stem is attached to the plug and passes through a packed bonnet to a handle or other operating mechanism.
3. When the stem is all the way in, the plug is in complete contact with the seat and flow stops. As the stem rises, an annular space opens for flow such that the annular area enlarges as the distance of the tapered plug from the seat increases.
4. Globe valves are well suited for throttling flows and for on-off service when they are combined with a regulator.
5. They are comparatively inexpensive but because of the tortuous fluid-flow path exhibit the greatest pressure drop of all common valves but needle valves. The Y-pattern valve has less drop than the angle valve, which has a lower drop than the standard globe valve.

Ball valves

1. A ball valve consists of a solid sphere which has a cylindrical port through an axis. The sphere is fitted into a socket in the valve housing so that the axis of the port is in line with the housing's inlet and discharge nozzles.
2. A stem rotates the ball 90°, from fully open to fully closed.

---

*Some recent references are R. H. Warring, *Handbook of Valves, Piping and Pipelines*, Gulf Publishing Company, Houston, 1982; and J. L. Lyons and C. L. Ashland, *Lyons' Encyclopedia of Valves*, Van Nostrand Reinhold Company, New York, 1975.

†*Chemical Engineering, Power Engineering,* and *Power* periodically feature such reports. One report appeared in *Power* in February 1983.

3. Suitable seal rings prevent leakage of fluid around the sphere, and O rings block fluids from escaping at the stem.
4. Ball valves are suited to a wide span of applications, from frequent on-off service to pressure regulation as well as flow control. They may be used with slurries and highly viscous liquids in addition to normally clean fluids.
5. They are comparable in price to gate valves and have the advantage of being repairable while the valve body is in a line.
6. They are available in full-port (the diameter of the port is the same as the line size) or reduced-port (70 to 85 percent of line size) types. The pressure drop of a full-port valve is slightly less than that of a gate valve, while the drop for a reduced-port (standard) ball valve is about 3 times as great.

Plug valves

1. These are similar to ball valves except that they use a tapered-plug closure mechanism instead of a sphere.
2. The port is usually oblong and is smaller in cross-sectional area than a circular port in a ball valve of the same nominal size.
3. The precision fit of the tapered plug into its housing and an O ring on a rotating stem prevents leakage out. Plugs may sometimes be lubricated to prevent galling, but lubrication must be compatible with the fluid service and operating temperature.
4. Plug valves are suited to on-off and throttling service and are excellent in vacuum conditions.
5. They are somewhat more expensive than ball valves in a comparable material, but the tapered plug can be lined with a polymeric or special material, resulting in an inexpensive valve with the sealing characteristics of a lubricated valve suitable for many corrosive services.
6. The pressure drop across the valve is greater by an order of magnitude than that of a gate valve.

Multiport ball and plug valves

1. Ball valves and plug valves can be constructed so that a fluid can enter at one nozzle and leave at one or more of several exit nozzles. The arrangements are known as three-, four-, and five-way valves.
2. A four-way double-ported valve can be used to alternate two streams to and from two sets of equipment.
3. Small directional arrows should be entered with multiport-valve representations on a flow diagram so that the correct configuration of the valve can be specified.

Butterfly valves

1. A disk (flapper) rotates on a pivotal stem within the valve so that flow is stopped completely when the disk is perpendicular to the flow path and is maximum when it is parallel.
2. Butterfly valves are often used in a wafer form which fits between flanges in a piping section; they are also available with flanged ends.
3. They are well suited for throttling as well as on-off operation. Seats are required for bubbletight shutoff.
4. Butterfly valves are the least expensive valves. They are available in lined material for corrosive service.
5. Their pressure drop is somewhat greater than that of gate valves.

Diaphragm valves

1. Diaphragm valves are usually constructed with a cusp or weir in a housing against which a flexible membrane, or diaphragm, impinges to form a seal. The diaphragm is connected to a stem which passes through the bonnet of the valve.
2. These valves are also constructed in a straight-through weirless design for special slurry applications.
3. They are well suited for throttling and on-off services with slurries.
4. They are available in lined bodies, which are considerably less expensive than high-alloy gate valves for corrosive service.
5. They have a high pressure drop.

Flush-bottom valves

1. This type of valve is designed so that there is no fluid in the body or the connecting nozzle when the valve is closed. It is well suited as a draw-off valve for slurries where packing of a dead space can occur or for a solution in which unwanted reactions might take place if the solution were unagitated.
2. There are two basic types; the receding-piston, or ram, valve is excellent for slurries, while the lifted-plug type is used with liquids.
3. Small piston or plunger types of valves are used for sampling slurries or liquids when it is important that there be no holdup in the connection to the sampling point. When the valve is closed, the piston is flush with the interior surface of the vessel or pipe.

Pinch or squeeze valves

1. One type of pinch valve consists of a flexible tube between a set of solid rollers which, when brought together, squeeze the tube to regulate flow or stop it completely.
2. In another design, the flexible tube is attached internally to both ends of a metallic housing. Regulating the pressure of air admitted to the space between the housing and the tube achieves a throttling or shutoff action.
3. These valves are low-cost, low-pressure-drop units well suited for slurries or gels. Large valves can be used in water systems, as they can effect a shutoff against big solid objects such as rocks, logs, or trash.
4. Flexible membranes are susceptible to erosion with some slurries and may have to be replaced periodically.
5. These valves have limitations of low operating temperature and pressure. Flexible membranes may deteriorate in some solvents or corrosive fluids.
6. These valves cannot be used in actual or incipient vacuum conditions as they tend to close and choke. If they are used on a pump suction, there must be a positive suction head at the valve under all conditions.

Needle valves

1. Needle valves are similar to globe valves except that the plug is an elongated needle which fits into a precisely tapered seat.
2. They are used to obtain precise throttling of low flows of clean fluids.

Slide valves

1. Slide valves are a form of gate valve consisting of a thin rectangular blade that moves in and out of a housing.

2. They are used to regulate or block the flow of solids.
3. Elastomeric seals on which the blade can impinge are provided within the housing to prevent gas or vapor leakage when a tight shutoff is required.

Spiral sock valves

1. These valves consist of a twistable fabric or elastomeric sock or tube.
2. They are used to control the flow of powders.

Check valves

1. Several types of check valves are available to prevent reversal of flow in a piping run.
2. Most common for horizontal runs is a swing check which consists of a hinged disk inside a housing. The disk opens when the flow is in the desired direction but is forced closed against a seat if the flow tries to go the other way. If a spring is used at the pivot, the valve may also be installed in a vertical run.
3. A lift check valve has a plug which is lifted when the flow is in the correct path but drops and reseats when the flow is reversed. It usually depends on gravity for closure, but a spring may be included to assist return of the plug (the valve is then known as a piston check valve).
4. A ball check valve is similar to a lift check valve except that a spherical element is forced from its seat under a positive pressure differential and returns to stop the flow. It may be used with some slurries since the spherical surface is self-cleaning.
5. A stop check valve not only prevents flow reversal but can be adjusted to vary the maximum rate of the forward flow.
6. Check valves have high pressure drops owing to a minimum differential imposed by the inertia of the moving elements and to friction losses resulting from the tortuous paths of the fluid.
7. Most check valves are suited to liquid and high-pressure gas service. A split flapper valve having a thin divided disk pivoting on a hinge, with or without a light spring, is required for low-pressure or vacuum service in which only a nominal pressure drop can be allowed.

Piping runs. Several aspects of the piping itself or its representation should be considered.

Bypasses for control valves

1. On-off valves, i.e., blocking valves, are usually placed immediately upstream and downstream of an automated control valve in continuous or critical service along with a bypass line containing a manual throttling valve about the group. Such a configuration permits continued operation or orderly and safe shutdown of a process in the event of a controller or control-valve malfunction.
2. The block valves and bypass line and valve are usually of line size for main lines up to $2\frac{1}{2}$-in diameter. For 3-in main lines and larger, it is good practice to have the bypass line and valve one line size smaller.
3. Several examples of bypass applications about control valves can be seen on the P&ID shown in Fig. 3.2. Some self-regulating valves in clean service, such as air-pressure regulators, are seldom provided with a bypass line.

Vent and drain sizes

1. Vent lines on a vessel should be of at least the size of the largest incoming non-recycled line.

TABLE 3.6 Suggested Minimum Vessel Vent and Drain Sizes

Vessel capacity, gal	Minimum vent size, in	Minimum drain size, in
Less than 350	$\frac{3}{4}$	1
350–1500	1	$1\frac{1}{2}$
1501–4500	1	2
4501–18,750	$1\frac{1}{2}$	3
Over 18,750	2	3

2. In addition, suggested minimum vent and drain sizes for vessels of various capacities are given in Table 3.6; these should permit normal filling or draining of an atmospheric vessel in a reasonable period without overpressure or collapse.

Condensate traps or drains on vents

1. Vent lines or relief-valve discharge lines which carry steam or condensable water vapor and have a horizontal section that can form a seal leg should be provided with a drain hole, if permissible, or with a suitable condensate trap at low points to prevent excessive pressure drops or the development of excessive and dangerous velocities of trapped liquids.
2. A manual drain is usually suitable for intermittent operations.

Protective screens

1. Atmospheric lines to the suction of a blower or compressor as well as pump intake lines from natural bodies of water or from sumps should be protected by screens from the entrance of foreign bodies.

Jacketing and bundling.   (See Fig. 3.39.)

1. An outer pipe, known as a jacket, is sometimes placed around a process line. An appropriate heating or cooling utility stream flows in the annulus to maintain the process fluid at a desired temperature.

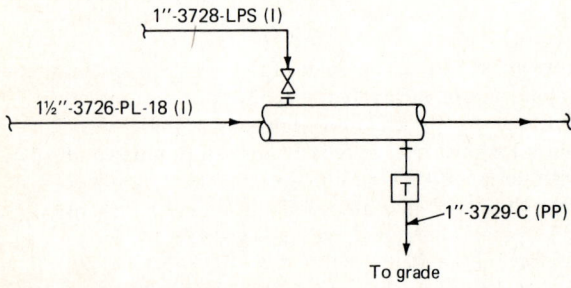

Figure 3.39 Jacketing and bundling.

2. Jacketing is shown on the body of the flow diagram, and the word "jacketed" is entered in the tracing column of the line tabulation.
3. A line is "bundled" when it is placed next to a hot or refrigerated line, and both lines are then insulated together. The word "bundled" is placed in the tracing column of the line tabulation.

Boundary definitions. (See Fig. 3.40.)

1. While building outlines per se are seldom defined on engineering flow diagrams, it may be necessary to locate a change point through which a pipeline passes for tracing, insulation, piping classification, or safety reasons.
2. The transition between classifications is marked by a short indicator perpendicular to the piping line. The boundary conditions are delineated.

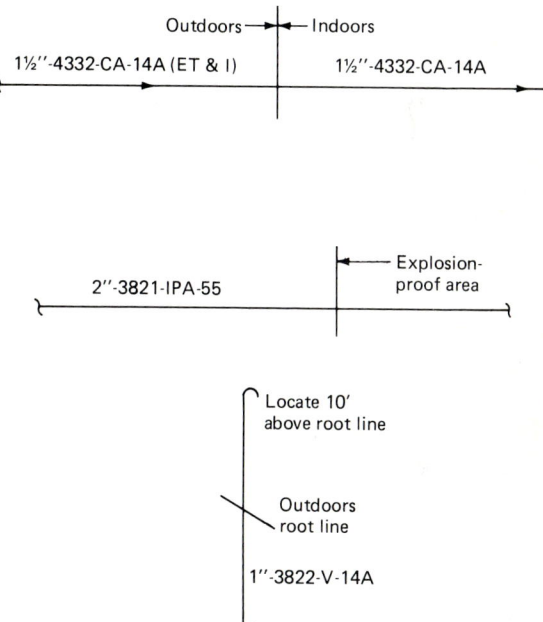

Figure 3.40 Boundary definitions.

Instrumentation and safety. The process engineer enters all sensing instrumentation, control valves, and safety devices on the P&ID or engineering flow diagram and designates whether instrument measurements are to local, or remotely indicated, or as recorded points. The instrument engineer completes the control loops on the P&ID or on a separate instrumentation flow diagram.

The following subsections discuss some of the more common instruments and safety devices to be considered for application on a flow diagram. Other specialized items should be provided as required.

Relief valves

1. A relief valve, also known as a pressure safety valve, protects equipment or piping from pressures beyond its design maximum allowable working pressure (MAWP).

2. Typical situations for relief-valve application are:
   a. Any vessel, exchanger, column, or other equipment that can be completely isolated by valving must be protected from an external fire or runaway exothermic heats of reaction.
   b. Vessels with an open vent or overflow that is of such size or length that excessive pressure could be generated in the event of fire or reaction excursion must be protected.
   c. Protection is also needed when the maximum discharge pressure of a pump or compressor feeding a piece of equipment or piping is greater than its MWAP and the pump or compressor can be deadheaded through the equipment or piping.
   d. Exchangers need protection when a liquid "cold" side is isolated and expanded owing to the continued flow of the "hot" side.
   e. A relief valve is usually placed after a pressure-reduction valve to ensure that subsequent equipment is protected in the event of a malfunction of the reduction valve.
   f. If a relief valve may not reseat completely owing to fouling by solids or gummy materials, a pair of relief valves is used in parallel and two ganged three-way valves are incorporated in the piping so that there is always one relief valve functioning with the equipment while the other is being cleaned.
3. Table 3.7 lists several of the basic codes that govern the sizing and installation of relief valves.
4. The sizes of relief-valve inlet and discharge lines are shown on the flow diagrams. The discharge side of most vapor safety valves is usually larger than that of the inlet, while those in liquid service usually have equal inlet and discharge connections.

TABLE 3.7 Codes Governing the Sizing and Installation of Relief Valves and Rupture Disks

Code*	Title
API RP 521	Guide for Pressure Relief and Depressuring Systems
API RP 526	Flanged Steel Safety Relief Valves for Use in Petroleum Refining
API RP 520	Recommended Practice for the Design and Installation of Pressure Relieving Systems in Refineries Part I: Design Part II: Installation
ASME Sec. VIII	Par. UG-125 General Par. UG-126 Safety and Relief Valves Par. UG-127 Rupture Disks Par. UG-128 Liquid Relief Par. UG-132 Determination of Pressure Relieving Requirements Par. UG-133 Pressure Setting Par. UG-134 Installation Par. UG-230 Capacity Calculation

*API: American Petroleum Institute, Washington, D.C. 20037; ASME: American Society of Mechanical Engineers, New York, N.Y. 10017.

Rupture disks

1. A rupture disk is an alternative to a relief valve when it is acceptable to allow pressure in equipment or piping to fall and to lose material until atmospheric pressure is reached.
2. It consists of a frangible wafer of composite materials or a thin metallic element which is shattered or ripped apart at a predetermined pressure.
3. The material of the wafer and the metallic element must be compatible with the fluid in the equipment or piping. A thin membrane of Teflon or other polymeric material is used to prevent chemical attack of the disk and results in an economical construction.
4. A rupture disk can be placed ahead of a relief valve to protect the relief valve from plugging or to permit the valve to be constructed of more economical materials. A pressure gauge is usually placed between the relief valve and the rupture disk to indicate the integrity of the disk.

Vacuum breakers

1. A vacuum breaker permits atmospheric air or a compatible gas or vapor to enter a vessel at a determined vacuum level to prevent collapse of the unit at its maximum vacuum design pressure.
2. Typical situations that require a vacuum breaker are the withdrawal of liquid with insufficient replacement of gas or vapor, the condensation of vapors in an isolated piece of equipment whereby the pressure in the vessel is reduced, and the connection of vessels to a powered vent system.

Conservation vents

1. A conservation vent is a combination of a relief valve and a vacuum breaker.
2. It normally allows storage vessels containing volatile fluids to float within a limited pressure range so that the loss of vapors is reduced.

Flame arresters

1. The vent on any unit containing an inflammable fluid should be provided with a flame arrester to prevent backflashing if the discharge vapors become ignited.
2. Conservation vents may be purchased with integral flame arresters.

Pressure sensors

1. Pressure connections to equipment should be made, where possible, in the gas or vapor space to eliminate corrections for hydraulic head.
2. Units such as filters, baghouses, heat exchangers, or distillation columns which induce considerable pressure drops often have a differential-pressure measurement across the unit in addition to an absolute- or gauge-pressure sensor at the inlet or discharge.
3. A pressure gauge on the discharge of a pump provides information regarding the operation of the pump; a permanent gauge (PI) may be installed, or a valved provisional tap (PP) may be used instead.
4. Pressure gauges should be isolated from equipment or piping with a suitable valve or chemical seal. The latter is used with corrosive fluids, toxic fluids, slurries, or fluids that would solidify in the gauge. All isolating valves and chemical seals are shown on the flow diagrams.

Temperature sensors

1. Whenever applicable, thermowells are located in the liquid portion of a two-phase system to give quicker responses.
2. Thermowells, with or without indicators, are frequently provided upstream and downstream of a heat exchanger.
3. It is good practice to provide a redundant temperature measurement to activate an alarm function in addition to the normal temperature measurement used to control an exothermic operation.

Sight glasses and lights

1. A viewing port is often included in agitated vessels or on distillation columns especially if there may be foaming problems.
2. If caustic or hydrogen fluoride is present, glass ports must be protected by a suitable transparent insert between them and the fluid.
3. Sight glasses are available in multiple sizes and configurations for use in piping. They are employed to indicate the presence of liquid or an interface, whether the liquid is stationary or flowing (the turning of a wheel), and the approximate flow rate (the lifting of a flapper).
4. A sight glass is often accompanied by a sight light; all sight glasses and lights are shown on the flow diagrams.

Level indicators. (See Fig. 3.41.)

1. All process vessels containing liquids require level measurement.
2. The simplest indicators are a shadowed liquid level through the translucent wall of a plastic tank with a calibration on the outside of the tank or a calibrated dipstick for use as an intermittent indicator in noncritical, nonhazardous service.
3. A simple continuous-indicating device is a window or a series of windows in the straight side or a vertical cylindrical glass tube along the side of a vessel. A series of tubes is required to determine an interfacial level.
4. A series of hollow external metal columns provided with translucent reflux faces is used when the application of glass windows or cylinders is limited by the design pressure of the vessel. This system is not suitable for showing interfaces.
5. Means must be provided to drain, flush, and clean external gauges.
6. A pressure-sensing device may be used at or near the bottom of a vessel as an indication of level. A differential-pressure unit with one leg connected to the vapor space is required if the tank is not open or if its pressure in the vapor space is greater than a few inches of water column.
7. A chemical seal is used to isolate the pressure sensor from liquids which are corrosive or toxic, could freeze at ambient temperatures, or are slurries.
8. A common level indicator uses a top-entering dip tube through which a small metered flow of compressed air, nitrogen, or other compatible gas* is metered and bubbles through the liquid. The pressure of the gas, corrected for liquid density, indicates the height of the liquid above the end of the dip tube. A differential-type pressure indicator is required for pressurized vessels.
9. Bubble-type indicators should not be used with saturated or nearly saturated solutions since evaporation may leave deposits at the end of the tube and lead to obstructions, causing erroneous level readings.

---

*Air should not be bubbled into flammable liquids.

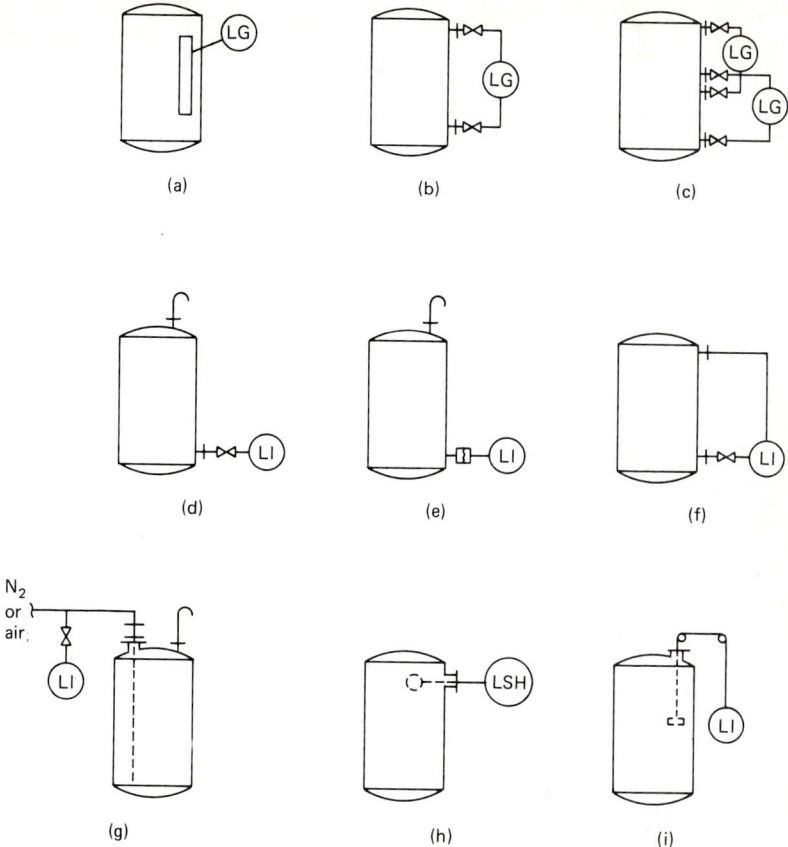

Figure 3.41 Typical level gauges. (*a*) Window-type level gauge. (*b*) External level gauge. (*c*) Interfacial level gauges. (*d*) Pressure gauge as level indicator. (*e*) Pressure gauge with chemical seal as level indicator. (*f*) Differential-pressure gauge as level indicator. (*g*) Bubbler-type level indicator. (*h*) Ball-float level switch. (*i*) Float-type level indicator.

10. A guided float sensor gives a continuous level reading of the surface of a liquid or a pile of solids. It is also used to indicate a narrow-range, "either-or" local condition.
11. Paddles, flappers, capacitance and sonic probes, and radiation units are some of the devices employed for specialized level indications. Strain gauges are used to determine weight. Notes on the flow diagrams are used to denote these special level indicators.

Flexible connections or hoses

1. Equipment or piping made of ceramics, glass, and certain resinous materials is isolated by flexible connections or hoses from damaging vibrations caused by mechanical equipment, especially reciprocating or high-speed types of gas movers (see the discharge of the air blower B-1741 in Fig. 3.2).
2. Flexible connectors are frequently used to ensure that liquid lines of 8-in diameter

or larger do not overstress inlet or discharge nozzles on pumps, filters, or similar equipment.
3. Flexible connections serve to isolate a weigh tank from its fixed entrance, discharge, or vent lines to permit unimpeded movement of the vessel.

Expansion joints. (See Fig. 3.42.)

1. An expansion joint is used in piping in lieu of bends and loops to account for linear growth due to changes in temperature.
2. Differential linear growth between the two sides of a heat exchanger is usually relieved by placing an expansion joint in the shell unless the tubes are coils or U tubes or the unit has a floating, pull-through, or packed-head configuration.
3. Consideration should be given to the extremes of start-up or emergency conditions as well as normal operating temperature in determining whether or not an expansion joint is required.

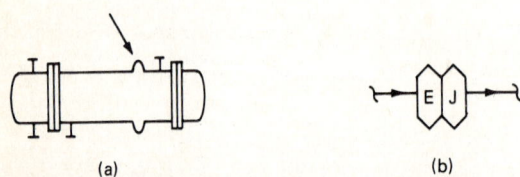

Figure 3.42 Typical expansion joints. (*a*) Fixed-tube-sheet exchanger with expansion joint on shell. (*b*) Piping run with expansion joint.

Fire valves

1. These are ball or plug valves which maintain their closure by a built-in spring mechanism but are held open in normal operation by a link mechanism containing a fusible section.
2. They are recommended for installation at draw-off nozzles on vessels with flammable contents.

Spring-closure valves. (See. Fig. 3.43.)

1. These valves are used when it is important that a manually operated valve return to a closed, open, or partially open position after being used.

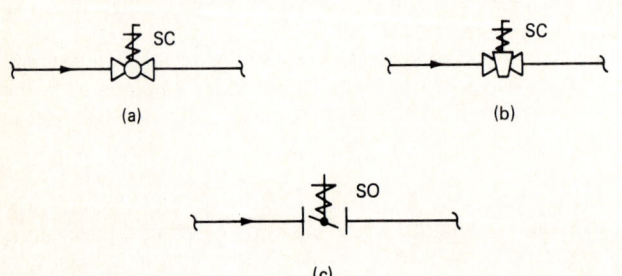

Figure 3.43 (*a*) Spring-closed ball valve. (*b*) Spring-closed plug valve. (*c*) Spring-opened butterfly valve.

2. They are often used instead of a restriction orifice in services which tend to clog openings. If the valve begins to clog, it can be opened fully and then returned to a stop for its partially open position.

Locked valves

1. Certain valves should remain in an open or a closed position until a change is authorized.
2. The abbreviation LO or LC next to a valve on a flow diagram notes that a valve is to be locked open or locked closed.

Limit switches

1. Limit switches are required when the position of a process or instrument valve must be known before a particular operational step can be taken or as part of an interlocked permissive sequence.
2. Information is provided by small limit switches on the valve housing. This is signified by using a Y as the final identification letter in an instrument bubble.
3. Only one switch is required to know whether a valve is fully open or fully closed. Two switches are needed to know if the valve is either fully open or fully closed.

Automatic switch-over of pumps

1. Automatic switch-over is required in critical services when a pumping operation must be maintained despite mechanical or electrical problems with the pump currently running.
2. Isolating valves about the spare pump are kept open and check valves provided in the discharge line from each pump.
3. A pressure-sensing device or flow switch in a common discharge line detects a pump failure, sounds an alarm, and automatically starts the reserve pump. Hold Tank Pumps P-1715 and P-1716 in Fig. 3.2 illustrate the procedure with indicator EA-1947 being electrically common to both.
4. Pumps with severe but intermittent duties are frequently placed on alternate start-up, whereby the pump started automatically is the one that had not previously been operating.

Failure alarms

1. Current to motors of critical pumps, compressors, agitators, or similar equipment is sometimes monitored to indicate the status of the process or of the equipment itself.
2. Engineering flow diagrams should reflect the presence of current indicators and accompanying low or high alarms. Agitator A-1715 in Fig. 3.2 illustrates such an indicator.

Interlocks

1. Often the operation of a piece of equipment depends on another function in the process before it can begin, continue, or terminate its operation.
2. The permissive conditions are covered by the appropriate instrumentation or notes on the engineering flow diagrams. Complex interlocks are indicated by an appropriate symbol such as ◇I◇ to show that a separate logic diagram prepared by the instrument engineer covers the instrumentation and sequence of operations.

Miscellanous and specialty items. Flow diagrams show many miscellaneous and specialty items to complete a project. Some of the more frequently encountered features are reviewed here.

Expansion tanks

1. Expansion tanks are required to relieve thermal expansion when liquid-filled sections of piping runs can be isolated and subsequently heated.
2. They may not be needed with water, as developed pressure is relieved by minor distortions of piping plus a small leakage at gaskets or by the use of a thermal relief valve.
3. They can be used to reduce water hammer or induced shock waves for any fluid.
4. Units are shown on a flow diagram with their dimensions or as a note giving a make and model number if they are to be purchased units.

Backflow preventers and air gaps

1. Backflow preventers are required by authorities supplying municipal water to prevent contamination of the water supply.
2. Backflow preventers or air gaps are used within a plant to ensure that contaminants do not enter the potable, locker-room, or safety-shower water systems.
3. A backflow preventer is a commercial item consisting of a series of special check valves* and pressure-differential chambers to ensure the absence of backflow. An air gap is an arrangement whereby the line bringing incipient process or plant water from the municipal supply is terminated several inches above the tank that is to supply the process or plant water.

Safety showers and eyewashes

1. These are required whenever hazardous liquids or solids are being unloaded or handled.
2. They are usually purchased as a single combination unit.
3. Units are sized to give about 30 gal/min to the shower portion and 5 gal/min to the eyewash for a water pressure of 30 psig. If the pressure is less, it must be boosted, and if it is too great, units must be supplied with the requisite restriction orifices.
4. The units should be connected directly to the potable-water supply. If they are taken from a pumped process- or plant-water supply, their feed lines should be taken immediately after the discharge of the process-water pump. A backflow preventer isolates the shower-eyewash-water takeoff point from any process or plant users.
5. Units located outdoors must be protected from freezing. This is accomplished by having an underground water-supply assembly designed so that water remains below the freezing level until the unit is activated and residual water at the end of use passes into a prepared drain at the base. Aboveground supply piping is usually protected by electric self-limiting tracing or electric induction heating. The shower or eyewash unit itself is then supplied with tracing or induction heating.
6. Safety-shower and eyewash units are shown on engineering flow diagrams near the equipment with which they are needed with a reference to the utility flow diagram containing the distribution header.

---

*Standard check valves may not be used as substitutes unless they are approved by local ordinances.

7. Symbolic containers showing dilute acetic acid or carbonate solutions are often shown next to an eyewash to denote their presence for swabbing purposes.

Utility service stations

1. It is good engineering practice to place utility hose stations at convenient locations throughout a plant. There should be air for such tasks as blowing dirt or water from equipment, unplugging or cleaning out lines, and operating pneumatically driven mechanical equipment or tools; water for housekeeping and emergency cooling; and steam to thaw or clean lines and equipment. Steam and water can be brought together to form a separate hot-water station.
2. Stations are sited on equipment-arrangement drawings and given identifying numbers. Each station is shown on the respective utility flow diagram and identified by the assigned number.
3. Air, steam, and water pressure for normal usage is limited to 30 psig by pressure reducers to prevent a pressure hazard caused by high-velocity fluid impinging on the skin or by the whipping action of hoses. High-pressure air required to drive pneumatic motors or tools should have special connectors.
4. Hoses are typically 30 ft long. If a series of hose stations is required, they can be located 50 to 60 ft apart.

Fire protection

1. Fire protection is required in operating and storage areas if flammable ingredients are used in the process or if gaseous or liquid fuels are present.
2. It is provided for control rooms, laboratories, maintenance shops, and office areas.
3. Although chemical canister units are provided, fire hydrants with fire-water pumps and underground fire-water mains are frequently included. The number and location of fire hydrants are set in accordance with local ordinances. The National Fire Protection Association (NFPA), Quincy, Massachusetts, provides standards for fire-protection equipment.
4. A method to reduce danger from a fire within a large oil storage tank is to sparge steam into the vapor space to displace air and snuff out the fire by smothering it.

Seal fluid systems

1. Reference is made at appropriate seal or packing locations on engineering flow diagrams to the seal or flushing fluid system required at the respective application point.
2. A notation lists the flow diagram which presents the distribution header and the line-identification codes of lines bringing and returning the fluid.
3. A separate section on a utility flow diagram should illustrate a schematic of the typical instrumentation and piping required at each application point for the various systems, list the various instruments and line numbers, and show the distribution systems with pertinent information in the line table.

Notes. The notes area on a P&ID or an engineering flow diagram can be used by the process engineer to clarify, explain, or give specific instructions regarding items on the drawings. Some typical categories covered by notes are:

1. Designate reference drawings for symbols and legends.
2. Add to or emphasize aspects of piping specifications for special conditions.
3. Specify minimum or maximum distances for process reasons.

4. Designate reference drawings for seal and bearing fluids.
5. Designate reference drawings for purges.
6. List special materials of construction.
7. Specify special locations for equipment, piping, or fittings.
8. Delineate sequences for interlocks.
9. Present instructions for the use of special spool pieces.
10. Present instructions for environmental compliance, sampling, and data reporting.
11. List special requirements for tracing and insulation.
12. Record special conditions for equipment furnished by others.
13. Designate equipment model numbers for specialized equipment.
14. Indicate special properties or hazards of process materials.

Chapter 4

Logistics of Flow Sheets and Flow Diagrams

Piping and instrumentation drawings (P&IDs) and/or engineering flow diagrams are not only complete road maps which enable the mechanical squad to prepare its routing of piping but also are vital links that connect all the disciplines to the basic project objectives. It is, therefore, of the utmost importance that engineering flow diagrams always reflect current information on the status of the project and that all engineers and designers associated with the project be aware of their current status.

The updating of engineering flow diagrams, like their preparation, is under the jurisdiction of the process engineers. Theirs is the responsibility of initiating and supervising the preparation of the various issues of the flow diagrams, distributing copies of the issues to the several disciplines and to the client, soliciting comments and changes, and maintaining a set of "master prints" which reflect the current status of the project.

Types of Drawing Issues

There are normally three stages in the preparation of the process flow sheets and the P&IDs. Various design groups may have different names or symbols for these stages. Those employed in the following discussions are the nomenclature in general use by engineering design companies and can be reconciled to other designations through the basic definitions. These are (1) the preliminary issues, (2) the zero issue or issued-for-construction drawings, and (3) revisions. A fourth and important type of issue for internal use is the interim issue.

The documentation of P&IDs or engineering flow diagrams as they pass through their various stages is given in the following subsections. The process flow sheets with the material and energy balances are treated similarly but do not normally have as great a distribution.

Preliminary issues

As was discussed in the subsection "Informational Basis of Engineering Flow Diagrams" in Chapter 3, engineering flow diagrams usually evolve from the process flow sheets. While only the more basic processing equipment is shown on the process flow sheets, the engineering flow diagrams show not only this equipment but all the auxiliary and subsidiary equipment for the processing sections, all the utility and specialty requirements, and a complete representation of the piping needed. Should an engineering flow diagram be intended as a P&ID, it would eventually contain all pertinent elements of the instrumentation required for the project. Should a separate instrumentation application diagram be used, only those elements of the instrumentation such as the basic sensors, control valves, or orifices would be included on the engineering flow diagrams.

Since the engineering flow diagrams are based on the process flow sheets, they generally go through several stages of development. The initial drawings in this evolution are usually termed "preliminary" issues to denote that developments for the given engineering flow diagram may not have been completed or approved by the client. The first set of engineering drawings is therefore labeled P-1 or P-A in the revision section of the title block, which is usually located in the lower right-hand corner of a drawing (see Fig. 4.1).

At this early stage in the preparation of the engineering flow diagram, many areas may well contain uncertainties with regard to the process itself or the equipment that will be used. In that case, the process engineer should enter on the flow diagram the most likely representation for the equipment or piping which is in doubt. To indicate that this is only a tentative solution to the problem the area to be resolved is enclosed with a heavy line and the word "HOLD" is printed within the enclosure. The hold is maintained until the problem has been resolved. Other disciplines are advised by

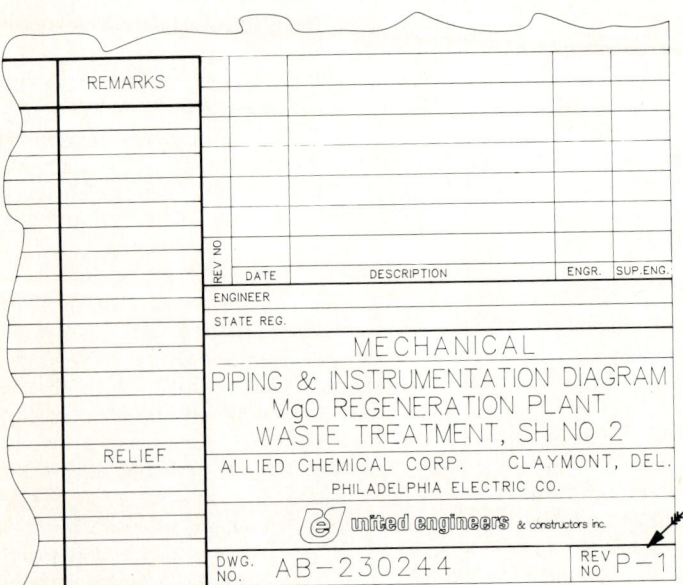

Figure 4.1 Title block, P-1 issue.

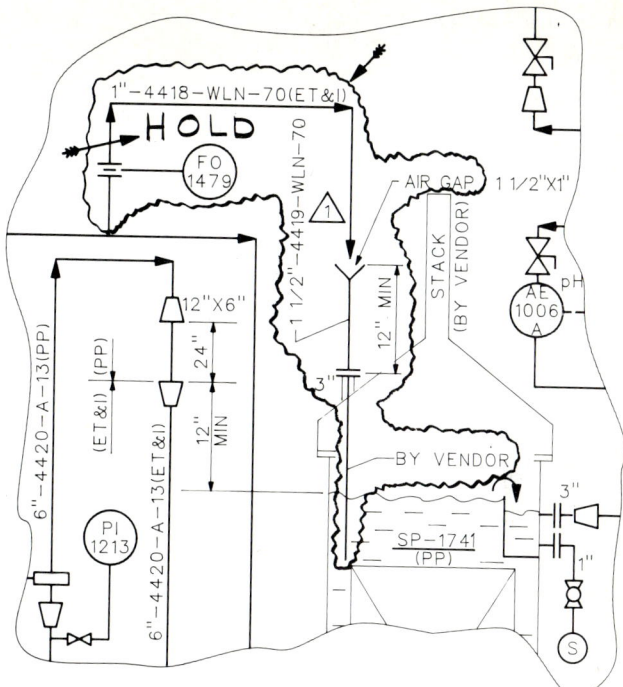

Figure 4.2 "HOLD" on design work.

this means not to act definitively on the work represented inside the enclosure. A typical hold situation is shown in Fig. 4.2.

At this point, process engineers request a meeting with all the discipline engineers and lead designers. The engineering flow diagrams should be presented to them and the purpose of each piece of equipment and line explained in detail. Any special nuances in the design of equipment should be noted and thoroughly reviewed. This type of conference enables the other disciplines to appreciate the scope of the project as seen from the viewpoint of the process engineers and helps them to make better-informed comments and suggestions on this and subsequent drawing issues.

A conference at an early stage of the project also helps the various disciplines to be aware of any special conditions or problems so that their initial efforts reflect the requirements of the process and do not have to be corrected or redone at a later time. A good example on Fig. 3.2 is that of line 4414-WLN entering line 4352-WLA with respect to line 4339-CA. Frequently, it makes no difference in which order two lines enter a third one. In this case, however, line 4414-WLN is intended not only to serve as a continuous recycle for pumps P-1741 and P-1742 but also to provide a continuous flow to all sections of line 4352-WLA downstream of the point where caustic from line 4339-CA enters it in the event that flow from V-1713 stops. In actuality, it would have been more convenient from a physical standpoint to have line 4414-WLN enter line 4352-WLA about 25 ft before line 4339-CA. However, the lead piping designer, having understood the intention for line 4414-WLN through discussion with the process engineer, routed the pipe correctly.

Prints of the set of preliminary P&IDs or engineering flow diagrams are made and distributed to all the discipline engineers and the lead designers with requests that they be reviewed and comments relevant to the particular discipline be entered and returned to the process engineer. The instrumentation group must pay special attention to this initial set of preliminary drawings, or P-1 issue. The P-1 issue includes the rudimentary portions of the instrumentation, and it is the duty of the instrument engineers to complete all the other elements of instrumentation that are required. If the P&IDs are to show the instrumentation and control requirements, the instrument engineers enter their comments and additional instrumentation requirements on the P-1 issue for return to the process engineers for entry on the tracings.* If separate instrument application diagrams are being used, the instrument engineers comment on the primary instrumentation entered by the process engineers and use the preliminary engineering flow diagrams as the basis for their application diagrams.

After receiving and reviewing the comments from the various discipline engineers, the process engineers should have a separate conference with each of the discipline engineers and lead designers to ensure that there is mutual understanding of why their comments and suggestions can or cannot be accepted. The process engineers then update the flow diagrams on the basis of the discussions with the discipline engineers and also ensure that the remainder of instrumentation work prepared by the instrument group is incorporated in the engineering flow diagrams. When this is completed, the drawings may be forwarded to the client or to the sponsoring division as the P-1 issue with the request that they be reviewed and commented upon. Any hold areas on the drawings and alternative choices for these situations should be indicated to the client with a request that the hold areas be resolved as soon as possible.

Comments from the client should be returned as soon as possible as marked-up drawings and written commentary when required. A meeting should then be arranged with the project manager for the client plus the necessary pertinent discipline engineers from the client group. The process and instrument engineers and other required discipline engineers represent the design group at the meeting with the client.

After a discussion of the client's markings and comments and a review of the suggestions regarding the areas marked "HOLD," plus any additional inputs from the engineering design group, there should be a consensus regarding the preliminary versions of the engineering flow diagrams. The agreed-upon changes to the engineering flow diagrams should be made to their tracings for the P-2, or P-B, issue.

Frequently, some of the piping in the P-1 issue may have been left unsized. The sizing of these lines should be completed for the P-2 issue. Every effort should then be made to clear up areas marked "HOLD" for this revision.

The P-2 issue should be reviewed internally within the discipline design groups in a manner similar to that used for the P-1 issue. Prints of the revised tracings are then forwarded to the client for approval.

The zero issue

Usually two or three preliminary issues of the engineering flow diagrams are sufficient to resolve the major questions regarding the process, equipment, and piping

---

*A tracing is an original linen or Mylar sheet used for a permanent drawing. Prints can be made from a tracing by several techniques. The tracing is altered as required as the project progresses. Sepias are transparencies derived from tracings at a given stage so that prints can be obtained of that stage when later required.

necessary to accomplish the objectives of the process. When there is that agreement, the flow diagrams may be made the "zero issue." This is accomplished by entering a 0 as the suffix to the drawing number and placing it in the revision box or by conforming to whatever is the custom of the design group. At the same time the words "Issued for Construction" or "Approved" are written into the description section of the area reserved for enumerating the revisions or table of additions and changes as Rev. No. 0 and dated. The lead designer, process engineer, and project manager for the engineering design group initial the Rev. No. 0 tracing in the appropriate areas in the revision section as having respectively reviewed, checked, and approved the work. Space often is provided also for members of the client group to sign off. An example of a completed title block and revision area for a zero issue is shown in Fig. 4.3.

The approval and distribution of the zero issue of the engineering flow diagrams is a significant milestone in the progress of the project. It signifies not only that the preliminary stages of the design concept have been completed but also that there is sufficient confidence in the design to use the contents of the engineering flow diagrams as the basis for the preparation by other disciplines of the drawings and materials which will constitute the bid packages for the construction of the project and for specifying and purchasing equipment.

In some situations, the term "Issued for Construction" often means that the project is "frozen"; i.e., no major revisions are allowed, except for the correction of errors, without the approval of the client's management. Even without the project's being frozen, it is recommended that as few changes as possible be initiated by process engineers at this stage. Any changes to the engineering flow diagrams after the zero issue could affect some of the construction packages with costly back charges if the change were made after the contract has been awarded.

Similarly, by the time that the engineering flow diagrams have been issued for

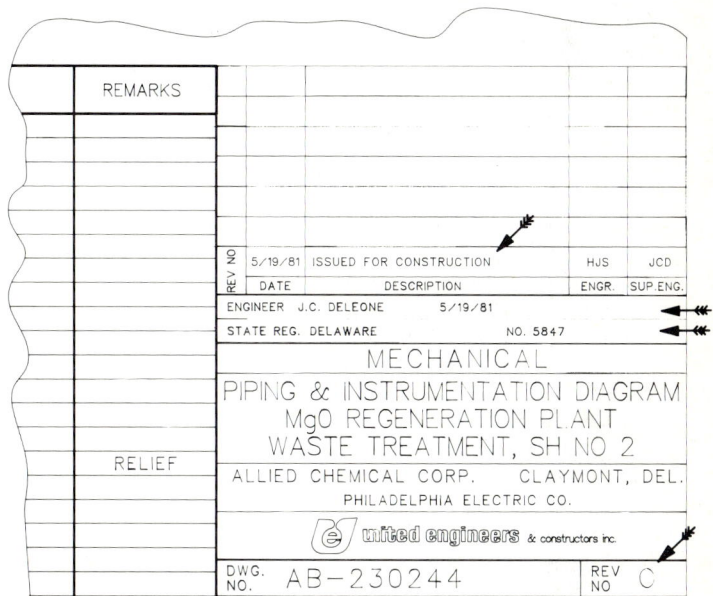

Figure 4.3 Title block, zero issue.

construction, most of the other disciplines should already have expended a considerable amount of effort in preparing their drawings, many of which are based on the preliminary flow diagrams. Any changes made to the flow diagrams can easily be magnified severalfold by the extra efforts that they induce for the other disciplines. Process engineers should bear this in mind when contemplating changes. There is no question that changes should be made and corrections instituted if they are necessary for the process to operate successfully or are required for safety or environmental-protection purposes. However, other types of changes should be weighed against the productivity of the design group and the effect on the project schedule before they are included in the project. Consultation with project management and the discipline engineers who would be affected by a proposed change aids in the decision whether or not to institute the change.

Revision issues

While "Issued for Construction" signifies the official beginning of the job, it does not mean that there will be no more changes to the project. It is rare that the process engineers and the client can foresee all eventualities by the time that the zero issue is made. There may be basic changes in the process due to further research on the part of the client or the impossibility of obtaining the required equipment for the process. Changes may also be incorporated if the equipment eventually purchased requires piping configurations or auxiliaries different from those upon which the process was originally predicated. It may also be discovered that the layout of equipment or the actual positioning of piping precludes following the original flow sheet exactly. Similarly, designers are often able to suggest less costly piping schemes than originally envisioned as their work progresses. New requirements of regulatory agencies can also cause additions or revisions to the process.

It is not unusual to have several revisions of the engineering flow diagrams after the diagrams have been issued for construction and the design stage of the project has been completed. Additional revisions may have to be made during the construction and start-up phases of the project to account for the changes that inevitably are required to correct faults that may be found in the design or to enable the plant to operate more efficiently.

The process engineers originate changes that are made for the above reasons. It is in the nature of these revisions or changes that they are encountered sporadically. In all cases, they must be discussed with the client's project manager and approval obtained before they are entered on the appropriate engineering flow diagram. On the other hand, many revisions or changes come directly from the client's project manager to the process engineers or through the design group's project manager. In these cases, the process engineers must fully understand the changes and be in agreement with them. Any objections must be explained to the client project manager and resolved before the changes are included in the project.

When sufficient revisions or changes have been collected or when the client requests it, a distribution is made of prints from the revised tracing. A revision is signified by entering the next sequential number in the revision box of the title block, the construction issue being 0. At the same time the designer enters a summary of the changes made since the zero issue or the previous revision issue into the appropriate area on the diagram. This description of changes should be checked by the responsible process engineer.

The title block for Fig. 3.2 illustrates that Revision 4 is the current issue of that drawing. The entries above the title block record the changes that had been made to the drawing in the various revision issues since the zero issue. The evolution of the drawing can be traced by a review of the information in the revision description area. It should be noted that Revisions 2, 3, and 4 delineate the exact changes that had been made in the drawings, whereas the word "general" was entered in the revision description box for Revision 1. In that case, the changes were so extensive and so pervaded the entire flow diagram that it would have been extremely difficult to summarize them. The user of the drawing must, therefore, refer to the revision labels on the diagram to distinguish the changes that had been made since this last revision.

All changes on a drawing after the zero issue must be marked in an appropriate manner to indicate that changes have been made. The usual procedure to delineate a change is to include it, and only this change, inside a continuous-wavy-line enclosure known as a "cloud," a "balloon," or a "circle" according to the custom of the individual design group. It is also mandatory to indicate the revision number for which the change was made. This is usually done by placing the revision number inside a small triangle* within the change enclosure. To avoid an excessive number of revision enclosures and revision triangles on the drawing at any one time, it is an accepted practice to include more than one change within an enclosure.

It should be noted that a deletion of a line, an instrument, or a piece of equipment is as much as a revision as an addition or a partial change in one of those categories. Thus, a complete erasure on a drawing must be indicated by an enclosure of the area where the item had existed and include the revision triangle and number.

The revision triangles and the numbers inside them are placed on the front of a tracing and remain there as part of the tracing. As with any other parts of the drawing, they should be moved if it becomes necessary to rearrange the drawing to make space for additional entries. In that event, the moved triangle and number should be placed as near as possible to their original position without being located within the confines of a piece of equipment or an instrument or crossing any line. The lines of subsequent revision enclosures should be drawn so that they do not include any prior revision triangles. The triangles and numbers may be relocated as required to avoid being included in subsequent enclosures.

Revision triangles are a permanent part of the tracing, but the revision enclosures are temporary since they are used to indicate the extent of the revisions for a particular issue only. Thus, when a revision issue has been made and work is started on the tracing for the next issue, one of the first things that should be done is to remove all revision enclosures. The only exceptions would be those enclosures which encircle holds that are not as yet resolved. Since all prior revision enclosures and holds must eventually be removed from a tracing, it is good practice to enter them on the back of the tracing, usually with a grease-film type of pencil.† Since this work must be done on the rear of the tracing, the designer should be careful and the process engineer should check that the enclosure line does not include any lines, instruments, or parts of equipment that are not part of the revision.

Interim issues

The preliminary, zero, and revision issues described previously are essentially formal, in that they are distributed officially to the client. At the same time, they are distrib-

---

*Other geometric symbols may be used in accordance with the practice of the client and design group, but the triangle seems to be the most widely used symbol.

†This procedure is carried out manually even with a computer-assisted design system.

uted internally among the design groups for their use and for reference when communicating with their counterparts in the client organization. It can be deduced from the date of the zero issue and those of the revisions that there may be considerable periods of time between formal issues. However, there occur during these intervals changes or revisions which should be communicated to the discipline engineers and their designers. In this case, the process engineers use their own discretion as to when there are sufficient changes to warrant making an interim issue.

Several interim issues may be made between formal issues. Although the convention of enclosing the revision number* is observed for an interim issue, no information regarding the changes is placed in the revision block at that time. Neither is a formal date assigned to the drawing in the title block. Instead, a phrase such as "Interim Issue" is written in a conspicuous area near the title block and the date of the interim placed next to it. The records of the interim revisions and their dates are removed when the next revision information is entered into the appropriate areas on the tracing immediately before the tracing is formally issued.

Internal Communication

One of the responsibilities of prime importance for process engineers is to ensure that all discipline engineers and designers who use the engineering flow diagrams are kept current with all changes or revisions with respect to the diagrams. A system that has proved satisfactory for accomplishing this task is the use of a "Master Set" of prints of the engineering flow diagrams and an organized method of identifying tracings as they are worked upon between formal revision issues.

Master sets

As soon as a formal distribution is made of a preliminary, zero, or revision issue, the process engineers should obtain a full-sized print of each of the engineering flow diagrams, including the lead sheet, and form a master set of them. This is done by labeling each of them as "Master Set" in large, wide letters, usually with a rubber stamp, in a distinctive color and placed in a conspicuous area on the print. The engineering flow diagrams are then assembled in numerical order and either clipped together or held together in a stick file clamp. The file is then classified as the "Process Engineering Master Set" or by a similar designation.

Any changes or revisions to the process which have been agreed upon by the client's project manager should be entered on the master set by the process engineer as soon as possible. These include changes in the hold category that the client wishes entered onto the next revision. Also entered on the master set at this stage are corrections of errors, oversights, or changes regarding pipe sizing, material specification, valves, renumbered equipment or instruments, or similar items. The client's representative should be advised that these changes are being made.

Since the master set is the sole responsibility of the process engineers, who are the most knowledgeable about all the markings on the master set, it is a general rule that normally only the process engineers or their designers make the changes on them. An exception occurs, of course, when P&IDs are used on the project instead of an

---

*Enclosures only are used for the interim issue when the drawing is in the preliminary stage since revision labels are normally not used for preliminary issues.

instrument application drawing as well as engineering flow diagrams. In that case, the instrument engineer shares the responsibility of keeping the instrumentation portion of the master set current. The process engineers use a distinctively colored crayon, pencil, or pen, e.g., a red one, to enter their markings, while the instrument engineers utilize a different color, e.g., dark purple or green, for their entries. The process and instrument engineers should inform one another as soon as they have made a change or revision to the master set which affects the other group.

Although the responsibility for the master sets of the engineering flow diagrams rests solely with the process engineers or is shared with the instrument engineers for P&IDs, other engineers or lead designers are permitted to make entries onto the master sets under certain circumstances. Normally, these persons bring any change or revision to the attention of the responsible process engineer, who then makes the appropriate entry if the change is warranted. If, however, the responsible process engineer is not or will not be available for a considerable period of time, the engineers or lead designers may make the change themselves on the master set if they deem it important to have it made at that time. However, they should initial and date their entry. In addition, they should send the process engineer a note outlining what change or revision has been made. This procedure is necessary to ensure that process engineers are always aware of any changes or revisions to the master set and can thus keep control of them.

While the file with the master set is consulted by all disciplines, it is most frequently used by the mechanical squad, and it should be kept on a designated reference table convenient to the lead mechanical designer. The process engineers should advise all discipline engineers and lead designers of any changes and revisions that affect them by indicating the marks on the master set and giving a full explanation for the changes. The lead designers, in turn, should pass the information on to the designers who will be doing the actual work. The designers, for their part, should frequently refer to the master set to determine any changes or revisions to the particular work area with which they are concerned. This will help assure that they have the information they need to keep themselves current with the process aspects of their work.

Provisional dating of tracings

Despite the efforts to disseminate the information regarding changes or revisions, many designers often depend only on an older print of an engineering flow diagram in their possession for their information. It should be stressed that designers should check the tracing in its storage location frequently to determine whether or not the print in their possession is the latest. It is, therefore, imperative that a method be devised and utilized to ensure that all users comparing a tracing and a print can immediately distinguish whether or not it is the same as the one in their possession and which version is more current.

A practical method to accomplish this is the use of a provisional dating system. A formal dating system should not be used since the formal date is entered in the title block only at the time of an issue. In essence, the tracings for a given formal issue are kept in a storage drawer or rack. After that the process engineer and the instrument engineer (in event of P&IDs) collect changes and revisions on the master set. When sufficient changes have accrued on a particular engineering flow diagram or set of diagrams, the process engineer authorizes that they be transferred to the respective tracings.

The designer or process engineer doing the correction work should do two things immediately upon removing a tracing from the place of storage. First, the current date should be entered as a provisional date on a clear area close to the title block; and, second, the current revision number, e.g., 1, should be erased from the revision box and the next revision sequential number, e.g., 2, entered in that box. To ensure that everyone realizes that the new number in the revision box does not signify that the tracing represents a formal issue, an X is placed through the number on the back of the tracing.

Once the provisional date has been placed near the title block and the X entered on the reverse side of the tracing behind the next revision number, all previous revision enclosures should be removed from the tracing. This is done so as not to confuse previous enclosures with those to be added for the current revision issue. Enclosures about any remaining hold items should be kept on the tracing.

After the changes have been made to the tracing for a given flow diagram, the process engineer should back-check a print and correct it as required. Another provisional date should be entered on the tracing near the title block if corrections are made to avoid the possibility, however remote, that the print from the corrected diagram could be confused with one from before the correction. If a correction is made on the same day that the tracing was originally worked on, a distinction should be made between the two date entries.

Interim-issue designation

After the corrections have been made and back-checked, the process engineers can elect to make an interim issue for internal use if the changes or revisions up to that point in time warrant it, or they can wait until more changes and revisions have been collected. Should the process engineer decide to wait before making an interim issue, marking of the original master set with changes and revisions should continue until there are sufficient changes for an interim issue. The same procedure of entering additional provisional dates on a tracing is followed whenever it is taken from its storage place or it has left a designer's board and is returned for additional work or this is done by the process engineer. Of course, no additional changes are made to the revision number box until after the next formal issue.

When an interim issue is to be made, the process engineer should write the words "Interim Issue," enter the date next to them, and delete the provisional dates. An internal distribution should be made, and the issue should also be sent to the client if copies are requested. The old print should be removed from the master file and a print of the interim issue, also stamped "Master Set," should replace it in the file.

After the interim issue, the designers should continue the procedure of entering provisional dates. The importance of the provisional dating system is to allow anyone with a print, be it an interim issue distributed by the process engineer or a "bootleg" print* which has been made at some time during the course of the project, to determine whether or not the print is current.

The procedure outlined above continues until it is decided by project management

---

*It is generally accepted that any engineer or designer can request and obtain a print of an engineering flow diagram from the tracing in storage whenever it is required for use in that person's work. Since this principle applies to all other drawings on the project, the provisional dating system can well serve as an identification system for those drawings also.

that a formal issue should be made for the particular revision. In that case, no additional changes or revisions should be made to the tracing, and the provisional dates, notification of interim issues, and the X behind the revision box are erased.

Archives

Few process engineers have perfect memories or total recall. Therefore, it is important that process engineers maintain a complete set of archives to document process decisions made on the project which affect its progress. This is required to refresh memories as to the chain of events leading to revisions and the sequence in which they have been made. It serves to fix such events in a time framework if that is required for engineering, commercial, or accounting purposes.

The three main elements that usually constitute a separate process archive are a record of the process calculations, a file of notes of conference, and a repository of all stages of the process flow sheets and of the engineering flow diagrams. Another set of pertinent information that is retained is a record of the markings added to the drawings received from equipment vendors. The file of the vendor drawing markups is usually maintained by the mechanical engineering group.

Calculation sheets

The sizing of all equipment and piping as well as all material and energy balances and the economic analysis used to select the most cost-effective processes or equipment is accompanied by calculations. These calculations are generally made on special sheets unique to the engineering group. Computer and calculator printouts are often part of the process engineering calculations record. The printouts should be identified properly and filed.

The calculations sheets and print sets are retained until the end of the project, when instructions are given as to their disposition. The usual procedure is that the client receives the original sheets plus several copies. The engineering group then keeps a copy for its files, and the process engineers each receive a set for their personal files.

Process engineers develop their own methods of organizing the calculations file. The final record file is usually presented with subdivisions of material balances, energy balances, equipment types, pipe sizing, insulation and tracing calculations, special studies, and miscellaneous work. Some engineers set up their main filing headings at the beginning of the project and place their calculations under them as the calculations are completed. New headings are added as required. The calculations are then in order for the final calculations file. Process engineers should separate those calculations which are void before completing the calculations file. The final file is indexed for quick reference. Other engineers keep their calculations in binders in chronological order. At the end of the project, the calculations are rearranged into the desired headings. The nonrelevant material is separated at that time and the final file indexed.

Notes of conference

It is general practice that all communications, in person or by telephone, between members of the engineering design group and the client representatives which result in the transfer of pertinent information or in decisions affecting the project are recorded in notes of conference. It is advisable that the notes be identified by a sequential number, the date of the conference, the location,* and the participants. The notes are usually prepared by an engineer from each discipline participating in the discussions and should contain a succinct summary of the material discussed.

The entire set of the notes becomes a part of the records for the project when the project is completed. However, to expedite understanding of the changes that have been made to the engineering flow diagrams during the course of the project, it is helpful for the process engineers to cull those notes of conference which impacted on the flow sheets and diagrams. Those sections dealing with process discussions or information should be identified by a suitable mark and included as an appendix to the calculation-sheets file.

Drawings

It is equally important that process engineers maintain a complete archive of the process flow sheets and engineering flow diagrams for which they are responsible. The archive should consist of the marked prints that were used to develop the formal issues of the process flow sheets and the engineering flow diagrams. These would essentially be all the master-set sheets, including interim issues, that were made between revisions. After back-checking any print, the process engineer should review the drawing and make the notations required as memory joggers for a future date and then deposit the print with the markings and notations in the drawings archive. The archive should be retained during the engineering phase of the project, through the start-up of the facility, and until the plant is accepted by the client.

---

*The location may be omitted for telephone conferences, but the title of the document can then be "Notes of Telephone Conference."

Section 2

Materials Selection and Piping Calculations

Chapter

5

Materials Selection

Once equipment has been selected, the materials for its construction must be established. Although a process engineer is not expected to be as knowledgeable as a metallurgist, the engineer should have a general idea of what materials are compatible with the process. Therefore, this chapter presents some general guidelines in the selection of materials for process equipment. As an introduction to the material-selection process, there is a brief discussion of the characteristic mechanical properties followed by a more detailed critique of the most generally used material groups: metals, plastics, ceramics, and glasses, respectively.

Mechanical Properties

The selection of materials for a specific application depends on their mechanical properties. Some of these properties are listed below. (See also Fig. 5.1.)

1. *Brittleness.* This is the tendency of a material to fracture, usually under low stress and without warning. Brittleness is the opposite of toughness.

2. *Creep.* The term "creep" is used to describe the rate of plastic deformation of a metal with time under a given stress load. Figure 5.2 shows the four basic stages of creep. First, there is an initial elongation when the load is applied, followed by a primary or transient creep stage. After the transient creep stage there is a stage in which the rate of deformation with time is relatively constant; this is referred to as the steady-state stage. The last stage prior to fracture is another transient creep stage called the tertiary creep stage.

For most ferrous alloys operating below 800°F creep is not important. However, support steel for poorly insulated furnaces and high-temperature reactors is susceptible to this type of failure. Lead used in radiation shielding or in reactors for manufacturing sulfuric acid needs support steel to reduce loading on the lead because it has a tendency to creep. Epoxies used in epoxy-bonded pipe and some polymeric bearings are also subject to creep failure.

3. *Ductility.* This is the ability to deform permanently by tensile stress without breaking. Ductility is the opposite of hardness. It is also referred to as the ability of a material to be drawn without breaking. Materials having this property are used in pressure components such as process tanks and pipes.

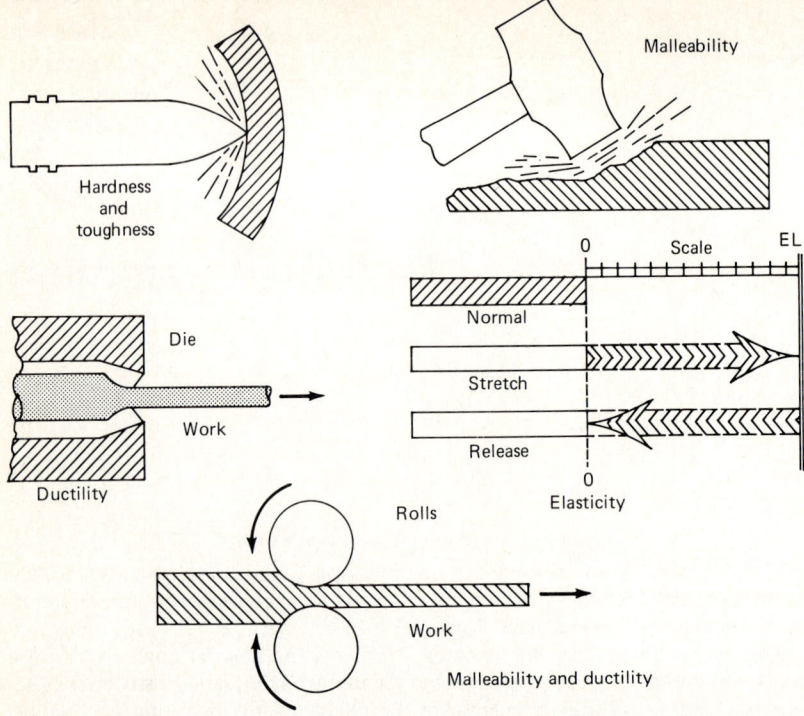

Figure 5.1 Mechanical properties of metals. (*Arthur M. Shrager,* Elementary Metallurgy and Metallography, *3d ed., Dover Publications, Inc., New York, 1969.*)

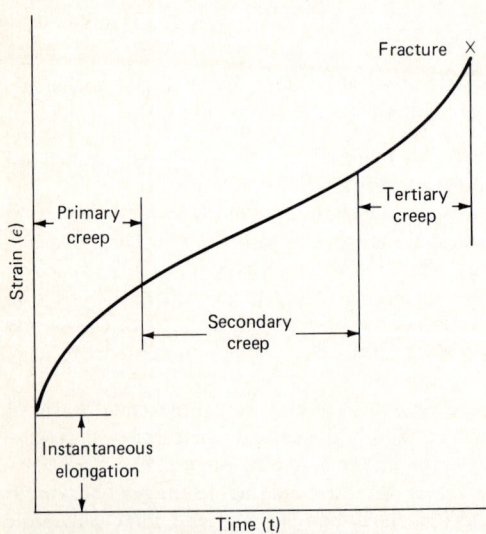

Figure 5.2 A characteristic creep curve showing the four elongation stages of a material as a function of time. *(William G. Moffatt, John Wulff, and H. W. Hayden,* The Structure and Properties of Materials, *vol. III: Mechanical Behavior, © John Wiley & Sons, Inc., New York, 1964.)*

4. *Fatigue.* Fatigue is the progressive fracturing of a material owing to cyclic loading. Cyclic loading causes a bending or cold working of the material, and the maximum stress that a material can take before failing is inversely proportional to the number of loading cycles it experiences during its lifetime. Ductile metals have considerable resistance to fatigue failure, while hard metals, ceramics, and most polymeric materials have a high likelihood of fatigue failure.

5. *Hardness.* The opposite of ductility, hardness is resistance to plastic deformation by bending, penetration, or scratching.

6. *Impact strength.* This is the ability of a metal to withstand shock loading without breaking. Wood and wrought iron because of their fibrous structure have high impact strength in relation to their unit weight.

7. *Malleability.* Malleability is a special case of ductility. It is a metal's ability to be permanently deformed by rolling, forging, or extruding without rupturing and without any noticeable increase in resistance to deformation. The degree of malleability is directly proportional to the temperature at which it occurs. Aluminum, copper, magnesium, and silver are typical malleable metals.

8. *Toughness.* Toughness is basically defined as the ability to withstand loading without breaking. It is also defined as the ability to resist shock loading. Therefore, toughness is a characteristic combining the properties of ductility and impact strength. It is the opposite of brittleness.

9. *Strength.* This is the ability of a material to withstand loading without failing. A given material has dynamic strength (impact strength) and static strength. Static strength may be further subdivided into compressive and tensile strength. Compressive strength is the ability of a material to be squeezed or pressed without failing, and tensile strength is the ability of a material to resist a pulling load without failing. The strength of a material is best described by stress-strain diagrams. These diagrams are the results of experiments performed on actual materials. A typical sample form used to test a material's strength is shown in Fig. 5.3.

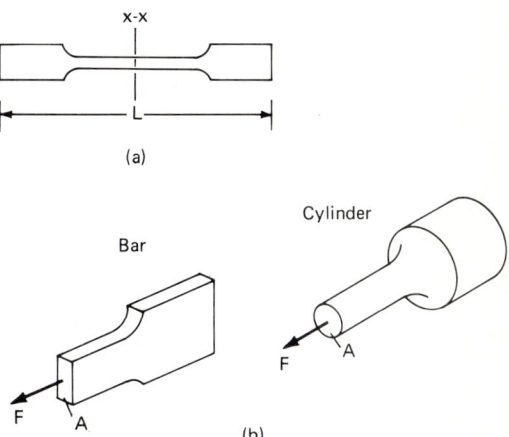

Figure 5.3 (*a*) Typical strength test sample shown in profile. The sample may be a bar or a cylinder. (*b*) The location at which stress is measured. L = length of sample; F = load (+ ⌢ tension; − ⌢ compression); A = area; $\sigma = F/A$ = stress.

When a load is applied to the material form, it causes the material to go into stress. Stress S is defined as the load F applied over a cross-sectional area A of the material; i.e.,

$$S = \frac{F}{A} \tag{5.1}$$

If the force is in pounds and area is in square inches, then stress is dimensionally pounds per square inch (psi).

When a material is under stress, it tends to deform. This is observed as a change in length. The measure of the degree of deformity is called strain ϵ and is defined by the formula

$$\epsilon = \frac{\Delta L}{L} \tag{5.2}$$

where ϵ = strain, dimensionless
L = original length of sample, ft
ΔL = change in sample length, ft

The resulting stresses and strains from applying various loads to the sample are plotted. Figures 5.4, 5.5, and 5.6 are typical curves for some engineering materials.

The slope of the stress-strain curve is called the "modulus of elasticity (E)." If the modulus of elasticity over a given range is constant, i.e.,

$$E = \frac{\Delta S}{\Delta \epsilon} = \text{constant} \tag{5.3}$$

the process is considered elastic.

If for a given change in strain, there is little or no change in stress over a given range, i.e., the modulus of elasticity is equal to zero and when the sample is unloaded some amount of deformity remains, the sample is said to have gone into plastic deformation.

Ultimate deformation and, accordingly, ultimate stress and strain occur when the sample finally fails. This is represented by an x in Figs. 5.4, 5.5, and 5.6.

Not only do materials have these strength characteristics in varying degrees, but these strength properties vary with direction; i.e., a material has different strength characteristics depending on whether it is in tension or in compression. This difference in properties is called "anisotropy." Figure 5.6 contrasts the respective characteristics of gray cast iron and concrete in tension and compression.

A knowledge of material-strength characteristics allows the engineer to select the proper materials efficiently for process equipment and its structural supports. A good example is reinforced concrete. Concrete is inexpensive and possesses good compressive strength. Steel has excellent tensile strength but is much more expensive than concrete. When a foundation is needed for a storage tank, it would be prohibitively expensive under normal circumstances to construct a foundation of thick steel plate. If the foundation were constructed of concrete alone, it would be inexpensive, but when the tank and its contents were placed on the foundation, the foundation would start to crack from the bottom of the slab upward toward the tank until it was completely useless. This problem is caused by the bending of the slab under the weight of the tank. In bending, the upper portion of the concrete slab is in compression, which is acceptable, but the lower portion experiences tension, for which concrete

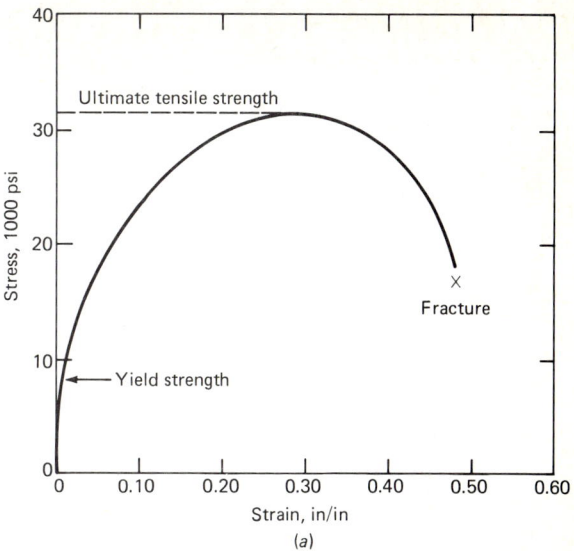

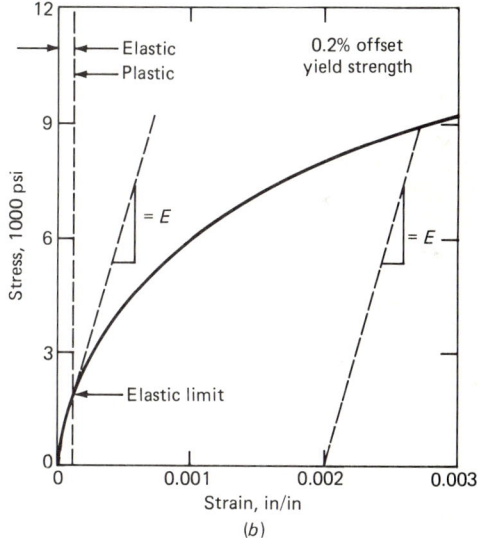

Figure 5.4 Engineering stress-strain diagram for polycrystalline copper. (*a*) Complete diagram. (*b*) Elastic region and initial plastic region showing 0.2 percent offset yield strength. (*William G. Moffatt, John Wulff, and H. W. Hayden,* The Structure and Properties of Materials, *vol. III:* Mechanical Behavior, © *John Wiley & Sons, Inc., New York, 1964.*)

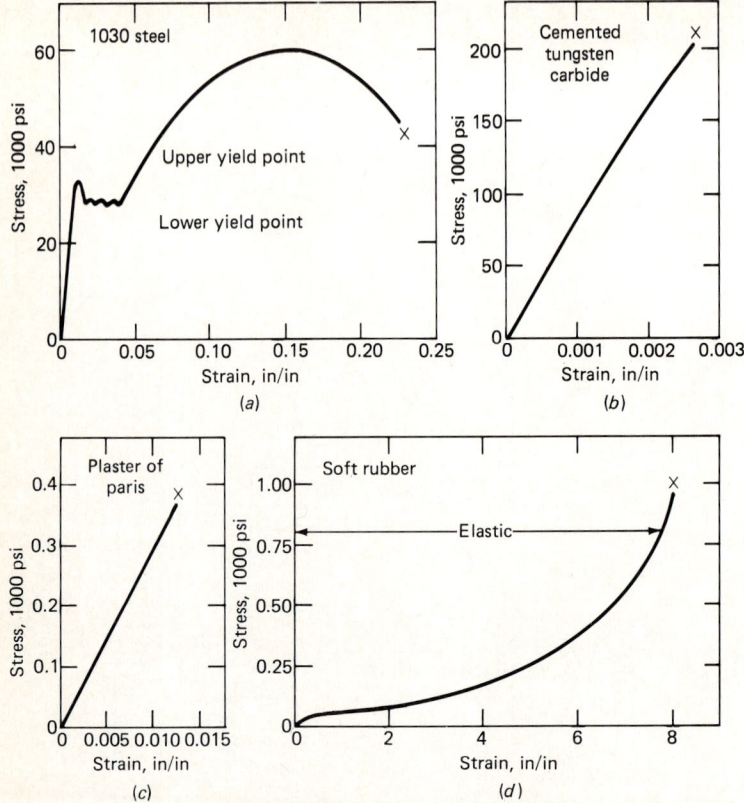

Figure 5.5 Engineering stress-strain curves for several engineering materials. (William G. Moffatt, John Wulff, and H. W. Hayden, The Structure and Properties of Materials, vol. III: Mechanical Behavior, © John Wiley & Sons, Inc., New York, 1964.)

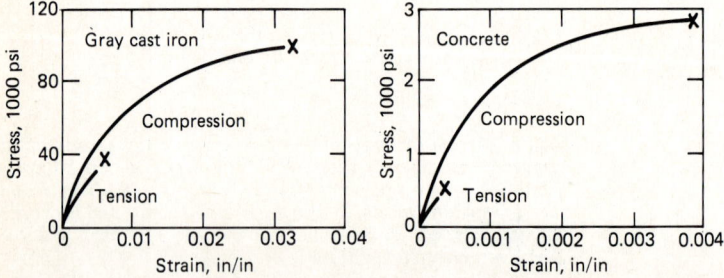

Figure 5.6 Tensile and compressive engineering stress-strain curves for gray cast iron and concrete. (William G. Moffatt, John Wulff, and H. W. Hayden, The Structure and Properties of Materials, vol. III: Mechanical Behavior, © John Wiley & Sons, Inc., New York, 1964.)

has a low tolerance. Placing steel bars through the lower portion of the concrete slab increases the tensile strength of the slab. The composite of steel and concrete is called reinforced concrete, which is stronger and more cost-effective than either concrete or steel alone.

The process industry uses composite materials, such as fiberglass-reinforced plastic (FRP) and clad steels, in the construction of corrosion-resistant equipment. The FRP materials are normally used for process equipment operating at temperatures below 200 to 300°F and from near-atmospheric pressure to about 50 psig. The clad steels are more expensive than FRPs but can be used over a wider range of operating pressures and temperatures.

Materials

There is a wide variety of materials used to construct process equipment. The type of material selected depends on its compatibility with process conditions and cost. The materials of choice can be divided into the following general categories: metals, polymeric materials, ceramics, and glasses. Each of these categories is described below together with representative materials used in the construction of process equipment. The materials' specific properties are described as well as their process compatibility.

Metals

The metals used in the process industry can be generally divided into two groups, ferrous and nonferrous. Ferrous metals are arbitrarily defined as those containing at least 50 percent iron. As a group they far exceed any other single group of metal alloys used in the process industry.

Ferrous alloys. The reason for the popularity of ferrous alloys is their relatively low cost, ready availability, and good workability. Ferrous alloys are subdivided into cast irons, carbon steels, low-alloy steels, and stainless steels. There are also tool steels, but these will not be discussed as they are of limited use in the process industry.

Cast iron. Cast irons are iron alloys that have greater than 1.5 percent carbon content (see Fig. 5.7). There are four types of cast iron: gray, white, ductile iron, and wrought iron.

Gray cast iron. This type is used in applications such as construction equipment requiring vibration dampening and wear resistance. Its gray appearance is caused by the graphite particles spread throughout its mass. Gray cast iron has a low tensile strength and is brittle, and it should not be used in high-pressure process applications. It is readily machinable but difficult to weld. If gray cast iron must be welded, a nickel-alloy filler metal must be used. Extreme care must be taken not to create excessive thermal gradients when welding cast iron; otherwise, the piece could crack because of the different thermal-expansion rates of the nickel alloy and gray cast iron.

White cast iron. White cast iron has a much lower silicon content than gray cast iron. No graphite particles exist in its microstructure, and any carbon in the cast iron is combined with iron in the form of iron carbide (Fe_3C). The metal is hard, extremely abrasive, brittle, and difficult if not impossible to machine. Because of these properties, white cast iron is not approved for use by pressure-vessel codes. It can be used, however, for grinding balls, dies, casings for slurry pumps, and car wheels.

Ductile cast iron. Ductile cast iron has the same constituents as gray cast iron, but it is manufactured differently. When ductile cast iron is manufactured, small spheri-

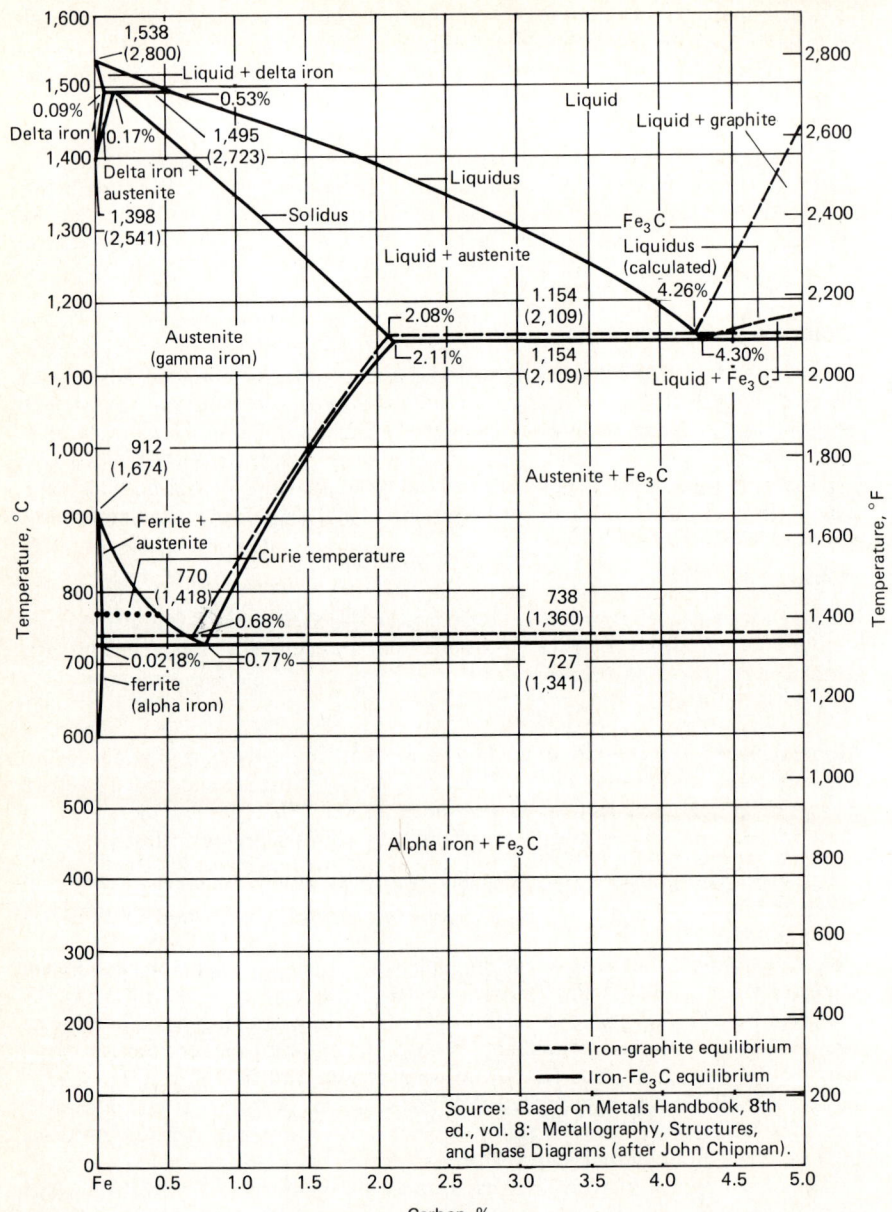

Figure 5.7 Iron-carbon equilibrium diagram. (*From* Metal Progress, © *American Society for Metals*, 1975.)

cal nodules of graphite are produced in its microstructure. This causes it to be ductile rather than brittle. Ductile cast iron is used for high-strength pipes, valve bodies, pump casings, compressor casings, crankshafts, and machine tools.

Wrought iron. Prior to the nineteenth century, wrought iron was the only form of iron used. It is basically pure iron low in carbon but with a few percent of slag remaining in the form of an iron silicate. The slag forms a fibrous structure within the wrought iron that gives it excellent shock, vibration, and corrosion resistance. Corrosion resistance apparently results from the fine fibrous structure of slag, which acts as a mechanical barrier limiting corrosion to the surface of the iron. Wrought iron can be used for water pipes, air-brake pipes, and engine bolts.

Silicon iron. Iron alloys with a relatively high silicon content (about 15 percent) are called silicon irons. They are available under such trade names as Duriron and Durichlor. Durichlor contains molybdenum for increased corrosion resistance. Both alloys are used where corrosion resistance to oxidizing and reducing acids (except hydrofluoric acid) is required or in piping handling slurries where erosion resistance is needed. These irons are available only in cast form, so they are very brittle and have low tensile strength. It is recommended that silicon iron pipe be used in systems operating below 50 psig. However, the material has been installed in processes experiencing temperatures up to 1400°F.

Carbon steel. The characteristic difference between cast iron and carbon steel is the carbon content. Carbon steel has less than 1.5 percent carbon. It is easy to fabricate and has better strength than cast iron. Depending on the type of heat treatment and alloying used, the steel can be given any degree of hardness or ductility. In addition, it is easier to weld than cast iron. These properties, in addition to abundance and low cost, make carbon steel the first choice to be considered for the design of equipment. The main disadvantage in using carbon steel is its susceptibility to corrosion.

There are a great number of carbon steel alloys on the market, and this variety can be overwhelming for a practicing engineer. The process engineer can improve the chance of selecting the proper steel for a given application by a little understanding of how alloying elements affect the properties of the steel. In addition to an understanding of the alloying process, design codes and standards can be used as a guide to material selection.

Alloying is to some degree empirical. Metallurgists have found from experience that certain elements, when added in small percentages, can change the properties of a steel. The more important alloying elements and their effects on steel are listed in Table 5.1. While the table shows the general effects of each element on alloying, these effects are by no means absolute. Sometimes a combination of various elements may create an alloy with properties entirely different from expectations. Therefore, the table should be used only as a general guide.

Low-alloy steel. Low-alloy steel was developed during the earlier years of the petroleum industry, when it was noticed that severe corrosion occurred in process equipment when high-sulfur crude oil was being processed. The carbon steel that was being used was replaced with a steel alloyed with small amounts of chromium. It was found that the chromium in the steel inhibited the formation of iron sulfides owing to the reaction of the sulfur in the crude with the iron in the steel. It was also discovered that the addition of chromium increased a steel's strength at high temperatures.

The basic difference between low-alloy steel and carbon steel is the chromium content. A carbon steel alloy is defined as one with less than 4 percent chromium. A low-

TABLE 5.1 Alloying Elements and Their Effect on Carbon Steels

Element	Effect
Aluminum (Al)	Used in concentrations of 2% or less; acts as a deoxidizer; restricts grain growth; aids in nitriling (a surface-hardening technique).
Chromium (Cr)	Increases strength, especially at high temperature; increases hardenability; reacts with carbon to form a wear-resistant structure; increases corrosion resistance in oxidation environments.
Cobalt (Co)	Used in cutting tools and tool steels. It allows the steel to get red-hot for long periods without breaking down the cutting edge.
Copper (Cu)	Used in concentrations of 0.5% or less; produces a tenacious self-sealing oxide film on the steel surface which resists atmospheric corrosion.
Manganese (Mn)	Used in concentrations of 2% or less; acts as a deoxidizer; increases hardenability; creates an austenitic structure to improve a steel's ductility and ease of forging; inhibits corrosion in sulfur environments.
Molybdenum (Mo)	Used in concentrations of 0.5% or less; increases high-temperature strength and hardenability by promoting a fine grain structure in the steel.
Nickel (Ni)	Used in concentrations of 5% or less; increases ductility, hardenability, and toughness by allowing the formation of an austenitic microstructure.
Silicon (Si)	Used in concentrations of 2.5% or less; increases toughness and hardness; acts as a deoxidizer. Killed steel is a steel in which oxygen is removed by the addition of aluminum or silicon.
Sulfur (S)	Used in concentrations of 0.5% or less; improves machinability but reduces ductility and weldability.
Tungsten (W)	Used in tool steels to increase hardness and strength at high temperatures.
Vanadium (V)	Used in concentrations of 0.3% or less; increases hardenability and wear resistance.

alloy steel is one with a chromium content between 4 and 9 percent. In addition to chromium, low-alloy steels contain molybdenum for increased high-temperature strength. Therefore, low-alloy steels used in the process industry include chromium and molybdenum in their composition.

In summary, low-alloy steels are useful in sulfur environments. They are also used in moderately oxidizing or acidic environments.

Stainless steel. Stainless steel was developed during the 1920s. Alloying experiments indicated that the corrosion resistance of steel was directly proportional to the amount of chromium in the steel (see Fig. 5.8). This led to the development of high-chromium stainless steels. Today, stainless steels are usually considered first for use in a corrosive and high-temperature environment.

A steel that has 12 percent or more chromium is considered a stainless steel. Another criterion defining a stainless steel is its passivity. Passivity is the ability of a metal to form an impervious surface coating which inhibits corrosion resulting from the electrochemical reaction of the metal with the surrounding environment. Stainless steels exhibit passivity in oxidizing environments. Although some tool steels such as the Type D high-chromium, high-carbon group contain 12 percent chromium,

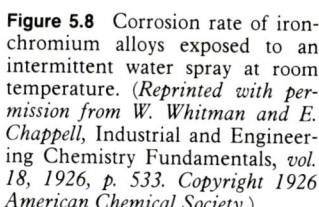
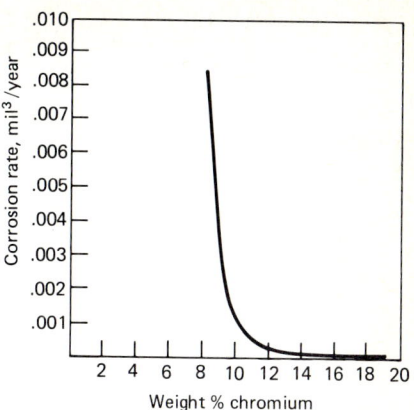

Figure 5.8 Corrosion rate of iron-chromium alloys exposed to an intermittent water spray at room temperature. (*Reprinted with permission from W. Whitman and E. Chappell,* Industrial and Engineering Chemistry Fundamentals, *vol. 18, 1926, p. 533. Copyright 1926 American Chemical Society.*)

they are not considered stainless steels because they do not exhibit passivity owing to their high carbon content (1.5 to 2.5 percent).

Figure 5.8 shows the corrosion rate as a function of chromium content in an iron-chromium alloy. It might be assumed that increasing the chromium content without limit would improve corrosion resistance. This does not happen. Figure 5.9 shows the phase diagram for iron and chromium in all proportions. With chromium contents between 20 and 70 percent, the sigma microstructure is formed. This microstructure is hard, brittle, and of poor corrosion resistance, which is contrary to what would be expected if Fig. 5.8 were considered by itself. Therefore, the commercially important stainless-steel alloys are formulated to minimize the formation as a sigma phase.

There are three general types of stainless steel of general interest to the process engineer: ferritic, austenitic, and martensitic stainless steels. The shaded areas in the phase diagrams of Figs. 5.10, 5.11, and 5.12 indicate the composition ranges where these alloys exist. In each case, the sigma region has been avoided.

Another point of interest presented by these phase diagrams is the formation of a stable austenitic stainless steel at room temperature. Normally, iron-chromium and

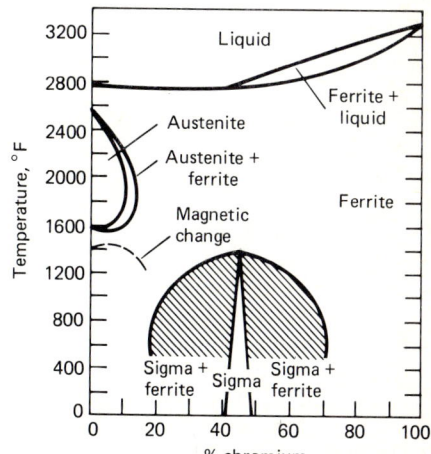

Figure 5.9 Iron-chromium phase diagram. (*By permission from A. W. Grosvenor,* Basic Metallurgy, © *American Society for Metals, 1962.*)

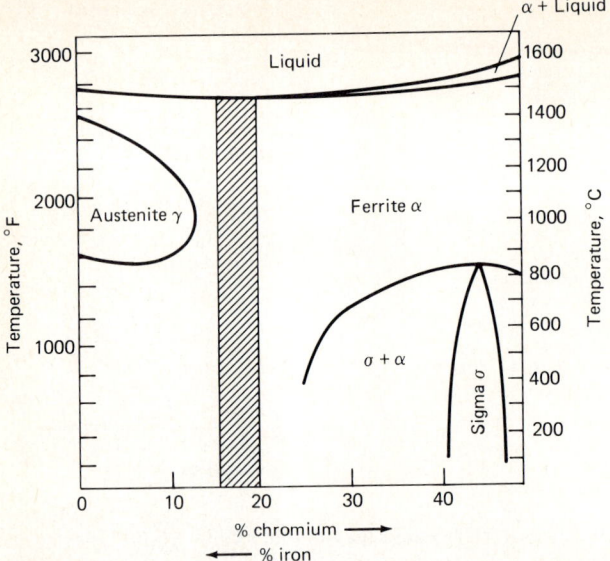

Figure 5.10 Iron-chromium phase diagram for alloys with a carbon content of about 0.2 percent. The shaded area shows the composition for ferritic strainless steels. (*K. Budinski,* Engineering Materials: Properties and Selection, *1978. Reprinted with permission of Reston Publishing Co., a Prentice-Hall Co., 11480 Sunset Hills Road, Reston, Va. 22090.*)

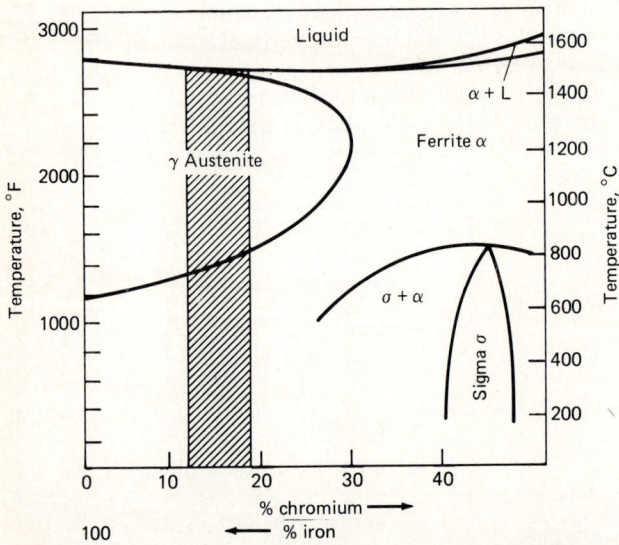

Figure 5.11 Iron-chromium phase diagram for alloys with about 1 percent carbon. The shaded area shows the composition for martensitic stainless steels. (*K. Budinski,* Engineering Materials: Properties and Selection, *1978. Reprinted with permission of Reston Publishing Co., a Prentice-Hall Co., 11480 Sunset Hills Road, Reston, Va. 22090.*)

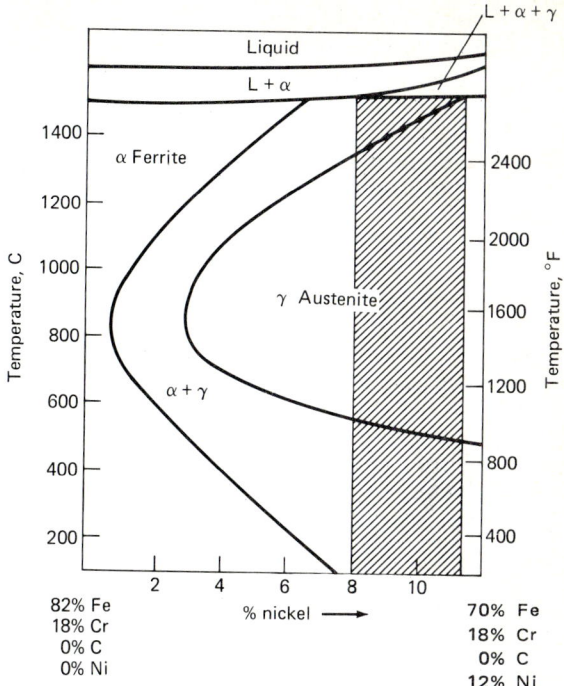

Figure 5.12 An Fe-Cr-Ni ternary diagram for an 18 percent chromium alloy with the austenitic-stainless-steel composition located within the shaded region. (*By permssion from C. A. Zapffe*, Stainless Steels, © *American Society for Metals, 1949, p. 136.*)

iron-chromium-carbon alloys cannot form an austenitic structure at room temperature. To force the formation of austenitic stainless steel, nickel must be added. Figure 5.12 shows how an increased nickel content increases an alloy's ability to form austenite, which improves the alloy's ductility.

This brief description of the stainless-steel alloying process is useful in understanding the behavior of this metal group. The specific characteristics of each of the three stainless-steel types can now be discussed in greater detail.

Ferritic stainless steel. Ferritic stainless steel has a carbon content of 0.2 percent or less and a chromium content between 11 and 18 percent. Although it cannot be heat-treated and has poor tensile and impact strength, it has better corrosion resistance than martensitic stainless steel and is suitable for use with strong oxidizing acids such as nitric acid.

Martensitic stainless steel. Martensitic stainless steel has a chromium content between 12 and 18 percent and a carbon content up to 1.2 percent. It has better hardenability and strength than does ferritic stainless steel. It is used as cladding to carbon steel for some process vessels. At the minimum chromium concentration it is resistant to water, steam, and mildly corrosive chemicals. Some forms tend to be magnetic.

Austenitic stainless steel. Austenitic stainless steel is a more complex material because the addition of nickel (3.5 to 22 percent) allows it to retain its austenitic microstructure at all temperatures. It has a high tensile strength and the best impact strength, ductility, and corrosion resistance of all the stainless steels over a very wide range of temperatures. Its corrosion resistance to sulfuric and organic acids is better than carbon steel, the low-alloy steels, or the ferritic or martensitic stainless steels.

All stainless steels are susceptible to pitting from exposure to high chloride concentrations. However, austenitic stainless steels with a high molybdenum content (1 to 5 percent) have improved resistance to pitting. Although austenitic stainless steels have excellent corrosion characteristics at temperatures up to 650°F, they experience problems with stress-corrosion cracking at high temperatures with high-pH (8 or more) waters, as in high-pressure boiler-feedwater systems and nuclear steam generators. Therefore, for severe corrodents, reducing media, or high-chloride environments, the process engineer must consider using nonferrous alloys.

Nonferrous alloys. As a general rule, nonferrous alloys are considered only after all possible ferrous alloys have been found unsuitable. The nonferrous alloys are secondary options because they are generally more expensive, are long-delivery items, and are difficult to machine or weld. Nonferrous alloys are normally selected for their superior corrosion resistance compared with ferrous alloys, but some are selected for other reasons. Aluminum is one such material.

Aluminum. Aluminum is a ductile material which is easily cast or machined. It has a high strength-to-weight ratio, is nonmagnetic, has good thermal and electrical conductivity, and will not rust when exposed to normal or salty air. Some alloys of aluminum are difficult to weld while others cannot be welded at all.

Aluminum is the material selected for constructing tank trucks if weight is a critical factor. Aluminum is used in the transportation, handling, and storage of high-purity water, various organic solutions, concentrated nitric acid, and aqueous solutions with a pH between 4.5 and 8.5. It is not used to handle the lower alcohols, organic halides, anhydrous organic acids, mercury, heavy metal salts, or steam. Aluminum is useful in cryogenic applications. However, care must be taken not to select an alloy which is susceptible to intergranular corrosion due to low-temperature aging.

A number of elements are used to alloy with aluminum. These are listed in Table 5.2.

The Aluminum Association has developed a four-digit designation system for the various alloys of aluminum, with the first digit on the left indicating the principal alloying element. This system is shown in Table 5.3.

The 2000 Series alloys, because of their brittleness, are not allowed by the pressure-vessel codes. They are acceptable, however, as bar stock and bolting, since these are not used as part of the pressure barrier.

Immediately following the four-digit alloy designation is a letter suffix which may be F, H, O, T, or W. The meanings of these suffixes are shown in Table 5.4.

The T suffix is usually followed by a series of numbers used to explain some of the details of the heat-treating process. Some of the most common number groups are listed in Table 5.5.

An aluminum alloy's usage for process equipment may be limited by its weldability. Weldable alloys require more careful preparation than a steel alloy prior to welding. First, the piece must be treated with a flux to remove any oxide present on its surface before welding. If the oxide is not removed, the aluminum cannot be welded since the oxide has a higher melting point than the aluminum itself, which prevents effective formation of a weld joint. After welding the flux must be washed off completely;

TABLE 5.2 Effects of Alloy Elements on Aluminum*

Alloying element	Effects
Iron	Naturally occurs as an impurity in aluminum ores. Small percentages increase the strength and hardness of some alloys and reduce hot-cracking tendencies in castings. Iron reduces pickup in die-casting cavities.
Manganese	Used in combination with iron to improve castability; alters the nature of the intermetallic compounds and reduces shrinkage. The effect on mechanical properties is improved ductility and impact strength.
Silicon	Increases fluidity in casting and welding alloys and reduces solidification and hot-cracking tendencies. Additions in excess of 13% make the alloy extremely difficult to machine. Silicon improves corrosion resistance.
Copper	Increases strength up to about 12%. Higher concentrations cause brittleness. Copper improves elevated-temperature properties and machinability. Concentrations over 5% reduce ability to hard-coat.
Magnesium	Improves strength by solid-solution strengthening. Alloys with over about 6% will precipitation-harden. Aluminum-magnesium alloys are difficult to cast because the molten alloy tends to "skin over" (dross) in contact with air.
Zinc	Lowers castability. High-zinc alloys are prone to hot cracking and high shrinkage. Percentages over 10% produce tendencies for stress-corrosion cracking. In combination with other elements, zinc promotes very high strength. Low concentrations in binary alloys (less than 3%) produce no useful effects.

*Adapted from K. Budinski, *Engineering Materials: Properties and Selection*, 1978. Reprinted with permission of Reston Publishing Co., a Prentice-Hall Co., 11480 Sunset Hills Road, Reston, Va. 22090.

TABLE 5.3 Aluminum-Alloy Designation System

Alloy	Designation
Commercially pure aluminum (99% minimum)	1XXX
Copper	2XXX
Manganese	3XXX
Silicon	4XXX
Magnesium	5XXX
Magnesium and silicon	6XXX
Zinc	7XXX

TABLE 5.4 Alloy-Designation Suffixes*

XXXX		
	F	As fabricated, no special controls
	W	Solution-heat-treated (used only on alloys that naturally age-harden)
	O	Annealed (wrought alloys only)
	H	Strain hardened (cold-worked to increase strength); wrought alloys only
	T	Thermally treated to produce effects other than F, O, or H

*K. Budinski, *Engineering Materials: Properties and Selection*, 1978. Reprinted with permission of Reston Publishing Co., a Prentice-Hall Co., 11480 Sunset Hills Road, Reston, Va. 22090.

TABLE 5.5   Treatment Method Used on an Aluminum Alloy*

XXXX-T1	Cooled from a hot-working temperature and naturally aged
XXXX-T2	Annealed (cast products only)
XXXX-T3	Solution-treated and cold-worked
XXXX-T4	Solution-treated and naturally aged
XXXX-T5	Cooled from a hot-working temperature and furnace-aged
XXXX-T6	Solution-treated and furnace-aged
XXXX-T7	Solution-treated and stabilized
XXXX-T8	Solution-treated, cold-worked, and furnace-aged
XXXX-T9	Solution-treated, furnace-aged, and cold-worked
XXXX-T10	Cooled from an elevated temperature, furnace-aged, and cold-worked
XXXX-T51	Stress-relieved by stretching
XXXX-T510	Stress-relieved by stretching with no further processing
XXXX-T511	Stress-relieved by stretching and minor straightening
XXXX-T52	Stress-relieved by compression
XXXX-T54	Stress-relieved by stretching and compression
XXXX-T42	Solution-treated from O or F temper and naturally aged
XXXX-T62	Solution-treated from O or F temper and furnace-aged

*K. Budinski, *Engineering Materials: Properties and Selection*, 1978. Reprinted with permission of Reston Publishing Co., a Prentice-Hall Co., 11480 Sunset Hills Road, Reston, Va. 22090.

otherwise, it will encourage corrosion around the weld joint. Once the joint has been welded, there may be a loss in strength at the weld, so that considerable care must be taken in the selection of an alloy and the welding procedure. Generally, 1000, 3000, and 5000 Series alloys are weldable, but alloys in the 2000 and 7000 Series are not recommended for gas welding. Table 5.6 lists typical aluminum alloys and recommended welding methods.

Alloy 20.   Alloy 20 is a generic name for a group of iron-chromium-nickel alloys. These are generally used in lieu of nickel alloys in severe corrosive service where stainless steel cannot be used. The Alloy 20s are less susceptible to intergranular corrosion, stress corrosion, and pitting in strongly oxidizing environments than are stainless steels. They are less sensitive to chlorides than stainless steels but more sensitive

TABLE 5.6   Weldability of Aluminum Alloys*

Alloy	Welding process			
	Gas	Electrode	TIG or MIG	Resistance
1100	R	R	R	R
2014	N	L	L	R
2017	N	L	L	R
2024	N	L	L	R
3003	R	R	R	R
5005	R	R	R	R
6061	R	R	R	R
6070	L	R	R	R
6071	R	R	R	R
7070	N	N	R	R
7075	N	N	L	R

*R = readily weldable; S = weldable, but may require special techniques; L = limited weldability; N = not recommended.

than nickel and copper alloys. A typical application for an Alloy 20 is in boiling sulfuric acid concentrations of 30 or 40 percent.

All Alloy 20s can be welded to either stainless-steel or nickel alloys, but the cast alloys have limited weldability owing to their lower strength. Alloy 20s can be welded by using the metal-arc (electrode) method, the tungsten–inert-gas (TIG) method, or the metal–inert-gas (MIG) method.

Copper alloys. Copper alloys are widely used where good corrosion resistance (especially to seawater) combined with machinability and electrical or thermal conductivity is required. They are also used where joining by soldering or brazing rather than welding is acceptable. Copper alloys are used for cookers (reactors) in the food-and-beverage industry because of their imputed ability to improve the flavor of the product. A number of copper alloys are used throughout a process plant. Each is presented below together with some general applications. In the galvanic series copper alloys generally exhibit noble behavior with respect to steel. Therefore, these alloys should not be joined to or be in contact with carbon steel because the resulting galvanic cell causes the steel to corrode.

Pure copper. Pure copper is not 100 percent pure copper but is made of weldable phosphorus-deoxidized copper. During production about 0.02 percent phosphorus is added to the molten copper to remove oxygen. This increases the copper's tensile strength but reduces its electrical conductivity. Copper is resistant to corrosion from seawater and fresh water as long as the velocities of the water are below 8 ft/s. However, if the water is slightly saline and is aerated, corrosion is noticeable at velocities above 4 ft/s. Copper is resistant to nonoxidizing acids such as dilute sulfuric acid and acetic acid, but it is not resistant to oxidizing acids such as concentrated sulfuric acid or to ammonium ions $(NH_4)^+$. Copper should not be used in process equipment handling fluids containing ammonium ions because copper reacts with ammonium ions to form $Cu(NH_3)_4^{++}$. This copper-ammonium complex is readily soluble in water, resulting in rapid corrosion of the copper.

Copper-beryllium alloys. These alloys, also referred to as beryllium bronze, contain between 0.3 and 3 percent beryllium, which increases the strength of the alloy to nearly that of steel. The beryllium also increases the electrical conductivity of the alloy. Copper-beryllium alloys are used instead of tool steels in services where sparking and ferromagnetic properties are a detriment. They are also used in plastic-injection-molding cavities and lubricated sliding systems. These alloys are also used in making springs because of their high resistance to fatigue.

Brasses. These are copper alloys containing 5 to 45 percent zinc. The zinc concentration has an offsetting effect on the corrosion resistance of the brass; i.e., increased zinc concentration increases susceptibility to stress-corrosion cracking but reduces susceptibility to impingement and erosion corrosion.

Muntz metal is a brass containing about 40 percent zinc. The brass has low ductility but high strength. It has fair resistance to salt water and is used to make tubes for inexpensive condensers.

Yellow brass contains about 30 percent zinc and has the best combination of strength and ductility of all the brasses. It is used when spinning, drawing, machining, or casting operations are required. However, yellow brass has a tendency to dezincify in salt and soft waters. Adding about 1 percent tin to the alloy produces admiralty brass, which has a much improved resistance to dezincification and is a more useful alloy for process-equipment construction.

Red brass contains about 15 percent zinc and is relatively immune to dezincification but is more susceptible to erosion corrosion.

Manganese bronze is actually a brass with an approximate composition of 56 percent copper, 41 percent zinc, 1 percent iron, 0.5 percent tin, 0.5 percent manganese, and 1 percent aluminum. It has high tensile strength and is used for pump casings and impellers, screens, and valve trims.

Bronzes. Bronzes have traditionally been defined as copper-tin alloys, but the current definition for a bronze encompasses all copper alloys that are not brasses. For example, there are other bronzes besides the copper-tin alloys, such as phosphorus bronzes, silicon bronzes, and aluminum bronzes. Bronzes are more costly but have higher strength and are more resistant to corrosion than brasses. They do not contain zinc and therefore do not corrode by dezincification, as do brasses.

Aluminum bronzes are copper alloys with 10 percent or less aluminum content. These bronzes are stronger, less ductile, and more wear-resistant than copper-tin bronzes and brasses. They are also the most corrosion-resistant of the copper alloys to high-temperature environments and to seawater. Of all the bronzes the aluminum bronzes have the highest resistance to acid and hydrogen sulfide corrosion.

Silicon bronzes contain up to 4 percent silicon, have high strength, are easily machinable, and are weldable. These bronzes are used in marine environments and environments containing hydrochloric and sulfuric acid. They are also resistant to erosion corrosion.

Copper-tin bronzes, the traditional bronzes, contain up to 12 percent tin and are the most ductile of the bronzes. They are resistant to erosion and impingement corrosion and are not susceptible to stress-corrosion cracking, as are the brasses. They are used in seawater and neutral-water environments.

Phosphorus bronzes are tin bronzes containing up to 0.35 percent phosphorus. Phosphorus increases the hardness of a bronze, thus improving its wear resistance. Because phosphorus bronzes are also resistant to fatigue, they are used in making springs and bearings.

Cupronickel alloys. Copper and nickel are soluble in each other in all proportions (see Fig. 5.13). During solidification they tend to form an austenitic structure and are generally ductile alloys. These alloys are more expensive than carbon steels and stainless steel but are used for corrosion-resistant process pressure vessels (see Table 5.7). Cupronickel alloys containing 50 percent or more copper are arbitrarily defined as copper alloys, while those containing 50 percent or more nickel are considered nickel alloys.

There are three commonly used cupronickel alloys: Cu/Ni = 90/10, 70/30, and 55/45. The 90/10 and 70/30 alloys are the most commonly used for process piping, tubes, and vessels. They are used for high-pressure steam fittings, condenser tubes, and process equipment exposed to seawater, brackish water, and neutral waters. They are also resistant to impingement and erosion corrosion due to high-velocity waters. The alloy containing 30 percent nickel is less susceptible to stress-corrosion cracking than the 10 percent nickel alloy. The 70/30 alloy may not be used for castings because of its high shrinkage rate.

The 55/45 alloys are popularly referred to as constantan. Constantan has high electric resistance and is used in the construction of electrical equipment such as electric process heaters. When constantan is joined to a dissimilar metal such as pure copper or iron, the combination is used as a thermocouple.

The cupronickel alloys are joined by welding, whereas bronzes and brasses are typically brazed or soldered. However, care must be taken during the welding process to prevent cracking from shrinkage.

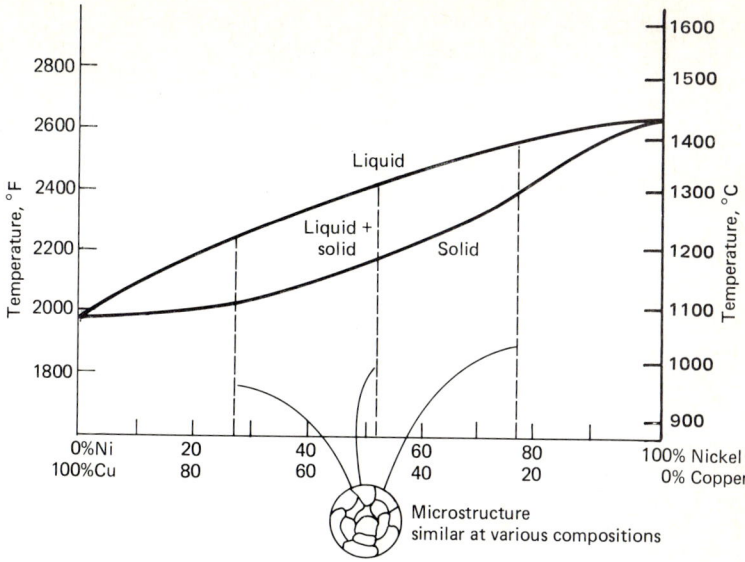

Figure 5.13 Temperature-composition diagram for copper-nickel alloys. Melting points: copper = 1981°F (1083°C); nickel = 2646°F (1452°C).

Nickel alloys. Nickel alloys have been arbitrarily defined as those alloys which contain nickel as the major constituent. They are usually the next choice after carbon steels and stainless steels for corrosive environments at elevated temperatures and pressures. Nickel alloys can be divided into five major groups:

1. Pure nickel alloys
2. Nickel-copper alloys called Monels
3. Nickel-chromium alloys called Inconels or Nichromes
4. Nickel-chromium-iron alloys called Incoloys
5. Nickel-molybdenum alloys called Hastelloys

TABLE 5.7 Copper-Nickel Alloys and Their Applications

Composition	Typical application
75–80% copper, balance nickel	Condenser tube; coins
70% copper, 30% nickel	High-pressure marine boilers and condensers because of its high ductility and strength and corrosion resistance
60% copper, 40% nickel	Constantan wire used in electrical heating equipment because of its high electrical resistance and low thermal coefficient of expansion; when joined to a dissimilar metal such as iron, used in thermocouples
40–70% nickel	Monels; used where corrosion resistance to some acids, neutral water, and seawaters is necessary
95% or more nickel	200 Series nickel alloys; used where corrosion resistance to strong caustics is necessary

Nickel and nickel alloys are used for process equipment because they form an austenitic microstructure, resulting in a ductile, malleable alloy. This makes nickel alloys most desirable for the construction of piping and pressure vessels despite their higher cost relative to carbon steel. Nickel alloys are readily weldable and can be readily joined to such dissimilar metals as copper alloys and steels by either welding or brazing techniques.

Pure nickels. Pure nickel alloys commercially available as nickel 200, 201, or 210 contain 95 to 99.5 percent nickel. Nickel 210 has the lowest nickel content and is used for castings. Nickel 200 and 201 are available in plate, sheet, piping, and tubing form. The essentially pure nickel alloys are used in corrosive environments at temperatures as high as 600°F. They are resistant to attack from strong caustic solutions, but they are not suitable for anhydrous ammonia or ammonium hydroxide at concentrations greater than 1 percent. The pure nickels are also used in handling phenols, foods, and synthetic fibers. Exposure to sulfur compounds at or about 600°F causes embrittlement of the alloy.

Monels. Monel is an International Nickel Corporation trademark for nickel-copper alloys which have better resistance to reducing environments than pure nickels and better resistance to oxidizing environments than copper alloys. Monels are usually TIG- or MIG-welded. They have good strength and ductility. Monels containing sulfur have enhanced machinability. Monels are resistant to seawater, neutral water, sulfuric acid, hydrochloric acid, lye, and deaerated hydrofluoric acid.

Inconels. Inconel is the trademark name for International Nickel Corporation nickel-chromium alloys having high corrosion resistance at elevated temperatures. Alloys 600 and 625 of the Inconels are the most often selected for process-equipment construction. The corrosion resistance of Inconel 600 is better at elevated temperatures than a pure nickel or a Monel. Inconel 600 can resist oxidation in a furnace up to a temperature of 2150°F. Inconel 625 is another alloy which has higher strength and toughness from cryogenic temperatures to 1800°F and exhibits high fatigue strength.

Inconels are used in constructing high-pressure boilers, nuclear-reactor equipment, and vessels for the food-and-beverage industry, pharmaceutical industry, and pulp-and-paper industry. In the pulp-and-paper industry Inconels have excellent resistance to the alkaline sulfur solutions used in the kraft-paper process. Unlike the stainless steels, Inconel 600 is immune to stress-corrosion cracking caused by aqueous solutions having high chloride concentrations. Therefore, it is compatible with hot salt water, magnesium chloride solutions, dry chlorine gas, and dry hydrogen chloride.

Incoloys. Incoloy is the trademark name for an International Nickel Corporation nickel-chromium-iron alloy having high corrosion resistance at elevated temperatures. Resistance to oxidation is slightly less than that of an Inconel. The Incoloys are resistant to pitting from reducing acids and stress-corrosion cracking caused by chlorides. They have been used for handling superheated steam up to a temperature of 1150°F. Incoloys 800 and 825 are the alloys generally used for process pressure vessels and piping.

Hastelloys. Hastelloy is the trademark name used by Cabot Corporation for its nickel-molybdenum alloys. These alloys have extremely good corrosion resistance at elevated temperatures. Some of the different types of Hastelloys are listed in Table 5.8 together with their compositions, forms, and code specifications.

Hastelloy B's and C's are normally used for corrosion-resistant piping. Hastelloy B's and N's are used in equipment handling molten salts at elevated temperatures such as breeder reactors.

TABLE 5.8 Characteristics of Hastelloys

	Hastelloy type			
	B-2	C-4	G	N
Composition				
Nickel	Balance	Balance	Balance	Balance
Molybdenum	28%	15.5%	6.5%	16.5%
Chromium	1% max.	16%	22%	7%
Cobalt	1% max.	2% max.	2.5% max.	0.2% max.
Iron	2% max.	3% max.	19.5%	5% max.
Silicon	0.1% max.	0.08% max.	1% max.	1% max.
Form and ASTM code				
Plate and sheets	B333	B575	B582	B434
Pipes and tubes	B619, B622, B626	B619, B622, B626	B619, B622, B626	
Rods	B335	B574	B581	B573
Fittings	B366	B366	B366	B366

Hastelloy B is resistant to hot hydrogen chloride, boiling hydrochloric and sulfuric acid (up to a concentration of 60 percent), and other reducing acids. Hastelloy C is resistant to wet chlorine gas, chlorine dioxide gas, ferric chloride solution, cupric chloride solutions, and other oxidizing chemicals. It is resistant to corrosion by pitting and stress-corrosion cracking. Hastelloy G has outstanding resistance to mixed acids, fluosilicic acid, sulfate compounds, contaminated nitric acid, and flue gases. It is used for flue-gas scrubbers, for pulp digesters, and in uranium-fuel-rod-reprocessing equipment.

Titanium. Titanium is used in the relatively pure state (90 to 99.9 percent titanium). It has excellent corrosion resistance but is one of the more costly and difficult alloys to weld. Titanium is not susceptible to biological action and is often used to solve corrosion problems in the food-and-beverage industry.

Titanium's corrosion resistance is due to the impervious oxide film that forms on its surface. Therefore, it is resistant to oxidizing environments but not to reducing environments. Titanium is used in seawater, with aqueous chloride solutions over a wide range of concentrations and temperatures, with wet chlorine gas, and with nitric acid (up to 90 percent concentration at 400°F). Titanium will corrode in dry chlorine gas, but the addition of moisture to the gas reduces the rate of corrosion.

Titanium is resistant to stress-corrosion cracking and erosion corrosion, but it is susceptible to crevice corrosion in stagnant chloride solutions. Titanium should not be used in handling hydrofluoric acid.

The difficulty in welding titanium is due to its high affinity for hydrogen, nitrogen, and oxygen in the molten state. Therefore, it must be welded by such inert-gas welding methods as the TIG or the MIG process. After a weld has been made, the inert-gas protection must be maintained until the welded joint cools below 1200°F; otherwise, the titanium will react with the oxygen, nitrogen, and moisture in the air, resulting in weld embrittlement.

Tantalum. Tantalum has excellent corrosion resistance, but its extremely high cost requires that it be used as metal cladding on vessels and pipes and only in extremely corrosive environments. It is resistant to most acids except hydrofluoric acid. It is also resistant to alkaline and salt solutions at room temperature but is attacked by boiling solutions. Tantalum is resistant to many oxidizing and reducing acids up to their boiling points.

Like titanium, tantalum must be welded in an inert gas, i.e., by TIG or MIG welding or by the plasma welding process. Vessels and piping clad with tantalum are normally connected by lap-joint flanges, with the tantalum extending over the flared end.

Zirconium. Zirconium has a structure and properties similar to those of titanium. Unlike tantalum and titanium, which are attacked by hot alkalies, zirconium is resistant to fused sodium hydroxide and alkalies in all concentrations up to their boiling points. It is resistant to nitric and hydrochloric acid in all concentrations but to sulfuric acid only up to 70 percent concentration. It is resistant to corrosion by hot water and steam up to about 800°F. Zirconium is not resistant to hydrofluoric acid, aqua regia, fluosilicic acid, wet chlorine, and oxidizing metal chlorides.

Like titanium, zirconium can absorb O_2, CO_2, H_2, and N_2 in the molten state, causing weld embrittlement. Therefore, zirconium must be welded and maintained under an inert-gas environment until it cools. Zirconium should therefore be welded by either the TIG, the MIG, or the plasma method or welded in an inert-gas closed chamber.

Codes and material selection. A knowledge of metals and their metallurgy does not guarantee that the process engineer will obtain the correct material for a process application. In fact, any equipment specification with a word description of the materials of construction to be used may be too vague or cumbersome for material suppliers to understand or adhere to. The use of standards and codes to define the type of material used and its required properties and composition greatly improves the probability of getting a satisfactory material.

American codes and standards. There are a number of codes and standards instituted by professional groups in the United States to ensure that the material used for fabricating process equipment meets certain minimum criteria for uniformity, safety, and reliability. Table 5.9 lists some of the organizations and the types of materials they are concerned with. For most process applications, the process engineer need refer only to American Society for Testing and Materials (ASTM) or American Society of Mechanical Engineers (ASME) designations to specify materials.

The first step in the material-selection process is definition of the process conditions: temperature, pressure, and corrosiveness of the environment. The next step requires the selection of a group of alloys which appear to be most compatible with these process conditions. The following step in material selection is a search through the proper codes or standards to see whether or not any of the desired materials are listed in the codes and what their designated numbers are.

The most frequently consulted source of material information is the ASTM standards. These standards consist of 48 volumes. Each volume, defined by the ASTM as a "part," is dedicated to a particular group of materials on the basis of their usage or qualification test. Table 5.10 lists some of the subjects covered by the ASTM standards.

Part 48, the index, is the most useful. Given a selection of materials and process conditions, the index is used as a guide to the standard number which meets these requirements. The index is organized both by ASTM standard number and by subject. Examples 5.1 and 5.2 show how the index can be used to select materials of construction and to check whether or not the proper material has been selected for a process application.

TABLE 5.9 Professional Organizations and Their Scope of Concern

Abbreviation	Organization	Scope of concern
ASTM	American Society for Testing and Materials 1916 Race Street Philadelphia, Pa. 19103	Provides testing procedures for a wide variety of materials in various applications to ensure material quality and uniformity defined by the standard.
ASME	American Society of Mechanical Engineers 345 East 47th Street New York, N.Y. 10017	Provides a list of recommended metals and welding rods, electrodes, and filler metals used in the fabrication of process pressure vessels, boilers, and nuclear power components. The society also provides quality-control standards for these metals.
AISI	American Iron and Steel Institute 1000 16th Street, N.W. Washington, D.C. 20036	Provides standard descriptions for carbon steels and alloy steels used in high-production items such as structural components, automobiles, and appliances and for high-strength steels.
SAE	Society of Automotive Engineers 400 Commonwealth Drive Warrendale, Pa. 15096	Provides a four-digit system for identifying carbon, alloy, stainless, and tool steels. This standard was developed in conjunction with AISI and is referred to as the AISI-SAE system.
CDA	Copper Development Association 405 Lexington Avenue New York, N.Y. 10174	Instituted a numerical system for identifying copper alloys.
ANSI	American National Standards Institute 1430 Broadway New York, N.Y. 10008	Provides safety and design criteria and test procedures for a wide variety of mechanical components and systems. Some standards, such as the piping standards, include a list of recommended materials.

Example 5.1 A chemical reactor is to be constructed of Inconel clad to carbon steel plate. How should the material be specified by using the ASTM standards?

solution

1. Consult the ASTM standards, specifically, *Annual Book of ASTM Standards,* Part 48, Index—Subject Index; Numeric List. Look in the subject index under "steel plate." Figure 5.14 shows a page of the index, and under steel plate there is an entry "Cladding, with nickel/nickel alloy, spec., (A265)01.03, 01.04." The A265 is the ASTM standard

TABLE 5.10 List of Sample Subjects Covered in ASTM Standards*

ASTM part number	Subject
1	Steel piping, tubing, and fittings
2	Ferrous castings; ferroalloys
3	Steel plate, sheet, strip, and wire
4	Structural steel; concrete-reinforcing steel; pressure-vessel plate and forgings; steel rails, wheels, and tires; fasteners
6	Copper and copper alloys
8	Nonferrous metals—nickel, lead, and tin alloys; precious metals; primary metals; reactive metals
16	Chemical-resistant nonmetallic materials—vitrified-clay and concrete pipe and tile; masonry mortars and units; asbestos-cement products
17	Refractories; glass and other ceramic materials; manufactured carbon and graphite products
34	Plastic pipe and building products
36	Plastics—materials, films, reinforced and cellular plastics; high-modulus fibers and their composites
37	Rubber, natural and synthetic—general test methods; carbon black
38	Rubber products, industrial—specifications and related test methods; gaskets; tires
48	Index—subject indexes, alphanumeric list

*This list is a sample of the ASTM volumes most often consulted by process engineers when selecting materials.

number which should be used when specifying steel-plate cladding. The 01.03 and 01.04 indicate the ASTM volume in which the standard is presented in detail.

2. If the reactor is a pressure vessel, then the ASTM number A265 is inadequate because pressure-vessel materials must come from the list of approved materials from the ASME pressure-vessel codes. For convenience, the ASME material number is usually the same as the ASTM number except that it is prefixed with an S. Therefore the ASME codes must be searched for under SA265, which is also identified as nickel/nickel-alloy-clad-steel plate. Fortunately such a designated material exists in the ASME codes.

A word of caution must be interjected at this point. ASME code materials may be used in applications requiring ASTM materials but not the reverse. ASTM materials may be of the same composition and properties as ASME materials but do not have the extensive quality-control requirements assigned to ASME materials.

Example 5.2 A reactor constructed of zirconium requires forged-zirconium flanges. The flange material specified by the manufacturer is SB495. Has the correct material been specified?

solution Because the material specification has an S prefix, the material must be one of the materials recommended by the ASME code. However, the code number less the S would be the standard for a similar material in the ASTM standards (see cautionary statement in Example 5.1). If the numeric list in the ASTM standards, Part 48, Index, is surveyed, the following listing is found for B495 (which is also reproduced in Fig.

Steel plate (*cont.*)
  carbon steel (*cont.*)
    sels), spec., **(A 299/A 299M) 01.04**
    manganese-titanium plate (for welded glass-lined pressure vessels), spec., **(A 562/A 562M) 01.04**
    for moderate-/lower-temperature service (in pressure vessels), spec., **(A 516/A 516M) 01.04**
    plates and bars, with 42,000 psi (290 MPa) minimum yield point, and ½ in. (12.7mm) maximum thickness, spec. for, **(A 529) 01.04**
    precipitation-hardening (maraging) 18 % nickel plate (for pressure vessels), spec., **(A 538/A 538M) 01.04**
  carbon steel (high-strength),
    for moderate/lower temperature service in welded pressure vessels, spec., **(A 612/A 612M) 01.04**
  carbon steel (high-strength, low-alloy),
    rolled floor plates, spec., **(A 786) 01.04**
  carbon steel (low),
    age-hardening nickel-copper-chromium-molybdenum-columbium/nickel-copper-columbium alloy steel plates/shapes/bars, spec.,
    (A 710) 01.04
  carbon steel (low-/intermediate-tensile strength),
    for fusion-welded pressure vessels, spec.,
    (A 285/A 285M) 01.04
    structural quality, for general purposes, spec.,
    (A 283) 01.04
  carbon steel (quenched and tempered),
    carbon steel (quenched and tempered), for structural applications, spec., **(A 678) 01.04**
    for welded layered pressure vessels, spec.,
    (A 724/A 724M) 01.04
  chromium/chromium-nickel/chromium-manganese-nickel, plate, sheet, and strip, for fusion-welded unfired pressure vessels, spec.,
    (A 240) 01.03, 01.04
  cladding,
    with nickel/nickel alloy, spec.,
    (A 265) 01.03, 01.04
  composite, tool-resisting, for security applications, spec., **(A 628) 01.03**
  copper/copper alloy clad, plate, spec., **(B 432) 02.01**
  corrosion-resisting chromium steel clad plate/sheet/strip, spec., **(A 263) 01.03, 01.04**
  corrugated structural, zinc-coated (galvanized), for pipe, pipe-arches, and arches, spec.,
    (A 761) 01.03, 15.08
  general requirements (for pressure vessels), spec.,
    (A 20/A 20M) 01.04
  general requirements (for rolled steel plates/shapes/sheet piling/bars), for structural use, spec.,
    (A 6) 01.04
  heat-resisting chromium/chromium-nickel stainless steel plate/sheet/strip (for fusion-welded unfired pressure vessels), spec.,
    (A 240) 01.03, 01.04
  high-strength/low alloy steel plate,
    high-strength/low-alloy columbium-vanadium steel shapes/plates/sheet piling/bars (for riveted/bolted/welded construction of bridges/buildings/other structures), spec.,
    (A 752) 01.04

    high-strength/low-alloy/hot-rolled structural steel plate, spec., **(A 656) 01.04**
    high-strength/low-alloy pressure vessel steel plates, spec., **(A 737/A 737M) 01.04**
    high-strength/low-alloy structural manganese vanadium steel shapes/plates/bars (for welded/riveted/bolted constructions in welded bridges and buildings), spec., **(A 441) 01.04**
    high-strength low-alloy structural steel shapes/plates/bars (for welded/riveted/bolted constructions), with enhanced atmospheric corrosion resistance, spec., **(A 242) 01.04**
    high-strength low-alloy structural steel shapes/plates/bars, with 50 ksi (345 MPa) minimum yield point to 4 in. (100 mm) thick, spec.,
    (A 588) 01.04
  industrial perforated, round opening series, spec. for openings, **(E 674) 01.03, 14.02**
  installing corrugated steel structural plate pipe (for sewers), practice, **(A 807) 01.06**
  nickel/nickel-base alloy clad steel plate, spec.,
    (A 265) 01.03, 01.04
  nil-ductility transition (NDT) temperature, drop-weight test, Δ **(E 208) 03.01**
  packaging/marking/loading methods (for steel products for domestic shipment), rec. practice,
    (A 700) 01.01, 01.03, 01.04, 01.05
  rolled floor plate,
    carbon, high-strength low-alloy/alloy steel, spec.,
    (A 786) 01.04
    stainless steel, spec., **(A 793) 01.03**
  structural,
    columbium-vanadium structural steel shapes/plates/sheet piling/bars, spec., **(A 572) 01.04**
    normalized high-strength low-alloy structural steel plates/shapes/bars (for welded/riveted/bolted construction), spec., **(A 633) 01.04**
    structural steel (carbon and high-strength, low-alloy steel) for bridges, spec., **(A 709) 01.04**
    structural steel shapes/plates/bars (for ship construction), spec., **(A 131) 01.04**
  through-thickness tension testing, **(A 770) 01.04**
  tie plates (low-carbon/high-carbon-hot-worked), spec., **(A 67) 01.04**
  ultrasonic examination,
    angle-beam ultrasonic examination (of steel plate for pressure vessels), spec.,
    (A 577/A 577M) 01.04
    straight-beam ultrasonic examination (of plain/clad steel plate for special applications), spec.,
    (A 578/A 578M) 01.04
    straight-beam ultrasonic examination (of steel plate for pressure vessels), spec.,
    (A 435/A 435M) 01.04
  wildcats (for ship anchor chains), spec.,
    (F 765) 01.02
  winding mesh,
    zinc-coated (galvanized) steel pipe winding mesh, spec., **(A 810) 01.06**
  zinc coatings on, by hot-dip galvanizing process, assuring high-quality coatings, rec. practice,
    (A 385) 01.06
  safeguarding against embrittlement/detection pro-

Figure 5.14 Subject index taken from the 1984 *Annual Book of ASTM Standards.*

Brass) Rod (Metric), **02.01**
B 454   Discontinued—Replaced by B 695 and B 696
B 455–83   Specification for Copper-Zinc-Lead Alloy (Leaded Brass) Extruded Shapes,   **02.01**
B 456–79   Specification for Electrodeposited Coatings of Copper Plus Nickel Plus Chromium and Nickel Plus Chromium,   **02.05**
B 457–67(1980)   Method for Measurement of Impedance of Anodic Coatings on Aluminum,   **02.02, 02.05**
B 458–83a   Specification for Sintered Nickel Silver Structural Parts,   **02.05**
B 459   Discontinued
B 460–68(1980)   Test Method for Dynamic Coefficient of Friction and Wear of Sintered Metal Friction Materials Under Dry Conditions,   **02.05**
B 461–67(1980)   Test Method for Frictional Characteristics of Sintered Metal Friction Materials Run in Lubricants,   **02.05**
B 462–82   Specification for Forged or Rolled Chromium-Nickel-Iron-Molybdenum-Copper-Columbium Stabilized Alloy (UNS N08020) Pipe Flanges, Forged Fittings, and Valves and Parts for Corrosive High-Temperature Ser   **02.04**
B 463–83   Specification for Chromium-Nickel-Iron-Molybdenum-Copper-Columbium Stabilized Alloy (UNS N08020) Plate, Sheet, and Strip,   **02.04**
B 464–83   Specification for Welded Chromium-Nickel-Iron-Molybdenum-Copper-Columbium Stabilized Alloy (UNS N08020) Pipe,   **02.04**
B 465–83   Specification for Copper-Iron Alloy Plate, Sheet, Strip, and Rolled Bar,   **02.01**
B 466–83a   Specification for Seamless Copper-Nickel Pipe and Tube,   **02.01**
B 466M–84   Specification for Seamless Copper-Nickel Pipe and Tube [Metric],   **02.01**
B 467–83   Specification for Welded Copper-Nickel Pipe,   **02.01**
B 468–83   Specification for Welded Chromium-Nickel-Iron-Molybdenum-Copper-Columbium Stabilized Alloy (UNS N08020) Tubes,   **02.04**
B 469–77   Specification for Seamless Copper Alloy Tubes for Pressure Applications,   **02.01**
B 470–68(1978)   Specification for Bonded Copper Conductors for Use in Hookup Wire for Electronic Equipment,   **02.03**
B 471–79   Specification for Chromium-Nickel-Iron-Molybdenum-Copper-Columbium Stabilized Alloy (UNS N08020) Round Spring Wire,   **02.04**
B 472–79   Specification for Chromium-Nickel-Iron-Molybdenum-Copper-Columbium Stabilized Alloy (UNS N08020) Billets and Bars for Reforging,   **02.04**
B 473–82   Specification for Chromium-Nickel-Iron-Molybdenum-Copper-Columbium Stabilized Alloy (UNS N08020) Bar and Wire,   **02.04**
B 474–82   Specification for Electric Fusion Welded Chromium-Nickel-Iron-Molybdenum-Copper-Columbium Stabilized Alloy (UNS N08020) Pipe,   **02.04**
B 475–79   Specification for Chromium-Nickel-Iron-Molybdenum-Copper-Columbium Alloy (UNS N08020) Round Weaving Wire,   **02.04**
B 476–72(1978)   Specification for General Requirements for Wrought Precious Metal Electrical Contact Materials,   **03.04**
B 477–82   Specification for Gold-Silver-Nickel Electrical Contact Alloy,   **03.04**
B 478–79   Test Method for Cross Curvature of Thermostat Metals,   **03.04**
B 479–82a   Specification for Annealed Aluminum Foil for Flexible Barrier Applications,   **02.02, 15.09**
B 480–68(1980)   Practice for Preparation of Magnesium and Magnesium Alloys for Electroplating,   **02.05**
B 481–68(1979)   Practice for Preparation of Titanium and Titanium Alloys for Electroplating,   **02.05**
B 482–68(1979)   Practice for Preparation of Tungsten and Tungsten Alloys for Electroplating,   **02.05**
B 483–83   Specification for Aluminum and Aluminum-Alloy Drawn Tubes for General Purpose Applications,   **02.02**
B 483M–83   Specification for Aluminum and Aluminum-Alloy Drawn Tubes for General-Purpose Applications [Metric],   **02.02**
B 484–83a   Specification for Sintered Nickel Steel Structural Parts,   **02.05**
B 485   Discontinued
B 486–74(1980)   Specification for Paste Solder,   **02.04**
B 487–79   Method for Measurement of Metal and Oxide Coating Thicknesses by Microscopical Examination of a Cross Section,   **02.05**
B 488–80   Specification for Electrodeposited Coatings of Gold for Engineering Uses,   **02.05, 03.04**
B 489–68(1979)   Practice for Bend Test for Ductility of Electrodeposited and Autocatalytically Deposited Metal Coatings on Metals,   **02.05**

B 490–68(1980)   Practice for Micrometer Bend Test for Ductility of Electrodeposits,   **02.05**
B 491–83   Specification for Aluminum-Alloy Extruded Round Tubes for General Purpose Applications,   **02.02**
B 491M–83   Specification for Aluminum and Aluminum-Alloy Extruded Round Tubes for General Purpose Applications [Metric],   **02.02**
B 492–82   Specification for Cast Copper-Nickel Ship Tailshaft Sleeves,   **02.01**
B 493–83   Specification for Zirconium and Zirconium Alloy Forgings,   **02.04**
B 494–79   Specification for Primary Zirconium,   **02.04**
B 495–79   Specification for Zirconium and Zirconium Alloy Ingots,   **02.04**
B 496–81   Specification for Compact Round Concentric-Lay-Stranded Copper Conductors,   **02.03**
B 497–79   Practice for Measuring Voltage Drop on Closed Arcing Contacts,   **03.04**
B 498–74(1979)   Specification for Zinc-Coated (Galvanized) Steel Core Wire for Aluminum Conductors, Steel Reinforced (ACSR),   **01.06, 02.03**
B 498M–83   Specification for Zinc-Coated (Galvanized) Steel Core Wire for Aluminum Conductors, Steel Reinforced (ACSR) [Metric],   **01.06, 02.03**
B 499–75(1980)   Method for Measurement of Coating Thicknesses by the Magnetic Method: Nonmagnetic Coatings on Magnetic Basis Metals,   **02.05**
B 500–72(1982)   Specification for Zinc-Coated (Galvanized) and Aluminum-Coated (Aluminized) Stranded Steel Core for Aluminum Conductors, Steel Reinforced (ACSR),   **02.03**
B 501–69(1983)   Specification for Silver-Coated Copper-Clad Steel Wire for Electronic Application,   **02.03**
B 502–70(1980)   Specification for Aluminum-Clad Steel Core Wire for Aluminum Conductors, Aluminum-Clad Steel Reinforced,   **02.03**
B 503–69(1984)   Recommended Practice for Use of Copper and Nickel Electroplating Solutions for Electroforming,   **02.05**
B 504–82   Method for Measurement of Thickness of Metallic Coatings by the Coulometric Method,   **02.05**
B 505–84   Specification for Copper-Base Alloy Continuous Castings,   **02.01**
B 506–81   Specification for Copper-Clad Stainless Steel Sheet and Strip for Building Construction,   **02.01**
B 507–70(1979)   Practice for Design of Articles to Be Electroplated on Racks with Nickel,   **02.05**
B 508–81   Specification for Copper Alloy Strip for Flexible Metal Hose,   **02.01**
B 509–77(1983)   Specification for Supplementary Requirements for Nickel Alloy Plate for Nuclear Applications,   **02.04, 12.01**
B 510–77(1983)   Specifications for Supplementary Requirements for Nickel Alloy Rod and Bar for Nuclear Applications,   **02.04, 12.01**
B 511–81   Specification for Nickel-Iron-Chromium-Silicon Alloy Bars and Shapes,   **02.04**
B 512–81   Specification for Nickel-Iron-Chromium-Silicon Alloy (UNS N08330) Billets and Bars,   **02.04**
B 513–79   Specifications for Supplementary Requirements for Nickel Alloy Seamless Pipe and Tube for Nuclear Applications,   **02.04, 12.01**
B 514–79   Specification for Welded Nickel-Iron-Chromium Alloy Pipe,   **02.04**
B 515–79   Specification for Welded Nickel-Iron-Chromium Alloy (UNS N08800) Tubes,   **02.04**
B 516–79   Specification for Welded Nickel-Chromium-Iron Alloy (UNS N06600) Tubes,   **02.04**
B 517–79   Specification for Welded Nickel-Chromium-Iron Alloy (UNS N06600) Pipe,   **02.04**
B 518–81   Specification for Nickel-Chromium-Iron-Columbium-Molybdenum-Tungsten Alloy (UNS N06102) Rod and Bar,   **02.04**
B 519–81   Specification for Nickel-Chromium-Iron-Columbium-Molybdenum-Tungsten Alloy (UNS N06102) Plate, Sheet, and Strip,   **02.04**
B 520–70(1980)   Specification for Tin-Coated Copper-Clad Steel Wire for Electronic Applications,   **02.03**
B 521–82   Specification for Tantalum and Tantalum Alloy Seamless and Welded Tube,   **12.01**
B 522–80   Specification for Gold-Silver-Platinum Electrical Contact Alloy,   **03.04**
B 523–79   Specification for Seamless and Welded Zirconium and Zirconium Alloy Tubes for Condensers and Heat Exchangers,   **02.04**
B 524–78   Specification for Concentric-Lay-Stranded Aluminum Conductors, Aluminum Alloy Reinforced (ACAR, 1350/6201),   **02.03**
B 524M–82   Specification for Concentric-Lay-Stranded Aluminum Conductors, Aluminum Alloy Reinforced (ACAR, 1350/6201) [Metric],   **02.03**

Figure 5.15   Alphanumeric index taken from the 1984 *Annual Book of ASTM Standards.*

5.15): "B494-79 Zirconium and Zirconium Alloy Ingots, 02.04." The -79 means that the standard was adopted in 1979. The 02.04 identifies the volume of the ASTM standards where the specification can be found.

Because the B495 is for ingots, this would make the material selection suspect. Further examination of the listing would produce the following listing: "B493-83 Zirconium and Zirconium Alloy Forgings, 02.04." This specification would be more applicable for the reactor flange material.

At this point the reactor manufacturer should be made aware of the discrepancy, and the material specification for the reactor flanges must be changed to SB493.

Using standard specifications can greatly improve the material-selection process. The standard number eliminates any ambiguity that may exist between the equipment manufacturer and the engineer when materials are being selected.

Codes and standards of other countries. Although American codes and standards are recognized and accepted throughout the world, there are certain circumstances in which material standards of other countries must be used in the construction of process equipment. This occurs when:

1. The equipment is designed in the United States but fabricated and installed in certain other countries.
2. The equipment is fabricated in one country, for example, in West Germany, and installed in another country whose codes allow only West German standards.

In cases such as these the process engineer needs a frame of reference to guide evaluation of the materials used in process-equipment construction. Table 5.11 is a list of equivalent materials prepared by K. K. Mahajan (see Bibliography) which cross-references the United States ASME code material designation with equivalent material designations for Belgium, Great Britain, France, West Germany, Italy, and Japan.

Polymeric materials

If corrosion resistance is an important equipment design factor and the process design temperature is less than 250°F, then polymeric materials such as plastics and rubbers are the most economical material choice. These polymeric materials may be used by themselves, in combination with fibrous materials for structural strength, or as a liner to carbon steel. When the polymeric material is combined with a fibrous material, usually fiberglass, the composite is called fiberglass-reinforced plastic (FRP), which can, with proper design, withstand pressures in excess of 150 psig. For higher pressures with good corrosion resistance the polymer may be used as a lining to carbon steel, which provides structural strength.

There is a variety of polymeric materials, most of which are organic compounds except for the silicones, which are silicon-based. Polymeric materials have been divided into two major groups, plastics and elastomers. Rubbers are placed in the latter category, but not all elastomers can be considered rubbers. Rubbers are defined as "substances that have at least 200 percent elongation in a tensile test and are capable of returning rapidly and forcibly to their original dimensions when the load is removed." Some elastomers may not be able to achieve this degree of elongation. Plastics, on the other hand, may be brittle, may exhibit permanent deformation under loading, or may have a tendency to creep under loading.

A disadvantage of polymers is their flammability; however, the halogenated poly-

TABLE 5.11 Equivalent Materials of Various Countries*

Composition	United States	France	Italy	Japan	Great Britain	Germany	Belgium
Plate							
Structural steels	SA–36	NF A35-501 E26-2	—	JIS G3106 SM 41A	BS 4360 Gr. 43B	DIN 17100 USt 42-2	NBN 631 AE26C
	SA–283 Gr. C	NF A35-501 E24-2	UNI 7070-72 Fe37B	JIS G3101 SS 41	BS 4360 Gr. 43A	DIN 17100 USt 37-2	NBN 631 AE24D
Vessel steel	SA–285 Gr. C	NF A36-205 A42-C1	UNI 5869-66 Fe42-1	JIS G3101 SB 42B	BS 1501-151 Gr. 26B	DIN 17155 HII	NBN 630 E42-2
Killed steels							
Higher temperature	SA–515 Gr. 70	NF A36-205 A52-C2	UNI 5869-66 Fe52-1	JIS G3115 SPV 32	BS 1501-221 Gr. 32B	DIN 17155 19 Mn 5	NBN 629 D52-2
Lower temperature	SA–516 Gr. 70	NF A36-205 A52-P2	UNI 5869-66 Fe52-2	JIS G3118 SGV 49	BS 1501-224 Gr. 32A	DIN 17135 A St 52	NBN 630 E52-2
Low-alloy steels							
C–½ Mo	SA–204 Gr. B	NF A36-206 18 MD 4.05	†	JIS G3103 SB 49M	†	16 Mo 5	
1 Cr–½ Mo	SA–387 Gr. 12 Cl. 1	†	UNI 5869-66 14 Cr Mo 3	†	BS 1501–620 Gr. 27A	DIN 17155 13 Cr Mo 44	NBN 629 14 Cr Mo 45
1¼ Cr–½ Mo-Si	SA–387 Gr. 11 Cl. 1	†	†	†	BS 1501–621 Gr. A	*	
2¼ Cr–1 Mo	SA–387 Gr. 22 Cl. 1	†	UNI 5869-66 12 Cr Mo 9 10	†	BS 1501–622 Gr. 31A	10 Cr Mo 9 10	NBN 629 12 Cr Mo 9 10
5 Cr–½ Mo	SA–387 Gr. 5 Cl. 1		...	†	†	12 Cr Mo 19 5	
High-alloy steels							
12 Cr–1 Al	SA–240 Gr. 405	NF A35-573 Z 6 CA 13	UNI 6902-71 X 6 Cr Al 13	JIS G4304 SUS 405	BS 1501 Gr. 405 S17	DIN 17440 X 7 Cr Al 13	
18 Cr–8 Ni	SA–240 Gr. 304	NF A35-573 Z 6 CN 18.09	UNI 6902-71 X 6 Cr Ni 18 10	JIS G4304 SUS 304	BS 1501 Gr. 304 S15	DIN 17440 X 5 Cr Ni 18 9	
18 Cr–10 Ni-Ti	SA–240 Gr. 321	NF A35-573 Z 6 CNT 18.11	UNI 6902-71 X 6 Cr Ni Ti 18 11	JIS G4304 SUS 321	BS 1501 Gr. 321 S12	DIN 17440 X 10 Cr Ni Ti 18 9	
18 Cr–10 Ni-Nb	SA–240 Gr. 347	NF A35-573 Z 6 CN Nb 18.11	UNI 6902-71 X 5 Cr Ni Nb 18 11	JIS G4304 SUS 347	BS 1501 Gr. 347 S17	DIN 17440 X 10 Cr Ni Nb 18 9	
16 Cr–12 Ni-2 Mo	SA–240 Gr. 316	NF A35-573 Z 6 CND 17.11	UNI 6902-71 X 5 Cr Ni Mo 17 12	JIS G4304 SUS 316	BS 1501 Gr. 316 S16	DIN 17440 X 5 Cr Ni Mo 18 10	

Pipe							
Carbon steel							
General	SA–53 Gr. B	†			BS 3601 Gr. 410	DIN 1629 St 45	NBN 630 E42-2
Killed steel	SA–106 Gr. B	†	UNI 5462-64 C 18	JIS G3454 STPG 42	BS 3602 Gr. 27S	DIN 17175 St 45.8	NBN 629 D42-2
	SA–155	†	†	JIS G3456 STPT 42	†	†	
Fusion-welded				†			
Low-alloy steels							
C-½ Mo	SA–335 Gr. P1	†	UNI 5462-64 16 Mo 5	JIS G3458 STPA 12	†	16 Mo 5	
1 Cr-½ Mo	SA–335 Gr. P12	†	UNI 5462-64 14 Cr Mo 3	JIS G3458 STPA 22	BS 3604 Gr. HF 620	DIN 17175 13 Cr Mo 44	
1¼ Cr-½ Mo-Si	SA–335 Gr. P11	12 CD 5 05	—	JIS G3458 STPA 23	BS 3604 Gr. HF 621	DIN 17175 13 Cr Mo 44	
2¼ Cr-1 Mo	SA–335 Gr. P22	12 CD 9 10	UNI 5462-64 12 Cr Mo 9 10	JIS G3458 STPA 24	BS 3604 Gr. HF 622	DIN 17175 10 Cr Mo 9 10	
5 Cr-½ Mo	SA–335 Gr. P5	Z 12 CD 5	—	JIS G3458 STPA 25	BS 3604 Gr. HF 625	12 Cr Mo 19 5	
High-alloy steels							
18 Cr-8 Ni	SA–312 TP 304	NF A35-573 Z 6 CN 18.09	UNI 6904-71 X 6 Cr Ni 18 10	JIS G3459 SUS 304 TP	BS 3605 Gr. 304 S18	DIN 17440 X 5 Cr Ni 18 9	
18 Cr-10 Ni-Ti	SA–312 TP 321	NF A35-573 Z 6 CNT 18.11	UNI 6904-71 X 6 Cr Ni Ti 18 11	JIS G3459 SUS 321 TP	BS 3605 Gr. 321 S18	DIN 17440 X 10 Cr Ni Ti 18 9	
18 Cr-10 Ni-Nb	SA–312 TP 347	NF A35-573 Z 6 CN Nb 18.11	UNI 6904-71 X 5 Cr Ni Nb 18 11	JIS G3459 SUS 347 TP	BS 3605 Gr. 347 S18	X 5 Cr Ni Nb 18 9	
16 Cr-12 Ni-2 Mo	SA–312 TP 316	NF A35-573 Z 6 CND 17.11	UNI 6904-71 X 5 Cr Ni Mo 17 12	JIS G3459 SUS 316 TP	BS 3605 Gr. 316 S18	DIN 17440 X 5 Cr Ni Mo 18 10	
Forgings							
Carbon steel							
General	SA–181 Gr. 1	NF A33-101 AF 42 2	UNI 3985 Aq 42	JIS G3201 SF 45	BS 1503–161 Gr. 28A	DIN 17100 R St 42-2	NBN 630 E37-1
Killed steel	SA–105	†	UNI 3986 Aq 50	JIS G3201 SF 50	BS 1503–161 Gr. 32A	DIN 17155 19 Mn 5	
Low-alloy steels							
C-½ Mo	SA–182 Gr. F1	‡	‡	‡	BS 1503–240	16 Mo 5	

TABLE 5.11 Equivalent Materials of Various Countries* *(Continued)*

Composition	United States	France	Italy	Japan	Great Britain	Germany	Belgium
1 Cr–½ Mo	SA–182 Gr. F12	15 CD 4.5	13 CD 4	†	BS 1503–621	DIN 17175 13 Cr Mo 44	
2¼ Cr–1 Mo	SA–182 Gr. F22	12 CD 9.10		†	BS 1503–622	DIN 17175 10 Cr Mo 9 10	
5 Cr–½ Mo	SA–182 Gr. F5	Z 12 CD 5		†	BS 1503–625	12 Cr Mo 19 5	
13 Cr	SA–182 Gr. F6	NF A 35-578 Z10 C13	†	JIS G4303 SUS 410	BS 1503 Gr. 410 S21	DIN 17440 X 10 Cr 13	
High-alloy steels							
18 Cr–8 Ni	SA–182 Gr. F304	NF A35-573 Z 6 CN 18.09	UNI 6901-71 X 6 Cr Ni 18 10	JIS G4303 SUS 304	BS 1503 Gr. 304 S40	DIN 17440 X 5 Cr Ni 18 9	
18 Cr–10 Ni–Ti	SA–182 Gr. F321	NF A35-573 Z 6 CNT 18.11	UNI 6901-71 X 6 Cr Ni Ti 18 11	JIS G4303 SUS 321	BS 1503–821 Gr. 321 S40	DIN 17440 X 10 Cr Ni Ti 18 9	
18 Cr–10 Ni–Nb	SA–182 Gr. F347	NF A35-573 Z 6 CN Nb 18.11	UNI 6901-71 X 5 Cr Ni Nb 18 11	JIS G4303 SUS 347	BS 1503 Gr. 347 S40	DIN 17440 X 10 Cr Ni Nb 18 9	
16 Cr–12 Ni–2 Mo	SA–182 Gr. F316	NF A35-573 Z 6 CND 17.12	UNI 6901-71 X 5 Cr Ni Mo 17 12	JIS G4303 SUS 316	BS 1503 Gr. 316 S40	DIN 17440 X 5 Cr Ni Mo 18 10	
Heat-exchanger tubes							
Carbon steel							
Seamless cold-drawn	SA–179	C 12 d	UNI 5462-64 C 14	JIS G3461 STB 33	BS 3059 Gr. 33	DIN 1629 St 35.4	NBN 629 D37–2
Welded	SA–214	†		JIS G3461 STB 35	BS 3059 Gr. 33	DIN 1626 St 37.2	NBN 629 D37–1
Alloy steel							
5 Cr–½ Mo	SA–213 Gr. T5	Z 12 CD 5	UNI 5462-64 12 Cr Mo 9 10	JIS G3462 STBA 25	†	12 Cr Mo 19 5	
2¼ Cr–1 Mo	SA–213 Gr. T22	12 CD 9.10		JIS G3462 STBA 24	BS 3059 Gr. 621	DIN 17175 10 Cr Mo 9 10	
18 Cr–10 Ni–Ti	SA–213 Gr. TP 321	NF A35-573 Z 6 CNT 18.11	UNI 6904-71 X 6 Cr Ni Ti 18 11	JIS G3463 SUS 321 TB	BS 3605 Gr. 321 S18	DIN 17440 X 10 Cr Ni Ti 18 9	
18 Cr–10 Ni–Nb	SA–213 Gr. TP 347	NF A35-573 Z 6 CN Nb 18.11	UNI 6904-71 X 5 Cr Ni Nb 18 11	JIS G3463 SUS 347 TB	BS 3605 Gr. 347 S18	DIN 17440 X 10 Cr Ni Nb 18 9	

Nonferrous							
Admiralty	SB-111 Alloy No. 443	‡			BS 1464 CZ 111	DIN 1785 Cu Zn 28 Sn	‡
Al-brass	SB-111 Alloy No. 687	‡			BS 1464 CA 110	DIN 1785 Cu Zn 20 Al	‡
Al-bronze	SB-111 Alloy No. 608	‡			BS 1464 CA 102	DIN 1785 Cu Al 15 As	‡
Cupronickel	SB-111 Alloy No. 715	‡			BS 1464 CN 107	DIN 1785 Cu Ni 30 Fe	‡
Monel	SB-163	‡			BS 3074 NA 13	DIN 17743 Ni Cu 30 Fe	‡
Bolts							
Carbon steel	SA-307 Gr. B	†		JIS G3101 SS 41	†	DIN 267 SD (St 5Q11)	
Alloy steel							
Cr-½ Mo	SA-193 Gr. B7	NF A35-559 42 CD 4	UNI 5332-64 40 Cr Mo 4	JIS G4105 SCM 4	BS 1506-621 Gr. A	DIN 17200 42 Cr Mo 4	NBN 253.02 42 Cr Mo 4
5 Cr-½ Mo	SA-193 Gr. B5	‡	‡	‡	‡	‡	‡
13 Cr	SA-193 Gr. B6 (410)	‡					
18 Cr-8 Ni	SA-193 Gr. B8	NF A35-559 Z 6 CN 18.09	UNI 6901-71 X 6 Cr Ni 18 10	JIS G4303 SUS 304	BS 1506-801 Gr. B	DIN 17440 X 5 Cr Ni 18 9	
Nuts							
Carbon steel							
General	SA-307 Gr. B	‡	‡	‡	‡	‡	‡
Elevated temperature	SA-194 Gr. 2H	NF A35-501 A60	UNI 5332-64 C45	JIS G4051 S 45 C	BS 1506-162	DIN 17240 C45	
Alloy steel							
C-Mo	SA-194 Gr. 4	40 D 2	...	†	†	†	‡
5 Cr-½ Mo	SA-194 Gr. 3	‡	‡	‡	‡	DIN 17240 24 Cr Mo 5	‡
12 Cr	SA-194 Gr. 6	‡	‡	‡	‡	‡	
18 Cr-8 Ni	SA-194 Gr. 8	NF A35-605 Z 6 CN 18.09	UNI 6901-71 X 6 Cr Ni 18 10	JIS G4303 SUS 304	BS 1506-801 Gr. B	DIN 17440 X 5 Cr Ni 18 9	‡

TABLE 5.11 Equivalent Materials of Various Countries* (Continued)

Composition	United States	France	Italy	Japan	Great Britain	Germany	Belgium
18 Cr-8 Ni-Nb	SA–194 Gr. 8C	‡	‡	‡	‡	‡	‡
18 Cr-8 Ni-Ti	SA–194 Gr. 8T	‡	‡	‡	‡	‡	‡
Castings							
Carbon steel	SA–216 Gr. WCA	†	∴	JIS G5151 SCPH 1	BS 1504–161 Gr. A	DIN 17245 GS–C25	
	SA–216 Gr. WCB	†	∴	JIS G5151 SCPH 2	BS 1504–161 Gr. B	DIN 17245 GS–C25	
Gray iron	SA–48	†	UNI 5007–69 G20	JIS G5501 FC20	BS 1452 Gr. 12	DIN 1691 GG20	
Ductile iron	SA–536	‡	‡	‡	BS 2789	DIN 1693	‡
Malleable iron	SA–197	‡	‡	‡	BS 310	DIN 1692	‡
Low-alloy steels							
C–½ Mo	SA–217 Gr. WC1	†	∴	JIS G5151 SCPH 11	BS 1504–240	DIN 17245 GS–22 Mo 4	
1¼Cr–½ Mo	SA–217 Gr. WC6	†	∴	JIS G5151 SCPH 21	BS 1504–621	DIN 17245 GS–22 Cr Mo 55	
2¼ Cr–1 Mo	SA–217 Gr. WC9	†	**	†	†	GS 12 Cr Mo 9 10	
5 Cr–½ Mo	SA–217 Gr. C5	‡	∴	JIS G5151 SCPH 61	BS 1504–625	GS 12 Cr Mo 19 5	
High-alloy steels							
13 Cr–½ Mo	SA–217 Gr. CA15	‡	†	†	BS 1504–713	†	
25 Cr–12 Ni	SA–351 Gr. CH8	‡	†	JIS G5122 SCH 13	†	G–X 35 Cr Ni Si 25 12	
25 Cr–20 Ni	SA–351 Gr. CK20	‡	†	JIS G5122 SCH 22	†	G–X 15 Cr Ni Si 25 20	
18 Cr–8 Ni	SA–351 Gr. CF8	Z 6 CN 9.10	X 6 Cr Ni 18 10	JIS G5121 SCS 13	BS 1504–801	DIN 17445 G X 6 Cr Ni 18 9	
18 Cr–9 Ni-2 Mo	SA–351 Gr. CF8M	Z 6 CND 18.12.2	X 6 Cr Ni Mo 17 12	JIS G5121 SCS 14	BS 1504–845 Gr. B	DIN 17445 G X 5 Cr Ni Mo 18 10	
18 Cr–10 Ni-Nb	SA–351 Gr. CF8c	‡	‡	‡	BS 1504–821	G–X 10 Cr Ni Nb 18 9	‡

*Kanti K. Mahajan, *Design of Process Equipment*, Pressure Vessel Handbook Publishing, Inc., Tulsa, Okla., 1975.
†Furnished in accordance with ASTM specifications.

mers and the silicones are less susceptible to burning. Table 5.12 lists a selected group of polymeric materials together with their characteristic physical properties.

The properties of plastics vary with the polymerization process and the individual monomers which make up the plastic. Generally, the longer the polymer chain, the higher the molecular weight, melting point, tensile strength, and compressive strength. If polymer chains are cross-linked to other polymer chains, there is an increase in stiffness and brittleness.

Plastics can be classified as either thermoplastic or thermosetting. The thermoplastics tend to melt at elevated temperatures and resolidify on cooling. The types of plastics which fall into this category are the polyolefins, vinyls, fluorocarbons, polyamides, polystyrene, polyimides, polyacetals, polycarbonates, and cellulosics. The thermosetting plastics tend to cross-link or set on heating and will not melt on reheating but will burn or sublime. The types of plastics which fall into this category are the polyurethanes, epoxies, phenolics, silicones, polyesters, and vinyl esters. However, depending on their monomers, polyurethanes, polyesters, and elastomers may be either thermoplastic or thermosetting.

Polyolefins. The polyolefins are produced by the polymerization of olefins such as ethylene, propylene, and butylene. The linkage occurs at the carbon-to-carbon double bond. Although the strength of a polyolefin is proportional to the molecular weight of its monomeric constituents, these polymers have the lowest strength of all polymer groups. However, they have excellent corrosion resistance to strong acids. Polyethylene is used for chemical-laboratory drain piping and in the construction of small tanks and containers for corrosive chemicals.

Polypropylene is slightly more expensive than polyethylene but has higher strength and corrosion resistance. It is easily formed into pipes and vessels.

TABLE 5.12 Physical Properties of Some Polymeric Plastics

Plastics	Common names	Maximum operating temperature, °F	Tensile strength, psi	Density, lb/ft^3
Acrylonitrile butadiene styrene	ABS	160	5,300–8,300	65.5
Chlorinated polyether	Penton	240	6,000	87.4
Epoxies		300	9,400	100–110
Phenolics		320	16,000	81–94
Polyacrylic (polymethyl methacrylate)	Plexiglas	140		74.9
Polyamide	Nylon	260	12,000	68.7
Polycarbonate		220	9,500	74.9
Polyethylene		140	2,800	57.4
Polyhexafluoropropylene	FEP	400	3,000	134.2
Polyimide		500	13,000	87.4
Polypropylene		190	4,900	56.2
Polystyrene	Styrofoam	160	7,500	64.9
Polytetrafluoroethylene	Teflon, TFE	450	2,500	137.3
Polytrifluorochloroethylene	TFCE	370	5,000	131.1
Polyvinyl chloride	Vinyl, PVC	140	7,000	81.1
Polyvinylidene chloride	Saran, PVDC	210	7,300	106.1
Polyvinylidene fluoride	Kynar	280	7,000	109.2

Polyethylene and polypropylene are both used as carbon steel liners in high-pressure process equipment.

Polyvinyl chloride (PVC). PVC is formed from the vinyl chloride (C_2H_3Cl) monomer. Its strength is comparable to that of the polyolefins, and it has excellent corrosion resistance. However, PVC may not be used with ketones, aromatics, or chlorinated hydrocarbons. PVC does not sustain combustion; it is self-extinguishing. It can be made into solid or lined piping and fittings.

Polyvinylidene chloride (PVDC). Polyvinylidene chloride is also recognized by the names PVDC and Saran™. It is formed from the vinylidene chloride ($C_2H_2Cl_2$) monomer. PVDC has slightly higher strength than PVC and has improved corrosion resistance to fluids at temperatures from 40 to 60°F above the limit for PVC. Because of its higher tolerance to temperatures it can be used in piping carrying hot corrosive liquids (less than 180°F) and in applications similar to those of PVC. PVDC plastic does not sustain combustion.

Fluorocarbons. Fluorocarbons are hydrocarbons whose hydrogen atoms have been totally or partially substituted by fluorine atoms. The most commonly used fluorocarbons are:

Acronym	Monomer	Formula
TFE, Teflon™	Tetrafluoroethylene	$(-CF_2-CF_2-)_n$
TFCE*	Trifluorochloroethylene	$(-CF_2-CFCL-)_n$
FEP	Hexafluoropropylene	$\left(-CF_2-\underset{\underset{CF_3}{\vert}}{CF}-\right)_n$
Kynar	Vinylidene fluoride	$(CH_2-CF_2-)_n$

*Although TFCE has a single chlorine atom in its monomer, it is considered a fluorocarbon.

The fluorocarbons are plastics with high-temperature resistance. Their maximum-operating-temperature limits range between 140 and 250°F. They also are characterized by their very high chemical resistance and very low coefficients of friction. These characteristics make them useful in constructing self-lubricating components in corrosive service. They are used as mechanical seals, gaskets, and bearings and as a lining for piping, fittings, and equipment employed in the food, pharmaceutical, and chemical-process industries. Fluorocarbons also have good dielectric properties and are used as electrical insulators.

Polystyrenes. Polystyrene is produced by a polymerization reaction on the ethylene group of the styrene molecule. This polymer is a low-strength plastic, sometimes referred to as the "cheap" plastic used to make toys that break easily. Polystyrene is used as a foam for insulation and as a sound deadener.

The strength of styrene is improved if it is copolymerized with acrylonitrile and butadiene. The resulting terpolymer, acrylonitrile butadiene styrene (ABS), is tough and has high impact strength. However, it has lower corrosion resistance and design strength than PVC. ABS ia typically used for machine and electrical-equipment housings and piping for waste lines, drainage, vents, potable-water systems, and gas-transmission systems.

Polyethers. The most widely used polyether is Penton™, the trade name for a chlorinated polyether made by Hercules Powder Co., Inc. It has excellent corrosion resistance to common acids, alkalies, and salts and is useful up to a temperature of 250°C. Although it costs about 2 to 3 times as much as PVC, it is normally used in services where PVC or ABS is not recommended. Penton is used as a liner for pumps, piping, fittings, and vessels in corrosive service.

Acrylics. Acrylics are clear plastics with moderate chemical resistance but with good tensile strength and impact resistance. The most commonly known acrylic polymer is polymethyl methacrylate, which is recognized by various trade names. These plastics have excellent optical properties and are used for safety glass in such applications as sight glasses and personnel guards on machining tools.

Polyamines. Polyamines are plastics formed as the condensate product of an amine with another organic compound. If the organic compound is an organic acid, the resulting polymer is a polyamide. If the reaction is between an acid dianhydride and an aromatic diamine, the polymer is known as a polyimide.

Nylons are the most commonly used polyamides. They are light, tough, and abrasion-resistant and have a low coefficient of friction. Nylon parts are used for unlubricated bearings, bushings, gears, and cams. They have higher strength than a polyolefin, PVC, or fluorocarbon but are not as strong as a metal.

Polyimides are used for ball-bearing separators, mechanical seals, and other sliding components.

Polyurethanes. The reaction of a diisocyanate with a low-molecular-weight alcohol produces polyurethane. Depending on the reaction process, the resulting polyurethane may behave like an elastomer or form a hard structure similar to a thermoset plastic. The hard-structured polyurethane is used to form insulation foam.

Although polyurethane is considered nonflammable or self-extinguishing, tests have shown that, when exposed to fire or excessive heat as in welding, the urethane may break down into toxic cyanide vapors and nitrogen oxides.

Polycarbonates. Organic esters reacted with a diphenyl carbonate form a polycarbonate. These are linear polyesters with properties of transparency, excellent impact strength, and high temperature resistance. They are used in applications similar to those of acrylics and have much greater impact strength but are more costly.

Phenolics. The most varied and versatile group of polymers consists of the phenolics. They are formed by the condensation-polymerization reaction between almost any phenolic compound and an aldehyde. Other types of phenolics are produced from a combination of phenol furfural and resorcinol formaldehyde. The phenolics are highly cross-linked thermoset plastics which are hard and possess great strength and impact resistance. They also have high chemical-corrosion resistance. Phenolics are used in electrical components such as insulators, switch plates, and enclosures. In mechanical applications they are used in corrosion-resistant piping and brake linings.

Epoxies. Thermoset plastics utilizing homologues of ethylene oxide are called epoxies. The polymer chains are shorter than those of other polymer types; however, when a hardener is added, the short chains become highly cross-linked, forming a hard, high-tensile-strength polymer with excellent temperature-resistant and corrosion-resistant properties. Thus, epoxies are used as corrosion-resistant surface coatings and also employed in FRP piping. Epoxies are also used to join metals to polyesters.

Polyesters. The condensation product of dimethyl terephthalate or terephthalic acid and ethylene glycol is polyethylene terephthalate, commonly referred to as a polyester. Polyesters may be either thermoplastic or thermosetting, depending on the acid and the alcohol combination. Polyesters have good strength and chemical resistance and are used in the construction of FRP tanks and piping. They are also used in tear-resistant fabrics and cords for belt drives and fire hoses.

Elastomers. Many types of elastomers are being manufactured. The most common elastomers along with some of their physical properties are listed in Table 5.13. Elastomers are normally used for sheeting, gaskets, and electrical insulation.

TABLE 5.13 Physical Properties of Some Polymeric Elastomers

Elastomers	Common names	Maximum operating temperature, °F	Tensile strength, psi	Density, lb/ft^3
Chlorosulfonated polyethylene	Hypalon	200	3000	70–80
Fluoroelastomer	Viton	400	2500	115
Polybutadiene acrylonitrile	Buna N	300	4000	
Polychloroprene	Neoprene	225	3500	76.8
Polyisobutylene isoprene	Butyl rubber	275	3000	
Polyisoprene	Natural rubber	160	4500	58.0
Polystyrene butadiene	Buna S	275	3500	
Silicone rubber		580	900	

Silicones. The term "silicones" refers to a group of chemical compounds using a silicon-oxygen framework in lieu of a carbon-to-carbon structure. Silicones are actually a hybrid structure of carbon-based and silicon-based molecules which forms a group of compounds called organosiloxanes. Silicones have excellent temperature stability and electrical-insulating properties. They are used chiefly in two forms, as oils and as elastomers. Low-molecular-weight silicones form oil-like compounds used as defoamers, high-temperature greases, and lubricants. Higher-molecular-weight silicones form elastomers used for heat-sealing devices and dielectrics and as high-temperature rubbers for gaskets and seats.

Ceramic materials

A ceramic material is usually employed when a process requires a material with high temperature and corrosion resistance. The term "ceramics" is applied to a group of inorganic brittle materials which include true ceramics, cermets, glass, and graphite. Table 5.14 presents the more frequently encountered ceramic materials together with their physical properties. The ceramics and cermets will be treated collectively, while the glasses and graphite will be treated in separate subsections.

TABLE 5.14 Physical Properties of Ceramic Materials

	Tensile strength, 10^3 psi	Modulus of elasticity, 10^6 psi	Thermal conductivity, Btu/(h·ft·°F)	Maximum temperature in oxidizing environments, °F
Ceramics				
Aluminum oxide	20–30	32–57	1.8	3540
Silicon carbide	15	55		3000
Cemented tungsten carbide	130	60–90		1200
Boron carbide	22.5	42		1000
Impervious graphite	2.6	2.3	85	400
Borosilicate glass	4–10	10	0.8–2	500
Glass, ceramic	20–30		1	

Ceramics and cermets. Ceramics are materials with high melting points and therefore have high heat resistance. They also have excellent corrosion resistance but tend to be brittle, resulting in low impact and tensile strength. Ceramics are also subject to thermal shock and will fracture if heated or cooled too rapidly.

Cermets are mixtures of metals and ceramics whose properties lie between those of a metal and those of a ceramic. The general characteristics of a cermet, as compared with a ceramic, are as follows:

Brittleness	Same
High temperature resistance	Same
Ductility	Greater
Mechanical- and thermal-shock resistance	Greater
Corrosion resistance	Same
Typical cermets	Al-Al$_2$O$_3$, TiC-Ni, B$_4$C–stainless steel

Table 5.15 lists some of the more important ceramics and their applications. Most carbides are used for wear-resistant or abrasive service, oxides as refractory materials and insulators, and sulfides as high-temperature dry lubricants.

Graphite

Pure allotropic graphite may be used as a dry lubricant. This form of graphite is unsuitable for use in process equipment and piping owing to its porosity. If allotropic graphite is impregnated with a phenolic or furfuryl alcohol resin, it forms what is called impervious graphite. The resultant composite material has better properties than either compound alone.

Although impervious graphite has low tensile strength, low impact strength, and poor abrasion resistance, it has high thermal conductivity and excellent corrosion resistance (see Table 5.14). A comparison of the thermal conductivity at room temperature of impervious graphite with that of metals is as follows:

	Thermal conductivity, Btu/[h·ft^2 (°F/in)]
Impervious graphite	1020
Carbon steel	325
Stainless steel	108

TABLE 5.15 Industrial Applications of Ceramics

Application	Ceramics
Chemical ceramics and porcelain	Mixed oxides $Al_2O_3 \cdot 2SiO_2 \cdot 2H_2O$ (kaolinite) $K_2O \cdot Al_2O_3 \cdot 6SiO_2$ (feldspar)
Wear-resistant coatings and materials	VC (vanadium carbide) TaC (tantalum carbide) TiC (titanium carbide) WC (tungsten carbide) Cr_3C_2 (chromium carbide) Cr_3O_2 (chromium oxide)
Abrasives	SiO_2 (silica) SiC (silicon carbide) B_4C (boron carbide)
Lubricants	MoS_2 (molybdenum disulfide) WS_2 (tungsten disulfide)
Insulators	Al_2O_3 (alumina) ZrO_2 (zirconia) MgO (magnesium oxide)
Cutting tools	WC (tungsten carbide)

Because of its high thermal conductivity, impervious graphite can be employed to construct corrosion-resistant heat exchangers. It can be used up to a maximum temperature of about 340°F, this temperature being limited by the impregnating resin and not by the graphite.

Impervious graphite is resistant to chlorides, fluorides, hydrofluoric acid (up to 60 percent), hot caustics, chlorinated hydrocarbons, and phosphoric acid but is not recommended for use with free halogens, chromic acid, and sulfur trioxide. Additional corrosion resistance of impervious graphite to specific chemicals can be found in the appendix to this chapter.

Impervious graphite is used for piping alone or as a lining. Piping made of graphite alone normally is constructed to withstand operating pressures up to 50 or 75 psig. Higher operating pressures can be achieved by using impervious graphite as a lining for steel.

Glasses

Glasses are considered by some as ceramics, but their properties are quite different from those of ceramics, metals, or plastics. A glass is an amorphous solid formed by subcooling its liquid phase. Glasses are hard, brittle materials with low thermal conductivity and good electrical-insulating properties (see Table 5.16). Glasses also possess excellent corrosion resistance to most chemicals.

The four types of glasses most frequently used for process-equipment constructions are fused silica, borosilicate glass, glass ceramic, and fiberglass. The general characteristics and applications of these glasses can be found in Table 5.16. Glasses are used whenever corrosion resistance at high temperatures (200°F or more) is required. The

TABLE 5.16 Glasses Used in the Process Industry

	Glass type			
	Fused silica	Borosilicate glass	Glass ceramic	Fiberglass
Approximate % composition				
SiO_2	96–99	80.8		52–56
Na_2O	0.2*	3.8		0–1
K_2O	0.2*	0.8		0–1
CaO	.....	0.2	See below.	16–25
MgO	.....	0.2		0–6
B_2O_5	2.9*	11.9		8–13
Al_2O_3	0.4*	2.0		12–16
Fe_2O_3	.....	0.1		
Application	High-temperature, high-corrosion resistance *Constituents of Vycor™ pure silica (quartz), 99% SiO_2	Good thermal-shock resistance; Pyrex™ used for chemical glassware and piping for pharmaceuticals and corrosive laboratory drains	Glasteel™ composition proprietary; used as lining for corrosion-resistant piping and equipment	Used to provide structural strength for polymeric materials employed in process piping and equipment

glasses are resistant to strong acids such as sulfuric, hydrochloric, nitric, and acetic acid, chlorinated hydrocarbons, brines, and hydrogen peroxide, but they are attacked by hydrofluoric acid, acidic fluoride compounds, and hot alkalies. Caustic solutions at room temperature can etch glass, affecting its transparency but not its strength.

If glass vessels and piping are to be used, a borosilicate-glass (Pyrex) type of glass should be selected. Depending on the size of the glass component, its joints, and whether or not it is armored, the glass component can be used within an operating-pressure range of 15 to 150 psig.

Glass clad to steel for structural strength is used when high operating pressures are expected and good corrosion resistance is required. A variety of process components such as reactors, piping, and pumps use a glass ceramic of proprietary composition bonded to the steel.

Corrosion

Corrosion is the deterioration of a material reacting with its environment. Selection of a material for equipment construction is partially based on its corrosion resistance in a given process application. The corrosion rate also depends on the mechanical stresses experienced by the equipment. Figure 5.16 shows the interdependence of environment, material, and mechanical design of the piece in which the material is

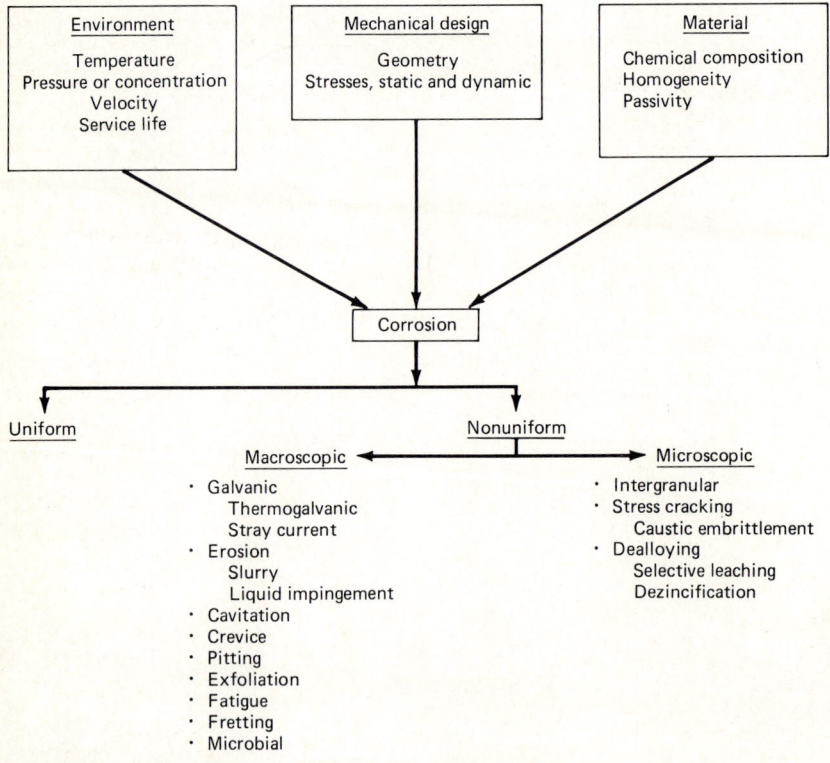

Figure 5.16 Factors affecting corrosion and types of corrosion.

used and how they all affect corrosion. The type and rate of corrosion depend on the combined effects of these factors. The subsequent discussion will explain how each type of corrosion may be caused by these factors and how to select materials to mitigate corrosion.

Factors affecting corrosion

Three factors affect the corrosion rate of a material: the material itself, the environment in contact with the material, and the mechanical design of the component. The material aspects have been covered in detail in the metal, polymeric-materials, and ceramics subsections of this chapter, but the other two factors need additional explanation.

Environment. The process conditions to which a material is exposed define its environment. The environmental parameters, as shown in Fig. 5.16, are:

1. Service life
2. Pressure or concentration
3. Temperature
4. Velocity

Given the service life and corrosion rate of a material, a piece of equipment can be designed with a corrosion allowance on its component parts sufficient to match the required service life. However, practical considerations of material costs and equipment tolerances would in many cases make this option prohibitive.

The concentration of a corrodent in an environment also affects the corrosion rate. In gaseous environments, the concentration of a corrodent is defined by its partial pressure or by total pressure if it is the only chemical constituent in the environment; e.g., pure hydrogen fluoride gas is not as corrosive as wet hydrogen fluoride gas. A similar phenomenon can exist for liquid corrodents. Figure 5.17 shows the corrosion rate of carbon steel as a function of sulfuric acid concentration and temperature.

All chemical reactions are affected by temperature, and since corrosion is an electrochemical reaction, it is no exception. Typically, the corrosion rate increases with increased temperature. Figure 5.17 shows that, for a given concentration of sulfuric acid, the corrosion rate of carbon steel increases with increased temperature.

Velocity can effect corrosion is three ways:

  1. It can bring fresh corrodent into contact with the material, thus accelerating the corrosion rate.
  2. It can wear away a material physically by the mechanical energy of high velocity. This is called erosion corrosion.
  3. The action of both mechanical energy and fresh corrodent can corrode the surface of a material which would otherwise be passive to stagnant corrodent. In this case, the erosive effect of high velocity scours off the passive surface formed by the reaction between the corrodent and the material, exposing subsurface material to corrosion by fresh corrodent in the stream.

Types of corrosion, causes, and cures

Different types of corrosion are caused by different circumstances. Understanding the characteristics of each type of corrosion and what causes it can reduce or eliminate

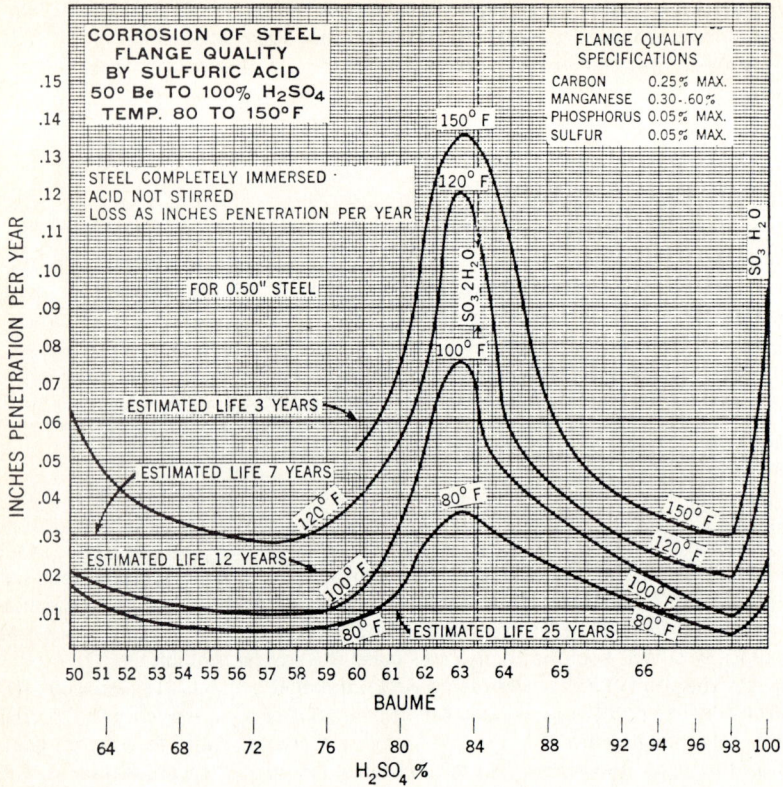

Figure 5.17 Steel corrosion by sulfuric acid. (*Reproduced with permission of Allied Corporation from product booklet entitled* Sulfuric Acid *by General Chemical Division, Allied Chemical Corporation, 1964.*)

the likelihood of corrosion. Described below are these 14 types of corrosion:

1. Uniform corrosion
2. Galvanic corrosion
3. Erosion corrosion
4. Cavitation
5. Pitting
6. Crevice corrosion
7. Exfoliation
8. Dealloying
9. Intergranular corrosion
10. Stress-corrosion cracking
11. Corrosion fatigue
12. Fretting
13. Hydrogen embrittlement
14. Microbial corrosion

Types 2 through 14 are generally characterized as nonuniform types of corrosion. These can further be classified as either macroscopic or microscopic. Macroscopic corrosions are observable by the naked eye, and their effects can be anticipated. Microscopic corrosion occurs within the microstructure of the material, its effects are not readily observable, and they are considered catastrophic.

Each corrosion type is discussed in this subsection along with the cause and the methods used to prevent it.

Uniform corrosion. Figure 5.18 shows the commonest and simplest form of corrosion. Uniform corrosion is characterized by a general loss of material over the whole surface exposed to a corrosive environment. This type of corrosion can occur in both wet and dry environments. Some examples of this type of corrosion are:

1. Metals in acid
2. Plastics soluble in organic solvents
3. High-temperature oxidation of metals in dry atmospheres

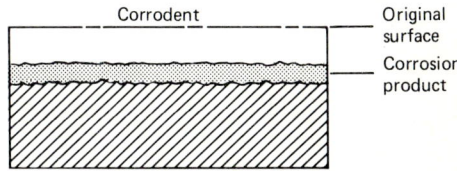

Figure 5.18 Uniform corrosion.

Uniform corrosion can be controlled by using a material resistant to the corrosive environment, changing the environment, applying a resistant coating to the material, adding a corrosion inhibitor to the environment, or applying cathodic protection. When a corrosion allowance is added to the wall thickness of a process vessel or process piping, the allowance thickness is based on the assumption of uniform corrosion.

Galvanic corrosion. Galvanic corrosion is caused by contacting two dissimilar metals with an electrolyte, creating an electrochemical cell. Figure 5.19 illustrates the galvanic-corrosion mechanism. One metal acts as a cathode, while the other acts as an anode. The metal acting as the anode is preferentially attacked, the rate of attack being determined by the relative position of the two metals in the galvanic series and the anode-to-cathode area ratio. (Table 5.17 shows the galvanic series for a selected group of metals in seawater.) For example, if an aluminum alloy is joined to a copper alloy, the galvanic series indicates that they are at the opposite ends of the series and will form a very reactive electrochemical cell when immersed in an electrolyte. As a result, the aluminum alloy, the anode, will corrode. However, if a copper alloy were joined to a nickel alloy, the anodic copper would be attacked, but the galvanic couple would be less reactive since the two alloys are closer in the galvanic series.

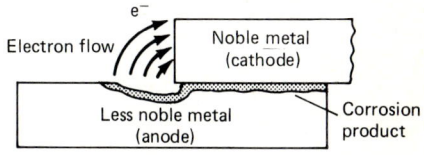

Figure 5.19 Galvanic corrosion.

Galvanic coupling should be avoided when feasible. The metals used in a process should be selected on the basis of close proximity to one another in the galvanic series. This recommendation includes not only the base metals of process equipment but also the metals used to join or fasten them, i.e., welding electrode, filler metals for welding, brazing, and soldering, bolts, and clips.

TABLE 5.17 Galvanic Series of Metals and Alloys in Seawater

Anodic or less noble metals (active) Cathodic or more noble metals (passive)	Magnesium and magnesium alloys Zinc Galvanized steel, galvanized wrought iron Aluminum and aluminum alloys Cadmium Low-carbon steel Alloy carbon steel Wrought iron Cast iron Stainless steel, Type 410 (active) Stainless steel, Type 430 (active) Stainless steel, Type 304 (active) Stainless steel, Type 316 (active) Lead Tin Muntz metal Yellow brass Admiralty brass Aluminum brass Red brass Copper Silicon bronze 90/10 copper-nickel 70/30 copper-nickel (low iron) 70/30 copper-nickel (high iron) Nickel Inconel 600 Silver Stainless steel, Type 410 (passive) Stainless steel, Type 430 (passive) Stainless steel, Type 304 (passive) Stainless steel, Type 316 (passive) Alloy 20 Monel Hastelloy, Alloy C Titanium Graphite Gold Platinum

Not all galvanic couples are harmful. Galvanized steel is steel coated with zinc. The zinc reacts with the iron to form a thin layer of iron-zinc alloy which is covered by a thicker layer of zinc. Although iron and zinc are dissimilar galvanically, the presence of zinc protects the iron from corrosion even where the continuity of the zinc coating is broken.

If dissimilar metals far apart in the galvanic series must be joined, there are basically three ways of reducing their galvanic effect on each other:

1. For relatively close joining, a dielectric insulator is inserted between the metals to inhibit the current flow between the galvanic pair. Figure 5.20 presents a bolted

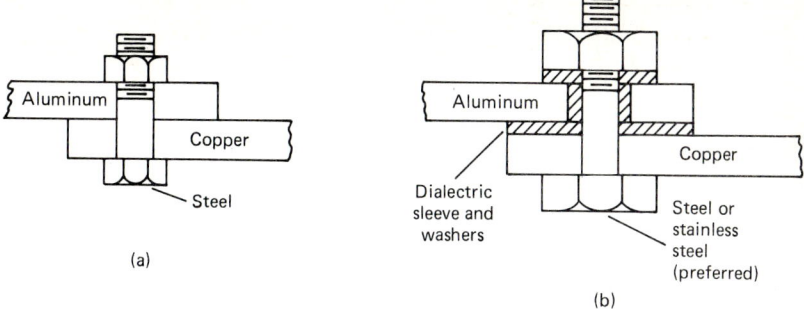

Figure 5.20 Joining two metals that are susceptible to galvanic corrosion. (*a*) Bad practice. This joint is highly susceptible to galvanic and crevice corrosion in strong electrolyte solutions such as seawater. (*b*) Better practice. This is a better joint; the galvanic-current flow is prevented by the insertion of a dielectric material between the dissimilar metals.

joint between aluminum and copper and shows how galvanic action between the metals can be reduced.

2. The anode-to-cathode area ratio must be kept as high as possible to minimize the corrosion current density from the anodic metal. The lower the current density, the slower the corrosion rate. Pumps and valves are examples of process equipment designed on this basis:

Equipment	Anode: cast steel or ductile iron	Cathode: stainless steel or bronze
Pump	Casing	Impeller
Valve	Body	Trim

The galvanic current is spread uniformly over the steel or iron surface. This reduces the corrosion rate and reduces the possibility of localized corrosion.

Other forms of galvanic corrosion. Galvanic corrosion can be produced by two other phenomena. These are described as thermogalvanic corrosion and stray-current corrosion.

Figure 5.21 schematically shows the mechanism of thermogalvanic corrosion. If a metal is exposed to a large thermal gradient, the microstructure of the metal changes and the hot part becomes more susceptible (anodic) to corrosion than the cold part. This type of corrosion can be prevented by properly insulating the metal part to distribute the heat evenly.

Stray-current corrosion is caused by the unintentional use of a metal as a direct-electric-current flow path (Fig. 5.22). Inducing current to flow through a metal causes it to become anodic. Even the noblest metals will be corroded by a stray current.

Stray-current corrosion is caused by poor grounding of electrical equipment. The leaking current will flow through any metal structures, piping, or mechanical equipment connected to the electrical equipment that provides a better path to the ground state. This is prevented by correcting the grounding fault or by providing cathodic protection for metal parts in contact with the electrical equipment. With cathodic protection, an electric current is intentionally applied to the metal so that it acts as a

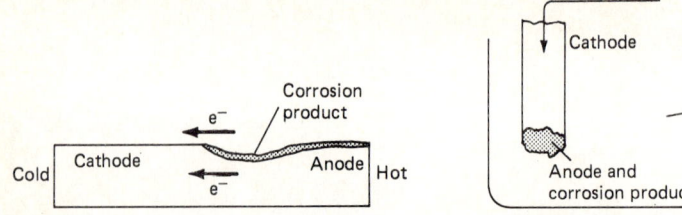

Figure 5.21 Thermogalvanic corrosion. **Figure 5.22** Stray-current corrosion.

cathode, while current of opposite polarity is directed to another piece of metal which acts as an anode. The latter metal is a sacrificial anode which is corroded in preference to the cathodically protected metal.

Erosion corrosion. The corrosion rate of a material can be accelerated by high-velocity corrodent or abrasive material flowing over materials surfaces. Figure 5.23 shows an example of erosion corrosion. This type of corrosion typically occurs to pump impellers and agitators and to elbows, tees, orifices, and valves experiencing very high pressure drops. Orifices and valves with high pressure drops also experience what is referred to as wire drawing.

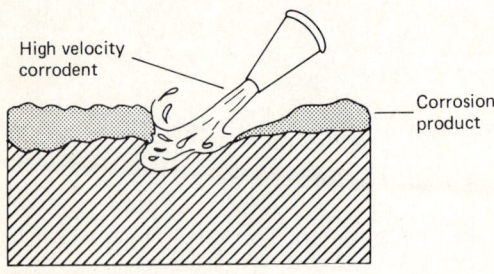

Figure 5.23 Erosion corrosion.

Erosion corrosion can be anticipated, but its rate is difficult to predict. It can be very rapid with galvanized or passivated metals since it removes the protective coating of surface metal and thus accelerates the corrosion process.

Erosion corrosion can be prevented by the following means:

1. Reduce the fluid velocity when possible. There are a number of empirical relationships for maximum flow rate in pipes. (See Chaps. 6, 7, and 9 for these relationships.)

2. Use a material more resistant to erosion, e.g., stainless steel or aluminum bronze.

3. Remove abrupt changes of direction in piping systems and replace them with more gradual changes; e.g., replace standard elbows with bends which have a bend radius that is at least 5 times the diameter of the pipe.

4. Reduce as much as possible surface roughness, which can contribute to erosion.

5. When high velocities cannot be eliminated, as in agitator blades, pump impellers, and impingement baffles, these parts should be designed for ease of installation and replacement.

Cavitation. Cavitation is sometimes misinterpreted as erosion corrosion. It has the same deterioration mechanism as erosion corrosion, but the cause is different. Cavitation is caused by the repeated collapse of vapor bubbles on a surface. When a vapor bubble collapses, the liquid rushes in to fill the space of the collapsed bubble. The resulting high velocity of the inrushing liquid "erodes" the surface, creating pits where the vapor bubbles collapse. This type of corrosion is presented in Fig. 5.24.

Cavitation is typically found in pumps carrying liquids operating at their flash points or in liquids with high concentrations of entrained gases. It can be reduced by analyzing the net positive suction head (NPSH) of the system and modifying system characteristics to improve the NPSH. In systems carrying entrained gases, either remove the gas prior to its entering the pump, if the gas is not needed in the downstream process, or increase the pressure or static head upstream of the pump, if the gas is required downstream of the pump. Specific details on NPSH calculations and cavitation can be found in the Chap. 11.

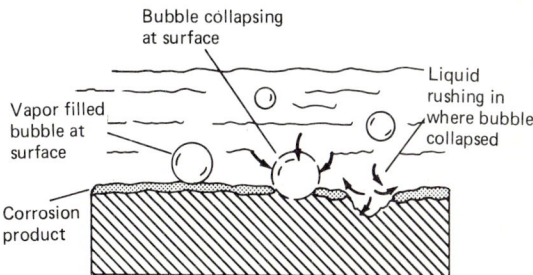

Figure 5.24 Cavitation.

Pitting. The formation of holes in an otherwise smooth metal surface is referred to as pitting corrosion. Figure 5.25 shows the general characteristics of pitting corrosion. Pitting is a local phenomenon producing surface holes of varying depths. The surface consists of a coating of corrosion product which acts as the cathode of a galvanic cell while the surface of each pit is the metal acting as an anode of a cell. Once the pit has been established, it continues to propagate down into the metal until it perforates the metal. The amount of metal consumed in pitting and the pits themselves are quite small and therefore difficult to detect in the early stages. So when the metal fails, it fails without warning.

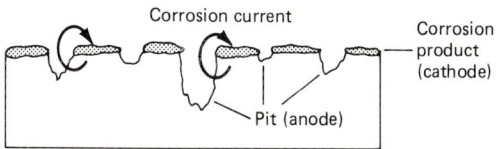

Figure 5.25 Pitting-type corrosion.

Stainless steels and aluminum alloys exposed to neutral-pH high-chloride or -fluoride-concentration solutions are susceptible to pitting.

Pitting corrosion can be prevented by selecting materials that are not susceptible to pitting corrosion and maintaining surface cleanliness and/or controlling the chemistry of the environment to reduce the chances of pitting.

Crevice corrosion. When two metal surfaces are in close contact with each other but are not sealed off from a corrosive environment, the space between the surfaces can be susceptible to crevice corrosion. Figure 5.26 shows how crevice corrosion propagates. When the corrosive environment squeezes between the surfaces, its concentration may change significantly from the environment in contact with the open surface. This results in a concentration gradient of either a metal-ion type or an oxygen concentration. This creates an electrochemical cell, resulting in crevice corrosion.

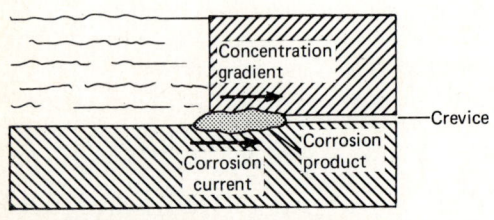

Figure 5.26 Crevice corrosion.

Crevice corrosion is a mechanical design problem; i.e., sleeves, bolted and riveted joints, gasket interfaces, and lap joints are susceptible to this type of corrosion. Metals that depend on their oxide coatings for corrosion protection, such as stainless steels, are also susceptible to crevice corrosion. The problem of crevice corrosion can be prevented by the following mechanical-design considerations:

1. Use continuous welds rather than bolted or riveted joints.
2. When bolting or riveting is required, fill crevices with nonabsorbent packing or gaskets.
3. Avoid stagnant areas in design. Provide free drainage in potentially stagnant areas.
4. Keep surfaces clean, and design the surface arrangement for ease of cleaning.

Exfoliation. If the metal corrodes by the falling away of layers (lamina) from the surface exposed to the corrosive environment, this process is called exfoliation. The surface (see Fig. 5.27) takes on a flaked or blistered appearance. The attack is directional, usually along the elongated grain boundaries of rolled or extruded metal surfaces. Aluminum and magnesium alloys are particularly susceptible to this type of corrosion. Exfoliation in these alloys is caused by exposure to high-chloride and -bromide-concentration solutions, acidic environments, or high temperatures. Intermittent wetting and drying can also cause exfoliation.

Exfoliation can be prevented by using alloys not susceptible to this type of attack, by proper heat treating of an alloy, or by providing a protective coating to the alloy to prevent exfoliation.

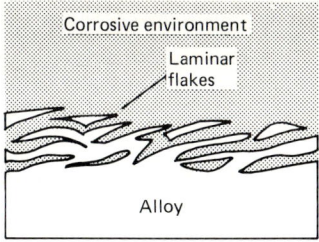

Figure 5.27 Exfoliation.

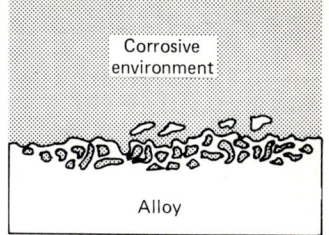

Figure 5.28 Dealloying (selective leaching, dezincification).

Dealloying. When one element of an alloy is singled out for corrosion attack, this type of corrosion is defined as dealloying or selective leaching (see Fig. 5.28). In the process, an alloying element is solubilized and leached out of the alloy by the corrosive environment. Dealuminification, dezincification, and degraphitization are special types of dealloying which by their names indicate the element attacked.

This type of corrosion can be prevented by selecting a material resistant to this type of corrosion or by changing the corrosiveness of the environment. For example, yellow brasses are susceptible to dezincification. By using an alloy with at least 85 percent copper, such as red brass, this type of attack can be eliminated.

Intergranular corrosion. A corrosion attack concentrating on the grain boundary of the alloy is defined as intergranular corrosion (see Fig. 5.29). The grains of the alloy's microstructure are preferentially attacked because they are high-energy areas of the crystal lattice and, therefore, more susceptible to corrosion. As corrosion progresses along the grain boundary, whole sections of metal become dislodged and appear sugary.

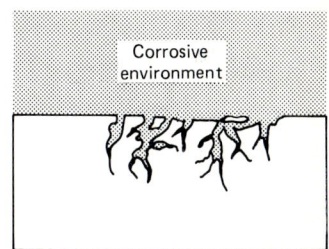

Figure 5.29 Intergranular corrosion.

Intergranular corrosion can result from the improper heat treatment of a metal. This is a typical problem when welding austenitic stainless steel. When stainless steel is heated in the range of 900 to 1500°F, the chromium carbides precipitate out at the grain boundaries, depleting the alloy's chromium content. These carbides are readily attacked by many corrosive environments. Table 5.18 lists chemical environments which can cause intergranular corrosion in austenitic stainless steel.

Intergranular corrosion can be prevented in the following ways:

1. Use low-carbon stainless steels (304L, 316L) which are less susceptible to carbide precipitation.

TABLE 5.18 Chemical Environments Causing Intergranular Corrosion in Austenitic Stainless Steel

Acetic acid at concentration of 35% or more and temperatures greater than 200°F
Boron trichloride
Monochlorobenzene
Nitric acid
Phosphoric acid
Sulfuric acid
Sodium chloride solutions

2. Use the proper heat treatment and/or welding procedures which reduce the chance of carbide precipitation.
3. Use stabilized stainless steels (Types 347 and T321) containing titanium or niobium as a stabilizing agent to minimize carbide precipitation.

Stress-corrosion cracking. The corrosion process can be accelerated by applying a static load on a metal. In Fig. 5.30 a static load is applied to a metal part exposed to a corrosive environment. As a result, the portion of the metal part experiencing a tensile stress will crack prematurely owing to exposure to the corrosive environment. This type of corrosion is called stress-corrosion cracking. A special type of stress-corrosion cracking is caustic embrittlement, which is caused by alkaline solutions reacting on steel. The exact mechanism for this type of corrosion is not well understood. A wide variety of metals is susceptible to this type of corrosion. Table 5.19 lists some metals and the type of environment in which they experience stress-corrosion cracking.

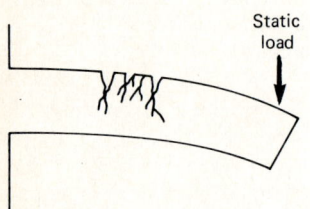

Figure 5.30 Stress-corrosion cracking.

TABLE 5.19 Some Metals and Types of Environment Which Cause Stress-Corrosion Cracking

Metal	Environment
Aluminum	Chloride solutions and steam
Copper	Ammonia and ammonium solutions
Stainless steel	Seawater and concentrated halide solutions; caustics
Carbon and low-alloy steels	Seawater; acid and caustic environments
Nickel	Sulfuric acid and caustics
Monel	Concentrated caustic; hydrofluoric and hydrochloric acids
Inconel	Concentrated caustic solutions
Titanium	Seawater and salt solutions in excess of 250°F; concentrated nitric acid

Stress-corrosion cracking can be prevented in the following ways:

1. Use a metal not susceptible to stress corrosion in a given corrosive environment.
2. Reduce tensile stresses by making the metal part experiencing the stress thicker or by distributing or reducing the load causing the high stress.
3. Stress-relieve to reduce stress concentration points in the metal.
4. Avoid riveting, bolting, and lap welding, which create higher stress concentrations than those of fillet or butt welds.

Corrosion fatigue. If a metal experiences changing stress patterns owing to cyclic mechanical loading and is exposed to a corrosive environment, it is susceptible to failure by corrosion fatigue (see Fig. 5.31). Fatigue normally causes a metal to fail at lower stress levels than expected. In addition, exposure to a corrosive environment can reduce a metal's stress tolerance even lower. The combined effect of cyclic loading and corrosive environment is usually greater than the damaging effects of either one separately.

The effect of corrosion fatigue can be reduced by:

1. Reducing or eliminating cyclic loading
2. Reducing stress loads by changing the geometry of the metal component
3. Altering the environment to reduce corrosion

Fretting. Fretting is a localized form of corrosion caused by a low-amplitude oscillatory motion between two metal surfaces. In Fig. 5.32 the metal parts continually rub against each other, causing metal attrition characterized by fret lines. Improperly balanced agitator and pump shafts experience fretting. Fretting can be reduced by reducing vibration, using harder metals, or improving lubrication to reduce friction and the resulting fretting.

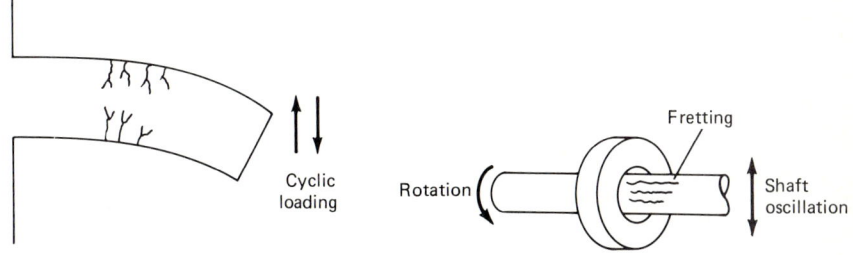

Figure 5.31 Corrosion fatigue.   **Figure 5.32** Fretting corrosion.

Hydrogen embrittlement. If hydrogen becomes entrapped in a metal, it can cause cracking and blistering. The cracking may seem to be stress-corrosion cracking, but the mechanism is different. The source of the hydrogen may be due to cathodic protection, welding, plating operations, pickling operations, or water vapor reacting with iron at high temperatures. Atomic hydrogen from these sources penetrates the metal, forming molecular hydrogen in the voids and metal imperfections. The buildup of molecular hydrogen in the metal lattice creates a high gas pressure, causing localized stress and metal disintegration.

Hydrogen embrittlement can be prevented or reduced by:

1. Using metals resistant to hydrogen embrittlement
2. Using low-hydrogen electrodes for welding
3. Using correct pickling procedures
4. Controlling cathodic protection
5. Using inhibitors in an environment, which could be a source of hydrogen

Microbial corrosion. Microbial interaction with organic and inorganic materials has been one of the least researched and understood types of corrosion. The compatibility of process materials with microorganisms must be considered when designing equipment for biochemical and pharmaceutical plants, wastewater-treatment plants, systems handling seawater or industrial cooling water, and systems using cooling towers. Table 5.20 lists typical bacteria and the corrosion problems these create.

Biological degradation occurs in aqueous solution at temperatures between 40 and 160°F. The rate of biological attack is directly proportional to temperature, but it is limited to a maximum temperature between 100 and 160°F, depending on the microorganism. At temperatures in excess of the maximum, the organism begins to die and biological attack does not occur.

Microbial attack can be reduced by injecting a biocide such as chlorine in the environment or by using materials or coatings resistant to biological attack.

Material selection for corrosion resistance

The preceding subsection presented the various types of corrosion that can be experienced, what causes them, and how they can be mitigated. While an understanding of the corrosion process improves the material-selection process, process engineers have access to empirically determined corrosion data for many materials in many process environments.

There are numerous sources of corrosion data for materials used in process equip-

TABLE 5.20 Typical Bacterial Groups and Their Corrosive Effects

Bacterial group	Active pH range	Corrosion problem
Sulfur-oxidizing bacteria (*Thiobacillus thiooxidans*)	1–6	Breaks down sulfur, sulfides, and sulfates to sulfuric acid; causes biodegradation of sulfur-containing materials such as vulcanized rubber, stone, and concrete. The sulfuric acid by-product can drop the environment's pH without harming the bacteria.
Sulfur-reducing bacteria (*Desulfovibrio desulfuricans*)	4–8	Produces a slime and generates a reducing environment of hydrogen sulfide as a by-product.
Iron-oxidizing bacteria (*Thiobacillus ferrooxidans*)	1–6	Oxidizes iron from the ferrous to the ferric state.

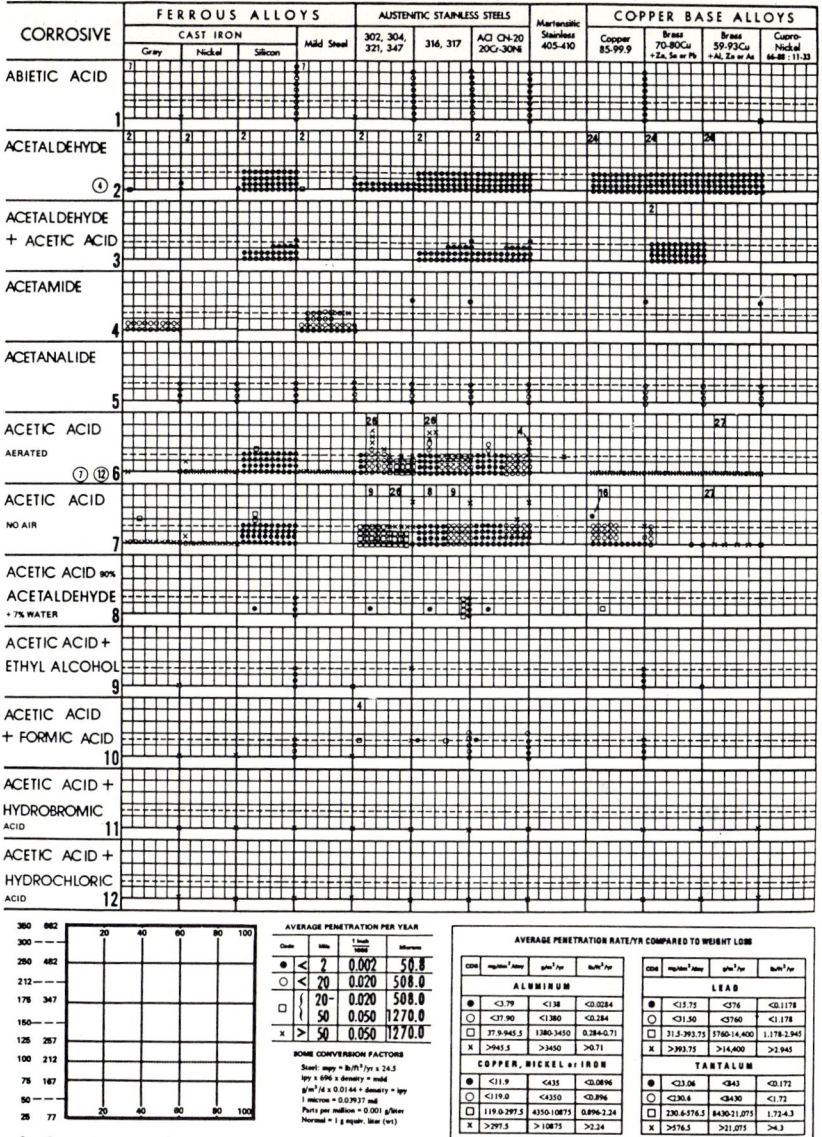

Figure 5.33a Page extracted from the *Corrosion Data Survey*, reproduced with permission of the National Association of Corrosion Engineers (NACE).

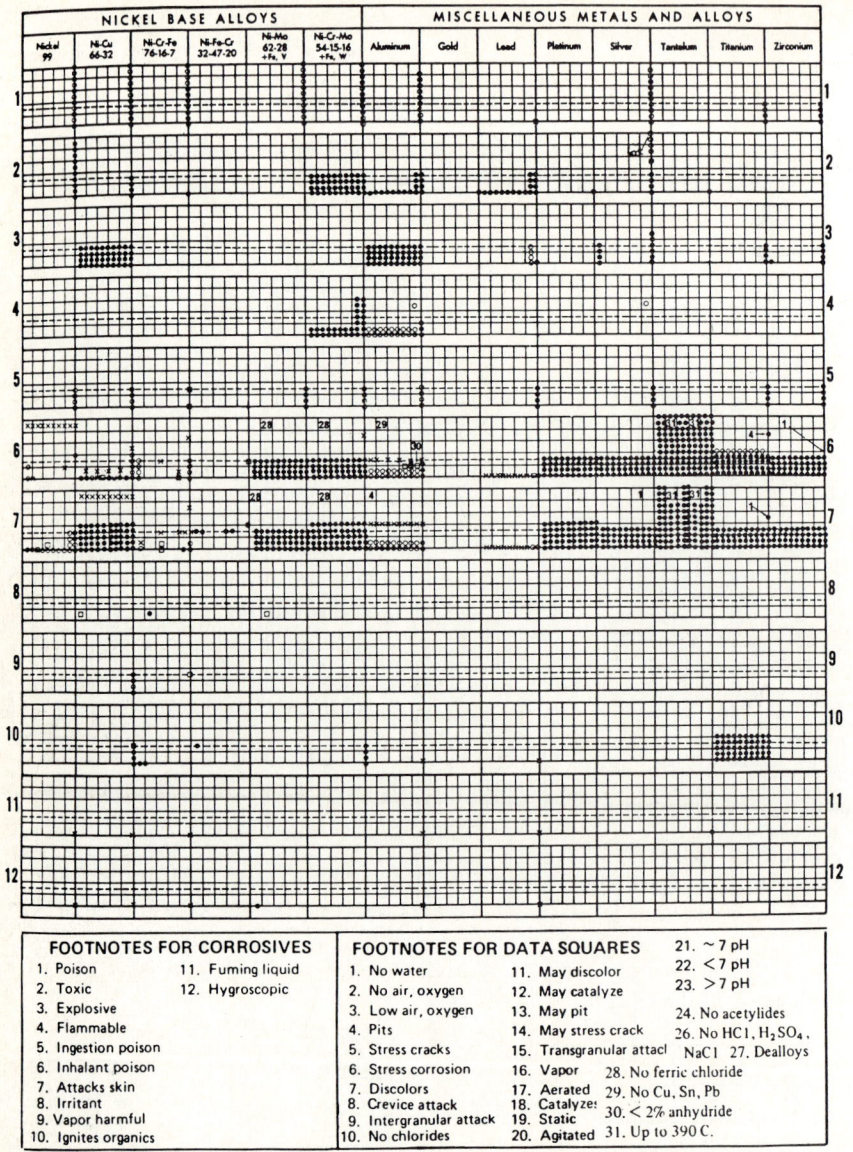

Figure 5.33b Page extracted from the *Corrosion Data Survey,* reproduced with permission of the National Association of Corrosion Engineers (NACE).

ment. The appendix to this chapter lists the corrosion resistance of some process materials to some selected process environments.

The most comprehensive source of corrosion data is the *Corrosion Data Survey*, published by the National Association of Corrosion Engineers (NACE). The *Survey* presents a table of graphs, each cross-referencing a given material to a corrosive environment. Each graph presents the corrosion rate of the material as a function of temperature and corrodent concentration in an aqueous solution. An excerpt from the *Corrosion Data Survey* has been reproduced in Figs. 5.33*a* and 5.33*b*. The bottom of Fig. 5.33*a* describes how corrosion information is represented in each small graph of the table. An inspection of Figs. 5.33*a* and 5.33*b* shows that footnote numbers have been placed either within a graph or in the box describing the corrodent. Definitions for these footnotes are found at the bottom of Fig. 5.33*b*.

BIBLIOGRAPHY

Budinski, Kenneth G.: *Engineering Materials: Properties and Selection*, Reston Publishing Company, Inc., Reston, Va., 1978.
Colangelo, V. J., and F. A. Heiser: *Analysis of Metallurgical Failures*, John Wiley & Sons, Inc., New York, 1974.
Flinn, Richard A., and Paul K. Trojan: *Engineering Materials and Their Applications*, Houghton Mifflin Company, Boston, 1975.
Giachino, J. W., W. Weeks, and G. S. Johnson: *Welding Technology*, 2d ed., American Technical Society, Chicago, 1968.
Henthorne, Michael: "Corrosion, Causes and Cures," CE Refresher Series, *Chemical Engineering*, May 17, 1971–Apr. 3, 1972.
Jefferson, T. B., and Gorham Woods: *Metals and How to Weld Them*, 2d ed., James F. Lincoln Arc Welding Foundation, Cleveland, 1962.
Kirby, Gary N.: "How to Select Materials," *Chemical Engineering*, Nov. 3, 1980, pp. 86–131.
Lindberg, Roy A.: *Processes and Materials of Manufacture*, 2d ed., Allyn and Bacon, Inc., Boston, 1977.
McNaughton, Kenneth J. (ed.): *Materials Engineering I—Selecting Materials for Process Equipment*, Chemical Engineering Magazine, McGraw-Hill, Inc., New York, 1980.
———: *Materials Engineering II—Controlling Corrosion in Process Equipment*, Chemical Engineering Magazine, McGraw-Hill, Inc., 1980.
Mahajan, Kanti K.: *Design of Process Equipment*, Pressure Vessel Handbook Publishing, Inc., Tulsa, Okla., 1979.
Megyesy, Eugene F.: *Pressure Vessel Handbook*, Pressure Vessel Handbook Publishing, Inc., Tulsa, Okla., 1975.
Moffatt, William G., John Wulff, and H. W. Hayden: *The Structure and Properties of Materials*, vol. III: *Mechanical Behavior*, John Wiley & Sons, Inc., New York, 1964.
———, ———, and George W. Pearsall, *The Structure and Properties of Materials*, vol. I: *Structure*, John Wiley & Sons, Inc., New York, 1964.
Nord, Melvin: *Textbook of Engineering Materials*, John Wiley & Sons, Inc., New York, 1952.
Pludek, V. Roger: *Design and Corrosion Control*, Halsted Press, John Wiley & Sons, Inc., New York, 1977.
Properties of Some Metals and Alloys, Bull. A-297A, International Nickel Company, Inc., New York, 1982.
Quick Reference Guide to High-Nickel Alloys, Bull. 73-202.1, Huntington Alloys, Inc., Huntington, W.Va., 1973.
Schweitzer, Philip A.: *Handbook of Corrosion Resistant Piping*, The Industrial Press, Inc., New York, 1969.
Shrager, Arthur M.: *Elementary Metallurgy and Metallography*, 3d ed., Dover Publications, Inc., New York, 1969.

Simons, Eric N.: *Guide to Uncommon Metals,* Hart Publishing Company, Inc., New York, 1967.
Stainless Steel, Bull. 5/81/10M—Carpenter 20Cb-3, Carpenter Technology Corporation, Reading, Pa., 1980.
Uhlig, Herbert H.: *Corrosion and Corrosion Control: An Introduction to Corrosion Science and Engineering,* John Wiley & Sons, Inc., New York, 1971.

Appendix
5.1
Corrosion Table*

	Brass	Bronze	Alloy 20	Hastelloy C	Monel	304 stainless steel	316 stainless steel	Titanium	Silicon iron	Tantalum	Copper	Aluminum	Carbon steel	Butyl rubber
Acetaldehyde			A	A			A	A	A	A			N	
Acetic acid, 20%	C	C	A	A	B	A	A	A	A	A	A	A	N	A
Acetic acid, 80%	C	C	A	A	B	A	A	A	A	A	A	A	N	A
Acetic acid, glacial	C	C	A	A	B		A	A	A	A	A	A	N	
Acetic anhydride	C	C	B	A		A	B	A		A	A	A	C	
Acetone	A	A	A	A	A	A	A	A	A	A	A	A	A	C
Aluminum chloride	C	C	C	A	N	N	C	C	C	A	C	C	N	A
Aluminum sulfate	C	C	C	A	C	C	C	A	A	A	C	C	N	
Ammonia, 10%	N	N	A	A	N	A	A	A			N	C	C	C
Ammonium chloride	N	N	C	A	C	C	C	A	C	A	N	C	C	
Ammonium nitrate	N	N	A	C	N	A	A	A	A	A	N	A	C	
Ammonium phosphate	N	N	A	A	N		A	A	A	A	C	C	C	
Ammonium sulfate	N	C	A	A	C	A	C	A	A	A	C	A	C	A
Amyl acetate			A	A	A		A	A		A				
Amyl alcohol	B		A	A			A	A	A	A	A	A		
Amyl chloride	B		A	A	A	A	B		A	A		N		
Aniline	N	N	A	A	A	A	A	A	C	A	N	N	A	
Aqua regia	N	N	N	C	N	N	N	A	N	A	N	N	N	
Arsenic acid	N		B	B	B	B	B	A	C			N		
Barium chloride	A	A	A	A	A	B	B	A	A	A	A	N		A
Barium sulfate	A	A	A	A	A	A	A	A	A	A	A	A		A
Beer			A	A		A	A	A			A	A	C	
Benzaldehyde	A	A	A	A	A	A	A	A			A	A		
Benzene	A		A	A	A	A	A	A	A		A	A	A	N
Benzoic acid		C	A	A		C	A	B	A					A
Borax	A	A	A	A	C	A	A	A	A		A	N		
Boric acid	N	A	A	A	A	A	A	A	A	A	C	A	N	A
Bromine water	C	C	N	A	C	N	N	A	N		C	N	N	
Butyl acetate	A		C	A	A	C	C	A	A	A	A	B		
Butyric acid	B	A	A	A	A	C	A	B	A	A	A	B		
Calcium bisulfite	N	A	C	A	N	C	C	C	N	A	A	A		A
Calcium chloride		A	A	A		C	A	A	A				C	A
Calcium hypochlorite	C	A	C	C	C	C	C	A	A	A	C	C	C	
Furfural	A	C	A	A	A	A	C	C	A	A	A	A	A	
Gasoline	A	A	A	A	A	A	A	A	N	A	A	A	C	N
Glycerine	A	A	A	A	A	A	A	A	A	A	A	A	A	A
Heptane	A	A	A	A	A	A	A	A	A	A	A	A	A	
Hexane			A	A	A	A	A	A				A		
Hydrobromic acid, 20%	N	N	N	A	N	N	N	A	N	A	N	N		
Hydrochloric acid, 0–25%	N	N	N	C	C	N	N	C	A	A	C	N	N	C

Epoxy	Hypalon	Natural rubber	Neoprene	Nitrile rubber	Nylon	Phenolic	Polyethylene	Polypropylene	PVC	Silicone elastomer	TFE	Ceramic, alumina	Graphite
N	N	A	A		N			B	N		A	A	A
A	A	C	C		N		C	A	A	A	A	A	A
B	A	A	C		N		C	A	C		A	A	A
B		N	C	A	N	A	C	B	N		A	A	A
	A	A	A		N	B	A	A	N		A	A	A
C	C	N	C	N	A	A	A	C	N	N	A	A	
A	A		A	N	N	A	A	A	A	C	A	A	A
A	A	C	A		C	A	A	A	A		A	A	A
C	A	N	A	N	A		A	A	A		A	A	A
A	A	N	A	N	N	A	A	A	A	A	A	A	
A	A	A	A		N	A	A	A	A		A	A	A
C	A	A	A		C	A	A	A	A		A	A	A
A	C	N	A	N	N	A	A	A	A	A	A	A	A
B	N	N	N		C		N	N	N		A	A	A
	C	B			A		A	A	C		A	A	A
C	N	B			C		N	N	N		A	A	A
B	B		N		N	B	N	B	N		A	A	A
N		N	N		N		B	N	N		A	A	A
		A	A		N		B	A	A		A	A	A
A	B	C	A	A	A	A	A	A	A	A	A	A	
A	A	C	A	N	A	A	A	A	A	A	A	A	
A	A	N	A		A		A	A	A		A	A	A
N	N	N	A		A	B	A	B	N	N	A	A	A
B	N	N	N	N	A	A	N	B	N	N	A	A	A
A	N	A	N	N	C	A	C	A	A	C	A	A	A
A	A	A	A		A	A	A	A	A		A	A	A
A	A	A	A	N	A	A	A	A	A	N	A	A	A
B	N		N		N		N	N	B		A	A	N
B	N	N			A	N	N	N	C		A	A	A
C	N	A	N		N		N	A	A		A	A	A
A	A	N	A	A	A	A	A	A	A	N	A	A	A
A	A	C	A	A	N	A	A	A	A	N	A	A	A
B	A	B	A		N		A	A	A		A	A	A
B		B	N		A	A	N	N	N		A	A	A
C	N	B	N	A	A	A	N	N	B	C	A	A	A
A	C	A	A	N	A	A	A	A	A	A	A	A	A
C	C	B	C			A	N	B	A		A	A	A
B	C	B	A		A		N	B	A		A	A	A
C	A	N	C		N		A	A	A		A	B	A
A	A	N	A	C	N	A	A	A	A	C	A	A	A

	Brass	Bronze	Alloy 20	Hastelloy C	Monel	304 stainless steel	316 stainless steel	Titanium	Silicon iron	Tantalum	Copper	Aluminum	Carbon steel	Butyl rubber
Hydrochloric acid, 25–37%	N	N	N	C	C	N	N	C	A	A	C	N	N	C
Hydrocyanic acid	N	N	A	A	A	A	A		A	A	N	N	N	
Hydrofluoric acid, 10%	N	N	B	A	A	B	B	N	N	N	N	N	N	A
Hydrofluoric acid, 30%	N	N	B	A	A	B	B	N	N	N	N	N	N	A
Hydrofluoric acid, 60%	N	N	B	A	A	B	B	N	N	N	N	N	N	A
Hydrogen peroxide, 30%	N	N	A	A	C	C	A	A	C	A	C	C	C	
Hydrogen peroxide, 50%	N	N	A	A	C	C	A	A	C	A	C	C	C	B
Hydrogen peroxide, 90%	N	N	A	A		C	A	A	C	A	C	C	C	N
Hydrogen sulfide, aqueous	N	N	A	A	N	C	A	A	A	A	N	A	C	A
Iodine in alcohol		N	A	A			A	N	N					N
Kerosene	A	A	A	A	A	A	A	A	A	A	A	A		
Lactic acid	A	A	A	A		B	B	A	A	A	A	A	N	
Lead acetate			A	A		A	A	A	A	A		A		
Magnesium chloride	C	C	A	A	A	A	A	A	A		C	C	C	A
Magnesium nitrate	A		A	A		A	A	A						
Magnesium sulfate	A		A	A	A	A	A	A			A	C	C	
Maleic acid	A		A	A		A	A	A	C					
Methanol	A	A	A	A	A	A	A	A	A	A	A	A		
Methyl chloride		C	A	A	A	A	A	A	A	A		N		N
Methyl ethyl ketone	A	A	A	A	A	A	A	A	A	A	A	A	A	A
Methylene chloride	N	N	A	A	N	A	A	A	A	A	N			
Napthalene	A		A	A	A	A	A	A	A	A		A		
Nickel chloride	N	N	A	A	A	A	A	A	A	A		N		A
Nickel sulfate		C	A	A	A	A	A	A				N		A
Nitric acid, 10%	N	N	A	A	N	A	A	A	A	A	N	B	N	A
Nitric acid, 20%	N	N	A	A	N	A	A	A	A	A	N	B	N	A
Nitric acid, 50%	N	N	A	A	N	C	C	A	A	A	N	B	N	N
Nitric acid, anhydrous	N		A	A	N	C	C	A	A	A	N	B	N	N
Oleic acid	A	A	A	A	A	A	A	A	A	A	C	A	C	
Oxalic acid	A	A	C	A	A	C	B	A	A	A	C	C	C	
Phenol	A		C	A	A	C	C	A	A	A	N	B	N	
Phosphoric acid, 0–50%	C	C	A	A	A	C	C	B	A	A	A	C	N	C
Phosphoric acid, 50–100%	C	C	A	A	A	C	C	B	A			C	N	C
Potassium bicarbonate			A	A	A	A	A	A						
Potassium bromide	A		A	A		A	A	A	A	A	A	N		
Potassium carbonate	A	A	A	A	A	A	A	A	A	N	A	N		A
Potassium chlorate	N		A	A	A	A	A	A	A	A		A		A
Potassium chloride	C	C	C	A	A	C	C	A	A	A	A	A		A
Potassium cyanide	N	N	A	A	A	A	A	A	A	A	N	N		
Potassium dichromate	N	N	A	A		A	A	A	A	A	N	A		

Epoxy	Hypalon	Natural rubber	Neoprene	Nitrile rubber	Nylon	Phenolic	Polyethylene	Polypropylene	PVC	Silicone elastomer	TFE	Ceramic, alumina	Graphite
A	A	N	A	C	N	A	A	A	A	N	A	B	A
A		A					A	A	A	A	A	B	A
A	A	A	A	N	N		A	A	B	A	A	N	A
A	A	A	A	N	N		A	A	B	C	A	A	A
A	C	A	N	N	N	A	A	A	C	N	A	N	A
B	A	N	C		N		A	B	A	A	A		A
	A	N	C		N		N	B	B	C	A		A
	C	N	N	A		A	B	B	A	C	A		A
A	A	N	A	A		A	A	A	B	B	A		A
	N	N	N	N			N	B	B	C	A	A	
A	N	N	N		A	N	C	A			A	A	A
A	A	A	A		C	A	A	A			A	A	A
A	N	A	N		A		A	A	A		A	A	A
A	A	A	A	N	A	A	A	A	A	C	A	A	
A	A	C	A		A		A	A	A		A	A	
A	A	A	A	N	A	A	A	A	A	C	A	A	
A	A						A	A	A		A	A	
A	A	A	A		A	B	B	A	B		A	A	A
	N	N	N	C		C	N	N	N	N	A	A	A
B	N		N		A	N	A	B	N		A	A	A
		N	N		N	N	N	N	N		A	A	A
A	N	N	N		A	A	B	N			A	A	A
A	A	A	A	N	A	A	A	A	A	C	A	A	A
A	A	A	A	N	A	A	A	A	A	C	A	A	A
	A	B	N	N		N	A	A	A		A	A	A
	A	B	N	N	C		N	A	A		A	A	A
	N	N	N	N	N		A	B	A	N	A	A	A
	N	N	N		N		A	N	A		A	A	A
	C	C	C		N		A	B	A		A	A	A
A	A	A			N	A	A	C	A		A	A	A
C	N	B	N		A			A	B	A	A	A	A
A	A	C	A	N	N	A	A	A	A		A	A	A
C	A	C	A	N	N		A	A	C	N	A	A	A
A	A	A	A		A		A	A	A	A	A	A	A
A	A	B	A		C		A	A	A		A	A	A
A	A	A	A	C	A	A	B	A	A	C	A	A	A
	A	C	A	N		A	A	A	A	C	A	A	
A	A	A	A		C	A	A	A	A	C	A	A	A
A	A	A	A		C		A	A	A		A	A	
A	A	A	A		N		A	A	A		A	A	A

	Brass	Bronze	Alloy 20	Hastelloy C	Monel	304 stainless steel	316 stainless steel	Titanium	Silicon iron	Tantalum	Copper	Aluminum	Carbon steel	Butyl rubber
Potassium hydroxide	N	N	A	A	A	C	C	C	N	N	N	N	C	A
Potassium nitrate	N	A	A	A	A	A	A	A	A	A	A	A	N	A
Potassium permanganate	A	A	C	A	A	C	C	A	A	A	A	A		
Potassium sulfate	C	C	A	A	A	A	A	A	A	A	A	A	C	A
Propyl alcohol	A	A	A	A		A	A	A			A	A	A	
Sodium acetate			A	A	A	A	A	A	A	A	A	A		
Sodium bicarbonate	A	A	A	A	A	A	A	A	A	A	A	A		N
Sodium bisulfate	N	C	A	A	A	A	A	A	A	N	C	C	N	A
Sodium bisulfite	A	A	C	A	A	C	C	A	N	A	A	A		
Sodium carbonate	A	A	A	A	A	A	A	A	N	A	N	A	A	N
Sodium chlorate	A		C	C		C	C	A	A					A
Sodium chloride			C	C		C	C	A	A				C	
Sodium cyanide	N	N	A	A	A	A	A	A			N	N	A	
Sodium hydroxide, 20%	N	C	A	A	A	A	A	A	N	N	C	N	A	A
Sodium hydroxide, 50%	N	C	A	A	A	A	A	A	N	N	C	N	A	A
Sodium hypochlorite	N	C	C	C	C	C	C	A	N	A	C	N	N	A
Sodium nitrate	C	C	A	A	A	A	A	A	A	A	A	A	A	A
Sodium silicate	A	A	A	A	A	A	A	A	A	A	A	A		
Sodium sulfate	A	A	A	A	A	A	A	A	C	A	A	A	C	A
Sodium sulfide	N	N	A	A	C	C	C	A	A	N	C	N	C	A
Stannic chloride	N		B	C	C	N	N	A	N	A		N	N	
Stearic acid	A	A	A	A	B	A	A	A	A		A	A		A
Sulfuric acid, 0–10%	N	A	A	A	C	N	N	B		A	N	N	N	N
Sulfuric acid, 10–75%	N	A	A	A	C	N	N	C	N	A	N	N	N	N
Sulfuric acid, 75–100%	N	A	A	C	C	N	N	N	A	A	N	A	N	
Tannic acid	A	A	A	A	A	C	C	A	A	A	A	N		
Tartaric acid	C	C	A	A	C	A	A	A	A	A	A	A		
Tetrahydrofurane			A	A		A	A							
Toluene	A	N	A	A	A	A	A	A	A	A	A	A	A	N
Trichloroethylene, dry	A	N	A	A	A	A	A	A	A	A	A	A	A	N
Turpentine	C	C	A	A	A	A	A	A	A	A	C	A	C	N
Urea			A	A		A	A	A				A		
Xylene			A	A		A	A	A						
Zinc chloride	N	N	A	A	A	A	A	A	A	N	N	N	A	
Zinc sulfate	C	A	A	A	A	A	A	A	A	A	C		C	A

*A = acceptable; B = acceptable up to 80°F; C = caution; use under limited conditions; N = not recommended; blank space = effect unknown. The corrosion data are a composite compilation from Michael Henthorne, "Corrosion, Causes and Cures," CE Refresher Series, *Chemical Engineering,* May 17, 1971–Apr. 3, 1972; Eugene F. Megyesy, *Pressure Vessel Handbook,* Pressure Vessel Handbook Publishing, Inc., Tulsa, Okla., 1975; Philip A. Schweitzer, *Handbook of Corrosion Resistant Piping,* The Industrial Press, Inc., New York, 1969; and *Stainless Steel,* Bull. 5/81/10M–Carpenter 20Cb-3, Carpenter Technology Corporation, Reading, Pa., 1980.

Epoxy	Hypalon	Natural rubber	Neoprene	Nitrile rubber	Nylon	Phenolic	Polyethylene	Polypropylene	PVC	Silicone elastomer	TFE	Ceramic, alumina	Graphite
A	C	C	A	C	A	A	A	A	A	A	A	N	A
A	A	A	A	N	A	A	A	A	A	A	A	A	A
A	A	A	A	N	C		A	A	A		A	A	A
A	A	A	A	N	C		A	A	A	A	A	A	A
	A		A					B	B		A	A	
A	A	A	A		A		A	A	A	A	A	A	A
A	A	C	A	N	A	A	A	A	A	N	A	A	A
	A	C	A	N	C	A	A	A	A	C	A	A	A
A	C	A	A		C		A	A	A		A	A	A
A	A	A	A	N		A	A	A	A	N	A	A	A
A	C	C	A	N	C	A	A	A	A	C	A	A	
A	A	A	A		A	A	A	A	A		A	A	A
A	A		A	N	A	A	A	A	A	C	A	A	A
A	A	C	A	A	A	C	A	A	A	A	A	N	A
A	C	C	A	A	A	C	A	A	A	A	A	N	A
	A	C	C	N	N	A	A	A	A	C	A	A	A
A	A	C	A	N	A		A	A	A	C	A	A	A
A	A	A	A		A	A	A	A	A	A	A	A	A
A	A	C	A	N	A	A	A	A	A	C	A	A	A
A	A	C	A	N	A	A	A	A	A	C	A	A	A
A	C	C	A	N			A	A	A	N	A	A	A
A	C	N	C				A	B	A		A	A	A
A	A	N	A	N	N	A	A	A	A		A	A	A
B	A	N	N	N	N	C	C	A	A	N	A	A	A
N	C	N	N	N	N	C	C	B	B	N	A	A	A
A	A	A	A	N		A	A	A		A	A	A	A
A	A	A	A			A	A	A		A	A	A	A
				A			A	B	A		A	A	
C	N	C	N	N	A	A	N	B	N	N	A	A	A
A	N	N	N	N	A	C	N	B	N	N	A	A	A
B	N	N	N	C	A	A	N	B	A	N	A	A	A
A	A		A		A		A	A	A		A	A	A
C	N		N		A	A	N	N	N		A	A	A
A	A	C	A	N	A	A	A	A	A	A	A	A	A
A	A	C	A	N	C	A	A	A	A	A	A	A	A

185

Chapter 6

Piping Calculations, Part 1: Limitations and Theoretical Considerations

One of the process engineer's most important tasks in the design of a facility is the sizing of the pipes and ducts that connect equipment or pipes with other equipment, pipes, or boundary points. In the MgO regeneration plant mentioned in Chap. 2 about 800 lines were provided to transport liquids, gases, or slurries, to provide equalizing lines, and to distribute the various utilities servicing the plant. Even allowing for duplication of some lines because of parallel usage and for others which were sized intuitively, it can be seen that the process engineers expended a large number of workhours to design the various piping systems properly. Only in this manner could excessive power consumption, piping costs, pipe erosion, water hammer, or plugging of lines be prevented or minimized.

The "proper design" of a piping system encompasses many practical and theoretical matters that process engineers must take into consideration. The subject of piping calculations is of such importance that four chapters are devoted to it. This chapter examines the constraints that limit the design of systems for the transport of flowing materials. It also provides a practical review of several theoretical concepts and basic relationships which must be understood to appreciate the correlations employed in the piping calculations themselves.

Chapter 7 develops the fundamentals between the various parameters that lead to a friction factor which relates friction loss to lengths of pipe or friction modulus for flowing fluids. Sources of information regarding friction moduli for various flow rates in different-sized piping are outlined for use with incompressible fluids. The affinity laws to be applied for their interpolation or extrapolation are described.

Chapter 8 presents methods of calculating compressible-fluid flows, while Chap. 9 reviews various two-phase-flow regimes. The two-phase-flow mixtures discussed are gas-liquid, solid-liquid, and solid-gas. Discussion of nonnewtonian fluids is limited to the basis of rheological methods of identification only and is given in this chapter.

Limitations on Piping

It is stating the obvious to say that the piping at any facility is limited to the piping available to the user. A review of the various sizes and physical characteristics of metallic piping that are generally used initiates discussion in this chapter. This will be followed by a short review of other conduits such as metal tubing, lined pipe, and glass pipe.

It is frequently useful to develop the economic basis for the selection of piping sizes to be used for the various streams in a process at the beginning of a project. Methods for accomplishing this selection are presented, and velocity and system constraints that could affect size selection are discussed.

Available piping

Standard commercial pipe sizes were originally defined by the American Standards Association (ASA)* and are known as iron pipe sizes (IPS). The pipes so defined are referred to by their nominal sizes, which in the smaller sizes consist of $\frac{1}{8}$, $\frac{1}{4}$, $\frac{3}{8}$, $\frac{1}{2}$, $\frac{3}{4}$, 1, $1\frac{1}{4}$, $1\frac{1}{2}$, 2, $2\frac{1}{2}$, 3, $3\frac{1}{2}$, 4, and 5 in. Progression then proceeds from 6 through 20 in by even integers. Starting at 24 in, diameters have 6-in increments through 60 in.

The ASA also defined the outside diameter (OD) for each of the nominal pipe sizes. The standard OD can be considerably greater than the nominal diameter at the lower end of the spectrum. For example, a $\frac{1}{8}$-in (0.125-in) nominal pipe has an OD of 0.405 in. For 1- through 12-in pipe, the OD is somewhat above the nominal size; i.e., an 8-in pipe has an OD of 8.625 in. However, at diameters of 14 in and larger, the OD is equal to the nominal size of the pipe.

The ASA also established several categories of pipe according to the pressure service to which a pipe may be subjected. These are termed "schedules" (Sch.), and they are listed as Sch. 10, Sch. 20, Sch. 30, Sch. 40, Sch. 60, Sch. 80, Sch. 100, Sch. 120, Sch. 140, and Sch. 160.† The OD for each nominal size is the same for all schedules. In sizes $\frac{1}{8}$ through 10 in, Sch. 40 is also referred to as "standard weight," while for sizes $\frac{1}{8}$ through 8 in, Sch. 80 is called "extra strong." In addition, the ASA defined a wall thickness for each schedule in the various nominal pipe sizes to meet its pressure rating. As would be expected, the wall thickness for a given schedule increases with the nominal size of the pipe.

The combination of OD and wall thickness leads to one of the more meaningful characteristics of a pipe, i.e., its internal diameter (ID). This parameter is used in calculating the various regimes, i.e., flow pattern, developed within a pipe and for determining important parameters for friction loss and heat-transfer calculations. The ID of Sch. 40 pipe (the most commonly used steel pipe) is slightly greater than the nominal pipe size for pipes between 1 and 10 in, while the ID is somewhat less than the nominal size for pipes of 12-in diameter and greater.

The ASA (ANSI) lists about 130 combinations of nominal sizes and schedules. In practice, however, a much smaller number of pipe sizes and schedules is typically

---

*Now known as the American National Standards Institute (ANSI), New York, N.Y. 10018.
†Not all schedules apply to all nominal pipe sizes.

used in the construction of chemical plants. Nominal sizes $2\frac{1}{2}$ and 5 in are seldom used in chemical plants although they are occasionally specified for boiler-house or other applications. Sizes $1\frac{1}{4}$ and $3\frac{1}{2}$ in are usually avoided in piping design. Should equipment be furnished with "odd-sized" couplings or nozzles, a reducing fitting or mating flange can be used to adapt to a common pipe size.

Table 6.1 presents the OD and ID for the more common nominal sizes and schedules.

Carbon steel pipe is typically limited in thickness to Sch. 40, Sch. 80, and Sch. 160. Schedule 40 is normally specified for steel-pipe sizes $1\frac{1}{2}$ in and larger when the pressure to be contained is 200 psig or less at ambient temperatures. However, for 1-in and smaller pipes, it is advisable to use Sch. 80 pipe. This presents less of a support problem and permits stronger joints with screwed and welded connections. Schedule 80 is used with carbon steel piping up to about 450 psig and Sch. 160 to about 900 psig at ambient temperatures.

The above-listed pressure ratings are useful guidelines if relatively little corrosion of the piping material is expected and a corrosion allowance of $\frac{1}{8}$ in or less is acceptable. For installations at higher temperatures or for which corrosion may be considerable, the required thickness of the pipe should be calculated and then checked against the equivalent thickness for the given diameter and schedule under consideration. The relationships between pressure, schedule, corrosion allowance, and wall thickness are expressed by the following equations:

$$\text{Schedule no.} = \frac{1000P}{S} \quad (6.1)$$

and

$$t = \left(\frac{P}{S} \times \frac{d}{2}\right) + C \quad (6.2)$$

where P = maximum internal working pressure, psig
S = allowable working stress at design temperature, psig*
t = pipe thickness, in
d = outside pipe diameter, in
C = corrosion allowance, in

Equation (6.1) may be substituted into Eq. (6.2) to give a quick method of obtaining the wall thickness of a given pipe size for a given pipe schedule number. If $\frac{5}{32}$ in (0.156 in) is taken as the design corrosion allowance, then Eq. (6.2) becomes

$$t = \left(\frac{\text{Sch. no.}}{1000} \times \frac{d}{2}\right) + 0.156 \quad (6.3)$$

and is a shortcut method of obtaining the wall thickness. The values so determined are within 5 percent of actual wall thicknesses for Sch. 40 pipe sizes from 3 to 12 in and within 10 percent for sizes 14 in and above. Equation (6.3) cannot be used for sizes 2 in and smaller since the assumed corrosion allowance of 0.156 in is equal to or greater than the actual wall thickness.

---

*The allowable working stress S for carbon steel seamless pipe and tubing varies with its specification and grade, e.g., 12,000 psi for A 53GrA or 17,500 psi for A 106GrC. These values are constant from $-20°F$ to 650°F and then begin to fall rapidly. Alloys and other metals have different values for S and change at different rates starting at different temperatures.

TABLE 6.1 Common Pipe Sizes and Schedule Dimensions*

Nominal size, in	Outside diameter, in	Sch. 10†		Sch. 40		Sch. 80		Sch. 160	
		Thickness, in	Inside diameter, in	Thickness, in	Inside diameter, in	Thickness, in	Inside diameter, in	Thickness, in	Inside diameter, in
½	0.840	0.083	0.674	0.109	0.622	0.147	0.546	0.187	0.466
¾	1.050	0.083	0.884	0.113	0.824	0.154	0742	0.218	0.614
1	1.315	0.109	1.097	0.133	1.049	0.179	0.957	0.250	0.815
1½	1.900	0.109	1.682	0.145	1.610	0.200	1.500	0.281	1.338
2	2.375	0.109	2.157	0.154	2.065	0.218	1.939	0.343	1.689
3	3.50	0.120	3.260	0.216	3.068	0.300	2.900	0.438	2.624
4	4.50	0.120	4.260	0.237	4.026	0.337	3.826	0.531	3.438
6	6.625	0.134	6.357	0.280	6.065	0.432	5.761	0.718	5.189
8	8.625	0.148	8.329	0.322	7.981	0.500	7.625	0.906	6.813
10	10.75	0.165	10.420	0.365	10.02	0.593	9.564	1.125	8.500
12	12.75	0.180	12.390	0.406	11.938	0.687	11.376	1.312	10.128
14	14.0	0.250	13.5	0.438	13.124	0.750	12.500	1.406	11.188
16	16.0	0.250	15.5	0.500	15.000	0.843	14.314	1.593	12.814
18	18.0	0.250	17.5	0.562	16.876	0.937	16.126	1.781	14.438
20	20.0	0.250	19.5	0.594	18.812	1.031	17.938	1.968	16.064
24	24.0	0.250	23.5	0.688	22.624	1.218	21.564	2.343	19.314

*Per ASA B36.10-1950.
†Add suffix S for stainless steel.

Corrosion allowances are not normally used in association with stainless-steel or alloy piping since these materials are chosen for their resistance to corrosion. An examination of Eq. (6.3) will show that the corrosion allowance of $\frac{5}{32}$ in for a 2-in carbon steel Sch. 40 pipe is about 70 percent of the calculated wall thickness and for a 6-in pipe is about 50 percent. However, if a corrosion allowance is not needed, a lower-schedule pipe for the same design pressure may be chosen. As a rule of thumb, the schedules for stainless and alloy piping can be about three schedule units less than that for the corresponding carbon steel piping. Thus, if Sch. 40 is indicated for carbon steel usage, Sch. 10S* would probably be satisfactory for stainless or alloy piping.

Caution must be exercised in making this transposition when stainless or alloy piping is to be used at higher temperatures. The internal strength of the material may then vary considerably from that of carbon steel, and the required thickness should be checked by a rigorous calculation.

Tubing and other flow conduits

A large variety of conduits besides pipe are available to process engineers for consideration to transport materials from one point to another. Tubing has mechanical characteristics different from those of pipe and can sometimes be used in place of it to great advantage. It is especially useful for runs of 1-in diameter and less when the flexibility of tubing compared with the greater rigidity of pipe is desired. Tubing is normally available in nominal sizes from $\frac{1}{8}$- to 12-in diameter and is used for applications such as reactor coils and heating coils in small vessels or large storage tanks.

Tubing is provided in a variety of stainless steels, alloys, and nonferrous materials such as copper, brass, nickel, and others. The most common tubing is made of copper or brass and is available in three series of wall thicknesses, known as Types K, L, and M. The actual outside dimensions and wall thicknesses for the various types and nominal diameters were defined by the Copper and Brass Research Association.†

A very important use for tubes is in heat exchangers. In this case, the OD of the tube is the same as its nominal size. The wall thicknesses are set by the standards of the Tubular Exchanger Manufacturers Association (TEMA)‡ in Birmingham wire-gauge units, a scale originally developed for wire diameters and also used for sheet metal thicknesses.

A specialty piping which has greatly increased in use during the past several decades is lined steel pipe. Linings can be provided from a variety of materials such as rubber (natural and synthetic), plastics (Saran, Kynar, Teflon, polypropylene, polyvinyl chloride, and others), cement, and glass. The linings are usually applied to or inserted into Sch. 40 or Sch. 80 steel pipe.

Nonmetallic solid piping is available in such diverse materials as hard rubber, graphite, plastics, resins, and glass as well as concrete, ceramic, and clay types of products.

Dimensional information for the more commonly used tubing and specialty piping is given in Table 6.2. Vendor literature should be consulted to ascertain wall and lining thicknesses and IDs for items not included in the table.

---

*Schedule numbers for stainless-steel and alloy piping use an S suffix. The wall thicknesses for S schedules are essentially the same as those for carbon steel piping schedules.

†Now known as the Copper Development Association, New York, N.Y. 10174.

‡Tarrytown, N.Y. 10591.

TABLE 6.2 Tubing Dimensions

Nominal size, in	Tubing outside diameter, in	Type K tubing		Type L tubing		Type M tubing		Copper and brass pipe		
		Thickness, in	Inside diameter, in	Thickness, in	Inside diameter, in	Thickness, in	Inside diameter, in	Outside diameter, in	Thickness, in	Inside diameter, in
3/8	0.50	0.049	0.402	0.035	0.430	0.0905	0.450			
1/2	0.625	0.049	0.527	0.040	0.545	0.028	0.569	0.675	0.0905	0.494
5/8	0.75	0.049	0.652	0.042	0.666	0.030	0.690	0.840	0.1075	0.625
3/4	0.875	0.065	0.745	0.045	0.785	0.032	0.811			
1	1.125	0.065	0.995	0.050	1.025	0.035	1.055	1.05	0.114	0.822
								1.315	0.1265	1.062
1½	1.625	0.072	1.481	0.060	1.505	0.049	1.527	1.900	0.150	1.600
2	2.125	0.083	1.959	0.070	1.985	0.058	2.009	2.375	0.1565	2.062
2½	2.625	0.095	2.435	0.080	2.465	0.065	2.495	2.875	0.1875	2.500
3	3.125	0.109	2.907	0.090	2.945	0.072	2.981	3.50	0.219	3.062
3½	3.625	0.120	3.385	0.100	3.425	0.083	3.459	4.00	0.250	3.500
4	4.125	0.134	3.857	0.110	3.905	0.095	3.935	4.50	0.250	4.000
5	5.125	0.160	4.805	0.125	4.875	0.109	4.907	5.563	0.250	5.063
6	6.125	0.192	5.741	0.140	5.845	0.122	5.881	6.625	0.250	6.125
8	8.125	0.271	7.583	0.200	7.725	0.170	7.785	8.625	0.3125	8.000

Materials Selection and Piping Calculations

Process engineers should also be aware that ductwork of various shapes is used for ventilation, heating, and certain special process applications. Likewise, open channels and trenches of various configurations are sometimes employed as convenient and economical means of conveying water and other fluids under gravity flow. The principles governing piping calculations in this and the following chapters can also be applied to ducts and open channels.

Economical sizing of pipe

Several criteria govern the choice of piping size. Of prime importance is the economics diagram that can be constructed for a given type of piping material. One such diagram is shown in Fig. 6.1, where annual power or operating costs are plotted against flow rate for a series of pipe sizes. Also given are the differences in annual fixed costs for adjacent pipe sizes in the series. Both fixed costs and operating costs are obtained for the same convenient length of pipe, say, 100 ft, and for 1 year's duration.

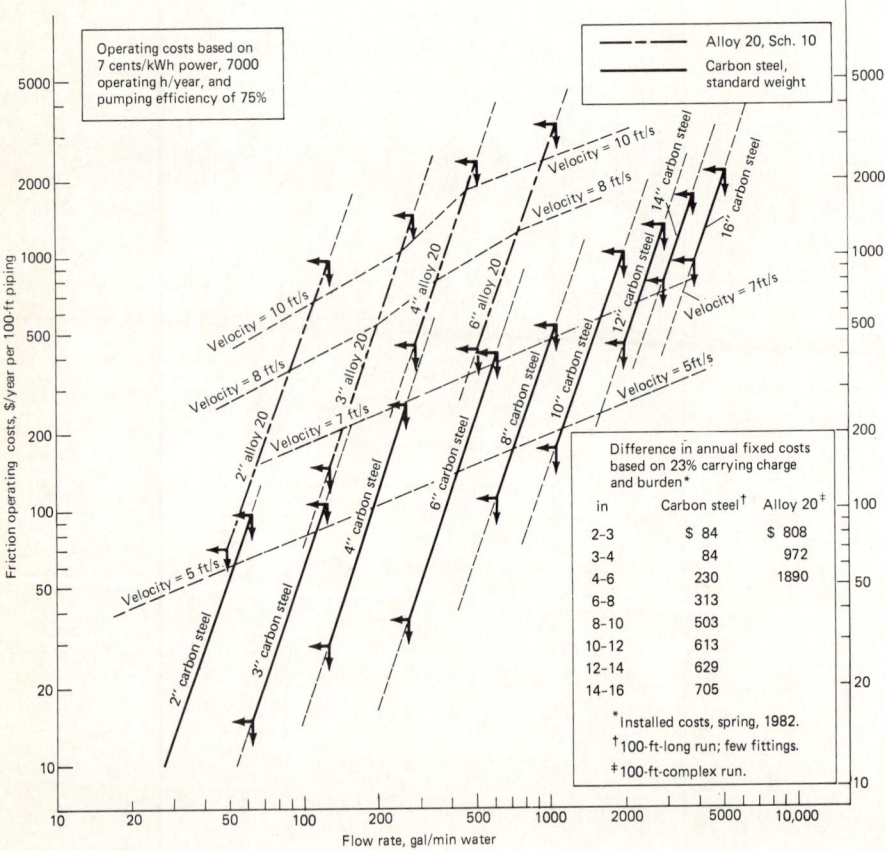

Figure 6.1 Economical selection of piping.

An excellent source of comparative information for the installed costs of a piping system is the Dow Chemical Company booklet *Comparing Installed Costs of Corrosion-Resistant Piping.** In it, the compilers list the material and total installed costs, including installation labor and materials, for 2-, 4-, and 6-in pipe sizes of various ferrous, nonferrous, and lined-steel piping.

Two types of piping systems are analyzed. One is for essentially long straight runs of flanged sections of piping with pipe supports and guides where required for outdoor installation. This would be typical of lines bringing utilities to or from the boundary limits of a plant or for their distribution through the plant. A second analysis is made of a moderately complex piping system and includes a considerably greater number of piping fittings and valves. A factor is also added for installation between closely placed pieces of equipment. This complex system is typical of a chemical plant or a boiler-house installation. Neither system includes "cost adders," which must be used to account for items such as insulation and tracing.

The examples used in the lower portion of Fig. 6.1 are a carbon steel straight-run system for cooling water. The upper section is typical of a complex piping system constructed of Alloy 20 to distribute 20 percent sulfuric acid to various reactors and to transfer their contents to other equipment. The total installed costs were converted to annual fixed costs by the use of a factor of 0.23 for the examples in Fig. 6.1.

In the subsection "General Pump Parameters" in Chap. 11, it will be demonstrated that the power consumed for a pumping operation is almost a direct function of the pressure change across the pump and the flow rate through it and an inverse function of the efficiency of the pump. The friction loss in 100 ft of pipe with its included fittings and valves can be treated as a pressure drop and expressed in terms of brake horsepower (bhp). It can easily be calculated that the cost of 1 bhp, when electricity costs 7 cents per kilowatthour with an operating schedule of 7000 h/year, is about $368 per year.

In the subsection "Friction-Loss Tables" in Chap. 7, it will be shown that the friction loss for a given length of pipe and its included fittings is a function of approximately the square of the flow rate for turbulent and nearly turbulent flows. It is a reasonable assumption that the flow-rate ranges being explored in Fig. 6.1 are in those regions. Thus, the power equivalent to friction loss in a given piping system is nearly proportional to the cube of the flow rate through the system.

Annual fixed costs for the 100 ft of a given size of pipe remain constant while operating costs of the motivating pump or compressor, on the other hand, are approximately proportional to the cube of the flow rate. Thus, only one flow rate need be used for each pipe size. For convenience when discussing liquids, this is taken as the flow rate for a velocity equal to 5 ft/s.†

Thus, the curves plotted on the diagram can be obtained by calculating the power costs for each of the pipe sizes in the service when the flow velocity is 5 ft/s. These isokinetic points, i.e., those representing the same velocity, are then entered on log-log paper‡, and a dashed line with a slope equal to 3 is entered through each point

---

*Dow Chemical Company, Midland, Mich. 48640.

†A simple and useful equation to bear in mind for any fluid is that velocity (ft/s) $\cong$ 0.4 X (gal/min)/d^2, where d = diameter, in. Conversely, gal/min $\cong d^2 \times$ velocity/0.4.

‡Two-cycle paper usually is sufficient. Three cycles are used for Fig. 6.1 to demonstrate the two ranges of annual costs represented on the figure.

and extended as far as possible in both directions. The absolute difference at the ordinate, between two adjacent lines, is an indication of the difference in annual power costs between two pipes for a given flow rate.

The graph is completed by examining each pair of adjacent lines in turn and noting for which flow rate annual operating costs nearly equal the difference in annual fixed costs. This flow rate approximates the upper limit for which the lower pipe size should be used and, in turn, is the approximate lower flow value for which the larger-sized pipe should be employed if a given piping system is to be near its economic optimum. For example, the difference in annual fixed costs between 100 ft of long-run 4- and 6-in carbon steel piping on Fig. 6.1 is given as $150. The operating power cost for 280 gal/min in 100 ft of 4-in pipe is about $171 per year, while the operating cost for that flow in 6-in pipe is only $21. At 280 gal/min in 4- and 6-in pipe, the difference in annual operating costs is $150, which equals the difference in fixed costs beween the two pipes. Thus, the upper economical point for 4-in-diameter pipe and the lower economical point for 6-in-diameter pipe are about 280 gal/min. Each succeeding pair of sizes may be similarly examined. The upper and lower economical limits have been noted for each pair.

The maximum flow rate to be serviced by any piping system is usually established before the plant is designed. The number of pipe sizes to be included in the graph should be about two nominal sizes greater than that needed for the estimated maximum flow rate. Experience has shown that economic evaluations for pipe sizes less than 2- or 3-in diameter are usually meaningless.

Once the diagram has been developed, it is best to shade in the sections of each pipe size that are included in the economical limits. In addition, it is helpful to have an indication of the approximate upper and lower velocities represented by the curves, as was done in Fig. 6.1. A tabulation similar to Table 6.3 is useful for quick reference.

While the above analysis for economical pipe sizing could be applied to any piping materials or fluids flowing,* a treatment of this kind is usually reserved for large flows, i.e., above 100 gal/min, and for relatively low-cost piping materials, i.e., carbon steel and plastic or resin pipes. In these cases annual operating costs quickly equal the

TABLE 6.3 Economical Pipe Sizes for Carbon Steel Sch. 40 Pipe

Flow-rate range, gal/min water	Nominal pipe size, in
64–130	3
130–280	4
280–660	6
660–1150	8
1150–2200	10
2200–3200	12
3200–4250	14
4250–5700	16

---

*Although the example used in Fig. 6.1 was that of a liquid, such an analysis could likewise be made for compressed air, blower gases, etc. In that event the flow rate should be expressed as standard cubic feet per minute (scfm) or pounds per hour (lb/h), stating temperature and pressure conditions.

annual-fixed-cost differential for the installed pipe at relatively modest velocities, approximately 6 to 7 ft/s.

Several variables may affect the analysis shown in Fig. 6.1. The addition of insulation and steam or electrical tracing to the piping increases the annual-fixed-cost differential for each pipe size. This increment is approximately a direct function of the diameter. The difference between annual fixed costs for two adjacent pipe sizes likewise increases. Thus, the differential in allowed operating costs beween two adjacent pipe sizes increases correspondingly. The result is that the upper-limit economical flow rate for a given pipe size shifts to a higher value. As a rough rule of thumb, the fractional increase in the economical flow rate is proportional to the cube root of the fractional increase in the differential annual fixed cost.

Another consideration is that the pipe may be used for fewer hours per year because of batch or intermittent operation. In that event operating costs per year are reduced for a given flow rate, and the flow range for the economical pipe size is shifted toward the higher flow rates. The revised flow rates are approximately inversely proportional to the cube root of the ratio of yearly operational hours.

A diagram prepared for a fluid of one density may be adapted to a fluid of another density by shifting the economical pipe size so that the revised flow rates are inversely proportional to the density ratios.

It was noted earlier that the original analysis shown in Fig. 6.1 was for uninsulated carbon steel pipe for which annual fixed costs were relatively small compared with potential operating costs. If the fixed costs, on the other hand, are much greater (as with stainless steels, high-alloy metals, and lined or glass piping) it can be seen that a much greater change in operating costs between two adjacent pipe sizes is required to make up for the incremental fixed costs between the two diameters. This is shown as the upper curves in Fig. 6.1. Other constraints then come into play in this case, so that it is rare that strictly economic considerations govern the choice of piping size for a given flow rate with these more expensive systems.

It will be noted that the power versus flow-rate curves for Alloy 20 pipes are not contiguous with those for the same-sized pipe in the lower carbon steel area. This results from the fact that friction is a function of the type of pipe and the characteristics of the fluids flowing, the equivalent length of pipe being dependent upon the number and type of valves and fittings included in the design, and the schedule of the pipe.

Velocity constraints

The existence of velocity constraints on fluids flowing in pipes must always be kept in mind. Each type of flow has its own peculiar and distinct velocity limits beyond which it should not be designed. Single-phase liquids such as water, hydrocarbons, solutions, and others can cause water hammer* or erosion of the piping if velocities are sufficiently great. A rough rule of thumb, based on engineering experience, for use with water in metallic pipes which are of 3-in diameter and greater and are in continuous service is that the velocity in feet per second should not exceed about 4 plus one-half of the nominal diameter in inches, i.e., $4 + d/2$. Thus, the velocity in

---

*"Water hammer" is the term given to the forces set up should the velocity of the flowing media be changed suddenly. At higher velocities, the forces can be destructive enough to damage piping or supports.

a 3-in pipe should normally be kept below $5\frac{1}{2}$ ft/s, and the velocity in a 16-in line could be as great as 12 ft/s, in the absence of other constraints. In no case should a pumped-liquid velocity be allowed to exceed 14 to 17 ft/s. Below 3-in diameter, velocities of 5 ft/s are acceptable for liquids.

The maximum allowable velocity is actually a function of the square root of the fluid density. However, most liquids encountered in refineries, chemical plants, and power stations will fall in the specific-gravity range of 0.6 (for light hydrocarbons) to 1.4 (for halogenated organics). When the square roots of these values are extracted, they have the value of 1 ± 0.2. This variation is sufficiently small so that the rule of thumb for the velocity of water in pipe can be applied to essentially all liquids.

Erosion of pipes can also be caused by high velocities of air, steam, combustion gases, or other fluids in the gaseous state. Rules of thumb are available for the allowable maximum flow of gases in pipes, but they are somewhat more complicated than the rules given for liquids since gas flows are apt to cover a much greater range of density than do liquid flows.

Another source of erosion is that caused by two-phase flows such as condensates and vapors, solids in liquids or slurries, solids in gases or pneumatic conveying, and liquids in gases or aerosols. Streams of two or more liquid phases need not be considered here since they are rarely encountered in normal engineering activity. For purposes of maximum allowable velocity emulsions may be considered as a single liquid in accordance with previous discussions.

It can easily be appreciated that the discontinuous solid particles in slurries or pneumatic-conveying streams or liquids in aerosols approach the speeds of the continuous phases. This circumstance can impart a high momentum to the particles, and excessive erosion will result if the velocity of the continuous phase is too great. The greatest degree of erosion will take place in elbows, tees, or other fittings where there is a change of direction. The turbulent action of the continuous medium also causes the particles to collide with the walls of the piping in straight runs, and erosion, albeit to a lesser degree, can thus occur along the straight pipe walls. If required, erosion can be dealt with either by lining the pipe with a relatively soft material such as rubber to absorb the impact energy of the particles, by employing an extra-hard material of construction for the pipe and fitting, or by using wear plates at the elbows.

It is difficult to give rules of thumb that will predict maximum velocities tolerable for slurries, pneumatic conveyors, or aerosols. Such values depend upon the densities of the continuous phase and of the particles being conveyed, the viscosity of the continuous phase, the size distribution of the particles, their spherosity, their hardness, the pipe diameter, and other variables. For practical purposes, the velocity of the continuous phase should be only somewhat greater than that necessary to suspend the particles.

While it is necessary to limit the upper velocity of a slurry, a pneumatically conveyed mixture, or an aerosol, it is important that the velocity of these streams be maintained above a minimum value to prevent the discontinuous phase from settling out and clogging the piping or causing excessive pressure drops. A minimum suspension velocity also depends upon the densities of the two phases, the viscosity of the continuous phase, the size of the particles, the spherosity of the particles, and the pipe diameter.

Both maximum and minimum flow velocities for slurries or other two-phase flows must be taken into account when constructing an economical pipe-size graph as in Fig. 6.1. In these cases, the loci of the maximum or minimum velocity points must be entered on the graph and then the flow rates for adjacent pipe sizes adjusted to account for these velocity constraints.

On rare occasions, an economical pipe size for a given regime will be limited as to both maximum and minimum velocities. In that case, flow-rate gaps would develop between the truncated flows for adjacent pipe sizes. To attempt to utilize either of the adjacent sizes for a flow rate falling in the gap would result in violation of either the upper or the lower limit of a given pipe size for that regime. One stratagem to avoid this dilemma is to employ a combination of pipes, preferably two of the same diameter, so that the reduced flow rate for each falls within acceptable ranges. An example of this type of problem is given in Fig. 6.2.

Here is a case in which the required flow rate is about 300 gal/min. In Fig. 6.2 it can be seen that a 4-in pipe should not be used, as the nearly 7.6-ft/s velocity exceeds the maximum allowable velocity for erosion. On the other hand, a 6-in pipe should not be used, since the velocity in it is only 3.4 ft/s, which is less than that required for suspension of the solid particles. In this case, consideration should be given to splitting the flow between two parallel 4-in lines. There would then be 150 gal/min in each, and their velocities would be about 3.8 ft/s, which in this case is satisfactory as to both maximum and minimum velocities.

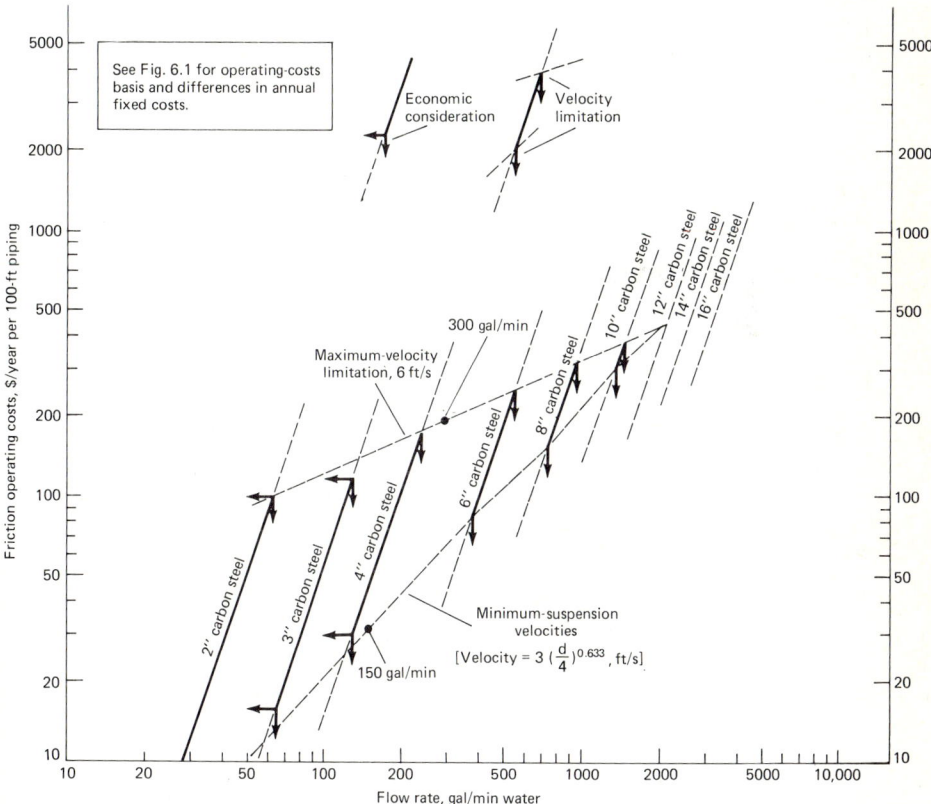

Figure 6.2 Economical selection of piping with maximum- and minimum-velocity limitations.

Friction losses

In addition to the various velocity constraints which must be respected, attention must often be paid to pressure losses for a given flow rate in different pipe sizes.

Effect of pumping head. Frequently it is necessary or desirable to utilize a pump with a given number of revolutions per minute (r/min) in certain services. In general, the common pump speeds result in certain maximum total dynamic heads (TDH) being developed by the pumps. These range from about 50 to 275 ft for pumps running at 3600 r/min, 25 to 150 ft for 1800-r/min pumps, and 25 to 80 ft for 1200-r/min pumps, depending upon the size of the pump.

The TDH is required primarily to overcome the pressure differential across equipment through which the flow passes, the allowed pressure drop across control valves, static changes, and friction in the piping system. Most of these items, except for friction losses in the piping, are essentially independent of the pipe sizings in the system. Thus, the TDH limitation imposed by a pump speed can set the maximum friction loss for the piping system after the other factors have been deducted from it. The pipe sizes must, therefore, be chosen so that friction losses do not exceed this maximum value.

Allowable pipe pressure. The material of construction for a piping system can also limit the pressure to which it may be subjected to a relatively low value. This would be the case for certain ceramic, plastic, or glass piping for which allowable pressure could be as low as 15 psig. Even metallic pipe, which normally is able to contain hundreds of pounds of pressure at ambient conditions, loses its strength at higher temperatures. The pressure limit for a 3-in Sch. 40 carbon steel pipe is reduced to less than 100 psig over 750°F. Thus, friction or pressure loss may well influence pipe size for a piping system involving low-strength materials.

Barometric legs. The pressure loss of fluids flowing in a piping system is of special interest in designing vacuum systems which have barometric legs. For example, a condenser under vacuum is usually located above a vessel containing condensate which is open to the atmosphere. A pipe runs from the condenser and is sealed from the atmosphere by ending it below the liquid level. When the vacuum is applied to the condenser, the liquid rises in the pipe. The condenser must be elevated sufficiently above the liquid level to prevent the liquid from entering it at the lowest pressure that can be developed by the vacuum-generating system. In addition, the condenser must be elevated slightly higher to account for the friction that is developed in the pipe when the condensate flows at its maximum rate. Thus, if the piping is too small in diameter, the friction developed could be so great that a costly supportive installation would be required to locate the condenser higher than necessary. A larger pipe size with a more reasonable friction loss reduces the extra elevation required to a reasonable length.

Vacuum service. Friction losses occurring in piping systems in vacuum service are of particular importance. In a system that is being run at 10 mmHg abs., for example, it is important to limit the pressure loss between the point where the vacuum is applied and the vacuum-inducing device to a reasonable amount. If not, the initial costs and the operating costs of the vacuum-generating equipment become excessive. In some instances, an economic appraisal of the pipe size may be necessary. In that event, the annual fixed costs of the vacuum equipment as well as the piping material must be taken into account along with the motive-steam or power costs.

Gravity flow. Besides induced flows by means of pumps, compressors, or vacuum devices, there can be gravity flow, or flow caused solely by differences in elevation or static pressure between two points in the system. In these cases, the total pressure or friction loss cannot be greater than the static head or pressure differential. A pipe of sufficiently large diameter must be chosen to maintain the friction loss at a reasonable value for the required flow. A properly sized diameter also prevents a high entrance loss or buildup of liquid due to constriction above the inlet to the piping system.

Inducing high pressure drops. In some instances, process engineers may want to induce a high pressure drop in a piping system rather than prevent one. This could be the case when there is need to drop a high pressure to a considerably lower one. If a control valve is included in the system, it may be desirable to limit the pressure drop across it. The excessive pressure can then be taken through a restrictive orifice or through a run of reduced-size piping. In the latter case, it is usually wise to use a higher-schedule or thicker-walled pipe.

At times it may be more desirable to use a run of smaller-sized pipe in place of a restrictive orifice. Although this is a more costly method of obtaining a given pressure drop, it is preferable in instances in which the orifice is subject to plugging. Of course, the pipe diameter would have to be large enough so that the pipe would not clog.

Types of Fluids

It is axiomatic that matter can assume one of three forms—solid, liquid, or gaseous—according to conditions of temperature and pressure. The latter two forms, liquid and gaseous, are termed "fluids" because, by definition,* they are substances which undergo continuous deformation when subjected to a shear stress. By means of a pressure differential, these fluids can be made to flow from one point to another in a closed channel. These fluid flows are most commonly encountered as a single-phase flow such as water, as a solution, as a hydrocarbon, as steam, as air, or as any other process components by themselves. However, combinations of the discrete forms must sometimes be transported or transferred, and these mixtures of forms are known as two-phase flow.

Single-phase flows

For pressure-drop calculations, fluids are divided into two categories: incompressible and compressible. Incompressible fluids are those whose densities remain relatively unchanged owing to changes in pressure. Thus, all liquids are incompressible fluids at most normally encountered pressures and, as such, lend themselves to rather simple analyses of their pressure losses due to friction.

Gases, on the other hand, are compressible since their densities vary essentially in direct proportion to their pressures for a given isothermal condition. For this reason, the analytical methods required for pressure-loss calculations with compressible fluids

---

*Committee on Definitions and Nomenclature of the Society of Rheology, Rheology Leaflet No. 7, November 1938, pp. 11–17.

are more difficult than those for incompressible fluids. The analysis for a gas is complicated by the fact that consideration must be given to changes in pressure, with or without a change in temperature, as a gas passes through the piping system. However, should the density of a gas be subject to less than a 10 percent change as it passes through the system, the gas may then be considered an incompressible fluid for the analysis of its pressure losses. The general relationships for determining friction losses for incompressible fluids are given in Chap. 7.

In a normal chemical plant, although most of the fluids transported in the piping systems are usually liquids, a considerable number of the fluids may be gases. In the example of the MgO regeneration plant cited earlier, many of the streams were concerned with gases from a calciner, while many others were process air, process steam, utility steam, compressed air, etc. The flow rates of these gases were such that, given the relatively short runs of pipe, the pressure losses of these gaseous streams were all considerably under 10 percent of their respective initial pressures. Thus, they could effectively be considered incompressible fluids. If a control valve were interposed in the gaseous piping, it would be necessary to analyze the piping on either side of the valve separately.

Chapter 8 is devoted to an explanation and analysis of pressure losses for compressible fluids. Such analyses are required for systems in which there are long runs of pipe or the pressure drops are relatively great with respect to the initial pressure. More important, it is necessary to understand the principles of pressure loss for compressible fluids in order to size control valves in gaseous service properly. Velocities in control valves can be extremely high if the valves are not sized properly, resulting in excessive pressure losses for the particular service.

Two-phase flows

One of the more common two-phase types of flow systems involves two phases of the same species. This is exemplified by the passage of "wet" steam, i.e., steam containing varying amounts of equilibrium water. This circumstance is due to saturated steam losing heat through the piping and insulation and condensate forming at the piping wall. At the other end of the spectrum, a high-pressure condensate passing through a steam trap to a lower pressure "flashes," and the result is a mixture of steam and condensate.

A similar phenomenon takes place in a refrigeration system in which high-pressure cooled liquid is flashed across a control valve to form a colder mixture of vapors and liquid. It is important to understand and appreciate the forms of flow that these two-phase single species may take in order to minimize pressure drop and avoid the water-hammer phenomenon. The relationships for two-phase systems are considered in Chap. 9.

Although solids are not considered fluids, it is obvious that they can be transported in piping as part of two-phase-flow systems. In the MgO regeneration facility, solids move in and out of storage bins; a portion leaves a calciner with its off gases, and some solids pass through the liquid-waste-treatment system.

Solids can flow in an upward direction only when they are aided by a mechanical device such as a conveyor or a belt or when they are carried along by a fluid. A mixture of solid particles suspended in a liquid is termed a "slurry." The movement of a suspension of solid particles in a gas is called "pneumatic conveying." Other types of mixed-phase transport are "aerosols," which are suspensions of liquid particles in a gaseous medium, and "foams," which are mixtures of liquids and gases with the liquid as the continuous phase. "Emulsions" are composed of two immiscible liquids, one of which forms an internal dispersal of liquid spheroids while the other is the continuous external phase.

Slurries, pneumatic suspensions of solids, and aerosols are generally unstable two-phase mixtures; i.e., they tend to separate under the influence of gravity. For a given two-phase mixture, a minimum velocity of the continuous phase is required to maintain the suspension of the discontinuous phase. The dependency of that velocity on various parameters of both phases is also developed in Chap. 9. The influence of the various factors on pressure drop or friction losses is included there.

The stability of emulsions and foams is greatly affected by such phenomena as surface tension and brownian movement rather than by gravity. Except for use in food-processing plants, the transport of emulsions is seldom encountered in chemical-plant applications. Movements of foams are very rare and require special equipment. For pressure-loss purposes, emulsions and foams may be considered as single-phase fluids having a volumetric average density but the viscosity of the continuous phase. These approximations for the relevant parameters generally yield pressure losses with acceptable accuracy.

Theoretical Considerations

The remainder of this chapter is devoted to several theoretical concepts which must be understood to appreciate better the several relationships that are needed for calculating friction losses. These concepts are treated from a practical engineering viewpoint in order to simplify their application.

Viscosity

An important physical property of a fluid that influences the type of flow pattern which will be developed in a conduit and, thus, the friction loss is the fluid's viscosity. It is intuitively known that it becomes more difficult to stir fluids and takes longer to pour or rearrange them as their viscosities increase. However, the theoretical concept of viscosity is often difficult to grasp. Since viscosity enters into so many process calculations, the following illustration is offered to help in understanding the basic concept.

Consider a fluid contained between a fixed horizontal plane of area A and another plane, of the same dimensions and area, placed at a distance d immediately above it. This is illustrated in Fig. 6.3a. Should a force F be applied to the upper plane, it

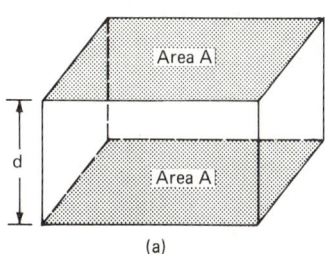
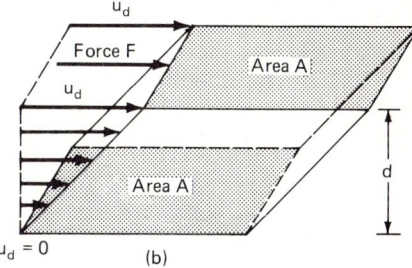

Figure 6.3 Viscosity concept. (a) Initial condition. (b) After application of force F; $F/A = \mu (u_d/d)$.

would move relative to the lower one with a velocity u, as in Fig. 6.3b. The difference in velocity from one layer of fluid to another which is at an incremental distance above it, between the two planes, is depicted at the left end of Fig. 6.3b. This shows that the velocity of a layer immediately adjacent to the upper plate is u and that the velocity of succeeding lower layers decreases linearly until it is zero immediately adjacent to the stationary lower plate. Experimental results show that, for all newtonian fluids, the slope of the velocity distribution line, i.e., u/d, correlates well with and is proportional to the shear rate or stress, i.e., F/A.

The proportionality factor between the two elements is termed the dynamic or absolute viscosity μ such that

$$\frac{F}{A} = \mu \frac{u}{d} \tag{6.4}$$

Thus, viscosity represents the resistance of the fluid to shear forces or stress. The greater the viscosity, the smaller the velocities attained by the nonfixed layers owing to applying a shear force. Conversely, as the viscosity becomes less, the shear force needed to produce a given velocity decreases. It is also interesting to note that, for a given viscosity, the shear force required to attain a certain velocity is inversely proportional to the distance between the point of application of the shear force and the fixed plane.

Equation (6.4) can be rearranged to express viscosity in terms of its dimensions as follows:

$$\mu = \frac{F}{A} \times \frac{d}{u} \tag{6.5}$$

Thus, the dimensions of absolute viscosity are

$$\frac{\text{Force} \times \text{time}}{\text{length}^2}$$

Most absolute viscosities are commonly referred to in terms of centipoise (cP), or one-hundredth of a poise (P), the metric unit for viscosity. One poise represents one dyne-second per square centimeter. In other words, one dyne of force applied to a square-centimeter plane which is located one centimeter from the stationary plane will maintain a velocity of one centimeter per second when the viscosity of the fluid is one poise. The English-measurement equivalent has the dimensions of pound-mass per foot-hour such that

$$\text{lbm}/(\text{ft} \cdot \text{h}) = 2.42 \text{ cP} \tag{6.6}$$

Frequently, measurements of viscosity are made when the fluid is flowing. The result so obtained is termed the "kinematic" viscosity z. Kinematic viscosity is related to absolute viscosity by dividing the latter by the mass density ρ of the fluid. The mass density, in turn, is the specific weight w of the fluid divided by the dimensional constant g_c. The dimensions of kinematic viscosity, therefore, are

$$\frac{\text{length}^2}{\text{time}}$$

In the metric system, kinematic viscosity is measured in square centimeters per second, or stokes (St). It is obtained by dividing the absolute viscosity in poises by the specific gravity s of the fluid, acceleration due to gravity being equal to 981 cm/s^2.

Similarly, centipoises divided by mass density gives a kinematic viscosity z in centistokes (cSt). This is the most commonly used expression for kinematic viscosity, and the relationship with absolute viscosity is shown by

$$z(\text{cSt}) = \frac{\mu(\text{cP})}{\rho(\text{g/cm}^3)} \approx \frac{\mu(\text{cP})}{s} \tag{6.7}$$

The kinematic viscosity in the English system in terms of square feet per second is obtained by dividing the equivalent absolute viscosity of the fluid lbm/(ft·s) by its mass density in pounds per cubic foot, one centipoise being equal to 2.42/3600, or 1/1488 lbm/(ft·s).

The concept of viscosity or resistance to shear stress and its values are sometimes difficult to visualize. While it is easy to comprehend a given object weighing 100 times a second one, a fluid flowing at 10 times the velocity of another, or a solid having 5 times the density of a second one, it is hard to realize what having 5, 10, or 100 times the viscosity means when comparing two fluids.

Many misconceptions regarding viscosity are prevalent, and it would be well to dispel some of them. It is often assumed that viscosity is a function of density or specific gravity; i.e., the greater the gravity, the higher the viscosity. While such a relationship holds for the absolute viscosity of gases, it is not true for liquids. This can be seen by an inspection of Table 6.4, which presents frequently encountered liquids along with their specific gravities and their absolute and kinematic viscosities. There is no direct or inverse correlation between specific gravity and either kind of viscosity.

TABLE 6.4 Viscosity of Selected Common Liquids

Liquid	Temperature, °F	Specific gravity	Viscosity cP	Viscosity cSt
Mercury	70	13.57	1.60	0.118
Freon-11	70	1.49	0.31	0.21
Gasoline	60	0.68	0.31	0.46
Water	60	1.00	1.13	1.13
Kerosene	68	0.80	2.17	2.71
Sulfuric acid (100%)	68	1.84	26.9	14.6
Ethylene glycol	70	1.12	19.9	17.8
Olive oil	100	0.90	38.9	43.2
Lard	100	0.95	59.0	62.1
Honey	100			73.6
Crankcase oil				
SAE-10	100	0.91	32.2–47.2	35.4–51.9
SAE-20	100	0.91	47.2–78.2	51.9–86.6
SAE-30	100	0.91	78.2–114.2	86.6–125.5
Fuel oils				
No. 2	70	0.82–0.95	2.7–7.0	3.0–7.4
No. 6	70	0.82–0.95	79.8–627	97.4–660
Printer's ink	100	1 10–1.4	550–3080	550–2200
Corn syrup	100	1.43	1570–157,000	1100–110,000
Blackstrap molasses	100	1.47	3900–82,000	2630–55,000
Road tar				
RT-2	122	1.07	46–69	43.2–64.9
RT-12	122	1.15	29,000–86,000	25,000–75,000

Liquids having a "slippery" feel don't necessarily have lower viscosities than those which are not slippery. Solutions of caustic or soap, although slippery, have viscosities nearly equal to or somewhat greater than that of water.

The easiest way to visualize or appreciate the measure of viscosity is to make comparisons with known common liquids, such as those given in Table 6.4, which can serve as mental standards.

Liquid viscosities can undergo a dramatic change in relative magnitude owing to a change in temperature. This is best shown in a graph of the viscosity index (Fig. 6.4). A specially constructed scale is used so that the change in kinematic viscosity for liquids appears as nearly a straight line as a function of temperature.* For illustrative purposes, only a few fluids are shown on Fig. 6.4. Standard references can be consulted for other fluids.

The graph is especially useful to obtain the viscosity of a liquid over a range of temperatures when the viscosity of the liquid is known at two temperature points. The viscosities at the known temperatures are entered on the graph, and a straight line between the points represents, in most cases, the viscosity-versus-temperature curve. The graph is also amenable to approximating the change in viscosity with temperature should viscosity information be available at only one temperature point. That point is entered on the graph, and a straight line is placed through it parallel to the viscosity-index line of another liquid which is close to it on the graph and has similar chemical properties. Process engineers should maintain their own collection of viscosity indices for special fluids as part of their files.†

The effect of temperature on the viscosity of a liquid is not easily reduced to a mathematical equation and is evaluated empirically as previously discussed. On the other hand, both the viscosity of gases adhering to the ideal gas laws and the effect of temperature on their viscosities are amenable to simple mathematical treatment as shown in Eqs. (6.8) and (6.9).

$$\mu^\circ \propto \frac{M^{1/2} T^{3/2}}{T_b(T + 1.47 T_b)} \tag{6.8}$$

$$\frac{\mu_2^\circ}{\mu_1^\circ} = \left(\frac{T_1 + 1.47 T_b}{T_2 + 1.47 T_b}\right)\left(\frac{T_2}{T_1}\right)^{3/2} \tag{6.9}$$

where μ° = gas viscosity at low pressure
M = molecular weight
T = absolute temperature, K
T_b = absolute temperature at normal boiling point

Variations of system pressure at a given temperature have a negligible effect on liquid viscosities, both absolute and kinematic, up to about 3000 atm. Beyond that, viscosities for all liquids with the exception of water increase at a rapid rate. At a given temperature, there is a slight increase (less than 10 percent) in the absolute viscosity of gases up to about 100 atm. Above 100 atm, the increase accelerates. On

---

*The American Society for Testing and Materials (ASTM), Philadelphia, Pa. 19103, has available a series of such viscosity-temperature grids.

†Technical handbooks such as *Perry's Chemical Engineers' Handbook* and others often contain alignment charts giving viscosities of common liquids and gases as a function of temperature.

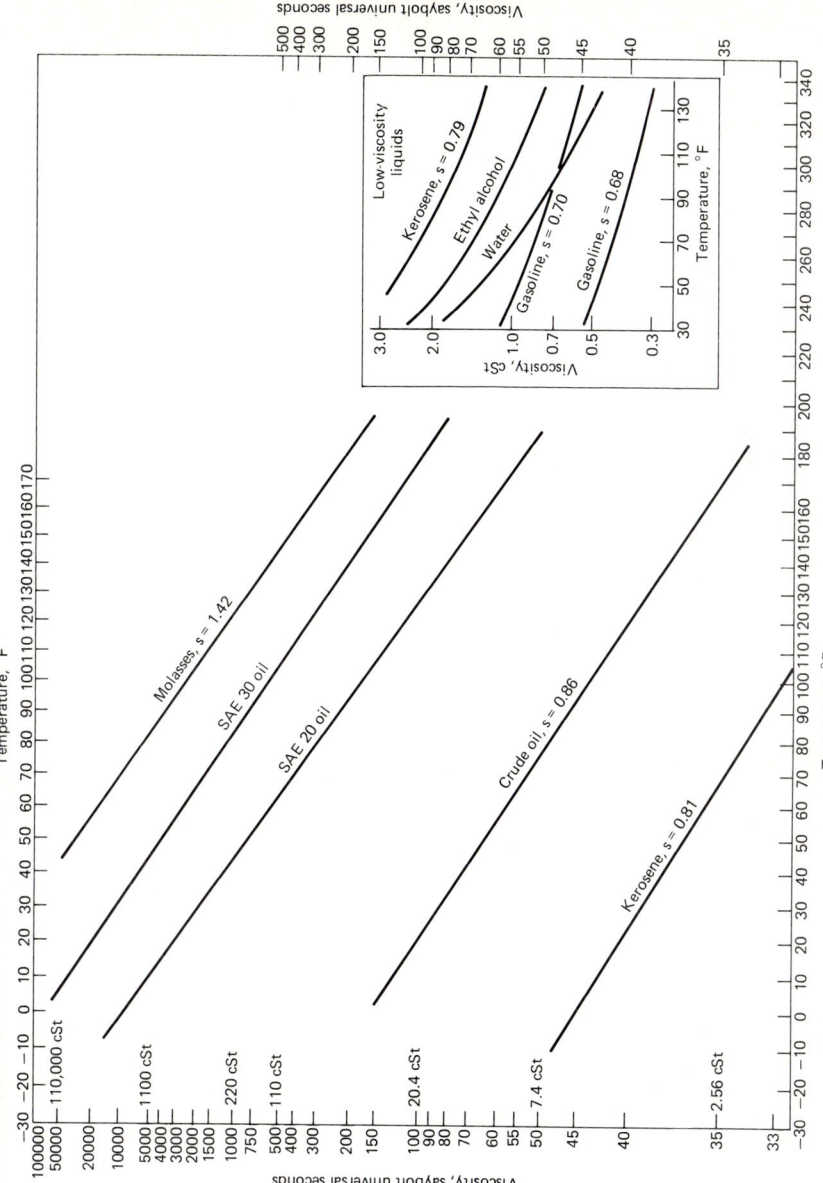

Figure 6.4 Effect of temperature on liquid viscosities. [*Modeled after ASTM Standard Viscosity-Temperature Charts for Liquid Petroleum Products (D341), Chart B, Saybolt Universal Viscosity, abridged.*]

the other hand, the kinematic viscosities of gases are nearly inversely proportional to their absolute pressures. Figure 6.5 shows the viscosities of several typical gases as functions of temperature and pressure.

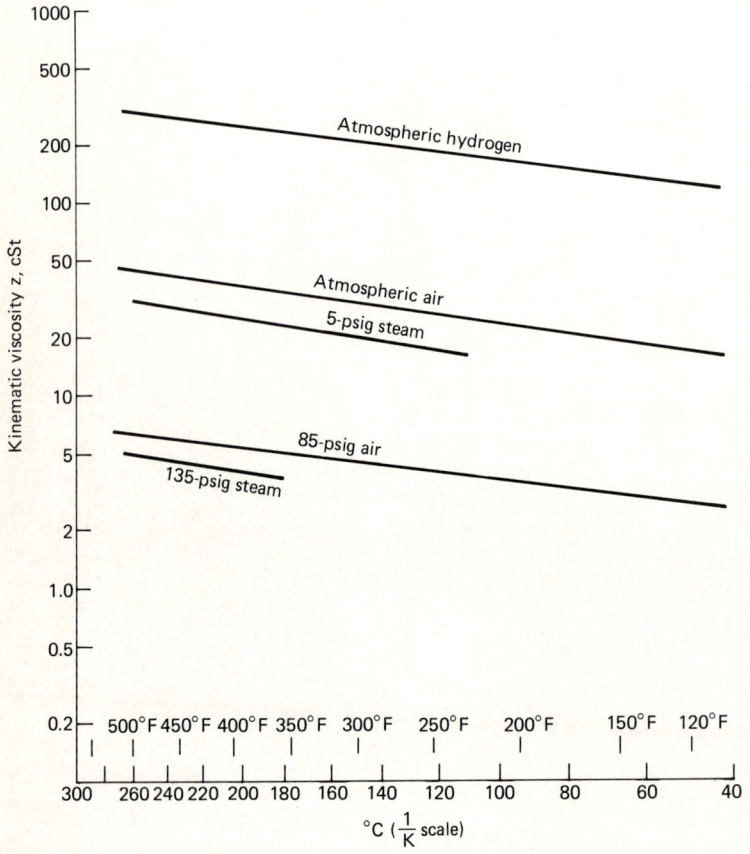

Figure 6.5 Gaseous viscosities.

It is often necessary to determine the resultant viscosity of a mixture of liquids of dissimilar viscosities. Mathematical expressions are available to calculate the viscosity of the resultant blend, but it is easier and more convenient to utilize a viscosity-blending chart as shown in Fig. 6.6. This chart uses the mole percent of each of the components in cartesian units as the abscissa while the viscosity scale on the ordinate is of a special construction. The viscosity of component A is entered at the 100 percent point on the right, and the viscosity of component B, at the same temperature, is located at the left. A straight line drawn between the respective points closely represents the viscosity of the various mole percent blends of the two components. In many instances the molecular weights of the components are unknown but may be close to one another. In that event the weight percentages may be used instead. Conversion factors from other viscosity units to Seconds Saybolt Universal are given in Table 6.5 on page 211.

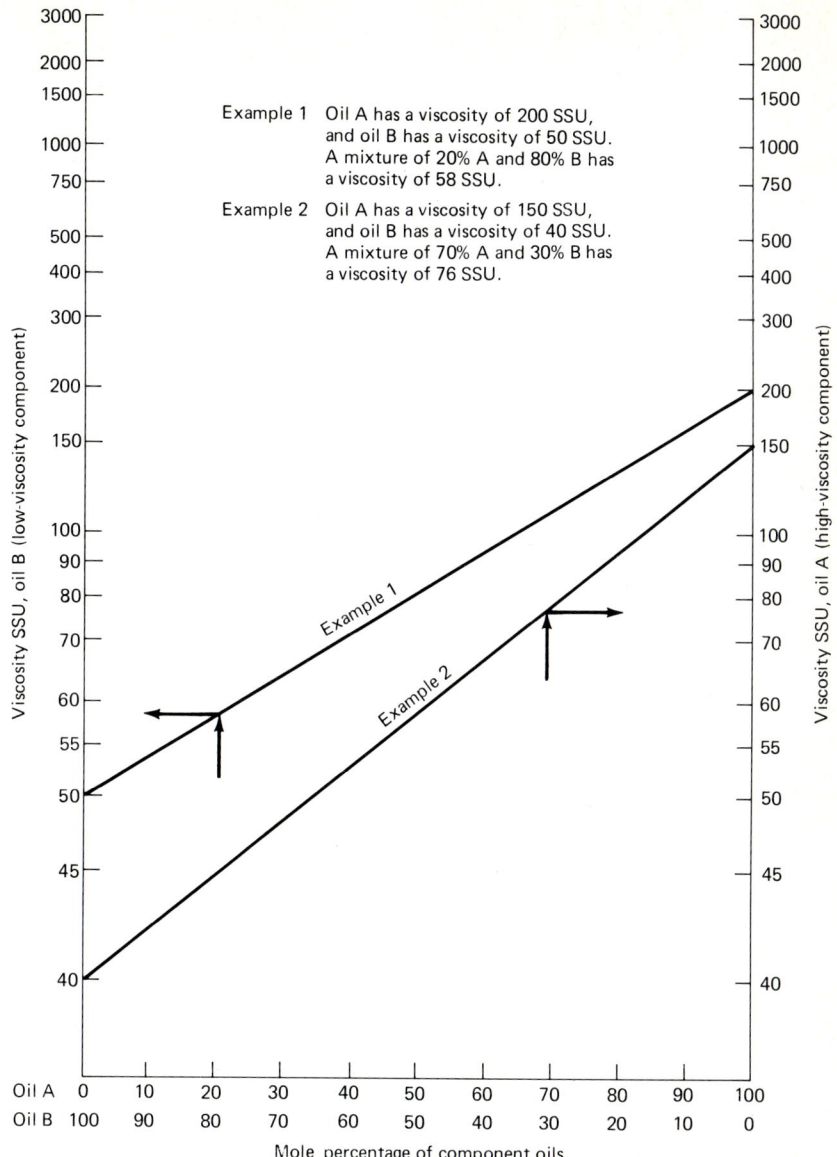

Figure 6.6 Viscosity-blending chart. (*Adapted from* Hydraulic Institute Engineering Data Book.)

Viscosity classifications. All gases and the great majority of liquids are known as true, simple, or newtonian fluids. The criterion for this classification is that the viscosity of the fluid be a point function of its temperature and pressure only and that it not be affected by the type or amplitude of motion to which it may be subjected or by time. Thus, for a newtonian liquid, when temperature and pressure remain constant, (1) the viscosity or ratio of shear stress to shear rate is a constant for all shear rates, (2) there is no shear rate only when there is no shear stress, and (3) the viscosity does not change with time. Nonnewtonian liquids obviously deviate from newtonian ones in that they do not follow one or more of these criteria.

Nonnewtonian fluids. Although the greater portion of a process engineer's work with fluids is with the newtonian type, it would be well to develop familiarity with the various categories of complex, or nonnewtonian, fluids. The several types of nonnewtonian liquids are plastic, dilatant, pseudoplastic, thixotropic, and rheopectic. They are best compared with a newtonian fluid and among themselves by examining the effect of shear stress on viscosity. This is shown in Fig. 6.7, which presents three curves for each viscosity classification. One shows the relationship between shear stress and shear rate, the slope of which is viscosity. The other two curves indicate the effects of shear stress and shear rate on viscosity.

The curves of the viscosity relationship for a newtonian fluid are the first ones, or Fig. 6.7a, and are seen to exhibit a constant viscosity value regardless of shear stress or shear rate.

Bingham plastic fluids. Plastic substances shown in Fig. 6.7b have a definite yield strength and do not flow until a certain degree of shear stress has been applied. It is as though the viscosity were infinite until the initial shear stress to induce flow is attained. Beyond that value of stress the material then has, as does a newtonian fluid, a constant viscosity defined by the slope of the curve for shear stress versus shear rate. Thus, Bingham plastic fluids are capable of being molded or given form. Oil paints are an example of this type of nonnewtonian fluid.

Dilatant fluids. Dilatant fluids exhibit an increase of viscosity with increasing rate of stress, as shown in Fig. 6.7c. This increase is also known as shear thickening, and the phenomenon can be accompanied by a change in volume. The changes in viscosity and volume are reversible and independent of time. Concentrated suspensions of titanium dioxide particles in water show both rheological and volume dilatancy. When the latter occurs, there is not sufficient water to fill the widened spaces between particles, and the surface appears to be dry as long as the shear stress is applied. Treading on moist sand at the seashore and having it appear to dry temporarily is a common example of dilatancy.

Pseudoplastic fluids. While a dilatant fluid exhibits an increase in viscosity as shear stress is increased, the viscosity of a pseudoplastic material decreases with increasing shear stress. This decrease, which is also known as shear thinning, is demonstrated in Fig. 6.7d. In contrast to plastic substances, there is no initial degree of shear stress that must be applied before flow is attained. Instead, many pseudoplastic fluids initially appear to be newtonian in behavior. They become nonnewtonian at a given shear stress and return to constant viscosity at a higher shear stress. The phenomenon is reversible as the shear stress is reduced. Pseudoplastics are the commonest of the nonnewtonian fluids. Many solutions of high-molecular-weight industrial polymers in organic solvents belong to this classification.

Thixotropic fluids. A second set of nonnewtonian fluids consists of those whose viscosity behavior is a function of time or the history by which the shear stress has been applied and may or may not be reversible. In this case, the structure of the fluid changes with time. Thixotropy is the property, exhibited by certain gels and emul-

Piping Calculations, Part 1: Limitations and Theoretical Considerations

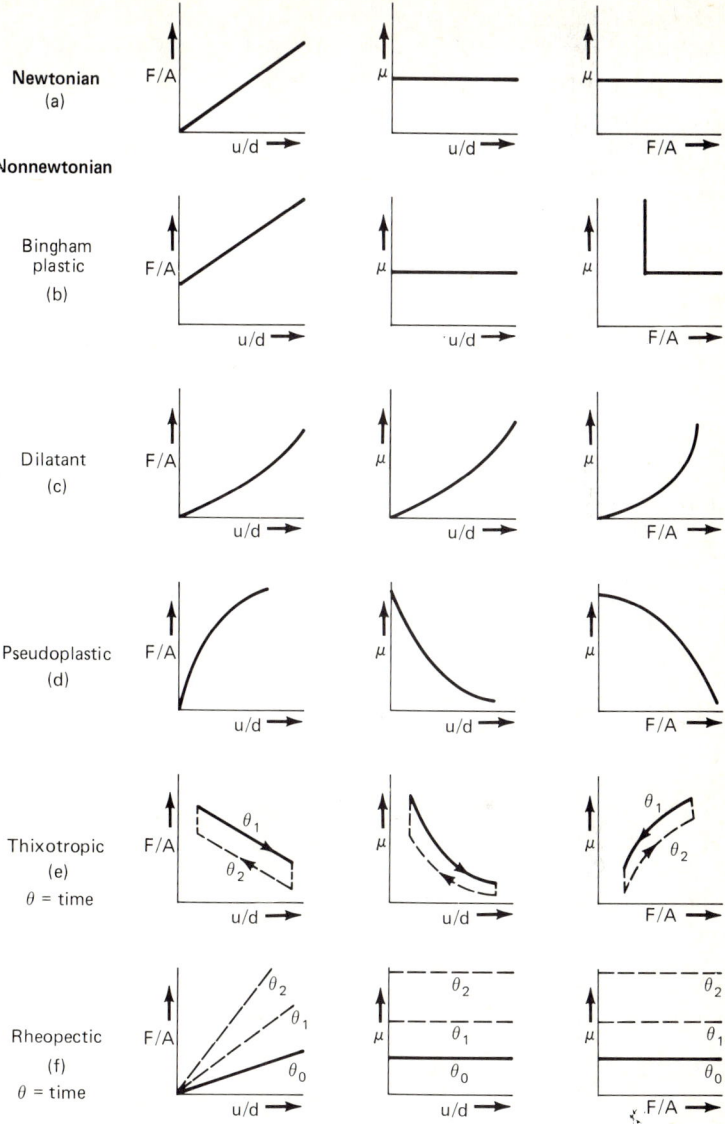

Figure 6.7 Comparison of viscosity relationships.

sions, of becoming more liquid when stirred or shaken. Thus, it can be seen in Fig. 6.7e that the viscosity of the fluid starts at a high value and decreases with both the shear rate and the shear stresss. However, there is a hysteresis effect, in that the viscosity does not follow the same shear-stress or shear-rate curves as these parameters are reduced and gradually returns to its original value only after a period of time. Many drilling muds, greases, paints, and inks, as well as common items such as mayonnaise and catsup, exhibit this property. The classic example of this phenomenon is the shaking of the catsup jar to cause the contents to flow more rapidly.

Rheopectic fluids. The viscosity of rheopectic fluids is also a function of time and the manner in which the shear stress was applied. Their viscosities remain constant at a given instant of time but are dependent on time. These fluids are described in section Fig. 6.7*f*, which shows that the slope of shear stress versus shear rate is constant for any shear rate but becomes greater after the initial application of stress. This property is exhibited by slowly gelling fluids, the most common of which are gelatins. In these cases, gelling proceeds more rapidly when the containing vessel is shaken.

Viscosity measurement. An approximate method of determining viscosity and one which gives the observer a physical feel for differences in viscosities is the use of a visual aid which consists of a series of liquids of known viscosities in separate similar tubes with a small air bubble of the same volume remaining in the stoppered mouth of each tube. A liquid of unknown viscosity could be placed in an identical tube and a similar-sized air bubble allowed to remain. When the tubes are inverted, the bubbles in the tubes with the known viscosities rise at different but discrete rates. The rate of ascent of the bubble for the unknown viscosity can be compared with the bubbles of the other samples, and its viscosity will be approximated by the sample whose bubble-rise rate it most nearly matches.

The bubble method can give only an approximate viscosity value. An instrument known as a viscosimeter (viscometer) is needed to determine the viscosity. Viscosimeters are of several basic types. The measurement of viscosities for newtonian fluids is usually made by allowing the fluid to flow through a restricted section of the viscosimeter and measuring the time for a given volume of fluid to pass through it. This procedure essentially measures the velocity through the calibrated restricted segment when acceleration is due to gravity. Viscosities of nonnewtonian fluids must be measured by means of an instrument which rotates in a stationary fluid. In this case, the torque due to the drag of the fluid is measured. This latter method may also be used with newtonian fluids, especially those with high viscosities.

Different types of viscosimeters are employed for various viscosity ranges. For low viscosities, a known volume of liquid is allowed to flow through one of several capillary tubes of specified diameter and length. For measurements of medium and higher viscosities, orifices of special construction and varied diameter are used in place of the capillaries. Each of the capillary tubes and orifices has its own factor by which effluent time is converted to viscosity units.

Instruments which utilize the capillary principle are known as Ostwald and Bingham viscosimeters for the men who devised them. The Saybolt instrument is the most common orifice-type viscosimeter used in the United States; the Saybolt Universal unit is used to measure medium viscosities and the Saybolt Furol, high viscosities. British equivalents of the respective Saybolt instruments are the Redwood and Redwood Admiralty. Most of the other European countries use the Engler viscosimeter for their measurements. Instruments used for determining much higher viscosities allow spheres of various diameter and density to fall through a fluid sample. Others, such as the Gardner mobilometer instrument, utilize a weighted perforated plate.

A whole series of instruments which employ a variety of shapes and configurations that rotate in liquids have been devised. The Stormer viscosimeter uses an inverted-horseshoe-shaped element. Shapes provided with the Brookfield viscosimeter, the most commonly used rotating unit in the United States, range from needles, to small cylinders, to large, flat disks. Each of the shapes finds application with a different type of fluid and viscosity range.

All viscosities determined by the viscosimeters discussed above are in centistokes or in special units unique to the fluid or to the industry which developed the method. The latter units can, in turn, be related to centistokes by means of conversion tables, charts, or alignment diagrams. Centistokes and viscosimeter readings are closely related, in the medium and high ranges of viscosity, by a direct multiplier or division factor. These are given for the more common viscosimeters in Table 6.5.

TABLE 6.5 Conversion Factors for Viscosimeter Readings to Centistokes (cSt)*

To convert		
Saybolt Universal	Divide seconds by	4.6
Saybolt Furol	Multiply seconds by	2.2
Redwood 1 (Standard)	Divide by	4.1
Redwood 2 (Admiralty)	Multiply by	2.4
Engler	Multiply degrees by	7.5
Dimmler No. 1	Multiply seconds by	3.2
Dimmler No. 10	Multiply seconds by	32
Stormer, 100 g	Multiply seconds by	2.8

*For liquids with viscosities greater than 70 cSt.

Types of flow patterns

Most of the problems encountered regarding pressure losses in piping relate to newtonian fluids. A review of some of the more important theoretical and practical aspects of fluid flow is presented here in preparation for the calculation of friction losses which follows in Chaps. 7, 8, and 9.

When a fluid flows within a conduit, be it a pipe, a noncircular duct, or an open channel, velocities represented in a cross-sectional traverse of the conduit are not equal. They take different profiles depending upon the type of flow pattern which prevails. Representations of these velocity profiles are easily and simply given for straight sections of pipe. In this instance a profile about the y axis also represents the profile along the z axis; i.e., the velocity profile in plane view z is the same as in elevation y, with the flow being in the x direction.

Figure 6.8 illustrates the typical linear-velocity profile development for the different types of flows that are encountered. Their average velocities, i.e., volume flows per unit of time divided by the cross-sectional area of the pipe, are represented by vertical lines u_a, u_b, u_c, and u_d. If the fluid were "ideal"* or the pipe were perfectly smooth, there would be no drag between the molecules themselves or between the fluid and the pipe. Thus, all molecules in the fluid would have the same linear veloc-

---

*A fluid is termed "ideal" if its viscosity is zero or no shear stress is required to deform it. This meaning is to be differentiated from that of the term "ideal" in thermodynamics, in which it is interchangeable with "perfect" for gases which conform to the gas laws.

ity, and the velocity profile from the wall on one side of the pipe to the wall opposite would be vertical lines as shown in Fig. 6.8.

However, there is no such state as an ideal fluid. All fluids exhibit shear stress. Even gases, in which molecules are at relatively great distances from one another, are subject to shear stress, albeit in small amounts. In addition, regardless of how smooth a pipe or a tube may appear to be, microscopic examination of its surface shows irregularities which impart a drag to the fluid flowing past it. Thus, the combination of viscosity and wall roughness results in the curved velocity profiles depicted in Fig. 6.8.

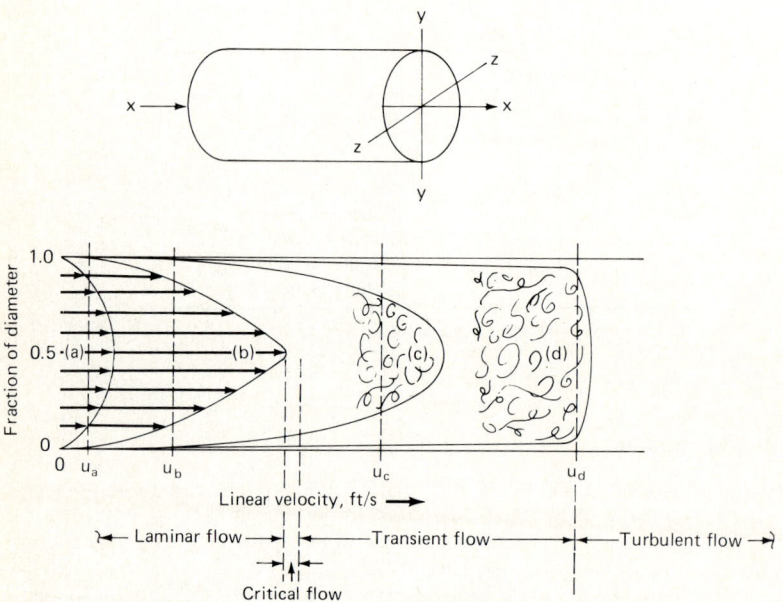

Figure 6.8 Linear-velocity profiles of various flow regimes.

The wall effect is evident in all the profiles, in that the velocity at the perimeter of the pipe is zero. This means that the molecules in contact with the pipe wall are essentially at rest. The molecules in the succeeding molecular layers away from the wall have to move past the more or less stationary molecules at the wall, and the viscosity of the fluid or shear-stress effect exerts a drag on that layer. Nevertheless, the fluid layers gradually accelerate as the distance from the pipe wall increases. The velocity in the traverse area increases until it reaches its maximum in the center of the pipe, the point where the wall effect is least. This is illustrated dramatically in curves *a* and *b*.

The mean linear velocity for curve *a* is relatively low, and the upper and lower halves of its profile approach a typical parabolic shape. At these low velocities or at high viscosities, the flow in the pipe can be thought to consist of a large number of annular rings or layers. Each ring or layer has a minute thickness, and all the molecules in it move at the same linear velocity in the *x* direction while having no significant velocity vectors in the y or z directions.* The annuli appear to be slipping past one another in the form of layers or laminae, and this type of flow is therefore termed "laminar." Since this effect is typical of highly viscous liquids, this type of flow is also called "viscous" flow.

The mean linear velocity $\bar{u}$ for a fluid in laminar flow is described by the equation

$$\bar{u} = \frac{2 \int_0^r u_r r\, dr}{r^2} \qquad (6.10)$$

with $u_r = f(r)$ and where u_r is the linear velocity at any point on the radius r.

When the profile relationship between the layer velocity and the layer radius approaches a parabola, the solution to Eq. (6.10) is such that the mean linear velocity is half of the maximum velocity† of the stream occurring at the center of the pipe.

Velocity profile b in Fig. 6.8 is similar to that for a except that its mean linear velocity for the same diameter is greater. All pipes have roughnesses or protrusions, no matter how minute, which cause perturbations in the fluid as it flows past them. At low flow rates, these disturbances are quickly dissipated. However, at certain higher rates the disturbances can be sustained and tend to be concentrated at the center of the pipe and cause the appearance of eddy currents.

For the sake of discussion, it can be assumed that the relationship between a fluid's mean linear velocity and the given pipe diameter is such that the maximum velocity shown in curve b is the greatest that could be attained for the given fluid without the commencement of sustained eddy currents at the center of the stream. Should sustained eddy currents form, molecules would migrate from the center section to the annular layer adjacent to it. There would no longer be the strict streamlining of the layers. When this occurs, laminar flow ends and critical flow begins.

The critical region of flow for a given fluid in a given pipe lasts for a relatively short span of mean-velocity increase. In this region the velocity profile is unstable, so that the flow is nearly laminar for short periods while at other times the incipient eddy currents appear intermittently. Thus, the transitional velocity will not appear constant at a given point along the length of the pipe. In addition, the transitional-velocity profile may appear to vary at different points along the length of the pipe at different instants. These nearly unpredictable variations in the appearance of the eddy currents make correlations of fluid-flow parameters difficult in the critical region.

Fortunately, the critical regime lasts for only a short span in the velocity spectrum as the flow regime is quickly transformed into the transient type of flow. The transient region also endures for a limited portion of the mean-velocity range; however, it does remain substantially longer than the critical regime as velocity increases.

The eddy currents which had appeared intermittently near the center core of the fluid in the pipe as the mean linear velocity increased now manifest themselves permanently in the transient-flow regime. As the mean linear velocity is further

---

*There is some y and z motion due to brownian movements in the fluid. However, the y and z vectors of any molecule are insignificant with respect to its x velocity and are not enough to cause appreciable numbers to transfer from one annulus to another or to affect the shape of the near-parabolic velocity curve.

†For derivation see W. H. King and E. F. Brater, *Handbook of Hydraulics,* McGraw-Hill Book Company, New York, 1976, pp. 6–7.

increased, the cross-sectional area encompassing the eddy currents spreads radially outward from the center of the pipe. The traverse profile of the linear velocities then deviates from a near parabola to become blunted, as in curve *c* in Fig. 6.8. The blunted portion delineates the region of the eddy currents.

Eddy currents have velocity vectors in all directions, in the *y* and *z* planes as well as in the *x* direction. All the velocity vectors occur in both plus and minus directions. However, at any given point along the pipe, the plus and minus vectors for the eddy currents in the respective *y* and *z* planes are essentially equal and cancel themselves out in each of these planes. Although some of the molecules in an eddy current may have a negative velocity vector in the *x* plane, others have larger positive vectors, so that the net result is a forward velocity of the eddy current. The forward-flow velocities of the various eddy currents for a given flow rate are all approximately equal. Thus, the traverse velocity profile of the space containing the eddy currents is nearly a straight vertical line, and the mean linear velocity for that section of the traverse area approaches the velocity of the eddy currents.

As the flow rate in a given pipe increases further, the traverse area enclosing the eddy currents spreads radially outward until it encompasses the entire cross section of the pipe except for a minute portion adjacent to the wall. The velocity profile is then essentially coincidental with the mean linear velocity, as in curve *d* of Fig. 6.8. Curve *d* also represents the end of the transient-flow regime and the beginning of turbulent flow.

Reynolds number

The Reynolds number is one of several dimensionless concepts that may be applied in correlating experimental data regarding the motion of fluids and in preparing calculations regarding these operations. It expresses the ratio of inertia forces (mass or velocity) of a moving fluid to its viscous forces and was developed in the late 1800s by Osborne Reynolds, an English scientist, who was one of the earlier investigators into the nature of fluid flow.

The Reynolds number (Re) is a key factor in determining the flow regime for a given flow rate in a given pipe. It can be defined by any of several expressions which relate flow rate, wetted perimeter or pipe diameter, and fluid velocity such that

$$\text{Re} = \frac{DG}{\mu} = \frac{Du\rho}{\mu} = \frac{Du}{z} = \frac{DW}{A\mu} = \frac{4W}{\pi D \mu} \qquad (6.11)$$

all in consistent units so that Re will be dimensionless. The normal units in the English system of measurements are:

D = inside diameter of pipe, ft
W = mass flow rate, lb/h
A = cross-sectional area of pipe, ft^2
G = mass velocity, lb/(h · ft^2)
u = mean linear velocity, ft/h
z = kinematic viscosity, ft^2/h
ρ = fluid density, lb/ft^3
μ = absolute viscosity, lbm/(ft · h)

Of the various expressions for Reynolds number given in Eq. (6.11) the simplest and easiest to remember are

$$\mathrm{Re} = \frac{DG}{\mu} = \frac{Du}{z}$$

Since mass velocity is usually stated in terms of pounds-mass per hour per square foot, absolute viscosity for the first equation should be given as pounds-mass per foot-hour. On the other hand, velocity is normally expressed as feet per second, and thus the dimension of kinematic viscosity in the second equation should be square feet per second.

Since absolute-viscosity data are usually listed in the metric units of centipoises, the conversion to English units may be obtained from the following equation:

$$1 \text{ lbm}/(\text{ft} \cdot \text{h}) = 2.42 \text{ cP} \tag{6.12}$$

or centipoises are multiplied by 2.42 to obtain pounds-mass per foot-hour. Should one wish to express mass velocity G in terms of pounds-mass per foot-second, one would divide the English units for the viscosity term by the number of seconds in an hour, i.e., 3600. Thus,

$$1 \text{ lbm}/(\text{ft} \cdot \text{s}) = \frac{1}{1488} \text{ cP} \tag{6.13}$$

In Fig. 6.8, it was noted that there are definite mean linear velocities in a given pipe or conduit which mark the boundary between one type of flow regime and another for a given fluid. Experimental results show that the Reynolds number, together with another dimensionless number, ϵ/D, relating the roughness of the pipe to its inside diameter, provides an excellent means for distinguishing the type of flow patterns. The *Hydraulic Institute Engineering Data Book** presents the basic information regarding the correlations for the transformations. These have been rearranged and are shown in Fig. 6.9. This figure clearly shows that all flows for which the Reynolds number is less than 2100 are laminar or streamline in nature. This can be interpreted as saying that those flows whose mass rates are low or are in relatively small-diameter pipe or whose fluid viscosities are high may well be in the laminar region.

The critical region in which eddy currents appear and disappear in the fluid has been found to occur in the narrow Reynolds-number range of 2100 to about 4000 regardless of pipe roughness or diameter. Beyond the critical region, the transient-flow area extends to at least a Reynolds number of 10,000 at high roughness-to-diameter ratios. The Reynolds number defining the boundary between the transient and turbulent regions then increases exponentially as the dimensionless roughness-to-diameter ϵ/D factor is reduced. Table 6.6 lists these transition points for various types of pipe or conduit.

A practical presentation of these relationships is to illustrate the boundaries between types of flow by plotting the flow regimes as a function of mean linear veloc-

---

*Hydraulic Institute, Cleveland, Ohio 44115.

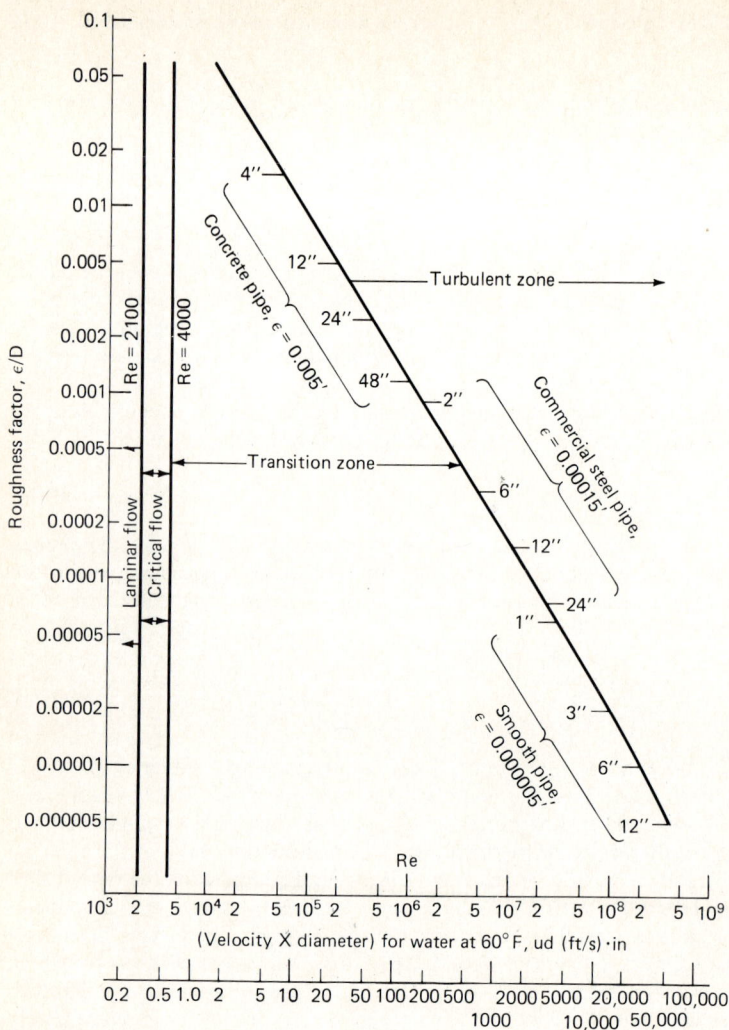

Figure 6.9 Types of flow as a function of Reynolds number and roughness parameter.

TABLE 6.6 Flow Regimes for Select Roughness-to-Diameter Ratios

Reynolds number	Water velocity-diameter product, (ft/s)·in	ϵ/D	Regime
<2100	<0.38	All	Laminar
2100–4000	0.38–0.78	All	Critical
4000–20,000	0.78–28	0.005	Transient
4000–2,600,000	0.78–500	0.0005	Transient
4000–10,000,000	0.78–1500	0.00015	Transient
>4000	>0.78	0.000005	Transient
>20,000	>28	0.005	Turbulent
>2,600,000	>500	0.0005	Turbulent
>10,000,000	>1500	0.00015	Turbulent

ity and pipe diameter only. This is done by fixing the viscosity and using roughness as the parameter for a series of curves and is shown in Fig. 6.10 for concrete (ϵ = 0.005 ft), commercial steel (ϵ = 0.00015 ft), and smooth (ϵ = 0.000005 ft) pipes with water at 60°F whose viscosity is 1.1 cP.

It would be well to consider Figs. 6.9 and 6.10 with respect to the effect of the various components of the Reynolds number on the boundaries between the various flow regions. The greater the roughness of a pipe, the more tendency there will be

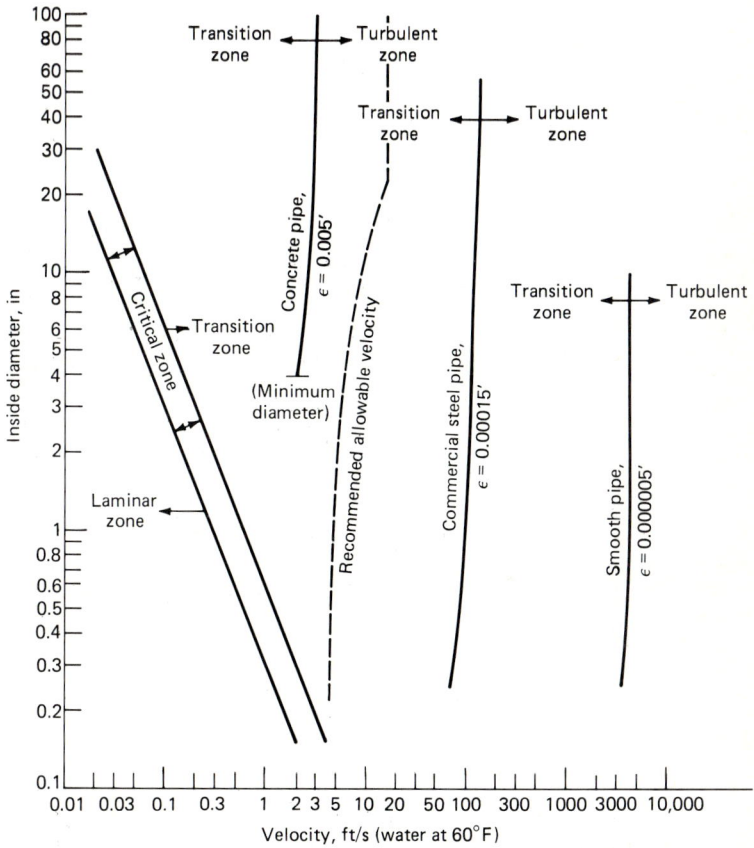

Figure 6.10 Types of water flow as a function of velocity and diameter.

that perturbations of greater amplitude will form at the wall more frequently. These are more readily propagated toward the center of the pipe and are thus more likely to form eddy currents there. This is borne out by the observation in both graphs that, for a given pipe diameter, the change from transient to turbulent flow takes place at lower Reynolds numbers or velocities as the roughness of the pipe increases.

It will be observed, however, that the flow regime in the pipe is always laminar as long as the Reynolds number is less than 2100 regardless of the roughness of a pipe or its diameter. No matter how great the mass velocity becomes, the flow will remain in the laminar region as long as the diameter is small enough or the viscosity great enough that the Reynolds number is less than 2100. Similarly, a pipe diameter may be large, yet a low flow rate or a high viscosity could maintain flow in the streamline region. Likewise, a fluid with a low viscosity may be flowing at such a low mass rate or in such a very small-diameter conduit that its Reynolds number is below 2100 and the flow would be in the viscous region also.

The critical region of flow is similarly little affected by the roughness of the conduit. Eddy currents may start more readily in pipe that is rougher, but they cannot be sustained in this flow region. Thus, as in laminar flow, the regime is a function of Reynolds number only.

Of great interest is the velocity at which the transformation between the transition zone and complete turbulence takes place. Superimposed on Fig. 6.10 are the recommended maximum allowable velocities for water in commercial pipe. It will be noted that in all cases the flows represented by these recommended allowable velocities may be turbulent in concrete pipe but are in the transient zone for commercial steel pipe. The viscosity of water is near the lower end of the range for liquids in most commercial operations. Therefore, other liquids having the same velocity as water in commercial pipe will be nearly as far as or farther from the transformation point as is water. It is rare indeed that one encounters completely turbulent flow for a liquid, especially in a pumped system.

Gaseous flows, on the other hand, are more likely to be in or near the turbulent zone for a larger portion of the time. It is only with very small-diameter pipe or low flow rates that they are in the transient area. They are rarely in the laminar region under normal process conditions.

Velocity head

A significant concept in all situations involving fluid flow is that of "velocity head." Basically, it is the energy required to impart a given velocity to a given mass. It is equivalent to the kinetic energy of the fluid and has the classic formula

$$h = \frac{u^2}{2g_c} \qquad (6.14)$$

and its corollary

$$u = \sqrt{2g_c h} \qquad (6.15)$$

where h = velocity head, ft fluid*
u = mean linear velocity, ft/s
g_c = dimensional constant, 32.17 (lbm/lbf)(ft/s^2)

It should be emphasized that the dimension of velocity head h is feet of the fluid that is flowing and it applies to any and all fluids: water or mercury, steam at 10 mmHg abs., and air at 100-atm pressure. All fluids that have the same velocity head will have the same linear velocity, and vice versa. This concept is relatively easy to comprehend for liquids by the use of an illustration which shows the effect of specific gravity on velocity for a given velocity head.

---

*Although h is conventionally given the dimension of feet fluid, it actually is an expression of pressure and has the implied dimension of foot pounds–force per pound-mass. This applies equally to subsequent discussions regarding static and friction heads.

Figure 6.11 depicts three columns of liquids in cylinders, each having a cross-sectional area of 1 ft². Near the base of each cylinder is an orifice of such a small cross-sectional area that, for all practical purposes, the fluid passing through it does not significantly affect the level in the column during the period of observation. It is also assumed that each orifice has the same cross-sectional area and is perfect,* i.e., offers

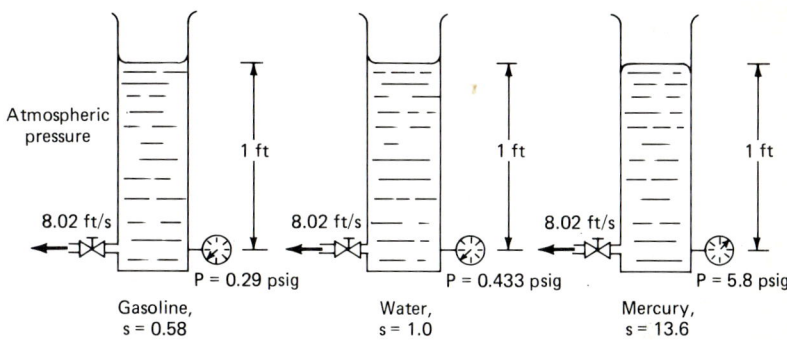

Figure 6.11 Effect of specific gravity on pressure for liquids. Cross section of all columns is 1 ft².

no resistance to flow. The flow from each column is collected in its own volumetrically calibrated receptacle.

Figure 6.11 shows the height of each liquid above its respective orifice to be 1 ft. Experimentation has shown that the volumes of liquid collected from each column during the same time interval are equal. When the volumes per unit of time are related to the cross-sectional areas of the orifices, it is seen that the velocity at the orifice is 8.02 ft/s regardless of fluid measured, or, making use of Eq. (6.15)

$$u = \sqrt{2g_c h} = \sqrt{2 \times 32.17 \times 1.0} = 8.02 \text{ ft/s}$$

The relationship in Eqs. (6.14) and (6.15) is such that velocities are proportional to the square root of the velocity head. Conversely, the velocity head required to provide a given velocity is in proportion to the square of the velocity.

Velocity head and pressure differential are related by a density factor. This may be seen by referring to Fig. 6.11, where it was noted that each column had a base of 1 ft². The liquid differential between the orifice level and the liquid level at the top was set at 1 ft. Thus, the respective volume of each column above the orifice is 1 ft³. Since density is defined as pounds per cubic foot, each column contains a weight above its orifice equal to its density. In the case of the water, there would be 62.4 lbf exerted over the 1 ft² of cross-sectional area. Since there are 144 in²/ft², it follows that the pressure of 1 ft of water over 1 in² is 62.4/144, or 0.433 lbf/in². Conversely, the reciprocal represents 1 lbf/in² as being equivalent to 2.31 ft of water.

The pressure represented by a 1-ft height of liquid is shown as a gauge pressure. This is valid since essentially 1 atm of pressure is exerted against the top of the column as well as at the level of the orifice. For the water column, the pressure gauge will read 0.433 psig. Since the pressure is a direct function of density, the pressure

---

*In actuality, only certain types of orifices are perfect. It is important to understand that many others do add resistance (see Table 7.10 in Chap. 7).

readings for the other two columns are obtained by multiplying the pressure exerted in the column with water by the respective specific gravities of the fluids in the other columns. This is shown by Eq. (6.16) and its corollary, Eq. (6.17):

$$\Delta P = \frac{\rho_f}{144} \times h = \frac{1}{2.31} \times \frac{\rho_t}{\rho_w} \times h \tag{6.16}$$

$$\Delta P = \frac{s}{2.31} \times h \tag{6.17}$$

where h = height of fluid, ft fluid
ρ_w = density of water at 60°F, 62.4 lb/ft^3
ρ_f = density of fluid, lb/ft^3
s = specific gravity of fluid (with respect to 60°F water)*
ΔP = pressure differential, psi

Now that the relationship between differential pressure and height of fluid for liquids [Eq. (6.17)] has been established, it is easier to examine the concept of velocity head for gases. The heights of the fluids in the example shown in Fig. 6.11 are readily established since liquids take the shape of the vessels that contain them. The height of the liquid in a container is simply a function of the geometry of the vessel and the volume of liquid to be contained. Gases, on the other hand, completely fill the whole container regardless of the geometry of the vessel or the initial volume of the gas.

A different mechanism is therefore required to demonstrate the relationship between fluid height and pressure differential for gases. In this case, we refer to Fig. 6.12, which shows the cylinders fitted with weightless and frictionless pistons instead of being open to the atmosphere. It will be assumed that the cross-sectional area of each cylinder is 1 ft^2. Each cylinder contains and is surrounded by its own gaseous medium at atmospheric pressure. For convenience, the temperature is assumed to be 60°F.

To start the exercise, the piston in each cylinder is placed at point B, and it is assumed that the contents of all the cylinders are at equilibrium with their surroundings. The valve at the orifice is closed, and on each piston is placed a weight equivalent in pounds to the density of the gas in the column, i.e., 0.0474 lb for methane, 0.0764 lb for air, and 0.1160 lb for carbon dioxide.

In each case the weight represents the mass of 1 ft^3 of gas or height of fluid equivalent to 1 ft of gas since the cross-sectional area of the column is 1 ft^2.† The weights will cause the piston to be lowered until the increased pressure in the cylinders sustains the respective weights. The pressures as read on the gauge-pressure instruments will be equal to the weight divided by 144. These will be 3.29×10^{-4} psi, 5.31×10^{-4} psi, and 8.05×10^{-4} psi, respectively. They represent the pressures equivalent to 1 ft of fluid for the different gas densities as calculated by Eqs. (6.16) and (6.17). The densities and weights should theoretically be recalculated to account for the

---

*It is assumed in Fig. 6.11 that the temperature of all the fluids was 60°F, so that $s_{(60°/60°)}$ was used. Had the temperatures and, thus, the specific gravities been different, the pressures shown by the gauges would also have been different. However, the discharge velocities would have remained the same.

†One should appreciate the great effect of density on height of fluid. Water, with a density of 62.4 lb/ft^3, needs but 2.31 ft of height to be equivalent to 1 psi, whereas air at atmospheric pressure and 60°F has a density of 0.0764 lb/ft^3 and requires a column more than 1880 ft high to be equivalent to 1 psi.

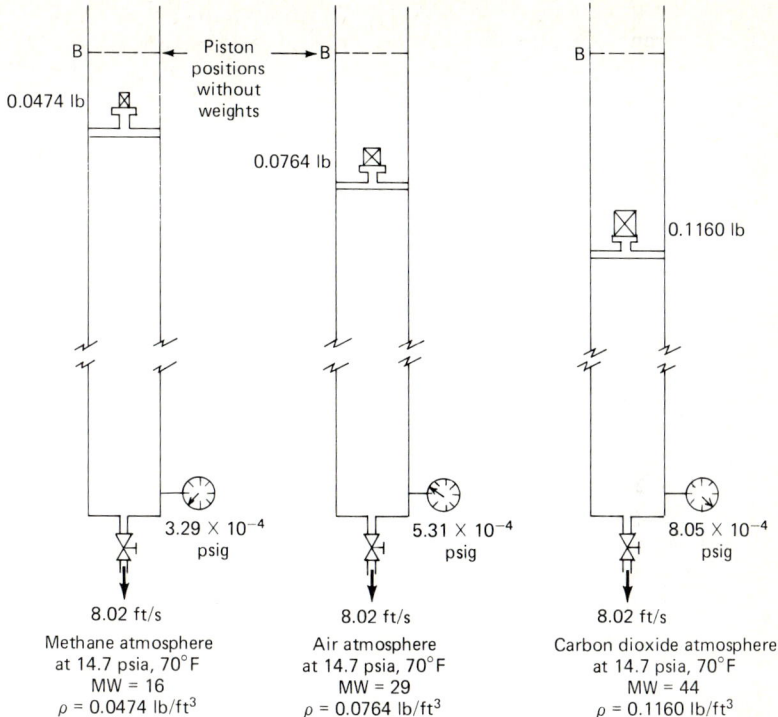

Figure 6.12 Effect of specific gravity on pressure for gases. Cross section of all columns is 1 ft².

increase of pressure in each cylinder due to the weight on the piston. However, in these cases in which gauge-pressure increases are so small with respect to atmospheric pressure, the relevant corrections would be insignificant. When the valves are opened, the velocities at each orifice are again 8.02 ft/s since there will be the equivalent of 1 ft of fluid differential at each orifice.

Consider duplicating the previous procedure but with each cylinder surrounded by its own atmosphere of gas at 150 psig and 60°F. In this case, the gases inside the cylinders will be at 150 psig originally, and their initial densities will be 11.3 times as great* as those shown in Fig. 6.12. If the weights shown in Fig. 6.12 are now added to the pistons, they will induce the same increases in pressure for the respective columns, i.e., 3.29×10^{-4} psi, 5.31×10^{-4} psi, and 8.05×10^{-4} psi, as they did previously. However, the weights now represent only 1/11.3 ft of fluid, and the velocity of the gases passing through the orifices will be reduced by a factor equivalent to the square root of the ratio of the atmospheric densities to the 150-psig densities. Thus, the velocities would be 2.39 ft/s instead of 8.02 ft/s.

To restore the velocity to 8.02 ft/s, which represents the velocity of 1 ft of fluid head, the weights required on the pistons would have to be increased by the same

---

*$p_2 \cong p_1 \left(\dfrac{P_2}{P_1} \right) \cong \dfrac{(150 + 14.7)}{14.7} \times p_1 = 11.3 p_1.$

factor of 11.3. The pressure differentials between the insides of the cylinders and the 150-psig atmospheres surrounding them would likewise be increased by the ratios of the absolute pressures. This is to be contrasted with the liquid columns represented in Fig. 6.11. Had the various liquid cylinders been placed in a 150-psig atmosphere of air, there would have been essentially no change in pressure differential between the insides of the cylinders and the surrounding atmosphere since the densities of the liquids would have been affected to only an insignificant degree.

It has been demonstrated that increasing the fluid head or pressure differential results in higher velocities in accordance with Eqs. (6.15) and (6.17). Normally there are no limitations to the extent to which liquid velocities may rise as a result of increased pressure drop as long as the change in pressure does not exceed the initial pressure minus the vapor pressure of the liquid. Gases, on the other hand, have definite limitations on the pressure drop to which they may be subjected because they are compressible fluids and are restricted by certain thermodynamic considerations which affect densities and velocities. This is discussed further in Chap. 8.

Chapter 7

Piping Calculations, Part 2: Friction Losses for Incompressible Fluids

The background for calculating friction loss in piping systems was given in Chap. 6 with the review of flow patterns and Reynolds numbers. The discussion continues in this chapter with a general method to determine the instantaneous friction factor for any fluid in motion in any type of conduit. Separate correlations are used for laminar flow and for transient or turbulent flows. Aids and shortcut methods in the form of special graphs, tables, slide rules, and mnemonic devices are then reviewed.

Most of the aids and shortcuts are based on standard clean steel pipe. Methods of conversion to surfaces with different roughness characteristics are given as well as means for converting various piping fittings and configurations into equivalent lengths of straight pipe. A convenient method for tabulating piping friction losses is shown. Included is a discussion of siphon effects and friction sealing of liquids in horizontal pipes and fittings.

All friction data presented are point functions and may be used for all fluids, although use of tabulated and nomograph information is normally limited to liquids. However, the information may also be used for gas flows when the total pressure drop in the system doesn't exceed 10 percent of the initial pressure.

Chap. 8 examines friction-loss calculations for compressible-fluid flow. Useful correlations are developed there for friction losses that are based on the initial instantaneous drop and can be applied to a large range of total-system pressure drops. Complex-flow-regime losses are reviewed in Chap. 9.

Friction Factors

In 1738, the famous Swiss mathematician and scientist Daniel Bernoulli published his most important work, *Hydrodynamica*. This classic was one of the first works to consider all aspects of fluids in motion and is still the basis of modern hydrodynamic work. Bernoulli's relationships for fluids in a simple piping or conduit system on which no work is being done, i.e., without the equivalent of a pump or compressor, are summarized as follows:

$$-\int_{P_1}^{P_2} \frac{dP}{\rho} = \frac{u_2^2 - u_1^2}{2g_c} + (h_1 - h_2) + F \tag{7.1}$$

where P = pressure of fluid, lbf/ft^2
ρ = density of fluid, lbm/ft^3
u = velocity of fluid, ft/s
g_c = dimensional constant, 32.17 (lbm/lbf)(ft/s^2)
h = elevation of fluid, ft
F = friction loss, (ft·lbf)/lbm
subscript 1 = condition at initial point
subscript 2 = condition at further point

All the terms of the equation have the dimensions of feet of fluid and relate the change in fluid pressure as it passes through a section of piping to the change in velocity, the change in elevation, and a third term known as friction. Friction is the head loss which occurs when "nonperfect" fluids, i.e., those with a finite viscosity, flow in channels that are not perfectly smooth. The drag effect caused by viscosity at the wall of the conduit in all flow regimes and the eddy currents that manifest themselves in the critical, transition, and turbulent zones all result in a net change of pressure for the fluid as it flows.

The friction that is developed in a section of conduit is a function of many variables, some of which are interrelated. They are:

1. Mean linear velocity and density or their product, mass velocity
2. Pipe diameter or hydraulic radius
3. Viscosity
4. Roughness of conduit
5. Length of pipe

The hydraulic radius is a convenient means of relating a noncircular conduit or any conduit that is not completely filled to an equivalent circular pipe. Hydraulic radius r_h equals the wetted area divided by the wetted perimeter, and equivalent diameter d_e equals 4 times the hydraulic radius, $4r_h$. It is noted that the equivalent diameter of an empty pipe is zero. The d_e increases as the height of the liquid rises until it is equal to the actual diameter when the pipe is half full. The equivalent diameter continues to increase until it reaches a maximum about 22 percent greater than the actual diameter when the height of liquid is about 81 percent of the actual diameter. As the height of liquid continues to rise, d_e diminishes until it is obviously equal to the actual diameter when the pipe is full (see Fig. 7.13 on page 272).

It was recognized very early that the friction head was related to the velocity head and the Reynolds number by means of a dimensionless value termed the "friction factor," or coefficient of friction. This relationship is shown in what is called Moody's friction equation:

$$f = \frac{F}{(L/D)(u^2/2g_c)} \tag{7.2}$$

or the Darcy equation:

$$F = f \times \frac{L}{D} \times \frac{u^2}{2g_c} = f \times \frac{L}{D} \times \frac{G^2}{\rho^2 g_c} \qquad (7.3)$$

where f = coefficient of friction, or friction factor
F = friction loss, (ft·lbf)/lbm
u = velocity, ft/s
G = mass velocity, lbm/(s·ft^2)
L = equivalent length of conduit, ft
D_e = equivalent diameter of conduit, ft
ρ = density, lbm/ft^3
g_c = dimensional constant, 32.17 (lbm/lbf)(ft/s^2)

The correlation between the Reynolds number and the friction factor has been the subject of studies by engineers, scientists, and mathematicians for over 100 years. Many practical experiments have been made, and the Hydraulic Institute publishes the latest valid data. These data are presented in a graphical form such as Fig. 7.1, which relates friction factor to Reynolds number for the various types of fluid flow for a series of roughness parameters. A companion piece to Fig. 7.1 is a graph, Fig. 7.2, which relates the relative roughness parameter (roughness divided by diameter) for the various types of surfaces that may be commonly encountered.

Roughness is a measure of the height of regular protrusions or unevennesses which extend from the surface of the conduit to disturb fluid flow. These protrusions can range from 0.001 to 0.01 ft (0.305 to 3.05 mm) for concrete pipe, 0.0005 ft (0.15 mm) for galvanized-iron pipe, and 0.00015 ft (0.05 mm) for commercial steel pipe. The least roughness is given as 0.000005 ft (1.5 μm) for smooth brass, lead, glass, or some lined pipe.

Laminar-flow friction

The combination of information derived from Figs. 7.1 and 7.2 makes it possible to solve any fluid-flow problem. The easiest flow to evaluate is laminar flow, also known as viscous flow. In this case roughness and, thus, the relative roughness parameter do not affect the friction factor, which is proportional only to the reciprocal of the Reynolds number. The slope of the relationship is such that $f_{\text{lam}} = 64/\text{Re}$. Using the definitions of Reynolds number from Eq. (6.11)

$$f_{\text{lam}} = \frac{64\mu}{\rho u D} = \frac{64\mu}{G D}$$

and substituting the resulting expressions for f_{lam} in Eq. (7.3) results in

$$f_{\text{lam}} = 32f \frac{Lu\mu}{D^2 \rho g_c} = 128 \frac{LW\mu}{\pi \rho D^4} \qquad (7.4)$$

where μ = absolute viscosity, lb/(ft·h) = 2.42 cP. Equation (7.4) is important because it presents the relevant relationships for friction loss when the flow is laminar. The magnitude of the friction loss, besides being directly proportional to the length of the pipe, is directly proportional to the weight flow rate and to viscosity but inversely proportional to the fourth power of the diameter. Thus, if the laminar friction drop is known for one fluid at a given flow rate in pipe of a certain diameter, the friction

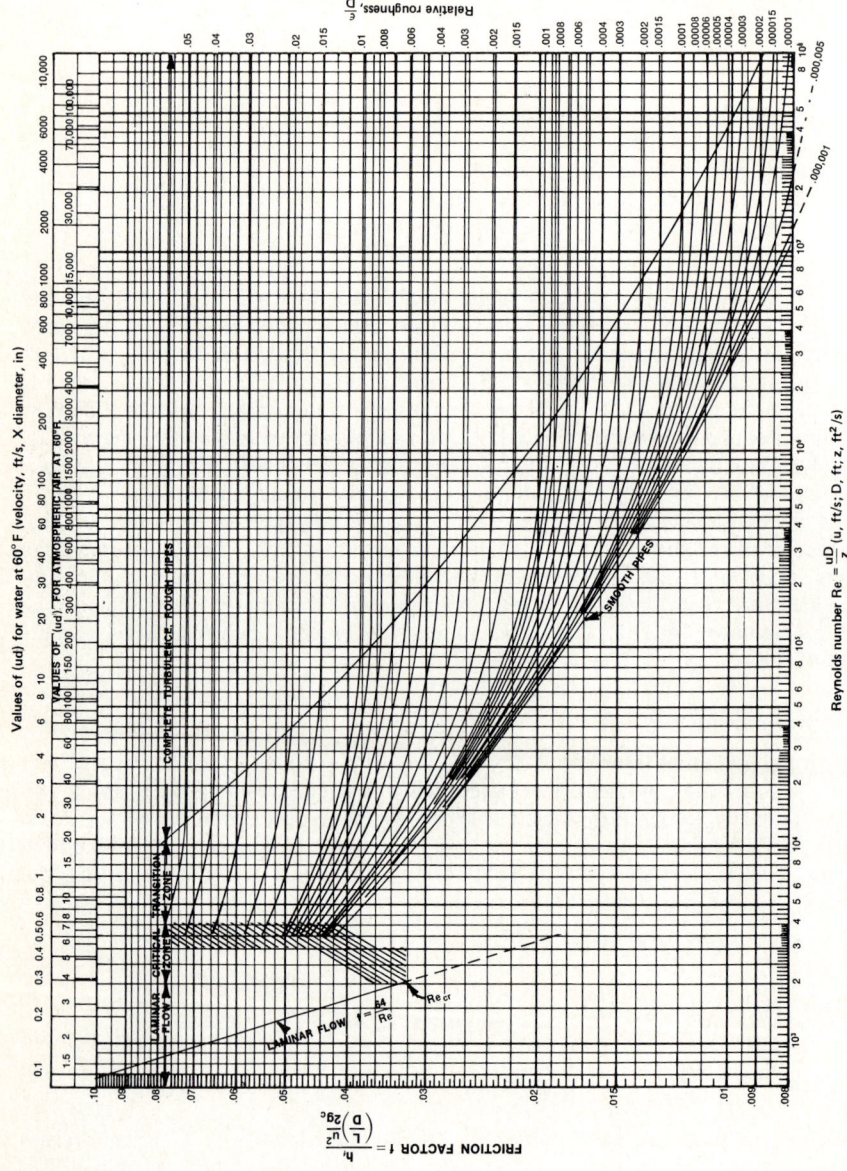

Figure 7.1 Moody chart (any roughness and size). *(The Hydraulic Institute, Cleveland, Ohio.)*

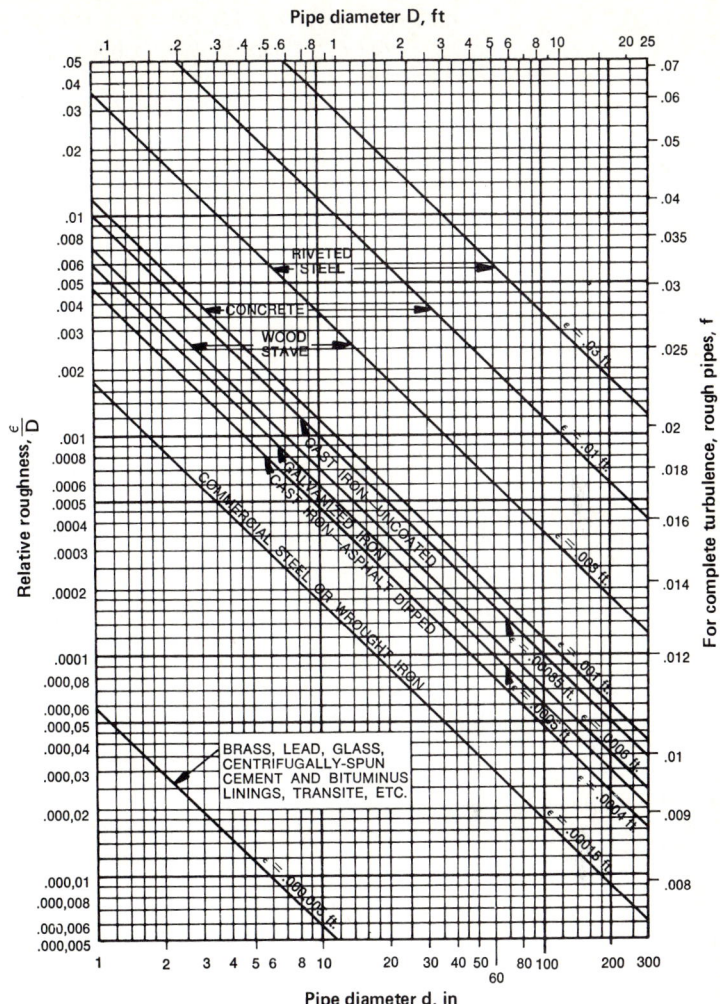

Figure 7.2 Relative roughness for clean pipe. *(The Hydraulic Institute, Cleveland, Ohio.)*

loss for a second fluid at a different flow rate or in another pipe size can easily be ascertained by using Eq. (7.4), provided the second flow also is laminar.

It must be emphasized that the preceding discussion expresses friction loss in terms of friction head measured in feet of fluid flowing as opposed to units of pressure drop. It is imperative that the distinction between friction loss in terms of head or feet of fluid flowing versus that of pressure drop be fully understood and appreciated in order to extrapolate from one set of conditions to another without having to resort to Fig. 7.1. This is shown by the following illustration. Two fluids having identical absolute viscosities are flowing in the laminar range at the same weight rate (pounds per hour) in identical piping systems. If the density of the second is one-half that of the first, the friction loss for the second will be twice that of the first in terms of head loss (feet of fluid). However, it is important to note that each fluid would have the same pressure drop (pounds per square inch).

Conversely, should the two fluids have been flowing at the same volume rate (gallons per minute), the friction loss for both in terms of fluid head would have been the same. But in this case the pressure change for the second, the fluid whose density is one-half that of the first, would have been one-half that of the first.

A word of caution must be given regarding the use of friction-factor diagrams such as that shown in Fig. 7.1. The Reynolds number is a dimensionless value and should remain the same as long as consistent units are used to calculate it. Although the Moody friction factor is dimensionless, its use gives friction loss expressed in various units, i.e., fact of fluid or pounds per square inch, depending on the constants used in the conversion equation. In addition, the Moody friction factor may have been scaled so as to accommodate a different coefficient in the equation for the friction factor. It is incumbent upon the user of any friction-factor diagram to understand completely the units of all terms, the units for the friction factor, and the units of all portions of the equation relating the friction factor to friction loss.

Transition- and turbulent-flow friction factors

Use of Fig. 7.1 becomes more complicated for the transition and turbulent zones of flow than for the laminar region. This complexity is due to the requirement that the roughness parameter of the pipe be known in addition to the Reynolds number in order to enter the chart to determine the Fanning friction factor.

Roughness parameter. In the critical region, between Reynolds numbers of 2000 and 4000, Moody friction factors take a sudden jump so that they are about 2 to 4 times as great as would be expected had the laminar-flow-regime relation been extended into that section. The larger the roughness parameter ϵ/D, the higher the friction factor. Thus, the greater the roughness of the pipe or the smaller its diameter, the greater the deviation of the Fanning friction factor from the extrapolated laminar friction factor.

As the curve of the friction factor for a given relative roughness parameter ϵ/D is traced with increasing Reynolds number, the Fanning friction factor is reduced at a decreasing rate until it becomes essentially constant. The point at which the friction factor becomes nearly constant represents the transformation from the transition- to the turbulent-flow regime. The loci of these transformation points are shown on Fig. 7.1. On the right side of the curve is complete turbulence, and on the left, down to the critical Reynolds number, is the transition area. Also included on Fig. 7.1 is a curve for smooth pipes which delineates the lower limit of friction factors in the transition zone. It should be noted that at extremely high Reynolds numbers the curve for smooth pipes parallels the lowest transition-zone curve. This means that flow in a smooth pipe will always be in the transition range at all Reynolds numbers above 4000.

It was previously shown that the Fanning friction factor in the laminar zone decreases as the first power of the Reynolds number. In the transition zone just beyond the critical zone the friction factor for pipes with lower roughness parameters is reduced approximately in proportion to the cube root of the increase in Reynolds number. However, for pipes with large roughness parameters, the influence for change in Reynolds number diminishes, and friction factors are reduced in proportion to as little as the one-tenth power of Reynolds-number change.

Mathematical models. An excellent mathematical model for calculating the Fanning friction factors depicted by the curves in the transition and turbulent zones in Fig. 7.1 has been made by C. F. Colebrook with his equation

$$\frac{1}{\sqrt{f}} = -2 \log_{10} \left(\frac{\epsilon}{3.7D} + \frac{2.51}{\text{Re}\sqrt{f}} \right) \tag{7.5}$$

The friction factors calculated by means of Eq. (7.5) require, at most, several iterations when starting with $f = 0.05$, the approximate average value for f in the chart. Equation (7.5) is amenable for use with a programmable calculator or in a computer program.

Other investigators later developed equations to approximate the friction factor directly without the need to perform iterations. Those by Moody and by Churchill are, respectively, as follows:

$$f = 0.0055 \left[1 + \left(20,000 \frac{\epsilon}{D} + \frac{10^6}{\text{Re}} \right)^{1/3} \right] \tag{7.6a}$$

and

$$\frac{1}{\sqrt{f}} = 6.96 \ln \left[\frac{1}{\frac{\epsilon}{3.7D} + \left(\frac{7}{\text{Re}}\right)^{0.9}} \right] \tag{7.6b}$$

Both these equations give friction factors within 5 percent of those derived from the Colebrook expression, Eq. (7.5). However, with the aid of only a scientific hand calculator, the iterations required by the Colebrook equation can be performed almost as expeditiously as the Moody or Churchill calculations for the friction factor directly.

An inspection of Fig. 7.1 shows that at complete turbulence the friction factor is essentially independent of the Reynolds number but is the inverse function of the roughness parameter ϵ/D. In the transition zone where most flows occur, the value of the friction factor is influenced by the Reynolds number itself as well as by the inside diameter of the pipe. It is not practical to assign specific exponents to which the diameter and the other Reynolds-number parameters should be raised in order to predict friction factors from one set of conditions to another in the transient zone. These exponents change continuously with the roughness-to-diameter ratio until the transformation to the turbulent zone and typical values are given in Table 7.1. It is generally accepted that f varies as the -0.18 power of flow and the 0.14 power of viscosity for most commercial pipes, and these values give reasonable results for most diameters and flow conditions normally encountered.

It is impossible to obtain a correlation of friction factor with Reynolds number and the roughness parameter for flows in the critical zone. This is true because the eddy currents in this region are transient and may appear and disappear at random. Thus, it is difficult to obtain repetitive experimental data. Readings taken in the bank of Reynolds numbers between 2100 and 4000 lie scattered in the crosshatched area shown on Fig. 7.1 and on Fig. 7.3 below.

Various researchers have proposed different-shaped curves to connect the terminus

TABLE 7.1 Exponents of Parameters Affecting Moody Friction Factor

ϵ/D	Position in transition zone	Re	Exponents			
			Diameter	Weight rate	Viscosity	Roughness
0.05	Beginning	4×10^3	-0.33	-0.07	0.07	0.40
0.05	End	1.3×10^4	-0.43	0	0	0.43
0.01	Beginning	4×10^3	$+0.01$	-0.18	0.18	0.17
0.01	End	9×10^4	-0.32	0	0	0.32
0.001	Beginning	4×10^3	$+0.26$	-0.28	0.28	0.02
0.001	End	1.3×10^6	-0.23	0	0	0.23
0.0001	Beginning	4×10^3	$+0.30$	-0.30	0.30	0.003
0.0001	End	1.5×10^7	-0.18	0	0	0.18
0.000,01	Beginning	4×10^3	$+0.30$	-0.30	0.30	0
0.000,01	End	4×10^8	-0.13	0	0	0.13
0.000,000,5	Beginning	4×10^3	$+0.30$	-0.30	0.30	0
0.000,000,5	End	4×10^9	-0.10	0	0	0.10

of the laminar line with the initial points of the friction-factor curves in the transition zone, but it is difficult to judge which curve is most suitable. Fortunately, the extent of Reynolds numbers in the critical zone is relatively small compared with the total normal Reynolds-number spectrum. Therefore, the most expeditious method to follow is to extrapolate either the laminar line forward to the Reynolds number of 4000 or the transition curve back to a Reynolds number of 2100 and then to use that Moody friction factor which results in the most conservative solution for the situation being studied. For example, if the problem involves head loss in a pipe in order to determine motive power required, then the friction factor from the extended transition curve gives the greater safety factor. On the other hand, if one had to prepare to receive the flow through a pipe under a gravity head, then the Moody friction factor derived from the projection of the laminar line represents the greatest flow.

Aids to Determine Head Losses

There are available to the process engineer several aids which eliminate most of the tedium involved in calculating friction losses in piping. These aids consist of nomographs, tables, and specialty items such as specifically constructed slide rules. They are available from a variety of sources including the basic reference, the *Hydraulic Institute Engineering Data Book,** manuals and specialty items prepared by vendors of equipment,† and specialized graphs which are found in standard engineering texts or which appear in journals from time to time.

All the shortcut methods for obtaining friction head loss should be used only in the context for which they are intended. It is important to know and understand their

---

*Published by The Hydraulic Institute, Cleveland, Ohio 44115.

†These may be distributed free or for a nominal fee.

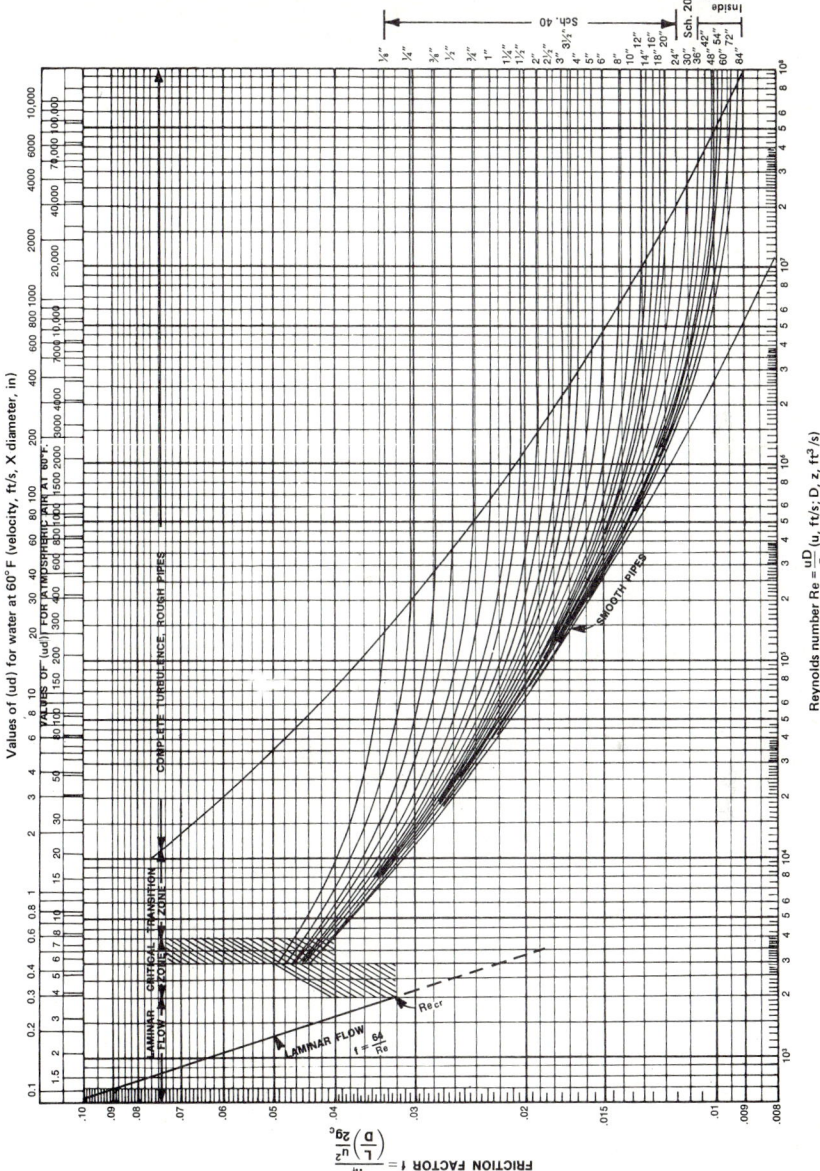

Figure 7.3 Moody chart (clean steel or wrought-iron pipe.) *(The Hydraulic Institute, Cleveland, Ohio.)*

limitations in the light of all the parameters included in the basic curves such as those shown as Fig. 7.1 or 7.3. However, if used within these constraints, the aids quickly and simply yield friction head losses well within normal engineering tolerances.

Simplified Moody charts

Although this is not a major simplification, the basic Moody-diagram curves in Fig. 7.1 can be modified by eliminating the roughness factor and limiting the curves to a series of diameters for pipe with a given roughness. This can be done for any of the types of surfaces shown in Fig. 7.2 or listed in Table 7.2.

TABLE 7.2 Values of Surface Roughness for Various Materials*

Material	Surface roughness	
	ϵ, ft	ϵ, in
Drawn tubing (brass, lead, glass, and the like)	0.000005	0.00006
Commercial steel or wrought iron	0.00015	0.0018
Asphalted cast iron	0.0004	0.0048
Galvanized iron	0.0005	0.006
Cast iron	0.00085	0.010
Wood stave	0.0006–0.003	0.0072–0.036
Concrete	0.001–0.01	0.012–0.12
Riveted steel	0.003–0.03	0.036–0.36

*L. F. Moody, *Transactions of the American Society of Mechanical Engineers,* vol. 66, 1944, pp. 671–684; and vol. 69, 1947, pp. 1005–1006. Additional values of ϵ for various types or conditions of concrete, wrought iron, welded steel, riveted steel, and corrugated metal pipes are given in H. W. King and E. F. Brater, *Handbook of Hydraulics,* 6th ed., McGraw-Hill Book Company, New York, 1976, pp. 6–12, 6–13.

A modified graph is given in Fig. 7.3, which presents the friction factor for new steel pipe, i.e., roughness equal to 0.00015 ft, as a function of Reynolds number. The diameters represented are the standard nominal sizes between $\frac{1}{4}$ and 36 in. The discussions pertaining to the Moody diagram given in Fig. 7.1 apply equally to Fig. 7.3 except that the roughness of the pipe is constant in the latter graph.

Diagrams similar to Fig. 7.3 can be and have been developed for surfaces other than new steel pipe when there is occasion for frequent analysis of flow over that type of surface. However, steel and wrought iron are the most frequently encountered flow surfaces, and most of the general aids for obtaining friction head loss are based on these materials.

Nomographs and alignment charts

Nomographs and alignment charts relating the various parameters to friction loss for a given length of pipe are available from several sources. One of the better known was prepared by Genereaux[1] and is given as Fig. 7.4. In this case friction loss is shown not as feet of fluid but as pounds per square inch or inches of water per foot of pipe.

Figure 7.4 Pipe flow chart (nomograph, clean steel). *(Excerpted by special permission from Chemical Engineering, vol. 44, 1937, p. 241. Copyright © 1984, McGraw-Hill, Inc., New York, N. Y. 10020.)*

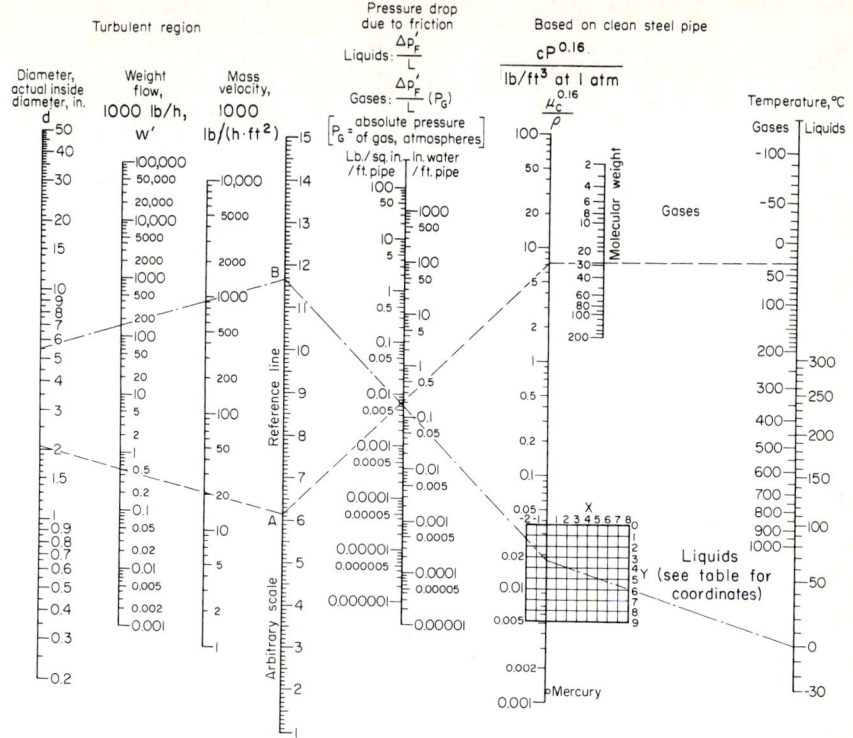

Coordinates for Liquids and Aqueous Solutions

	X	Y		X	Y
Acetaldehyde	−0.3	3.7	Glycerol, 100%	6.9	1.8
Acetic acid, 100%	1.0	4.0	Glycerol, 50%	3.0	3.7
Acetic acid, 77%	2.6	3.8	Hydrochloric acid,		
Acetic anhydride	0.7	4.3	31.5%	1.1	4.2
Acetone, 100%	0.9	3.4	Linseed oil, raw	3.4	1.8
Acetone, 35%	2.7	3.7	Mercury	See chart	
Ammonia, anhydrous	0.9	3.6	Methanol, 100%	0.8	3.3
Ammonia, 26%	1.9	3.6	Methanol, 40%	2.8	3.6
Aniline	2.5	3.4	Methyl acetate	0.0	4.2
Benzene	0.6	3.6	Methyl chloride	−0.8	4.3
Butanol	2.6	2.6	Nitric acid, 95%	0.8	5.8
Calcium chloride			Nitric acid, 60%	1.5	4.8
brine, 25%	2.6	4.2	Nitrobenzene	1.7	4.4
Carbon disulfide	0.0	5.6	Octane	0.4	2.7
Carbon tetrachloride	0.7	6.0	Phenol	2.4	3.4
Chloroform	0.0	6.0	Propionic acid	0.6	3.8
Chlorosulfonic acid	1.5	5.8	Sodium chloride		
Cyclohexanol	5.3	2.2	brine, 25%	2.1	4.4
Diphenyl	0.0	3.5	Sodium hydroxide,		
Ethyl acetate	0.2	3.9	50%	5.3	3.7
Ethyl alcohol, 95%	1.9	3.0	Sulfur dioxide	−0.2	6.1
Ethyl alcohol, 45%	3.6	3.4	Sulfuric acid, 110%	3.7	4.7
Ethyl chloride	0.2	4.3	Sulfuric acid, 98%	3.5	4.8
Ethyl ether	−0.3	3.2	Sulfuric acid, 78%	3.2	4.8
Ethylene glycol	3.5	2.9	Tetrachloroethylene	0.3	6.2
Fluorocarbon F-11	0.0	6.2	Toluene	0.4	3.6
Fluorocarbon F-12	−1.2	5.9	Trichlorethylene	0.1	5.9
Fluorocarbon F-21	−0.4	5.9	Turpentine	1.1	3.1
Fluorocarbon F-22	−1.7	5.5	Vinyl acetate	0.4	4.2
Fluorocarbon F-113	0.9	6.2	Water	2.0	4.2
Formic acid	1.5	4.5			

This nomograph is designed primarily for turbulent flow in clean steel pipe. Since most flows are in the transition region, friction losses from the nomograph are somewhat lower than those which could be calculated with the Fanning friction factors obtained directly from Fig. 7.3 for commercial steel and wrought-iron pipe. The effect of viscosity and density on friction loss is taken into account in the nomograph so that it may be used for both liquids and gases of varying species and conditions.

Slide rules

Tube Turns Inc.* markets a slide rule, shown in Fig. 7.5, which permits the determination of friction loss in pounds per square inch per 100 ft for laminar- and turbulent-flow regimes. Separate scales are used for liquids and for gases. The slide rule also gives velocity and Reynolds numbers when weight flow rate, pipe diameter, viscosity, and density or specific volume are entered as the variables. A scale is also included for correction coefficients that can be applied to friction losses shown in the turbulent region for Reynolds numbers that fall in the transition zone.

The Tube Turns slide rule uses the roughness of clean steel as the basis for the relationships, and the results obtained are well within accepted tolerances of good engineering practice. A scale relating equivalent lengths of various piping fittings to pipe diameter is provided as a convenience.

The rule can easily be used to extrapolate for pressure losses for gases or vapors whose specific volumes or flow rates are beyond the apparent scope of the rule. By exercising ingenuity in altering the scale of the input data and remembering the effect of these variables on the friction loss, the user can make calculations far beyond the rule's original range of flow rates, pressures, and pipe sizes. With some practice, many of the calculations can be accomplished by interpolation with the rule itself. The rule makes an important addition to the auxiliary tools normally used by the process engineer.

Friction-loss tables

The most rapid and accurate means of obtaining friction loss in terms of feet of fluid for liquids in clean steel pipe is the use of tables based on volume flow rates and pipe diameter.

Hydraulic Institute tables. The *Hydraulic Institute Engineering Data Book* includes a series of tables for all pipe sizes in the range of $\frac{1}{8}$ to 60 in for commonly used schedules of steel and asphalt-dipped cast-iron pipes. The tables relate flow in cubic feet per second and gallons per minute to velocity, velocity head, and friction loss as feet of fluid per 100 ft of pipe for water at 60°F.

Cameron tables. The *Cameron Hydraulic Data Book*† presents tables similar to those given by the Hydraulic Institute for Sch. 40 new steel and for asphalt-dipped cast-iron pipe. In addition, it includes data on Sch. 80 and Sch. 160 pipe for pipe diameters 12 in and smaller. Table 7.3 shows a sample of the type of information presented, in this instance for 4-in nominal-diameter pipe.

---

*Tube Turns Inc., Box 32160, Louisville, Ky. 40232.

†This book is published by the Ingersoll-Rand Company, Woodcliff Lake, N.J. 07675, and is available from it for a nominal fee.

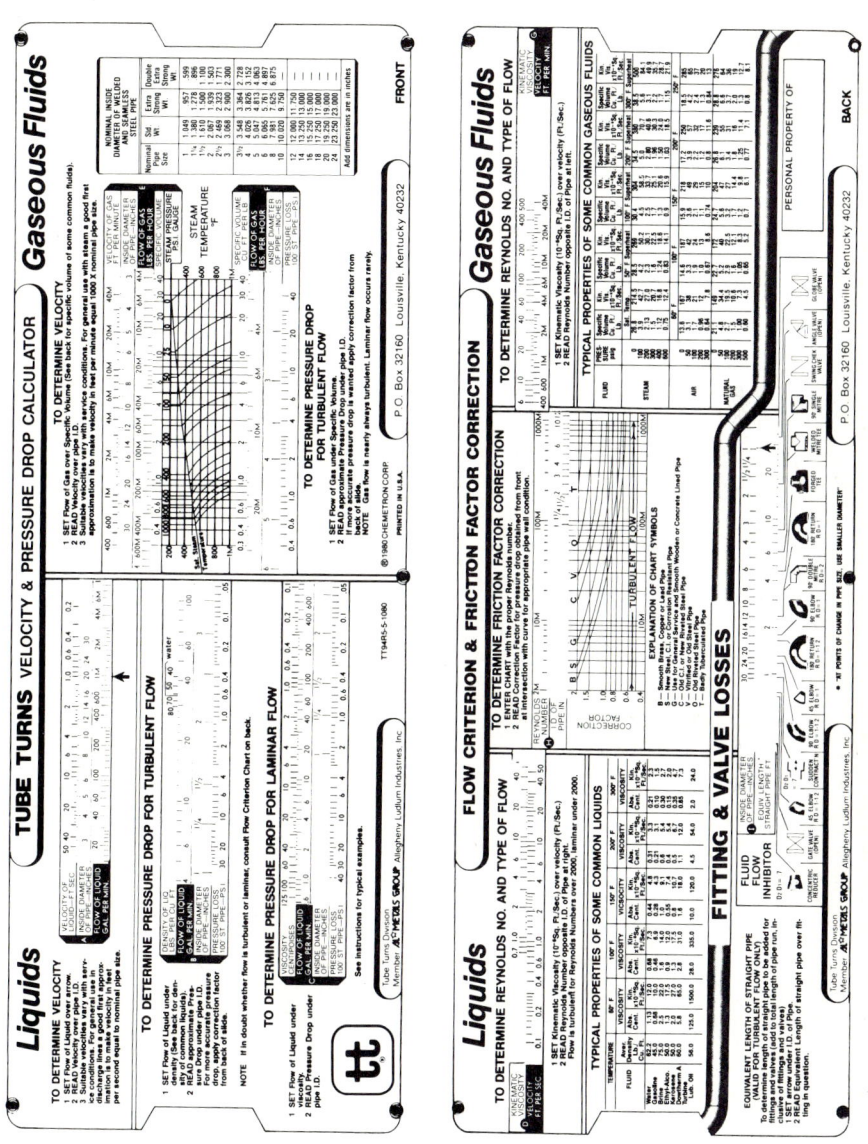

Figure 7.5 Pressure-drop slide rule. *(With permission from Tube Turns Inc., Louisville, Ky.)*

TABLE 7.3 Friction of Water: Asphalt-Dipped 4-in Cast-Iron and New Steel Pipe*

Flow, U.S. gal/min	Asphalt-dipped cast iron, 4 in ID			Standard-weight steel, Sch. 40, 4.026 in ID			Extra-strong steel, Sch. 80, 3.626 in ID			Sch. 160 steel, 3.438 in ID		
	Velocity, ft/s	Velocity head, ft	Head loss, ft/100 ft	Velocity, ft/s	Velocity head, ft	Head loss, ft/100 ft	Velocity, ft/s	Velocity head, ft	Head loss, ft/100 ft	Velocity, ft/s	Velocity head, ft	Head loss, ft/100 ft
20	.511	.004	.038	.504	.004	.035	.56	.00	.045	.691	.007	.074
30	.766	.009	.076	.756	.009	.072	.84	.01	.092	1.04	.017	.154
40	1.02	.016	.128	1.01	.016	.120	1.12	.02	.153	1.38	.030	.258
50	1.28	.025	.194	1.26	.025	.179	1.40	.03	.230	1.73	.046	.387
60	1.53	.037	.273	1.51	.036	.250	1.67	.04	.320	2.07	.067	.540
70	1.79	.050	.365	1.76	.048	.330	1.95	.06	.424	2.42	.091	.691
80	2.04	.065	.470	2.02	.063	.422	2.23	.08	.541	2.77	.119	.885
90	2.30	.082	.588	2.27	.080	.523	2.51	.10	.649	3.11	.150	1.10
100	2.55	.101	.719	2.52	.099	.613	2.79	.12	.789	3.46	.185	1.34
110	2.81	.123	.862	2.77	.119	.732	3.07	.15	.943	3.80	.224	1.61
120	3.06	.146	1.02	3.02	.142	.861	3.35	.17	1.11	4.15	.267	1.89
130	3.32	.171	1.19	3.28	.167	1.00	3.63	.20	1.29	4.49	.313	2.20
140	3.57	.199	1.37	3.53	.193	1.15	3.91	.24	1.48	4.84	.363	2.53
150	3.83	.228	1.57	3.78	.222	1.31	4.19	.27	1.69	5.18	.417	2.89
160	4.08	.259	1.77	4.03	.253	1.48	4.47	.31	1.91	5.53	.475	3.26
170	4.34	.293	1.99	4.28	.285	1.66	4.75	.35	2.14	5.88	.536	3.66
180	4.60	.328	2.23	4.54	.320	1.85	5.02	.39	2.38	6.22	.601	4.09
190	4.85	.368	2.47	4.79	.356	2.05	5.30	.44	2.64	6.57	.669	4.53
200	5.11	.406	2.73	5.04	.395	2.25	5.58	.48	2.91	6.91	.742	5.00
220	5.62	.490	3.29	5.54	.478	2.70	6.14	.59	3.49	7.60	.897	6.00
240	6.13	.583	3.90	6.05	.569	3.19	6.70	.70	4.13	8.30	1.07	7.09
260	6.64	.685	4.55	6.55	.667	3.72	7.26	.82	4.81	8.99	1.25	8.27
280	7.15	.794	5.26	7.06	.774	4.28	7.82	.95	5.54	9.68	1.45	9.55
300	7.66	.912	6.02	7.56	.888	4.89	8.38	1.09	6.33	10.37	1.67	10.9
320	8.17	1.04	6.84	8.06	1.01	5.53	8.94	1.24	7.17	11.06	1.90	12.4

340	8.68	1.17	7.70	8.57	1.14	6.22	9.50	1.40	8.06	11.75	2.14	13.9
360	9.19	1.31	8.61	9.07	1.28	6.94	10.0	1.6	9.00	12.44	2.40	15.5
380	9.70	1.46	9.58	9.58	1.43	7.71	10.6	1.7	9.99	13.13	2.68	17.3
400	10.2	1.62	10.6	10.1	1.58	8.51	11.2	1.9	11.0	13.82	2.97	19.1
420	10.7	1.79	11.6	10.6	1.74	9.35	11.7	2.1	12.1	14.52	3.27	21.0
440	11.2	1.96	12.8	11.1	1.91	10.2	12.3	2.3	13.3	15.21	3.59	22.9
460	11.7	2.14	13.9	11.6	2.09	11.2	12.8	2.5	14.5	15.90	3.92	25.0
480	12.3	2.33	15.2	12.1	2.27	12.1	13.4	2.8	15.7	16.59	4.27	27.2
500	12.8	2.53	16.4	12.6	2.47	13.1	14.0	3.0	17.0	17.28	4.64	29.5
550	14.0	3.06	19.8	13.9	2.99	15.8	15.3	3.6	20.5	19.00	5.61	35.5
600	15.3	3.65	23.6	15.1	3.55	18.7	16.7	4.3	24.3	20.74	6.67	42.1
650	16.6	4.28	27.6	16.4	4.17	21.7	18.1	5.1	28.4	22.46	7.83	49.2
700	17.9	4.96	32.0	17.6	4.84	25.3	19.5	5.9	32.8	24.19	9.08	57.0
750	19.1	5.70	36.6	18.9	5.55	28.9	20.9	6.8	37.6	25.92	10.4	65.2
800	20.4	6.48	41.6	20.2	6.32	32.8	22.3	7.7	42.7	27.65	11.7	74.1
850	21.7	7.32	46.9	21.4	7.13	37.0	23.7	8.7	48.1	29.38	13.4	83.4
900	23.0	8.20	52.6	22.7	8.00	41.4	25.1	9.8	53.8	31.10	15.0	93.4
950	24.3	9.14	58.5	23.9	8.91	46.0	26.5	10.9	59.8	32.83	16.7	104
1000	25.5	10.1	64.8	25.2	9.87	50.9	27.9	12.1	66.2	34.56	18.5	115
1100	28.1	12.3	78.3	27.7	11.9	61.4	30.7	14.6	79.8	38.02	22.4	139

*Based on Darcy's formula. Reprinted with permission from *Cameron Hydraulic Data Book*, copyright © Ingersoll-Rand Company, 1981.
NOTE: No allowance has been made for age, difference in diameter, or any abnormal condition of interior surface. Any factor of safety must be estimated from local conditions and the requirements of each particular installation. It is recommended that for most commercial design purposes a safety factor of 15 to 20% be added to the values in the table.

Cameron also includes a separate set of tables with similar information regarding the friction loss of water flows in smooth tubing (roughness equal to 0.000005 ft) and for copper, brass, plastic, and glass piping for the various standard inside diameters. Table 7.4, showing data for 4-in tubing and smooth pipe, is typical of the tables for nominal diameters which range from $\frac{3}{8}$ to 12 in.

A third and perhaps the most important set of tables contained in the *Cameron Hydraulic Data Book* consists of those entitled "Friction Loss for Viscous Liquids." An example, for new 4-in steel pipe, is shown as Table 7.5. The initial portion of these tables relates friction loss* to volume flow rate as a function of a series of kinematic viscosities ranging from 0.6 cSt (31 SSU) to 650 cSt (3000 SSU) for nominal pipe sizes spanning 1- through 48-in diameter. Sizes from 1 through 24 in are reported as Sch. 40, while Sch. 30 is used for 30 and 36 in and inside diameters for 42 and 48 in.

The term "viscous liquids" as used in these tables does not necessarily mean that the values reported are in the viscous, or laminar, zone of flow. On the contrary, a distinction is made in the tables between flows in the transition regime† and those in the laminar area by shading friction losses for combinations of flow and viscosity in the various pipe-size tables where the Reynolds number is 2100 or less. This differentiation between the two zones for any table permits easy interpolation between pairs of points on the table for conditions other than those represented on that table.

A supplementary pair of tables is given in the *Cameron Hydraulic Data Book* to cover the laminar-flow regime, i.e., viscosities from 4000 SSU (863 cSt) to 20,000 SSU (4315 cSt) for flow rates of 5 to 3000 gal/min in Sch. 40 steel pipe ranging from $1\frac{1}{4}$ to 18 in in diameter. An example of the table for the various flow and diameter ranges is given in Table 7.6. Because the friction-loss points represented in these two tables in the Cameron book lie essentially in the laminar zone, they all appear shaded.

In the shaded- or laminar-zone area in the tables represented by Tables 7.5 and 7.6, the interpolation for intermediate conditions of flow rate and viscosity can be eyeballed as an arithmetic proportion. A more exact friction loss can be obtained by correcting a given friction loss from a nearby point as follows:

$$\Delta H_2 = \Delta H_1 \times \frac{(\text{gal/min})_2}{(\text{gal/min})_1} \times \frac{z_2}{z_1} \tag{7.7}$$

where ΔH = friction loss, ft of fluid
 z = kinematic viscosity, cSt

For flow rates and viscosities that lie in the unshaded area but adjacent to a condition in the shaded area, one should first determine the Reynolds number to ascertain whether or not the new condition is still in the laminar region (Re $\leq$ 2100).

Shortcut equations. There are several simplifying equations which are shortcuts for calculating fluid velocities and obtaining Reynolds numbers from the various param-

---

*Most nomographs and tables, including the others in the *Cameron Hydraulic Data Book*, report friction loss in terms of 100 ft of pipe length. However, attention must be paid to the fact that the Cameron viscous-liquids tables use 1000 ft as the basic length.

†None of the points in the viscous-liquids tables, of which Table 7.5 is an example, is in the turbulent region with steel pipe. Similarly, none of the points in the friction-loss tables for water of which Tables 7.3 and 7.4 are examples is in the laminar or turbulent zone.

eters that are commonly used and are generally available from the literature. These equations with the appropriate coefficients for the dimensions used are as follows:

$$u = \frac{0.408 \times (\text{gal/min})}{d^2} \tag{7.8}$$

and

$$Re = \frac{3162 \times (\text{gal/min}) \times s}{d \times \mu} \tag{7.9a}$$

or

$$Re = \frac{3162 \times (\text{gal/min})}{d \times z} \tag{7.9b}$$

where u = velocity, ft/s
Re = Reynolds number, dimensionless
d = inside or equivalent diameter, in
μ = absolute viscosity, cP
z = kinematic viscosity, cSt
s = specific gravity (compared with water at 60°F)

Interpolations and extrapolations. Interpolations between adjacent points in the transition areas or unshaded portions of the Cameron viscous-liquids tables can also be eyeballed, but special care must be taken. In the transition zone, the interpolation factors are such that the flow rates are raised to the 1.82 power while the viscosity is raised only to the 0.14 power. This means that the eyeballed interpolations should be skewed toward the lower flow-rate point when flow is the only variable; i.e., there is a changed rate for the same fluid. When a viscosity change is involved, there is essentially no skewing between adjacent viscosity points.

Interpolation by calculation in the transient zone can be obtained by the following relationship:

$$\Delta H_2 = \Delta H_1 \times \left[\frac{(\text{gal/min})_2}{(\text{gal/min})_1}\right]^{1.82} \times \left(\frac{z_2}{z_1}\right)^{0.14} \tag{7.10}$$

It should be pointed out that the interpolation can be approximated by using the squares of the flow rates instead of the 1.82 power if there is no change or only a small change in viscosity. The friction loss so determined deviates by no more than 3 percent of its actual value. This deviation is well within normal engineering tolerances, and the procedure is often worth the convenience afforded.

Using liquid tables for gases. The Cameron viscous-liquids tables are easily adapted for use with gases since the units of friction loss are in feet of fluid. This measurement can be related to volume flow rate and fluid kinematic viscosity regardless of whether the fluid is a liquid or a gas. Gas flows are normally expressed in terms of standard cubic feet per minute (scfm) or in pounds per hour.* In the former case, standard

---

*A convenient engineering number to remember is that there are 379 standard cubic feet in a pound-mole of any gas at 1-atm pressure and 60°F.

TABLE 7.4 Friction of Water: 4-in Copper Tubing and Standard-Pipe-Size Copper and Brass Pipe*

Flow, U.S. gal/min	Type K tubing 3.857-in ID; 0.134-in wall thickness		Type L tubing 3.905-in ID; 0.110-in wall thickness		Type M tubing 3.935-in ID; 0.095-in wall thickness		Pipe 4.000-in ID; 0.250-in wall thickness	
	Velocity, ft/s	Head loss, ft/100 ft	Velocity, ft/s	Head loss, ft/100 ft	Velocity, ft/s	Head loss, ft/100 ft	Velocity, ft/s	Head loss, ft/100 ft
100	2.74	0.72	2.68	0.68	2.64	0.65	2.55	0.60
110	3.02	0.85	2.94	0.80	2.90	0.77	2.81	0.71
120	3.29	0.99	3.21	0.94	3.16	0.90	3.06	0.83
130	3.57	1.15	3.48	1.08	3.42	1.04	3.31	0.96
140	3.84	1.31	3.74	1.23	3.69	1.19	3.57	1.10
150	4.11	1.48	4.01	1.40	3.95	1.35	3.83	1.25
160	4.39	1.67	4.28	1.57	4.21	1.51	4.08	1.39
170	4.66	1.86	4.55	1.75	4.48	1.69	4.33	1.56
180	4.94	2.06	4.81	1.94	4.74	1.87	4.58	1.73
190	5.21	2.27	5.08	2.14	5.00	2.06	4.84	1.91
200	5.49	2.49	5.35	2.35	5.27	2.26	5.10	2.09
220	6.04	2.96	5.89	2.79	5.80	2.68	5.61	2.48
240	6.59	3.46	6.42	3.26	6.32	3.14	6.12	2.90
260	7.14	4.00	6.95	3.77	6.85	3.63	6.63	3.36
280	7.69	4.57	7.49	4.31	7.38	4.15	7.14	3.84

300	8.24	5.18	8.02	4.88	7.90	4.70	7.65	4.35
350	9.60	6.85	9.36	6.46	9.22	6.22	8.92	5.75
400	11.0	8.74	10.7	8.23	10.5	7.93	10.2	7.33
450	12.4	10.83	12.0	10.20	11.9	9.83	11.5	9.08
500	13.7	13.12	13.4	12.36	13.2	11.91	12.8	11.00
550	15.1	15.61	14.7	14.71	14.5	14.17	14.1	13.09
600	16.5	18.31	16.0	17.24	15.8	16.61	15.3	15.35
650	17.9	21.19	17.4	19.96	17.1	19.23	16.6	17.77
700	19.2	24.28	18.7	22.86	18.4	22.03	17.9	20.35
750	20.6	27.55	20.1	25.95	19.8	25.00	19.1	23.09
800	22.0	31.01	21.4	29.21	21.1	28.14	20.4	25.99
850	23.3	34.67	22.8	32.65	22.4	31.46	21.7	29.05
900	24.7	38.51	24.1	36.27	23.7	34.94	23.0	32.27
950	26.1	42.54	25.4	40.06	25.0	38.60	24.2	35.64
1000	27.4	46.76	26.8	44.03	26.4	42.42	25.5	39.17
1100	30.2	55.74	29.4	52.48	29.0	50.56	28.1	46.69
1200	32.9	65.45	32.1	61.62	31.6	59.37	30.6	54.82
1300	35.7	75.89	34.8	71.45	34.2	68.83	33.1	63.55
1400	38.4	87.05	37.4	81.95	36.9	78.95	35.7	72.89
1500	41.1	98.23	40.1	93.13	39.5	89.71	38.3	82.82
1600	43.9	111.5	42.8	105.0	42.1	101.1	40.8	93.34
1800	49.4	138.8	48.1	130.6	47.4	125.8	45.8	116.1
2000	54.9	168.9	53.5	158.9	52.7	153.1	51.0	141.3
2200	60.4	201.7	58.9	189.8	58.0	182.8	56.1	168.7

*Based on Darcy's formula. Reprinted with permission from *Cameron Hydraulic Data Book*, copyright © Ingersoll-Rand Company, 1981.

NOTE: No allowance has been made for age, difference in diameter, or any abnormal condition of interior surface. Any factor of safety must be estimated from local conditions and the requirements of each particular installation. It is recommended that for most commercial design purposes a safety factor of 15 to 20% be added to the values in the table.

TABLE 7.5 Friction Loss for Viscous Liquids: Loss in Feet of Liquid per 1000 Ft of Pipe; 4-in (4.026-in-ID) Sch. 40 New Steel Pipe*

Flow		Kinematic viscosity, cSt								
U.S. gal/min	Bbl/h (42 gal)	0.6	1.1	2.1	2.7	4.3	7.4	10.3	13.1	15.7
					Approximate SSU viscosity					
			31.5	33	35	40	50	60	70	80
20	28.6	.30	.34	.40	.43	.49	.57	.50	.63	.75
30	42.9	.62	.70	.82	.87	.98	1.14	1.25	.95	1.13
40	57.1	1.05	1.18	1.35	1.44	1.62	1.86	2.04	2.20	1.51
50	71.4	1.58	1.76	2.02	2.13	2.37	2.75	3.00	3.21	3.40
60	85.7	2.22	2.44	2.80	2.93	3.27	3.77	4.12	3.29	4.62
70	100	2.96	3.24	3.69	3.88	4.31	4.93	5.39	5.72	6.03
80	114	3.79	4.16	4.67	4.93	5.44	6.20	6.78	7.23	7.55
90	129	4.72	5.15	5.77	6.08	6.72	7.63	8.32	8.87	9.29
100	143	5.77	6.27	6.91	7.33	8.12	9.15	9.97	10.6	11.2
120	171	8.09	8.81	9.66	10.2	11.2	12.7	13.6	14.6	15.3
140	200	10.8	11.7	12.9	13.4	14.8	16.6	17.9	19.0	20.0
160	228	13.9	15.0	16.4	17.1	18.8	21.1	22.7	24.0	25.2
180	257	17.4	18.7	20.4	21.5	23.2	26.0	28.1	29.6	30.8
200	286	21.4	22.7	24.9	25.9	28.0	31.4	33.7	35.7	36.9
220	314	25.6	27.2	29.8	30.8	33.2	37.3	40.2	42.3	44.0
240	343	30.3	32.0	34.9	36.1	38.8	43.4	46.8	49.1	51.4
260	371	35.4	37.2	40.4	42.0	45.0	50.1	53.9	56.7	59.3
280	400	40.8	42.7	46.4	48.2	51.7	57.4	61.6	65.0	67.4
300	429	46.6	48.7	53.0	54.8	58.8	64.9	69.6	73.3	76.0
350	500	62.7	65.6	70.6	72.8	77.9	85.4	91.2	96.2	101
400	571	81.4	84.7	90.4	93.7	99.8	109	116	122	127
450	643	102	106	113	117	124	135	144	151	157
500	714	125	130	137	142	151	164	174	182	189
550	766	151	157	165	170	180	195	206	216	224
600	857	179	185	195	200	212	229	242	253	263
650	929	209	216	228	231	246	266	280	291	303
700	1000	242	249	260	267	283	306	322	333	346
750	1070	276	285	298	305	321	348	366	377	391
800	1140	314	324	337	345	362	392	411	426	439
850	1215	355	364	378	387	406	438	459	476	489
900	1285	396	408	424	434	453	486	510	531	543
950	1360	441	451	470	481	502	536	563	584	600
1000	1430	488	500	521	527	550	591	621	641	662
1100	1570	587	602	627	634	659	708	740	765	790
1200	1715	699	712	741	754	780	835	869	898	924

*Based on Darcy's formula. Reprinted with permission from *Cameron Hydraulic Data Book*, copyright © Ingersoll-Rand Company, 1981.

NOTE: For this pipe size, $u = 0.0252 \times$ gal/min; velocity head $= 9.858 \times 10^{-6}$ (gal/min)^2. Figures in shaded area are laminar (viscous) flow. Loss in lb/in^2 = 0.433 (sp gr) (figures from table). No allowance has been made for age,

cubic feet per minute can be converted to actual flow in cubic feet per minute (acfm) by employing the appropriate pressure and absolute-temperature ratios. Gas flow in a nonstandard form, i.e., gallons per minute, can be obtained by multiplying actual cubic feet per minute by the factor 7.48, the number of gallons in a cubic foot.

				Kinematic viscosity, cSt					
20.6	26.4	32.0	43.2	65.0	108.4	162.3	216.5	325	435	650
				Approximate SSU viscosity					
100	125	150	200	300	500	750	1000	1500	2000	3000
.99	.95	1.15	1.55	2.34	3.91	5.85	7.80	11.7	15.7	23.4
1.48	1.27	1.54	2.07	3.12	5.21	7.80	10.4	15.6	20.9	31.2
1.96	1.90	2.30	3.11	4.68	7.82	11.7	15.6	23.4	31.3	46.8
2.47	2.54	3.08	4.15	6.25	10.4	15.6	20.8	31.2	41.8	62.5
5.01	3.17	3.84	5.18	7.81	13.0	19.5	26.0	39.0	52.2	78.1
6.53	3.80	4.61	6.22	9.37	15.6	23.4	31.2	46.8	62.7	93.7
8.17	4.44	5.38	7.25	10.9	18.2	27.3	36.4	54.6	73.2	109
10.0	8.81	6.15	8.29	12.5	20.8	31.2	41.6	62.4	83.6	125
12.0	10.8	11.3	9.33	14.1	23.4	35.1	46.8	70.2	94.1	141
16.5	12.9	13.7	10.4	15.6	26.0	39.0	52.0	78.0	105	156
21.6	17.6	18.6	20.3	18.8	31.2	46.8	62.4	93.7	125	187
27.2	22.9	24.3	26.5	21.9	36.4	54.6	72.8	109	146	218
33.3	29.0	30.3	33.2	25.0	41.7	62.4	83.2	125	167	250
40.0	35.5	37.4	40.7	45.7	46.9	70.2	93.6	140	188	281
47.0	42.6	45.0	48.7	54.8	52.1	78.0	104	156	209	312
54.6	50.3	53.0	57.1	64.7	57.3	85.8	114	172	230	343
62.7	58.5	61.5	65.1	74.7	62.5	93.6	125	187	251	375
71.3	67.2	70.8	76.8	85.7	67.7	101	135	203	272	406
81.0	76.4	80.5	87.2	97.3	73.0	109	146	218	292	437
107	85.8	90.8	98.5	110	127	117	156	234	313	468
135	98.5	104	113	125	146	127	169	254	340	508
167	112	118	128	143	166	136	182	273	366	547
201	127	133	145	161	187	146	195	293	392	585
239	143	149	162	180	208	156	208	312	418	625
278	178	184	198	222	254	285	234	351	470	703
319	213	221	237	265	305	343	260	390	523	781
365	252	263	280	313	360	404	286	429	575	860
414	296	305	328	364	417	467	507	468	627	937
465	338	353	378	419	480	528	583	507	680	
519	386	402	433	474	546	608	663	546	732	
574	437	455	488	533	616	685	745	585	784	
632	490	510	546	570	687	764	830	624	836	
694	544	570	608	663	763	848	920	663	889	
822	603	629	674	739	844	939			941	
966	666	696	743	813	927				993	

difference in diameter, or any abnormal conditon of interior surface. Any factor of safety must be estimated from local conditions and the requirements of each particular installation. It is recommended that for most commercial design purposes a safety factor of 15 to 20% be added to the values in the table.

The velocity of any gas flow is analogous to the velocity for liquid flow [Eq. (7.8)] and can be calculated from the expression

$$u = \frac{3.05 \times \text{acfm}}{d^2} \quad (7.11)$$

TABLE 7.6 Friction Loss for Viscous Liquids (4000 to 20,000 SSU); Loss in Feet of Liquid per 1000 ft of Pipe, 1¼- to 6-in Pipe Sizes, Sch. 40*

Flow, U.S. gal/min	Nominal pipe size, in	Velocity of flow, ft/s	Kinematic viscosity, cSt								
			863	1079	1295	1726	2158	2589	3021	3452	4315
			Approximate SSU viscosity								
			4000	5000	6000	8000	10000	12000	14000	16000	20000
5	1¼	1.07	749	937	1125	1499	1874	2248	2623	2998	3747
	1½	0.79	405	506	607	809	1011	1214	1416	1618	2022
	2	0.48	149	186	223	298	372	447	521	596	744
10	1¼	2.15	1499	1874	2249	2998	3748	4496	5247	5995	7494
	1½	1.58	809	1011	1214	1618	2023	2427	2832	3236	4045
	2	0.95	298	372	447	595	745	893	1042	1191	1489
15	1½	2.36	1213	1517	1821	2427	3035	3641	4248	4854	6068
	2	1.43	447	558	670	893	1117	1340	1564	1787	2233
	2½	1.01	219	274	329	439	549	658	768	877	1097
20	1½	3.15	1618	2023	2428	3236	4046	4854	5664	6472	8090
	2	1.91	596	745	894	1191	1489	1787	2085	2382	2978
	2½	1.34	292	366	440	585	731	878	1024	1170	1463
25	2	2.39	744	931	1117	1489	1862	2233	2606	2978	3722
	2½	1.68	367	458	549	731	914	1098	1280	1463	1828
	3	1.08	153	192	230	307	383	460	537	613	767
30	2	2.87	893	1117	1341	1787	2234	2680	3127	3573	4467
	2½	2.01	440	549	659	879	1095	1315	1536	1755	2194
	3	1.30	184	230	276	368	461	552	643	737	920
40	2	3.82	1191	1489	1787	2382	2979	3573	4170	4764	5956
	2½	2.68	585	732	878	1170	1463	1755	2048	2340	2926
	3	1.74	245	307	368	491	613	736	859	981	1227

50	2	4.78	1489	1862	2234	2978	3723	4467	5212	5956	7445
	2½	3.35	731	914	1098	1463	1829	2194	2560	2926	3657
	3	2.17	307	384	460	614	767	920	1073	1227	1534
60	2	5.73	1786	2234	2681	3573	4468	5360	6255	7147	8934
	2½	4.02	878	1097	1317	1755	2195	2633	3072	3511	4388
	3	2.60	368	460	552	736	921	1104	1289	1473	1840
70	2½	4.69	1024	1280	1537	2048	2560	3072	3584	4096	5120
	3	3.04	429	537	644	859	1074	1288	1503	1718	2147
	4	1.76	144	181	217	290	362	435	507	579	724
80	2½	5.36	1170	1463	1756	2340	2926	3511	4097	4681	5851
	3	3.47	491	614	737	982	1227	1473	1718	1963	2454
	4	2.02	166	207	248	331	414	497	579	662	828
90	2½	6.03	1317	1646	1976	2633	3292	3950	4609	5266	6583
	3	3.90	552	690	829	1104	1381	1657	1933	2209	2761
	4	2.27	186	233	279	372	466	559	652	745	931
100	2½	6.70	1462	1829	2195	2926	3658	4388	5121	5851	7314
	3	4.34	614	767	921	1227	1534	1841	2148	2454	3068
	4	2.52	207	259	310	414	517	621	724	828	1035
125	3	5.42	766	959	1151	1534	1918	2301	2685	3068	3835
	4	3.15	259	323	388	517	647	776	905	1035	1293
	6	1.39	50	63	75	100	126	151	176	201	251
150	3	6.51	920	1151	1381	1841	2301	2761	3222	3681	4602
	4	3.78	310	388	466	621	776	931	1086	1241	1552
	6	1.67	60	75	90	121	151	181	211	241	301

*Based on Darcy's formula; laminar flow–figures suitable for any interior roughness. Reprinted with permission from *Cameron Hydraulic Data Book*, copyright © Ingersoll-Rand Company, 1981.

NOTE: No allowance has been made for age, difference in diameter, or any abnormal condition of interior surface. Any factor of safety must be estimated from local conditions and the requirements of each particular installation. It is recommended that for most commercial design purposes a safety factor of 15 to 20% be added to the values in the table.

Similarly, the Reynolds number for gas flow can be expressed as

$$\text{Re} = \frac{23{,}650 \times \text{acfm} \times \rho}{d \times \mu \times 62.4} = \frac{379 \times \text{acfm} \times \rho}{d \times \mu} \qquad (7.12)$$

where ρ = density, lb/ft³.

Although gases have low absolute viscosities, their kinematic viscosities (absolute viscosity in centipoises divided by specific gravity) are comparable to those of liquids. For example, the kinematic viscosity of 100-psig air at 60°F is 2.2 cSt, that of saturated steam at 20 psig is 15 cSt, and that of hydrogen at ambient conditions of temperature and pressure is 110 cSt. It is important to remember that, like liquid viscosities, gas viscosities are a function of temperature; however, they increase directly with temperature increases, while those of liquids are reduced.

The gallons per minute that represents gas flow will usually be of an order of magnitude or more greater than the maximum liquid gallons per minute reported in the Cameron tables of friction loss for viscous liquids for any given pipe size. Nevertheless, a close approximation of the friction loss for the gaseous flow can be made by noting the head loss for the largest value of gallons per minute in the liquid table at the kinematic viscosity nearest that of the gas and then scaling up for the gas value of gallons per minute. Since the Reynolds numbers for these gas-flow points approach the turbulent zone, it is best to adjust liquid friction head loss in the transient zone by squaring the flow rates and raising the viscosity-ratio correction to the 0.14 power in accordance with the equation

$$\Delta H_G = \Delta H_L \left(\frac{7.49 \times \text{acfm}_G}{(\text{gal/min})_L} \right)^2 \left(\frac{z_G}{z_L} \right)^{0.14} \qquad (7.13)$$

Crane tables for air. Technical Paper 410, prepared by the Crane Company,* contains separate pressure-drop tables for water and air. The table for water is similar to the Cameron friction-loss tables, except that Crane uses only Sch. 40 pipe and presents loss as a pressure drop (pounds per square inch) per 100 ft of pipe. The air table shown as Table 7.7 not only is useful by itself but also can serve as the initial point for calculating friction losses for other gases. The Crane air table presents pressure drops in pounds per square inch per 100 ft of length for 100-psig air at 60°F at discrete flows from 1 to 30,000 acfm in Sch. 40 new steel piping ranging from $\tfrac{1}{8}$- to 12- in in diameter. The interpolation between flow points is easily accomplished by using the square of the flow ratios.

To extrapolate to air at other conditions of temperature and pressure, it is necessary to correct for the change in the specific volume or density of the air. This is required since the friction loss in the Crane table is given in terms of pressure drop rather than in fluid head loss. In addition, a change in temperature means a change in absolute viscosity, while a change in pressure always is reflected as a change in kinematic viscosity unless there is a compensating change in temperature. Thus, the equation for the extrapolation of an airflow pressure drop in which the weight rate remains constant (i.e., the standard cubic feet per minute is the same but temperature and pres-

---

Flow of Fluids through Valves, Fittings and Pipe, Technical Paper 410, 1982, Crane Company, Chicago, Ill. 60632.

sure conditions differ) in the same-sized pipe is as follows:

$$\Delta P_2 = \Delta P_1 \left(\frac{114.7}{P_2}\right) \left(\frac{t_2 + 460}{520}\right) \left(\frac{z_2}{0.224}\right)^{0.14} \quad (7.14)$$

where ΔP_2 = extrapolated pressure drop, psi/100 ft
ΔP_1 = initial pressure drop, psi/100 ft
P_2 = new pressure, psia
t_2 = new temperature, °F
z_2 = new viscosity, cSt

However, should one wish to represent an entirely different actual volume flow rate or express the original flow in terms of actual volume flow rate, then the relationship is expressed as

$$\Delta P_2 \cong \Delta P_1 \left(\frac{\text{acfm}_2}{\text{acfm}_1}\right)^2 \left(\frac{114.7}{P_2}\right) \left(\frac{t_2 + 460}{520}\right) \left(\frac{z_2}{0.224}\right)^{0.14} \quad (7.15)$$

where acfm_2 = actual volume rate at new conditions
acfm_1 = actual volume rate at initial conditions

Should the weight flow rate be used as the variable, Eq. (7.15) would then read

$$\Delta P_2 \cong \Delta P_1 \left(\frac{W_2}{W_1}\right)^2 \left(\frac{114.7}{P_2}\right) \left(\frac{t_2 + 460}{520}\right) \left(\frac{z_2}{0.224}\right)^{0.14} \quad (7.16)$$

where W_2 = new weight flow rate, lb/h
W_1 = initial weight flow rate, lb/h

Equations (7.15) and (7.16) can be adapted for use with other gases by inserting the multiplier MW/29 in the temperature-pressure term of Eq. (7.15) and 29/MW in that of Eq. (7.16), where MW is the molecular weight of the second gas and 29 is the molecular weight of air.

Pressure drops for both liquids and gases for pipe schedules other than Sch. 40 are approximated by the relationship

$$\Delta P_2 = \Delta P_1 \left(\frac{d_{\text{Sch. 40}}}{d_{\text{Sch.}}}\right)^{4.82} \quad (7.17)$$

Effect of specific gravity on friction loss. It should be recalled that the friction loss of a liquid, when expressed in terms of height of fluid, is a function of weight or volume flow rate to the 1.82 power and of kinematic viscosity to the 0.14 power. Specific gravity enters into consideration only when converting from liquid weight to liquid volume rates and from absolute viscosity to kinematic viscosity.

When the fluid is a gas, however, and the flow is given in terms of a weight rate, extrapolation of friction loss as height of fluid from one gas condition to another will be a function of the combination of weight ratio, absolute-temperature ratio, and inverse-pressure ratio squared along with the kinematic-viscosity ratio to the 0.14 power. If volume flow rates are being compared, the specific-gravity relationship between the two conditions as well as the pressure and temperature ratios may be ignored if friction loss is expressed as height of fluid.

On the other hand, specific gravity is an important consideration for extrapolations in terms of pressure drop since friction loss in height of fluid is converted to pressure

TABLE 7.7 Flow of Air through Sch. 40 Steel Pipe*

Pressure drop of air, psi/100 ft of Sch. 40 pipe for air at 100 psig and 60°F

Free air, q'_m, ft³/min at 60°F and 14.7 psia	Compressed air, ft³/min at 60°F and 100 psig	1/8"	1/4"	3/8"	1/2"	3/4"	1"	1 1/4"	1 1/2"	2"	2 1/2"	3"
1	0.128	0.361										
2	0.256	1.31										
3	0.384	3.06	0.083	0.018	0.020							
4	0.513	4.83	0.285	0.064	0.042							
5	0.641	7.45	0.605	0.133	0.071	0.027						
6	0.769	10.6	1.04	0.226	0.106	0.037						
8	1.025	18.6	1.58	0.343	0.148	0.062	0.019					
10	1.282	28.7	2.23	0.408	0.255	0.094	0.029					
15	1.922		3.89	0.848	0.356	0.201	0.062					
20	2.563		5.96	1.26	0.834	0.345	0.102	0.026				
25	3.204		13.0	2.73	1.43	0.526	0.156	0.039	0.019			
30	3.845		22.8	4.76	2.21	0.748	0.219	0.053	0.025			
35	4.486		35.6	7.34	3.15	1.00	0.293	0.073	0.035			
40	5.126			10.5	4.24	1.30	0.379	0.095	0.044			
45	5.767			14.2	5.49	1.62	0.474	0.116	0.055			
50	6.408			18.4	6.90	1.99	0.578	0.149	0.067	0.019		
60	7.690			23.1	8.49	2.85	0.819	0.200	0.094	0.027		
70	8.971			28.5	12.2	3.83	1.10	0.270	0.126	0.036		
80	10.25			40.7	16.5	4.96	1.43	0.350	0.162	0.046	0.019	
90	11.63				21.4	6.25	1.80	0.437	0.203	0.058	0.023	
100	12.82				27.0	7.69	2.21	0.534	0.247	0.070	0.029	
125	16.02				33.2	11.9	3.39	0.825	0.380	0.107	0.044	
150	19.22					17.0	4.87	1.17	0.537	0.151	0.062	0.021
175	22.43					23.1	6.60	1.58	0.727	0.205	0.083	0.028
200	25.63					30.0	8.54	2.05	0.937	0.264	0.107	0.036

Size				3½"	4"	5"		6"			
225	28.84	0.134	0.045	0.022			10.8		2.59	1.19	0.331
250	32.04	0.164	0.055	0.027			13.3		3.18	1.45	0.404
275	35.24	0.191	0.066	0.032			16.0		3.83	1.75	0.484
300	38.45	0.232	0.078	0.037			19.0		4.56	2.07	0.573
325	41.65	0.270	0.090	0.043			22.3		5.32	2.42	0.673
350	44.87	0.313	0.104	0.050	0.030		25.8		6.17	2.80	0.776
375	48.06	0.356	0.119	0.057	0.034		29.6		7.05	3.20	0.887
400	51.26	0.402	0.134	0.064	0.038		33.6		8.02	3.64	1.00
425	54.47	0.452	0.151	0.072	0.042	37.9	37.9		9.01	4.09	1.13
450	57.67	0.507	0.168	0.081	0.042	⋮	⋮		10.2	4.59	1.26
475	60.88	0.562	0.187	0.089	0.047	⋮	⋮		11.3	5.09	1.40
500	64.08	0.623	0.206	0.099	0.052	⋮	⋮		12.5	5.61	1.55
550	70.49	0.749	0.248	0.118	0.062	⋮	⋮		15.1	6.79	1.87
600	76.90	0.887	0.293	0.139	0.073	⋮	⋮		18.0	8.04	2.21
650	83.30	1.04	0.342	0.163	0.086	⋮	⋮		21.1	9.43	2.60
700	89.71	1.19	0.395	0.188	0.099	0.032	24.3		24.3	10.9	3.00
750	96.12	1.36	0.451	0.214	0.113	0.036	27.9		27.9	12.6	3.44
800	102.5	1.55	0.513	0.244	0.127	0.041	31.8		31.8	14.2	3.90
850	108.9	1.74	0.576	0.274	0.144	0.046	35.9		35.9	16.0	4.40
900	115.3	1.95	0.642	0.305	0.160	0.051	40.2		40.2	18.0	4.91
950	121.8	2.18	0.715	0.340	0.178	0.057		0.023	⋮	20.0	5.47
1,000	128.2	2.40	0.788	0.375	0.197	0.063		0.025	⋮	22.1	6.06
1,100	141.0	2.89	0.948	0.451	0.236	0.075		0.030	⋮	26.7	7.29
1,200	153.8	3.44	1.13	0.533	0.279	0.089		0.035	⋮	31.8	8.63
1,300	166.6	4.01	1.32	0.626	0.327	0.103		0.041	⋮	37.3	10.1
1,400	179.4	4.65	1.52	0.718	0.377	0.119		0.047			11.8
1,500	192.2	5.31	1.74	0.824	0.431	0.136		0.054			13.5
1,600	205.1	6.04	1.97	0.932	0.490	0.154		0.061			15.3
1,800	230.7	7.65	2.50	1.18	0.616	0.193		0.075			19.3
2,000	256.3	9.44	3.06	1.45	0.757	0.237		0.094	0.023		23.9

TABLE 7.7 Flow of Air through Sch. 40 Steel Pipe* *(continued)*

Free air, q'_m, ft³/min at 60°F and 14.7 psia	Compressed air, ft³/min at 60°F and 100 psig	\multicolumn{9}{c}{Pressure drop of air, psi/100 ft of Sch. 40 pipe for air at 100 psig and 60°F}
		2½"	3"	3½"	4"	5"	6"	8"	10"	12"
2,500	320.4	14.7	4.76	2.25	1.17	0.366	0.143	0.035	0.016	
3,000	384.5	21.1	6.82	3.20	1.67	0.524	0.204	0.051	0.022	
3,500	448.6	28.8	9.23	4.33	2.26	0.709	0.276	0.068	0.028	
4,000	512.6	37.6	12.1	5.66	2.94	0.919	0.358	0.088	0.035	
4,500	576.7	47.6	15.3	7.16	3.69	1.16	0.450	0.111		
5,000	640.8		18.8	8.85	4.56	1.42	0.552	0.136	0.043	0.018
6,000	769.0		27.1	12.7	6.57	2.03	0.794	0.195	0.061	0.025
7,000	897.1		36.9	17.2	8.94	2.76	1.07	0.262	0.082	0.034
8,000	1025			22.5	11.7	3.59	1.39	0.339	0.107	0.044
9,000	1153			28.5	14.9	4.54	1.76	0.427	0.134	0.055
10,000	1282			35.2	18.4	5.60	2.16	0.526	0.164	0.067
11,000	1410				22.2	6.78	2.62	0.633	0.197	0.081
12,000	1538				26.4	8.07	3.09	0.753	0.234	0.096
13,000	1666				31.0	9.47	3.63	0.884	0.273	0.112
14,000	1794				36.0	11.0	4.21	1.02	0.316	0.129
15,000	1922					12.6	4.84	1.17	0.364	0.148
16,000	2051					14.3	5.50	1.33	0.411	0.167
18,000	2307					18.2	6.96	1.68	0.520	0.213
20,000	2563					22.4	8.60	2.01	0.642	0.260
22,000	2820					27.1	10.4	2.50	0.771	0.314
24,000	3076					32.3	12.4	2.97	0.918	0.371
26,000	3332					37.9	14.5	3.49	1.12	0.435
28,000	3588						16.9	4.04	1.25	0.505
30,000	3845						19.3	4.64	1.42	0.520

*With permission from Crane Company, New York.
†Values in the table represent instantaneous pressure loss dP/dL at stated flow conditions. Values may be applied to length of line as long as the total pressure drop does not exceed 10% of initial pressure. Otherwise, pressure drop must be analyzed according to methods given in Chap. 8.

drop by multiplying by a density factor. Thus, interpolations or extrapolations utilizing weight flow rates for gases require that the weight ratio be raised to the 1.82 power, multiplied by the combination of the pressure ratio and the inverse ratio of the absolute temperatures and by the kinematic-viscosity ratio raised to the 0.14 power.

Pipe flow diagrams

The information contained in the series of friction-loss tables such as Table 7.3 for water in clean steel pipes is often condensed into a graphical representation as a pipe flow diagram such as Fig. 7.6. The chart is entered at the volume flow rate, and the friction loss usually is found in terms of pressure drop per 100 ft of pipe as a function of pipe diameter. Curves correlating velocity with flow rate are frequently included on the graphs.

Manufacturers of relevant equipment often include diagrams similar to Fig. 7.6 for water and standard clean-steel-pipe sizes in the engineering-data sections of their catalogs. Producers of specialty piping such as glass, plastic, rubber, flexible, or lined pipe often utilize pipe flow diagrams for presenting pressure-drop-modulus information on their products in their literature. In most cases there is only one inside diameter for any nominal-size pipe in a particular product line. If a product is available in a series of inside diameters, the manufacturer will usually place a series of curves for each nominal size on the graph, present separate graphs, or advise that the diameter-affinity relationship be used to make the necessary correction to the modulus.

Flow charts which relate the friction-loss modulus to volume flow rate for a range of liquids other than water are sometimes available. In this instance, a separate graph is used for each nominal pipe size, usually for the standard schedules, while the parameter for the series of curves on each graph is fluid kinematic viscosity. Figure 7.7 is representative of this type of diagram. In this case, a clean steel Sch. 40 4-in pipe is the basis for friction loss expressed in feet of fluid per 100 ft of pipe for flow rates in gallons per minute. The viscosity parameter is given both in seconds Saybolt Universal and in centistokes.

The graph clearly delineates the laminar- and transient-flow zones. The narrow band of steep viscosity curves on the right-hand side of Fig. 7.7 represents the transient-flow area. One curve in this section has a viscosity equivalent to water at 60°F and is so identified. The curves with the lesser slopes on the left side of the graph represent flows in the laminar region, while the curve separating the laminar and transient zones essentially represents the critical-flow area.

Mnemonic devices

Frequently complex relationships can be reduced to a simple analogy to allow a speedy approximation of the required value. Two such mnemonic devices for fluid flow and pressure drop are presented here.

Approximate flow-diameter correlations. It was shown earlier that the pressure drop in a conduit varies approximately in proportion to the square of the velocity and inversely as the fifth power of the diameter. Economical flow rates in carbon steel lines and headers were calculated and summarized in Table 6.5. If a plot were to be made of these flow rates Q in gallons per minute versus nominal diameter d in inches

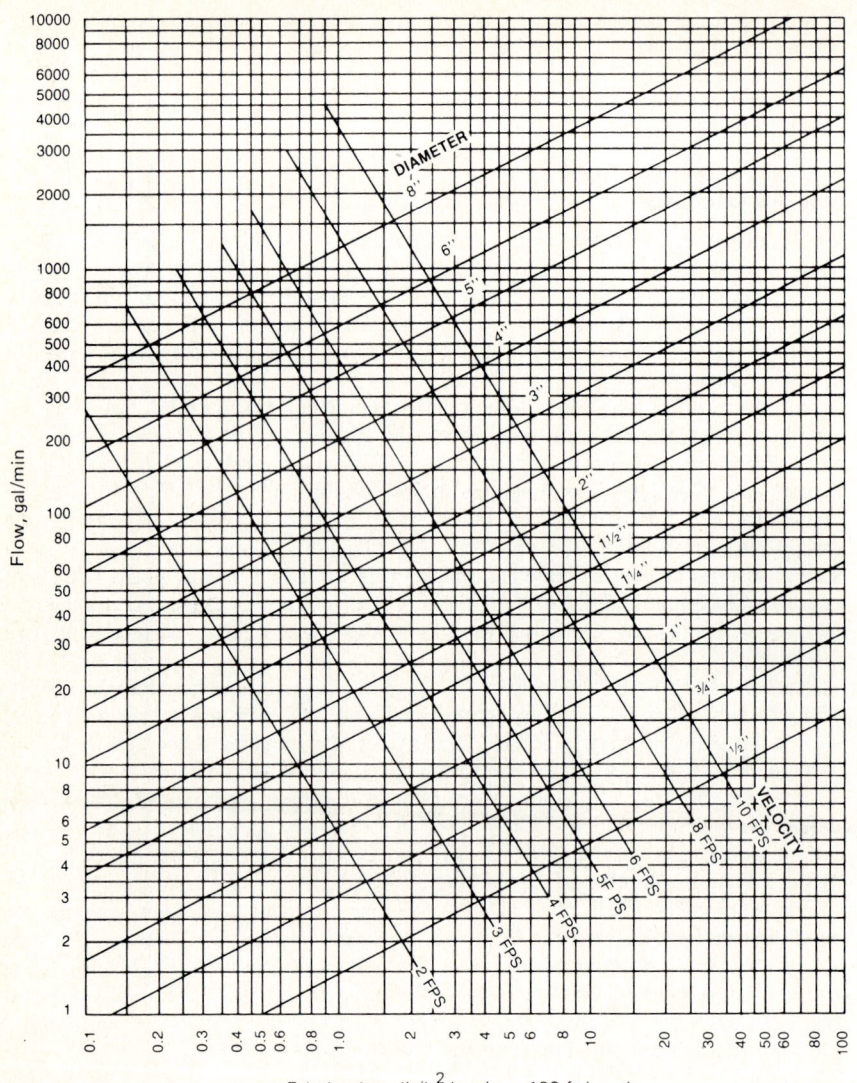

Figure 7.6 Flow diagram for water in clean steel pipe, $C_f = 140$. *(With permission of American Iron and Steel Institute, Washington, D.C.)*

to the 0.4, or 2/5, power, the slope of the curve could be about one-seventh. Thus the simple relationships

$$Q = 7d^{5/2} \tag{7.18}$$

and

$$d = \left(\frac{Q}{7}\right)^{2/5} \tag{7.19}$$

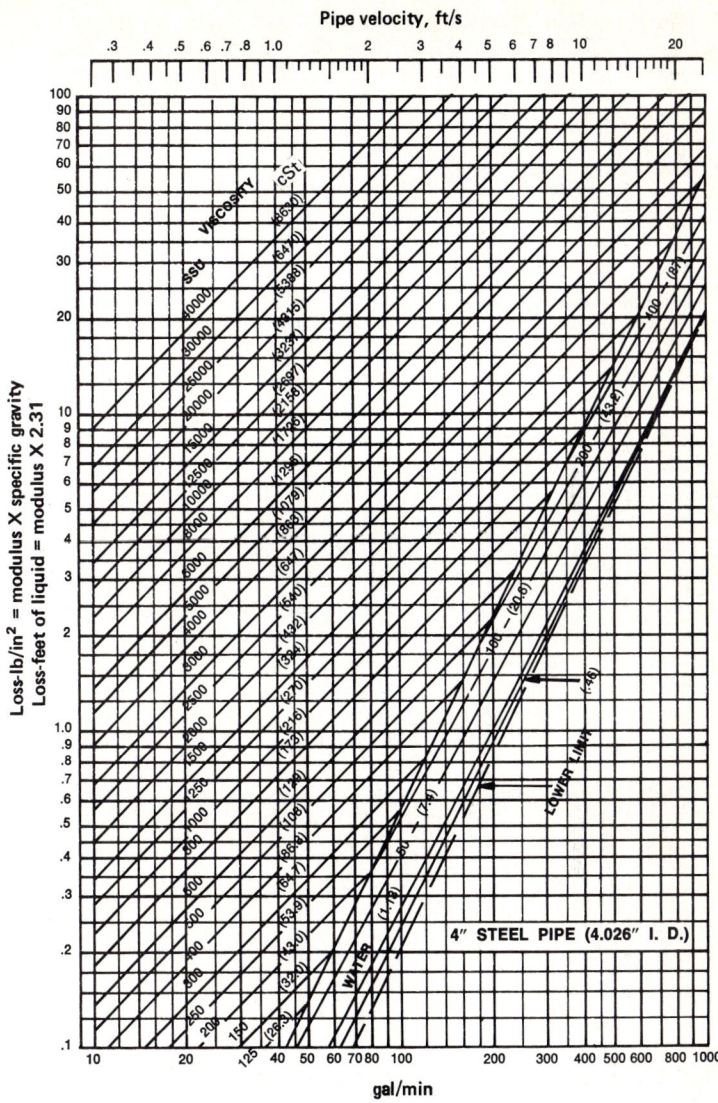

Figure 7.7 Friction-loss modulus for 100 ft of pipe. *(The Hydraulic Institute, Cleveland, Ohio.)*

can be used for quick approximations of liquid flows in various-sized headers or as the first guess for header size for a known flow rate of liquid.

The values obtained by means of Eqs. (7.18) and (7.19) correspond to water-pressure drops of about 2 psi/100 ft of commercial, somewhat-rusted steel pipe ($C_f = 100$). Flows in clean commercial pipe ($C_f = 140$) are about 40 percent greater than in somewhat-rusted pipe ($C_f = 100$) for the same pressure drop per unit length. Therefore, use of 10 instead of 7 in Eqs. (7.18) and (7.19) gives the corresponding flow rates and diameters for clean commercial pipe ($C_f = 140$).

A number to the 2.5 power is equal to the product of its square root and its square, while a number to the 0.4 power closely equals the square root of the product of the number's square root and its cube root. These relationships are useful if a calculator is not available.

The Rule of Fours for pressure drop. Using friction-loss tables and specialized slide rules is more convenient than finding pressure loss in an alignment pipe flow chart or determining the Reynolds number, finding the friction factor on a Moody chart, and then calculating friction loss by the Darcy relationship. Regardless of the ease of using the tables, there may be occasions when the tables are not available to the engineer. This might occur when the engineer is away from the necessary supports and is required to make an estimate of pressure drop on the spot without the use of a calculating aid. Most engineers would find it difficult, if not impossible, to make such a determination. However, if the engineer recognizes that friction loss is essentially a point function, it is necessary only to recall the pressure loss for a given set of conditions. The pressure loss for almost any other set of conditions can then be obtained by simple arithmetic calculations or approximations involving the friction-loss affinity laws.

A starting condition and one that is easily memorized is the so-called Rule of Fours: "The pressure drop of *four* hundred gallons per minute of water in a *four*-inch clean steel pipe ($C_f = 140$) is almost *four* pounds per square inch per hundred feet." When applied to any normal set of flow rates and pipe diameters, this set of conditions will give pressure drops within 10 percent accuracy when corrected by the affinity correlations developed earlier, as follows:

$$\Delta P_2 = 4 \left[\frac{(\text{gal/min})_2}{400} \right]^2 \left(\frac{4}{d_2} \right)^5 \left(\frac{s_2}{1.0} \right) \tag{7.20}$$

The 10 percent accuracy afforded by Eq. (7.20) can be improved considerably by applying a fourth four and redefining the initial flow rate as 400 plus *four* (4) percent, i.e., 416 gal/min.

Friction Factors for Other Than Clean Steel Pipe

It is evident from the earlier theoretical discussions that the degree of roughness of the wetted pipe or conduit surface greatly influences the friction loss in the transient- and turbulent-regime areas. Regardless of the surface of the pipe, eddy currents that induce inefficiencies do not manifest themselves until the Reynolds number approaches the range of 2100 to 4000. Once eddy currents are started, however, the greater the roughness of the surface, the more inefficiency or friction loss will be induced in the flow pattern.

Effect of roughness parameter

The effect of surface roughness on friction loss is easily demonstrated by comparing the Fanning friction factors in Fig. 7.1 at any Reynolds number above 4000 and assuming that the diameter portion of the roughness parameter (ϵ/D) remains constant. The minimum friction factor is that equivalent to smooth pipe, and it increases as the degree of roughness, as expressed by ϵ, increases.* The relationship between roughness and the Fanning friction factor is

$$\frac{f_2}{f_1} = \left(\frac{\epsilon_2}{\epsilon_1}\right)^{0.18} \tag{7.21}$$

The exponent of 0.18 is a reasonable value obtained from Table 7.1 which applies to most normal Reynolds numbers for the roughnesses generally encountered.

For materials of construction and for a fluid where no interaction between the two is expected, it is satisfactory to use the clean-roughness factor in developing the friction factors. In other instances, however, the roughness of a pipe can increase with time depending upon the use to which it has been put or on whether or not there has been interaction between the surface material and the fluids passing over it. For example, a hydrocarbon flowing in a steel pipe does not affect the roughness of the surface material, nor is there any change with water flowing through a stainless-steel or glass pipe. However, when water is pumped through steel, wrought-iron, or cast-iron pipe, sufficient corrosion or rusting takes place to alter the roughness of the surface. The roughness factor of slightly rusty steel pipe has been measured at 0.00044 ft, while that for old rusty pipe can be as great as 0.0030 to 0.0095 ft compared with 0.00015 ft for clean steel pipe. The friction loss for slightly rusty pipe is thus increased by about one-third, while that for old rusty pipe is almost 2 to 3 times as great as that for new clean pipe.

Roughness friction factor

Many engineers find remembering roughness values in terms of hundredths of a thousandth of a foot difficult and the arithmetic manipulation of these values awkward. Fortunately, there is an alternative index of roughness which facilitates considering this parameter when dealing with friction-loss tables or diagrams based on clean pipe in order to account for surfaces with different roughness characteristics. The index uses a series of numbers known as the Hazen and Williams roughness-friction-factor coefficients C_f to represent the various surfaces that are normally listed for piping or conduits.

The index is a direct representation of the relative flows in pipes of varying roughness with equal inside diameters and having the same pressure loss per given length. It uses an arbitrary value of 100 for steel pipe that has been in water service for a reasonable period of time. Comparisons are then made with other material surfaces such that the relationship between the Moody friction factors and the Hazen and Williams coefficients assigned to those materials is as follows:

$$\frac{f_2}{f_1} = \left(\frac{C_{f,1}}{C_{f,2}}\right)^{1.85} \tag{7.22}$$

The *Cameron Hydraulic Data Book* presents a listing of the C_f range, the average C_f value for clean pipe, and a recommended design number for commonly encountered materials as well as for the correction ratio $(100/C_f)^{1.85}$. This is presented as Table 7.8.

---

*For all practical purposes, the maximum roughness in piping is about 0.03 ft in riveted steel pipes. Thus, a roughness parameter equal to 0.05 for a 6-in riveted pipe is a realistic upper limit to the series of ϵ/D curves.

TABLE 7.8 Hazen and Williams Friction Factor C_f*

Type of pipe	Range: high = best, smooth, well laid; low = poor or corroded	Average value for clean new pipe	Commonly used value for design purposes
Cement, asbestos	160–140	150	140
Fiber	....	150	140
Bitumastic-enamel-lined iron or steel centrifugally applied	160–130	148	140
Cement-lined iron or steel centrifugally applied	....	150	140
Copper, brass, lead, tin, or glass pipe and tubing	150–120	140	130
Wood stave	145–110	120	110
Welded and seamless steel	150–80	130	100
Interior-riveted steel (no projecting rivets)	....	139	100
Wrought iron, cast iron	150–80	130	100
Tar-coated cast iron	145–50	130	100
Girth-riveted steel (projecting rivets in girth seams only)	....	130	100
Concrete	152–85	120	100
Full-riveted steel (projecting rivets in girth and horizontal seams)	....	110	100
Vitrified, spiral-riveted steel (flow with lap)	....	115	100
Spiral-riveted steel (flow against lap)	....	100	90
Corrugated steel	....	60	60

Values of C_f	150	140	130	120	110	100	90	80	70	60
Multipler (basis: $C_f = 100$)	0.47	0.54	0.62	0.71	0.84	1.0	1.22	1.50	1.93	2.57

*Reprinted with permission from *Cameron Hydraulic Data Book,* Copyright © Ingersoll-Rand Company, 1981.

It will be seen from Table 7.8 that the Hazen and Williams friction coefficient for clean steel pipe is 140 and the value of $(100/C_f)^{1.85}$ equals 0.537. Corrections may easily be made to the friction-loss moduli ΔH_f and ΔP for clean steel pipe that are found in the various diagrams, nomographs, slide rules, and tables that have been discussed previously according to the expression

$$\Delta H_{f,2} = \frac{1}{0.537}\left(\frac{100}{C_{f,2}}\right)^{1.85} \times \Delta H_{f,1} = 1.86 \left(\frac{100}{C_{f,2}}\right)^{1.85} \times \Delta H_{f,1} \qquad (7.23a)$$

and

$$\Delta P_2 = \frac{1}{0.537}\left(\frac{100}{C_{f,2}}\right)^{1.85} \times \Delta P_1 = 1.86\left(\frac{100}{C_{f,2}}\right)^{1.85} \times \Delta P_1 \qquad (7.23b)$$

with $(100/C_{f,2})^{1.85}$ being obtained by calculation or directly from Table 7.8.

Friction Loss in Fittings and Piping Components

When the Cameron, Crane, or similar tables of pipe flow diagrams are used, the friction-loss moduli are given in terms of discrete lengths of pipe, e.g., 100 or 1000 ft. Therefore, to calculate total friction loss in a given piping system it is necessary to determine the length of piping. The total length of straight pipe is relatively easy to determine from a piping takeoff or from an approximation of the routing of the pipe. However, in addition to the length of straight pipe, it is necessary to take into account the friction losses of the fluid as it passes through valves and pipe fittings, encounters sudden contractions and enlargements, or is subjected to entrance and exit losses. An accounting for these losses is especially required for nonlaminar flows. Friction losses through valves and fittings are usually much greater than would appear from their nominal length. This can be attributed to the excessive amounts of turbulence and, thus, inefficiency incurred when changes of directions or variations in velocity are imparted to the fluid. The following discussions permit prediction of the equivalent lengths of various components.

Resistance coefficient K

The friction loss caused by fittings or by the reduction or enlargement of pipe size can be determined by two methods. Experimental analysis has been used to relate friction loss ΔH_f for a fitting to velocity head $u^2/2g_c$, and, as a result, a resistance coefficient K has been developed for the various common categories of fittings and reducers or increasers such that

$$\Delta H_f = K\left(\frac{u^2}{2g_c}\right) \qquad (7.24a)$$

The values of K for the various sizes and types of fittings are given in the *Hydraulic Institute Engineering Data Book* based on data from Crane Company Technical Paper 410. The Ingersoll-Rand *Cameron Hydraulic Data Book* presents these materials in graphical form.

Equivalent lengths of fittings

Instead of obtaining the friction loss directly and separately for each fitting at each pipe size, the equivalent length of a fitting may be found by equating Eq. (7.24a) with Eq. (7.3), the Darcy friction-loss formula for fluids in nonlaminar flow, such that

$$\Delta H_f = K\left(\frac{u^2}{2g_c}\right) = f\left(\frac{L}{D}\right)\left(\frac{u^2}{2g_c}\right) \qquad (7.24b)$$

Thus, it can be seen that K, the resistance coefficient for a given type of fitting, divided by the Moody friction factor f is equal to the ratio of the "length" of the fitting divided by its inside diameter, or

$$\left(\frac{L}{D}\right) = \left(\frac{K}{f}\right) \tag{7.25}$$

The K for a given type of fitting is a direct function of the friction factor, so that L/D for that type of fitting is the same dimensionless constant for all pipe diameters.

The right ordinate of Fig. 7.2 gives the Moody friction factor for complete turbulent flow as a function of the inside pipe diameter for clean steel pipe. These factors are used with the K values found in the Hydraulic Institute or Crane data to determine L/D. The L/D value multiplied by the inside diameter* of the pipe† gives the so-called equivalent length of pipe corresponding to the resistance induced by the fitting.

The Cameron tables present the L/D constant in addition to the K values for a wide variety of fittings for common pipe sizes. The fittings are shown as pictorial representations to assist the user in selecting the fitting most nearly corresponding to the actual one to be incorporated in the piping.

Table 7.9 summarizes the L/D constants for the various fittings discussed in the *Cameron Hydraulic Data Book.* The constants for valves are based on full-port openings in clean steel pipe.

An interesting feature of the Cameron tables is the delineation of minimum flow velocity required by the various check valves and foot valves to lift their disks. These velocities, in feet per second, are appropriate constants divided by the square root of the fluid density in pounds per cubic foot. Should the velocity in the pipe exceed that value, the L/D constant may be applied directly in calculating the equivalent length of the fitting. If, on the other hand, the velocity in the pipe were to be less than the minimum velocity specified in the table, the L/D constant for the fitting would have to be corrected by a ratio such that

$$\left(\frac{L}{D}\right)_{corr} = \left(\frac{L}{D}\right)_{table} \left(\frac{\text{minimum velocity in table}}{\text{actual velocity in pipe}}\right)^2 \tag{7.26}$$

Several of the fittings have constrictions in their flow paths such that there is a minimum internal diameter of a reduced-flow area compared with the pipe diameter or cross-sectional area. In that event, the L/D constant must be corrected by a ratio of the respective areas squared such that

$$\left(\frac{L}{D}\right)_{corr} = \left(\frac{L}{D}\right)_{table} \left(\frac{a}{a_i}\right)^2 \tag{7.27a}$$

or

$$\left(\frac{L}{D}\right)_{corr} = \left(\frac{L}{D}\right)_{table} \left(\frac{d}{d_i}\right)^4 \tag{7.27b}$$

---

*The diameter for the conversion is usually taken as feet, but if K is divided by 12, then L/D has the dimension of equivalent feet per inch of diameter.

†Use of nominal pipe size instead of actual inside diameter is usually sufficient for most engineering purposes.

L/D values in laminar region

The L/D values included in Table 7.9 and those found in the Cameron and Crane literature are valid for flows in the transient and turbulent regions. Some authors suggest that these L/D values can be applied to the entire spectrum of Reynolds numbers. However, this practice could lead to erroneous conclusions, especially when calculating friction for a highly laminar flow in the suction line to a pump. It has been found good practice to modify L/D somewhat for use in the laminar region where Re < 1000. In this area it is suggested that equivalent length can be calculated according

TABLE 7.9 Equivalent Length-to-Diameter Ratios for Fittings*

Fitting (fully open)	L, equivalent length, ft		Remarks
	D, ft L/D	d, in L/d	
Gate valve	8	0.67	Correct for constriction
Globe valve	340	28	Correct for constriction
Angle valve	55	4.6	Plug or Y type; correct for constriction
Angle valve	150	12.5	Globe type; correct for constriction
Ball valve	3	0.25	Correct for constriction
Butterfly valve	45	3.8	Correct for constriction
Plug valve	18	1.5	Correct for constriction
Three-way plug valve	30	2.5	Through flow; correct for constriction
Three-way plug valve	90	7.5	Branch flow; correct for constriction
Standard tee	20	1.7	Through flow
Standard tee	60	5.0	Branch flow
Standard 45° ell	16	1.3	
Standard 90° ell	30	2.5	$r/d = 0.5$
Long-radius 90° ell	16	1.3	$r/d = 1.0$
90° bend	20	1.7	$r/d = 1.0$
90° bend	12	1.0	$r/d = 2.0$
90° bend	14	1.2	$r/d = 4.0$
90° bend	30	2.5	$r/d = 10$
90° bend	50	4.2	$r/d = 20$
30° miter bend	8	0.67	
45° miter bend	15	1.25	
60° miter bend	25	2.1	
90° miter bend	60	5.0	
Close return bend	50	1.2	
Stop check valve (vertical disk rise, straight flow)	400	33	Minimum velocity = $55/\sqrt{\rho}$; correct for constriction
Stop check valve (vertical disk rise, right-angle flow)	200	16.7	Minimum velocity = $75/\sqrt{\rho}$; correct for constriction
Stop check valve (disk at 45°, right-angle flow)	350	25	Minimum velocity = $60/\sqrt{\rho}$; correct for constriction
Swing check valve	50	12.2	Minimum velocity = $50/\sqrt{\rho}$; correct for constriction

*Adapted from *Hydraulic Institute Engineering Data Book.*

to the expression

$$\left(\frac{L}{D}\right)_{corr} = \frac{Re}{1000}\left(\frac{L}{D}\right)_{table} \quad (7.28)$$

However, in no case should the equivalent length so calculated be less than the actual length of the fitting.

Effect of surface roughness on *L/D* values

Theoretically, the L/D factors for the fittings in Table 7.9 should be corrected for variations in surface roughness for other than clean steel pipe. In practice, this is not normally required since the correction is usually relatively minor and the resistance of these fittings is often only a small part of the total "length" of the piping system. However, for large-diameter pipe or conduit with a roughness considerably different from that of clean steel, it is advisable to make use of the Hazen and Williams friction-factor coefficient C_f (see Table 7.8) for the correction. If a friction-loss modulus for clean steel is being used for the remainder of the piping, the correction is

$$\left(\frac{L}{D}\right)_{corr} = \left(\frac{L}{D}\right)_{table} \times \left(\frac{140}{C_f}\right)^{1.85} \quad (7.29)$$

Entrance, discharge, and pipe-size-change losses

The friction losses which occur at the entrance and discharge of vessels and other equipment, as well as in piping reductions or enlargements or in orifices, must be included in the design of any piping system. In these instances the friction loss is mainly a function of the fluid velocity and the configuration of the nozzle or pipe and, to a lesser extent, of the diameter of the piping. For nozzles and orifices, the friction loss is defined by Eq. (7.24a) and modified by a function which is the ratio of the diameter of the aperture in the nozzle or orifice to that of the pipe diameter:

$$\Delta H_f = K\left(\frac{u^2}{2g_c}\right)\left[1 - \left(\frac{d_1}{d_2}\right)^4\right] \quad (7.30)$$

In this case, K is $1/C_d^2$, where C_d is the discharge coefficient representing inefficiencies found in the basic expression $u = C_d\sqrt{2g_c h}$. Diameter d_1 is the constricted diameter, while d_2 is the vessel "diameter" for use with nozzles or the pipe diameter when working with orifices.

The friction losses caused by a sudden contraction (entrance) or enlargement (discharge) are determined respectively by the expressions

$$\Delta H_f = K\left(\frac{u^2}{2g_c}\right)\left[1 - \left(\frac{d_1}{d_2}\right)^2\right] \quad (7.31a)$$

$$\Delta H_f = K\left(\frac{u^2}{2g_c}\right)\left[1 - \left(\frac{d_1}{d_2}\right)^2\right]^2 \quad (7.31b)$$

for which $K = 0.5$ for a contraction and $K = 1.0$ for an enlargement. Equations (7.31a) and (7.31b) may also be used if there is a gradual change in pipe size, as in a reducer or a diffuser. However, in these instances the value of K for use in these equations is given by the following relationships:

Gradual contraction

$$K = 0.8 \sin \theta/2 \qquad \theta < 45° \tag{7.32a}$$
$$K = 0.5 \sqrt{\sin \theta/2} \qquad 45° < \theta < 180° \tag{7.32b}$$

Gradual enlargement

$$K = 2.6 \sin \theta/2 \qquad \theta < 45° \tag{7.32c}$$
$$K = 1.0 \qquad 45° < \theta < 180° \tag{7.32d}$$

where θ is the angle of contraction or enlargement included between the sides of the reducer or the diffuser.

An identification of the various orifices, nozzles, and the more common pipe-size-change devices is shown pictorially in Table 7.10 along with the appropriate C_d and K values. Also given is the equation necessary for a diameter correction factor D_c if a correction is required. Figure 7.8 may be used in conjunction with Table 7.10 as an expedient in determining the diameter correction factor D_c.

The $K(D_c)$ values from Table 7.10 may be converted directly into friction loss in terms of feet of fluid by multiplying by the velocity head $u^2/2g_c$. However, should it be desirable to have the resistance of the orifice, nozzle, or size-change device in terms of equivalent length of pipe, then L/D may be obtained by the expression

$$\frac{L}{D} = \frac{\Delta H_f(D_c)}{f} = \frac{K(D_c)\left(\dfrac{u^2}{2g_c}\right)}{f} \tag{7.33}$$

where K = fitting resistance factor*
D_c = diameter correction factor
f = Moody friction factor

Figure 7.3 may be used to determine the Moody friction factor f with respect to the diameter of a pipe. Although Fig. 7.3 is based on Sch. 40 pipe, the Moody friction factor is not sensitive to small changes in inside diameter, and the use of Fig. 7.3 for other schedules gives results well within acceptable engineering tolerances.

It should be stressed that, in working with orifices, nozzles, and size-changing devices, the velocities used in the various equations are based on the smaller of the diameters. Therefore, calculations which result in equivalent length of pipe are also based on the smaller diameter. To obtain an equivalent length of pipe in the larger size, one would have to apply Eq. (7.27a).

Valve coefficient C_v

A convenient concept which is used to express the capacity of valves is that of the valve coefficient C_v. This parameter can be applied with conventional valves and is especially helpful in describing control valves. Control valves are usually conventional valves which may be provided with reduced trim or smaller internal dimensions to

---

*Dividing K by 12 allows the use of diameter in inches.

TABLE 7.10 Friction-Loss Coefficients for Orifices, Nozzles, and Changements*

$$u = C_d\sqrt{2g_c H_f}$$

$$H_f = \frac{1}{C_d^2} \times \frac{u^2}{2g_c} = K(Dc)\left(\frac{u^2}{2g_c}\right)$$

Device	C_d	K	D_c
Reentrant tube (length = ½:1 diameter)	0.52	3.7	$1 - \left(\dfrac{d_1}{d_2}\right)^4$
Sharp-edged orifice	0.61	2.7	$1 - \left(\dfrac{d_1}{d_2}\right)^4$
Square-edged nozzle (stream clears sides)	0.61	2.7	$1 - \left(\dfrac{d_1}{d_2}\right)^4$
Reentrant tube (length = 2½ diameters)	0.73	1.88	$1 - \left(\dfrac{d_1}{d_2}\right)^4$
Square-edged nozzle (tube flows full)	0.82	1.49	$1 - \left(\dfrac{d_1}{d_2}\right)^4$
Well-rounded orifice	0.98	1.04	$1 - \left(\dfrac{d_1}{d_2}\right)^4$
Pipe exit	....	1.0	1.0
Pipe entrance (inward projection)	....	0.78	1.0

*With permission of Ingersoll-Rand Company and Crane Company.

TABLE 7.10 (continued)

$$u = C_d\sqrt{2g_c H_f}$$
$$H_f = \frac{1}{C_d^2} \times \frac{u^2}{2g_c} = K(D_c)\left(\frac{u^2}{2g_c}\right)$$

Device	C_d	K	D_c
Pipe entrance (flush)			
Sharp-edged	....	0.5	1.0
$r/d = 0.02$	....	0.28	1.0
$r/d = 0.04$	....	0.24	1.0
$r/d = 0.06$	....	0.15	1.0
$r/d = 0.10$	....	0.09	1.0
$r/d = 0.15$ and up	....	0.04	1.0
Sudden enlargement	....	1.0	$\left[1 - \left(\frac{d_1}{d_2}\right)^2\right]^2$
Sudden contraction	....	0.5	$1 - \left(\frac{d_1}{d_2}\right)^2$

alter their configuration for specific applications and also have electric or pneumatic operators to control their position.

The term C_v represents the flow of 60°F water in gallons per minute that will pass through the valve at a differential pressure across the valve of 1 psi. It follows that the water flow rate for any other pressure differential is

$$\text{gal/min} = C_v\sqrt{\Delta P} \qquad (7.34a)$$

and the pressure drop for any other flow rate is

$$\Delta P = \left(\frac{\text{gal/min}}{C_v}\right)^2 \qquad (7.34b)$$

where gal/min = flow rate of 60°F water
ΔP = pressure drop across valve, psi
C_v = valve coefficient, (gal/min)/(psi)$^{1/2}$

The use of the C_v concept can be extended to any fluid by converting ΔP from pounds per square inch to feet of flowing fluid by the relationship 1 psi equals (2.31/s) ft. Then, Eqs. (7.34a) and (7.34b) become

$$\text{gal/min} = \frac{C_v}{1.52}\sqrt{\Delta H_f \times s} \qquad (7.35a)$$

and

$$\Delta H_f = 2.31 \, s \left(\frac{Q}{C_v}\right)^2 \qquad (7.35b)$$

where Q = flow rate of any fluid, gal/min
ΔH_f = friction loss, ft of fluid
s = specific gravity of fluid with respect to 60°F water
C_v = valve coefficient, (gal/min water)/(psi)$^{1/2}$

It then follows that the flow rate of any fluid in terms of valve coefficient and pressure drop is:

$$\text{gal/min} = C_v \sqrt{\Delta P/s} \qquad (7.36a)$$

$$\Delta P = s \left(\frac{\text{gal/min}}{C_v}\right)^2 \qquad (7.36b)$$

An interesting and useful relationship can be developed between Eqs. (6.15), (7.8), and (7.35a) to permit the calculation of the inner diameter in inches of a valve from its C_v. The combination of these equations will lead to the expression

$$d^2 = \frac{C_v}{29.89} \cong \frac{C_v}{30} \qquad (7.37)$$

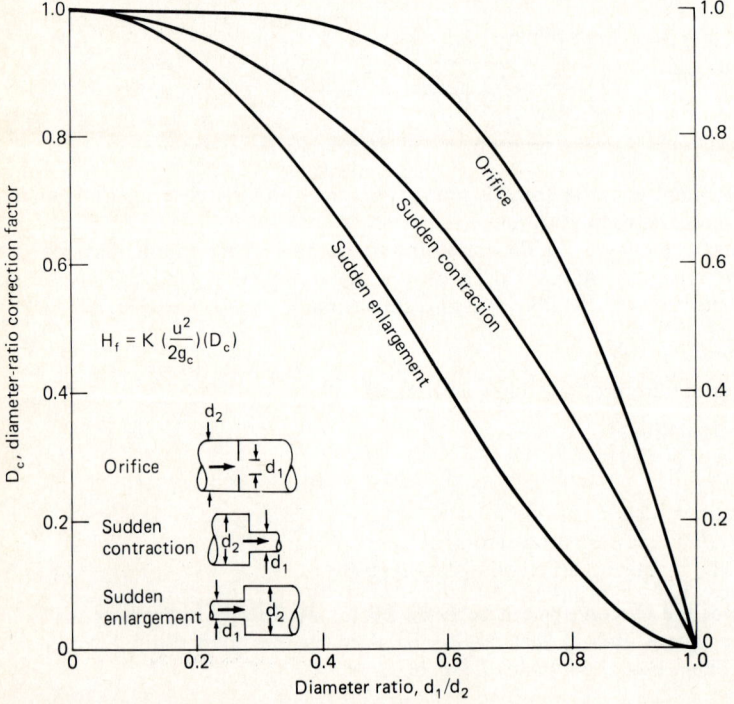

Figure 7.8 Values of diameter-ratio correction factors.

When it is necessary to know the effect of the partial closing of a valve on its flow rate or pressure drop, this can be presented as the percent change in C_v or percent of fully open flow as a function of valve opening. Each valve type has its own characteristic curve. Several of these patterns are given in Fig. 7.9. The curves for butterfly, plug, and ball valves are reported in terms of degrees open, while those for various types of control valves are given in fractions of rated travel.

A note of caution regarding pressure drop across control valves must be given. A calculated pressure drop for a given flow and C_v is valid only when the valve is fully open. Normally, an automated control valve will be operated at less than its fully open

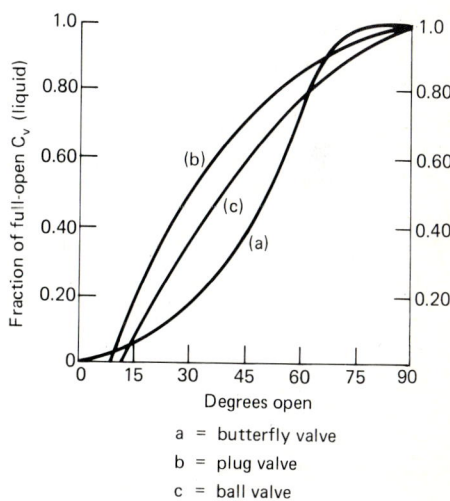

a = butterfly valve
b = plug valve
c = ball valve

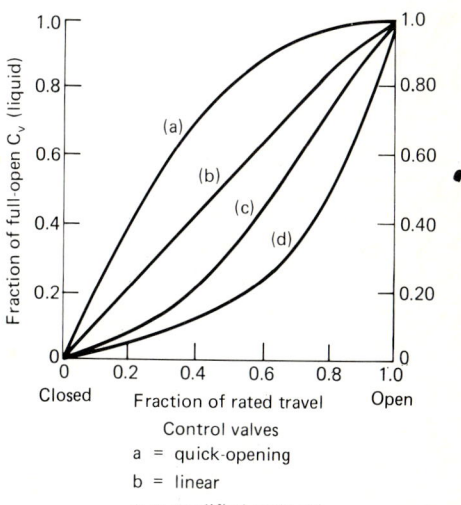

Control valves
a = quick-opening
b = linear
c = modified parabolic
d = equal percentage

Figure 7.9 Typical valve characteristic curves.

position; i.e., it is throttled. To throttle the flow effectively, the friction loss across the valve will be several orders of magnitude greater than when the valve is fully open.

Even though equations could be developed to predict the equivalent lengths of control valves, they are generally not assigned to them for inclusion in the summations of equivalent piping lengths for the calculation of friction loss. Instead, a discrete allowable pressure drop is assigned to each control component in accordance with the dynamics of the piping system and is considered separately from piping and fitting friction losses.

Cumulative Head Losses in Piping Systems

The pressure change or head loss in a piping system is a summation of one or more of the following: difference in absolute pressure between the initial and final points of the system, change in static head due to differences in elevation, inlet- and discharge-velocity head losses, pressure drops imposed upon the system by flow through equipment or control valves, and total friction loss through pipes and fittings. A convenient method for enumerating these factors is a tabulation similar to that given in Table 7.11. This form is ostensibly designed for use in calculating the head requirements for pumps, and thus several of the terms are more appropriate to the design of pumped systems. However, with certain adaptations Table 7.11 can be used with gas services such as fans, blowers, compressors, etc.

Typical evaluations

It is important when evaluating pumping systems to segregate suction and discharge piping in order to analyze the conditions at the entrance of the pump. Therefore, the tabulation in Table 7.11 is divided into two sections, one pertaining to the piping on the suction side of the pump and the other for the discharge side.* Either section would suffice to analyze a piping system which does not contain a pump, blower, or compressor. However, the format for discharge piping is somewhat better suited to the evaluation and thus is used for the following examples.

Several items should be kept in mind when using the tabulation. Since the discharge tabulation is being used for the entire piping system, velocity entrance losses h_c as well as exit losses h_e should be included in the row for discharge losses. The equivalent length for any fitting can be taken from convenient tables or calculated by multiplying the L/d of the fitting in Table 7.9 by the nominal size of the pipe.

It will be noted that several columns are provided for calculations. They are used for trial evaluations of several adjacent pipe sizes, as shown in Table 7.12 for the piping system depicted in Fig. 7.10. This permits selection of a final pipe size for the run in accordance with sound engineering judgment for the information developed. If piping sizes are fixed for several branches in a piping system, the several columns in the tabulations can then be used to find the losses in each of the sections and the sum of the friction losses totaled on the sheet itself.

Siphon effects

While all the terms in Table 7.11 are defined on the sheet, further explanation of the meaning of the elevation term h_d is warranted. The actual path of a piping system is

---

*Use will be made of this form for calculating pump heads in the subsection "General Performance Characteristics of Pumps" in Chap. 11.

TABLE 7.11 Pump Tabulation Sheet*

FORM 5053

united engineers & constructors inc.

PUMP CALCULATION SHEET

PROJECT _____ DESIGN GPM _____ DESIGN TDH _____ PUMP NO. _____

PUMP SERVICE _____ DATE _____

		SUCTION LINE		DISCHARGE LINE		
PUMP NO.						VISCOSITY @ MIN P.T. _____ cpl μ)
LBS/HR		SCH. & PIPE MATL. _____				SPEC.GRAV. @ MIN P.T. _____ (SG)
NORMAL GPM		FLUID PUMPED _____				KINEMATIC VISCOSITY _____ cst z)
DESIGN GPM		TEMP., MAX _____ MIN _____				VAPOR PRESS. @ MAX P.T. _____ mmHg(VP)
REFERENCE DRAWINGS		FROM _____ TO _____ F				
LINE NO.						
LINE SIZE, INCHES I.D.	d			d		
VELOCITY, FT/SEC	V			V		
FRICT. LOSS, FT.FLUID/100 FT	F			F		
REYNOLDS NUMBER	Re			Re		
LINEAL FEET OF PIPE				ⓐ		
GATE VALVES	ⓐ			ⓑ		
GLOBE VALVES	ⓑ			ⓒ		
ELLS	ⓒ			ⓓ		FORMULAE:
TEES, THRU BRANCH	ⓓ			ⓔ		1 psi = (2.31/SG) = _____ FT. FLUID
TEES, THRU RUN	ⓔ			ⓕ		$V = .408\ GPM/d^2$
TOTAL EQUIVALENT LENGTH	EL			EL		$d = \sqrt{\dfrac{GPM}{2.45v}}$
LINE FRICT.(F x EL/100)1.15	F_s			F_d		$z = \mu/SG$
VESSEL PRES.psig x 2.31/SG	P_s^{**}	+		P_d^{**}	+	$h_c = V^2/128.8$
LIQUID ELEV.ABOVE PUMP.FT.	h_s^*	+		h_d	+	$h_e = v^2/64.4$
TANK CONTR.OR ENL. LOSS	h_c			h_e	+	Re= 3162 x GPM x SG / (d x μ)
ORIFICE PL.,psi x 2.31/SG					+	
CONTR.VALVE,psi x 2.31/SG					+	
EXCHANGER,psi x 2.31/SG					+	
ALG.SUM = SUCT.OR DISCH.HEAD	Hs			Hd		
TOTAL DYNAMIC HEAD,Hd–Hs	TDH	FT.l	" SUCT..	" DISCH.)		
NPSH AVAILABLE:						
ABS. PRESSURE HEAD	P_s^{**} + 34/SG	+				
STATIC HEAD	h_s^*	+				
FRICTION HEAD	$h_c + F_s$	–				
VAP. PRES. HEAD	.045 x VP/SG	–				
ALG.SUM = NPSH,FT.FLUID						

*MINUS FOR SUCTION LIFT. **MINUS FOR PRESSURE LESS THAN ATMOSPHERIC. FOR ELEVATIONS ABOVE SEA LEVEL, DECREASE 34 BY 1.1 FT FOR EACH 1000 FT OF ALTITUDE ABOVE SEA LEVEL.

*Used with permission of United Engineers & Constructors, Philadelphia.

TABLE 7.12 Summary of Piping Calculations*

FORM 5053

united engineers & constructors inc.

PUMP CALCULATION SHEET

PROJECT: M₂O R-gen. DESIGN GPM: 140 DESIGN TDH: _____ PUMP NO. P-1515
PUMP SERVICE: DILUTE H₂SO₄ TO COOLING TOWER DATE: 3-27-81

		SUCTION LINE				DISCHARGE LINE		
PUMP NO.	P-1515	SCH. & PIPE MATL.	10S, 316SS		VISCOSITY @ MIN P.T.	0.80	cpl μ	
LBS/HR	61,200	FLUID PUMPED	3% H₂SO₄		SPEC.GRAV. @ MIN P.T.	1.02	(SG)	
NORMAL GPM	120	TEMP., MAX	140 MIN 70		KINEMATIC VISCOSITY	0.78	cst z	
DESIGN GPM	140	FROM V-1515 TO V-1525		F	VAPOR PRESS. @ MAX P.T.	1.48	mmHg(VP)	
REFERENCE DRAWINGS								
LINE NO.		1527				1529/1930	1529/1930	1529/1930
LINE SIZE, INCHES I.D.	d	3"(3.260)	4"(4.260)		d	2"(2.157)	3"(3.260)	4"(4.260)
VELOCITY, FT/SEC	V	5.37	3.15		V	12.2	5.37	3.15
FRICT. LOSS, FT.FLUID/100 FT	F	3.25	0.83		F	26.9	3.25	0.83
REYNOLDS NUMBER	Re	172,000	132,000		Re	261,000	173,000	132,000
LINEAL FEET OF PIPE		35	35			191	191	191
GATE VALVES	2 @	1.6 = 3	2.1 = 4		4 @	1.1 = 7	1.6 = 7	2.1 = 9
GLOBE VALVES					7 @	5.2 = 36	7.7 = 54	10.2 = 71
ELLS	4 @	7.7 = 31	10.2 = 41		4 @	3.5 = 14	5.2 = 21	6.8 = 27
TEES, THRU BRANCH	1 @	15.5 = 16	20.3 = 20		1 @	(35) = 35	(35) = 35	(139) = 139
TEES, THRU RUN						134 = 13		
TOTAL EQUIVALENT LENGTH	EL	85	100		EL	254	308	437
LINE FRICT.,IF × EL/100 1.15	F_s				F_d			
VESSEL PRES.psig × 2.31/SG	P_s**	+	+		P_d**	+	+	+
LIQUID ELEV.ABOVE PUMP.FT.	h_s^*	+	+	−	h_d	+	+	+
TANK CONTR.OR ENL. LOSS	h_c	+	+		h_e	+	+	+
ORIFICE PL.,psi × 2.31/SG								
CONTR.VALVE,psi × 2.31/SG								
EXCHANGER,psi × 2.31/SG								
ALG.SUM = SUCT.OR DISCH.HEAD	Hs				Hd			
TOTAL DYNAMIC HEAD,Hd−Hs	TDH	FT.I	" SUCT.,	" DISCH.)	FORMULAE: 1 psi = (2.31/SG)	2.26 FT. FLUID		
NPSH AVAILABLE:					$V = .408 \text{ GPM}/d^2$			
ABS. PRESSURE HEAD	P_s** + 34/SG	+	+		$d = \sqrt{\dfrac{GPM}{2.45v}}$			
STATIC HEAD	h_s^*	+	+		$z = \mu/SG$	$F = \left(\dfrac{d_{SCH 40}}{d_{SCH 10}}\right)^5 F_{SCH 40}$		
FRICTION HEAD	$h_c + F_s$	−	−		$h_c = V^2/128.8$			
VAP. PRES. HEAD	.045 × VP/SG	−	−		$h_e = V^2/64.4$			
ALG.SUM = NPSH,FT.FLUID					$Re = 3162 \times \dfrac{GPM \times SG}{d \times \mu}$			

*MINUS FOR SUCTION LIFT. **MINUS FOR PRESSURE LESS THAN ATMOSPHERIC. FOR ELEVATIONS ABOVE SEA LEVEL, DECREASE 34 BY 1.1 FT FOR EACH 1000 FT OF ALTITUDE ABOVE SEA LEVEL.

*Used with permission of United Engineers & Constructors, Philadelphia.

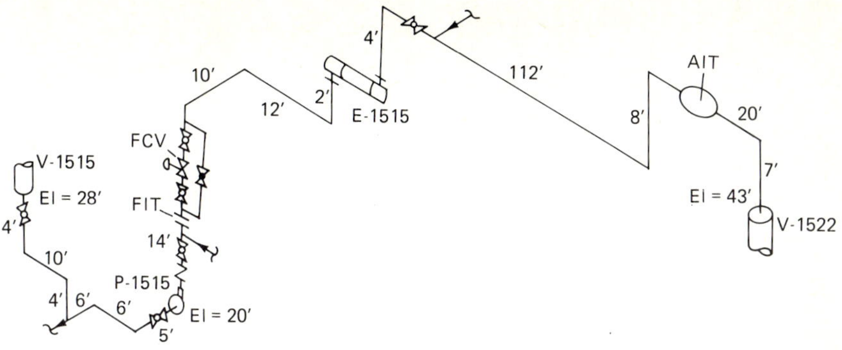

Piping Calculations, Part 2: Friction Losses for Incompressible Fluids   269

Figure 7.10 Typical pipe run.

rarely a straight line for its total length but usually takes the form of a series of horizontal and vertical steps connected by elbows or other fittings. The vertical steps may easily be a combination of upward and downward lines to meet the convenience of the piping runs in the space allocated to piping in the project.

Figure 7.11a depicts a typical run of pipe in isometric form. The discharge of the pump is at elevation 10 ft, and the line rises to elevation 40 ft and then drops into the vessel nozzle at elevation 23 ft. At first glance, it would appear that there will be

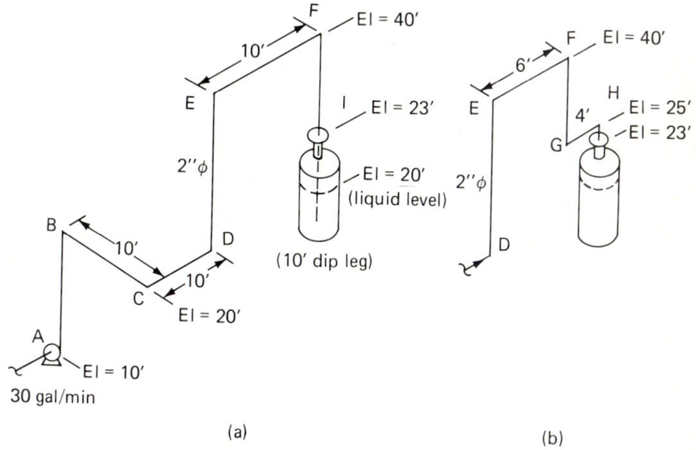

Figure 7.11 Pipe runs with siphon effect.

friction due to 77 ft of pipe plus five elbows and that the pump will have a static head h_d of 13 ft (23 ft at the vessel nozzle minus 10 ft at the pump discharge). These conditions will be valid only if the line is liquid-sealed from the discharge of the pump to the point at which it enters the vessel. It is obvious that the line is filled with liquid from the discharge point of pump A to the high point of the line at E since there is a continuous rise in the line and any gas (usually air) in the line would have been expelled as the liquid displaced it. Whether or not the last horizontal portion of the line E-F and the vertical drop F-I are filled completely or only partially with liquid depends upon the manner in which the drop F-I terminates in the vessel.

If the nozzle is provided with a dip leg that reaches below the surface of the liquid and the leg does not have a weep hole or a siphon break, there will be a liquid seal for the entire pipe. At the beginning of the pumping operation, air remains in the drop section F-I and possibly in a portion of the horizontal line E-F. However, over a period of time the air is expelled into the liquid owing to a combination of factors including solubility, turbulence in the horizontal section, and the splash effect in the vertical leg. As the trapped air is gradually removed by attrition from the lines, the piping run becomes full of liquid. The original premise that the total run of pipe is 77 ft and the static-elevation difference is 13 ft would be nearly correct, except that the amount of piping would have to be increased by the length of the dip leg, i.e., 10 ft, while the static head would be decreased by 3 ft, the distance of the liquid level below the nozzle. Table 7.13 presents a simplified tabulation of total head requirements, in this case as Col. A.

On the other hand, if there were a siphon break or weep hole in the dip leg at the vessel nozzle or the vertical downcomer were to terminate at the nozzle itself, there would be no liquid seal. In this case, unless the flow was so large that the friction loss was 1 ft or more per foot of pipe, there would be a continuous air or gas channel from the liquid level to the top of the vertical downcomer at point F. This means that the total pressure at point F is essentially the same as it is in the vessel. For all intents and purposes, the piping system can be considered terminated at the high point F. The static-head differential to be used in the calculation is then the difference between the discharge elevation of the pump and that at point F, i.e., 30 ft, as opposed to 13 ft should the line end at the nozzle. Since point F is now considered the termination of the line, the portion beyond it need not be accounted for in calculating friction loss; i.e., friction is calculated for 60 ft of pipe instead of 75 ft, as it would have been if the line had ended at the nozzle. Column B in Table 7.13 reflects total head requirements for this case.

At times there may be horizontal sections of piping and fittings in the vertical drop from the high point F, as in Fig. 7.11b. If the flow rate is great enough or there is sufficient equivalent length of piping and fittings in the lower horizontal section, a liquid seal can be formed at that point. This happens because the liquid flow from a partially filled horizontal section to the atmosphere or, in this case, to the vertical leg at the nozzle simulates flow over a weir. Obviously, if the flow is stopped, all fluid eventually runs out of the horizontal section and the height of the chord is zero. Likewise, as with a weir, an increase in flow causes the height of the chord to increase. If

TABLE 7.13 Head Requirements for Conditions in Fig. 7.11

	A	B	C
	Dip leg without weep hole	Dip leg without weep hole	Pipe with siphon seal
Length of pipe, ft	87	60	75
(Number of elbows); equivalent length, ft	(4) 21	(5) 26	(6) 32
Total equivalent length, ft	108	86	107
Friction loss (50 gal/min), ft	4.1	3.3	4.1
Static-head difference, ft	10	13	15
Total friction and static head, ft	14.1	16.3	19.1

the flow is great enough, the height of the chord equals the diameter of the pipe and the horizontal pipe can be said to be "running full" or is sealed at its discharge point. This precludes the possibility of air or a gas entering the pipe at that point and getting beyond the liquid seal. The head requirements for this case are given as Col. C in Table 7.13.

The relationship of the height of water as it exits various-shaped open conduits is thoroughly reported by King and Brater.[2] Figure 7.12 is an interpretation of their correlation of the ratio of height of fluid to diameter for a circular duct (pipe) as a function of flow rate in gallons per minute divided by diameter in inches to the 2.5 power. An extrapolation of these data to the point where h/d is 1.0 shows that $(gal/min)/d^{2.5}$ is about 10.2. At the flow rate represented by this relationship for a given pipe size, a horizontal pipe will be sealed at its end by the flow of liquid in the pipe itself. These minimum sealing values for typical nominal pipe sizes are shown in Table 7.14.

It is unlikely that the velocity in a pumped system, especially for pipe sizes of 2-in diameter and greater, will attain the minimum sealing velocity. Thus, the liquid level at the end of the horizontal run will not "reach" the top tangent point of the inside diameter and an air gap will exist at that location to provide a continuous path of equal pressure along the horizontal conduit. Should this path continue back to the point where there is another vertical drop, as at point G in Fig. 7.11b, the vertical leg F-G will not be sealed either and credit may not be taken in the static balance for the height of the leg F-G. If, however, there were sufficient friction loss in the horizontal section G-H, the level of the liquid in the upstream portion of the horizontal section could become great enough to make contact with the top of the pipe and thus create a liquid seal in the horizontal section even if there were none at its end.

The liquid level h inside a horizontal run as it terminates in an open system or at an elbow or tee connecting to a vertical drop can be found by means of Fig. 7.12. The "air" gap at the end of the horizontal section is then d minus h. A liquid seal takes place if there is sufficient friction loss in the horizontal section that it is equal to or greater than the air-gap distance.

A rigorous determination of friction loss entails the use of the hydraulic-radius and equivalent-diameter concepts since flow in the air-gap section is, in reality, flow in an open channel. Figure 7.13 presents the relationship between the fractional equivalent

TABLE 7.14 Sealing Flow Rates in Horizontal Pipe

Nominal pipe size, in	Minimum sealing flow rate, gal/min	Minimum seal velocity, ft/s
$\frac{1}{2}$	3.2	3.3
$\frac{3}{4}$	6.3	3.8
1	11.5	4.3
$1\frac{1}{2}$	35	5.3
2	63	6.0
3	169	7.3
4	340	8.4
6	930	10.3
8	1840	11.8
10	3300	13.2
12	5100	14.5

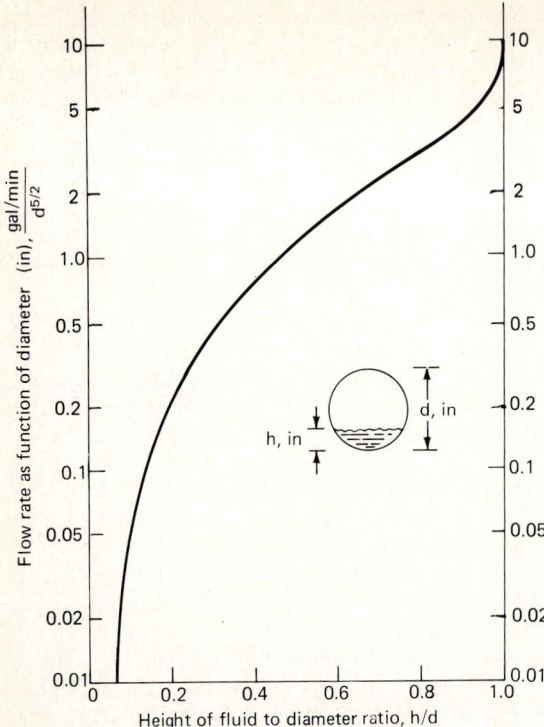

Figure 7.12 Flow in partially filled pipes.

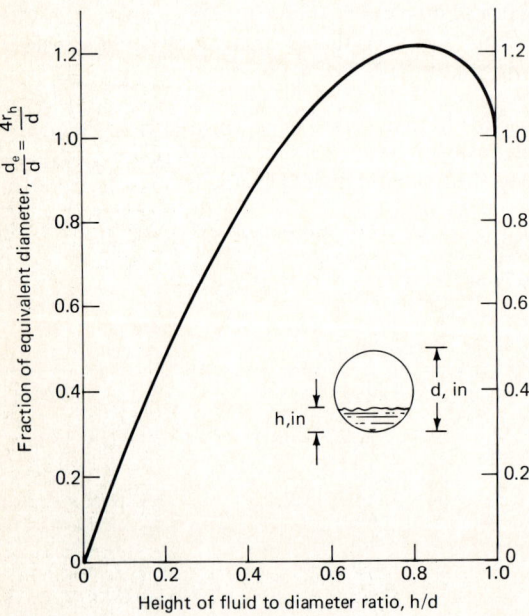

Figure 7.13 Fractional equivalent diameter versus liquid height.

diameter and the chord height. The relationship exhibits an anomaly in that a portion of the curve shows equivalent diameters greater than the pipe diameter itself. The rigorous analysis for friction loss in the air-gap section is complicated by the fact that the height of the liquid in the horizontal section is continuously changing. A simplification of the problem is to use the friction modulus for a fully flowing pipe and to include a reasonable safety factor on the length required. It is necessary to perform a rigorous analysis only if the horizontal length of pipe available appears to be marginal. Alteration of the piping configuration with horizontal ells can frequently add sufficient equivalent length to provide the liquid seal.

REFERENCES

1. R. P. Genereaux, *Chemical and Metallurgical Engineering,* vol. 44, 1937, pp. 241–248.
2. H. W. King and E. F. Brater, *Handbook of Hydraulics,* 6th ed., McGraw-Hill Book Company, New York, 1976.

Chapter 8

Piping Calculations, Part 3: Compressible Fluids

The great and obvious difference between an incompressible fluid and a compressible fluid is the effect of a change in pressure on the specific volume or density of the fluid. As implied by their names, there is essentially no change in the specific volume of an incompressible fluid if the pressure on the system is varied, while the density of a compressible fluid varies almost proportionately with the system pressure. A variation in system temperature also affects incompressible and compressible fluids differently. Whereas the densities of most liquids vary little or only moderately with a change in temperature, the densities of gases at constant pressure are, for most practical purposes, inversely proportional to the system temperature. It can thus be recognized that liquids at ordinary temperatures and pressures are incompressible while gases are compressible fluids.

While the molar volume or density of a liquid remains essentially constant over large pressure ranges and moderate temperature spans, these parameters for gases which are so-called perfect or are nearly *ideal* vary in accordance with

$$V = \frac{RT}{P} \quad \text{or} \quad PV = RT \tag{8.1a}$$

$$\text{or} \quad \rho = \frac{PM}{RT} \tag{8.1b}$$

where V = molar volume, ft³/(lb·mol)
ρ = density, lb/ft³
M = molecular weight, lb/(lb·mol)
P = pressure, lbf/ft²
T = temperature, °R
R = universal gas constant, 1545 (ft·lbf)/[°R·(lb·mol)]

The effect of a change in pressure or temperature on the density of a gas causes the determination of pressure loss experienced by flowing compressible fluids in conduits, orifices, and nozzles to differ from that for incompressible fluids. Reference should be made to the Bernoulli equation, (7.1), in Chapter 7 to evaluate pressure change in any fluid flowing in a conduit. The equation is shown here in its differential form:

$$-dh = \frac{du^2}{2g_c} + dz + dF \tag{8.2}$$

where h = head exerted by fluid, ft of fluid
u = fluid velocity, ft/s
z = elevation of fluid, ft
F = friction loss, ft of fluid
g_c = dimensional constant, 32.17 (lbm/lbf)(ft/s^2)

The friction term dF from Eq. (8.2) can be written as

$$dF = dh_F = f \times \frac{u^2}{2g_c} \times \frac{dL}{D_e} \tag{8.3}$$

where h_F = head loss due to friction, ft of fluid
f = Moody friction factor
D_e = equivalent diameter, ft
L = length of conduit, ft

It will be noted that, for a liquid flowing in a pipe of constant cross-sectional area, velocity remains essentially constant regardless of the pressure in the system. Thus, total head loss is a function of the change in static height between the initial and final points of the system and total friction loss for the system. The friction term dh_F/dL for an incompressible fluid remains essentially constant throughout the length of the system. Except for inlet- and discharge-velocity effects, there is no contribution of the velocity term $du^2/2g_c$ as the liquid flows from the inlet to the discharge point.

The effect of pressure change on the terms in Eqs. (8.2) and (8.3) is quite different for a compressible fluid. These equations are written as absolute-pressure losses in pounds-force per square inch by expressing linear velocity as a function of mass velocity G and density ρ and then multiplying all the head terms by $\rho/144$ so that they become

$$-dP = -\left(\frac{G^2}{144 \times 2g_c} \frac{d\rho}{\rho^2}\right) + \left(\frac{\rho}{144} dz\right) + \left(\frac{f}{144 D_e} \times \frac{G^2}{2g_c \rho} dL\right) \tag{8.4}$$

where G = mass flow rate, lb/(ft$^2 \cdot$s).

Each of the terms in the pressure-loss equation now contains a density term. However, the flow of a gas in a differential length of conduit of constant cross-sectional area results in a friction loss which causes the pressure of the system to drop. The result is a reduction in density with an attendant increase in velocity. If the static-elevation term were to be taken into account, the effect of changes in density along the length of the flow due to elevation variations would have to be recognized.

It can thus be seen that, in contrast to liquid flow, there is an interaction among the three right-hand terms for gases in Eqs. (8.2) and (8.4) and that exact solutions of the equations are very complex. The pressure-drop relationships are further complicated for those gases which deviate from the perfect gas law as given by Eqs. (8.1a) and (8.1b).

Fortunately, a change in elevation in a gas-flow situation normally has little effect on the total pressure drop of the system since gas densities are usually very low. While the pressure-drop effect of the velocity-difference term should be taken into account, the velocity loss is normally only a relatively small portion of the friction loss in the conduit unless the final velocity is excessively large. If, for the sake of discussion, the static- and velocity-loss terms were to be ignored, the effect of change in pressure on friction and thus on total pressure drop in a system could easily be demonstrated, as shown in Fig. 8.1.

In Fig. 8.1 it is assumed that an incompressible fluid and a compressible fluid are flowing in an isothermal manner or an adiabatic manner and that both fluids start at the same pressure and with the same differential pressure drop per unit length of

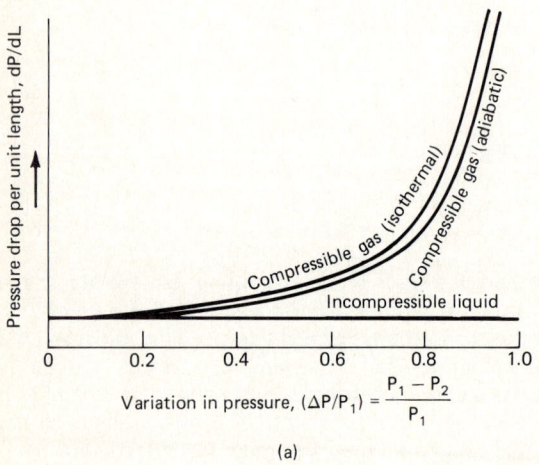

(a)

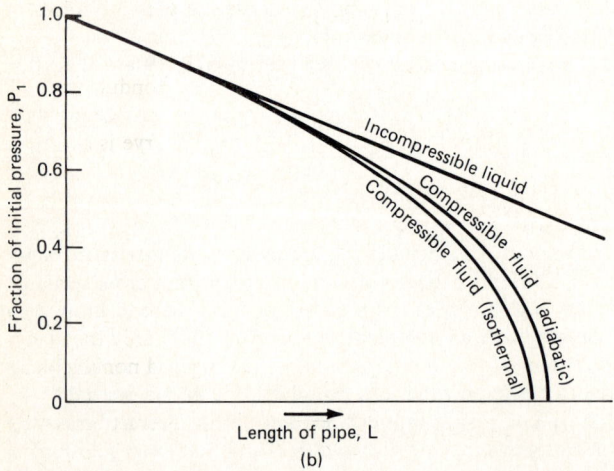

(b)

Figure 8.1 Pressure characteristics of compressible and incompressible fluid flow. (*a*) Change in pressure drop per unit length as a function of pressure loss. (*b*) Change in pressure as a function of length of pipe. All fluids have the same initial dP/dL.

conduit. In Fig. 8.1a, it can be seen that the incompressible fluid retains the same dP/dL regardless of the fraction of original pressure lost. Obviously the liquid fraction pressure loss cannot be more than its original pressure unless there has been a loss of static head at the same time. In actuality, the liquid dP/dL due to friction remains constant until the system pressure becomes less than the vapor pressure of the liquid, at which point a portion of the liquid flashes into vapor, forming a two-phase system in which the vapor phase is a compressible fluid.

The dP/dL for the gas systems (isothermal and adiabatic) begins to increase as soon as there is a friction loss. The value of dP/dL at any point on the isothermal curve is essentially

$$\left(\frac{dP}{dL}\right)_2 = \frac{P_1}{P_2}\left(\frac{dP}{dL}\right)_1 \tag{8.5}$$

where P = system pressure
L = length of conduit
subscript 1 = initial point in system
subscript 2 = point in system L distant from initial point

Thus, the rate of pressure loss per unit length of conduit length for a given equivalent diameter increases at an accelerating rate as the gas passes along the length of the conduit. A similar change in dP/dL takes place for a system flowing adiabatically. However, the acceleration of the increase is somewhat less for the adiabatic system than that for the isothermal system since the adiabatic system experiences a temperature drop and, for any pressure point downstream, the density of the adiabatic gas system is greater than and its dP/dL is less than the respective value for an isothermal gas system.

Although the velocity of a flowing gas accelerates as it loses pressure owing to friction, there is a maximum-velocity limit for a gas flowing through a conduit of constant cross-sectional area. This limit is dictated by the laws of thermodynamics and is equal to the sonic velocity of the gas for adiabatic-flow conditions. It is slightly less than the sonic velocity for isothermal conditions. Thus, the locus of points of a gas-flow dP/dL curve reaches a high point at some low-pressure value.

Figure 8.1b contrasts the change in system pressure for liquid and gaseous (both isothermal and adiabatic) systems as a function of the length of the conduit when all systems have the same initial dP/dL. Since the dP/dL for a liquid remains essentially constant for the length of the conduit, the pressure-versus-length curve is a straight line until it approaches a low-pressure value equal to the vapor pressure of the liquid. The curves for the gaseous systems are asymptotic to the incompressible curve in the initial portion of the conduit. However, the deviation of the gas pressure curves from those of the liquid system becomes greater and greater so that, if extrapolated, they would approach zero pressure in about one-half of the length that it would take for an incompressible fluid with negligible vapor pressure to approach zero total pressure.

The properties of compressible fluids as they affect pressure drop are discussed in the next subsection. They include the distinction between perfect and nonideal gases and the thermodynamic effect on various gas properties of a change in pressure. Subsequent subsections present several methods to calculate friction pressure loss.

Properties of Compressible Fluids

Equations (8.1a) and (8.1b) showed that, in contrast to liquids, the densities of gases and vapors cannot be considered constant during flow through pipes. In addition, there are a number of other factors which must be considered in the analysis of compressible-fluid-flow problems. They are:

1. Corrections for nonideal behavior
2. Distinction between abiabatic and isothermal flow
3. Effect of sonic velocity on compressible-fluid flow

Each of these factors increases the complexity of the flow analysis. However, an understanding of their effects aids the engineer in deciding whether to consider or ignore them in analyzing a given system.

Nonideal behavior

There are many situations in which the ideal gas law may be applied in analyzing compressible-fluid flow. However, under certain conditions some gases, especially condensable vapors, show considerable deviation from ideal behavior. The easiest method to take nonideality into account is to use a modified form of the ideal gas law which introduces another term in the ideal-gas-law equation. This is the so-called compressiblity factor to account for deviations from ideality. The modified ideal-gas-law equation then takes the form

$$PV = ZRT \tag{8.6}$$

with Z being defined as the compressibility factor. The compressibility factor can be found in thermodynamic tables, calculated directly from experimental data, or obtained from functional correlations between reduced temperature T_r and reduced pressure P_r. This generalized relationship is presented by the chart shown in Fig. 8.2.

The reduced properties are defined as

$$\text{Reduced temperature } (T_r) = \frac{\text{actual absolute temperature}}{\text{critical absolute temperature}} = \frac{T}{T_c} \tag{8.7}$$

$$\text{Reduced pressure } (P_r) = \frac{\text{actual absolute pressure}}{\text{critical absolute pressure}} = \frac{P}{P_c} \tag{8.8}$$

$$\text{Reduced volume } (V_r) = \frac{\text{actual mole volume}}{\text{critical mole volume}} = \frac{V}{V_c} \tag{8.9}$$

Critical temperature T_c is that temperature above which the substance cannot be liquefied, with critical pressure P_c being the minimum pressure required to liquefy the substance at the critical temperature. The volume of 1 mol of the substance at its critical temperature and pressure is the critical volume V_c. P_c, T_c, and V_c or Z_c values for many substances can be found in thermodynamic texts or in standard mechanical or chemical engineering handbooks or references.

Thermodynamic effects

The flow of a gas is actually a thermodynamic process and should be analyzed within the context of the laws of thermodynamics, most of which were developed for ideal gases. The concepts of a polytropic process and of compressible-fluid flow with friction were developed to explain actual rather than theoretical fluid-dynamics processes in a practical manner. However, explanations for these processes have always utilized classic thermodynamic theories based on reversible processes and true equilibrium

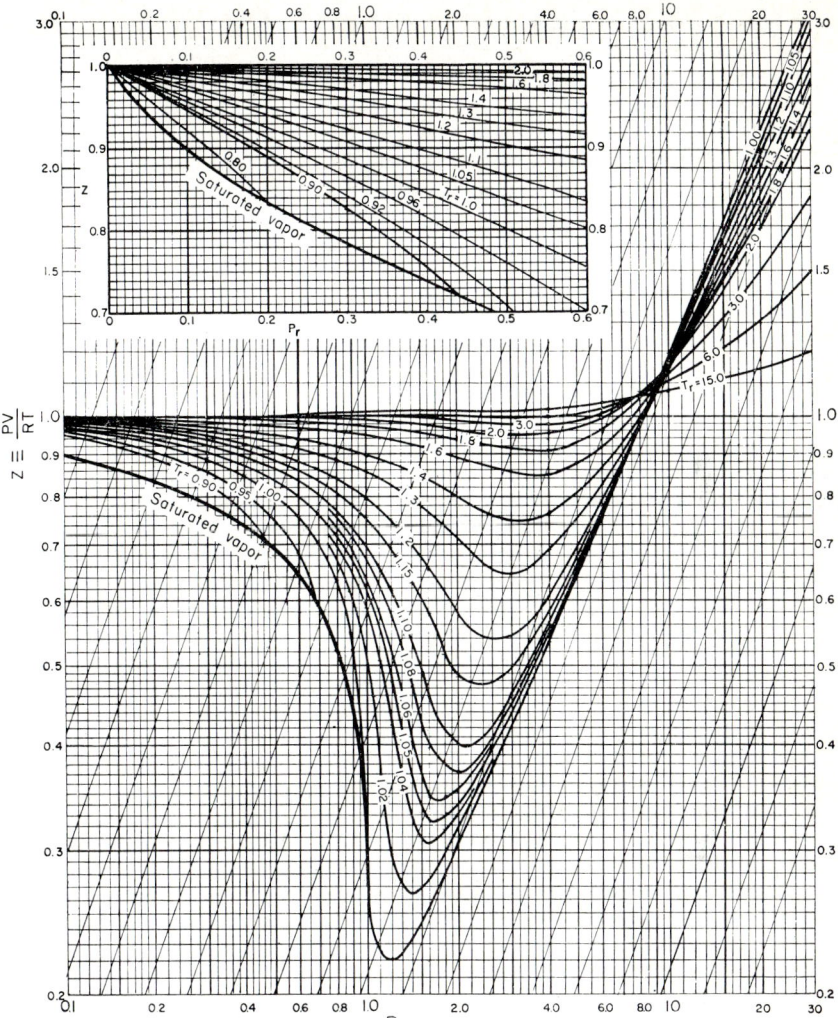

Figure 8.2 Generalized compressibility-factor diagram. (*Based on data compiled by A. L. Lyderson, R. A. Greenkorn, and O. A. Hougen,* Generalized Properties of Pure Fluids, *University of Wisconsin Engineering Experiment Station Rep. 4, Madison, 1955; and Perry et al. (eds.),* Chemical Engineers' Handbook, *4th ed., McGraw-Hill Book Company, New York, 1963, by permission.*)

states. While the concept of reversibility assumes that a physical process can be represented by a time-dependent equation, the solution of this equation is invariant with regard to the algebraic sign affixed to the variable time. In other words, the same amount of work is done on a system while compressing it (forward time) as would expected to be obtained from a system if it were allowed to expand back to its original state (reverse time).

Because friction actually exists regardless of whether compression or expansion occurs, more work is done on the system during compression and less work is obtained from it when expanding than are theoretically possible by assuming reversibility. Thus, actual work compressor operations and gas flows through a pipeline are irreversible processes. The standard equations representing these operations, however, assume that classic reversible thermodynamics can be used and the effects of irreversibility can be compensated for by the addition of an empirical correction factor, such as an efficiency factor or a discharge coefficient, or by other similar conventions. Fortunately, the accuracy of these methods is satisfactory for most practical applications and is the authors' justification for analyzing irreversible compressible-fluid-flow systems as if they were reversible in this chapter and in the explanation of compressor operations and design in Chap. 12.

The first step in analyzing the flow of a compressible fluid is to determine which type of polytropic process is being considered and particularly whether it is a special case, such as an isothermal or an adiabatic process. In an isothermal process, the gas flows at a constant temperature through a pipe and an ideal gas follows the relationship

$$PV = \text{constant} \tag{8.10}$$

To attain a constant temperature, the gas must exchange heat with its surroundings, which act as a uniform-temperature heat sink. The isothermal-flow assumption typically can be applied to most uninsulated lines carrying gases flowing at a low velocity. The lack of insulation and the low velocity allow the gas to approach thermal equilibrium with its surroundings. Most plant air and gas supply lines may be treated as isothermal fluids. However, this is not the case for gas traveling at high velocities and or in insulated pipelines.

Gases and vapors flowing through orifices, through pipes at high velocities, or in insulated pipelines cannot maintain a constant temperature and, therefore, should not be treated as isothermal processes. The effect of velocity acceleration due to changes in flow cross-sectional area and the effect of friction create changes in temperature. High gas velocities and insulation prevent the compressible fluid from coming to thermal equilibrium with its surroundings. When heat cannot be exchanged with the surroundings, compressible-flow situations may be assumed to be essentially adiabatic processes, and ideal fluids then follow the relationship

$$PV^k = \text{constant} \tag{8.11}$$

where k = specific-heat ratio = C_p/C_v = $C_p/(C_p - R)$
C_p = specific heat at constant pressure
C_v = specific heat at constant volume
R = universal gas constant

When Eq. (8.10) is compared with Eq. (8.11), it can be seen that the isothermal process is equivalent to having an exponent k of unity for the volume term. The isothermal and adiabatic processes are special cases of the so-called polytropic fluid-flow process. The isothermal process represents thermal equilibrium with the environment, while the adiabatic process exemplifies no heat exchange with the environment. Actual flow processes can often be approximated by these extremes, but there are other situations which can be represented by the general relationship

$$PV^n = \text{constant} \tag{8.12}$$

where n = polytropic coefficient. The exponent n can be substituted for k in Eq. (8.11) and can have a value between zero and infinity, depending on the amount of heat added or removed from the flowing system. In the special polytropic process known as the adiabatic case, heat is neither added nor taken away, so that $n = k$. In the special case, the isothermal process, sufficient heat is added to a flowing system to maintain it at a constant temperature and n is unity. For other so-called polytropic processes, the value of n is commonly considered to be in the range

(Isothermal) $1 < n < k$ (adiabatic)

Table 8.1 lists several important relationships for the various thermodynamic processes which are used in analyzing expansion due to pressure drop. The expressions in the table are valid for perfect or nearly ideal gases only. However, if the deviation from the ideal is not too great, i.e., less than about 5 percent, or the change in Z value is not greater than about 5 percent owing to change in pressure, the relationships in Table 8.1 may be used to obtain results within acceptable engineering limits. The values for the polytropic and adiabatic friction-loss ratios are slightly lower than shown in the table since a reduction in temperature also reduces the viscosity of the gas, resulting in an increase in its Reynolds number and an accompanying reduction of the Moody friction factor. Since the friction losses are converted to a small degree of heat input to the gas, the actual values for the polytropic and adiabatic temperature ratios could be slightly greater than those predicted by the expansion equations alone. While the resulting temperature changes affect the density, velocity, and friction-loss ratios, this friction-loss energy-conversion effect can easily be ignored in most normal cases.

Effect of specific-heat ratio. As can be seen in Table 8.1, the value of the specific-heat ratio k can have a significant effect on the temperature, density, velocity, and incremental friction-loss ratios owing to pressure change. For example, for a final pressure equal to one-half of the original, the absolute temperature at the lowered pressure is 0.76 of the original for k equal to 1.67 but 0.94 when k equals 1.10. Similarly, the velocity and friction-loss pressure ratios for a 2:1 pressure ratio are 1.51 of

TABLE 8.1 Gas-Law Thermodynamic Relationships for a Pressure Change

Pressure ratio = $\dfrac{P_2}{P_1}$

Property	Isothermal process	Polytropic process	Adiabatic process
Temperature ratio	$\left(\dfrac{T_2}{T_1}\right)_I = 1$	$\left(\dfrac{T_2}{T_1}\right)_P = \left(\dfrac{P_2}{P_1}\right)^{(n-1)/n}$	$\left(\dfrac{T_2}{T_1}\right)_A = \left(\dfrac{P_2}{P_1}\right)^{(k-1)/k}$
Density ratio	$\left(\dfrac{\rho_2}{\rho_1}\right)_I = \dfrac{P_2}{P_1}$	$\left(\dfrac{\rho_2}{\rho_1}\right)_P = \left(\dfrac{P_2}{P_1}\right)^{1/n}$	$\left(\dfrac{\rho_2}{\rho_1}\right)_A = \left(\dfrac{P_2}{P_1}\right)^{1/k}$
Velocity ratio	$\left(\dfrac{u_2}{u_1}\right)_I = \dfrac{P_1}{P_2}$	$\left(\dfrac{u_2}{u_1}\right)_P = \left(\dfrac{P_1}{P_2}\right)^{1/n}$	$\left(\dfrac{u_2}{u_1}\right)_A = \left(\dfrac{P_1}{P_2}\right)^{1/k}$
Friction-loss ratio (head)	$\left[\dfrac{(dh/dL)_2}{(dh/dL)_1}\right]_I = \left(\dfrac{P_1}{P_2}\right)^2$	$\left[\dfrac{(dh/dL)_2}{(dh/dL)_1}\right]_P = \left(\dfrac{P_1}{P_2}\right)^{2/n}$	$\left[\dfrac{(dh/dL)_2}{(dh/dL)_1}\right]_A = \left(\dfrac{P_1}{P_2}\right)^{2/k}$
Friction-loss ratio (pressure)	$\left[\dfrac{(dP/dL)_2}{(dP/dL)_1}\right]_I = \dfrac{P_1}{P_2}$	$\left[\dfrac{(dP/dL)_2}{(dP/dL)_1}\right]_P = \left(\dfrac{P_1}{P_2}\right)^{1/n}$	$\left[\dfrac{(dP/dL)_2}{(dP/dL)_1}\right]_A = \left(\dfrac{P_1}{P_2}\right)^{1/k}$

NOTE: $k = C_p/C_v$; $1 < n < k$.

those of the original values at the higher pressure if k is equal to 1.67, as opposed to 1.77 for k equal to 1.10. The deviations from the isothermal expressions and between the higher and lower k values become more pronounced as the ratio of initial to final pressure becomes greater.

The specific-heat ratio of the so-called noble gases, i.e., those such as helium, neon, and other gases and vapors having monatomic molecules, are the highest of the gases at normal temperatures and pressures. They have k values in the range of 1.62 to 1.67 at normal ambient conditions. Diatomic gases such as hydrogen, carbon monoxide, and air have ratios of about 1.40, while those of the higher-molecular-weight diatomic gases such as bromine are as low as 1.32. The value of the specific-heat ratio continues to decrease primarily with the number of atoms in the molecules and secondarily with the structure and molecular weight of the gas. Methane (MW = 16) with 5 atoms has a k value of 1.31, while chloroform (MW = 134), also with 5 atoms, has a value of 1.15. Benzene (12 atoms) has a value of 1.10, while n-hexane (20 atoms) has a value of 1.08.

As the number of atoms in a molecule of gas increases, its heat capacities both at constant pressure and at constant volume become larger so that the k-ratio value approaches unity or nearly isothermal behavior. Thus, friction-loss problems with organic vapors which have 15 or more atoms in a molecule can often be treated as if the vapors were isothermal regardless of k value. It is always important, however, to consider k values when determining the maximum flow rate through a pipe, a restrictive orifice, a control valve, or a relief valve.

Maximum flow and choking. The mass flow rate of a gas in a given piping system has a finite maximum value which can be determined by using the gas-law relationships and the laws of thermodynamics. The flow of a compressible fluid in a pipe of constant cross-sectional area results in a series of continuous infinitesimal pressure changes with accompanying decreases in fluid density along the flow path. These changes, for high-velocity flows, may be considered as an adiabatic process and, under certain conditions, as isentropic. The partial differential for such an isentropic process defines a maximum limit and is represented as

$$g_c \left(\frac{\partial P}{\partial \rho} \right)_S = c^2 \qquad (8.13)$$

where P = pressure of fluid
ρ = density of fluid
c = constant (in terms of velocity)
g_c = dimensional constant
subscript S = constant entropy

In actuality, the value of c in Equation (8.13) is the velocity of sound (also known as sonic or acoustic velocity) in the compressible-fluid field. It is also the maximum linear velocity that the fluid can attain in a conduit with a constant cross-sectional area when its initial velocity is less than sonic. By substituting Eqs. (8.6) and (8.11) into Eq. (8.13), the velocity of sound in any gas can be determined as

$$c = \sqrt{kg_c \left(\frac{ZRT}{M} \right)} \qquad (8.14)$$

where c = velocity of sound, ft/s
k = constant-pressure-to-constant-volume specific-heat ratio
g_c = dimensional constant, 32.17 (lbm/lbf)(ft/s^2)
Z = compressibility factor
R = gas constant, 1545 (ft·lbf)/[(lb·mol)·°R]
T = temperature at flow conditions, °R
M = molecular weight

While the flow temperature T at sonic velocity in an adiabatic process is different from that for an isothermal process, the sonic pressure and temperature or those of the reservoir are usually known for these processes. A useful relationship is then given in the form of an expression for the maximum mass flow of the gas at sonic conditions:

$$\left(\frac{W}{A}\right)_{max} = G_{max} = P_0 \sqrt{\left(\frac{kg_c M}{ZRT_0}\right)\left(\frac{2}{k+1}\right)^{(k+1)/(k-1)}} \qquad (8.15)$$

where W = mass flow rate, lb/s
A = cross-sectional flow area, ft^2
G = mass flux, lb/(s·ft^2)
P_0 = pressure at source condition, psia
T_0 = temperature at source condition, °R
k = specific-heat-ratio constant

The derivation of Eq. (8.15) involves complex interrelationships between Eqs. (8.11) through (8.14) with the mass flow rate W defined as

$$W = uA\rho \qquad (8.16a)$$

where W = mass flow rate, lb/s
u = linear velocity, ft/s
A = cross-sectional flow area, ft^2
ρ = density, lb/ft^3

The mass flow rate at sonic velocity c takes the form

$$W_{max} = cA\rho\star \qquad (8.16b)$$

where $\rho\star$ = density of fluid at sonic-velocity conditions, lb/ft^3. The validity of Eq. (8.16b) requires complex derivations that may be found in standard thermodynamic texts on gas flow such as those by A. H. Shapiro,[1] J. E. A. John,[2] or M. A. Saad.[3]

Since sonic velocity is the maximum rate of motion that can be attained in most configurations,† Eq. (8.15) is used to determine the maximum mass flow that can pass through a given pipe, orifice, or valve. Attempts to increase the mass flow rate above that value result in a choking effect. The maximum mass flow rate can be made greater, as shown by Eq. (8.15), by increasing the source pressure, by lowering the source temperature, by providing a larger cross-sectional flow area, or by a combination of those changes.

---

†Velocities above sonic can be obtained in units with venturis or any sudden contraction followed by an expansion.

For most operating conditions, the maximum mass operating flow rate is not important since the mass flow rate of the gas is usually considerably lower than that at its sonic velocity. However, important exceptions for which there could be choking or maximum-flow-rate constraints are systems with safety relief valves or those containing other restrictions. These conditions are considered below in the subsection "Compressible-Fluid Flow: A Rigorous Approach."

Typical Evaluation of Compressible-Fluid Friction Loss

The quantitative analysis of pressure changes for a perfect gas in an isothermal-flow situation can be derived easily from the Bernoulli expression shown in Eq. (8.4) by including the Darcy pressure-change-versus-length relationship, eliminating the head change due to elevation variation ρdz and then incorporating the isothermal density-pressure correlations from Table 8.1 and the weight-flow-rate relationship in Eq. (8.16a). The expression thus obtained usually relates the weight flow of a gas to a given length of pipe as a function of two pressures along the line and is as follows:

$$G = \sqrt{\left[\frac{144 g_c}{\overline{V}_1 \left(f \frac{L}{D} + 2 \ln \frac{P_1}{P_2}\right)}\right] \times \left[\frac{P_1^2 - P_2^2}{P_1}\right]} \quad (8.17)$$

where G = mass flux, lb/(s·ft²)
g_c = dimensional constant, 32.17 (lbm/lbf)(ft/s²)
f = Moody friction factor
L = length of pipe from point 1 to point 2, ft
D = diameter of pipe, ft
$\overline{V}_1$ = specific volume of gas at point 1, ft³/lb
P_1 = pressure at point 1, psia
P_2 = pressure at point 2, psia

It is difficult to obtain a simple expression for pressure drop ΔP from Eq. (8.17). Thus, when G is given, it is necessary to solve Eq. (8.17) by trial and error with several iterative steps to obtain ΔP. However, if the pressure drop due to velocity acceleration is relatively small compared with the friction drop, the natural-log term in Eq. (8.17) may be eliminated, and the equation then reads:

$$G = \sqrt{\left(\frac{144 g_c}{\overline{V}_1 f \frac{L}{D}}\right) \times \left(\frac{P_1^2 - P_2^2}{P_1}\right)} \quad (8.18)$$

Since $P_2 = P_1 - \Delta P$, Eq. (8.18) can be solved algebraically so that

$$\Delta P \cong P_1 - \sqrt{P_1^2 - \frac{P_1 \overline{V}_1 f \frac{L}{D}}{144 g_c} G^2} \quad (8.19)$$

Equations (8.18) and (8.19) are usually applicable with sufficient accuracy when the total pressure drop is less than about 25 to 40 percent of the pressure of the system at its initial point. Above that, Eq. (8.17) should be used only for isothermal-flow processes and for gases which are nearly ideal at their initial conditions. However, when the pressure drop in a system is below 10 percent of the initial system pressure, not only is the contribution toward pressure drop due to velocity acceleration usually negligible but the variations due to a deviation from the ideal gas laws or due to adiabatic or polytropic flow are minor. Thus, Eqs. (8.18) and (8.19) can be revised to read

$$G = \sqrt{\frac{288 g_c \Delta P}{\overline{V}_{1,2} f \frac{L}{D}}} \qquad (8.20a)$$

and $\Delta P = \dfrac{\overline{V}_{1,2} f \dfrac{L}{D} G^2}{288 g_c}$ \qquad (8.20b)

where ΔP = pressure drop between points 1 and 2, psia
$\overline{V}_{1,2}$ = average specific volume for conditions at points 1 and 2, using temperature at point 1, ft³/lb

Various graphical methods have been proposed to evaluate compressible-fluid flow for adiabatic and polytropic processes for wide ranges of relative pressure change. Typical procedures are those given by Crane[4] and in Perry.[5] These particular methods have several disadvantages, in that they (1) require complex mathematical analysis, (2) ΔP must be calculated indirectly, (3) head loss due to velocity acceleration is not taken into account, and (4) the relationships are not suitable for fluids which deviate from ideal behavior. The next subsection presents a graphical method which is useful for determining pressure-drop values for most normal fluid-flow conditions encountered regardless of flow regime or thermodynamic process. The major limitations are that the final velocity must be less than about one-half of the sonic value and that the pressure of the gas at its initial condition should be less than about 0.8 of its critical value.

The analytical methods described in the last subsection of the chapter are needed when analyzing flow situations in which high velocities are encountered and for supersonic flows, particularly when shock waves are formed.

Compressible-Fluid Flow: Shortcut Methods

A convenient and acceptable means (within certain limits) of analyzing pressure losses for compressible-fluid flow is to consider the terms in the Bernoulli expression given in Eq. (8.2) as being independent of one another. This equation shows that the change in head for the fluid is equal to the change in kinetic energy due to velocity acceleration, the change in potential energy due to variation in elevation, and friction loss in the pipe. The kinetic-energy and elevation-difference terms may both be considered as point functions relating to the conditions at the initial and final locations along the conduit or pipe and as essentially independent of the intervening path of the pipe. Friction loss, on the other hand, is a function of the equivalent length of the pipe or conduit between the initial and final locations.

The potential-energy term in Eq. (8.2) must always be considered in evaluating liquid pressure changes. However, the densities of most common gases and vapors are sufficiently low that differences in potential energy for normal piping configurations and flow conditions are negligible with respect to kinetic-energy changes and friction losses and thus usually need not be included in the evaluation. However, consideration should be given to the potential-energy term should the density of the fluid due to high pressure or high molecular weight be greater than 5 lb/ft³ with a difference in elevation of about 50 ft or more.

Review of loss due to kinetic-energy change

It should be recalled that the kinetic energy of any fluid is represented in the height of a column of fluid by the relationship

$$h_K = \frac{u^2}{2g_c} \tag{8.21a}$$

or in terms of pressure by

$$\Delta P_K = \frac{\rho h_K}{144} = \frac{\rho u^2}{288 g_c} \tag{8.21b}$$

where h_K = kinetic-energy head component, ft of fluid
ΔP_K = kinetic-energy pressure component, psi
u = velocity of fluid, ft/s
g_c = dimensional constant, 32.17 (lbm/lbf)(ft/s²)
ρ = density of fluid, lb/ft³

Thus, the variation in pressure due to a kinetic-energy change between two locations in a flowing system is

$$\Delta P_{K,I} = -\frac{\rho_2 u_2^2 - \rho_1 u_1^2}{288 g_c} \tag{8.22}$$

where $\Delta P_{K,I}$ = change in kinetic energy for isothermal flow, psi
u_1, u_2 = velocity at upstream and downstream locations, respectively, ft/s
ρ_1, ρ_2 = density at upstream and downstream locations, respectively, lb/ft³

Isothermal flow of nearly ideal gases. The density of a liquid under adiabatic or isothermal conditions remains essentially unchanged for normal pressure variations. Thus, for liquid flow in a pipe or conduit of uniform cross-sectional area, the velocity along the length of the conduit remains essentially the same. The kinetic-energy-change term for liquids is thus insignificant for flow within the pipe and may be disregarded except for entrance and exit losses. For gases, however, the density is a function of its pressure, as seen in Table 8.1. Velocities in a conduit of uniform cross section are inversely proportional to the density of the gas for isothermal flow.

Many gases, in most practical situations, have nearly ideal behavior. If the density- and velocity-ratio expressions in Table 8.1 for the isothermal-flow process are substituted into Eq. (8.22), the pressure drop due to kinetic-energy change for a nearly ideal gas in isothermal flow may be expressed in terms of the initial velocity and density as a function of the initial and final pressures:

$$\Delta P_{K,I} = -\frac{\left(\rho_1 \frac{P_2}{P_1}\right)\left(u_1 \frac{P_1}{P_2}\right)^2 - \rho_1 u_1^2}{288 g_c} = -\frac{\rho_1 u_1^2 \left(\frac{P_1}{P_2} - 1\right)}{288 g_c} \tag{8.23}$$

where P_1, P_2 = pressure at upstream and downstream locations, respectively, psia.

In other words, the change in kinetic energy for a gas flowing in a conduit of uniform cross-sectional area is its initial kinetic pressure $\rho_1 u_1^2/(288 g_c)$ times a correction factor. For nearly ideal gases in isothermal flow, the correction factor is $[(P_1/P_2) - 1]$. In a subsequent discussion on friction losses, a method is given to predict P_2 based on initial flow conditions and length of flow. When this is applied to Eq. (8.23), it is possible to determine pressure loss due to kinetic-energy change based on initial flow conditions and length of flow.

Isothermal flow of nonideal gases. The change in density and velocity with a change in pressure for nonideal gases must take into account the initial and final compressibility factors. Thus, the ideal relationship in Table 8.1 becomes

$$\rho_1 = \rho_2 \times \frac{Z_1 P_2}{Z_2 P_1} \tag{8.24}$$

and
$$u_2 = u_1 \times \frac{Z_2 P_1}{Z_1 P_2} \tag{8.25}$$

where Z_1, Z_2 = compressibility factor at upstream and downstream locations, respectively.

Equation (8.23) then becomes

$$\Delta P_{K,I} = \frac{\rho_1 u_1^2}{288 g_c} \times \left(\frac{Z_2 P_1}{Z_1 P_2} - 1\right) \tag{8.26}$$

Inspection of the compressibility-factor diagram (Fig. 8.2) in the low-pressure range shows that, for reduced pressures below 0.8, the reduced-temperature lines are all nearly straight or that, if curved, the maximum relative deviation of the compressibility factor from its values had the T_r line been straight would be within acceptable limits. Thus, the relationship between compressibility factors at different pressure levels is defined by the slope of the T_r line from the initial T_r and P_r points to the condition where Z is equal to 1 when P_r equals 0. Thus,

$$Z_{2,I} = \left[1 - \frac{P_2}{P_1}(1 - Z_1)\right]_{P_r < 0.8} \tag{8.27}$$

Equation (8.26) can then be interpreted as showing that the change in pressure due to kinetic-energy variation is the kinetic-energy pressure of the gas at its initial point times a correction factor $K_{K,I}$ which is a function of the initial and final pressures, so that

$$K_{K,I} = \frac{1}{Z_1}\left(\frac{P_1}{P_2} - 1\right) \tag{8.28}$$

where $K_{K,I}$ = correction factor to initial kinetic-energy pressure to account for kinetic-energy changes under isothermal flow. Thus,

$$\Delta P_{K,I} = \frac{\rho_1 u_1^2}{288 g_c} \times K_{K,I} \tag{8.29}$$

Values of $K_{K,I}$ are shown graphically in Fig. 8.3a for various ratios of P_2/P_1 as a function of initial compressibility factors.

Adiabatic flow of nearly ideal gases. It can be seen from Table 8.1 that the density of a gas going from one pressure to another is greater if the flow is adiabatic rather

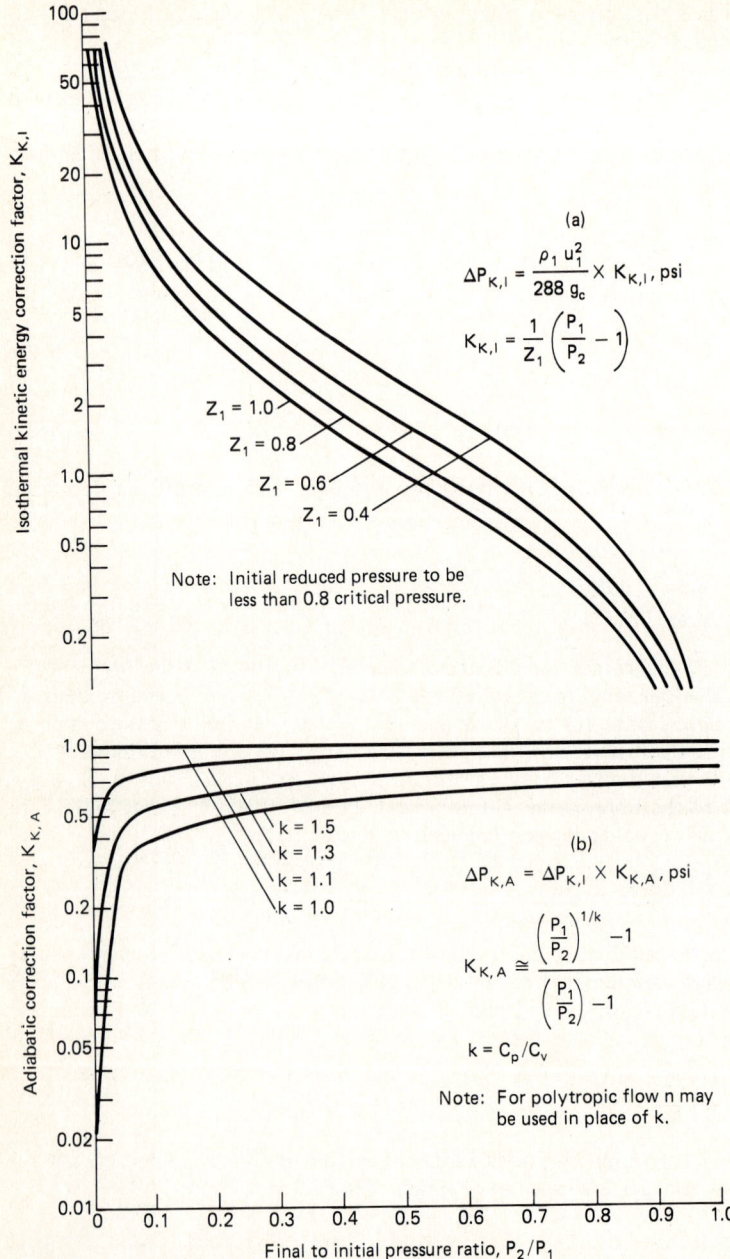

Figure 8.3 Correction factors for kinetic-energy losses.

than isothermal. As a result, the velocity at the downstream pressure is less for an adiabatic flow that for an isothermal flow. Consequently, the pressure loss due to kinetic-energy change, as expressed by Eq. (8.30) for adiabatic flow, is less than that given by Eq. (8.23) for isothermal flow for the same flow length:

$$\Delta P_{K,A} = \frac{\rho_1 u_1^2}{288 g_c} \left[\left(\frac{P_1}{P_2}\right)^{1/k} - 1 \right] \qquad (8.30)$$

where $\Delta P_{K,A}$ = change in kinetic energy for adiabatic flow, psi
k = ratio of constant-pressure and constant-volume specific heats

Adiabatic flow of nonideal gases. If the relationship between Z values shown in Eq. (8.27) and the adiabatic-expansion relationships given in Table 8.1 are used, it can be seen that the pressure change due to a variation in kinetic energy is

$$\Delta P_{K,A} = \frac{\rho_1 u_1^2}{288 g_c} \left[\left(\frac{Z_2}{Z_1} \times \frac{P_1}{P_2}\right)^{1/k} - 1 \right] \qquad (8.31)$$

Equation (8.31) for kinetic-energy loss in adiabatic flow can then be represented as a correction factor $K_{K,A}$ times Eq. (8.30) for isothermal flow such that

$$\Delta P_{K,A} = \frac{\rho_1 u_1^2}{288 g_c} \times K_{K,I} \times K_{K,A} \qquad (8.32)$$

and $K_{K,A} = \dfrac{\left(\dfrac{Z_2}{Z_1} \times \dfrac{P_1}{P_2}\right)^{1/k} - 1}{\left(\dfrac{Z_2}{Z_1} \times \dfrac{P_1}{P_2}\right) - 1} \qquad (8.33)$

where $K_{K,A}$ = correction factor to convert isothermal-flow kinetic-energy pressure drop to adiabatic-flow kinetic-energy pressure drop.

The expression for Z_2/Z_1 as a function of P_2/P_1 is complex; however, for the ranges of Z_1 and k to which Eq. (8.33) applies

$$K_{K,A} \cong \frac{\left(\dfrac{P_1}{P_2}\right)^{1/k} - 1}{\dfrac{P_1}{P_2} - 1} \qquad (8.34)$$

and approximate values for the correction factor $K_{K,A}$ are given in Fig. 8.3b for various pressure ratios as a function of the specific ratio k of the gas.

Polytropic flow of gases. The determination of kinetic-energy changes during polytropic flow of gases is similar to that of adiabatic flow except that the polytropic ratio n is used in place of the specific-heat ratio k in the various adiabatic relationships. Figure 8.3b takes note of this. Thus, the subsection on adiabatic applications can be used for polytropic applications as well. It should be recalled that the value of the polytropic ratio normally lies between $n = 1$ for isothermal flow and $n = k$ for true adiabatic flow.

Relative importance of pressure loss due to kinetic-energy change. In many instances of gas flow in a pipe of uniform cross-sectional area, the pressure loss due to kinetic-energy change is only a minor fraction of the pressure loss due to friction and can be left out of consideration for all practical purposes. However, should the initial density of the gas be sufficiently great, its initial velocity be high enough, or there were to be a large total relative change in pressure, then the kinetic-energy loss would be a significant portion of the total pressure drop and must be taken into account.

Figure 8.4 is presented as an aid in determining whether or not the kinetic-energy loss need be considered when analyzing friction loss. The figure is valid for isothermal flow of a nearly ideal gas. The relative pressure losses for a gas that deviates from ideality would be slightly greater than that shown in Fig. 8.4, while those for adiabatic or polytropic flow of ideal or real gases would in general be somewhat less than that shown in the figure.

Friction loss: the incompressible analogy

Equation (8.1b) can be substituted into the friction portion of Eq. (8.4), and a differential equation for $-dP_F/dL$ expressing frictional pressure drop for an ideal gas as a function of conduit length is obtained:

$$-\frac{dP_F}{dL} = \left[\left(\frac{f}{288 g_c} \times \frac{G^2}{D_e} \right) \left(\frac{RT}{M} \right) \right] \times \frac{1}{P} \qquad (8.35)$$

Isothermal flow. When integrated for a fixed amount of gas flowing isothermally in a pipe of uniform cross-sectional area and equivalent diameter for the pressures at two different positions 1 and 2 of length, Eq. (8.35) becomes

$$\Delta L = \left[\left(\frac{288 g_c}{f} \times \frac{D_e}{G^2} \right) \left(\frac{M}{RT} \right) \right] \left(\frac{P_1^2 - P_2^2}{2} \right) \qquad (8.36)$$

If, for expediency, an intermediate pressure P_{int} is used as a constant value for P in Eq. (8.35) to give the same results as Eq. (8.36), then

$$\Delta L = \left[\left(\frac{288 g_c}{f} \times \frac{D_e}{G^2} \right) \left(\frac{M}{RT} \right) \right] (P_{int})(P_1 - P_2) \qquad (8.37)$$

By equating Eqs. (8.36) and (8.37), clearing terms, and factoring, it can be seen that the assumed intermediate pressure which yields a $-dP_L/dL$ value valid for use over the entire range of pressure drop is

$$P_{int} = \frac{P_1^2 - P_2^2}{2} \times \frac{1}{(P_1 - P_2)} = \frac{P_1 + P_2}{2} = P_{avg} \qquad (8.38)$$

Thus, for a gas flowing at isothermal conditions, pressure drop due to friction can be obtained by applying the friction portion of the basic flow relationship of Eq. (8.4) and using the average pressure for the system to obtain the velocity or density values. The friction portion of Eq. (8.4) may then be rewritten and integrated as

$$\Delta P_{F,I} = \frac{f}{288} \times \frac{G^2}{g_c \rho_{avg}} \times \frac{L}{D_e} \qquad (8.39)$$

where $\Delta P_{F,I}$ = pressure drop due to friction for an isothermal flow, psi
ρ_{avg} = arithmetic average of densities at initial and final position, or $\frac{\rho_1 + \rho_2}{2}$, lb/ft^3

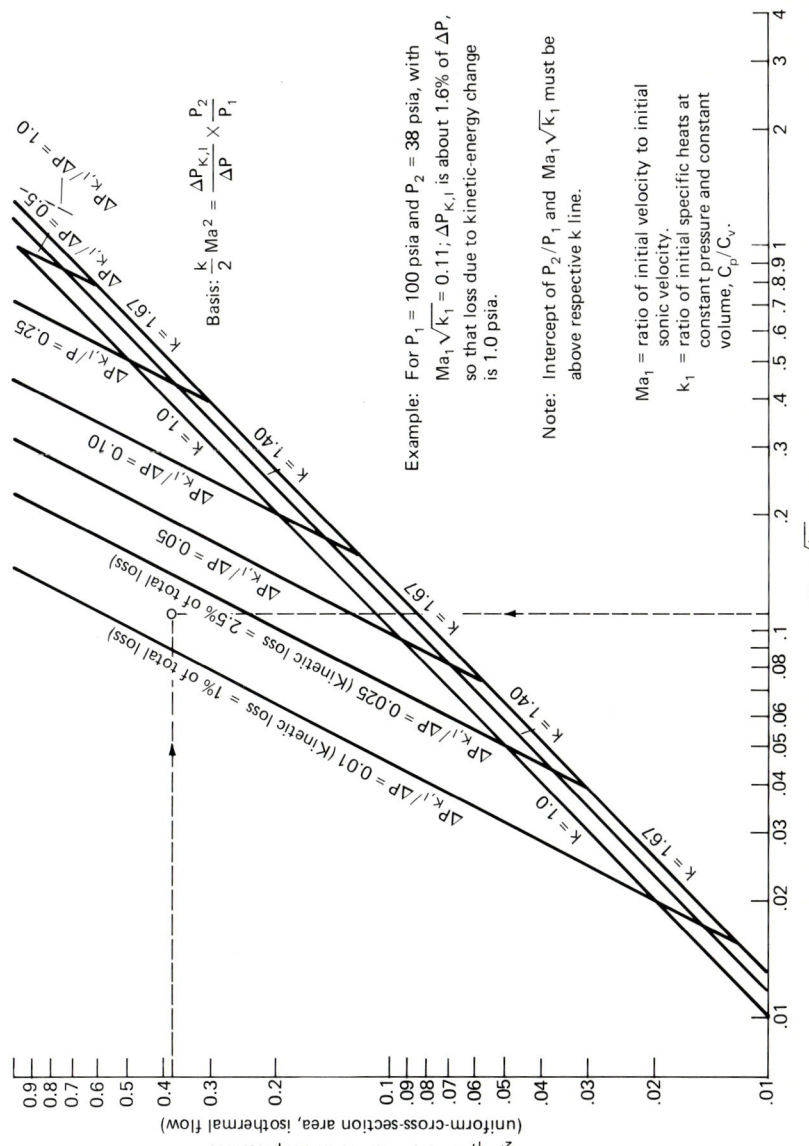

Figure 8.4 Relative importance of kinetic-energy change to total pressure change in systems.

The thermodynamic relationships given in Table 8.1 plus the density–pressure–compressibility-factor relationship* in Eq. (8.24) are used to show that

$$\Delta P_{F,I} = \left(\frac{f}{288} \times \frac{G^2}{g_c \rho_1} \times \frac{L}{D_e}\right) \left[\frac{2}{1 + \left(\frac{P_2}{P_1}\right)\left(\frac{Z_1}{Z_2}\right)}\right] \quad (8.40)$$

It can be recognized that

$$\left(\frac{f}{288} \times \frac{G^2}{g_c D_e} \times \frac{1}{\rho_1}\right) = \left(-\frac{dP}{dL}\right)_1 \quad (8.41)$$

where $(-dP/dL)_1$ = instantaneous pressure loss per unit conduit length at inlet conditions, psi/ft, so that Eq. (8.40) becomes

$$\Delta P_{F,I} = L \left(\frac{dP}{dL}\right)_1 \left[\frac{2}{1 + \left(\frac{P_2}{P_1}\right)\left(\frac{Z_1}{Z_2}\right)}\right] \quad (8.42)$$

Thus, the pressure drop due to friction for a compressible fluid flowing in an isothermal manner is the length times the initial instantaneous pressure drop and a correction factor, or

$$\Delta P_{F,I} = K_{F,I} \left(\frac{dP}{dL}\right)_1 L \quad (8.43)$$

The correction factor $K_{F,I}$ is thus seen to be

$$K_{F,I} = \frac{\rho_1}{\rho_{\text{avg}}} = \frac{Z_1 P_1}{(ZP)_{\text{avg}}} = \frac{2}{1 + \left(\frac{Z_1}{Z_2}\right)\left(\frac{P_2}{P_1}\right)} = \frac{Z\left(\frac{P_1}{P_2} - 1 + Z_1\right)}{\left(\frac{P_1}{P_2} - 1 + Z_1\right)} \quad (8.44)$$

The correction factor is a function of the initial and final pressures and, for those cases in which the initial reduced pressure is less than 0.8, of the initial compressibility factor [see Eq. (8.27)].

Shortcut pressure-drop analysis. Values for the isothermal friction-loss correction factors $K_{F,I}$ are given on the left side of Fig. 8.5 for use in converting friction loss in a system considered as incompressible to a compressible loss. The factor in this case is given as a function of the ratio of final pressure to initial pressure with typical initial compressibility factors as parameters.

The correction factor $K_{F,I}$ can also be expressed as a function of $\left[L\left(\frac{dP}{dL}\right)_1 \Big/ P_1\right]$.

This is represented in the right-hand portion of Fig. 8.5 with typical initial compressibility factors as parameters.

---

*Use of the relationship in Eq. (8.24) assumes that the Z for a given T_r is essentially linear with P_r. It was shown that the assumption is valid for reduced pressures up to 0.8.

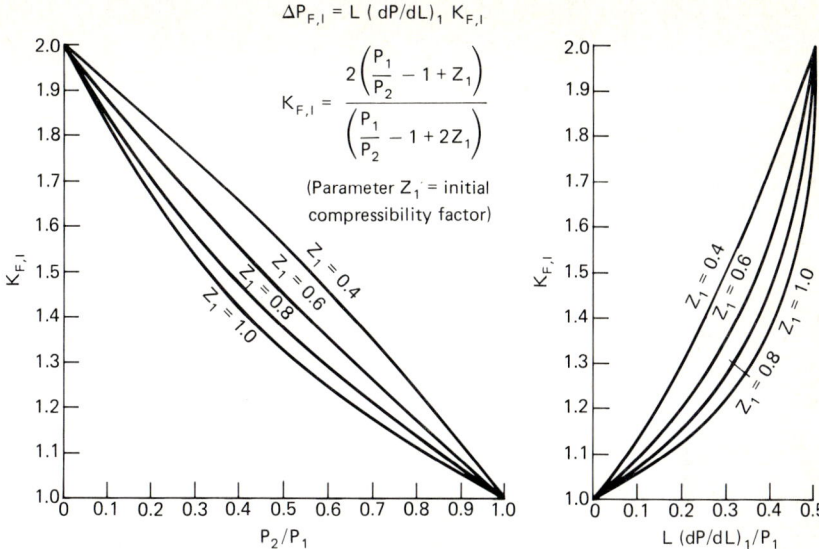

Figure 8.5 Correction factor $K_{F,I}$ to incompressible friction loss for compressible isothermal loss.

The use of both sections of Fig. 8.5 facilitates a quick evaluation of total pressure drop in a system. The value of $(dP/dL)_1/P_1$ is calculated by a convenient means as in the subsection "Friction-Loss Tables" in Chap. 7, multiplied by the equivalent length of the piping section under consideration and then divided by the initial pressure of the system. By using this value as the abscissa of the right graph, the correction factor $K_{F,I}$ is then determined for the initial compressibility factor. The $K_{F,I}$ ordinate on the left graph leads to the P_2/P_1 ratio on the left abscissa, yielding the downstream pressure. The friction pressure drop, or $\Delta P_{F,I}$, may then be calculated by means of Eq. (8.43).

Total pressure drop. The above shortcut method for determining pressure drop due to friction yields acceptable results directly if the contribution due to kinetic-energy change is minimal, i.e., less than 2 to 4 percent of the friction loss. Figure 8.4 can be used to predict the influence of kinetic-energy changes. An iterative method is required for those cases in which the pressure drop due to kinetic-energy change is significant. The total pressure drop $\Delta P_{T,I}$ must be determined for isothermal flow such that

$$\Delta P_{T,I} = \Delta P_{K,I} + \Delta P_{F,I} \tag{8.45}$$

where $\Delta P_{T,I}$ = total pressure drop in a compressible-fluid system flowing isothermally, psi.

Equation (8.45) is amenable to solution by the iterative method, as follows:

1. Enter Fig. 8.5 with $L(dP/dL)_1/P_1$ and determine $K_{F,I}$ for Z_1.
2. Calculate $\Delta P_{F,I}$ by means of Eq. (8.43) and determine the approximate values of P_2 and P_2/P_1.

3. Enter Fig. 8.3a to determine $K_{K,I}$ and enter Fig. 8.5 to determine revised $K_{F,I}$, using Z_1 in both cases.

4. Calculate $\Delta P_{K,I}$ and revised $\Delta P_{F,I}$ by means of Eqs. (8.29) and (8.43), respectively. Determine $\Delta P_{T,I}$ from Eq. (8.45).

5. If the value of $\Delta P_{T,I}$ is within several percent of $\Delta P_{F,I}$ calculated in Step 2, $\Delta P_{T,I}$ may be accepted as the solution. If the relative difference between $\Delta P_{T,I}$ and $\Delta P_{F,I}$ is too great, complete another iteration.

6. Calculate P_2/P_1 based on the last-calculated $\Delta P_{T,I}$.

7. Repeat Steps 3 and 4.

8. If the value of recalculated $\Delta P_{T,I}$ is within several percent of the previously calculated $\Delta P_{T,I}$, the value may be taken as the solution. If not, a subsequent iteration should be made.

At the conclusion of the iterations, the downstream velocity of the gas should be checked. If it exceeds the sonic velocity for the final conditions, the obtained pressure drop should be rejected as being moot. The original premise of flow rate, initial conditions, or pipe size would have to be reexamined. Because of the change in internal energy which takes place at very high velocities but which is not accounted for in the derivation of the shortcut analysis methods, it is best to limit application of the shortcut methods to those instances in which the downstream velocity is less than half of the sonic value. When the velocity in any portion of the system is half of the sonic velocity or more, flow analysis of the system should be accomplished by means of the rigorous approaches described below in the subsection "Frictional Flow in Pipes and Ducts."

Adiabatic or polytropic flow. If the flow condition is adiabatic or polytropic rather than isothermal, there will be a reduction in the temperature of the gas along the flow path. This results in greater average density than if the flow had been isothermal, so that the adiabatic or polytropic pressure loss for a given flow rate and length of pipe is consistently less than the isothermal loss. The frictional pressure drop for adiabatic or polytropic flow can thus be easily obtained by multiplying the system isothermal friction loss by a correction factor $K_{F,A}$ equivalent to the ratio of the average density for the isothermal flow to the average density for adiabatic or polytropic flow. The frictional pressure drop for adiabatic or polytropic flow $\Delta P_{F,A}$ is then given by

$$\Delta P_{F,A} = L \left(\frac{dP}{dL}\right)_1 \times K_{F,I} \times K_{F,A} \tag{8.46}$$

where $Z_{2,I}$ = downstream compressibility factor for isothermal-flow condition
$Z_{2,A}$ = downstream compressibility factor for adiabatic- or polytropic-flow condition

By applying the thermodynamic relationships in Table 8.1 and correcting for any change in the compressibility factor, it can be shown that the value of the adiabatic or polytropic friction correction factor is

$$K_{F,A} = \frac{(\rho_{avg})_I}{(\rho_{avg})_A} = \frac{1 + \left(\frac{Z_1}{Z_2} \times \frac{P_2}{P_1}\right)}{1 + \left(\frac{Z_1}{Z_2} \times \frac{P_2}{P_1}\right)^{1/n}} \tag{8.47}$$

It should be noted that n is the polytropic exponent such that its value is normally between 1.0 for isothermal conditions and k, the ratio of constant-pressure specific heat to constant-volume specific heat for truly adiabatic conditions.

Solving Eq. (8.47) exactly is complex, as there is no simple relationship to predict $Z_{2,A}$ for a given pressure change. However, in actuality the ratio of $Z_{2,I}$ to $Z_{2,A}$ is always somewhat greater than 1.0, and assuming it to be 1.0 provides a conservative solution and one within acceptable engineering limits. The correction factor can then be defined as

$$K_{F,A} \cong \frac{1 + \dfrac{P_2}{P_1}}{1 + \left(\dfrac{P_2}{P_1}\right)^{1/n}} \tag{8.48}$$

Values for the correction factor $K_{F,A}$ are given in Fig. 8.6.

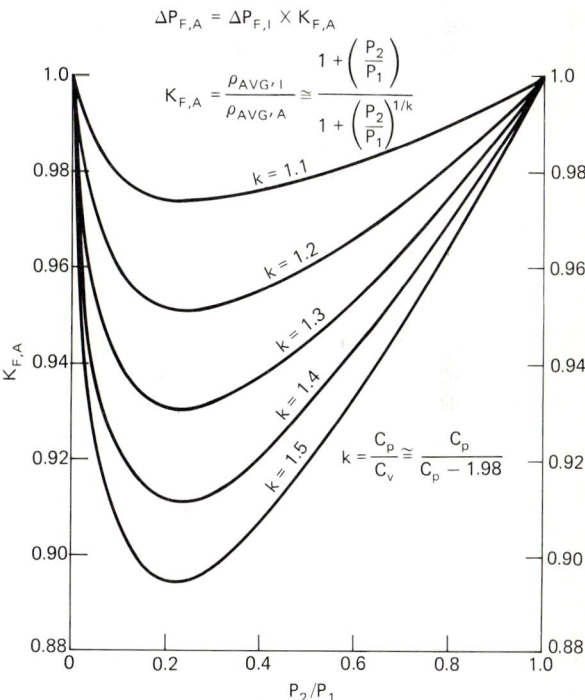

Figure 8.6 Correction factor $K_{F,A}$ to compressible isothermal friction loss for adiabatic fluid flow.

Total pressure drop. The total pressure drop $\Delta P_{T,A}$ for an adiabatic- or polytropic-flow condition is obtained by

$$\Delta P_{T,A} = \Delta P_{K,A} + \Delta P_{F,A} \tag{8.49}$$

Essentially all adiabatic- or polytropic-flow systems require an iterative means for solving for total pressure loss. The method is the same as outlined above for solving for total pressure loss in an isothermal system, with the exception that $\Delta P_{K,I}$ and $\Delta P_{F,I}$ are multiplied by $K_{K,A}$ (Fig. 8.3b) and by $K_{F,A}$ (Fig. 8.6), respectively.

Typical compressible-fluid-flow problem. The following illustrates a typical application of the shortcut method for solving a compressible-fluid-flow problem.

Example 8.1 Ethyl ether gas at 150 psig and 300°F is to be sent to a process vessel from a generating reactor located about 1800 equivalent ft from where it is to be used. A maximum of 6100 lb/h is required. A pressure of 50 psig is required at a control valve just ahead of the reactor. What will be the pressure of the gas if a 2-in Sch. 40 insulated steel line is used?

Preliminary calculations show the following initial conditions:

Mol wt = 74.1
Pressure P_1 = 150 psig
Temperature T_1 = 300°F
Reduced pressure $(P_r)_1$ = 0.316
Reduced temperature $(T_r)_1$ = 0.905
Compressibility factor (Z_1) = 0.82
Density ρ_1 = 1.826 lb/ft³

Velocity u_1 = 40.0 ft/s
Sonic velocity C_1 = 677 ft/s
Viscosity μ_1 = 0.011 cP
Reynolds number Re_1 = 1,790,000
Friction loss $(dP/dL)_1$ = 3.56 psi/100 ft
$C_p/C_v, (k_1)$ = 1.095
Equivalent length L = 1800 ft

solution Since the line is to be insulated, it is assumed that the flow is adiabatic, so that n has the value of k in determining correction factors. Perform an iterative analysis until the $\Delta P_{T,A}$ values between successive iterations are within a few percent of one another.

a. *Initial evaluation*

$(dP/dL)_1$ = 3.56 psi/100 ft
$L(dP/dL)/P_1$ = 0.39
$K_{F,I}$ (Fig. 8.5) = 1.43 P_2/P_1 (Fig. 8.5) = 0.43
∴ P_2 = 70.8 psia, or 56 psig

b. *Iterative analysis*

	First iteration	Second iteration	Third iteration
P_2/P_1	0.43	0.44	0.44
$K_{K,I}$ (Fig. 8.3a)	1.62	1.61	
$K_{K,A}$ (Fig. 8.3b)	0.88	0.89	
$K_{F,I}$ (Fig. 8.5)	1.43	1.425	
$K_{F,A}$ (Fig. 8.6)	0.98	0.98	
$\Delta P_{K,A}$ (Eq. 8.32), psi	0.44	0.43	
$\Delta P_{F,A}$ (Eq. 8.46), psi	91.60	91.28	
$\Delta P_{T,A}$ (Eq. 8.49), psi	92.04	91.71	
P_2, psia	72.6	73.0	
P_2, psig	57.9	58.3	

Since the values of $\Delta P_{T,A}$ and P_2 for the second iteration are within 2 percent of those for the first, a third iteration is not required. The pressure just ahead of the control valve is about 73.0 psia, or 58.3 psig. The result meets the criterion of 50 psig before the control value. The end gas velocity is 85 ft/s, which is considerably less than the sonic velocity of the gas.

It should be noted in Example 8.1 that an assumption that the gas was always at its initial density would have resulted in an estimation that the pressure just upstream of the control valve would have been 85.9 psig. Use of that value instead of the 58.3 psig determined when taking expansion into account could result in an error when sizing the control valve.

Discontinuity. It should be recalled that the *dP/dL* for a given *liquid* flow rate in a uniformly cross-sectioned pipe remains essentially constant along the length of the pipe. If there is a sudden pressure drop in the path due to a throttling valve, orifice, or other restrictions, the *dP/dL* downstream of the discontinuity is essentially the same as the upstream value. Thus, the total friction loss for liquid flow can be obtained by multiplying the initial *dP/dL* by the total length regardless of where the discontinuity is located.

Unfortunately, solving gas-flow problems is not as simple. The sudden pressure drop caused by a restriction is to be contrasted to the continual gradual change that takes place with friction loss in gas flow. The sudden change in pressure, in turn, results in an abrupt change in *dP/dL* downstream of the restriction point compared with its value just upstream of the point. Thus, if an abrupt pressure change is to be imposed on a gas flowing in a uniform pipe, it is important to note the exact restriction point with respect to both ends of the pipe, and the lengths of both discrete portions should be analyzed separately by the methods just outlined. For the first portion, the initial *dP/dL* is used with the length from the starting point to the restriction point and corrected, as required, by the factors from Figs. 8.3, 8.5, and 8.6. This provides the pressure just upstream of the restriction point. The drop across the valve or orifice can be determined by standard methods and the system pressure, temperature, and gas density just downstream of the restriction point calculated. A *dP/dL* just downstream of the restriction point can then be determined and used to analyze the pressure drop in the remaining length of the pipe. A change in pipe size can be treated similarly.

Preliminary evaluation of pipe sizes. The pipe size required for compressible-fluid flow when the initial and final pressure conditions are set for a given weight flow rate through a given uniform, continuous length of pipe may be approximated by means of Fig. 8.5 and Eq. (8.43). In this case, the ratio P_2/P_1 is determined and $K_{F,I}$ at the initial compressibility factor of the gas or vapor is obtained in the left-hand relationship of Fig. 8.5. This value can then be substituted in Eq. (8.43), which can be rearranged to read

$$\left(\frac{dP}{dL}\right)_1 < \frac{P_2 - P_1}{L \cdot K_{F,I}} \tag{8.50}$$

The inequality sign recognizes that piping is available only in discrete sizes and that there is probably only a small kinetic-energy loss. The size for which $(dP/dL)_1$ most closely approaches the right-hand value of Eq. (8.50) should be chosen for first inspection. An aid in choosing the pipe size is to express *dP/dL* in psi/100 ft of pipe and to use a tool such as the Tube Turns slide rule, as described in the subsection "Slide Rules" in Chap. 7. Another expedient is to determine $(dP/dL)_1$ in feet of fluid at the initial conditions per 100 ft of pipe and by adapting the Cameron tables for use with gases, as explained in the subsection "Friction-Loss Tables" in Chap. 7.

Maximum equivalent length and flow rates

Equation (8.43) and Fig. 8.5 reveal several significant points about compressible-fluid flow. Of great importance is the fact that $K_{F,I}$ tends to a value of 2.0 as $\dfrac{L(dP/dL)_1}{P_1}$

approaches 0.5. With $K_{F,I}$ approaching 2.0 and $L(dP/dL)_1$ approaching 0.5 P_1, Eq. (8.43) shows that $\Delta P_{F,I}$ approaches P_1. This can happen only as the absolute discharge pressure approaches zero. Thus, 0.5 is the upper limit for $\dfrac{L(dP/dL)_1}{P_1}$, so that the maximum equivalent length of a line for an ideal compressible fluid with a given weight flow rate at isothermal conditions in a uniform line is less than one-half of the comparable length if the gas had been incompressible and essentially all the pressure were transformed into friction. This is represented by the expression

$$L_{max} \cong \frac{P_1}{2}\left[\frac{1}{(dP/dL)_1}\right] \tag{8.51}$$

The quantity L_{max} represents the choke point for a given gas weight flow rate in a certain-sized line with an initial pressure of P_1. If the line is actually longer than L_{max}, the flow is choked or diminished in order to reduce $(dP/dL)_1$ to satisfy Eq. (8.42) or Eq. (8.43) so that L_{max} equals the actual length of the line. If, on the other hand, it is necessary to maintain the original weight flow rate, the initial instantaneous pressure $(dP/dL)_1$ must be reduced by increasing the diameter of the line.

Isothermal maximum lengths and flow rates. The relationship between choke length and flow rate can be considered from another aspect. It is obvious that the pressure of a compressible gas cannot become zero in value since, according to the relationships in Table 8.1, the velocity in the line would be infinite. In fact, it is not possible for the velocity in a uniform continuous line to exceed its choking value. Choking velocity in isothermal flow for an ideal gas occurs when the ratio of inlet pressure to final pressure times the initial velocity in the pipe is slightly less than sonic velocity. Conversely, the minimum pressure that can be attained in isothermal flow in a continuous uniform line is the inlet pressure times a ratio of the initial velocity divided by the sonic velocity c_1 so that

$$(P_2)_{min} = P_1 \frac{u_1}{c_1} \sqrt{k} \tag{8.52}$$

Thus, the above-mentioned choke point will occur if P_2 [calculated by means of the friction-pressure-drop relationship of Eq. (8.40) plus the kinetic drop from Eq. (8.26)] is equal to or less than $(P_2)_{min}$ from Eq. (8.52). To reduce $(P_2)_{min}$ for the actual line for a given inlet pressure, the inlet velocity must also be reduced. This, in turn, can be done only by reducing the weight flow rate through the pipe or by using a larger-sized pipe to maintain the original weight flow rate.

Adiabatic and polytropic maximum lengths and flow rates. If the flow is adiabatic or polytropic, the choke point for a given mass rate in a given line occurs somewhat farther from the inlet point than for isothermal flow. The pressure at the choke point will also be somewhat less than that at isothermal flow. It can be seen from the adiabatic-expansion expressions in Table 8.1 that the temperature of the gas drops as the pressure is reduced. Thus, the density of the gas for a given pressure after the inlet point in adiabatic flow is greater than that at the comparable pressure in isothermal flow. It follows that the dP/dL value for adiabatic flow at any given pressure is less than that for isothermal flow. A lowered dP/dL gives a longer L_{max} and a smaller P_{min}.

The effect of a smaller dP/dL is somewhat diminished by the reduction of the sonic velocity due to lowered temperature. However, in most cases, the deviation of adiabatic or polytropic flow from isothermal flow is minor.

Compressible-Fluid Flow: A Rigorous Approach

The rigorous approach to analyzing compressible-fluid flow is used only for those applications in which the linear velocity of the fluid approaches sonic velocity, as in the case with choking, when the linear velocity is supersonic, or when keeping track of temperatures, pressures, and velocities during the flow process is important. Since the rigorous approach is not needed for calculating flow characteristics in the majority of compressible-flow situations, this subsection will only highlight some of its more important aspects.

Compressible-fluid-flow parameters

The rigorous method for analyzing compressible-fluid flow integrates fluid-flow theory with the thermodynamic properties of the fluid. The subsection "Compressible-Fluid Flow: Shortcut Methods" presented an approach which superimposed the thermodynamic properties of the fluid onto the Bernoulli equation. The rigorous method integrates the two into a more unified approach.

There are other sets of parameters, in addition to those listed in Table 8.1, which are used to characterize and simplify the calculations of the flow properties in the rigorous method. These are the stagnation state, the critical state, and Mach number.

Stagnation state. The stagnation state is defined as the state in which the compressible fluid is at rest, i.e., has zero linear velocity. The thermodynamic properties of the fluid at this state are defined by a 0 subscript to indicate a zero-velocity condition; i.e.,

P_0 = absolute pressure at stagnation state
T_0 = absolute temperature at stagnation state
ρ_0 = density at stagnation state
H_0 = enthalpy at stagnation state
u_0 = linear velocity of the fluid at stagnation = 0

The stagnation state is an important concept in compressible-fluid flow since it defines the total potential energy of the fluid which can be transformed into kinetic energy as the fluid flows. However, the amount of energy transformed into velocity is limited by the enthalpy of the fluid in its stagnation state. This can be expressed by the equation

$$H_0 = H + \alpha \left(\frac{u^2}{2g_c} \right) \tag{8.53}$$

where α = conversion factor (kinetic energy to enthalpy).

Equation (8.53) shows that the fluid enthalpy equals the stagnation enthalpy at zero velocity. As the fluid begins to flow, a portion of the enthalpy is transformed into velocity or kinetic energy with the result that the enthalpy of the flowing fluid is reduced. For a thermodynamically reversible system which is frictionless, all the enthalpy can be recovered as the fluid decelerates back to zero velocity.

The enthalpy of the system can be changed only if work is done on or by the fluid by means of a compressor or turbine or if heat is exchanged with the fluid. In an adiabatic process, however, the stagnation enthalpy cannot be changed. If an adiabatic process is reversible, it is also an isentropic process, which means that all the properties of the flowing gas can be determined by knowing its velocity and stagnation enthalpy.

Example 8.2 Given superheated steam at 90 psia and 1000°F, what adiabatic-flow condition reduces the steam pressure to 75 psia?

solution The stagnation conditions for the steam are 90 psia and 1000°F. Its stagnation enthalpy and entropy are about 1532 Btu/lb and 1.93 Btu/(lb·°F), respectively. If adiabatic and nearly frictionless flow conditions are assumed, the process may be considered as isentropic. A steam chart or table shows the enthalpy at 75 psia, and an entropy of 1.93 Btu/(lb·°F) is about 1504 Btu/lb. The difference between the two enthalpies (28 Btu/lb) is the amount of static energy converted to velocity head. This procedure is represented in Fig. 8.7.

The 28 Btu/lb can be converted into velocity units by using Eq. (8.53), with the conversion factor α being equal to 1/778 Btu/(ft·lbf), to show that increasing the fluid velocity to 1184 ft/s in adiabatic frictionless flow reduces the steam pressure from a stagnation value of 90 psia to a flowing value of 75 psia.

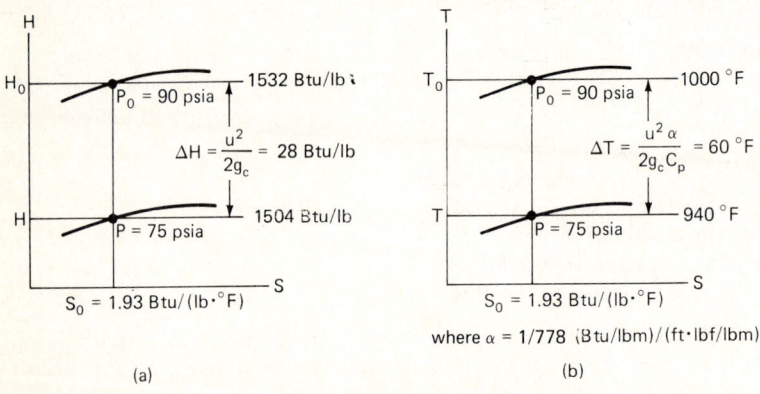

Figure 8.7 Graphical representation for Examples 8.2 and 8.3.

Although the assumption of frictionless flow may not be specifically applicable, it is often used to estimate conditions due to the throttling of gases through nozzles, orifices, and valves. In these cases isentropic analysis gives an acceptable estimate of the properties of the fluid at the throats or points of constriction.

In an isentropic process there is a temperature change in addition to pressure and enthalpy changes. The temperature change can be found by using the relationship

$$\Delta H = H_0 - H = C_p(T_0 - T) \tag{8.54}$$

By substituting Eq. (8.53) into Eq. (8.54) it can be seen that the change in fluid temperature is related directly to the linear velocity of the fluid by the relationship

$$\Delta H = C_p(T_0 - T) = \alpha \left(\frac{u^2}{2g_c}\right) \quad (8.55)$$

Equation (8.55) can be rearranged and made dimensionally consistent to show the temperature-velocity relationship as

$$u = \sqrt{778 \times 2g_c C_p(T_0 - T)} = \sqrt{1556 g_c C_p(T_0 - T)} \quad (8.56a)$$

where u = fluid velocity, ft/s
 C_p = specific heat, Btu/(lb·°R)
 T = temperature, °R
 g_c = dimensional constant, 32.17 (lbm/lbf)(ft/s^2)
 778 = conversion factor, (ft·lbf)/Btu

Example 8.3 What is the attendant change in temperature in Example 8.2?

solution The resulting temperature may be calculated by means of Eq. (8.54) or Eq. (8.55). The heat capacity of steam at the stagnation and final conditions is about 0.52 Btu/(lb·°R), so that the final temperature is 946°F. The results of this example are also shown graphically in Fig. 8.7.

In the absence of an outside influence such as a compressor, a heat exchanger, or friction to cause thermodynamic irreversibility, the maximum available energy of the fluid is defined by its stagnation properties. Were there to be irreversibility, there would be a change in the entropy of the flowing fluid.

It would appear from consideration of Eqs. (8.53) through (8.56a) that all the stagnation enthalpy of the fluid could be converted to velocity. However, this can happen only if the temperature of the flowing fluid reaches absolute zero so that Eq. (8.56a) reduces to

$$u_{max} = \sqrt{1556 g_c C_p T_0} \quad (8.56b)$$

The u_{max} thus determined appears to be a handy reference velocity, but there is a lower, limiting velocity which is more practical for use in analyzing compressible-fluid flow. It is based on the critical-velocity state for the fluid.

Critical state. The subsection "Thermodynamic Effects" reviewed some of the thermodynamic effects of compressible-fluid flow. In that subsection, the maximum flow was determined to be fixed by the sonic velocity c for the fluid and was calculated by Eq. (8.14). When the linear velocity of the compressible fluid is equal to its sonic velocity, the fluid is said to be at its critical state and the various properties of the fluid in that condition are symbolized by an asterisk; i.e.,

 P^* = fluid pressure at the critical state (absolute)
 T^* = fluid temperature at the critical state (absolute)
 u^* = fluid linear velocity at the critical state (equals c, the velocity of sound)

The critical state is an important reference state since it defines the maximum limit of weight flow in a compressible-fluid-flow system. This limit is known as the choking flow rate. The concept of Mach number is used to identify the critical state and to perform calculations which highlight the importance of the state.

Mach number. The Mach number is a dimensionless number which references the actual linear velocity of the fluid to its sonic velocity, i.e.,

$$\text{Ma} = \frac{u}{c} \tag{8.57}$$

where Ma = Mach number
u = linear velocity of compressible fluid, ft/s
c = sonic velocity [as determined by Eq. (8.14)], ft/s

If the linear velocity is less that the sonic velocity, the flow is termed "subsonic" and the Mach number is less than unity. If the linear velocity is greater than the sonic velocity, the flow is "supersonic." However, it was previously postulated that mass flow is limited to sonic velocity. Thus supersonic velocities are the result of conditions other than routine flow in a uniform conduit. These velocities are produced under certain operating conditions in the throttling processes. When the compressible fluid experiences a reduction of flow cross-sectional area, the linear velocity increases. If the velocity upstream of the constriction is sufficiently high or the change in cross-sectional flow area is sufficiently reduced, sonic velocity (a Mach number of unity) is produced at the point of minimum cross-sectional area. Typically, the point of minimum area is at the throat of a nozzle, orifice, or valve. Downstream of the throat, the area usually increases and the compressible fluid expands. However, if the downstream back pressure is lower than the pressure at the throat, a supersonic velocity can be produced.

The rigorous technique for pressure-drop analysis uses Mach numbers to determine the temperature, pressure, density, and other fluid properties at various points in a compressible-fluid-flow system. In addition, the Mach number when used with stagnation pressure and temperature can be employed to calculate the mass flow of gas as follows:

$$\frac{W}{A} = G = 144\, P_0 \text{Ma} \sqrt{\left(\frac{k g_c M}{Z_0 R T_0}\right)\left(1 + \frac{k-1}{2}\text{Ma}^2\right)^{(k+1)/(1-k)}} \tag{8.58}$$

Equation (8.58) is the general form of Eq. (8.15), and when Ma = 1, it can be used to calculate the maximum mass flow rate through a given cross-sectional area. The subsequent portions of this subsection demonstrate methods which utilize Mach numbers for analyzing flow in piping systems, nozzles, and orifices.

Property ratios. The rigorous method uses the specific-heat ratio k and the Mach number (Ma) to calculate all the properties of a flowing compressible fluid. The respective properties, thus determined, are also represented as dimensionless ratios. The ratio defines a given property relative to either its stagnation state or its critical state. For example, stagnation-state ratios are

P/P_0 for pressure
T/T_0 for temperature
ρ/ρ_0 for density

The critical-state ratios are

$P/P^\star$ for pressure
$T/T^\star$ for temperature
$\rho/\rho^\star$ for density

The ratio of cross-sectional area to critical flow area $(A/A^\star)$, although not a ratio of physical properties of a gas, may be determined from k and Ma and is useful in analyzing a constriction in a conduit. Additional important parameters may also be calculated from the Mach number and specific-heat ratio. They are the maximum resistance to flow and entropy difference. While they are not property ratios either, these parameters are important for frictional-flow calculations and are discussed in greater detail later.

The ratio-representation convention is a convenient way of generalizing compressible-fluid-flow properties. It allows an engineer to calculate properties at a given flow point without calculating properties of the fluid at intermediate points. For example, the temperature, pressure, gas density, and Mach number may be known at one point in a flow stream but only the Mach number known at some other point. The temperature, pressure, and density at the second point can then be calculated by the following steps:

1. Given the first Mach number Ma_1, determine the stagnation pressure, temperature, and density ratios for this point, $(P/P_0)_1$, $(T/T_0)_1$, and $(\rho/\rho_0)_1$, respectively.

2. Determine the stagnation pressure, temperature, and density ratios for the second point, $(P/P_0)_2$, $(T/T_0)_2$, and $(\rho/\rho_0)_2$, respectively, for the second Mach number Ma_2.

3. Calculate the actual pressure, temperature, and density values at the second point by using the first-point values and the ratios as follows:

$$P_2 = \frac{(P/P_0)_2}{(P/P_0)_1} P_1 \qquad P_2 = \frac{(P/P^\star)_2}{(P/P^\star)_1} P_1$$

$$T_2 = \frac{(T/T_0)_2}{(T/T_0)_1} T_1 \quad \text{and} \quad T_2 = \frac{(T/T^\star)_2}{(T/T^\star)_1} T_1$$

$$\rho_2 = \frac{(\rho/\rho_0)_2}{(\rho/\rho_0)_1} \rho_1 \qquad \rho_2 = \frac{(\rho/\rho^\star)_2}{(\rho/\rho^\star)_1} \rho_1$$

The stagnation-state ratios were used in the above example. However, the critical-state ratios could also have been used to calculate the fluid properties at the second point had they been known originally.

The specific equations for calculating the necessary ratios for rigorous compressible-fluid-flow analysis are presented in Table 8.2. The development of these equations is beyond the scope of this book. Detailed explanations and derivations of the equations may be found in a text on compressible fluid flow such as those by Shapiro,[6] John,[7] and Saad.[8]

Tables 8.3 through 8.6 give numerical solutions for relevant property ratios and frictional-flow characteristics as a function of typical Mach numbers in accordance with the equations in Table 8.2. Exact solutions according to the equations or interpolated values from the listings in the tables may be used to analyze the various types of compressible-fluid problems that are discussed in the following subsections.

The equations and relationships in Table 8.2 were originally developed for ideal gases; i.e., the compressibility factor of the gas at all conditions is 1.0, and the k value remains constant. Shapiro[9] has shown that the equations and relationships in Table 8.2 may also be used with real gases to give results within acceptable limits of accuracy. This would be expected for flow conditions involving sonic or supersonic velocities since pressures at these conditions are usually low, so that the compressibility factor approaches 1.0 and k is essentially constant.

TABLE 8.2 Thermodynamic Relationships for Rigorous Flow Calculations for Perfect Gases

	Isentropic flow	Isothermal friction flow
Mach number	$\text{Ma}\star = 1$	$\text{Ma}\star^t = \dfrac{1}{\sqrt{k}}$
Temperature	$\dfrac{T_0}{T} = 1 + \dfrac{k-1}{2}\text{Ma}^2$	$\dfrac{T_0}{T_0\star^t} = \dfrac{2k}{3k-1}\left(1 + \dfrac{k-1}{2}\text{Ma}^2\right)$
Pressure	$\dfrac{p_0}{p} = \left(1 + \dfrac{k-1}{2}\text{Ma}^2\right)^{\frac{k}{k-1}}$	$\dfrac{p_0}{p_0\star^t} = \dfrac{1}{\sqrt{k}}\left(\dfrac{2k}{3k-1}\right)^{\frac{k}{k-1}} \dfrac{\left(1+\dfrac{k-1}{2}\text{Ma}^2\right)^{\frac{k}{k-1}}}{\text{Ma}}$
		$\dfrac{p}{p\star^t} = \dfrac{1}{\text{Ma}\sqrt{k}}$
Density	$\dfrac{\rho_0}{\rho} = \left(1 + \dfrac{k-1}{2}\text{Ma}^2\right)^{\frac{1}{k-1}}$	$\dfrac{\rho}{\rho\star^t} = \dfrac{1}{\text{Ma}\sqrt{k}}$
Area	$\dfrac{A}{A\star} = \dfrac{1}{\text{Ma}}\left[\left(\dfrac{2}{k+1}\right)\left(1 + \dfrac{k-1}{2}\text{Ma}^2\right)\right]^{\frac{k+1}{2(k-1)}}$	$\dfrac{A}{A\star^t} = 1$
Friction function	$f\dfrac{L_{\max}}{D} = 0$	$f\dfrac{L_{\max}}{D} = \dfrac{1-k\text{Ma}^2}{k\text{Ma}^2} + \ln k\text{Ma}^2$
Entropy function	$\dfrac{s-s\star}{c_p} = 0$	$\dfrac{s-s\star^t}{c_p} = \dfrac{k-1}{2}\left[\dfrac{k\text{Ma}^2-1}{k} - \ln(k\text{Ma}^2)\right]$

NOTE:
 $k = c_p/c_v$.
 $(\ )_0$ = stagnation state.
 $(\ )\star$ = state at which Ma = 1 for adiabatic constant-area flow.
 $(\ )\star^t$ = state at which Ma = $1/\sqrt{k}$ for isothermal constant-area flow.
 $(\ )_x$ = conditions upstream of a normal shock wave.
 $(\ )_y$ = conditions downstream of a normal shock wave.
 f = friction factor from Moody chart, Fig. 7.1.

Adiabatic friction flow (Fanno line)	Flow with normal shock wave
$Ma^* = 1$	$Ma_y^2 = \dfrac{Ma_x^2 + \dfrac{2}{k-1}}{\dfrac{2k}{k-1}Ma_x^2 - 1}$
$\dfrac{T}{T^*} = \dfrac{k+1}{2\left(1 + \dfrac{k-1}{2}Ma^2\right)}$	$\dfrac{T_y}{T_x} = \dfrac{\left(1 + \dfrac{k-1}{2}Ma_x^2\right)\left(\dfrac{2k}{k-1}Ma_x^2 - 1\right)}{\dfrac{(k+1)^2}{2(k-1)}Ma_x^2}$
$\dfrac{p}{p^*} = \dfrac{1}{Ma}\sqrt{\dfrac{k+1}{2\left(1 + \dfrac{k-1}{2}Ma^2\right)}}$	$\dfrac{p_y}{p_x} = \dfrac{2k}{k+1}Ma_x^2 - \dfrac{k-1}{k+1}$
$\dfrac{p_0}{p_0^*} = \dfrac{1}{Ma}\sqrt{\left[\dfrac{2\left(1+\dfrac{k-1}{2}Ma^2\right)}{k+1}\right]^{\frac{k+1}{k-1}}}$	$\dfrac{p_{0y}}{p_{0x}} = \dfrac{\left[\dfrac{k+1}{2}Ma_x^2\right]^{\frac{k}{k-1}}}{\left[\dfrac{2k}{k+1}Ma_x^2 - \dfrac{k-1}{k+1}\right]^{\frac{1}{k-1}}}$
$\dfrac{\rho}{\rho^*} = \dfrac{1}{Ma}\sqrt{\dfrac{2\left(1+\dfrac{k-1}{2}Ma^2\right)}{k+1}}$	$\dfrac{\rho_y}{\rho_x} = \dfrac{p_y}{p_x} \Big/ \dfrac{T_y}{T_x}$
$\dfrac{A}{A^*} = 1$	$\dfrac{A_y}{A_x} = 1$
$\bar{f}\dfrac{L_{max}}{D} = \dfrac{1 - Ma^2}{kMa^2} + \dfrac{k+1}{2k}\ln\dfrac{(k+1)Ma^2}{2\left(1+\dfrac{k-1}{2}Ma^2\right)}$	$f\dfrac{L_{max}}{D} \to 0$
$\dfrac{s - s^*}{c_p} = \ln Ma^2 \sqrt{\left[\dfrac{k+1}{2Ma^2\left(1+\dfrac{k-1}{2}Ma^2\right)}\right]^{\frac{k+1}{k}}}$	$\dfrac{s_y - s_x}{R} = -\ln\dfrac{p_{0y}}{p_{0x}}$

TABLE 8.3 Isothermal-Friction-Flow Values ($T = T^{*t}$)

	Ma	T_0/T_0^{*t}	P/P^{*t}	P_0/P_0^{*t}	ρ/ρ^{*t}	fL_{max}/D
$k = 1.02$	0.00	0.9903	∞	∞	∞	∞
	0.01	0.9903	99.014	60.205	99.014	9793.7
	0.05	0.9903	19.803	12.055	19.803	385.185
	0.10	0.9904	9.9015	6.0510	9.9015	92.453
	0.15	0.9905	6.6010	4.0598	6.6010	38.798
	0.20	0.9907	4.9507	3.0721	4.9507	20.310
	0.40	0.9919	2.4754	1.6329	2.4754	3.3147
	0.60	0.9939	1.6502	1.2052	1.6502	0.7215
	0.80	0.9966	1.2377	1.0419	1.2377	0.1054
	0.90	0.9983	1.1002	1.0094	1.1002	0.0194
	1.00	1.0002	0.9901	1.0000	0.9901	0.0002
	1.10	1.0023	0.9001	1.0107	0.9001	0.0207
	1.20	1.0046	0.8251	1.0402	0.8251	0.0653
	1.40	1.0097	0.7072	1.1572	0.7072	0.1929
	2.00	1.0299	0.4951	2.2247	0.4951	0.6512
	3.00	1.0794	0.3300	16.264	0.3300	1.3260
	4.00	1.1487	0.2475	291.68	0.2475	1.8537
	7.00	1.4755	0.1414	5.85E+07	0.1414	2.9316
	10.00	1.9806	0.0990	1.36E+14	0.0990	3.6348
	∞	∞	0.0000	∞	0.0000	∞

	Ma	T_0/T_0^{*t}	P/P^{*t}	P_0/P_0^{*t}	ρ/ρ^{*t}	fL_{max}/D
$k = 1.40$	0.00	0.8750	∞	∞	∞	∞
	0.01	0.8750	84.515	52.965	84.515	7132.9
	0.05	0.8754	16.903	10.610	16.903	279.05
	0.10	0.8768	8.4515	5.3334	8.4515	66.159
	0.15	0.8789	5.6344	3.5867	5.6344	27.288
	0.20	0.8820	4.2258	2.7230	4.2258	13.974
	0.40	0.9030	2.1129	1.4784	2.1129	1.9682
	0.60	0.9380	1.4086	1.1259	1.4086	0.2989
	0.80	0.9870	1.0564	1.0092	1.0564	0.0063
	0.90	1.0168	0.9391	0.9953	0.9391	0.0076
	1.00	1.0500	0.8452	1.0025	0.8452	0.0508
	1.10	1.0868	0.7683	1.0280	0.7683	0.1174
	1.20	1.1270	0.7043	1.0703	0.7043	0.1971
	1.40	1.2180	0.6037	1.2039	0.6037	0.3738
	2.00	1.5750	0.4226	2.0720	0.4226	0.9013
	3.00	2.4500	0.2817	6.4848	0.2817	1.6131
	4.00	3.6750	0.2113	20.103	0.2113	2.1537
	7.00	9.4500	0.1207	313.21	0.1207	3.2429
	10.00	18.375	0.0845	2247.6	0.0845	3.9488
	∞	∞	0.0000	∞	0.0000	∞

$k = 1.20$							$k = 1.67$					
0.00	0.9231	∞	∞	∞	∞	∞	0.00	0.8329	∞	∞	∞	∞
0.01	0.9231	91.287	56.475	91.287	8323.3	5978.3	0.01	0.8329	77.382	49.065	77.382	233.04
0.05	0.9233	18.257	11.311	18.257	326.52	233.04	0.05	0.8336	15.476	9.8328	15.476	233.04
0.10	0.9240	9.1287	5.6812	9.1287	77.910	54.787	0.10	0.8357	7.7382	4.9472	7.7382	54.787
0.15	0.9252	6.0858	3.8159	6.0858	32.425	22.332	0.15	0.8392	5.1588	3.3326	5.1588	22.332
0.20	0.9268	4.5644	2.8921	4.5644	16.796	11.264	0.20	0.8441	3.8691	2.5358	3.8691	11.264
0.40	0.9378	2.2822	1.5529	2.2822	2.5581	1.4228	0.40	0.8776	1.9346	1.3970	1.9346	1.4228
0.60	0.9563	1.5215	1.1637	1.5215	0.4755	0.1545	0.60	0.9334	1.2897	1.0860	1.2897	0.1545
0.80	0.9822	1.1411	1.0242	1.1411	0.0381	0.0022	0.80	1.0115	0.9673	0.9952	0.9673	0.0022
0.90	0.9978	1.0143	1.0013	1.0143	0.0004	0.0414	0.90	1.0589	0.8598	0.9917	0.8598	0.0414
1.00	1.0154	0.9129	1.0004	0.9129	0.0157	0.1116	1.00	1.1119	0.7738	1.0081	0.7738	0.1116
1.10	1.0348	0.8299	1.0188	0.8299	0.0616	0.1983	1.10	1.1705	0.7035	1.0416	0.7035	0.1983
1.20	1.0560	0.7607	1.0549	0.7607	0.1257	0.2933	1.20	1.2347	0.6449	1.0907	0.6449	0.2933
1.40	1.1040	0.6521	1.1806	0.6521	0.2804	0.4913	1.40	1.3798	0.5527	1.2332	0.5527	0.4913
2.00	1.2923	0.4564	2.1261	0.4564	0.7769	1.0488	2.00	1.9490	0.3869	2.0417	0.3869	1.0488
3.00	1.7538	0.3043	8.8560	0.3043	1.4721	1.7766	3.00	3.3442	0.2579	5.2279	0.2579	1.7766
4.00	2.4000	0.2282	43.613	0.2282	2.0070	2.3228	4.00	5.2974	0.1935	12.340	0.1935	2.3228
7.00	5.4462	0.1304	3402.91	0.1304	3.0911	3.4169	7.00	14.505	0.1105	86.833	0.1105	3.4169
10.00	10.1538	0.0913	1.00E+05	0.0913	3.7958	4.1240	10.00	28.735	0.0774	334.04	0.0774	4.1240
∞	∞	0.0000	∞	0.0000	∞	∞	∞	∞	0.0000	∞	0.0000	∞

TABLE 8.4 Adiabatic-Friction-Flow Values

$k = 1.02$

Ma	$T/T^\star$	$P/P^\star$	$P_0/P_0^\star$	$\rho/\rho^\star$	fL_{max}/D
0.00	1.0100	∞	∞	∞	∞
0.01	1.0100	100.498	60.505	99.503	9793.8
0.05	1.0100	20.099	12.1157	19.901	385.25
0.10	1.0099	10.0494	6.0808	9.9509	92.508
0.15	1.0098	6.6992	4.0796	6.6343	38.845
0.20	1.0096	5.0239	3.0868	4.9762	20.351
0.40	1.0084	2.5105	1.6397	2.4896	3.3407
0.60	1.0064	1.6720	1.2090	1.6614	0.7376
0.80	1.0036	1.2522	1.0438	1.2478	0.1131
0.90	1.0019	1.1122	1.0103	1.1101	0.0232
1.00	1.0000	1.0000	1.0000	1.0000	0.0000
1.10	0.9979	0.9081	1.0096	0.9100	0.0165
1.20	0.9957	0.8315	1.0379	0.8351	0.0572
1.40	0.9906	0.7109	1.1517	0.7177	0.1768
2.00	0.9712	0.4927	2.1924	0.5074	0.6084
3.00	0.9266	0.3209	15.656	0.3463	1.2287
4.00	0.8707	0.2333	272.169	0.2679	1.6892
7.00	0.6779	0.1176	4.81E+07	0.1735	2.5083
10.00	0.5050	0.0711	9.63E+13	0.1407	2.9129
∞	0.0000	0.0000	∞	0.0995	3.5894

$k = 1.40$

Ma	$T/T^\star$	$P/P^\star$	$P_0/P_0^\star$	$\rho/\rho^\star$	fL_{max}/D
0.00	1.2000	∞	∞	∞	∞
0.01	1.2000	109.543	57.873	91.288	7134.4
0.05	1.1994	21.903	11.5914	18.2620	280.02
0.10	1.1976	10.9435	5.8218	9.1378	66.921
0.15	1.1946	7.2866	3.9103	6.0995	27.932
0.20	1.1905	5.4554	2.9635	4.5826	14.5333
0.40	1.1628	2.6958	1.5901	2.3184	2.3085
0.60	1.1194	1.7634	1.1882	1.5753	0.4908
0.80	1.0638	1.2893	1.0382	1.2119	0.0723
0.90	1.0327	1.1291	1.0089	1.0934	0.0145
1.00	1.0000	1.0000	1.0000	1.0000	0.0000
1.10	0.9662	0.8936	1.0079	0.9249	0.0099
1.20	0.9317	0.8044	1.0304	0.8633	0.0336
1.40	0.8621	0.6632	1.1149	0.7693	0.0997
2.00	0.6667	0.4082	1.6875	0.6124	0.3050
3.00	0.4286	0.2182	4.2346	0.5092	0.5222
4.00	0.2857	0.1336	10.7188	0.4677	0.6331
7.00	0.1111	0.0476	104.142	0.4286	0.7528
10.00	0.0571	0.0239	535.93	0.4183	0.7868
∞	0.0000	0.0000	∞	0.4083	0.8215

$k = 1.20$

0.00	1.1000	∞	∞	∞	∞	∞
0.01	1.1000	104.880	59.205	95.346	8324.1	5980.2
0.05	1.0997	20.973	11.8568	19.071	327.09	234.36
0.10	1.0989	10.4828	5.9529	9.5394	78.365	55.828
0.15	1.0975	6.9842	3.9959	6.3636	32.811	23.206
0.20	1.0956	5.2336	3.0258	4.7768	17.133	12.0184
0.40	1.0827	2.6013	1.6151	2.4027	2.7680	1.8680
0.60	1.0618	1.7174	1.1986	1.6175	0.5999	0.3878
0.80	1.0338	1.2710	1.0409	1.2294	0.0902	0.0558
0.90	1.0176	1.1208	1.0096	1.1015	0.0183	0.0111
1.00	1.0000	1.0000	1.0000	1.0000	0.0000	0.0000
1.10	0.9813	0.9005	1.0087	0.9177	0.0128	0.0074
1.20	0.9615	0.8172	1.0340	0.8498	0.0437	0.0248
1.40	0.9197	0.6850	1.1317	0.7448	0.1320	0.0721
2.00	0.7857	0.4432	1.8837	0.5641	0.4247	0.2105
3.00	0.5789	0.2536	6.7354	0.4381	0.7724	0.3440
4.00	0.4231	0.1626	28.355	0.3844	0.9718	0.4071
7.00	0.1864	0.0617	1.47E+03	0.3309	1.2115	0.4714
10.00	0.1000	0.0316	3.16E+04	0.3162	1.2857	0.4889
∞	0.0000	0.0000	∞	0.3015	1.3647	0.5064

$k = 1.67$

0.00	1.3350	∞	∞	∞	∞	
0.01	1.3350	115.540	56.234	86.549		
0.05	1.3339	23.098	11.2649	17.3169		
0.10	1.3305	11.5349	5.6607	8.6693		
0.15	1.3250	7.6739	3.8052	5.7916		
0.20	1.3173	5.7388	2.8871	4.3563		
0.40	1.2671	2.8141	1.5599	2.2209		
0.60	1.1913	1.8191	1.1759	1.5270		
0.80	1.0993	1.3106	1.0351	1.1922		
0.90	1.0501	1.1386	1.0081	1.0843		
1.00	1.0000	1.0000	1.0000	1.0000		
1.10	0.9499	0.8860	1.0070	0.9327		
1.20	0.9006	0.7908	1.0267	0.8781		
1.40	0.8059	0.6412	1.0981	0.7957		
2.00	0.5705	0.3777	1.5297	0.6620		
3.00	0.3325	0.1922	2.9903	0.5781		
4.00	0.2099	0.1145	5.6083	0.5457		
7.00	0.0767	0.0396	23.848	0.5160		
10.00	0.0387	0.0197	65.183	0.5084		
∞	0.0000	0.0000	∞	0.5009		

309

TABLE 8.5 Isentropic-Flow Values

	Ma	A/A^*	P/P_0	ρ/ρ_0	T/T_0		Ma	A/A^*	P/P_0	ρ/ρ_0	T/T_0
$k = 1.02$	0.00	∞	1.0000	1.0000	1.0000	$k = 1.40$	0.00	∞	1.0000	1.0000	1.0000
	0.01	60.505	0.9999	1.0000	1.0000		0.01	57.873	0.9999	1.0000	1.0000
	0.05	12.115	0.9987	0.9988	1.0000		0.05	11.591	0.9983	0.9988	0.9995
	0.10	6.0808	0.9949	0.9950	0.9999		0.10	5.8218	0.9930	0.9950	0.9980
	0.15	4.0796	0.9886	0.9888	0.9998		0.15	3.9103	0.9844	0.9888	0.9955
	0.20	3.0868	0.9798	0.9802	0.9996		0.20	2.9635	0.9725	0.9803	0.9921
	0.40	1.6397	0.9217	0.9232	0.9984		0.40	1.5901	0.8956	0.9243	0.9690
	0.60	1.2090	0.8325	0.8355	0.9964		0.60	1.1882	0.7840	0.8405	0.9328
	0.80	1.0438	0.7223	0.7269	0.9936		0.80	1.0382	0.6560	0.7400	0.8865
	0.90	1.0103	0.6627	0.6681	0.9920		0.90	1.0089	0.5913	0.6870	0.8606
	1.00	1.0000	0.6020	0.6080	0.9901		1.00	1.0000	0.5283	0.6339	0.8333
	1.10	1.0096	0.5415	0.5481	0.9880		1.10	1.0079	0.4684	0.5817	0.8052
	1.20	1.0379	0.4823	0.4893	0.9858		1.20	1.0304	0.4124	0.5311	0.7764
	1.40	1.1517	0.3716	0.3789	0.9808		1.40	1.1149	0.3142	0.4374	0.7184
	2.00	2.1924	0.1353	0.1407	0.9615		2.00	1.6875	0.1278	0.2300	0.5556
	3.00	15.656	0.0123	0.0134	0.9174		3.00	4.2346	0.0272	0.0762	0.3571
	4.00	272.16	0.0005	0.0006	0.8621		4.00	10.7188	0.0066	0.0277	0.2381
	7.00	4.81E+07	1.47E−09	2.19E−09	0.6711		7.00	1.04E+02	2.42E−04	2.61E−03	0.0926
	10.00	9.63E+13	4.44E−16	8.88E−16	0.5000		10.00	5.36E+02	2.36E−05	4.95E−04	0.0476
	∞	∞	0.0000	0.0000	0.0000		∞	∞	0.0000	0.0000	0.0000

$k = 1.20$						$k = 1.67$				
0.00	∞	1.0000	1.0000	1.0000	1.0000	0.00	∞	1.0000	1.0000	1.0000
0.01	59.205	0.9999	1.0000	1.0000	1.0000	0.01	56.234	0.9999	1.0000	1.0000
0.05	11.856	0.9985	0.9988	0.9998	0.9988	0.05	11.264	0.9979	0.9988	0.9992
0.10	5.9529	0.9940	0.9950	0.9990	0.9950	0.10	5.6607	0.9917	0.9950	0.9967
0.15	3.9959	0.9866	0.9888	0.9978	0.9889	0.15	3.8052	0.9815	0.9889	0.9925
0.20	3.0258	0.9763	0.9802	0.9960	0.9803	0.20	2.8871	0.9674	0.9803	0.9868
0.40	1.6151	0.9092	0.9237	0.9843	0.9250	0.40	1.5599	0.8780	0.9250	0.9491
0.60	1.1986	0.8088	0.8379	0.9653	0.8437	0.60	1.1759	0.7529	0.8437	0.8924
0.80	1.0409	0.6892	0.7333	0.9398	0.7483	0.80	1.0351	0.6162	0.7483	0.8235
0.90	1.0096	0.6267	0.6774	0.9251	0.6988	0.90	1.0081	0.5497	0.6988	0.7866
1.00	1.0000	0.5645	0.6209	0.9091	0.6497	1.00	1.0000	0.4867	0.6497	0.7491
1.10	1.0087	0.5039	0.5649	0.8921	0.6018	1.10	1.0070	0.4282	0.6018	0.7116
1.20	1.0340	0.4461	0.5104	0.8741	0.5557	1.20	1.0267	0.3749	0.5557	0.6746
1.40	1.1317	0.3417	0.4086	0.8361	0.4708	1.40	1.0981	0.2842	0.4708	0.6036
2.00	1.8837	0.1328	0.1859	0.7143	0.2811	2.00	1.5297	0.1201	0.2811	0.4274
3.00	6.7354	0.0213	0.0404	0.5263	0.1256	3.00	2.9903	0.0313	0.1256	0.2491
4.00	28.355	0.0032	0.0084	0.3846	0.0632	4.00	5.6083	0.0099	0.0632	0.1572
7.00	1.47E+03	2.37E−05	1.40E−04	0.1695	0.0141	7.00	23.848	0.0008	0.0141	0.0574
10.00	3.16E+04	5.64E−07	6.21E−06	0.0909	0.0051	10.00	65.183	0.0001	0.0051	0.0290
∞	∞	0.0000	0.0000	0.0000	0.0000	∞	∞	0.0000	0.0000	0.0000

TABLE 8.6 Normal-Shock-Wave Values

	Ma_x	Ma_y	P_y/P_x	P_{0y}/P_{0x}	ρ_y/ρ_x	T_y/T_x
$k = 1.02$	1.00	1.000	1.0000	1.0000	1.0000	1.0000
	1.05	0.952	1.1035	0.9998	1.1014	1.0019
	1.10	0.909	1.2121	0.9988	1.2075	1.0038
	1.15	0.870	1.3257	0.9964	1.3183	1.0056
	1.20	0.834	1.4444	0.9919	1.4338	1.0074
	1.30	0.770	1.6968	0.9761	1.6785	1.0109
	1.40	0.716	1.9695	0.9499	1.9415	1.0144
	1.70	0.592	2.9087	0.8130	2.8369	1.0253
	2.00	0.505	4.0297	0.6210	3.8846	1.0373
	2.50	0.409	6.3020	0.3116	5.9412	1.0607
	3.00	0.345	9.0792	0.1190	8.3394	1.0887
	3.50	0.300	12.3614	0.0357	11.0223	1.1215
	4.00	0.267	16.148	0.0086	13.9310	1.1592
	4.50	0.241	20.440	0.0017	17.008	1.2018
	5.00	0.221	25.237	0.0003	20.200	1.2494
	6.00	0.192	36.346	0.0000	26.735	1.3595
	7.00	0.173	49.475	0.0000	33.214	1.4896
	8.00	0.159	64.623	0.0000	39.414	1.6396
	10.00	0.140	100.980	0.0000	50.500	1.9996
	∞	0.099	∞	0.0000	101.000	∞

	Ma_x	Ma_y	P_y/P_x	P_{0y}/P_{0x}	ρ_y/ρ_x	T_y/T_x
$k = 1.40$	1.00	1.000	1.0000	1.0000	1.0000	1.0000
	1.05	0.953	1.1196	0.9999	1.0840	1.0328
	1.10	0.912	1.2450	0.9989	1.1691	1.0649
	1.15	0.875	1.3762	0.9967	1.2550	1.0966
	1.20	0.842	1.5133	0.9928	1.3416	1.1280
	1.30	0.786	1.8050	0.9794	1.5157	1.1909
	1.40	0.740	2.1200	0.9582	1.6897	1.2547
	1.70	0.641	3.2050	0.8557	2.1977	1.4583
	2.00	0.577	4.5000	0.7209	2.6667	1.6875
	2.50	0.513	7.1250	0.4990	3.3333	2.1375
	3.00	0.475	10.3333	0.3283	3.8571	2.6790
	3.50	0.451	14.1250	0.2129	4.2609	3.3151
	4.00	0.435	18.500	0.1388	4.5714	4.0469
	4.50	0.424	23.458	0.0917	4.8119	4.8751
	5.00	0.415	29.000	0.0617	5.0000	5.8000
	6.00	0.404	41.833	0.0297	5.2683	7.9406
	7.00	0.397	57.000	0.0154	5.4444	10.4694
	8.00	0.393	74.500	0.0085	5.5652	13.3867
	10.00	0.388	116.500	0.0030	5.7143	20.387
	∞	0.378	∞	0.0000	6.0000	∞

$k = 1.20$							$k = 1.67$					
1.00	1.000	1.0000	1.0000	1.0000	1.0000	1.0000	1.00	1.000	1.0000	1.0000	1.0000	1.0000
1.05	0.953	1.1118	0.9998	1.0923	1.0178	1.0497	1.05	0.953	1.1282	0.9999	1.0749	1.0497
1.10	0.911	1.2291	0.9989	1.1873	1.0352	1.0985	1.10	0.913	1.2627	0.9990	1.1494	1.0985
1.15	0.873	1.3518	0.9965	1.2848	1.0521	1.1471	1.15	0.878	1.4034	0.9969	1.2235	1.1471
1.20	0.838	1.4800	0.9924	1.3846	1.0689	1.1956	1.20	0.846	1.5504	0.9934	1.2968	1.1956
1.30	0.779	1.7527	0.9777	1.5902	1.1022	1.2933	1.30	0.794	1.8631	0.9813	1.4406	1.2933
1.40	0.729	2.0473	0.9542	1.8027	1.1357	1.3934	1.40	0.751	2.2009	0.9627	1.5795	1.3934
1.70	0.619	3.0618	0.8362	2.4663	1.2415	1.7162	1.70	0.662	3.3643	0.8754	1.9603	1.7162
2.00	0.546	4.2727	0.6767	3.1429	1.3595	2.0827	2.00	0.607	4.7528	0.7634	2.2821	2.0827
2.50	0.469	6.7273	0.4162	4.2308	1.5901	2.8059	2.50	0.553	7.5674	0.5782	2.6970	2.8059
3.00	0.421	9.7273	0.2298	5.2105	1.8669	3.6783	3.00	0.523	11.0075	0.4283	2.9925	3.6783
3.50	0.390	13.2727	0.1198	6.0562	2.1916	4.7041	3.50	0.504	15.073	0.3177	3.2043	4.7041
4.00	0.369	17.363	0.0610	6.7692	2.5651	5.8848	4.00	0.491	19.764	0.2384	3.3585	5.8848
4.50	0.354	22.000	0.0309	7.3636	2.9877	7.2214	4.50	0.482	25.080	0.1816	3.4731	7.2214
5.00	0.342	27.181	0.0159	7.8571	3.4595	8.7142	5.00	0.476	31.022	0.1406	3.5600	8.7142
6.00	0.327	39.181	0.0044	8.6087	4.5514	12.1694	6.00	0.467	44.782	0.0883	3.6799	12.1694
7.00	0.317	53.363	0.0013	9.1356	5.8413	16.251	7.00	0.462	61.044	0.0585	3.7562	16.251
8.00	0.311	69.727	0.0004	9.5135	7.3293	20.961	8.00	0.459	79.809	0.0406	3.8075	20.961
10.00	0.303	109.000	0.0001	10.0000	10.9000	32.262	10.00	0.455	124.842	0.0217	3.8696	32.262
∞	0.289	∞	0.0000	11.0000	∞	∞	∞	0.448	∞	0.0000	3.9850	∞

Frictional flow in pipes and ducts

For most process piping systems, the methods for analyzing frictional losses presented in the subsections "Typical Evaluation of Compressible-Fluid Friction Loss" and "Compressible-Fluid Flow: Shortcut Methods" usually give acceptable results. However, a more rigorous approach is needed when analyzing systems approaching sonic velocity or in the supersonic condition. These conditions are typically found in piping connected to the discharge of a safety relief valve. Back pressures can sometimes affect the performance of the valve.

The method presented in this subsection includes the concept of generating entropy. This thermodynamic framework has no significant effect at low subsonic velocities but must be considered here.

Friction causes a thermodynamic process to become irreversible. This irreversibility increases the entropy of the fluid in the system and reduces the usable energy of the fluid. The result is a reduction in the stagnation enthalpy of the fluid. This phenomenon, as it applies to frictional flow through pipes and ducts, can be expressed as

$$H'_0 = H_0 - \Sigma F = \left(H + \alpha \frac{u^2}{2g_c}\right) - \Sigma F \tag{8.59}$$

where H'_0 = reduced stagnation enthalpy due to friction losses.

A decrease in stagnation enthalpy or usable energy and an increase in entropy result from friction along the length of a pipe. The ensuing changes in thermodynamic properties cannot proceed indefinitely. Therefore, there is a finite limit to the amount of friction that can be generated in a given compressible-fluid-flow system. This limit is expressed in the form of a resistance coefficient K which takes the form

$$K = \bar{f} \frac{L_{max}}{D} \tag{8.60}$$

where K = maximum flow resistance factor in a pipe for choking velocity
D = uniform diameter, ft
$\bar{f}$ = average Moody friction factor along the length of the pipe
L_{max} = maximum length of the pipe before choking occurs, ft

The maximum-length and maximum-resistance coefficients can be calculated directly from the specific-heat ratio k and the Mach number of the fluid at the inlet to the duct. In addition, the maximum length thus calculated depends upon whether the flow is adiabatic, polytropic, or isothermal. The difference in choking length between an adiabatic- or polytropic-flow process and an isothermal process is relatively small for vapors having low k values but can be significant with noble and diatomic gases.

The maximum-length and the maximum-entropy production for an adiabatic-flow process occurs when the Mach number at the outlet of the duct is unity. The isothermal-flow process, however, has a shorter maximum length for the same flow rate and an exit Mach number of less than unity. Table 8.2 shows that the exit Mach number for an isothermal-flow process can be calculated by the following equation:

$$\text{Ma}^{\star^i} = \frac{1}{\sqrt{k}} \tag{8.61}$$

where $\text{Ma}^{\star^i}$ = critical (choking) Mach number for isothermal flow.

It was noted earlier that the specific-heat ratios of high-molecular-weight gases approach unity. Thus, for all practical purposes, the isothermal- and adiabatic-flow properties for these vapors are nearly identical.

The working equations used in calculating frictional flow in pipes and ducts are given in Table 8.2. Most of the fluid properties required, with the exception of the maximum-resistance coefficient, are expressed as ratios of inlet to discharge properties. The discharge properties are based on choking or critical flow which occurs at the exit. It may appear that such ratios are not useful for calculating flows in pipes where the exit velocity is less than sonic. However, the ratios are very helpful since they are directly related to the maximum-resistance coefficient.

The maximum-choking-resistance coefficients at any two points in a flow system are a function of the actual flow resistance coefficients between the two points as shown by the relationship

$$\left(\frac{fL_{max}}{D}\right)_{Ma_2} = \left(\frac{fL_{max}}{D}\right)_{Ma_1} - \left(\frac{fL}{D}\right)_{actual} \tag{8.62}$$

The inlet resistance coefficient can be determined from the inlet Mach number Ma_1. The relationships in Eq. (8.62) are shown graphically in Fig. 8.8. Once the exit resistance $(fL_{max}/D)_{Ma_2}$ has been determined, the discharge Mach number can be cal-

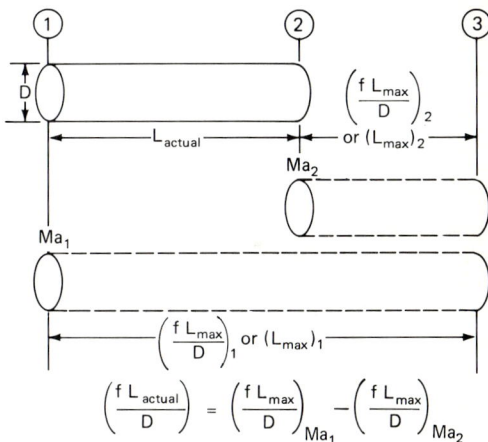

Figure 8.8 Graphical representation of the calculation of frictional compressible-fluid flow in a pipe by using the rigorous method.

culated by the appropriate friction function in Table 8.2 or be interpolated from Table 8.3 or Table 8.4. The exit Mach number can then be used to determine the discharge fluid property ratios by means of the equations, or they can be interpolated from Table 8.3 or Table 8.4 according to the respective flow process. A correlation of adiabatic friction functions with Mach numbers most applicable for this type of analysis is given in Fig. 8.9. The isothermal friction function values and stagnation pressure ratios are somewhat lower for the equivalent Mach numbers.

A number of sources[10,11,12,13] cite examples using this method for sizing gas piping. While Lappie[14] and others[15,16] have reduced the working equations to graphs and nomographs as an aid with manual calculations, these graphical techniques are lim-

316   Materials Selection and Piping Calculations

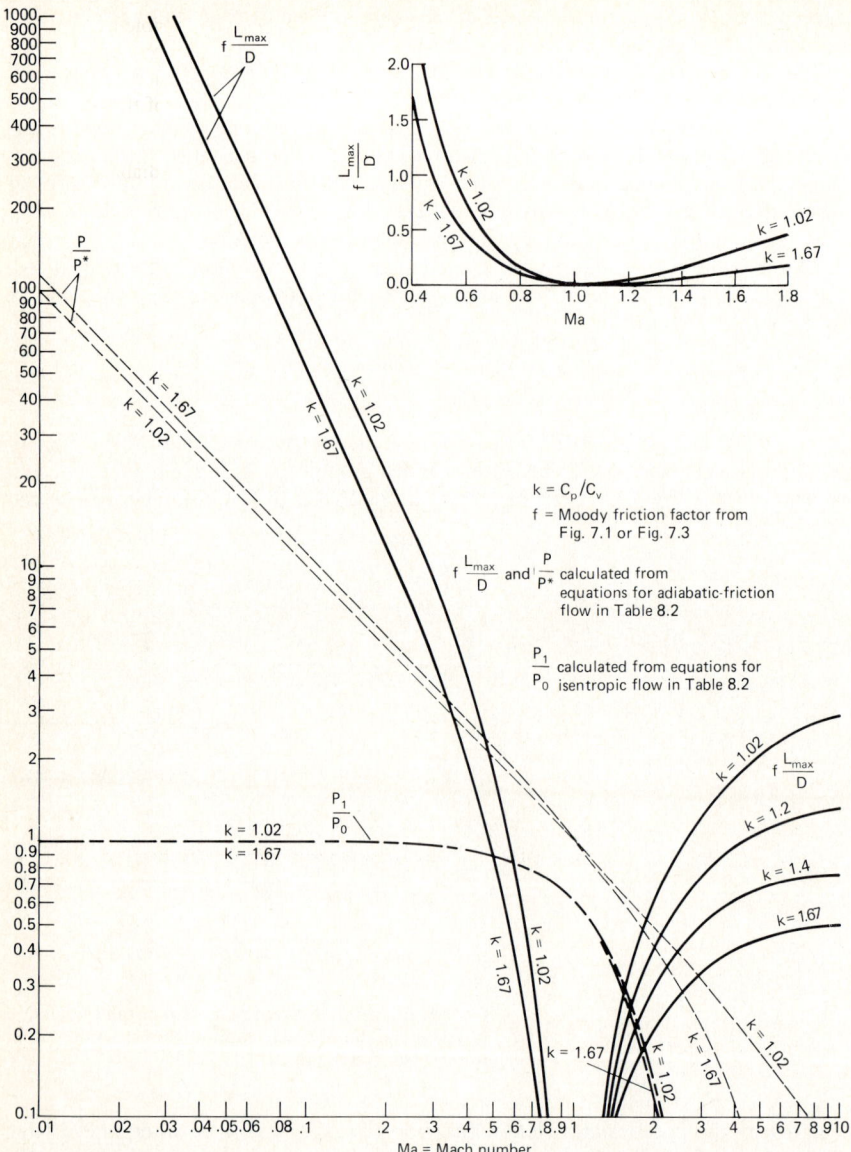

Figure 8.9 Adiabatic friction function and pressure-ratio relations versus Mach number. $f(L_{max}/D)$ = friction function; P/P^* = flow pressure relative to pressure at sonic velocity; P_1/P_0 = initial flow pressure relative to stagnation pressure.

ited to subsonic flow. Example 8.4 shows the application of the adiabatic-frictional-flow working equations under subsonic and supersonic conditions.

Example 8.4 Superheated steam having stagnation-state conditions of 250 psia and 600°F flows through an insulated 24-in Sch. 80 commercial steel line from a boiler to a process vessel. The line is a straight run of pipe 50 equivalent ft long and has no constrictions. If it is assumed that the inlet and discharge losses are negligible and that the steam behaves as a nearly ideal gas, what is the mass flow of the steam and what are the properties of the steam entering the process vessel when the Mach number of the steam entering the line is 0.20? Define the properties at the inlet and outlet of the pipe when the inlet steam is at Ma = 1.5.

solution Since the pipe is insulated, the steam flow is assumed to be adiabatic, and the adiabatic equations in Table 8.2 or interpolated values from Table 8.5 may be used.

Steam properties	Pipe characteristics
$P_0 = 250$ psia	$D_i = 1.797$ ft
$T_0 = 600°F$	$\epsilon = 0.00015$ ft
$\rho_0 = 0.412$ lb/ft^3	$A_i = 2.535$ ft^2
$C_p = 9.63$ Btu/[(lb·mol)·°F]	
$k = C_p/(C_p - 1.98) = 1.26$	
$\mu = 0.0206$ cP	

Part a. Mach number = 0.20 at the inlet.

(1) Given the above steam properties and Ma = 0.20, the steam mass flow rate from Eq. (8.58) is 2.63 lb/s.
(2) The inlet temperature and pressure are 584.6°F and 243.8 psia as calculated from the ratios for T/T_0 and P/P_0 at Ma = 0.20.
(3) The $(fL_{max}/D)_i$ for inlet conditions is calculated from the adiabatic relationships to be 16.27 at Ma = 0.20.
(4) The friction factor for the steam flow is 0.017 for Re = 134,000 in Fig. 7.1.
(5) The maximum length of pipe before choking occurs for Ma = 0.20 and $(fL_{max}/D)_i$ = 16.27 would be 1680 ft.
(6) Since the maximum length is considerably greater than the actual 50-ft length of pipe, choking does not occur and the exit Mach number is less the unity.
(7) To find the exit Mach number, the fL_{actual}/D is calculated to be 0.740. When this is subtracted from fL_{max}/D at the inlet, the $(fL_{max}/D)_e$ at the end of 50 ft of pipe is determined. The difference between $(fL_{max}/D)_i$ and fL_{actual}/D is 15.53.
(8) A Mach number equal to 0.204 gives $(fL_{max}/D)_e$ = 15.53 at the pipe exit.
(9) The exit temperature and pressure of 584.4°F and 238.9 psia are calculated from the relationships:

$$T_{exit} = \left[\frac{(T/T^\star) \text{ exit}}{(T/T^\star) \text{ inlet}}\right] \times T_{inlet}$$

$$P_{exit} = \left[\frac{(P/P^\star) \text{ exit}}{(P/P^\star) \text{ inlet}}\right] \times P_{inlet}$$

Part b. Mach number = 1.5 at the entrance. Since the Mach number is greater than unity, the flow is supersonic. It should be noted in Table 8.5 and Fig. 8.9 that the friction function fL_{max}/D for subsonic flows decreases from infinity to zero as the Mach number increases from stagnation (Ma = 0) to sonic velocity (Ma = 1). However, when there is supersonic flow (Ma > 1), the fL_{max}/D increases as the Mach number becomes greater and becomes asymptotic to a finite value as the Mach number approaches infinity. Thus, a gas at a subsonic velocity accelerates toward a Mach number of 1 owing to friction, while a gas at supersonic velocity decelerates toward a Mach number of 1 because of friction. It should be noted that, regardless of whether the initial velocity is subsonic or supersonic, transposition from one regime to the other cannot be made owing to friction alone since the absolute value of fL_{max}/D approaches 0 at Ma = 1.

(1) The mass flow rate calculated by Eq. (8.58) for the inlet Mach number = 1.0 is 6.61 lb/s.

(2) The entrance temperature and pressure are calculated to be 360.1°F and 72.1 psia from the ratios T/T_0 and P/P_0 at Ma = 1.5.
(3) The $(fL_{max}/D)_i$ is calculated as in Step 3 of Part a but is equal to 0.1659.
(4) From the inlet properties, the friction factor for Re = 338,000 in Fig. 7.1 is 0.015.
(5) The maximum length, given Ma = 1.5, is calculated to be about 20 ft. This is considerably shorter than the actual length of 50 ft and indicates that choking would occur. On the basis of the limited information available, an entrance Mach number of 1.5 is not possible in this example.

A Mach number of 1.5 could only have existed in Example 8.4 had a shock wave formed within the pipe. The shock-wave phenomenon and its influence on the results of Part b of Example 8.4 are reviewed below in the subsection "Shock-Wave Formation."

Analyzing nozzles and orifices

Nozzles and orifices present restrictions to the flow of a fluid. The resulting reduction in cross-sectional area can cause significant changes in the pressure and temperature of a compressible fluid as it passes through the constriction. The magnitude of the change is a function of the linear velocity of the fluid or, more specifically, of the Mach number at the constriction. Figure 8.10 depicts an idealized frictionless nozzle and also shows pressure profiles for different linear velocities and mass flow rates through the nozzle.

Curve A in Fig. 8.10 represents the pressure distribution in the nozzle when the flow remains subsonic throughout. It is typical of most orifices and venturis that are used to measure flow rates. Curve B represents a higher velocity but one which is still below its sonic value. Curve C shows the pressure distribution for a flowing system with sonic velocity at the throat of the restriction but which still has subsonic veloc-

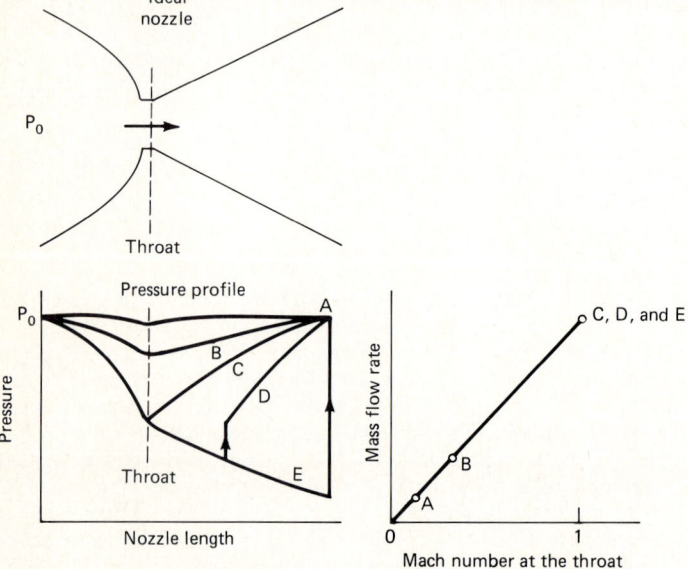

Figure 8.10 Compressible-fluid flow in a nozzle.

ities on both sides of the throat. This curve represents the point at which choking begins. At that condition, the velocity at the throat is the maximum possible for a given mass flow rate with the pressure at the throat being at its minimum value.

Pressure distributions downstream of the constriction are affected by (1) the Mach number at the throat and (2) the back pressure downstream of the nozzle. For the idealized nozzle in Fig. 8.10, all the pressure loss through the nozzle is recovered downstream of the nozzle as long as the velocity remains subsonic. In this condition the back pressure is equal to the pressure upstream of the nozzle. If sonic velocity and choking occur at the throat, as in curves C, D, and E, the downstream velocity could be subsonic, supersonic, or a combination of both. Curve C represents the critical mass flow through the system when the pressure cannot be recovered. Curve E represents a system having a back pressure lower than the pressure at the throat. In this case the pressure difference between the throat and the nozzle exit can be maintained only by supersonic velocities. The abrupt change in pressure in curve D is caused by a shock wave in the nozzle. The shock wave is a phenomenon in compressible-fluid flow which results in the fluid making a transition from a supersonic to a subsonic velocity.

The concepts of shock-wave formation, supersonic velocity, and choking cannot be predicted by using only incompressible fluid-flow relationships; however, the incompressibility assumption can be used for certain flow-constriction conditions.

Incompressible treatment of flow through nozzles and orifices. When compressible-fluid-flow velocities at the throat of a nozzle or at the constriction of an orifice are well below their sonic values, flow conditions may be analyzed by relationships similar to these for incompressible-fluid flow. These relationships (see Crane[17]) may be used as long as the pressure drop across the restriction relative to its inlet pressure is small.

For $\Delta P/P_1 \leq 0.1$,

$$W = 1891 d_o^2 C_n \sqrt{\frac{\Delta P}{\overline{V}_1}} \tag{8.63}$$

For $0.25 \geq \Delta P/P_1 > 0.1$,

$$W = 1891 Y d_o^2 C_n \sqrt{\frac{\Delta P}{\overline{V}_1}} \tag{8.64}$$

where W = mass flow of gas, lb/h
d_o = diameter of throat or orifice, in
C_n = flow coefficient (from Fig. 8.11)
Y = net expansion factor (from Fig. 8.12)
P_1 = upstream pressure, psia
ΔP = pressure drop across throat or orifice, psi
$\overline{V}_1$ = specific volume of inlet gas, ft³/lb

Equations (8.63) and (8.64) assume that friction is the dominant factor affecting pressure drop and flow rate. Figures 8.11 and 8.12, used in conjunction with the equations, represent empirically determined correlations to account for frictional effects and expansion.

It should be noted that Eq. (8.63) is applicable to incompressible fluids as well as

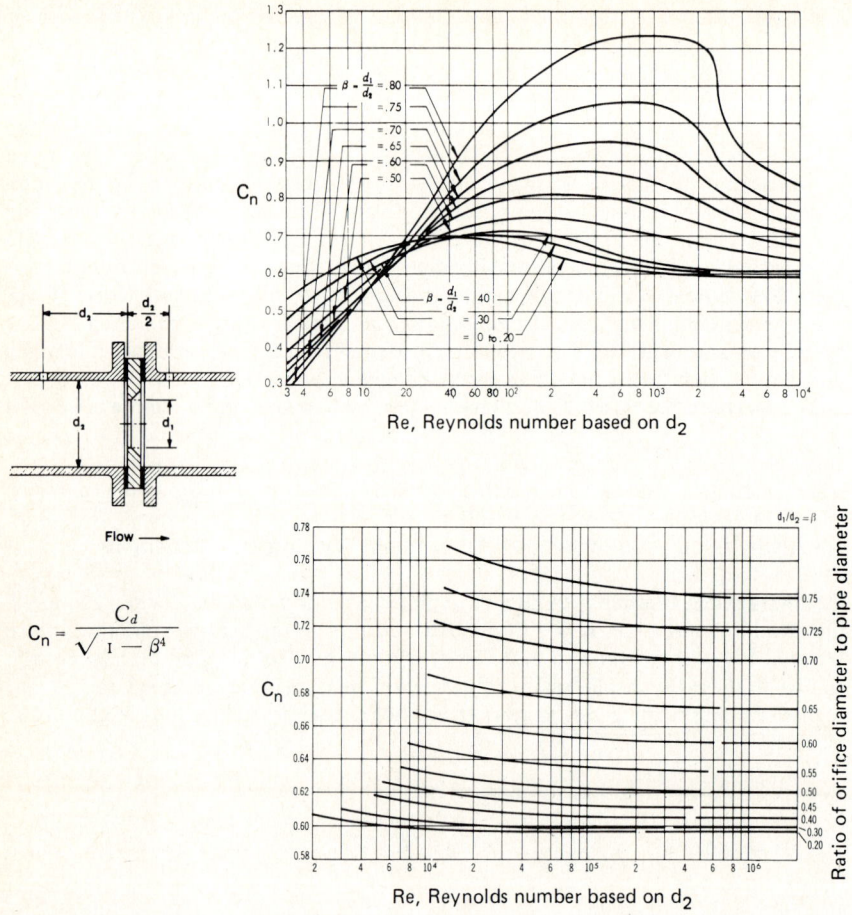

Figure 8.11 Discharge coefficients for a square-edged orifice. *(With permission from Crane Company.)*

compressible fluids as long as flashing does not occur before or in the constriction. Equation (8.64), on the other hand, should be used only with compressible fluids. A comparison of the results obtained by applying both equations to a given set of conditions is given in Example 8.5.

Example 8.5 Ethane vapor in Tank 1 at 100 psig and 60°F flows at the rate of 13,500 lb/h through a 3-in Sch. 40 pipe into Tank 2. There is a restrictive orifice in the line to limit vapor flow between the tanks. The orifice is 2.0 in in diameter and is located 100 equivalent ft after Tank 1 and 600 equivalent ft before Tank 2 (see Fig. 8.13). The pressure at Tank 2 is unknown, but it is known that pressure in the tank is maintained constant by the ethane-vapor line entering it. Find the pressure in Tank 2, assuming (*a*) incompressible-fluid flow and (*b*) incompressible-fluid flow with corrections for compressibility as required.

solution

Stagnation physical properties
$M = 30$ lb/(lb·mol)
$\mu = 0.01$ cP
$\rho_0 = 1/V = 0.656$ lb/ft³
$C_p = 13.5$ Btu/[(lb·mol)·°F]
$k = C_p/(C_p - 1.98) = 1.172$

Pipe relationships
$d = 3.068$ in
$\epsilon = 0.00015$ ft
$A = 0.0513$ ft²

Part a. Assume incompressibility; i.e., the density does not change.

(1) Calculate Reynolds number (Re) to be 2,780,000.
(2) Determine the relative pipe roughness ϵ/D or $12\epsilon/d$ to be 0.00059.
(3) Determine the friction factor f from Re and ϵ/D in Fig. 7.1 as 0.0176.
(4) Determine the friction pressure drop by means of the Darcy equation or the shortcut methods for the total 700 ft of pipe to be 42.1 psi.
(5) The pressure drop across the orifice, using Eq. (8.63), is 5.9 psi.

$k = 1.3$ approximately

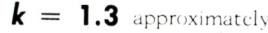

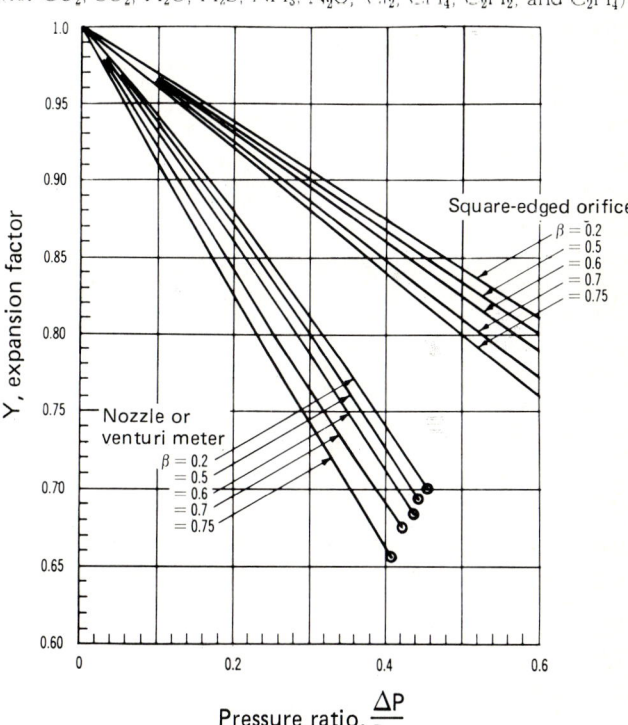

Figure 8.12 Net expansion factor Y for compressible flow through nozzles and orifices. Note that the expansion factor for other k values may be interpolated between or extrapolated linearly from $k = 1.3$ and $k = 1.4$, given on page 322. (*With permission from Crane Company.*)

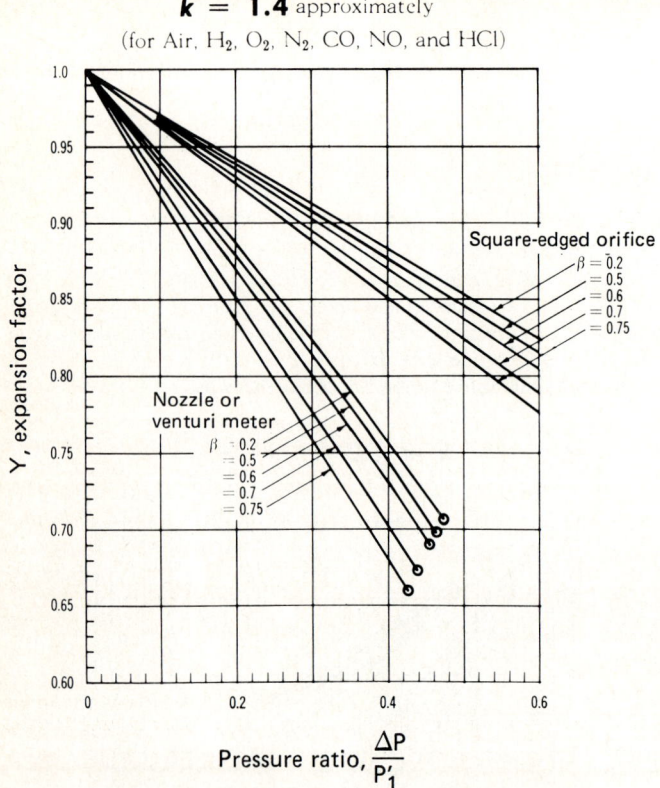

Figure 8.12 (*Continued*) Net expansion factor Y for compressible flow through nozzles and orifices. Note that the expansion factor for other k values may be interpolated between or extrapolated linearly from $k = 1.3$ and $k = 1.4$. (*With permission from Crane Company.*)

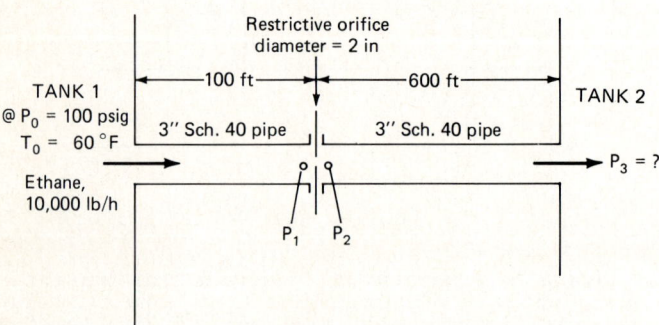

Figure 8.13 Diagram of Example 8.5.

(6) The total pressure drop of 52.9 psi is subtracted from the inlet 100 psig in Tank 1 to give 47.1 psig as the pressure in Tank 2.

Part b. Assume incompressibility with corrections for compressibility as required.

(1) Break the problem into three separate pressure-drop calculations, i.e., pressure drop upstream of the orifice, pressure drop across the orifice, and pressure drop downstream of the orifice.
(2) The upstream Re and f are the same as in Part a: 2,780,000 and 0.0176, respectively.
(3) By use of the Darcy equation, the upstream friction pressure drop for the first 100 ft is 6.0 psi. No compressibility correction is needed since the friction drop is less than 10 percent of the upstream absolute pressure and the density difference is negligible.
(4) The pressure immediately upstream of the orifice is 94.0 psig and is employed to calculate the orifice pressure drop by using Eq. (8.63).
(5) The ethane density, assuming ideality, upstream of the orifice is 0.622 lb/ft³ at 94.0 psig and 60°F.
(6) The ratio of orifice diameter to pipe diameter is 0.65.
(7) From Fig. 8.11 the flow coefficient is 0.67.
(8) According to Eq. (8.63), the pressure drop across the orifice is 12.4 psi. The pressure drop across the orifice is greater than 10% of the absolute pressure (108.7 psia) just upstream of the orifice. Therefore, a correction must be made for compressibility.
(9) By use of Eq. (8.64) and obtaining the net expansion factor Y from Fig. 8.12 by means of an iteration for $k = 1.172$ ($Y = 0.96$), the ΔP across the orifice is calculated to be about 12.4 psia.
(10) The pressure immediately downstream of the orifice is 81.6 psig, and the density of the gas is 0.554 lb/ft³.
(11) The incompressible friction loss in the 600-ft line after the orifice is now found to be 44.1 psi. This is greater than 10 percent of the upstream pressure of 81.6 psig, so that a correction for compressibility must be used when calculating the downstream friction loss.
(12) Recalculation of the line friction loss by Eq. (8.17) or by the shortcut or rigorous methods shows the pressure drop in the downstream leg to be 57.3 psi for a total drop in the system of 75.7 psi. Thus Tank 2 is at about 24.3 psig if compressibility is taken into account, as against 47.1 psig if it is ignored.

Example 8.5 shows that significant errors can occur if the effect of compressibility is not compensated for when required. However, the equations used in solving the example cannot predict such phenomena as choking, the effects of adiabatic or isothermal flow, isentropic expansion or compression, or the creation of shock waves in piping systems carrying gases and vapor. This can be done by using the rigorous approach to analyzing nozzles and orifices when the pressure drop across the unit exceeds 25 percent of the upstream pressure value.

Rigorous analysis of nozzles and orifices. The rigorous method undertakes the analysis of nozzles, orifices, and other cross-sectional restrictions by first determining the Mach number at the throat to see if choking occurs. Equation (8.15) is used to calculate the mass flow rate when there is choking at the throat. If the actual mass flow is less than the maximum flow calculated by Eq. (8.15), no choking occurs. However, if the expected actual flow exceeds the maximum calculated flow, the actual flow must be reduced to the maximum calculated flow value or the throat area increased. The pressure and temperature profiles can then be determined by means of the isentropic-flow equations in Table 8.2 or interpolated from the values in Table 8.5. The equations and table values indicate that the key reference point is the stagnation state except for the cross-sectional area, which uses the area at the critical state as its reference. The resulting property ratios are used to calculate fluid properties at any desired point.

Example 8.6 A converging-diverging nozzle is placed in a 14-in Sch. 40 pipeline and is used to measure the flow rate of ethane which is at 100 psia and 480°F (see Fig. 8.14). The nozzle is assumed to be frictionless and isentropic with the ethane behaving as an ideal gas. If the nozzle throat has a cross-sectional area of 0.1065 ft^2, what are the Mach numbers, temperatures, and pressures of the ethane at the throat inlet and nozzle exit when the ethane mass flows are (a) 9000 lb/h, (b) 90,000 lb/h, and (c) 900,000 lb/h?

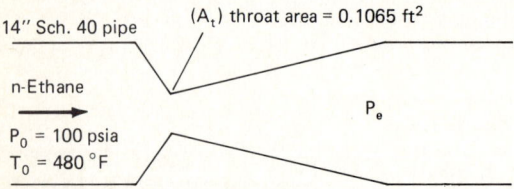

Figure 8.14 Diagram of Example 8.6.

solution The stagnation-state properties for ethane are 100 psia, 480°F, and 0.297 lb/ft^3. The specific-heat ratio is assumed to be constant at 1.12. The sonic velocity for the ethane, by Eq. (8.14), is 1320 ft/s. The cross-sectional area for a 14-in standard steel pipe is 0.940 ft^2. The isentropic-flow equations in Table 8.2 and the isentropic-flow values in Table 8.5 are used in analyzing this example.

Part a. Flow at 9000 lb/h.

(1) The inlet linear velocity is 8.96 ft/s.
(2) The Mach number, calculated by dividing the linear velocity by the sonic velocity, is 0.0068. At low Mach numbers, T/T_0 and P/P_0 approach unity. Therefore, the temperature and pressure at the nozzle inlet are essentially equal to the stagnation-state values of 100 psia and 480°F.
(3) The throat cross-section which results in choking is calculated from the A/A^* ratio, which is found to be 143. Thus, the throat A^* which causes choke is 0.0065 ft^2.
(4) Since the actual throat area A_t is 0.1065 ft^2, it is larger than A^* and no choking occurs at the throat.
(5) The actual Mach number at the throat is found by determining the ratio A_t/A^*, which is 16.23. The ratio A_t/A^* is then substituted into the area-ratio equation, and the actual Mach number at the throat is found, by successive iteration, to be about 0.06.
(6) From the temperature-ratio equation, the T/T_0 value is about 0.998 and T at the throat equals 479.8°F.
(7) From the pressure-ratio equation, P/P_0 equals 0.998 and P at the throat equals 99.8 psia.

Because velocity at the throat is subsonic, the velocity in the diverging section is subsonic. Since the nozzle exit has the same cross-sectional area as the entrance with the flow being frictionless and isentropic, there is a 100 percent pressure recovery and the exit conditions are the same as the inlet conditions, so that

$$T_{inlet} = T_{exit} = 480°F$$
$$P_{inlet} = P_{exit} = 100 \text{ psia}$$

Part b. Flow at 90,000 lb/h.

(1) Proceed as in Part a, Steps 1 through 4, for the following results:

V_i = 89.6 ft/s

Ma_i = 0.0679

T_i = 479.7°F

P_i = 99.7 psia

$A^\star$ = 0.0658 ft² $<$ A_{throat} = 0.1052 ft²

Since the throat area (0.0658 ft²) required for choking is less than the actual area (0.1052 ft²), choking does not occur.

(2) Proceed as in Part a, Steps 5 through 7, resulting in

$A_i/A^\star$ = 1.627

Ma_{throat} = 0.97

T/T_0 = 0.948 and T = 431°F

P/P_0 = 0.615 and P = 61.5 psia

Because throat velocity is subsonic, as in Part a, the temperature and pressure are the same at the exit and the entrance.

Part c. Flow at 900,000 lb/h.

(1) By inspection, the Mach number at the nozzle entrance appears to be greater than 0.5. Therefore, the nozzle-entrance Mach number must be found by substituting 900,000 lb/h into Eq. (8.58) and iterating. The Mach number then calculates to be 0.452.

(2) The entrance and nozzle properties are calculated as in Part a, Steps 2 through 4; i.e.,

T_i = 468°F

P_i = 89.3 psia

$A^\star$ = 0.486 ft²

Since the throat area (0.486 ft²) for choking is greater than the actual area (0.1052 ft²), choking (Ma = 1) occurs at the throat and a flow of 900,000 lb/h is impossible.

(3) By use of Eq. (8.58), maximum flow through the nozzle is found to be 90,100 lb/h.

(4) Since Ma = 1 at the throat, the pressure and the temperature at the nozzle throat can be calculated by the appropriate equations to be 58.1 psia and 428°F, respectively.

Because the throat is at the critical state, the gas continues to accelerate in the divergent section to create supersonic velocities.

(5) The supersonic Mach number at the exit is found by setting $A^\star$ = A_{throat} and calculating the Mach number from the area equation. The exit Mach number is 2.691.

(6) The exit temperature and pressure are then calculated to be 200.3°F and 3.43 psia, respectively.

Example 8.6 illustrates the characteristics of an ideal gas flowing through an ideal pipeline discontinuity (variable cross-sectional area) when there is no back pressure influencing the exit conditions of the nozzle. But nonideal gases flowing through a nonideal discontinuity such as a square-edged orifice behave differently.

Actual compressible flow through nozzles and orifices. The above analysis assumed frictionless flow or no change in the entropy of the fluid so that the flow was isentropic. Whenever a gas flowing through an actual piping system experiences a change in cross-sectional area, the change cannot be approximated accurately by the assumption of isentropic flow. Cross-sectional-area discontinuities in valves, orifices, and other fittings have attendant frictional effects. These departures from isentropic flow are accounted for by using nozzle efficiencies and discharge coefficients. These coefficients can be derived from the thermodynamic relationship for a difference in fluid enthalpy. The difference in enthalpy between a flowing gas and its stagnation state was defined in the subsection "Compressible-Fluid-Flow Parameters" by Eqs. (8.53) and (8.54). The resulting combination is

$$H_0 - H = \alpha \frac{u^2}{2g_c} = C_p(T_0 - T) \tag{8.65}$$

However, for frictional nonisentropic flows, Eq. (8.65) requires an additional factor so that it takes the form

$$H_0 = H + \alpha \frac{u^2}{2g_c} + \Sigma F \tag{8.66}$$

where ΣF = frictional effects.

The frictional forces cause a reduction in the stagnation enthalpy so that

$$H'_0 = H_0 - \Sigma F = H + \alpha \frac{u^2}{2g_c} \tag{8.67}$$

where H'_0 = reduced stagnation enthalpy due to frictional effects.

The new reduced stagnation enthalpy has less energy to be transformed into kinetic energy of gas velocity, but its entropy is greater. For instance, Fig. 8.15 shows a gas

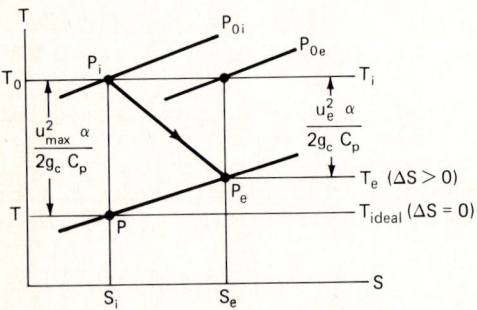

where $\alpha = 1/778$ (Btu/lbm)/(ft·lbf/lbm)

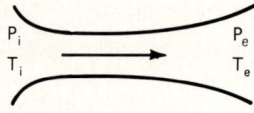

Nozzle efficiency: $\eta = \dfrac{T_i - T_e}{T_i - T_{ideal}}$

Figure 8.15 Temperature-entropy diagram showing the effect of nozzle efficiency on isentropic flow.

entering a nozzle having properties defined as inlet pressure (P_i), inlet temperature (T_i), inlet velocity (u_i), inlet enthalpy (H_i), and inlet entropy (S_i).

The inlet stagnation enthalpy is represented by Eq. (8.53) as

$$H_{0i} = H_i + \alpha \frac{u_i^2}{2g_c} \tag{8.68}$$

and the stagnation entropy for isentropic conditions is, by definition,

$$S_{0i} = S_i \tag{8.69}$$

The stagnation or inlet enthalpy and entropy can be used to define the stagnation temperature T_{0i} and stagnation pressure P_{0i} of the gas.

The pressure of the gas leaving the nozzle is defined as P_e, and the exit temperature of an adiabatic, isentropic process is defined as T_{ideal}. This temperature can be substituted into Eq. (8.56), forming

$$u_{\max} = \sqrt{2g_c C_p (T_i - T_{\text{ideal}})} \tag{8.70}$$

Although there are frictional effects in the nozzle, the exit pressure P_e is essentially the same as that for isentropic conditions. However, the exit entropy increases to S_e. This causes a change in stagnation properties at the exit* and an increased exit temperature T_e above the exit temperature T_{ideal} had the process been isentropic.

The velocity for the nonisentropic-flow nozzle is calculated from

$$u_e = \sqrt{2g_c C_p (T_i - T_e)} \tag{8.71}$$

These relations are represented graphically in Fig. 8.15.

The deviations from ideality are expressed as nozzle efficiency η and are defined as

$$\eta = \frac{T_i - T_e}{T_i - T_{\text{ideal}}} = \frac{H_{\text{actual}}}{H_{\text{isentropic}}} \tag{8.72}$$

If Eqs. (8.70) and (8.71) are substituted into Eq. (8.72), the relationship is

$$\eta = \left(\frac{u_e}{u_{\max}}\right)^2 = \left(\frac{u_{\text{exit,actual}}}{u_{\text{exit,isentropic}}}\right)^2 \tag{8.73}$$

The nozzle efficiency is used to determine discharge coefficients such as shown in Fig. 8-11. The relationship between the nozzle efficiency and the discharge coefficient C_n is defined as

$$C_n = \sqrt{\eta} \tag{8.74}$$

The mass flow rate for nonisentropic changes due to a reduced area can be calculated by substituting Eq. (8.74) into Eq. (8.58):

$$G = \frac{W}{A} = 144 \, C_n P_0 \text{Ma} \sqrt{\left(\frac{kg_c M}{ZRT_0}\right)\left(1 + \frac{k-1}{2} \text{Ma}^2\right)^{(k+1)/(1-k)}} \tag{8.75}$$

---

*Changes in stagnation properties due to frictional flow were reviewed in the preceding portion of this subsection.

Equation (8.58) was developed for a Mach number of 1.0. It was revised for use with any Mach number in Eq. 8.75. Thus, for critical flow (Ma = 1)

$$G = \frac{W}{A^*} = 144\, C_n P_0 \sqrt{\left(\frac{kg_c M}{ZRT_0}\right)\left(\frac{2}{k+1}\right)^{(k+1)/(k-1)}} \qquad (8.76)$$

Equations (8.75) and (8.76) are the basic equations used to develop the safety-relief-valve-sizing formula such as that presented by the ASME:[18]

$$W = CKAP \sqrt{\frac{M}{ZT}} \qquad (8.77)$$

The variables in Eqs. (8.76) and (8.77) are similar, as can be seen by the following comparison:

Eq. (8.76)	Eq. (8.77)
W	W
C_n	K
$\sqrt{\left(\dfrac{kg_c}{R}\right)\left(\dfrac{2}{k+1}\right)^{(k+1)/(k-1)}}$	C
P_0	P
M	M
T_0	T

Example 8.7 Ammonia gas (MW = 17) at 120°F and 286.4 psia (Z = 0.84) flows through a square-edged orifice, as shown in Fig. 8.16. The duct is 14-in Sch. 40 pipe, and the orifice has an area of 0.1065 ft². What is the flow through the orifice when the pressure downstream of the orifice is 100 psia, assuming (*a*) the orifice behaves as an ideal isentropic nozzle and (*b*) the nonisentropic behavior of the orifice can be approximated by using Fig. 8.11?

solution

Part *a*. The pressure change from 186.4 to 100 psia is caused by the transformation of a portion of the ammonia stagnation-state enthalpy into kinetic energy. Initially, calculations would be made to determine whether sonic velocity (Ma = 1) is achieved at the

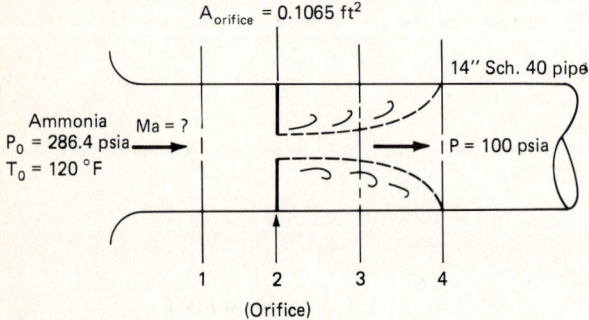

Figure 8.16 Diagram of the compressible-flow process in Example 8.7.

orifice. The isentropic-flow equations in Table 8.2 or interpolated values from Table 8.5 may be used in solving this example.

(1) The ratio of pipe cross-sectional area to orifice area is 8.82.
(2) From the area-ratio equation for isentropic flow, the Mach number for the ammonia upstream of the orifice is found to be 0.0676.
(3) Based on $P/P_0 = 0.572$ at the orifice and assuming Ma = 1 there, the pressure at the orifice is 163.9 psia.
(4) From the isentropic-flow relationships of Table 8.2 or Table 8.5, the temperature and pressure of the ammonia upstream of the orifice are calculated to be 119.8°F and 285.6 psia, respectively.

The pressure at the throat is greater than the stated downstream pressure of 100 psia. Thus, there is no pressure recovery downstream of the orifice, and the downstream flow is supersonic when the Mach number at the orifice is 1.

(5) By substituting Ma = 0.0676 into Eq. (8.58), the mass flow is determined as 0.483 lb/s. According to the law of the conservation of mass, the mass flow rate is constant throughout the system during steady-state flow conditions.
(6) The temperature and pressure of the ammonia at the orifice (point 2, Fig. 8.16), where the Mach number is unity, are 77°F and 163.9 psia, respectively.
(7) Because supersonic-flow conditions exist downstream of the orifice, the properties of ammonia at point 4 are not the same as at point 1 even though the gas is allowed to expand to the original pipe diameter. From the equation for A/A^* for isentropic flow, the Mach number at point 4 is 3.086 for an A/A^* ratio of 8.821. It should be noted that the same equation was used in Step 2 to find the Mach number at point 1. There are always two solutions for a single A/A^*, one subsonic (Ma < 1) and one supersonic (Ma > 1). However, the equation converges to a single solution at Ma = 1.
(8) The temperature and pressure at point 4 for Ma = 3.086 are -130.8°F and 4.72 psia, respectively.
(9) The Mach number at point 3 is found by using the isentropic-flow equation for P/P_0, where $P = 100$ psia and $P_0 = 286.4$ psia. The Mach number thus calculated is 1.397.
(10) The Mach number at point 3 is used to calculate the temperature at point 3 by the relationship for T/T_0. T/T_0 is found to be 1.156, and the temperature is 41.65°F.

The following table summarizes the properties along the length of the pipe when frictionless isentropic flow is assumed:

Point (Fig. 8.16)	1	2 (orifice)	3	4
Mass flow, lb/s	0.483	0.483	0.483	0.483
Mach number	0.0676	1.0	1.397	3.086
Temperature, °F	119.8	77.0	41.65	-130.8
Pressure, psia	285.6	163.9	100.0	4.72

Part b. Nonisentropic behavior is assumed. Portions of the calculations for Part a are used in solving Part b. However, the discharge coefficient must be calculated in order to determine the nonisentropic behavior of the orifice.

(1)–(4) See corresponding steps in Part a.
 (5) The flow regime for ammonia is nearly turbulent. From Fig. 8.11, the discharge coefficient C_n for the orifice is found to be 0.6 for an orifice-diameter-to-pipe-diameter (d_0/d_i) ratio of 0.337 at higher Reynolds numbers.

(6) From Eq. (8.58), the nonideal mass flow is calculated to be 0.27 lb/s with Z_0 assumed to be unity. However, Z_0 is actually 0.84, so that the calculated nonideal mass flow is 0.320 lb/s.

(7) The nonisentropic behavior causes an increase in ammonia temperature over that of isentropic flow. From Eq. (8.72), with η equal to 0.36, as calculated by Eq. (8.74), the temperatures at points 2, 3, and 4 are now found to be 105°F, 92°F, and 30°F, respectively.

The properties of the ammonia, after correction for nonisentropic behavior of the gas and the orifice, are approximated as follows:

Point	1	2 (orifice)	3	4
Mass flow, lb/s	0.32	0.32	0.32	0.32
Compressibility factor, Z	0.84	0.88	0.91	0.92
Heat capacity ratio, k	1.16	1.18	1.19	1.36
Mach number, Ma	0.07	1.00	1.4	3.1
Temperature, °F	119.8	103	87	−13
Pressure, psia	285.8	163	100	6.5

Frictional effects in compressible-flow systems result in higher gas temperatures than in frictionless systems. In addition, the presence of friction causes a change in the stagnation properties of the gas. The relationship between friction and changes in stagnation properties was presented in a preceding subsection on adiabatic and isothermal frictional flow in ducts of constant cross-sectional area.

The concept of shock waves which accompany supersonic flows and the creation of abrupt changes in the properties of a flowing gas are reviewed next.

Shock-wave formation

Whenever supersonic velocities occur in nozzles, orifices, or pipelines, there is a likelihood that a shock wave will form. A shock wave is a natural phenomenon in a compressible-fluid-flow system and causes an abrupt change in the Mach number and the properties of the fluid. A shock wave can occur only if the Mach number of the fluid upstream of the shock point is greater than unity; i.e., flow is supersonic. Shock waves cannot occur with subsonic velocities since that would be a violation of the laws of thermodynamics. A shock wave occurs in systems having a back pressure greater than the pressure wave being produced in the supersonic-flow stream. The fluid adjusts its properties internally to resolve this difference and create a "normal shock wave."* This results in an increase of entropy.

Figure 8.17 is a schematic representation of a normal shock wave and the changes that take place across it. By convention, a subscript x defines the properties ahead of the shock wave, while a subscript y denotes properties after it. The mechanics of normal-shock-wave formation is rightfully the subject for a gas dynamics text, e.g., Shapiro,[18] John,[19] or Saad.[20] The purpose of this discussion is to note its existence and review its impact on the properties of a flowing fluid.

---

*Shock waves have small but finite thickness. A normal shock wave is oriented perpendicularly to the direction of flow; oblique shock waves are not considered here.

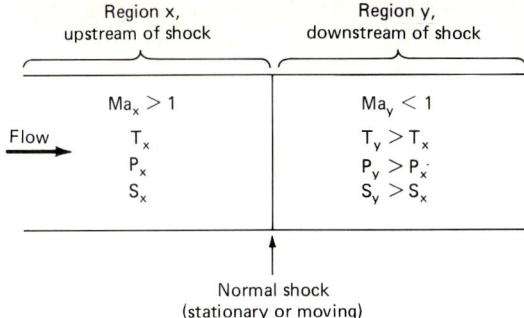

Figure 8.17 Schematic diagram of a normal shock wave.

The changes in fluid properties across a normal shock wave can be calculated. The equations for normal shock waves in Table 8.2 can be used to evaluate these changes. The values listed in Table 8.6 for select Mach numbers may be interpolated to facilitate calculations. Examples of the application of the equations or table are now presented.

Example 8.8 The ethane vapor in Example 8.6 passes through a converging-diverging nozzle and is expected to experience varying back pressures P_e at the exit of the nozzle (see Fig. 8.16). Assume that ethane behaves as an ideal gas and that choking exists at the nozzle throat. Determine the pressure and temperature at the exit of the nozzle and at the upstream and downstream side of the shock wave when (*a*) the normal shock occurs at the nozzle throat, (*b*) the normal shock occurs where the nozzle discharge area is 0.5 ft², and (*c*) the normal shock occurs at the junction of the nozzle discharge and the 14-in Sch. 40 pipe.

solution

Part *a*. Normal shock at the throat.

(1) Example 8.6 determined that the mass flow for ethane with a Mach number equal to unity at the throat is 90,100 lb/h at a temperature of 428°F and a pressure of 58.1 psia.
(2) From the Prandtl equation, $Ma_x = Ma_y = 1$, and the gas velocity remains subsonic downstream of the throat.
(3) The Mach number at the nozzle exit is calculated to be 0.068 by using the A/A^* relationship between the nozzle-exit cross-sectional area and the throat area.
(4) The Mach number is used to calculate an exit temperature and pressure of 479.7°F and 99.7 psia, respectively.

Part *b*. The normal shock occurs where the nozzle downstream area is 0.5 ft².

(1) Refer to Step 1 in Part *a*.
(2) The Mach number at $A = 0.5$ ft² is calculated as in Step 3 in Part *a*. The Mach number must be supersonic and is found to be 2.60. This is the Mach number upstream of the shock wave (Ma_x).
(3) The temperature and pressure are calculated from T/T_0 and P/P_0 for Ma_x, respectively, and are found to be 213.5°F and 4.14 psia.
(4) The Mach number downstream of the shock wave (Ma_y) is calculated for Ma_y as a function of Ma_x and is found to be 0.431.

(5) The temperature and pressure at Ma_y are 469.9°F and 29.36 psia, respectively, as calculated for the ratios T_y/T_x and P_y/P_x.
(6) Because a shock wave results in an increase in entropy, it also causes a change in stagnation state. The new stagnation temperature and pressure are 480°F and 32.6 psia, respectively. These are calculated from $T_{0y}/T_{0x} = 1$ and $P_{0y}/P_{0x} = 0.326$ at Ma = 2.60. The new stagnation state is used to calculate the nozzle-exit conditions.
(7) Since the exit area is known, an A^* must be calculated to determine the exit Mach number. At $Ma_y = 0.431$, $A/A^* = 1.528$. A_y (area at the shock) is given as 0.5 ft². Therefore, A^*, the choking area, is calculated from $(A^*/A) \times A_y$ to be 0.327 ft².
(8) The ratio of exit area to the choking area is calculated to be 2.87. From this area ratio, a Mach number at the exit is determined as 0.213.
(9) The nozzle-exit temperature and pressure are calculated to be 477.5°F and 31.7 psia, respectively.

Part c. Normal shock at the nozzle exit.

(1) The Mach number, temperature, and pressure values upstream of the shock wave are the same as those calculated in Example 8.6, Part c: $Ma_x = 2.691$, $T_x = 200.3°F$, and $P_x = 3.43$ psia.
(2) The downstream Mach number (Ma_y) at the nozzle exit is calculated as in Step 4 in Part b, so that Ma_y is 0.421.
(3) The temperature and pressure downstream of the shock wave when it is at the exit are 470°F and 26.0 psia, respectively. These are calculated from the ratios T_y/T_x and P_y/P_x and $Ma_x = 2.691$.

Example 8.9 In Example 8.4, Part b, it was determined that choking occurred and that a Mach number of 1.5 at the pipe inlet could not be maintained if friction alone were considered. Determine the effect of shock-wave formation in the pipe and the resulting gas properties at the pipe discharge.

solution A Mach number of 1.5 in Example 8.4, Part b, could possibly occur if, as noted earlier, sufficient back pressure at the pipe discharge existed to cause the formation of a shock wave within the pipe. The analysis of a shock wave requires an iterative solution using an estimate of where the shock wave is located, calculating the frictional losses in the supersonic regime, calculating the property changes across the normal shock wave, and finally calculating the frictional losses in the subsonic regime. The steps involved in this analysis are as follows:

(1) Establish the location of the normal shock wave in the pipe. The first estimate should be at the pipe inlet ($L = 0$ ft).
(2) Calculate $(fL_{max}/D)_x$ just before the shock wave by the difference between fL_{max}/D at the pipe inlet and fL_x/D, where L_x is the distance of the shock wave from the inlet.
(3) Calculate the Mach number (Ma_x) from the $(fL_{max}/D)_x$ determined in Step 2. Use this Mach number to calculate the gas properties P_x, T_x, P_{0x}, T_{0x} ahead of the shock wave.
(4) Calculate the Mach number (Ma_y) after the shock wave by using the equations in Table 8.2. Use this Mach number to calculate the change in properties across the shock wave, i.e., P_y, T_y, and P_x. Note that $T_{0y} = T_{0x}$.
(5) Calculate $(fL_{max}/D)_y$ just after the shock wave from Ma_y.
(6) Calculate fL_{max}/D at the pipe discharge by the difference between $(fL_{max}/D)_y$ after the shock wave and fL_y/D, where L_y is the distance between the shock wave and the pipe discharge.
(7) Calculate the Mach number at the pipe outlet by using fL_{max}/D calculated in Step 6.
(8) Calculate the discharge properties by using the discharge Mach number. If the dis-

charge pressure is greater than the actual back pressure, go back to Step 1 and try a shock-wave location farther down the pipe. If the discharge pressure is lesser in value, try a location closer to the inlet. If the trial location is at the inlet, the Mach number at the inlet cannot exist.

Table 8.7 shows the gas discharge properties for a normal shock wave located at five different points in Example 8.9. It should be noted that choking occurs when the normal shock wave is 4.7 ft from the inlet and that a Mach number of 1.5 cannot be maintained at the pipe inlet if the normal shock wave moves farther down the pipe or the back pressure drops below 89 psia.

TABLE 8.7 Effect of Normal Shock-Wave Location on Gas Properties (Example 8.9)

Distance of normal shock wave from pipe inlet	0 ft	2.0 ft	4.0 ft	4.5 ft	4.7 ft*
Properties ahead of shock wave					
Ma_x	1.5	1.434	1.366	1.348	1.341
T_x, °R	820	836	853	857	859
P_x, psia	72	76	81	82	82.6
T_{0x}, °R	1060	1060	1060	1060	1060
P_{0x}, psia	250	240	231	229	228
$(fL_{max}/D)_x$	0.166	0.1359	0.1058	0.0984	0.0950
Properties after shock wave					
Ma_y	0.478	0.515	0.560	0.572	0.578
T_y, °R	998	993	988	987	985
P_y, psia	172.5	166	159	157	156
T_{0y}, °R	1027	1028	1028.3	1028.6	1028.7
P_{0y}, psia	199	195	192.5	192	191
$(fL_{max}/D)_y$	1.442	1.087	0.773	0.702	0.670
Properties at pipe discharge					
Ma	0.572	0.649	0.791	0.873	1.0
T, °R	986	974	951	936	915
P, psia	143	130	110	100	89
T_0, °R	1027.6	1027.7	1028.3	1028.6	1028.7
P_0, psia	175	169	161	158	156
fL_{max}/D	0.702	0.377	0.0934	0.0291	0.0
Mass flow rate, lb/s	655	680	709	718	721

*Choking occurs beyond this point.

REFERENCES

1. A. H. Shapiro, *The Dynamics and Thermodynamics of Compressible Fluid Flow*, vol. I, The Ronald Press Company, New York, 1953.
2. J. E. A. John, *Gas Dynamics*, Allyn and Bacon, Inc., Newton, Mass., 1969.
3. M. A. Saad, *Compressible Fluid Flow*, Prentice-Hall, Inc., Englewood Cliffs, N.J., 1985.
4. *Flow of Fluids through Valves, Fittings and Pipe*, Technical Paper 410, Crane Company, New York, 1982, pp. 3-3–3-5.
5. R. H. Perry, D. W. Green, and J. O. Maloney (eds.), *Perry's Chemical Engineers' Handbook*, 6th ed., McGraw-Hill Book Company, New York, 1984, pp. 5-28–5-32.
6. Shapiro, op. cit., pp. 58, 59.
7. John, op. cit., chap. 6.

8. Saad, op. cit., chap. 3.
9. Shapiro, op. cit., pp. 95–97.
10. Ibid., chap. 6.
11. D. M. Kirkpatrick, "Simpler Sizing of Gas Piping," *Hydrocarbon Processing,* October 1969, pp. 135–138.
12. John, op. cit., chap. 9.
13. Saad, op. cit., chap. 5.
14. C. E. Lapple, *Transactions of the American Institute of Chemical Engineers,* vol. 39, 1943, pp. 385–432.
15. S. H. Tan, "Simplified Flare System Sizing," *Hydrocarbon Processing,* October 1967, pp. 149–154.
16. M. B. Loeb, "New Graphs for Solving Compressible Flow Problems," *Chemical Engineering,* May 19, 1969, pp. 170–184.
17. Crane, op. cit., pp. A-20, A-21.
18. American Society of Mechanical Engineers, ASME Boiler and Pressure Vessel Code, sec. VIII, div. 1, par. UG-131, New York, 1983.
19. Shapiro, op. cit., chap. 16.
20. John, op. cit., chap. 6.
21. Saad, op. cit., chap. 4.

Chapter 9

Piping Calculations, Part 4: Complex Fluids

In addition to sizing lines for gases or liquids by themselves, the process engineer on occasion must deal with lines carrying two-phase fluids. This chapter presents the more often used methods for analyzing these complex fluids.

The analysis of two-phase-flow piping is more complex and less understood than that of incompressible- or compressible-fluid flow. Many correlations are used in analyzing two-phase flow. Govier and Aziz[1] have compiled the most exhaustive listing of these correlations. Different methods are used, depending on the two-phase type, whether gas-liquid, gas-solid, or liquid-solid, and on the orientation of flows, i.e., horizontal or vertical. The methods presented in this chapter were selected on the basis of simplicity, ability to provide acceptable results, and conformity to empirical observations.

Gas-liquid systems are presented first because they are the most widely encountered of the two-phase systems and are the best correlated. The gas-solid systems and liquid-solid systems are then discussed. These systems are more difficult to analyze and, therefore, less accurately correlated.

Gas-Liquid Systems

This subsection presents analysis and design methods used for gas-liquid and vapor-liquid two-phase flow systems. Both gases and vapors correlate similarly in two-phase systems except for certain conditions of continuous vapor condensation or liquid flashing in the flowing system.

Gas-liquid flow regimes

The type of flow regime of a gas-liquid system must be determined before any system analysis can be performed. Figure 9.1 shows the types of flow regimes that can exist in two-phase gas-liquid systems. Table 9.1 lists the characteristic linear velocities of the gas and liquid phases in each flow regime.

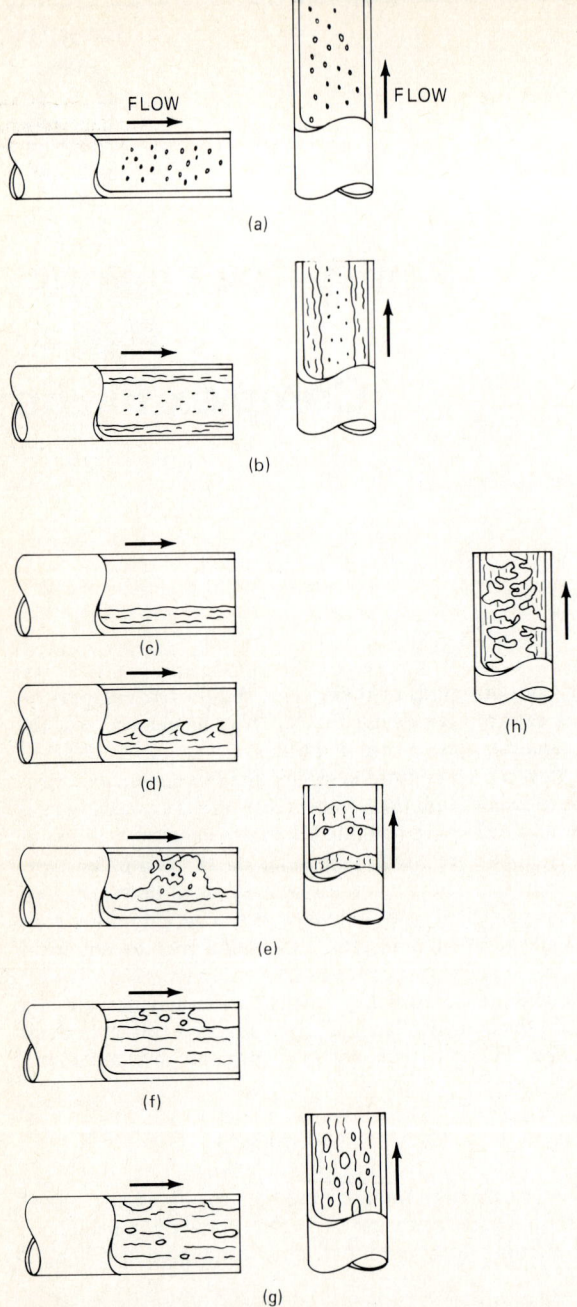

Figure 9.1 Two-phase gas-liquid flow regime. (*a*) Dispersed (spray flow). (*b*) Annular flow. (*c*) Stratified. (*d*) Wavy. (*e*) Slug. (*f*) Plug. (*g*) Bubble. (*h*) Froth (both phases dispersed in vertical flow).

TABLE 9.1 Two-Phase-Flow Regimes and Characteristic Linear Velocities

Regime	Liquid phase, ft/s	Vapor phase, ft/s
Dispersed	Close to vapor velocity	>200
Annular	<0.5	>20
Stratified	<0.5	0.5–10
Slug	15 (but less than vapor velocity)	3–50
Plug	2	<4
Bubble	5–15	0.5–2

Dispersed flow, also referred to as spray or mist flow, occurs at very high gas velocities with the liquid phase dispersed as droplets throughout the gas phase. The liquid-droplet velocity approaches the gas-phase velocity in this flow regime because the droplet terminal velocity is negligible and the slip velocity, i.e., the velocity difference between the gas- and liquid-phase velocities, approaches zero.

Annular flow occurs at relatively lower gas velocities than dispersed flow. The liquid phase forms an annulus about the circumference of the pipe with the gas flowing through the central core. There is significantly more slip with annular flow than with dispersed flow.

Stratified flow occurs only in horizontal pipes when the gas-phase velocity is not great enough to maintain an annulus of liquid about the circumference of the pipe. One form of stratified flow, called "wavy flow," is characterized by the formation of waves on the surfaces of the liquid phase. Wavy flow is formed close to the transition point where stratified flow can be transformed into slug flow with a further increase in gas velocity.

Slug flow is characterized by an intermittent pattern of alternating liquid phases and gas phases along the length of the line. The entire pipe cross-sectional area can be occupied by a "slug" of either liquid or gas at different points along the flow path.

Plug flow occurs when the liquid phase forms a nearly continuous phase with large elongated bubbles, "plugs" of gas located within the liquid phase.

Bubble, or froth, flow, like plug flow, has a dominant liquid phase, but the liquid phase in bubble flow is at a higher velocity than the liquid phase in plug flow. This higher velocity causes the vapor phase to disperse into many smaller bubbles within the liquid phase.

Each flow regime behaves differently, having its own set of empirical correlations for predicting flow behavior. Before flow behavior can be calculated, the type of flow regime must be predicted. Prediction of the various two-flow regimes is very difficult, and many methods have been proposed for their prediction. The most often used method to determine flow regimes is the Baker plot,[2] which has been reproduced in Fig. 9.2. Its horizontal axis is defined as

$$X = \left(\frac{W_L}{W_g}\right) \lambda \psi \tag{9.1}$$

where W_L = liquid mass flow rate, lb/h
W_g = gas mass flow rate, lb/h

$$\lambda = 0.463 \sqrt{\rho_g \rho_L} \tag{9.2}$$

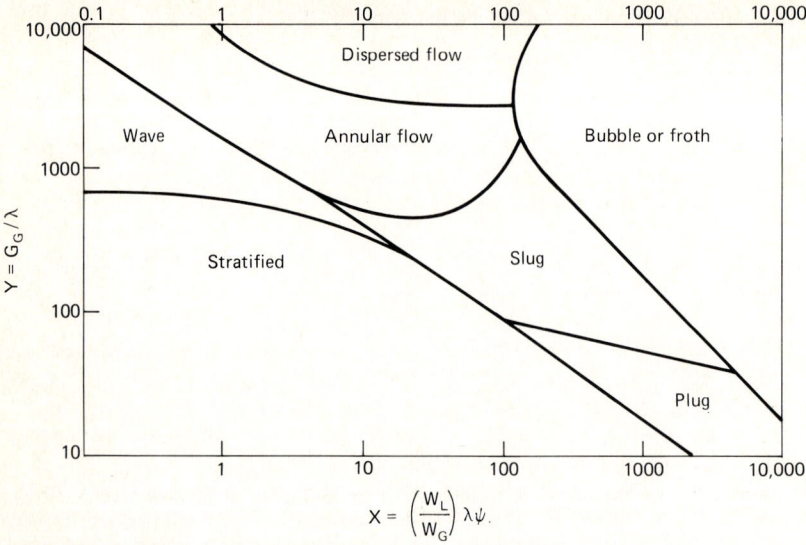

Figure 9.2 Baker plot for a two-phase-flow-regime correlation.

and $\psi = \dfrac{1147}{\sigma}[\mu_L/\rho_L^2]^{1/3}$ (9.3)

where μ_L = liquid viscosity, lb/(ft·h)
ρ_g = gas density, lb/ft^3
ρ_L = liquid density, lb/ft^3
σ = surface tension, dyn/cm

The vertical axis in Fig. 9.2 is defined as

$Y = G_g/\lambda$ (9.4)

where $G_g = (W_g/A)$ = superficial mass flux of vapor phase, lb/(ft^2·h)
A = total internal cross-sectional area of pipe, ft^2

Once the values of X and Y have been calculated, their point of intersection on the Baker plot determines the expected two-phase flow regime. However, the transition boundaries between regimes are not as sharp as the Baker plot indicates. Therefore, some overlap can occur, and the user should be cautious when applying this plot.

Although the Baker plot is based on horizontal flow, it has been used with limited success in estimating vertical flow regimes. Most published correlations for vertical upward flow are for special cases of fluids or flow regimes. Govier and Aziz[3] have proposed a generalized correlation for vertical two-phase flow based on data they compiled from the findings of other investigators.

Pressure-drop calculations for gas-liquid systems

The calculation of pressure drop in a two-phase gas-liquid flow system is more complicated and less accurate than calculating pressure drops in single-phase gas or liquid systems. In the absence of any correlation, a very rough approximation for the two-phase-flow frictional pressure drop can be calculated in either one of two ways.

If the gas phase and the liquid phase are expected to be well mixed and evenly distributed, the mixture can be considered to be pseudo-homogeneous and the pressure drop can be approximated by using the methods described in Chap. 7 for single-phase fluid. The weighted average physical properties are used to calculate the pressure drops for these pseudo-homogeneous systems.

More accurate methods for two-phase flow use different correlations for each flow regime. The most commonly used method was developed by Lockhart and Martinelli.[4] This is one of the simplest and most effective methods for calculating two-phase-flow pressure drop due to friction.

This method requires the calculation of X, the Martinelli two-phase-flow modulus, defined as

$$X^2 = (\Delta P_{100,L})/(\Delta P_{100,g}) \tag{9.5}$$

where $\Delta P_{100,L}$ = pressure drop per 100 ft of pipe, only liquid being assumed to flow through the pipe, psi/100 ft
$\Delta P_{100,g}$ = pressure drop per 100 ft of pipe, only gas being assumed to flow through the pipe, psi/100 ft

An alternative form of calculating the Martinelli modulus by using the Darcy relationship is

$$X^2 = (W_L/W_g)(\rho_g/\rho_L)(f_L/f_g) \tag{9.6}$$

The Martinelli modulus is then used to calculate the two-phase-flow modulus ϕ from a relationship taking the general form

$$\phi = aX^b \tag{9.7}$$

where a and b are empirical constants for the specific flow regime. Equation (9.7) for a specific flow regime is correlated as follows:

1. *Annular flow*

$$\phi = aX^b \tag{9.8}$$

where $a = 4.8 - 0.3125d$
$b = 0.343 - 0.021d$
d = inside pipe diameter, in (If $d > 10$ in, set $d = 10$ in the correlation.)

2. *Bubble flow*

$$\phi = 14.2\, X^{0.75}/(W_L/A)^{0.1} \tag{9.9}$$

3. *Stratified flow*

$$\phi = 15{,}400\, X/(W_L/A)^{0.8} \tag{9.10}$$

4. *Slug flow*

$$\phi = 1190\, X^{0.815}/(W_L/A)^{0.5} \tag{9.11}$$

5. *Plug flow*

$$\phi = 27.315\, X^{0.855}/(W_L/A)^{0.17} \tag{9.12}$$

After the two-phase flow modulus has been calculated, it is used to determine the two-phase pressure drop by

$$\Delta P_{100,\text{TP}} = \Delta P_{100,g}\, \phi^2 \qquad (9.13a)$$

so that $\quad \Delta P_{\text{TP}} = \Delta P_g\, \phi^2 \qquad (9.13b)$

where $\Delta P_{100,\text{TP}}$ = two-phase-flow pressure drop per 100 ft of pipe, psi/100 ft.

Dispersed-flow pressure drop. The Lockhart-Martinelli correlation for dispersed flow requires the calculation of a two-phase-flow modulus ϕ to be used in conjunction with either the liquid-phase pressure drop or the gas-phase pressure drop to calculate the two-phase pressure drop in a pipe; i.e.,

$$\Delta P_{100,\text{TP}} = \Delta P_{100,g}\, \phi_g^2 \qquad (9.14)$$

$$\Delta P_{100,\text{TP}} = \Delta P_{100,L}\, \phi_L^2 \qquad (9.15)$$

The two-phase flow modulus is also dependent on whether each phase is laminar (viscous) or turbulent.* Table 9.2 shows the modulus symbol used in relation to the phase type and flow characteristics.

TABLE 9.2 Martinelli Symbols Used for Two-Phase-Flow Modulus

Symbol	Flow characteristics	
	Liquid phase	Gas phase
1. Liquid-phase moduli		
ϕ	Turbulent	Turbulent
$\phi_{L,tv}$	Turbulent	Viscous
$\phi_{L,vt}$	Viscous	Turbulent
$\phi_{L,vv}$	Viscous	Viscous
2. Gas-phase moduli		
$\phi_{g,tt}$	Turbulent	Turbulent
$\phi_{g,tv}$	Turbulent	Viscous
$\phi_{g,vt}$	Viscous	Turbulent
$\phi_{g,vv}$	Viscous	Viscous

The flow regime of each phase is determined along with their individual pressure drops. The X value and flow regime are used for finding the Martinelli two-phase-flow modulus. Table 9.3 shows the correlation between X and the various possible combinations of laminar and turbulent flow in the gas and liquid phases.

Wave-flow pressure drop. When wave flow is encountered, the pressure drop is calculated by using the Huntington correlation. The Huntington correlation calculates a two-phase-flow friction factor by the following relationship:

$$f_{\text{TP}} = 0.0044 \left(\frac{W_L \mu_L}{W_g \mu_g} \right)^{0.216} \qquad (9.16)$$

---

*The Martinelli correlation considers transitional flow as turbulent flow.

TABLE 9.3 Two-Phase-Flow Moduli as a Function of Martinelli Moduli X

X	$\phi_{L,tt}$	$\phi_{g,tt}$	$\phi_{L,vt}$	$\phi_{g,vt}$	$\phi_{L,tv}$	$\phi_{g,tv}$	$\phi_{L,vv}$	$\phi_{g,vv}$
0.01	(128)	(1.28)	(120)	(1.20)	(112)	(1.12)	(105)	(1.05)
0.02	(64.8)	(1.37)	(64)	(1.28)	(58)	(1.16)	(53.5)	(1.02)
0.04	38.5	1.54	(34)	(1.36)	(31)	(1.24)	(28.0)	(1.12)
0.07	24.4	1.71	20.7	1.45	(19.3)	(1.35)	(17.0)	(1.19)
0.10	18.5	1.85	15.2	1.52	(14.5)	(1.45)	(12.4)	(1.24)
0.20	11.2	2.23	8.9	1.78	(8.7)	(1.74)	(7.0)	(1.40)
0.4	7.05	2.83	5.62	2.25	(5.5)	(2.20)	4.25	1.7
0.7	5.04	3.53	4.07	2.85	(4.07)	(2.85)	3.08	2.16
1.0	4.20	4.20	3.48	3.48	(3.48)	(3.48)	2.61	2.61
2.0	3.1	6.20	2.62	5.25	(2.62)	(5.24)	2.06	4.12
4.0	2.38	9.5	2.05	8.20	(2.15)	(8.60)	1.76	7.0
7.0	1.96	13.7	1.73	12.0	(1.83)	(12.8)	1.60	11.2
10.0	1.75	17.5	1.59	15.9	(1.66)	(16.6)	1.50	15.0
20.0	1.48	29.5	(1.40)	(28.0)	(1.44)	(28.8)	1.36	27.3
40.0	1.29	51.5	(1.25)	(50.0)	(1.25)	(50.0)	1.25	50.0
70.0	1.17	82.0	(1.17)	(82.0)	(1.17)	(82.0)	(1.17)	(82.0)
100.0	1.11	111.0	(1.11)	(111.0)	(1.11)	(111.0)	(1.11)	(111.0)

NOTE: Parentheses indicate area of limited data.

where f_{TP} = two-phase-flow friction factor
W_L = liquid mass flow, lb/h
W_g = gas mass flow, lb/h
μ_L = liquid viscosity, cP
μ_g = gas viscosity, cP

The friction factor is used to calculate the pressure drop by the following equation:

$$\Delta P_{100,TP} = 3.33 \times 10^{-9} \frac{f_{TP} G_g^2}{D \rho_g} \qquad (9.17)$$

where $G_g = W_g/A$ = gas mass flux, lb/(ft²·h)
D = internal pipe diameter, ft
A = internal cross-sectional area of pipe, ft²
ρ_g = density of gas, lb/ft³
$\Delta P_{100,TP}$ = two-phase-flow pressure drop per 100 ft of pipe, psi/100 ft

Effect of hydrostatic head. The contribution of static-head losses can be significant in two-phase flow systems. The significance of static head depends on the relative volume fraction of the liquid phase. The greater the liquid-volume fraction, the greater the static-head loss. This can be explained by a modified version of the static-head contribution of the Bernoulli equation:

$$\Delta P_{static} = (\rho_M z_{static})/144 \qquad (9.18)$$

where ΔP_{static} = pressure difference due to elevation difference, psi
z_{static} = elevation difference relative to some reference, ft

Two-phase mixtures with high liquid-volume fractions have high average densities ρ_M and significant static heads. Therefore, bubble-flow mixtures with high liquid-vol-

volume fractions have significant static heads, while dispersed-flow mixtures with high gas-phase fractions have minor static-head contributions. These two flow regimes are pseudo-homogeneous mixtures, and reasonable approximations of static head can be made by using Eq. (9.18). However, the other flow regimes have more complex heterogeneous mixtures, and their static heads cannot be estimated by using Eq. (9.18).

Baker has proposed a modification to Eq. (9.18) for two-phase systems of the form

$$\Delta P_{\text{static}} = \rho_L z_{\text{static}} E_H / 144 \tag{9.19}$$

where E_H = liquid head factor calculated by

$$E_H = 1.61(u_{sg})^{-0.7} \tag{9.20}$$

where u_{sg} = superficial gas velocity, ft/s.

The correlation has an error of about ±15 percent, probably owing to the fact that it neglects the slip between phases. Slip is the relative linear-velocity difference between the gas and liquid phases and is defined as

$$S = u_g - u_L \tag{9.21}$$

where S = slip velocity, ft/s
u_g = gas-phase linear velocity, ft/s
u_L = liquid-phase linear velocity, ft/s

Because of the relatively high slip between phases in plug, slug, froth, and annular flow regimes, a correlation other than Eq. (9.19) must be used. Govier and Aziz[6] present the most exhaustive analysis on vertical gas-liquid flow and the effects of static head; however, DeGance and Atherton[7] present a summary of the more useful two-phase correlations for vertical flow. Most of the correlations are rather complex and include some relationship for slip. A survey of these relationships is beyond the scope of this book, but they generally take the form of Eq. (9.19), with the slip correction being factored into E_H. For most practical applications, the slip effect is adequately provided for by Eq. (9.20).

Another limitation of vertical two-phase-flow correlations is that they are based on upward-flowing fluids. There has been very little research in correlating downward-flowing mixtures, but the consensus indicates that downward flow does not pose as many problems as upward flow.

When calculating pressure gradients in two-phase flow systems, the hydrostatic-head contribution is treated differently than a single-phase liquid system. In liquid systems, the upward- and downward-flowing lines are added algebraically to determine the net static-head contribution in the piping system. In two-phase flow systems, only the static heads in upward-flowing lines are added together to get the total static head. All downward-flowing lines are assumed to have no static-head contribution. Figure 9.3 shows an arbitrary piping configuration carrying a two-phase mixture. Only the legs z_2, z_4 and z_6 contribute to static head.

Pressure drop through orifices. Murdock's correlation[8] is the easiest to use for calculating pressure drop through an orifice. Murdock tested a wide range of two-phase fluids, liquid mass fractions, and Reynolds numbers. He found the standard deviation for the correlation to be 0.75 percent. The first step of his correlation requires calculation of the expected pressure drop through the orifice as if the liquid and gas

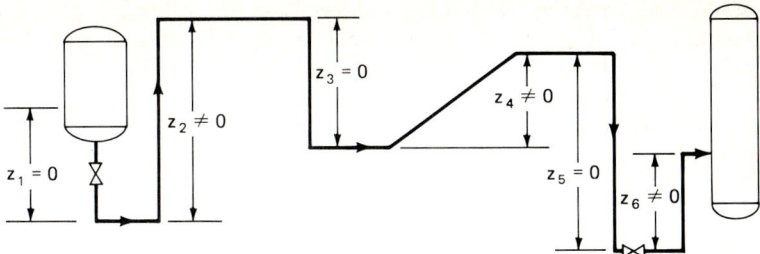

Figure 9.3 A hypothetical piping run between two process vessels in which two-phase flow exists. The drawing shows which vertical runs are used to calculate the total-static-head contribution for the piping system. Total static head = $z_2 + z_4 + z_6$ (upward-flowing runs only), ft. All downward-flowing runs are assumed to have a static head equal to zero.

phases alone pass through the orifice by using the orifice equations

$$W_L = 2407 A_o C \sqrt{\rho_L \Delta P_L} \qquad (9.22)$$

$$W_g = 2407 A_o C \sqrt{\rho_g \Delta P_g} \qquad (9.23)$$

where A_o = orifice-flow area, in²
C = orifice coefficient
W_L = liquid mass flow, lb/h
W_g = gas mass flow, lb/h

The calculated pressure drops for the gas phase alone and the liquid phase alone are then used to calculate the two-phase pressure drop by the equation

$$\sqrt{\frac{\Delta P_{TP}}{\Delta P_g}} = 1 + 1.26 \sqrt{\frac{\Delta P_L}{\Delta P_g}} \qquad (9.24)$$

where ΔP_{TP} = two-phase-flow pressure drop in the dimensions of ΔP_g.

Design criteria for two-phase (gas-liquid) flow systems

As a rule, two-phase (gas-liquid) flow systems should be avoided if possible. If a two-phase mixture is expected to occur in a system, the two phases should be separated into individual streams of nearly homogeneous gas and liquid. Air and steam traps are used to separate liquids from gas streams. Large two-phase flows require the use of flash tanks. The increase in cross-sectional area of the flash tank causes the liquid phase to disengage from the gas phase. The gas exits the top of the tank, while the liquid phase is pumped out of the tank bottom. The two streams continue their flow downstream in two separate pipelines.

However, some process conditions such as condensate in return lines flashing into steam, two-phase feed lines entering distillation columns, or process-plant refrigerant-return lines which must maintain a specific ratio of vapor- to liquid-flow rate for proper operation require or inevitably produce two phases. These two-phase systems must have their piping sized to minimize operating problems. The two problems that must be addressed are the proper flow regime for a given operating condition and the minimization of erosion corrosion.

Mitigating erosion. Depending on the flow regime, the liquid in a two-phase flow system can be accelerated to velocities approaching or exceeding vapor velocities (see Table 9.1). In some cases these velocities are higher than desirable for a process piping system. Such high velocities can cause a phenomenon known as "erosion corrosion" in equipment and piping systems. There are no general correlations that predict the rate of erosion corrosion in piping systems, but Coulson et al.[9] have proposed an index based on velocity head to determine the range of mixture densities and velocities below which erosion corrosion should not occur. The index takes the form

$$\rho_M u_M^2 \leq 10,000 \tag{9.25}$$

If the product of mixture density ρ_M and mixture velocity u_M is below 10,000, erosion corrosion should not be a problem. Mixture density is calculated by

$$\rho_M = \frac{W_L + W_g}{\left(\dfrac{W_L}{\rho_L}\right) + \left(\dfrac{W_g}{\rho_g}\right)} \tag{9.26}$$

and mixture velocity (u_M) is calculated by

$$u_M = u'_g + u'_L = \text{mean velocity, ft/s} \tag{9.27}$$

where u'_g = superficial linear gas velocity, ft/s
u'_L = superficial linear liquid velocity, ft/s
W_g = gas mass flow rate, lb/h
W_L = liquid mass flow rate, lb/h

Example 9.1 A 6-in Sch. 40 pipe has a two-phase mixture flowing through it. The flow rate and density of each phase are as follows:

	Liquid	Vapor
Mass flow rate, lb/h	7200	21,600
Density, lb/ft^3	58	0.7

Determine whether the two-phase flow will experience erosion corrosion.

solution

  a. Mixture density calculated by Eq. (9.26) is 0.930 lb/ft^3.
  b. Given that the pipe cross-sectional area is 0.20 ft^2, the superficial liquid velocity is calculated to be 0.172 ft/s and the superficial gas velocity is 42.86 ft/s.
  c. The mean velocity calculated by Eq. (9.27) is 43.03 ft/s.
  d. The index calculated by Eq. (9.25) is found to be 1722.
Since 1722 < 10,000, erosion is not expected to be a problem.

Maintaining the proper flow regime. When designing piping systems for two-phase flow, it is best to avoid slug flow. Slug flow poses mechanical and process problems, such as the following:

1. Water hammer is caused by the slug of liquid impinging on pipe and equipment walls at every change of flow direction. If slug flow is not eliminated, there may be equipment damage due to erosion corrosion and/or overstressing of joints due to cyclic loading.

2. If slug flow enters a distillation column, the alternating composition and density of the gas and liquid slugs cause cycling of composition and pressure gradients along the length of the column. This cycling causes problems with product quality and process control. The problem of pressure gradients is particularly disturbing in vacuum columns.

Although slug flow is the main flow regime to be avoided, dispersed flow may also be undesirable. Dispersed flow is nearly a homogeneous mixture of the liquid phase in the gas phase. This makes its behavior very similar to that of a compressible fluid. While dispersed flow is an ideal flow regime in piping systems, it causes phase-disengagement problems in flash tanks and distillation columns.

Example 9.2 An uninsulated n-butane line runs between the storage and the processing areas. The butane is transferred at the rate of 25,000 lb/h. The vapor pressure of butane is such that in winter at 0°F it is expected that 5 weight percent of the butane will be vapor, while in summer at 100°F the vapor fraction increases to 30 percent. What size of line should be used to transfer the butane?

solution Physical properties of n-butane (molecular weight = 58.1, T_c = 306°F, P_c = 37.4 atm) at the operating temperatures of 0°F and 100°F:

	$T = 0°F$	$T = 100°F$
Density, lb/ft^3		
ρ_L	38.6	34.8
ρ_v	0.0863	0.552
Viscosity, cP		
μ_L	0.247	0.145
μ_v	0.00655	0.00789
P_{vp} (vapor pressure), atm abs	0.49	3.5
σ (surface tension), dyn/cm	17.2	10.5

Calculation of Baker-correlation parameters by Eqs. (9.2) and (9.3):

$$\lambda = 0.463\sqrt{\rho_v \rho_L} = \begin{cases} 0.463\sqrt{(0.0863)(38.6)} = 0.843 \text{ at } 0°F \\ 0.463\sqrt{(0.552)(34.8)} = 2.028 \text{ at } 100°F \end{cases}$$

$$\psi = \left(\frac{73}{\sigma}\right)\left[\mu_L \left(\frac{62.3}{\rho_L}\right)^2\right]^{1/3} = \begin{cases} \left(\dfrac{73}{17.2}\right)\left[0.247\left(\dfrac{62.3}{38.56}\right)^2\right]^{1/3} = 3.66 \text{ at } 0°F \\ \left(\dfrac{73}{10.5}\right)\left[0.145\left(\dfrac{62.3}{34.8}\right)^2\right]^{1/3} = 5.38 \text{ at } 100°F \end{cases}$$

346 Materials Selection and Piping Calculations

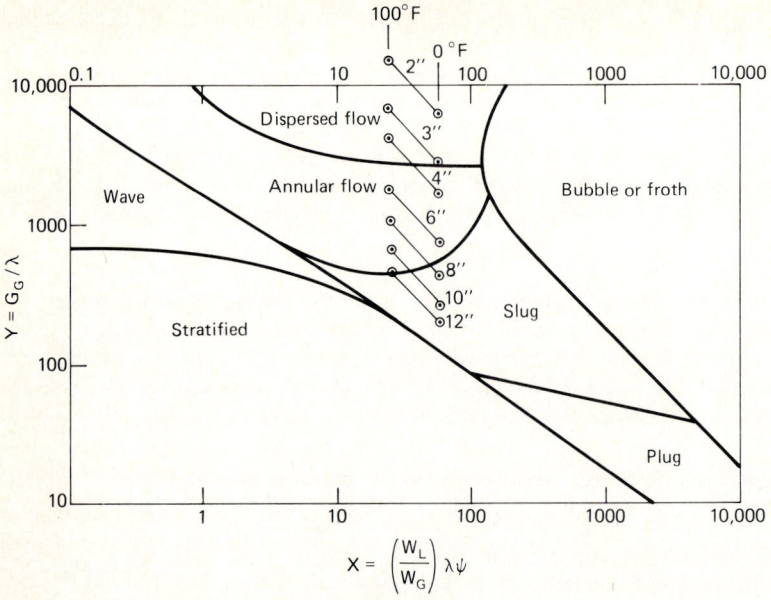

Figure 9.4 Baker plot of the solution to Example 9.2.

Calculation of mass flow rates, $G = W/A$:

	$T = 0°F$	$T = 100°F$
Liquid rate W_L, lb/h	23,750	17,500
Vapor rate W_v, lb/h	1,250	7,500
Total flow rate, lb/h	25,000	25,000
$\dfrac{W_L}{W_V} = \dfrac{W_{L/A}}{W_{y/A}} = \dfrac{G_L}{G_V} =$	19.0	2.3333
$X \text{ (coordinate)} = \left(\dfrac{G_L}{G_V}\right)\lambda\psi =$	58.79	25.49

See Baker correlation.

Calculating the Y coordinate for the Baker correlation for various pipe sizes:

Nominal pipe size	Cross-sectional area, ft²	$T = 0°F$			$T = 100°F$		
		G_v, lb/(ft²·h)	$Y = G_v/\lambda$	Regime	G_v, lb/(ft²·h)	$Y = G_v/\lambda$	Regime
2 in Sch. 40	0.02333	53,600	63,500	Dispersed	320,000	158,600	Dispersed
3 in Sch. 40	0.0513	24,300	28,800	Dispersed	146,000	72,100	Dispersed
4 in Sch. 40	0.0884	14,140	16,750	Annular	84,800	41,800	Dispersed

Piping Calculations, Part 4: Complex Fluids

Nominal pipe size	Cross-sectional area, ft²	T = 0°F			T = 100°F		
		G_v, lb/(ft²·h)	$Y = G_v/\lambda$	Regime	G_v, lb/(ft²·h)	$Y = G_v/\lambda$	Regime
6 in Sch. 40	0.200	6,230	7,380	Annular	37,300	18,420	Annular
8 in Sch. 40	0.347	3,590	4,260	Slug	21,500	10,640	Annular
10 in Sch. 40	0.547	2,280	2,700	Slug	13,700	6,750	Annular
12 in Sch. 40	0.777	1,608	1,905	Slug	9,650	4,750	Annular*

*Borderline with slug-flow regime.

See the above data plotted on Baker correlation chart shown in Fig. 9.4.

Conclusion. To operate out of the slug-flow regime at all times, the pipe size must be 6 in or smaller. For additional conservatism, set the size criteria at 4 in or smaller.

Calculation of two-phase-flow pressure drop in 4-in and 6-in lines:

Line size	4-in Sch. 40	6-in Sch. 40
Inside pipe diameter, in	4.026	6.065
Liquid ΔP calculation		
W at 0°F, lb/h	23,750	23,750
W at 100°F, lb/h	17,500	17,500
N_{Re} at 0°F	150,700	100,000
N_{Re} at 100°F	189,100	125,500
f at 0°F	0.0199	0.0199
f at 100°F	0.0194	0.0193
P_{100}, psi/100 ft at 0°F	0.0919	0.01188
P_{100}, psi/100 ft at 100°F	0.0542	0.00694
Vapor ΔP calculation		
W at 0°F, lb/h	1,250	1,250
W at 100°F, lb/h	7,500	7,500
N_{Re} at 0°F	299,000	198,500
N_{Re} at 100°F	1,489,000	988,000
f at 0°F	0.1881	0.01833
f at 100°F	0.01771	0.01651
P_{100}, psi/100 ft at 0°F	0.1079	0.01355
P_{100}, psi/100 ft at 100°F	0.570	0.0686
$X^2 = \Delta P_{100}^{liq}/\Delta P_{100}^{vap}$ at 0°F	0.852	0.876
$X^2 = \Delta P_{100}^{liq}/\Delta P_{100}^{vap}$ at 100°F	0.905	0.1011
ϕ^2 at 0°F	3.46 (annular correlation)*	2.86 (annular correlation)*
ϕ^2 at 100°F	2.61 (dispersed correlation)†	2.26 (annular correlation)*
ϕ^2 at 0°F	12.038	8.20
ϕ^2 at 100°F	6.8	5.15
P (two-phase) $= P_{100}^{vap} \phi^2$ at 0°F	1.299 psi/100 ft	0.111 psi/100 ft
P (two-phase) $= P_{100}^{vap} \phi^2$ at 100°F	3.88 psi/100 ft	0.353 psi/100 ft

NOTE: f is based on a roughness $\epsilon = 0.0002$.
*Calculated by using Eq. (9.7).
†Calculated by using Eq. (9.8).

Summary

For short lines, < 100 ft, suggest the use of 4-in-diameter pipe with expected two-phase pressure drop ranging from 1.3 to 3.9 psi/100 ft.

For long lines, > 100 ft, suggest the use of 6-in-diameter pipe with expected two-phase pressure drop ranging from 0.11 to 0.35 psi/100 ft.

The 4-in lines may have either dispersed or annular two-phase flow, while 6-in lines have annular two-phase flow.

Fluid-Solid Systems

Fluid-solid systems such as found in slurry lines and in pneumatic conveying piping pose problems that differ from those of other two-phase systems. A solid phase does not readily flow through a pipe unless the pipeline is inclined downward at an angle greater than the angle of repose of the solid. In horizontal lines and vertically rising lines, solid flow must be induced by a fluid phase traveling at a linear velocity sufficient to fluidize the solid phase. The minimum fluid linear velocity required to induce solid fluidization depends on characteristic parameters of both the fluid and the solid phases. One of the more important parameters is the terminal velocity of the solid phase in a given fluid.

Terminal velocity

When a solid particle is dropped in a vacuum, the force of gravity F_g causes the solid to accelerate as it falls. However, if a solid particle is dropped into a stationary fluid, the viscous properties of the fluid will cause a drag force F_D on the solid counteracting the force of gravity. This results in a deceleration of the solid until the drag force is in equilibrium with the force of gravity; i.e.,

$$F_g = F_D \tag{9.28}$$

At this point, the solid particle will fall at a constant velocity. This constant velocity is called the terminal velocity of the solid particle u_t.

The terminal velocity of a solid spherical particle can be calculated by recognizing that the gravitational force can be equated to

$$F_g = \left(\frac{\pi D_s^3}{6}\right) g(\rho_s - \rho_f) \tag{9.29}$$

where D_s = solid-particle diameter, ft
g = gravitational acceleration, 32.17 ft/s^2
ρ_s = solid density, lb/ft^3
ρ_f = fluid density, lb/ft^3

The drag-force contribution can be expressed as

$$F_D = \frac{\pi}{8} D_s^2 C_D \rho_f u_t^2 / g_c \tag{9.30}$$

where C_D = drag coefficient
u_t = terminal velocity, ft/s

For nonspherical particles, the term $\left(\frac{\pi}{8} D_s^2\right)$ can be replaced by the more generalized parameter A_p divided by 2. A_p is the projected area of the solid normal to flow.

The terminal velocity can be calculated by substituting Eqs. (9.29) and (9.30) into Eq. (9.28) and rearranging it to form

$$u_t = \sqrt{\frac{4}{3}\frac{(\rho_s - \rho_f)g}{\rho_f}\left(\frac{D_s}{C_D}\right)} \tag{9.31}$$

In Eq. (9.31) each of the variables except the drag coefficient C_D can be determined directly from the physical properties of the solid and fluid phases. The drag coefficient is a nonlinear function of the particle Reynolds number Re_p, defined as

$$\text{Re}_P = \frac{D_s u_t \rho_f}{\mu_f} \tag{9.32}$$

where μ_f = absolute viscosity of the fluid, lbm/(ft·s).

Figure 9.5 shows the relationship between Reynolds number and drag coefficient. The curves in Fig. 9.5 show that the drag coefficient is also a function of the sphericity ψ of the solid particle, which is defined as

$$\psi = \frac{\text{surface area of a sphere whose volume equals that of the particle}}{\text{surface area of particle}} \tag{9.33}$$

Sphericity is an adjustment factor for the geometry of a solid particle. Table 9.4 lists the sphericity for some typical process materials and geometric shapes.

Normally, the calculation of terminal velocity is an iterative process involving the calculation of the solid-particles[1] Reynolds number and determining whether the flow region is (1) laminar, $\text{Re}_p < 1$; (2) transitional, $1 < \text{Re}_p < 1000$; or (3) turbulent, $\text{Re}_p > 1000$.

The first step requires an assumption of terminal velocity used to calculate the first trial Reynolds number. The Reynolds number is used to find from Fig. 9.5 a C_D which is then used to calculate a new terminal velocity for the next iteration. Iteration continues until the terminal velocity used in the Reynolds number matches the terminal velocity calculated from the drag coefficient.

McCabe and Smith[10] have developed a convenient test for determining the flow region of the falling solid which in turn can be used to calculate drag coefficient and terminal velocity. Their test requires the calculation of a K factor by the equation

$$K = D_s \left[\frac{g\rho_f(\rho_s - \rho_f)}{\mu_f^2}\right]^{1/3} \tag{9.34}$$

The K factor is used directly to determine the flow region. Table 9.5 shows the relationship between the K factor and the flow region and the formulas used to calculate drag coefficient and terminal velocity for spherical particles.

Effect of solids concentration on terminal velocity. The above equations for calculating terminal velocity were derived from experimental data on single particles or particles in fluid streams at low concentrations. As the solids loading increases the predicted terminal velocity, drag coefficient and solid velocity deviate from the actual values. Yang[11] has proposed a correlation to compensate for this deviation. The pro-

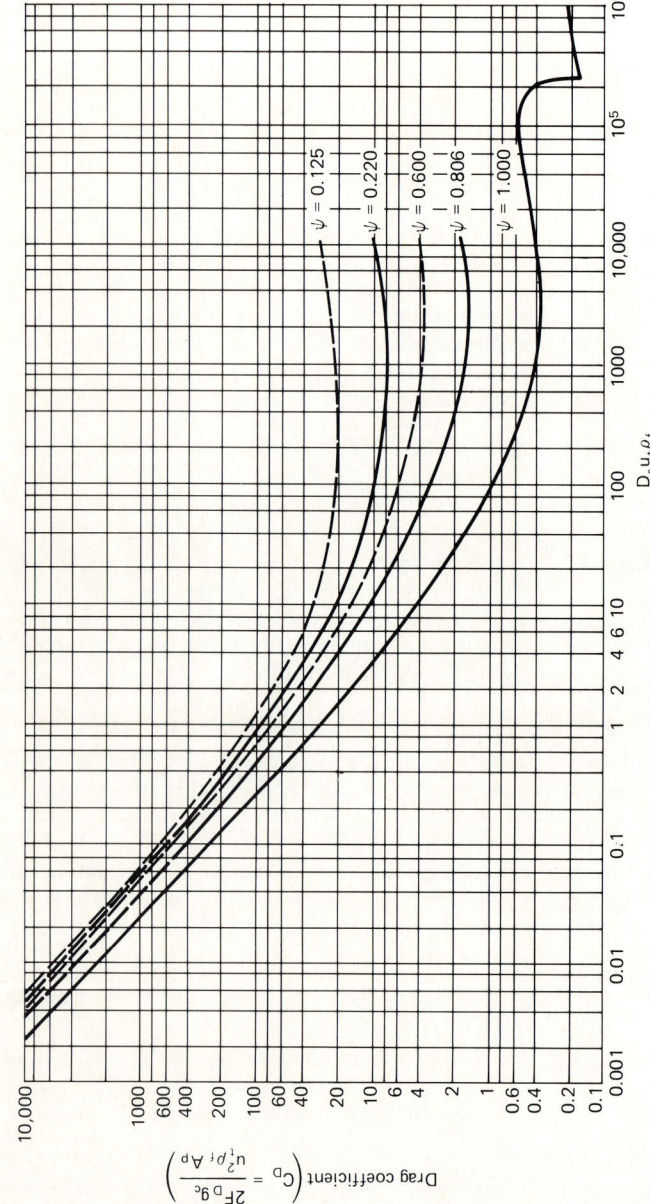

Figure 9.5 Drag coefficient as a function of Reynolds number. (G. G. Brown and associates, Unit Operations, John Wiley & Sons, Inc, New York, 1950.)

Piping Calculations, Part 4: Complex Fluids   351

TABLE 9.4 Typical Sphericity Values

	ψ
Materials	
Anthracite coal	0.63*
Fischer-Tropsch catalyst	0.58*
Sand, round	0.86*
Sand, sharp	0.67*
Iron catalyst	0.578†
Pulverized coal	0.696†
Geometric shapes	
Sphere	1.00
Cube	0.806
Cylinders	
$h = 3r$	0.86
$h = 10r$	0.691
$h = 20r$	0.580
Disks	
$h = r$	0.827
$h = r/3$	0.594
$h = r/10$	0.323
$h = r/15$	0.254

*O. Levenspiel and D. Kunii, *Fluidization Engineering,* John Wiley and Sons, Inc., New York, 1969, chap. 3, Table 3, p. 72.
†Ibid., chap. 3, Table 2, p. 65.

posed drag-coefficient correlation is a function of voidage:

$$C_{D,c} = C_D(\epsilon)^{-4.7} \qquad (9.35)$$

where C_D = drag coefficient as calculated from Table 9.5
$C_{D,c}$ = drag coefficient at high solids concentration
$\epsilon = 1 - e_s$ = voidage
e_s = volume fraction of the solid phase

This modified drag coefficient can be used in Eq. (9.31) to determine the terminal velocity corrected for solids concentration.

TABLE 9.5 Drag-Coefficient and Terminal-Velocity Equations Using the K Factor

K-factor range	Reynolds-number regime	C_D	u_t
$K < 3.3$	Laminar	$24/\text{Re}_p$	$\dfrac{g(\rho_s - \rho_f)D_s^2}{18\mu_f}$
$3.3 < K \leq 43.6$	Intermediate	$10/\sqrt{\text{Re}_p}$	$D_s\sqrt{\left(\dfrac{4}{225}\right)\dfrac{(\rho_s - \rho_f)g^2}{\rho_f\mu_f}}$
$43.6 < K \leq 2360$	Turbulent	0.43	$\sqrt{\dfrac{3.1g(\rho_s - \rho_f)D_s}{\rho_f}}$

Fluid-solid-system flow regimes

No generalized plot such as the Baker plot for gas-liquid systems has been developed as yet for fluid-solid systems. However, it is acknowledged that fluid-solid flow regimes fall into four general categories: symmetrical flow, asymmetrical flow, moving bed, and stationary bed. In symmetrical flow, the solid phase is homogeneously distributed throughout the fluid phase. Asymmetrical flow has an irregular solids distribution within the system; for example, in horizontal flow systems the lower portion carries a higher concentration of solids than the upper portion. In the moving-bed regime, the solid particles tend to settle out in a bed along the length of a horizontal pipe but continues to move at a much slower rate than the fluid. A moving-bed regime may be considered as asymmetric flow with a more pronounced density gradient. At lower fluid velocities the solid particles settle out, and the dense solid phase starts to resist any movement. At this point a stationary-bed regime develops. The fluid in a horizontal pipe with a stationary bed flows above the bed, and as it passes over the bed, it may continue to drop out solid particles onto the bed and/or reentrain smaller solid particles into the flow stream. This phenomenon can be a steady-state, transient, or oscillatory process.

A steady-state fixed bed can exist only where solid-particle diameters are all of about the same size. If particle diameters vary over a wide range, the fixed bed will continue to increase in size until it finally plugs up the line, preventing the flow of fluid. At some intermediate particle-diameter distribution range, the stationary bed builds up and breaks down on a cyclic basis, creating dunes or slugs of solids along the length of the line. This cyclic process is referred to as "saltation" or "slugging."

There are no generalized correlation methods for predicting any of these flow regimes in either horizontal or vertical lines. However, the fluid-solid flow relationship can be plotted as mixture linear velocity u_M versus pressure gradient. Figure 9.6a shows the general form of the pressure gradient in a horizontal line as a function of mixture velocity, and Fig. 9.6b shows the relationship for flow in a vertical line. Each figure also shows the pressure-gradient line for the pure fluid alone for a given linear velocity as well as the boundaries between symmetric, asymmetric, moving-bed, and stationary-bed flow regimes. The figure also shows approximate velocities in the transition zones between the various regimes. The four boundary velocities are designated as

u_{M1} = average mixture velocity at symmetric-asymmetric boundary
u_{M2} = average mixture velocity at asymmetric–moving-bed boundary
u_{M3} = average mixture velocity at moving-bed–stationary-bed boundary
u_{M4} = mixture velocity at which choking or complete solids blockage occurs

When a fluid-solid system is designed, the terminal velocity provides a convenient basis for establishing the minimum mixture transport velocity $u_{M,\min}$.[12] One method often used requires the minimum mixture velocity to be twice the solids terminal velocity, i.e.,

$$u_{M,\min} = 2u_t \tag{9.36a}$$

or $$u_{M,\min} = 2(u_{f,\min} + u_s) \tag{9.36b}$$

where $u_{M,\min}$ = minimum mixture transport velocity, ft/s
$u_{f,\min}$ = minimum superficial fluid velocity, ft/s
u_s = superficial solid velocity, ft/s

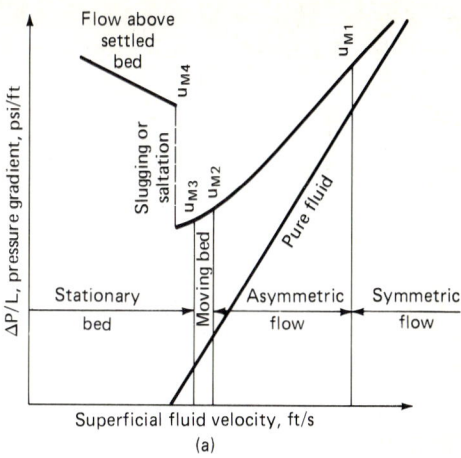

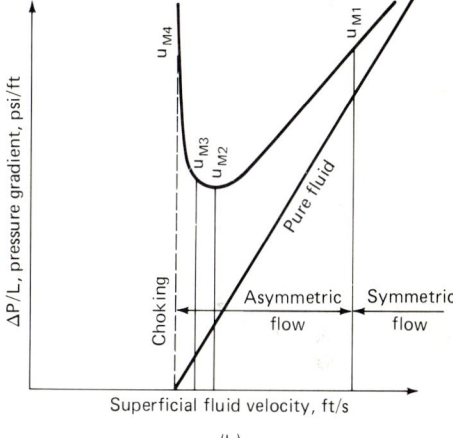

Figure 9.6 Generalized flow–pressure-gradient curves for fluid-solid systems. (a) For horizontal flow. (b) For vertical flow.

However, in some applications it may not be practical to achieve this velocity. Therefore, the engineer must decide whether an alternative mode of solid transport such as a mechanical conveyor should be considered or if a lower velocity can be tolerated. A decision to accept the latter must be based on a knowledge of the type of flow regimes expected at the lower velocities. Another design criterion for fluid-solid systems based on a lower linear velocity than calculated by Eq. (9.36a) is

$u_M \geq u_{M1}$ desirable

$u_M \geq u_{M2}$ minimum requirement

This criterion simply states that the mixture linear velocity should be greater than the velocity bounding asymmetric and moving-bed flow regimes but that the more desirable minimum velocity would be the boundary velocity between the symmetrical and asymmetrical flow regimes.

Finding a correlation for predicting these boundary velocities is difficult. Govier and Aziz[13] refer to work done by Spell suggesting the following boundary-velocity equations for liquid-solid systems:

$$u_{M1} = 134 C_D^{0.816} D^{0.633} u_t^{1.63} \tag{9.37}$$

$$u_{M2} = 54.4 C_D^{0.815} D^{0.633} u_t^{1.63} \tag{9.38}$$

where D = pipe diameter, ft.

However, Govier and Aziz did not recommend a correlation for calculating u_{M3} and u_{M4}. This should not be of concern, since these velocities are well below u_{M2}. If the mixture velocity is below u_{M2}, solid-transport problems exist and should be corrected by increasing the fluid velocity so that the mixture velocity exceeds u_{M2}.

Unlike liquid-solid systems, gas-solid systems have no effective correlation of u_{M2}, u_{M3}, and u_{M4}. However, Knowlton and Bachovchin[14] have developed a correlation for predicting choking velocity for vertical gas-solid systems which agrees within a range of 10 percent with experimental data for coal at high pressures. The correlation is

$$\frac{u_{\text{choked}}}{\sqrt{gD_s}} = 9.07 \left(\frac{\rho_s}{\rho_g}\right)^{0.347} \left(\frac{W_s D_s}{\mu_g}\right)^{0.214} \left(\frac{D_s}{D}\right)^{0.246} \tag{9.39}$$

where μ_g = absolute gas viscosity, lbm/(ft·s)
u_{choked} = superficial gas velocity at choking conditions, ft/s
D_s = solid diameter, ft
D = pipe diameter, ft

From the work of Zenz and Othmer[15] on air and rapeseed mixtures, it appears that u_{choked} can be used to estimate u_{M4} by

$$u_{M4} \approx u_s + u_{\text{choked}} \tag{9.40}$$

where u_s approaches zero.

The velocities calculated by Eqs. (9.37), (9.38), and (9.39) can be used to diagnose solids-conveying systems to determine whether the fluid velocities are high enough to prevent plugging or choking in piping systems.

Solid velocity. The previous correlations tacitly assume that the solid velocity is known. However, the solid velocity is not easily determined. This subsection presents the various methods used in calculating solid velocities. The Rose and Duckworth correlation, which will be discussed later, also includes a method for calculating the minimum transport velocity. Solid velocity is defined as the difference between the mixture velocity u_M and the terminal velocity u_t of the solid; i.e.,

$$u_s' = u_M - u_t \tag{9.41}$$

where u_s' = actual solid velocity, ft/s.

The difference between actual fluid velocity and solid velocity is known as the slip; i.e.,

$$S = u_f' - u_s' \tag{9.42}$$

where S = average slip between phases, ft/s
u_f' = actual fluid velocity, ft/s
u_s' = actual solid velocity, ft/s

The slip relationship assumes that the two phases are evenly distributed throughout a piping system. Although this is generally a valid assumption, in some situations it may not be true.

Rose and Duckworth[16,17] have developed a general correlation for calculating solid velocity. The equation takes the form

$$u_s = 0.63 u_f \phi_4 \tag{9.43}$$

The value of the empirical constant ϕ_4 is a function of

$$(u_f^2/g) D_s s^2$$

where $s = \rho_s/\rho_f$
$g = 32.17$ ft/s^2
$D_s =$ solid-particle diameter, ft

The relationship shown in Fig. 9.7 has been correlated over these ranges:

$960 < s < 9800$

$0.03 < D_p/D < 0.101$

$0.0001 < (u_f^2/gD_s s^2) < 0.1$

Therefore, caution should be exercised if the correlation is used to calculate solid velocities for operating conditions outside any of these ranges.

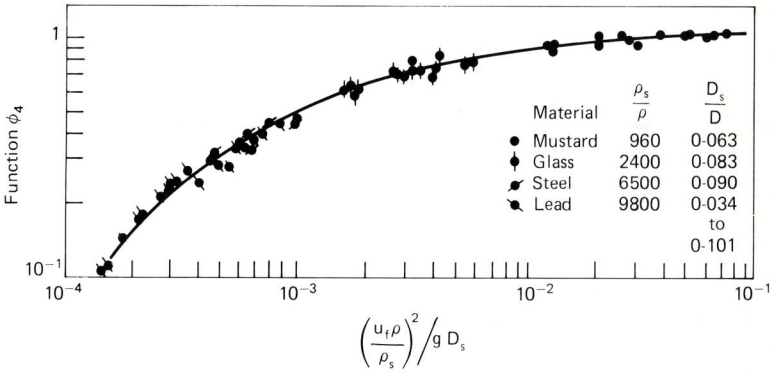

Figure 9.7 Rose and Duckworth function ϕ_4. (*H. E. Rose and R. A. Duckworth, The Engineer, vol. 227, no. 5904, 1969, p. 430. Morgan-Grampian Ltd.*)

McLeman[18] developed a correlation for calculating solids velocity in pneumatic conveying systems. The correlation agrees with experimental data within a range of ± 10 percent.

$$u_a - u_s = \frac{u_t}{0.212 + 3.28 \sqrt{u_t/\rho_s}} \tag{9.44}$$

where $u_a =$ superficial air velocity, ft/s.

Pressure gradients for fluid-solid systems

The total pressure gradient for fluid-solid systems is composed of three components; i.e.,

$$\Delta P_{total} = \Delta P_{acceleration} + \Delta P_{static\ head} + \Delta P_{friction} \qquad (9.45)$$

Pressure-gradient components in a fluid-solid system are difficult to determine and often require complex calculations. If the symmetric flow regime is expected to exist for the fluid-solid mixture, the two-phase flow may sometimes be treated as a pseudo-homogeneous mixture. For this situation, the Bernoulli and Darcy equations are used with adjustments for density and viscosity. For example, the average density for the mixture is used in these equations; however, the mixture viscosity (μ_M) is more difficult to estimate. Einstein[19] proposed the following equations for estimating mixture viscosity with spherical solids:

1. For a solid-volume fraction of less than 2 percent,

$$\mu_M = \mu(1 + 2.5e_s) \qquad (9.46)$$

2. For solid-volume fractions up to 20 percent,

$$\mu_M = \mu(1 + 2.5e_s + 14.1e_s^2) \qquad (9.47)$$

where μ = pure-fluid absolute viscosity.

Another correlation for mixture viscosity was proposed by Guth and Simha[20] for water slurries in either transition or pure turbulent flow and with solids between 10 and 100 μm in diameter. The relationship is

$$\mu_M = \mu \left/ \left[1 - \frac{e_s}{e_{s,max}}\right]^{2.5} \right. \qquad (9.48)$$

where $e_{s,max}$ = maximum possible solid-volume fraction. This would occur when the solids are arranged in the piping similarly to a packed bed.

The above treatment of fluid-solid systems as a pseudo-homogeneous mixture provides reasonable estimates of pressure gradients at dilute solids concentrations, i.e., solids-volume fractions of 2 percent or less. For higher solids concentrations or fluid mixtures that are not symmetrically distributed, more complex calculations must be performed. Govier and Aziz[1] provide the most exhaustive survey of methods used to calculate fluid-solid pressure gradients. Klinzing[10] lists the various methods used for calculating pressure gradients for gas-solid systems. In addition, Klinzing includes a number of examples comparing the pressure gradients calculated by the different methods.

The most general correlation for fluid-solid pressure gradients has been developed by Rose and Duckworth.[17] This correlation will be presented in greater detail in the following subsection. A method proposed by McLeman[18] for frictional pressure drop in pneumatic conveying systems is a simpler correlation and also is presented.

Calculation of static head. The static-head component is represented by

$$\Delta P_{static\ head} = [\rho_s e_s + \rho_f(1 - e_s)]z/144 \qquad (9.49)$$

where z = vertical height of two-phase column, ft.

For most pneumatic conveying systems the fluid-density group is usually negligible since the fluid is a gas and may be ignored. Therefore, Eq. (9.46) can be reduced to

$$\Delta P_{\text{static head}} = \rho_s e_s z / 144 \tag{9.50}$$

Calculation of acceleration pressure drop and length. When solids enter a fluid stream, they must be accelerated to the steady-state solids velocity for the system. Therefore, a pressure drop exists for the change in kinetic energy of the solids. The Rose and Duckworth correlation for this pressure drop is

$$\Delta P_{\text{acceleration}} = \frac{\rho_f u_f^2}{2g_c} + \frac{0.56}{g_c} \rho_f u_f^2 R \phi_4 \phi_5 \quad \text{lbf/ft}^2 \tag{9.51}$$

where g_c = dimensional constant, 32.17(lbm/lbf)·ft/s^2
u_f = superficial fluid velocity, ft/s
R = mass ratio, lb of solids/lb of fluid
ρ_f = fluid density, lb/ft^3
ϕ_4 = empirical constant obtained from Fig. 9.7
ϕ_5 = empirical constant obtained from Fig. 9.8

and the correlation for the length of pipe required to fully accelerate the solids is

$$L_{\text{acceleration}} = 6D \left[\frac{w_s}{\rho_f} \sqrt{\frac{\rho_s}{gD^4 D_s \rho_f}} \right]^{1/3} \tag{9.52}$$

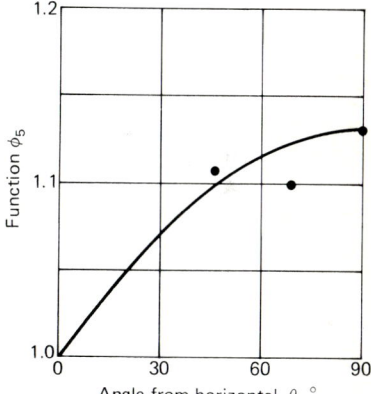

Figure 9.8 Rose and Duckworth function ϕ_5. (H. E. Rose and R. A. Duckworth, The Engineer, vol. 227, no. 5904, 1969, p. 430. Morgan-Grampian Ltd.)

Calculation of frictional loss. The pressure-gradient contribution due to friction is treated in one of two ways for fluid-solid systems. One method adds an excess pressure drop to the pressure drop expected from the fluid alone. Another method adds a factor to the fluid friction factor.

The excess-pressure-drop method has been used by McLeman[18] for a pneumatic conveying system. The McLeman correlation defines frictional pressure drop as

$$\Delta P_{\text{friction}} = \Delta P_x + \Delta P_a \tag{9.53}$$

where ΔP_a = pressure drop for air alone, psi
ΔP_x = excess pressure drop due to the solids, psi

The excess pressure drop is calculated by

$$\Delta P_x = \Delta P_a \frac{KF}{u_t u_s^2} \qquad (9.54)$$

where F = solid mass flow rate, lb/h
$K = 47{,}470 \exp(-0.135d)$
d = pipe diameter, in

This correlation as well as other more complex correlations for gas-solid systems[21] tends to predict excessively high pressure drops for small particle sizes. Therefore, for systems carrying 50-μm or smaller solids, experimentation may be required to confirm calculated pressure drops.

For determining the pressure drop in liquid slurries, Govier and Aziz[22] recommend the use of the Rose and Duckworth correlation.[17] However, liquid slurries of fine solids (less than 10 μm) usually exhibit nonnewtonian behavior and cannot be analyzed by this method.

Calculation of frictional and static-head loss. The Rose and Duckworth correlation has a combined expression for frictional and static-head loss. However, an inspection of the equation shows that the contribution of each component can be segregated. The frictional contribution can be defined as

$$\Delta P_{\text{friction}} = (f_f + \psi_s)\rho_f \left(\frac{L}{D}\right)\frac{u_f^2}{2g_c} \quad \text{lbf/ft}^2 \qquad (9.55)$$

where f_f = Moody friction factor based on superficial fluid velocity
ϵ_r = coefficient of restitution, where $\epsilon_r = 0.0$ for inelastic-particle collisions and $\epsilon_r = 1$ for perfectly elastic-particle collision (e.g., for coal particles, $\epsilon_r = 0.63$)
D = pipe diameter, ft
g_c = dimensional constant, 32.17(lbm/lbf)·ft/s^2
L = pipe length, ft
R = solids loading, lb of solids/lb of fluid
u_f = superficial fluid velocity, ft/s
ρ_f = fluid density, lb/ft^3
$\psi_s = \psi_1\psi_2\psi_3\psi_4\psi_5\psi_6$
$\psi_1 = f(R)$ (see Fig. 9.9)
$\psi_2 = f(D_s/D)$ (see Fig. 9.10)
$\psi_3 = f(\epsilon_r)$ (see Fig. 9.11)
$\psi_4 = f(s)$ (see Fig. 9.12)
$\psi_5 = f(\theta)$ (see Fig. 9.13)
$\psi_6 = f(u_f^2/gD)$ (see Fig. 9.14)
θ = angle from the horizontal, °
$s = \rho_s/\rho_f$

while the contribution for hydrostatic-head loss is given by

$$\Delta P_{\text{static head}} = \rho_f L \sin\theta \left[1 + R\left(\frac{u_f}{u_s}\right)\left(1 - \frac{1}{s}\right)\right] \quad \text{lbf/ft}^2 \qquad (9.56)$$

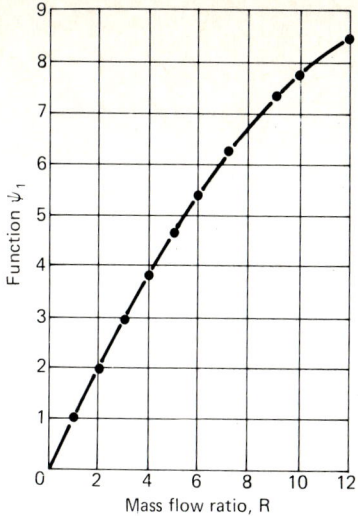

Figure 9.9 Rose and Duckworth function ψ_1. (*H. E. Rose and R. A. Duckworth, The Engineer, vol. 227, no. 5905, 1969, p. 478. Morgan-Grampian Ltd.*)

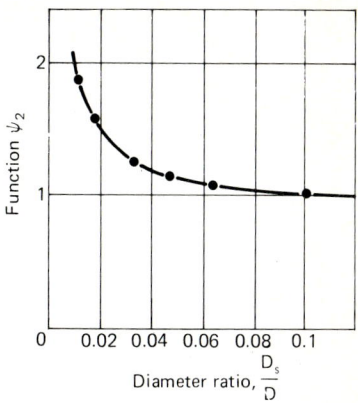

Figure 9.10 Rose and Duckworth function ψ_2. (*H. E. Rose and R. A. Duckworth, The Engineer, vol. 227, no. 5905, 1969, p. 478. Morgan-Grampian Ltd.*)

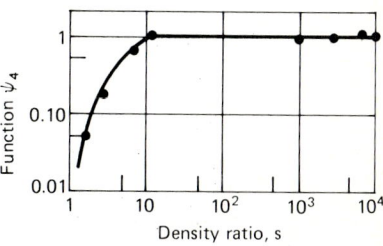

Figure 9.12 Rose and Duckworth function ψ_4. (*H. E. Rose and R. A. Duckworth, The Engineer, vol. 227, no. 5905, 1969, p. 478. Morgan-Grampian Ltd.*)

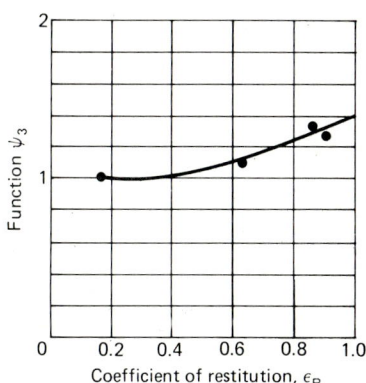

Figure 9.11 Rose and Duckworth function ψ_3. $\epsilon_R = 1$, perfectly elastic collision of solid particles; or $\epsilon_R = 0$, inelastic collision. ϵ_R, coal ≈ 0.63. (*H. E. Rose and R. A. Duckworth, The Engineer, vol. 227, no. 5905, 1969, p. 478. Morgan-Grampian Ltd.*)

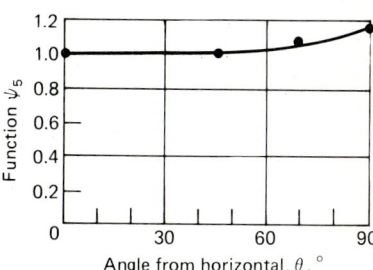

Figure 9.13 Rose and Duckworth function ψ_5. (*H. E. Rose and R. A. Duckworth, The Engineer, vol. 227, no. 5905, 1969, p. 478. Morgan-Grampian Ltd.*)

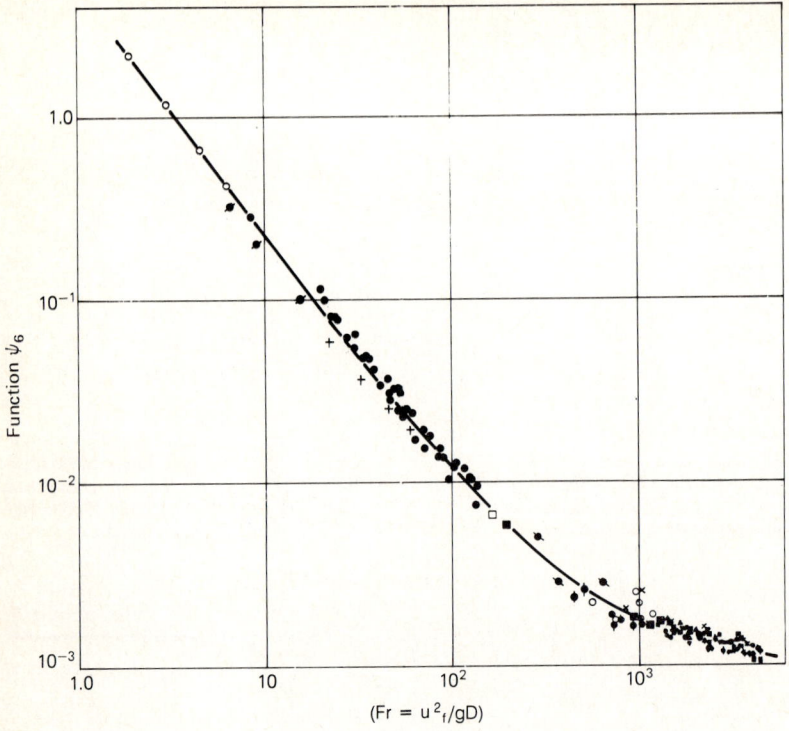

Symbol	Material	Fluid	d, in	D, in	ρ, lbf/ft^3
	Coal	Water	0.5	3.0	94
	Sand	Water	0.118	5.9	165
	Steel	Water	0.125	1.265	460
	Lead	Water	0.095	1.265	695
	Blaud's pills	Air	0.28	2.0	114
	Blaud's pills	Air	0.28	3.0	114
	Mustard	Air	0.08	1.265	73
	Lead	Air	0.128	1.265	695
	Lead	Air	0.08	1.265	695
	Lead	Air	0.06	1.265	695
	Lead	Air	0.043	1.265	695
	Lead	Air	0.022	1.265	695
	Lead	Air	0.015	1.265	695
	Steel	Air	0.125	1.265	450
	Glass	Air	0.118	1.265	187
	Limestone	Air	1.25	6.0	89
	Coke	Air	1.0	6.0	32
	Iron ore	Air	0.125	6.0	200
	Wheat	Air	0.20	16.0	87
	Wheat	Air		12.0	87
	Wheat	Air		10.0	87

Figure 9.14 Rose and Duckworth function ψ_6. (*H. E. Rose and R. A. Duckworth,* The Engineer, *vol. 227, no. 5905, 1969, p. 478. Morgan-Grampian Ltd.*)

Minimum transport velocity. Rose and Duckworth also developed a correlation for minimum transport velocity $u_{M,\min}$:

$$u_{M,\min} = 3.2u_t \left[R^{0.2} \left(\frac{D}{D_p}\right)^{0.5} \left(\frac{1}{s}\right)^{0.7} \left(\frac{u_c^2}{gD}\right)^{0.25} \right] \qquad (9.57)$$

where u_c = terminal velocity corrected for solids concentration by using Eq. (9.35). Equation (9.57) corresponds to u_{M2} for horizontal flow and u_{choked} for vertical flow.

Design criteria for fluid-solid systems

When designing or troubleshooting a fluid-solid system, the engineer should consider the fluid velocity carefully. This is done to determine whether the velocity is high enough to prevent solids setting, which could lead to plugging and choking. The procedure consists of the following steps:

1. Determine the properties of the solid and the fluid, i.e., flow rates, densities, particle diameters, and fluid viscosity.
2. Calculate the drag coefficient and terminal velocity of the solids by using the equations in Table 9.5. Correct the drag coefficient for solids concentration by using Eq. (9.35) if necessary.
3. Calculate the solids velocity by using either Eq. (9.41) or Eq. (9.44), whichever matches the situation.
4. Determine the mixture velocity and compare it with the velocity limits for choking at the inception of a moving-bed flow regime, i.e., Eqs. (9.37), (9.38), (9.39), or (9.40). If the calculated mixture velocity is at or near the limiting velocity, one or more of the following measures should be taken to increase mixture velocity.
    a. Reduce solids loading R.
    b. Reduce pipe diameter to increase velocity.
    c. Increase pump head or blower discharge pressure to increase fluid velocity beyond the limiting velocity.
5. Determine the resulting pressure gradient of the system by using the appropriate Eqs. (9.45) and (9.49) through (9.56).

One should exercise caution when dealing with fluid-solid systems. The above methods assume that phenomena such as electrostatic charges in pneumatic systems or nonnewtonian behavior in slurries do not exist. No effective method has been developed to predict electrostatic behavior in pneumatic systems. However, Klinzing[23] provides a very good discussion of the theoretical aspects of electrostatic behavior along with some experimental data.

REFERENCES

1. G. W. Govier and K. Aziz, *The Flow of Complex Mixtures in Pipes*, Van Nostrand Reinhold Company, New York, 1972.
2. Ovid Baker, "Simultaneous Flow of Oil and Gas," *Oil and Gas Journal*, vol. 53, 1954, pp. 185–190.
3. Govier and Aziz, op. cit., chap. 8, Fig. 8.9, pp. 335–337.
4. A. E. DeGance and R. W. Atherton, "Chemical Engineering Aspects of Two-Phase Flow," part 4, *Chemical Engineering*, Apr. 20, 1970, pp. 96, 97.
5. R. Kern, "How to Size Process Piping for Two-Phase Flow," *Hydrocarbon Processing*, October 1969, p. 115.

6. Govier and Aziz, op. cit., chap. 8.
7. DeGance and Atherton, op. cit., part 6, *Chemical Engineering,* Oct. 5, 1970, pp. 87–94.
8. L. S. Tong, *Boiling Heat Transfer and Two-Phase Flow,* John Wiley & Sons, Inc., New York, 1965, pp. 102–103.
9. J. M. Coulson et al., *Chemical Engineering,* vol. 1, 3d ed., Pergamon Press, New York, 1978, pp. 91, 92.
10. G. E. Klinzing, *Gas-Solid Transport,* McGraw-Hill Book Company, New York, 1981, p. 58.
11. Ibid., chap. 4, pp. 82–83.
12. Govier and Aziz, op. cit., chap. 9, p. 468.
13. Ibid., chap. 11, pp. 642–657.
14. Klinzing, op. cit., chap. 4, p. 99, Eq. (4-22).
15. F. A. Zenz and D. F. Othmer, *Fluidization and Fluid-Particle Systems,* Reinhold Publishing Corporation, New York, 1960, chap. 10, pp. 313–350.
16. Govier and Aziz, op. cit., chap. 9, pp. 474, 475.
17. H. E. Rose and R. A. Duckworth, *The Engineer,* vol. 227, no. 5903, 1969, p. 392; vol. 227, no. 5904, 1969, p. 430; vol. 227, no. 5905, 1969, p. 478.
18. J. M. Coulson et al., *Chemical Engineering,* vol. 2, 3d ed., Pergamon Press, New York, 1978, pp. 283–288.
19. A. Einstein, *Annalen der Physik,* vol. 17, 1905, p. 459; vol. 19, 1906, p. 289; vol. 34, 1911, p. 591.
20. E. Guth and R. Simha, *Kolloid-Zeitschrift,* vol. 74, 1936, p. 266.
21. Klinzing, op. cit., chap. 4, pp. 87–93.
22. Govier and Aziz, op. cit., chap. 9, pp. 497–500; chap. 11, pp. 704–708.
23. Klinzing, op. cit., chap. 6, pp. 122–149.

Section 3

Equipment Selection, Sizing, and Related Subjects

Chapter 10

Vessels

An understanding of the design of process vessels is important for process engineers. Vessels are used both for storing and for processing materials. As a result, different design criteria may be required depending on the vessel's function. In most cases the design of a vessel is regulated by government or quasi-governmental agencies.

This chapter outlines the selection of vessel geometry, the design of a vessel, the codes that regulate a vessel's construction, and how a vessel is inspected after construction. Many vessel appurtenances were reviewed in Chap. 3, with respect to their presentation on flow diagrams. A discussion of the design of the vessel itself and its accessories is given in this chapter.

It should be emphasized that although various codes used for the design of vessels are presented and data and equations are excerpted from these codes, this chapter should not be used in lieu of a code document for the actual design of a process vessel. The intent of this chapter is to acquaint the process engineer with the regulations used to design vessels and how to determine which regulations apply.

Vessel-Design Criteria

The purpose of a vessel in a process plant is to contain process fluids safely for storage or during processing. The most basic requirement is that the vessel will not leak or fail catastrophically under normal operating conditions. Therefore, the vessel design selected involves investigation of the process materials handled and the temperatures and pressures that the vessel will experience during operation. Selection of materials compatible with these process conditions and determination of vessel geometry must also be considered. The final design must use an economy of material but not at the expense of overstressing the vessel to the point of failure. Material selection was covered in Chap. 5; the importance of vessel stress will be reviewed in this chapter.

Vessel stresses

Chapter 5 showed that applying a load to a material causes it to deform, exhibiting a "strain." The load, represented as a force applied per unit cross-sectional area, is called the "stress." As the magnitude of the load changes, it will cause a characteristic stress-and-strain relationship for the material. If stress is directly proportional to strain, the material is said to behave elastically. However, if there appears to be little or no relationship between stress and strain, the material's behavior is plastic. Some materials, such as ductile steels, have a distinct elastic and plastic deformation pattern, while other materials such as cast iron have no distinct elastic or plastic deformation pattern but a combination of the two. An idealized elastic-plastic deformation pattern for ductile steel is shown in Fig. 10.1a.

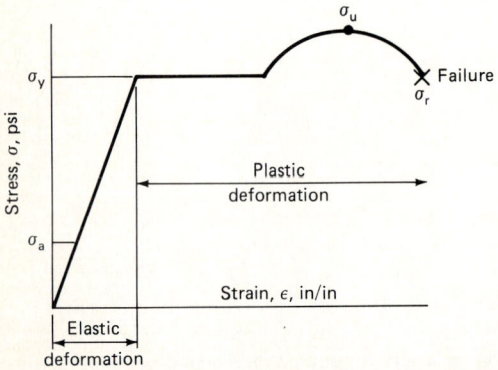

Figure 10.1a Idealized stress-strain curve for a ductile steel. σ_r = rupture stress; σ_u = ultimate stress; σ_y = yield stress; σ_a = allowable stress.

The curve shows four important stress points to be considered when selecting a material for use in constructing a process vessel. The allowable stress σ_a is a stress recommended by a recognized applicable vessel-design code as the maximum stress that should be applied on the material. The yield stress σ_y is the stress point at which the material goes through a transition from elastic to plastic deformation behavior. The ultimate stress σ_u is the maximum stress that the material can withstand, and the rupture stress σ_r is the stress at which the material fails.

Not all materials have distinct stress points. Extremely hard and brittle materials can be perfectly elastic over their stress-strain curve and show no yield but fail at some stress at the end of their elastic ranges. Therefore, ultimate stress and rupture stress for these materials will be the same. Materials such as concrete and cast iron, have no elastic region at all and can fail without warning because their ultimate stress and rupture stress, as in the case of hard and brittle materials, are the same. Therefore, ductile materials are recommended for use in process-vessel design. Even if the allowable stress is inadvertently exceeded, the ductile material will go through plastic deformation before failure. This plastic deformation manifests itself as a bump, dent, or elongation before the vessel material ruptures.

Once a material has been selected for vessel construction, the expected actual stresses must be calculated. These stresses depend on the type of loading that the vessel experiences under various conditions. They also depend on vessel geometry and orientation. Table 10.1 lists some of the operating conditions for a vessel and the loadings that may be applied individually or in combination with each other.

TABLE 10.1 Possible Stresses Applied to Vessels

Normal operation stresses	
Deadweight	σ_d, vessel and process-fluid weight
Internal pressure	σ_i
External pressure	σ_e
Seismic (vibrational) loading	σ_s
Wind loading	σ_w
Eccentric loading	σ_x
Thermal loading	σ_t
Field-erection stresses	
Deadweight	σ_d, vessel only
Seismic loading	σ_s
Wind loading	σ_w
Hydrotesting	
Testing pressure	σ_h

The stress on a vessel wall is defined by the general equation

$$\sigma = F/A \tag{10.1}$$

where σ = stress, psi
     F = force or load applied, lbf
     A = cross-sectional area over which the force is being applied, in^2

Stresses can be either compressive or tensile. Compressive stresses are indicated by a negative sign. Figure 10.1b shows the orientation of compressive and tensile stresses on some simple shapes.

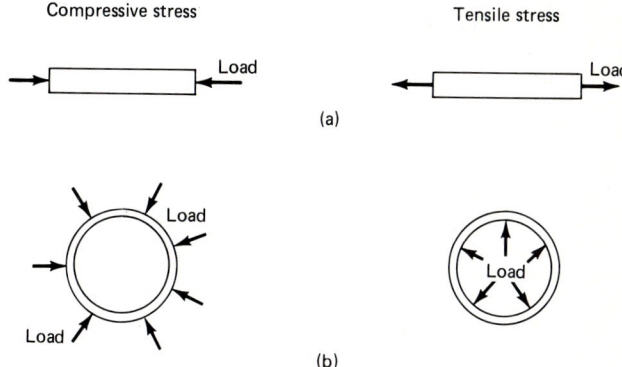

Figure 10.1b Tensile and compressive stress orientation on some common shapes. (a) Bars. (b) Thin shells.

If more than one load is being applied simultaneously to a point on the vessel, the stresses of the loads are simply added together or added according to a prescribed formula of a vessel-design regulation; i.e.,

$$\sigma_t = K_1\sigma_1 + K_2\sigma_2 + \cdots + K_N\sigma_N \tag{10.2}$$

where σ_t = total stress
σ_j = stress contribution for load condition $j = 1, 2, \ldots N$
K_j = regulation multiplier or safety factor for stress condition $j = 1, 2, \ldots N$. If no regulation applies, K is usually $+1$ or -1, depending on whether the load is tensile or compressive.

The total stress during normal operations must be less than the material's allowable stress; however, total stress may be permitted to exceed allowable stress by 20 to 100 percent, depending on the applicable regulation, but only during erection or testing. The principal contributors to total stress are discussed in detail in the following subsections.

Deadweight stresses. The actual weight of a vessel creates a compressive stress on itself. This is particularly important in tall vessels, such as distillation columns, where, for a constant shell-wall thickness, compressive stress increases directly in proportion to the distance from the top of the vessel. Compressive stress can be calculated by

$$\sigma_d = \frac{W}{\pi D t} \tag{10.3}$$

where D = vessel diameter, in
t = vessel-wall thickness, in
W = weight of vessel above point where stress is being calculated, lb

For extremely tall columns, wall thicknesses are usually thinner at the top than at the bottom of the vessel. However, most vertical vessels under 12 ft usually have a uniform shell-wall thickness.

Horizontal tanks on saddles. The weight of a horizontal tank on saddles and its contents causes bending stresses on the shell. A typical horizontal tank with two saddles is presented in Fig. 10.2 together with a diagram showing the bending moments produced by the weight of the empty horizontal tank.

The calculation of stresses in a horizontal tank requires the solution of formulas more complex than Eq. (10.3), which is used for a vertical tank. References 1, 2, and 3 provide the necessary equations to calculate rigorously deadweight stresses on horizontal tanks and their supports. These equations are beyond the scope of this book, but Fig. 10.3 shows a simplified but conservative method for selecting saddles and locating them by assuming an average volume density of 42 lb/ft^3. This is roughly equivalent to a tank two-thirds full of water.

Pressure stresses. When a process vessel contains a gas or a vapor, the fluid applies a uniform pressure to the walls of the vessel. If the fluid's pressure is greater than the atmospheric pressure on the outer walls of the vessel, then the walls experience a tensile stress and the vessel experiences internal-pressure stresses. If the pressure is lower on the inside of the vessel, the vessel walls are in compression and the vessel experiences external-pressure stresses.

Internal pressure. Internal-pressure stresses are the most frequently encountered stresses in process vessels. The basic theory for their calculation is based on the general relationship given by Eq. (10.1). Thus for a spherical thin-walled vessel internal-

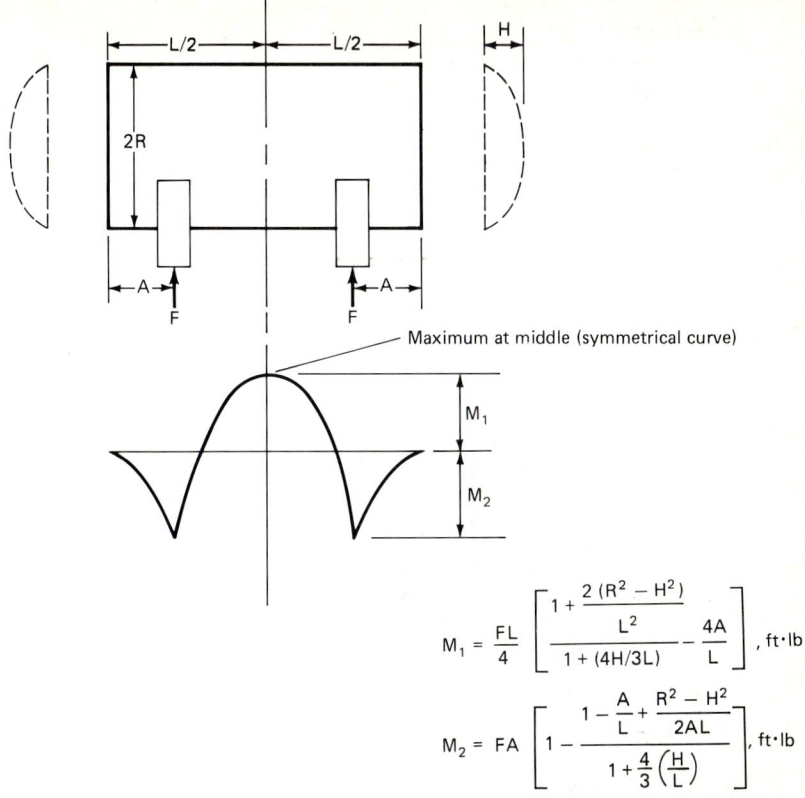

Figure 10.2 Bending stresses in an empty horizontal cylindrical vessel supported on two saddles.

pressure stress can be calculated by the equation

$$\sigma_{i,r} = \frac{PD}{4t} \tag{10.4}$$

where $\sigma_{i,r}$ = radial or hoop stress due to internal pressure, psi
P = internal pressure, psi
D = sphere diameter, in
t = wall thickness, in

This stress is the same no matter where it is measured on the sphere's surface. This stress equation is also applicable to hemispherical heads, but it does not completely represent the stresses on a cylindrical shell.

Cylindrical shells by their geometry experience a different stress in radial and longitudinal directions. Figure 10.4 shows schematically the stresses applied to a cylindrical shell.

If pressure, vessel diameter, and wall thickness are substituted into Eq. (10.1), the equation for calculating radial stress, also called circumferential or hoop stress, takes

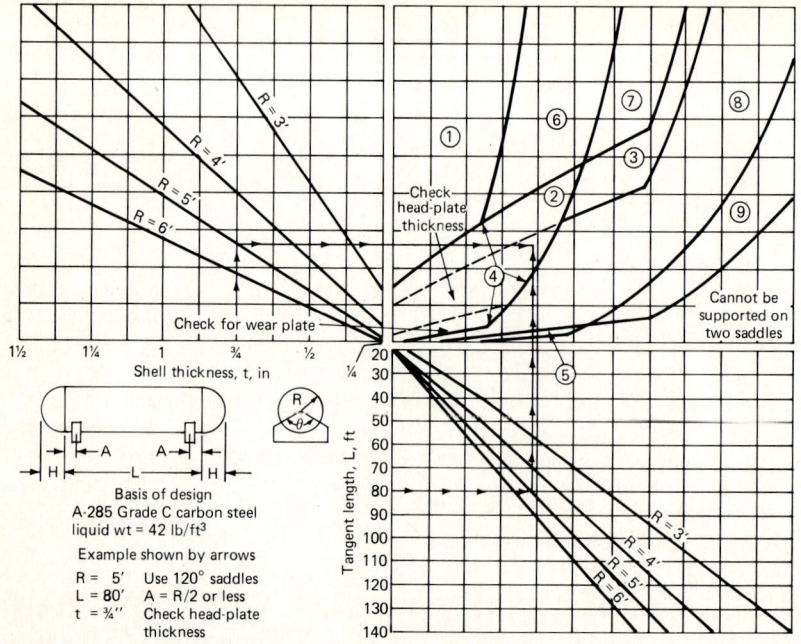

Figure 10.3 Location and type of supports for a horizontal vessel.
① θ = 120°, 0 ≦ A/L ≦ 0.2.
② θ = 150°, 0.5 R ≦ A ≦ 0.16L.
③ θ = 150°, A/L = 0.11.
④ θ = 120°, 0 ≦ A/R ≦ 0.50.
⑤ Add rings at supports, A/L = 0.25.
⑥ θ = 120°, 0 ≦ A/L ≦ 0.123.
⑦ θ = 120°, A/L = 0.09.
⑧ Add rings at supports, 0.17 ≦ A/L ≦ 0.24.
⑨ Add rings at supports, A/L = 0.1933.

(L. P. Zick, "Stresses in Large Horizontal Pressure Vessels on Two Saddles," Welding Journal (N.Y.), Research Supplement 30, 1951, p. 435-S. Courtesy of American Welding Society.)

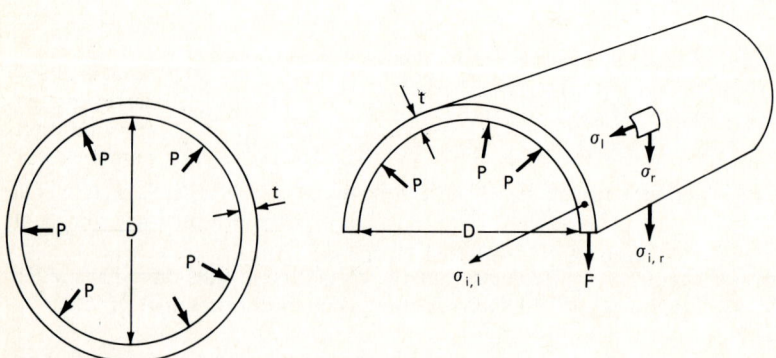

Figure 10.4 Stresses on a cylindrical shell.

the form

$$\sigma_{i,r} = \frac{PD}{2t} \tag{10.5}$$

Equation (10.5) is similar to Eq. (10.4), used to calculate internal-pressure stress for a sphere. However, the longitudinal-stress formula is

$$\sigma_{i,l} = \frac{PD}{4t} \tag{10.6}$$

and is identical to Eq. (10.4).

Comparing Eq. (10.5) with Eq. (10.6) shows that a cylinder experiences greater stress in the radial direction than in the longitudinal direction. This also explains the observation that cylindrical vessels, pipes, and garden hoses split along their longitudinal seam rather than pull apart.

Practical internal-stress equations. Experience has shown that the above equations tend to underestimate stress and thus the necessary wall thicknesses for pressure vessels. Therefore, various agencies concerned with the safe design of pressure vessels have modified these equations to provide a more conservative design. For example, the ASME code, Sec. VIII, for unfired pressure vessels uses the following general form to calculate vessel-wall thicknesses:

$$t = \frac{PDa_1}{SE - a_0 P} \tag{10.7}$$

where S = maximum allowable stress, psi
E = weld-joint efficiency, a percentage, expressed as a fraction, which can range from 70 percent for no examination to 100 percent for full radiographic inspection

For cylindrical vessels the coefficients a_0 and a_1 are

	a_0	a_1
Longitudinal stress	−0.2	0.25
Radial stress	0.6	0.50

Note that Eq. (10.7) becomes identical to Eqs. (10.4) and (10.5) when S is the actual stress, E equals 1, and a_0 equals zero.

Internal-pressure stresses of vessel heads. The theory behind the calculation of internal-pressure stresses on vessel heads is even more complicated. Harvey[4] presents the theoretical analysis for deriving stress equations for a number of vessel-head configurations. The equations used by the ASME employ the following numerical coefficients in Eq. (10.7) to calculate head-wall thickness:

Head type	Figure no.	a_0	a_1
Spherical	10.5	0.1	0.25
Ellipsoidal*	10.6	0.1	0.50
Flanged and dished heads (torispherical)*	10.7	0.1	0.88
Conical	10.8	0.6	$1/(2 \cos \theta)$, where $\theta \leq 30°$

*Standard head design.

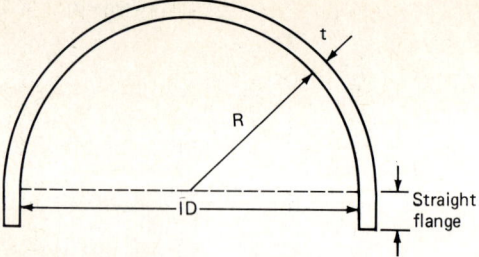

Figure 10.5 Hemispherical head.

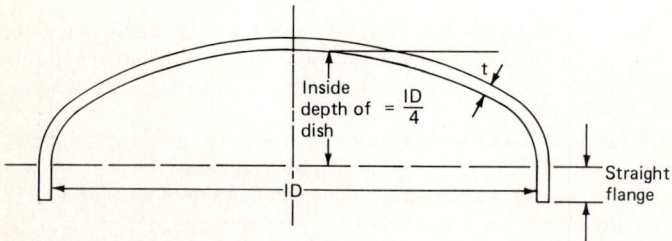

Figure 10.6 Ellipsoidal dished head. Standard ratio of major axis to minor axis = 2:1.

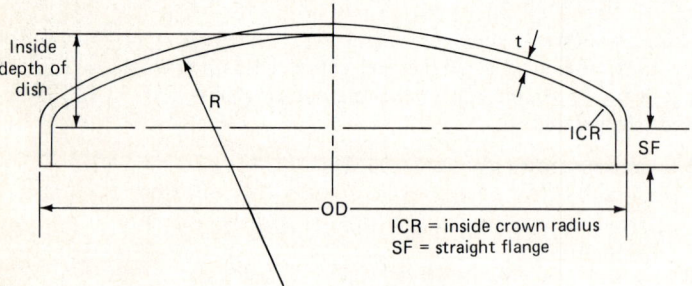

ICR = inside crown radius
SF = straight flange

Figure 10.7 ASME code flanged and dished (torispherical) head.

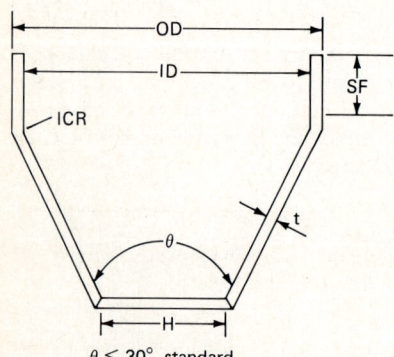

$\theta \leq 30°$, standard

Figure 10.8 Conical dished head.

External pressure. External-pressure stresses are more difficult to calculate than internal-pressure stresses. External pressures cause compressive stresses, which in turn cause the phenomenon of buckling, an effect difficult to predict. If external pressure were given a negative sign and substituted into Eq. (10.6), the resulting wall thickness would be thinner than needed to maintain the structural stability of the vessel.

A vessel's structural stability can be estimated from its collapsing pressure P_c. Calculation of collapsing pressure is based on vessel geometry and takes one of two forms:

$$P_c = 2.2E(t/D_o)^3 \qquad (10.8)$$

for very long cylinders ($L \gg 1.1D_o(D_o/t)^{1/2}$).

$$P_c = 2.80E(t/D_o)^{2.5}/(L/D_o) \qquad (10.9)$$

for short cylinders. E is the modulus of elasticity, which for most pressure-vessel steels is about 27 million psi, D_o is the outside diameter of the cylinder, and L is the length of the cylinder, all in dimensionally consistent units. The above equations assume a Poisson ratio* of about 0.3, which is valid for most pressure-vessel steels.

The ASME Boiler and Pressure Vessel Code publishes a series of charts, such as the one shown in Fig. 10.9, which are used to calculate external pressure. The method is an iterative procedure. The first step requires an assumption of a wall thickness which is substituted in an equation. The results of the equation are plotted on the chart to find a working pressure which must be greater than the design external pressure.

Wind loading. Vessels that are not located in buildings are subject to wind loading. Most horizontal vessels have negligible wind-loading stresses, and analysis for this type of stress may normally be omitted for horizontal vessels. However, wind loading is significant for tall vessels such as distillation columns.

The method most commonly used to calculate wind loading assumes that the vessel can be treated as a uniformly cantilevered beam with the wind pressure P_w being applied uniformly over the surface exposed to the wind. This is related to the wind velocity by the equation

$$P_w = 0.0025 u_w^2 \qquad \text{lbf/ft}^2 \qquad (10.10)$$

where u_w = wind velocity, mi/h.

This equation is based on a rectangular surface facing the wind. The calculated P_w must be multiplied by 0.6 for cylindrical or ellipsoidal cross sections. The wind pressure is then used to calculate stresses along the length of the vessel.

Another, more complicated method, which includes correction factors for terrain and wind gusting, is given by ANSI† in its standard A58.1-1972.

Seismic loading. Vessels that are located in geographical areas subject to earthquakes should be analyzed for seismic-loading stresses. Earthquakes cause a sudden

---

*The Poisson ratio is a ratio of contracting strain to elongation strain. Because structural shapes are not unidimensional, their deformation depends on where forces are applied. For example, if a force is applied to the corners of a rectangle shown in Fig. 10.10, the load will create an elongation strain and a contracting strain, creating a parallelogram configuration.

†American National Standards Institute (ANSI), 1430 Broadway, New York, N.Y. 10018.

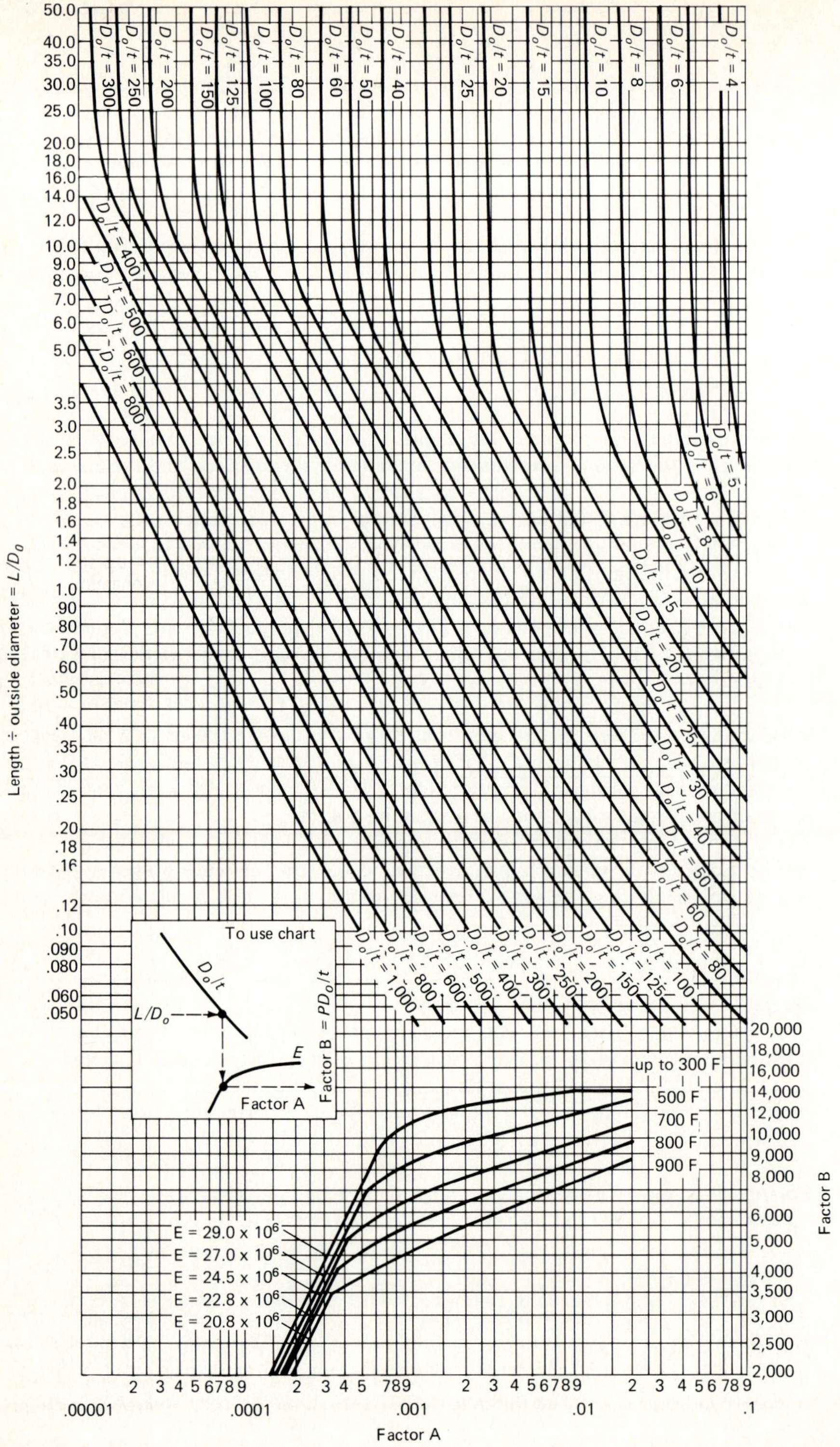

Length ÷ outside diameter = L/D_o
Factor A
Factor B
D_o/t = 300
D_o/t = 250
D_o/t = 200
D_o/t = 150
D_o/t = 125
D_o/t = 100
D_o/t = 80
D_o/t = 60
D_o/t = 50
D_o/t = 40
D_o/t = 25
D_o/t = 20
D_o/t = 15
D_o/t = 10
D_o/t = 8
D_o/t = 6
D_o/t = 4
D_o/t = 400
D_o/t = 500
D_o/t = 600
D_o/t = 800
D_o/t = 5
D_o/t = 30
D_o/t = 1,000
To use chart
L/D_o
E
Factor B = PD_o/t
up to 300 F
500 F
700 F
800 F
900 F
E = 29.0 x 10^6
E = 27.0 x 10^6
E = 24.5 x 10^6
E = 22.8 x 10^6
E = 20.8 x 10^6
.00001
.0001
.001
.01
.1

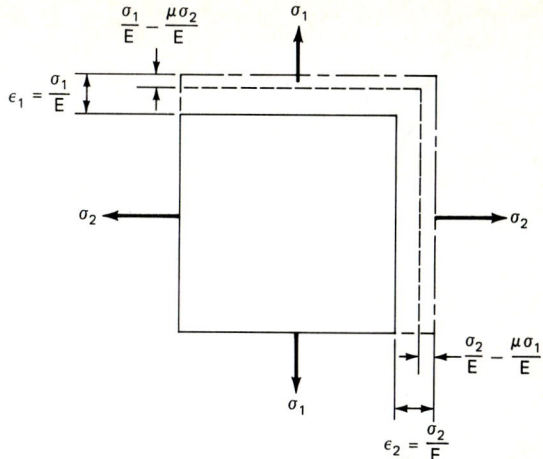

Figure 10.10 Determination of the Poisson ratio μ for a rectangular plate under stress. $\mu = \epsilon_c/\epsilon_e \approx 0.3$ for carbon steel. ϵ_c = lateral contraction; ϵ_e = axial elongation.

erratic vibration to the ground. Because of their erratic nature their effects on a pressure vessel are difficult to predict accurately. However, spectral response curves charting seismic accelerations and durations for many areas in the United States have been compiled. The seismic accelerations a are reported as multiples of gravitational acceleration g; i.e.,

$$c = \frac{a}{g} = \text{empirically determined seismic coefficient} \tag{10.11}$$

The seismic force F_s for a given seismic coefficient is calculated by the equation

$$F_s = cW \quad \text{lbf} \tag{10.12}$$

where W = operating weight of the vessel during an earthquake, lbf.

If this force is applied to the center of gravity of the vessel and multiplied by the height h of the center of gravity from grade, the overturning moment M_s of the vessel is calculated. This can be used to calculate the seismic stress by the equation

$$\sigma_s = \frac{M_s}{Z} = \frac{F_s h}{Z} = \frac{cWh}{Z} \tag{10.13}$$

where Z = section modulus for the vessel. For cylindrical vessels Z is calculated by

$$Z = \left(\frac{\pi}{32}\right)\frac{D_o^4 - D_i^4}{D_o} \quad \text{ft}^3 \tag{10.14}$$

Figure 10.9 Chart for determining thickness of carbon steel vessels under external pressure. The factor B equals PD_o/t, where P is the external pressure, psi; D_o is the shell outside diameter, in; and t is the shell wall thickness, in. (*Adapted by permission of the American Society of Mechanical Engineers from Figs. 5-UGO-28.0 and 5-UCS-28.1 from Sec. VIII, Div. 1, of the 1983 edition of the ASME Boiler and Pressure Vessel Code.*)

where D_o = outside diameter, ft
D_i = inside diameter, ft

References 5, 6, and 7 present a more detailed explanation of the calculation of seismic stresses.

Combining stresses. A number of stress conditions can occur simultaneously during normal operation of a process vessel. The standard procedure is to sum them together algebraically to determine the total stress on the vessel shell. Table 10.2 lists how these stresses are added.

Once the total stress for each case listed in Table 10.2 has been calculated, the total stress with the largest magnitude is the maximum expected stress applied to the vessel. This maximum total stress must be less than the maximum allowable stress for material used in constructing the vessel. If it exceeds the maximum allowable stress, either a different material must be used or the design of the vessel changed to reduce the total stress.

Welded joints

A welded joint could be the strongest or the weakest part of a pressure vessel. The welding rods or electrodes used to join two metal parts of a vessel are selected to have a higher allowable tensile stress than the metal being jointed. Therefore, the weld material should not, in theory, fail before the vessel material does. However, practical experience has shown that most vessel failures occur on or around a weld. This is caused by either poor workmanship or poor heat treatment of the welded area.

To compensate for weld imperfections, vessel codes provide weld-efficiency factors which are used to adjust the allowable stress value for the weld material. Weld efficiency is based on the type of weld and the degree of weld inspection. Table 10.3 lists some typical weld-joint efficiencies.

Vessel inspection

Part of the construction of any process vessel includes inspection of the workmanship applied in its construction. Often a visual inspection for leaktightness is sufficient for vessels not used in hazardous applications. However, most process vessels as well as process piping require a more thorough inspection. Table 10.4 lists various inspection techniques together with their advantages and disadvantages.

TABLE 10.2 Total-Stress Formulas

	Windward or tensile loading	Leeward or compressive loading
Internal pressure	$\sigma_t = \sigma_w + \sigma_i - \sigma_d$ $\sigma_t = \sigma_s + \sigma_i - \sigma_d$	$\sigma_t = -\sigma_w + \sigma_i - \sigma_d$ $\sigma_t = -\sigma_s + \sigma_i - \sigma_d$
External pressure	$\sigma_t = \sigma_w - \sigma_e - \sigma_d$ $\sigma_t = \sigma_s - \sigma_e - \sigma_d$	$\sigma_t = -\sigma_w - \sigma_e - \sigma_d$ $\sigma_t = -\sigma_s - \sigma_e - \sigma_d$

NOTE: Seismic- and wind-loading stresses are not added together because the likelihood of an earthquake and a windstorm occurring simultaneously is remote.

σ_t = total stress; σ_d = deadweight stress; σ_e = external-pressure stress; σ_i = internal-pressure stress; σ_s = seismic-loading stress; σ_w = wind-loading stress.

TABLE 10.3 Typical Weld-Joint Efficiencies

Weld type	Radiographic examination, %		
	Full	Spot (partial)	None
Butt			
Single	90	85	65
Double	100	85	70
Fillet	80	...	60
Plug	80	...	60

TABLE 10.4 Process-Piping and Vessel Inspection Techniques

Technique	Advantage	Disadvantage
Radiographic: Used to examine the internal soundness of weldments and metals by bombarding the piece with x-rays or gamma rays.	1. Sharp picture of any defects that may exist. 2. Rays penetrating the piece being examined impinge on a sensitized film, which provides a permanent record of the examination.	1. Special protection and training must be provided for the inspection personnel. 2. Two sides of the piece being examined must be accessible: one side for the radiographic equipment and the other for the film.
Ultrasonic: Uses high-frequency sound waves to locate defects.	1. Very sensitive; can detect very fine surface and subsurface cracks. 2. Equipment is portable. 3. Only one side need be accessible.	1. Personnel must be trained in interpreting the equipment responses. 2. Not effective on rough-surface materials or welds with backing rings. 3. No permanent record provided.
Magnetic-particle: Used to detect surface defects by applying a liquid suspension of fine particles that flow into fine cracks. When a strong magnetic field is applied to the inspection area, the particles concentrate themselves in the area of the defect, highlighting its size and shape.	1. Useful for showing fine cracks that are not noticeable in radiographic examination. 2. Useful in showing where and what material must be removed for weld repair.	1. Cannot be used on nonmagnetic material. 2. Detects surface cracks only. 3. Cannot detect defects parallel to the magnetic field.
Dye-penetrant: Used to detect surface defects. A liquid dye is applied to a clean, dry surface and allowed to penetrate surface cracks and dry. Once it is dry, a developer is put over the surface, causing the dye in the surface imperfection to outline the defect clearly.	1. Useful for nonmagnetic materials. 2. Can be used on nozzles and surfaces difficult to inspect radiographically.	1. Detects surface defects only. 2. Not practical on rough surfaces.

Vessel-design codes

The function of design codes for process vessels is to provide the engineer with a procedure for the design, material selection, and inspection procedure for such vessels. The codes list minimal requirements to reduce the chances of process-vessel failure. Many different codes and standards are available with which to design process vessels. Selection of the proper code depends on the process conditions that the vessels will experience.

United States codes and standards. There are various codes and standards in the United States that are used for designing process vessels. However, individual states and municipalities may have additional regulations whose requirements must also be met. The choice of the required code is generally not made by process engineers but by the vessel designer and the plant manager. If there are questions as to code jurisdiction, the proper state or local government agency must be consulted.

The most frequently used codes for designing process vessels in the United States are listed in Table 10.5. Of the codes listed in the table the one most often referenced is the ASME code, Sec. VIII, for unfired pressure vessels. This code gives extensive coverage to all aspects of pressure-vessel design. Figure 10.11 shows a composite drawing of a pressure vessel with annotations regarding applicable sections of the ASME code for designing various portions of the vessel.

TABLE 10.5 Typical United States Codes and Standards Used for Vessel Fabrication

Applicable range	Code or standard*
Operating pressure, 0–0.5 psig (storage tanks)	1. Underwriters Laboratories Inc., standards 2. American Petroleum Institute (API) standards 650 and 12A, 12B, 12D, and 12F.
Operating pressure, 0.5–15 psig (storage tanks)	API standard 620†
Operating pressure, 15 psig (process and storage vessels)	ASME code, Sec. VIII

*The addresses of these agencies are:

American Petroleum Institute (API)
1271 Avenue of the Americas
New York, N.Y. 10020

The American Society of Mechanical Engineers (ASME)
345 East Forty-Seventh Street
New York, N.Y. 10017

Underwriters Laboratories Inc.
333-T Pfingsten Road
Northbrook, Ill. 60062

†Although not applicable, the ASME code, Sec. VIII, is sometimes used if a more conservative design is desired.

Codes of other countries. Although many of the vessel codes and standards used in the United States are accepted in numerous other countries, some countries have their own vessel-design regulations. Table 10.6 lists the names and addresses of agencies responsible for the regulation of pressure-vessel design in several countries.*

---

*Composite drawings similar to Fig. 10.13 for codes of various other countries were published in *Hydrocarbon Processing*, December 1978.

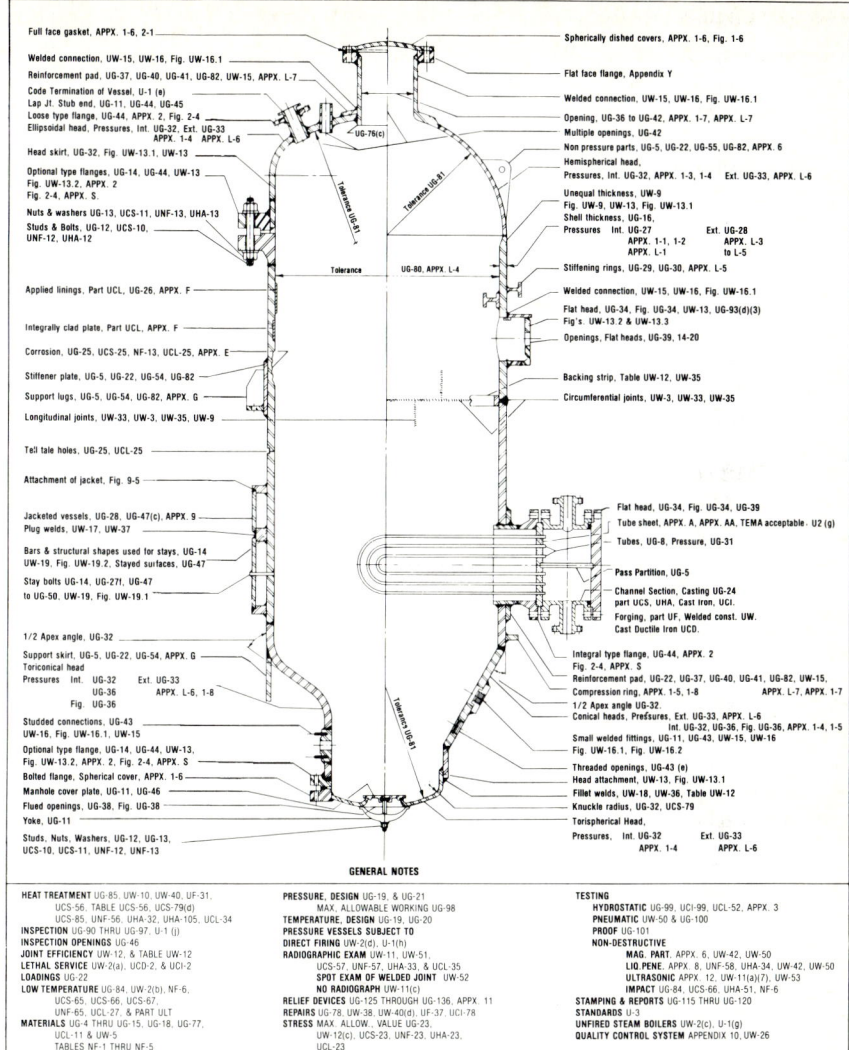

Figure 10.11 Guide to ASME Sec. VIII, Div. 1, Pressure Vessel Code. (*Courtesy of Missouri Boiler & Tank Inc., St. Louis.*)

Vessel Selection

An important decision to be made before designing a vessel is the vessel's general geometry, which, in turn, is dictated by the vessel's application. The most basic consideration is whether the vessel will be used as a storage vessel or as a process vessel.

Storage vessels

Many vessels in a process plant are used exclusively for storing raw materials, intermediate products, or finished products. The type of vessel used depends on the pro-

TABLE 10.6 Agencies Regulating the Design of Pressure Vessels in Various Countries

Belgium
    Code: NBN 121, Code de Bonne Pratique pour la Construction des Appareils Soumis à Pression
    Agency: Institut Belge de Normalisation, Avenue de la Brabançonne 29, Bruxelles 4
France
    Code: Réglementation des Appareils à Pression de Gaz, AOAVE
    Agency: SADAVE, 34 Rue St-Lazare, 75, Paris 9^e
West Germany
    Code: UVV (Unfall Verhütungsvorschrift Druckbehälter)
    Published by: Carl Heymanns Verlag KG, Köln 1, Gereonstrasse 1832
Italy
    Code: ANCC Pressure Vessel Code
    Published by: G. Pirola, Via Comelico 24, 20135 Milan
Sweden
    Code: Pressure Vessel Code—Calculation of the Strength of Pressure Vessel
    Agency: Tryckkarls Commissionem, IVA, P.O. Box 5073, S-10242, Stockholm 5
United Kingdom
    Codes: Fusion Welded Pressure Vessels, BS 1500 and BS 5500
    Agency: British Standards Institution, 2 Park Street, London W1

cess material being stored. Liquids and gases require storage vessels different from those of solids. The type of storage tank used for liquids and gases depends to a degree on the storage pressure or vapor pressure of the fluid within the range of storage temperatures. Table 10.7 lists possible storage-tank designs used at different fluid-storage pressures.

If solids are being stored, the vessels usually do not need watertight or gastight construction unless the solid is in a slurry or the contents create hazardous conditions when exposed to the weather. Solids are generally stored in vertical tanks with conical bottoms for ease of solids discharge. The angle of the bottom cone should exceed the angle of repose of the solid to prevent solid choking. These tanks may also contain internal conical disks and baffles above the discharge to prevent bridging of solids.

TABLE 10.7 Storage-Tank Types versus Storage Pressure

Storage pressure, psig	Tank type
1	1. Standard API tanks, vented and fixed roof. 2. If air inhalation is undesirable, use API tanks with a floating head or a fixed roof. An API tank has an inert-gas blanket for the vapor space.
1–20	1. Vertical or horizontal tanks with dished heads. 2. Sphere, sometimes used.
20–100	1. Sphere. 2. Long horizontal cylinder.
>100	1. Small high-pressure cylinder or group of cylinders manifolded to a common intake and discharge header.

Process vessels

The selection of a process vessel is dictated by process requirements. It is difficult to apply simple rules to cover the selection of all process vessels, but there are some vessel-design details that should be considered when specifying the construction of process vessels. Many of these details have been presented in the subsection "Checklists" in Chap. 3. Additional ones are presented here.

Orientation. There is always some question as to whether a vessel is to be horizontal or vertical. Table 10.7 presented one criterion for storage-tank orientation. Another criterion for selecting a vessel's orientation is based on physical space limitations. Horizontal tanks are used in areas of low headroom, while vertical tanks would be used where plant area is limited.

Vertical vessels are practical in processes in which the force of gravity is important, such as settling tanks, distillation columns, and gas scrubbers. However, some settling processes and liquid-extraction systems operate more efficiently in horizontal vessels.

Vessel supports. The method by which a vessel is supported depends on its size and orientation. Most horizontal vessels are mounted on saddles; exceptions to this support method are underground horizontal tanks and tanks supported on lugs. Lugs, legs, and skirts are used to support vertical vessels.

Lugs. Lugs shown in Fig. 10.12 are used to suspend vessels between floors of a building or from a steel-frame structure. The operating weight of the vessel must be

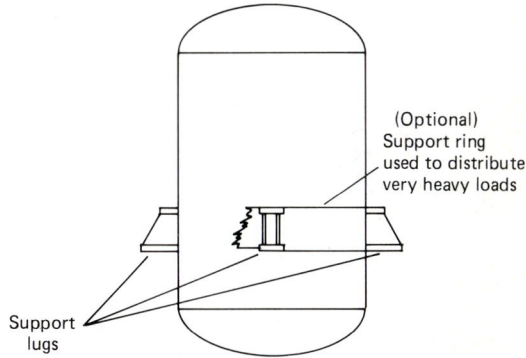

Figure 10.12 A process vessel supported by lugs.

considered when designing lug supports. If the operating weight is too great, a stiffener plate or support ring should be used to distribute the weight rather than concentrate load stresses at each lug.

Vertical vessels can also be supported by legs or by a skirt. Vertical process vessels which are 6 ft in diameter or less are usually suppported on legs. For larger-diameter vessels, a skirt should be considered.

Legs. Legs are generally used on process vessels that are less than 6 ft in diameter. The number of legs used depends on the deadweight load stress carried by each leg. Usually 3 legs suffice for vessels that are about $3\frac{1}{2}$ ft in diameter; larger-diameter vessels usually have 4 legs. If deadweight stresses exceed the allowable stresses for each leg, up to 10 legs have been used. If it appears that 8 or 10 legs are needed, a skirt may be a more practical alternative.

Skirts. A skirt is a cylindrical plate slightly larger in diameter than the process vessel. It is attached by a continuous weld to the base of the straight side of the vessel extending down to the ground, where it is bolted to a concrete pad. As a result, the deadweight load of the vessel is evenly distributed about the skirt rather than being concentrated only at certain points, as would be the case when legs are used.

Skirts are designed with openings to pass process piping through, provide access for maintenance and inspection, and furnish openings for weep holes at the skirt base in order to drain rainwater or condensate.

Holdup. The required volume of a vessel varies with the application. A distillation column's volume depends on the required volume of the packing or on the number of trays and the tray spacing. The volume of a reactor vessel or a neutralizing tank depends on the required residence time for the reaction. The volume of surge tanks and accumulators for compressed-air or gas systems depends on the volume needed to reduce pressure variations in the system or as backup during emergency periods.

Day tanks. A day tank is an intermediate storage tank. It is used when the main storage tanks are remote from the process facilities or when the source of supply can be intermittent or interrupted. The day tank provides a continuous supply of raw materials to the process and allows an orderly shutdown of the process. They are also used for product testing prior to release or recycle.

The name "day tank" implies a 24-h capacity. This is not necessarily true, since some day tanks could have volumes equivalent to as little as 10 min or as much as 1 week. The actual volume depends on what is economically the best volume to prevent interrupted service and allow orderly shutdown of the process system.

Reflux drums. A reflux drum collects liquid from a condenser as well as acting as a liquid seal on the condenser. It also is a distribution point for reflux and distillate. The volumes of these vessels usually have a liquid holdup of 5 to 10 min with a normal freeboard, i.e., vapor space volume, of about 25 to 50 percent. The flow rates of recycle or reflux streams are not taken into account for holdup volume unless it is possible that such streams may not be returned immediately.

Elevated above grade. With the exception of API-type storage tanks (see Table 10.5) and underground storage tanks, the bottom of a tank must be elevated a minimum distance above grade level. This distance accommodates the piping and equipment required beneath the tank and provides ease of access for maintenance and inspection as required. An elevation of 3 to 5 ft is adequate for most process vessels. However, if there are height limitations, the elevation may be scaled down to a minimum dimension that can fit the vessel's discharge-nozzle length plus one standard-radius elbow. For a vessel located in a multistoried structure, the elevation can be reduced to little more than the length from the vessel shell to the bolting flange of the discharge nozzle. This is acceptable only if the discharge piping can run through the floor to the next level and there is physical access to the flange bolts and valving.

Design versus operating conditions. Since there can be periodic upset conditions which temporarily increase temperatures and pressures in the vessel, it is good engineering practice to design a process vessel or a storage vessel for other than normal operating conditions. Therefore, design pressures and temperatures for vessels are selected to meet the extremes that would be expected in the process.

Design temperatures are based on worst-case operating conditions. Vessels normally operating at ambient conditions are designed for highest summer temperatures plus 20 to 50°F, depending on the degree of conservatism desired. Vessels used in high-temperature service usually have design temperatures that are 5 to 10 percent greater than normal operating temperatures. The design must also account for the maximum heating or cooling temperatures resulting from a control-system failure.

Design pressures are determined on the basis of the vessel type. API-type storage tanks operate at near atmospheric pressure. While the side plates at the bottom of the vessel are designed to withstand the hydrostatic pressure of the stored liquid, their roof and upper shell are subject to lower stresses. Therefore, their design pressures range from 10 to 20 in of water column for tanks that are less than 20 ft in diameter and 2 to 5 in of water column for larger-diameter tanks.

Process vessels with dished heads which normally operate at pressures between atmospheric and 5 psig are usually designed for 10 psig. A good rule of thumb for selecting design pressures for other pressure ranges is as follows:

Operating-pressure range, psig	Design pressure as a percent of operating pressure
10–100	130–150
100–500	115–130
500–1000	110–115
Over 1000	105–110

REFERENCES

1. David S. Azbel and Nicholas P. Cheremisinoff, *Chemical and Process Equipment Design, Vessel Design and Selection,* Ann Arbor Science Publishers, Ann Arbor, Mich., 1982, pp. 726–742.
2. Lloyd E. Brownell and Edwin H. Young, *Process Equipment Design,* John Wiley & Sons, Inc., New York, 1959, chap. 11.
3. Eugene F. Megyesy, *Pressure Vessel Handbook,* 3d ed., Pressure Vessel Handbook Publishing, Inc., Tulsa, Okla., 1975, pp. 72–85.
4. John F. Harvey, *Theory and Design of Modern Pressure Vessels,* 2d ed., Van Nostrand Reinhold Company, New York, 1974, pp. 41–47.
5. Henry H. Bednor, *Pressure Vessel Design Handbook,* Van Nostrand Reinhold Company, New York, 1981, pp. 13–21.
6. Brownell and Young, op. cit., pp. 163–168.
7. Megyesy, op. cit., pp. 49–51.

BIBLIOGRAPHY

Chuse, Robert: *Unified Pressure Vessels: The ASME Code Simplified,* McGraw-Hill Book Company, New York, 1960.

Mahajan, Kanti K.: *Design of Process Equipment: Selected Topics,* Pressure Vessel Handbook Publishing, Inc., Tulsa, Okla., 1979.

Mallinson, John H.: *Chemical Plant Design with Reinforced Plastics,* McGraw-Hill Book Company, New York, 1969.

Chapter 11

Pumps

The Pump

Pumps are used to provide sufficient kinetic energy to a liquid to cause it to flow through a piping system. Pumps come in a variety of types and configurations. The type used depends on the nature of the fluid and the hydraulics of the piping system. This chapter presents the various classifications and types of pumps, typical applications, and methods of installing pumps in a piping system.

Pump Classification and Types

Pumps can be divided into three categories: reciprocating pumps, rotary pumps, and centrifugal pumps. The pumps in each category can be subdivided into different types on the basis of their mechanical design. Figure 11.1 presents a graphical relationship of pump categories and types.

Reciprocating pumps

Reciprocating pumps are positive-displacement pumps which consist of a chamber or group of chambers which alternately draws in a fixed volume of liquid during the suction cycle of operation and expels a pressurized liquid during the discharge cycle. Reciprocating pumps are divided into three general types based on the device used to create compression. These are piston pumps, plunger pumps, and diaphragm pumps.

Reciprocating pumps are specified according to their orientation, the number of discharge strokes per cycle, the number of pumping chambers, and their characterization as a power or a direct-acting pump. Pump orientation is usually either horizontal or vertical as defined by the orientation of the centerline of the pump chamber. If a pump is single-acting, it has only one discharge stroke per cycle. When a pump has two discharge strokes per cycle, it is called double-acting; i.e., liquid is being pumped during both the forward and the reverse strokes. Figure 11.2 illustrates the characteristic curves produced by single-acting and double-acting pumps.

If a pump has only a single pumping chamber to pressurize the liquid, it is a simplex pump. If it has two chambers, it is a duplex pump. Pumps can also be triplex (with three chambers) and multiplex (with many chambers). As can be seen in Fig.

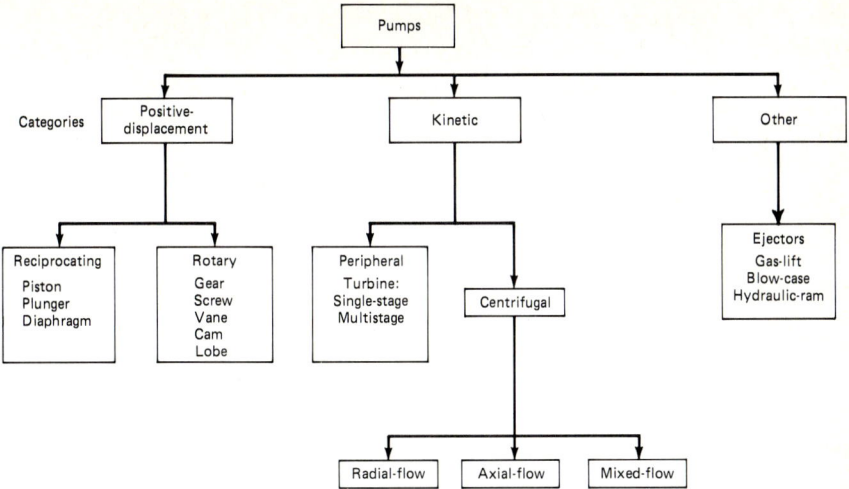

Figure 11.1 Pump categories and types.

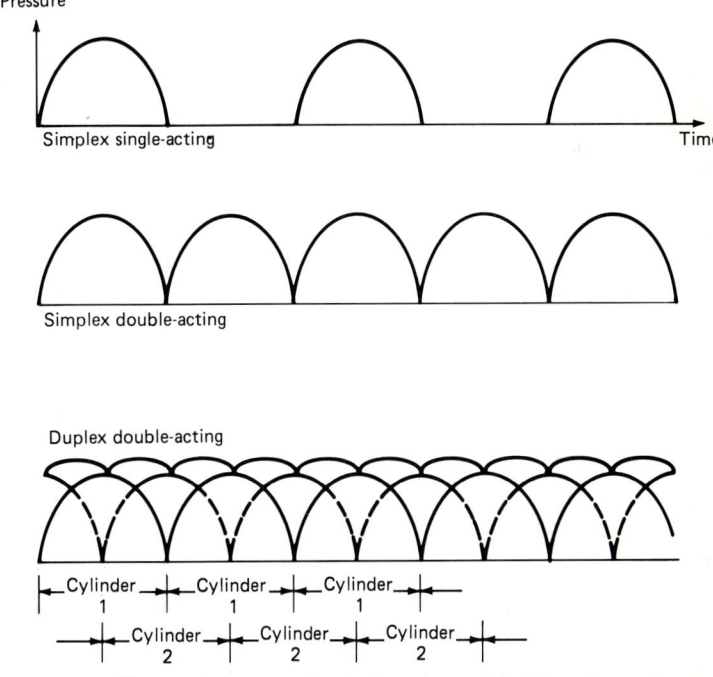

Figure 11.2 Characteristic curves for single-acting and double-acting reciprocating pumps.

11.2, multiplexing a reciprocating pump reduces the ripples in the discharge flow curve. However, the cost of a pump increases with the number of pumping chambers.

The ripple factor of a reciprocating pump can also be reduced by placing an air-pressurized surge tank downstream of the pump. The surge tank is an economical alternative to multiplexing when discharge pressures are not high or when absorption of air from the surge tank into the liquid is acceptable. As a general rule, simplex pumps require a surge volume of 6 to 8 times the volume of a pump chamber, while duplex pumps require a surge volume equal to only 3 to 4 times their chamber volumes.

Reciprocating pumps are also described as being either power or direct-acting. A direct-acting pump uses a fluid such as a gas, steam, or air to provide the pumping action, while a power pump is motor-operated. Power pumps are more efficient, but direct-acting pumps are more practical for remote locations.

Piston and plunger pumps. Piston and plunger types of pumps have a moving member which pressurizes the liquid as it moves through a chamber. The difference between a piston pump and a plunger pump is the method by which each is sealed. The piston-pump seal is located on the piston itself, while the plunger-pump seal is stationary and is mounted in the plunger shaft.

Both piston and plunger pumps are normally used in high-pressure applications (400 to 10,000 psig). In addition, these pumps normally have high efficiencies, ranging from 85 to 95 percent. Some plunger pumps are provided with variable-speed drives and are used for metering liquid flow. Applications for piston and plunger pumps include hydraulic systems, high-pressure cleaning, hydrostatic testing of piping and vessels, deep-well pumping, homogenizing food products, boiler-feedwater pumping, and pumping volatile or viscous fluids.

While piston and plunger pumps have higher initial and maintenance costs than centrifugal pumps, they have considerably greater efficiencies at high pressures. These pumps are not used for solids and abrasive slurries and require overpressure protection. They are not recommended for systems in which flow pulsations are unacceptable.

Diaphragm pumps. A diaphragm pump consists of a flexible membrane in a housing. As the diaphragm is flexed, liquid is drawn in and out of the chamber containing the diaphragm. Liquid passage through the pump is made unidirectional by the placement of check valves in the suction and discharge portions of the pump.

Diaphragm pumps have operating characteristics similar to those of piston and plunger pumps, but they do not have packing or seal leakage problems. This makes them desirable for services handling corrosive, toxic, or expensive materials.

Most diaphragm pumps operate at pressures below 150 psig, but high-pressure operations can be achieved by selection of the diaphragm material; i.e.,

Diaphragm material	*Maximum pressure, psi*
Elastomeric	750
Teflon	1,500
Metallic	45,000

Teflon and elastomeric materials are limited to operating temperatures of 280 and 212°F, respectively.

While diaphragms are subject to flex-fatigue failure, a properly designed diaphragm pump can have a long service life. For those applications handling toxic or hazardous materials, a diaphragm failure is unacceptable. Therefore, pumps with double diaphragms are used. A hydraulic oil is placed between the two diaphragms to maintain mechanical continuity between the liquid being pumped and the driver.

Diaphragm pumps can be either motor-driven or driven by the direct action of compressed air, steam, or a hydraulic fluid. Diaphragm pumps used for metering flow are primarily motor-driven, while those having high capacities (over 10 gal/min) usually have direct-action drives. The capacity of metering-type diaphragm pumps may be controlled by adjusting the stroke of the plunger which moves the diaphragm.

Diaphragm pumps, unlike other positive-displacement pumps, are used to handle slurries and abrasive materials. This is possible because they have no seals or packing that can deteriorate owing to solids buildup and use self-cleaning ball check valves.

Rotary pumps

Rotary pumps are positive-displacement pumps which force a fixed volume of liquid through the pump by the rotating action of an internal component. This component can be either gears, lobes, screws, or vanes. Rotary pumps are constant-speed and, therefore, constant-capacity pumps. However, their capacities can drop off at high discharge pressures. This is true because rotary pumps do not operate at the close tolerances required for piston and plunger pumps and some slipping of liquid back into the suction side of a pump can occur.

Rotary pumps have certain advantages over other positive-displacement pumps, these being lower initial cost, fewer component parts requiring maintenance, need for less space, and ability to handle entrained gases and vapors.

The disadvantages of rotary pumps are that, owing to their higher slip, they cannot produce the high pressures that plunger and piston pumps can. They are also susceptible to overpressure damage and require relief valves, as do all positive-displacement pumps. Rotary pumps are generally used only with self-lubricating liquids. They cannot, with certain exceptions, handle abrasive solids.

The basic types of rotary pumps are cam, gear, lobe, screw, and vane pumps.

Cam pumps. Cam pumps consist of a cam-and-piston type of arrangement. The eccentric motion produced by the cam causes the hollow piston to trap liquid and discharge it into the high-pressure side of the pump casing. Rotation of the cam causes a nearly continuous flow of liquid through the pump.

Lobe pumps. Lobe pumps are similar to external gear pumps in operation. The impeller in a lobe pump is a wheel with two, three, or four lobes as shown in Fig. 11.3. The two impellers in a lobe pump have their rotation synchronized by an exter-

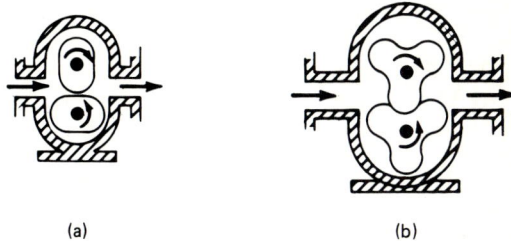

(a)   (b)

Figure 11.3 Lobe pump. (*a*) Two-lobe. (*b*) Three-lobe.

nal drive chain. Impeller synchronization is necessary because the lobe pump pushes fewer but larger-volume units of liquid through it during each rotation than does a gear pump. The synchronization maintains a constant flow through the pump. The more lobes on an impeller, the less pulsation developed by the pump; however, considerably fewer pulsations are produced by a gear pump than by a lobed pump.

Gear pumps. Because of their compact and simple design, gear pumps, such as the one shown in Fig. 11.4, are probably the most often used rotary pumps. The liquid enters the pump suction and gets trapped in the space between the teeth of the gear. The liquid in the space is circulated around the outer position of the casing space and out through the discharge port.

Figure 11.4 Gear pump. *(Roper Pump Company, Commerce, Ga.)*

Gear pumps have two gears with the driver connected to only one of them. The other gear, called the idler gear, moves in response to the driven gear. Gear pumps come in two arrangements, an external gear arrangement and an internal gear arrangement. These are shown in Fig. 11.5.

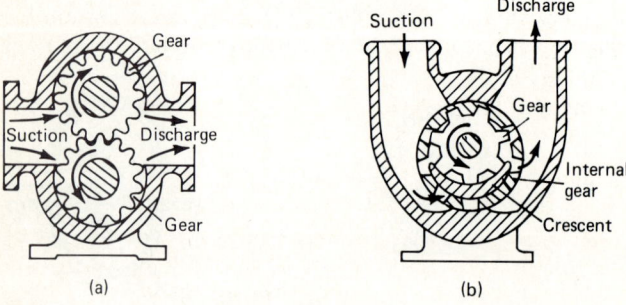

Figure 11.5 Cutaway view of two types of gear pumps. (*a*) External gear arrangement. (*b*) Internal gear arrangement.

Gear pumps are often used to pump such viscous fluids as hydraulic oil. They are also used to pump liquids with small amounts of solid impurities. However, pumps in this service require a special driver-train arrangement; i.e., the idler gear does not idle but must be synchronized to the rotation of the other gear by a drive chain. This reduces the chance of solids buildup jamming the gears.

The external gear pump is limited to about 200 gal/min., while internal-gear-pump capacities can range up to about 1100 gal/min. These pumps can handle liquids with viscosities up to 5 million SSU.

Screw pumps. A screw pump consists of one, two, or three threaded screws in a pump housing. Such pumps are used for handling liquids with very high viscosities (up to about 1 million SSU) and solid slurries. They are self-priming, have low shear, and generally have longer operating lives than other positive-displacement pumps because they operate at low revolutions per minute. Figure 11.6 shows a typical screw pump.

Figure 11.6 Twin-screw pump. *(Roper Pump Company, Commerce, Ga.)*

If a screw pump is used to pump solids and slurries, then a single-screw pump such as the one shown in Fig. 11.7 is used. The casing consists of a helical cavity which matches the threads of the screw. The rotation of the screw moves the liquid slurry along the length of the cavity and out the discharge.

Vane pumps. A vane pump consists of a rotor placed slightly off center in a casing. Embedded in the rotor is a series of sliding or swinging blades as shown in Fig. 11.8. As the pump rotates, centrifugal action presses the blades along the inner casing wall, trapping liquid between blades and forcing it out the discharge.

Vane pumps have negligible pulsations, and the vanes are self-compensating for wear. Vane pumps are not recommended for handling abrasive fluids, but they are self-priming and can pump liquids containing entrained gases or vapors.

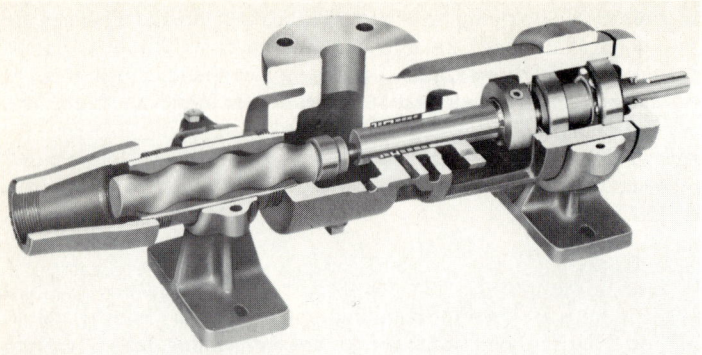

Figure 11.7 Single-screw pump used to pump solids and slurries. *(Roper Pump Company, Commerce, Ga.)*

Figure 11.8 Vane pump. *(Dover Corporation, Blackmer Pump Division, Grand Rapids, Mich.)*

Centrifugal pumps

Centrifugal pumps are the workhorses of the process industries. They are the most common type of pump in most process plants. Centrifugal pumps transfer the kinetic energy of rotation to the liquid, thrusting it out of the pump casing tangentially to the rotation at a higher pressure than when it entered. Centrifugal pumps have a relatively simple, compact construction and require little maintenance compared with other pump types. These characteristics make centrifugal pumps the economical choice for most applications. Because they do not require close clearances, as do positive-displacement pumps, they can be used with liquids containing a variety of solids and abrasive materials.

Another advantage of centrifugal pumps is that they dynamically respond to changes in system characteristics. The discharge pressure of centrifugal pumps is roughly inversely proportional to the square of the flow capacity over a large portion of their operating curves.

Centrifugal pumps, however, have some disadvantages. They cannot achieve the high pressures that positive-displacement pumps can produce without going to a multistage design, and their efficiencies drop appreciably with increasing fluid viscosity.

Centrifugal pumps are available in a number of casing arrangements and impeller designs. Each is tailored to different application needs.

Casing design. Centrifugal pumps may be either horizontal or vertical. Vertical pumps such as the one shown in Fig. 11.9 are typically used for wells and cooling-tower sumps, while the pump shown in Fig. 11.10 is a vertical in-line pump used in certain petroleum services to produce very high discharge pressures.

The horizontal configuration is the most frequently selected centrifugal-pump orientation. In most cases, centrifugal pumps have a vertical split-case design, as shown in Fig. 11.11. Large-capacity, high-head centrifugal pumps usually have a horizontal split-case design, as shown in Fig. 11.12. The latter casing is also used for multistage pumps.

A propeller pump, such as the one shown in Fig. 11.13, is selected for large capacity but relatively low head.

Impeller design. There are many types of impellers, but they can generally be classified as axial-flow, mixed-flow, or radial-flow impellers.

With axial-type impellers, the flow is parallel to the axis and head is produced by the vortexing action and lift imparted by the impeller blades. These pumps are also called propeller pumps. They are used in applications requiring large flows and low to moderate heads. Typical applications include cooling-water recirculation and evaporator or crystallizer circulation.

Radial-type impellers accelerate entering liquid radially outward from the center of the impeller. Head is produced by centrifugal action. Most centrifugal pumps with low to moderate flows are of this design. Radial impellers can be subdivided into open impellers (Fig. 11.14), semiclosed impellers (Fig. 11.15), and closed impellers (Fig. 11.16). In general, closed impellers are used to pump clean liquids, while open impellers are used to pump liquids containing solids. Open impellers are less efficient than closed impellers but are less susceptible to clogging by solids. A compromise between clogging and efficiency is found in semiclosed impellers. These impellers have a single shroud rather than the two shrouds on closed impellers. Many nonclogging pump impellers are a variation of semiclosed-impeller design.

Mixed-flow impellers are a compromise between axial-flow and radial-flow impellers. The mixed flow, as the name implies, develops head by a combination of blade-lift and centrifugal actions.

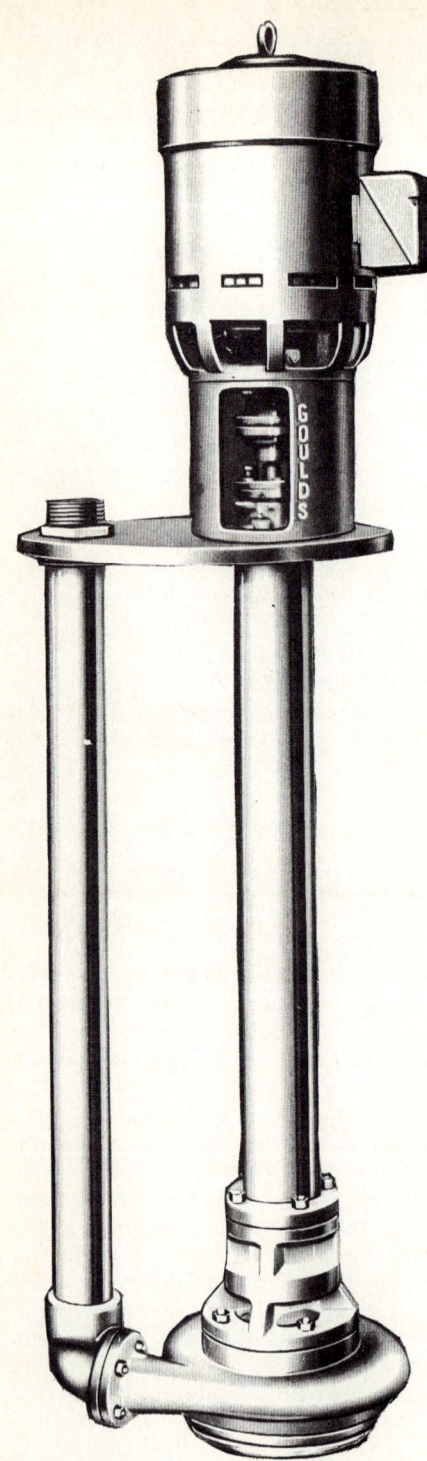

Figure 11.9 Vertical centrifugal pump used in wells and sumps. *(With permission from Goulds Pumps, Inc., Seneca Falls, N.Y.)*

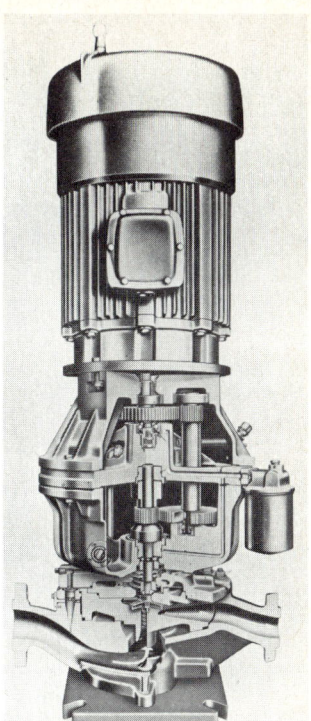

Figure 11.10 Vertical in-line centrifugal pump used for high-pressure-process and petroleum service. *(Courtesy of Sundstrand Fluid Handling, Arvada, Colo.)*

Figure 11.11 Horizontal centrifugal pump with a vertical split-case design. The pump shown is used for heavy-duty-process service such as pulp and paper processing, slurry handling, food slurries, and waste treatment. *(With permission from Goulds Pumps, Inc., Seneca Falls, N.Y.)*

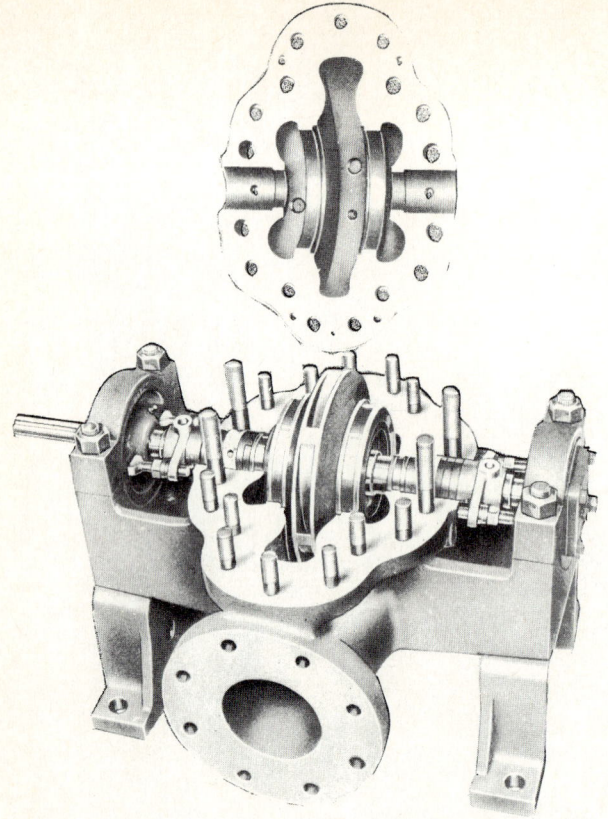

Figure 11.12 Horizontal centrifugal pump with a horizontal split-case design. *(With permission from Goulds Pumps, Inc., Seneca Falls, N.Y.)*

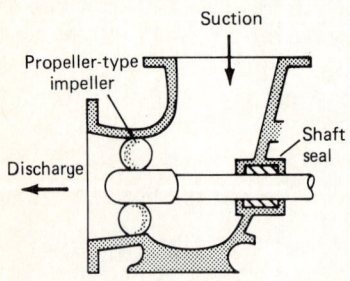

Figure 11.13 Propeller-type centrifugal pump used for large-capacity, low-head applications.

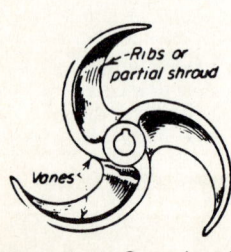

Figure 11.14 Open impeller.

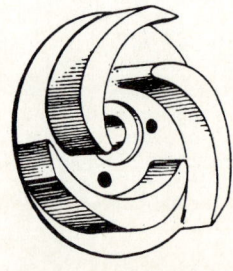

Figure 11.15 Semiclosed impeller.

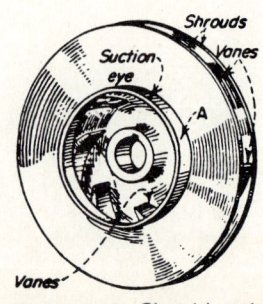

Figure 11.16 Closed impeller.

TABLE 11.1 General Characteristics of Centrifugal Pumps by Impeller Type

	Impeller type		
	Axial	Radial	Mixed
Head	Low	High	Intermediate
Capacity	High	Low	Intermediate
Power curve	Rises as shutoff is approached	Level or drops as shutoff is approached	Intermediate

Table 11.1 lists the general operating characteristics of the three impeller designs. In practice, there is no marked delineation between the types. Pump manufacturers market a variety of pump designs that cover the spectrum between axial- and radial-type impellers. The process engineer who knows the pump application should consult pump manufacturers to select the pump best fitted to the application.

Turbine pumps

Turbine pumps, also called regenerative pumps, have a series of small vanes placed along the periphery of the impeller (see Fig. 11.17). Liquid is drawn into the pump and circulates along the periphery in the space between the impeller vanes. The rotat-

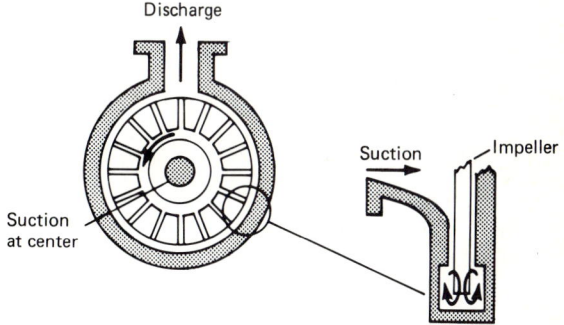

Figure 11.17 Turbine pump.

ing vanes impart a velocity to the liquid which is converted into pressure head in the annular space between the impeller and the pump casing.

Turbine pumps have very close clearances between the impeller and the pump casing but not as close as those of rotary pumps. Although turbine pumps resemble rotary pumps, their behavior is more like that of centrifugal pumps in that their efficiencies drop off with flow and fluid viscosity.

These pumps are similar to rotary pumps in that their close clearance allows them to handle vapors. Therefore, they are self-priming and are used for small boiler feeds and for condensate return systems.

Regenerative or turbine pumps should not be confused with vertical turbine pumps used for deep wells and circulating-water applications. Vertical turbine pumps are centrifugal pumps with one or more mixed-flow impellers in series.

Pump-Performance Characteristics

There are several important parameters that must be considered when evaluating a pump to be used for a given application. They are throughput, total dynamic head (TDH), available net positive suction head (NPSHA), required net positive suction head (NPSHR), hydraulic horsepower (hhp), efficiency, and brake horsepower (bhp). These terms must be understood to analyze the graphs that describe the performance of the various classes of pumps. The various parameters will be reviewed, and their application to the unique characteristics exhibited by the various types of pumps will be described.

General pump parameters

Simplified explanations of the various important characteristics which describe pump performance are given for use in the practical selection of pumping equipment.

Flow rate. The flow rate or capacity Q of a pump is the actual flow of material that enters at the suction and exits at the discharge nozzle of the pump. It does not include internal recirculation. However, if a pump includes a relief mechanism, as do some reciprocating and rotary pumps, the relieved flow is part of the throughput. Flow rate is normally rated in terms of gallons-per-minute (gal/min) flow, while many pump manufacturers now include the rating in cubic meters per hour (m^3/h) also.

Total dynamic head. Total dynamic head is a practical application of Bernoulli's equation in determining the design of the pump. It is the difference between the discharge head h_d and the suction head h_s of the system downstream and upstream of a pump, respectively; i.e.,

$$\text{TDH} = h_d - h_s \tag{11.1}$$

TDH is usually expressed in terms of pressure differential (pounds per square inch, or psi) for reciprocating pumps and as feet of fluid for centrifugal pumps. The TDH of rotary pumps may be given as pounds per square inch or feet of fluid, depending upon the practice of the pump manufacturer. Head of fluid now is often included in pump catalogs in meters as well as in feet.

The suction and discharge heads can be broken down into component individual terms which include:

1. Line friction contributions to head for the suction side F_s and for the discharge side F_d are important constituents. They should be based on the design flow rates shown on the process flow sheet.

2. Pressure contributions to head at the source P_s and at the destination P_d must be accounted for. These values are usually based on psig. If either is a vacuum, the difference between 14.7 psia and the respective psia should be considered as(—) psig. The value for the suction head should be taken as the lowest value expected at the suction source, while that of the discharge head should be the greatest pressure expected there. The values should be adjusted so that they reflect the maximum difference expected during pump operation.

3. The liquid-level contribution at the source h_s and the destination h_d relative to the centerline of the pump must be analyzed. These level contributions are often of special interest and are given special names. If the liquid level is above the pump, h_s

is termed the static suction head and is given a positive sign. When the liquid level is below the pump, h_s is termed the static suction lift and is assigned a negative value. The discharge-level contribution is called the static discharge head, and h_d may be similarly positive or negative, depending on its position relative to the centerline of the pump.

The suction level should be taken as the lowest possible level from which the pump will ever have to transfer, regardless of the normal operating level. Similarly, the discharge head must be taken as the highest possible liquid level at the destination regardless of the normal operating level at that point.

4. Contributions due to velocity changes in the contraction in the piping at the source h_c and at the enlargement at the destination h_e must be included.

5. Friction loss through equipment or sensing instrumentation located in the suction or discharge piping must be taken into account.

6. The pressure drop across a control valve is usually assigned in conjunction with the instrument engineer, but as a rough rule of thumb it should be equivalent to a minimum of 5 to 10 psi where possible. For optimum operation of most control valves in liquid service, however, the drop at the design flow rate is usually about 30 to 50 percent of the total friction losses in the system including those across equipment and other instrumentation as well as in the piping.

To keep the sign conventions for each contribution correct and to ensure that no items are forgotten in the calculation of the TDH, many groups have pump calculation sheets similar to that shown in Table 11.2. This type of calculation sheet reduces pump calculations to a simple accounting problem with suction on one side of the balance sheet and discharge on the other. Once the suction and discharge heads have been calculated, the TDH is determined by the difference. If the difference as defined by Eq. (11.1) is negative, the suction head is greater than the discharge head and the liquid could flow between the source and the destination of the piping system without the aid of a pump. In this instance, the process engineer should review the path of the piping to determine whether or not any intervening point is above the assumed discharge point. If there is a piping elevation above the assumed discharge level, the height of the highest point should be employed in setting the minimum TDH to be used so that flow is initiated in the system. The discussion regarding siphon effects in piping in the subsection "Siphon Effects" in Chap. 7 should be considered when assigning the contribution of static discharge head h_d to the TDH.

When the TDH is positive, a pump is required and the magnitude of the TDH is the minimum head of the pump for the given flow rate. It is good practice to provide an allowance of about 10 percent of all frictional losses for piping and equipment which should be added to the calculated TDH; the sum is then rounded to the next greater multiple of 5 ft. The result is designated as the "design TDH."

Available net positive suction head. Once the TDH of the pump has been calculated, the available net positive suction head must be determined. The NPSHA is defined as the difference between the suction head h_s of the piping system and the vapor pressure head h_{vp} of the liquid, both expressed as feet of fluid, at the suction of the pump so that

$$\text{NPSHA} = h_s - h_{vp} \qquad (11.2)$$

As a liquid passes through a piping system into the suction of the pump, the total pressure loss due to friction continuously increases and the static head of the fluid

TABLE 11.2 Pump Calculation Sheet*

FORM 5053

united engineers & constructors inc.

PUMP CALCULATION SHEET

PROJECT _____ DESIGN GPM _____ DESIGN TDH _____ PUMP NO. _____
PUMP SERVICE _____ DATE _____

PUMP NO.	SCH. & PIPE MATL.
LBS/HR	FLUID PUMPED
NORMAL GPM	TEMP., MAX _____ MIN _____ F
DESIGN GPM	FROM _____ TO _____
REFERENCE DRAWINGS	
	VISCOSITY @ MIN P.T. _____ cp (μ)
	SPEC.GRAV. @ MIN P.T. _____ (SG)
	KINEMATIC VISCOSITY _____ cst (z)
	VAPOR PRESS. @ MAX P.T. _____ mmHg(VP)

SUCTION LINE / **DISCHARGE LINE**

	SUCTION LINE	DISCHARGE LINE
LINE NO.		
LINE SIZE, INCHES, I.D.	d	d
VELOCITY, FT/SEC	V	V
FRICT. LOSS, FT.FLUID/100 FT	F	F
REYNOLDS NUMBER	Re	Re
LINEAL FEET OF PIPE	@	@
GATE VALVES	@	@
GLOBE VALVES	@	@
ELLS	@	@
TEES, THRU BRANCH	@	@
TEES, THRU RUN	@	@
TOTAL EQUIVALENT LENGTH	EL	EL
LINE FRICT.,IF x EL/100 1.15	F_s	F_d
VESSEL PRES.,psig x 2.31/SG	P_s^{**}	P_d^{**}
LIQUID ELEV.ABOVE PUMP,FT.	h_s^*	h_d
TANK CONTR.OR ENL. LOSS ORIFICE PL.,psi x 2.31/SG CONTR.VALVE,psi x 2.31/SG EXCHANGER,psi x 2.31/SG	h_c	h_e
ALG.SUM = SUCT.OR DISCH.HEAD	Hs	Hd

TOTAL DYNAMIC HEAD,Hd–Hs _____ TDH _____ FT.(" SUCT.," " DISCH.)

NPSH AVAILABLE:
ABS. PRESSURE HEAD P_s^{**} + 34/SG +
STATIC HEAD h_s^* + F_s +
FRICTION HEAD h_c + F_s –
VAP. PRES. HEAD .045 x VP/SG –

ALG.SUM = NPSH,FT.FLUID _____ FT. FLUID

*MINUS FOR SUCTION LIFT. **MINUS FOR PRESSURE LESS THAN ATMOSPHERIC. FOR ELEVATIONS ABOVE SEA LEVEL, DECREASE 34 BY 1.1 FT FOR EACH 1000 FT OF ALTITUDE ABOVE SEA LEVEL.

FORMULAE:
$$1 \text{ psi} = (2.31/SG)$$
$$V = .408 \text{ GPM}/d^2$$
$$d = \sqrt{\frac{GPM}{2.45v}}$$
$$z = \mu/SG$$
$$h_c = V^2/128.8$$
$$h_e = V^2/64.4$$
$$Re = \frac{3162 \times GPM \times SG}{d \times \mu}$$

*With permission of United Engineers & Constructors Inc., Philadelphia.

varies with the geometry of the system. However, once the fluid enters the pump, it is pressurized and the pressure gradient rises sharply across the pump.

If the pressure at the inlet to the pump is equal to or less than the vapor pressure of the liquid, there will be vapor present as the liquid enters the pump. Once inside the pump, the action of the pump increases the pressure on the vapor-liquid system and vapor bubbles formed at the pump suction collapse. This phenomenon is called "cavitation." Cavitation is undesirable for two reasons: the capacity of the pump is reduced, and there could be material damage to the pump. The destructive effects of cavitation are discussed in the subsection "Types of Corrosion, Causes, and Cures" in Chap. 5. The reduction in pump capacity is approximately equal to the volume of flashed vapors. For example, if the rated capacity of a pump is 100 gal/min of 150°F water (h_{vp} = 8.6 ft) and 0.01 percent by weight of the water is flashed during cavitation, the vapor occupies 37.3 percent of the volume at the suction. Because a pump at a given speed maintains a given volumetric flow rate, the pump will still have a suction of 100 gal/min but only 62.7 percent by volume will be liquid. Therefore, the pump's liquid capacity will drop to 62.7 percent of its original rating without cavitation.

NPSHA is a useful index for determining whether or not the suction piping system can cause cavitation since each pump has a minimum NPSH value below which cavitation occurs. Therefore, the more positive the NPSHA, the less likely are the possibilities for cavitation. The reason for this behavior can be explained by an expanded form of Eq. (11.2):

$$\text{NPSHA} = (P_s + h_s - F_s) - h_{vp} \qquad (11.3)$$

The three terms in parentheses are the factors which contribute to the suction head P_s.

At the bottom of the pump calculation sheet in Table 11.2 is a procedure to calculate NPSHA. If the calculated NPSHA of the system as designed is negative or close to zero, one or more of the parameters in the parentheses of Eq. (11.3) must be changed to increase the NPSHA. This can be done in any combination of the following changes:

1. Increasing the absolute-pressure head P_s at the source vessel
2. Increasing the liquid-level elevation h_s on the suction side of the pump
3. Decreasing friction in the piping F_s by reducing the equivalent length of the suction line, or by increasing the suction-line diameter, or by doing both

The procedure shown in Table 11.2 to calculate the NPSHA applies to centrifugal pumps and rotary pumps. However, a reciprocating pump has an additional factor which reduces its NPSHA slightly. This is the head loss due to acceleration h_a.

NPSHA for reciprocating pumps. Although positive-displacement pumps are self-priming, the NPSHA must still be determined for them. Equation (11.2) applies to all pumps. However, an additional term to those considered in Table 11.2 for centrifugal and rotary pumps must be included in the NPSHA formula for reciprocating pumps. This term accounts for head loss due to acceleration.

The instantaneous liquid flow rate in a reciprocating pump varies with time. Therefore, there are accelerations and decelerations during each stroke which, in turn, create instantaneous high pressure drops in the suction line during these surges. The momentary accelerations and associated head losses must be accounted for. This is termed the "acceleration head loss h_a" and can be calculated by

$$h_a = \frac{L \times l_s \times N^2 \times d_c^2}{70{,}700 \, d^2} \tag{11.4}$$

where d = suction-pipe diameter, in
d_c = pump-cylinder diameter, in
L = suction-pipe length, ft
l_s = suction-stroke length, in
N = pump speed, r/min

The acceleration head loss is treated in the same way as friction and is inserted into the available NPSH equation to give

$$\text{NPSH}_{\text{available}} = P_s + h_s - h_a - F_s - h_{vp} \tag{11.5}$$

where P_s = pressure-contribution head, ft
h_s = static-elevation head, ft
h_a = acceleration head loss, ft
F_s = suction-line frictional loss, ft
h_{vp} = vapor-pressure-contribution head, ft

Therefore, the available NPSH for a reciprocating pump is lower than that for a comparable centrifugal or rotary pump.

The Hydraulic Institute presents an alternative relationship for calculating the acceleration head for reciprocating pumps. The equation includes correction factors for the type of reciprocating pump and the nature of the fluid. The equation is similar to Eq. (11.4) but takes the form

$$h_a = \frac{L \times u \times N \times C}{K \times g} \tag{11.6}$$

where u = average velocity in suction line, ft/s
g = gravitational constant, 32.17 ft/s
C = constant for pump type such that
$C = 0.200$ for duplex single-acting
$C = 0.115$ for duplex double-acting
$C = 0.066$ for triplex single- or double-acting
$C = 0.040$ for quintuplex single- or double-acting
$C = 0.028$ for septuplex single- or double-acting
$C = 0.022$ for nonuplex single- or double-acting
K = factor to account for any compressibility of the type of fluid such that
$K = 2.5$ for hot oil
$K = 2.0$ for most hydrocarbons
$K = 1.5$ for water, glycols, and amines
$K = 1.4$ for deaerated water

Required net positive suction head. An ideal pump would not cavitate until the NPSHA of the system becomes negative. However, a real pump does cavitate within a range of positive NPSHA values. This is due to the fact that the physical geometries and operating characteristics of real pumps generate internal head losses at the intake section.

The additional internal losses vary from one type of pump to another and among a manufacturer's various models for the same type of pump. Therefore, these losses are not predictable. Fortunately, pump manufacturers recognize these differences and test their pumps to determine the minimum NPSHR allowed before the pump begins to cavitate. The results are usually plotted on the performance curves for the pump. The NPSHR depends primarily on flow rate and to a lesser extent on the TDH of the pump.

Some centrifugal pumps can be fitted with a flow inducer, a corkscrew-shaped device which literally screws in the incoming fluid and reduces the NPSHR of the pump somewhat.

When selecting a pump for a given application, the process engineer must make sure that the NPSHA of the system exceeds the NPSHR of the pump selected. As a practical rule, the difference between NPSHA and NPSHR should be at least 1 to 10 ft. The actual value depends on the confidence that the process engineer has in the calculated NPSHA and the behavior of NPSHR on the performance curves with respect to flow rate and TDH. Consideration must be given to the possibility that an excursion in fluid flow rate could decrease the NPSHA while increasing the NPSHR.

Hydraulic horsepower. Hydraulic horsepower is the theoretical work done by a pump to impart a given head to a certain flow rate of fluid from the point at which it enters the pump to the point at which it is discharged if the pump were operating at 100 percent efficiency. The hhp for a particular flow rate and head is simply the product of the weight rate in pounds per minute and the head in feet of fluid divided by 33,000 (ft·lbf)/hp, or

$$\text{hhp} = \frac{W \times h}{33,000} = \frac{8.33 \times Q \times h \times s}{33,000} = \frac{Q \times h \times s}{3961} \tag{11.7}$$

where hhp = theoretical or hydraulic horsepower
W = weight flow rate, lb/min
Q = volumetric flow rate, gal/min
h = total dynamic head, ft of fluid
s = specific gravity of fluid relative to water at 85°F

Brake horsepower and efficiency. The actual brake horsepower required for a given pumping operation is always greater than the calculated hhp because of internal losses due to friction caused by the packing or the mechanical seals in the pump's stuffing box and at the shaft bearings. Additional rotational losses are induced by short-circuiting and erratic paths of the liquid through the pump so that the actual bhp is greater than the hhp. The ratio of the two is known as the shaft efficiency η and is given as

$$\eta = \frac{\text{hhp}}{\text{bhp}} \tag{11.8}$$

where η = shaft efficiency.

The values of the respective parameters defining pump performance, i.e., head, efficiency, and NPSHR, are a function of capacity and revolutions per minute and are the subject of complex mathematical analysis. However, the pump manufacturer can determine them directly when performing a certified test on a given pump or represents their approximate values for the average pump of that design in the manufacturer's literature. Unless otherwise requested or specified, all the test work and the resulting pump performance curves are for clean water at a temperature of not more than 85°F. Performance for fluids with other specific gravities or viscosities must be adjusted accordingly.

General performance characteristics of pumps

Reciprocating pumps, rotary pumps, and centrifugal pumps each have unique capacity-head characteristics. Figure 11.18 shows a set of simplified head-capacity curves for the three types of pumps. Intersecting the three curves is a system friction curve. The intersection of a pump curve with the system curve is the operating point of the pump-system combination.

Reciprocating-pump characteristics. Since reciprocating pumps are constant-speed and essentially constant-capacity pumps, the ideal reciprocating pump should not be affected by any parameter other than speed. Since the pumping is performed by the reciprocating action of a piston or a plunger in a compression chamber, the only variable is discharge pressure. An ideal reciprocating pump would have zero clearance between the piston and the chamber walls and therefore would have no leakage. In addition, there would be no space between the piston head and the chamber head at the end of a compression cycle. Theoretically an infinite pressure could be developed by such a pump. This is represented by curve A in Fig. 11.18.

An infinite discharge pressure is a practical impossibility. One reason is that actual reciprocating pumps have a finite space at the end of a compression cycle. In addition, the discharge valve is partially open, so that fluid always moves out of the chamber. Also, the materials of which the pump is constructed would fail under excessive pressure stress. It is for this last-named reason that a well-designed positive-displacement pump is supplied with pressure-relief valves set to relieve discharge liquid back into the pump's suction on high pressure. The relief-valve set pressure is below the design pressure of the pump but well above its normal operating pressure. The relief valve also protects downstream piping and equipment. The pump just described is characterized by curve B in Fig. 11.18.

The closer the clearances of a reciprocating pump, the less leakage out of the compression chamber. However, this results in greater friction and high maintenance. Therefore, a compromise between leakage and friction must be made so that an acceptable leakage is designed into the pump. This leakage is proportional to the pump discharge pressure. As discharge pressure increases, leakage also increases and pump capacity drops off accordingly. This reduction in pump capacity is called "slip" and is represented by curve C in Fig. 11.18.

When a reciprocating pump is placed in a piping system, its operation depends on the flow-versus-head characteristics of the system curve. This can be explained by using a simple piping system with a throttling valve on the pump discharge as shown in the lower portion of Fig. 11.18.

The upper portion of the figure shows system curves A through E superimposed on the pump characteristic curves. The system curves are the head-versus-capacity curves for the discharge piping with the valve position ranging from 20 to 100 percent open. The point of intersection of a system curve with a pump curve is the operation point for a complete pump and piping system. It should be noted that as the valve is throttled from the 100 percent open position, the head or back pressure on the pump increases. The positive-displacement pump with no slip, represented by curves A and B, shows no change in capacity when the valve is 80 or 100 percent open. This is to

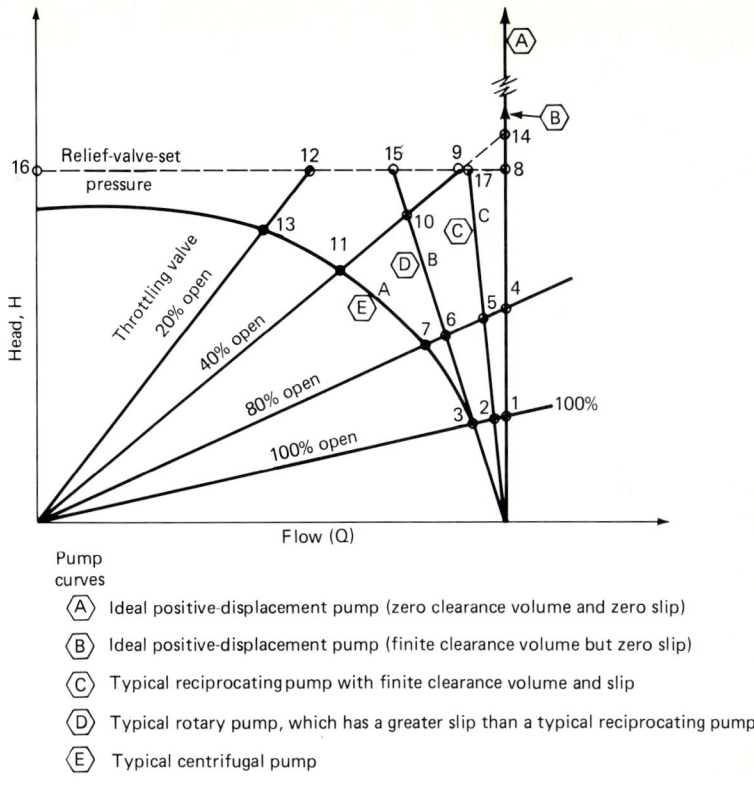

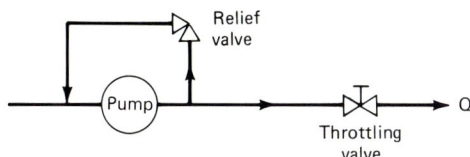

Figure 11.18 Generalized capacity-versus-head characteristic curves for reciprocating pumps. Points 1 through 17 are the various operating points for each pump with a given throttling-valve position.

be expected. However, when the valve is 40 percent open, the back pressure caused by throttling exceeds the set pressure of the relief valve on the pump. At point 8, the no-slip pump has a reduction in discharge pressure to the system equal to the relief-valve set point. The difference between the system flow (point 9) and the rated capacity of the pump (point 8) is the flow of liquid recirculating through the relief valve. At point 12, with the valve 20 percent open, the recirculation flow is even greater.

The other pump curve C, which has slip, will always have some reduction in capacity depending on the discharge pressure of the system. In a system with the valve 80 percent open, the pump operates at point 5, which is below the set pressure of the relief valve. If the valve is throttled to the 20 percent open position, the back pressure will cause the relief valve to open. This results in a reduction in system flow which is represented by point 12. The difference between points 12 and 17 is the amount of flow recycle through the relief valve. Note that point 16 on the chart is the condition at which the valve is fully closed and the discharge flow is zero. Therefore, the full capacity of each pump is recirculated through the relief valve.

Reciprocating pumps have high efficiencies. The loss of efficiency is due mainly to the residual volume at the end of the discharge stroke. However, it is evident from the discussion regarding Fig. 11.18 that the efficiency of such a pump is reduced with increasing pressures owing to higher slippage rates.

The NPSHR of a reciprocating pump is usually low and independent of the TDH. It increases with pump speed since the throughput goes up accordingly.

Rotary-pump characteristics. Rotary pumps are essentially positive-displacement pumps with much higher slip than reciprocating pumps. The high slip is the result of greater clearances between the rotary-pump impeller and its casing. The performance characteristics of a rotary pump are shown as curve D in Fig. 11.18. When a rotary pump is installed in a piping system, its operating points are 3, 6, 10, and 15, depending on the throttling-valve position. At point 15, the system flow is at point 12, and the recirculation flow through the relief valve is the difference between points 15 and 12.

Centrifugal-pump characteristics. A centrifugal type pump has a greater clearance between its impeller and its casing than does a rotary pump. However, unlike a positive-displacement pump, it converts kinetic energy to pressure head. Therefore, the centrifugal-pump characteristic curve is represented by the relationship

$$h = f\left(\frac{u^2}{2g_c}\right) \tag{11.9}$$

where h = pump head
u = linear velocity of the fluid in the pump
$f\left(\dfrac{u^2}{2g_c}\right)$ = functional relationship

Since linear velocity and volumetric flow are proportional, the head-flow characteristic curve is roughly approximated by a parabola, for which the head decreases with increasing throughput as shown by curve E in Fig. 11.18, points 3, 7, 11, and 13.

Centrifugal pumps respond dynamically to the head characteristics of a piping system. As a result, a pump's capacity can change considerably with changes in piping friction and control head. The dynamic responses of centrifugal pumps and their interaction with system curves are important to understand.

Performance curves. A given pump with its own characteristic casing-impeller configuration and inlet and discharge connections will have various combinations of flow rate, total dynamic head, required net positive suction head, and other important parameters, depending on the impeller size being employed and the speed at which the impeller turns.

Flow rate and head. The locus of the flow-rate and dynamic-head points is termed the "performance curve" or "pump curve." Figure 11.18 presented an idealized curve for a centrifugal pump. However, actual performance curves for pumps with radial-flow, axial-flow, and mixed-flow impellers are different from one another. Typical curve shapes for these impellers are given in Fig. 11.19.

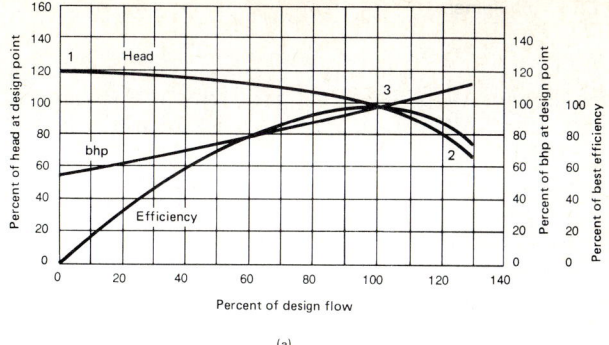

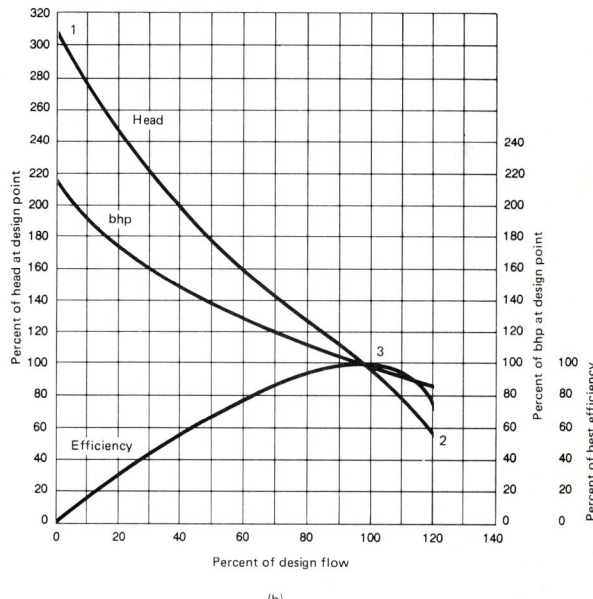

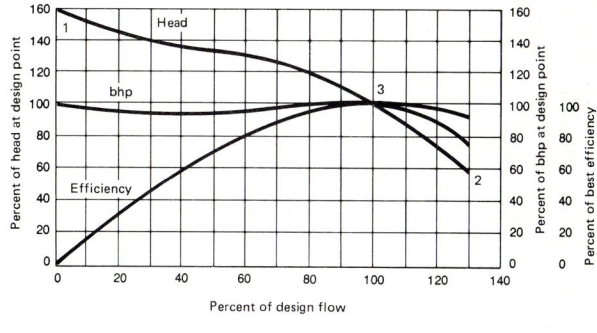

Figure 11.19 Typical performance curves for (*a*) radial-flow, (*b*) axial-flow, and (*c*) mixed-flow impellers of centrifugal pumps. System design point is taken at maximum efficiency. *(With permission from Goulds Pumps, Inc., Seneca Falls, N.Y.)*

There are several significant points on a performance curve. The initial point is the "end valve," or shutoff head, when there is no flow (i.e., $Q=0$), and is noted as points 1 on Fig. 11.19. It is easy to visualize the value of the head at this point, as it is the height of fluid that could be sustained by the pump in a vertical line sufficiently high so that there is no flow from its end. It is also the pressure, in feet of fluid, that would be read on an indicator placed on the pump discharge nozzle with a valve in the discharge line closed so as to deadhead the pump. The end value is essentially the static-head equivalent to the kinetic energy represented by the linear tip velocity of the impeller. The linear velocity at the tip of the impeller is a function of the impeller diameter and the angular velocity at which it revolves and is represented by the equation

$$u_{\text{lin}} = \frac{\pi DN}{60} \tag{11.10}$$

where u_{lin} = impeller tip speed, ft/s
D = impeller diameter, ft
N = pump speed, r/min

The approximate deadheaded height, or kinetic energy, at no flow is given by the equation

$$h_o = \frac{u_{\text{lin}}^2}{2g_c} = \frac{\pi^2}{2g_c}\left(\frac{DN}{60}\right)^2 \tag{11.11}$$

where h_o = end value of pump performance curve, ft.
g_c = dimensional constant, 32.17 (lbm/lbf)(ft/s^2)

Example 11.1 What is the approximate shutoff head of a pump with a 10-in impeller which is rotating at 1750 r/min?

solution The kinetic energy or head at no flow according to Eq. (11.11) is 90.5 ft.

The value of 90.5 ft given in the solution to Example 11.1 is somewhat lower than the end values for typical pumps with 10-in impellers in pump manufacturers' catalogs since actual impeller diameters are larger than their nominal values.

Once flow has been established through the pump, the kinetic head represented by the end value usually decreases in most cases owing to head losses within the pump. However, in some instances the hydraulic effects within the casing and nozzles may be such that the head remains constant or even rises slightly for a small span of flowrate increase before the head begins to drop.

It is not possible to estimate other points on the centrifugal-pump performance curve with the same facility as the end value. The performance at other points for a given impeller diameter and rotational speed is governed by the geometric shape of the impeller, the contours of the casing, and the configurations of inlet and discharge ports. The head values at flow rates other than no flow are subject to complex mathematical analysis. Manufacturers of pump equipment perform these analyses to obtain optimum design for their standard lines. However, the actual performance curve can vary somewhat even from pump to pump of the same model. A true pump curve can be obtained only by a precision test conducted for a fee by the manufacturer after the pump has been constructed. However, general curves are available in manufacturers' literature or engineering data, and these are usually sufficient for selecting and using most pumps.

Once the dynamic head of the pump begins to drop with increasing flow through the pump, it continues to be reduced, usually at an accelerating rate. At a certain flow rate internal losses are such that pump operation is no longer stable but becomes erratic owing to choking at the pump inlet or discharge points so that there is intermittent cavitation and surging within the pump casing. This is known as the "runoff point," and a given pump can never deliver more than the flow rate represented by that point. The runoff points are marked 2 on curves in Fig. 11.19. The TDH of the pump, were it to operate at the runoff flow rate, would fluctuate unpredictably between the endpoint head on the diagram and zero. It is good engineering practice to ensure that the planned flow rate for a proposed pump is such that it operates well below the runoff value so that the pump system remains predictable despite moderate increases in flow rates.

Efficiency. The efficiency for a centrifugal pump at cutoff is zero since the flow rate is nil. At the other extreme, the efficiency just beyond the runoff point is an indeterminate value approaching zero since the TDH varies from some low value to nil. Between these points there is a finite efficiency for each point on the performance curve, increasing from the cutoff point until there is a maximum value at a flow rate about one-half to three-quarters of the runoff flow-rate value.

Curves for the efficiency and brake-horsepower values for the various types of pumps are represented in Fig. 11.19. The maximum-shaft-efficiency point is marked 3 on the performance curves in Fig. 11.19. A well-designed centrifugal pump suitable for most uses in chemical plants, refineries, powerhouses, and related facilities has a maximum shaft efficiency between 65 and 80 percent. However, larger pumps frequently have maximum efficiencies up to 90 percent. It should be pointed out that the overall efficiency of a pump is less than its shaft efficiency since the efficiency of the driver (an electric motor, a steam turbine, a diesel or gasoline engine, an air motor, etc.) represents another small horsepower loss.

Brake horsepower. The large majority of centrifugal pumps utilized are the radial-flow type. The brake horsepower starts at a definite value at pump shutoff and rises continuously as the flow rate through the pump increases until the flow approaches the point at which internal cavitation occurs. At that condition, the brake horsepower levels off or may decrease slightly. The brake horsepower for an axial-flow pump starts at a high value and decreases at varying rates until cavitation begins.

Required net positive suction head. The internal friction losses generated by the physical geometry and operating characteristics of centrifugal pumps vary from one type of pump to another and from one manufacturer's model to another, so that these losses are not predictable. Pump manufacturers recognize these differences and test their pumps to determine the minimum NPSHR that a pump can tolerate before it begins to cavitate. The results are then plotted on the performance curves for the particular pump. The NPSHR depends primarily on the flow rate and to a lesser extent on the TDH of the pump. Since the most important element in setting the NPSHR for any pump is the friction losses of the fluid as it passes through the inlet nozzle and hub before it reaches the impeller, the NPSHR for a given pump casing is lowest at the shutoff point and increases as the net flow rate through the pump rises. The change in NPSHR may vary directly with the flow rate or could increase as the square of that value, depending on the design of the pump or the particular size of impeller in a given casing.

Presentation of performance curves. The performance curve for a centrifugal pump at a given motor speed can be given similarly to those in Fig. 11.19 if it is the result of test data obtained for only one impeller diameter. Curves based on engineering data are also sometimes reported for one diameter only. Most pump casings, however, can be fitted with impellers of varying diameter from a minimum necessary to accommodate the inlet volute opening to the maximum size that can fit in the casing. Impellers of any diameter, usually manufactured in increments of $\frac{1}{8}$ or $\frac{1}{16}$ in, may be specified between minimum and maximum for the pump. For convenience, most manufacturers include a series of performance curves for each pump model that they make. By showing performance curves in this way, the head and brake-horsepower efficiency and the NPSHR can be shown for diameters from the maximum to minimum sizes of impellers. It is necessary to eyeball the values between adjacent curves or to use the affinity relationships discussed below in the subsection "Affinity Relationships for Performance Curves" if more exact determinations are required.

A method used by some manufacturers to represent their performance data is given in Fig. 11.20. Here, the efficiency, brake horsepower, and NPSHR curves are shown separately as a function of flow rate with impeller diameter as the curve parameter. Other pump manufacturers use only one graph and present the respective data as curves of equal value of the particular parameter for the different head and flow-rate combinations. These are shown on the pump performance graphs for impeller diameters between maximum and minimum. Figure 11.21 is an example of this type of representation. The increments between lines of constant efficiency may vary from 1 to 10 percent, depending on the section of the graph and the clarity required. The lines of equal horsepower usually represent the available motor horsepowers that are applicable to operations covered by the graph. These curves assume that water at 85°F

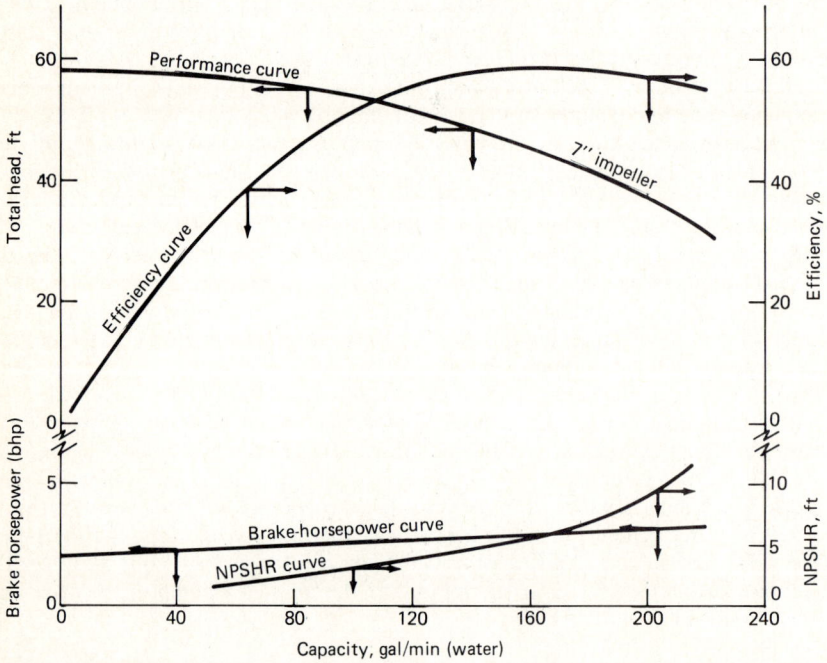

Figure 11.20 Typical performance curves supplied by a centrifugal-pump manufacturer for one impeller size. *(Based on Goulds model 3196, 2 × 3 − 10, 1750 = r/min pump.)*

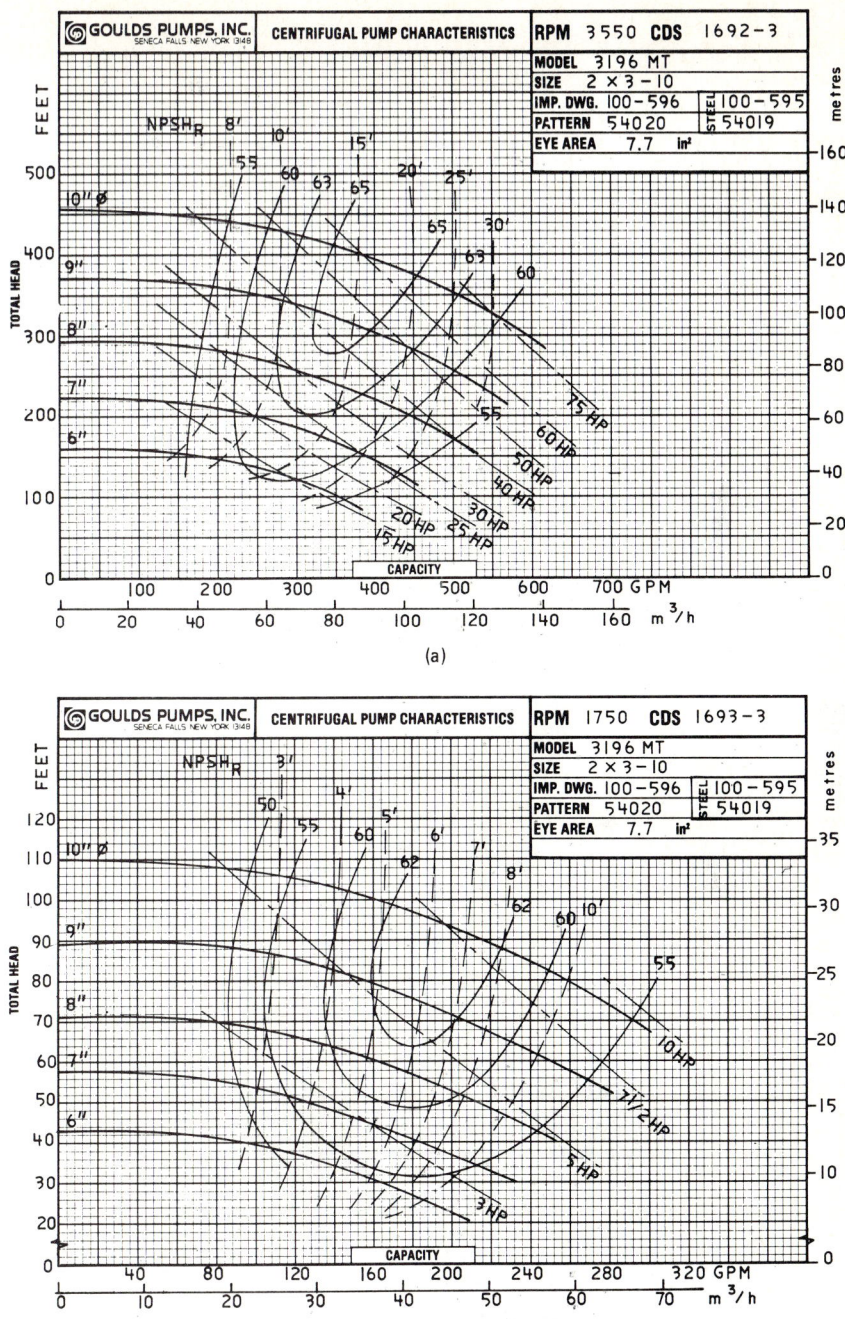

Figure 11.21 Typical performance curves with superimposed motor horsepower, efficiency, and NPSHR curves. (*a*) 3350 r/min. (*b*) 1750 r/min. (*With permission from Goulds Pumps, Inc., Seneca Falls, N.Y.*)

is the fluid being pumped. Separate sets of curves are required for different pump speeds.

The three most common electric-motor speeds for operating pumps are 1200, 1800, and 3600 r/min. However, pump performance curves are usually reported at speeds of 1150, 1750, and 3550 r/min to compensate for motor slip caused by the motor working on the fluid. Manufacturers, therefore, usually present a separate set of performance curves for several of these common speeds or other special speeds peculiar to the application of the particular pump.

The curves for a pump with a 3-in suction, a 2-in discharge, and a 10-in maximum impeller, as shown in Fig. 11.21, represent those of a given manufacturer. A comparison with equivalent curves for the same nominal-sized pump by a different vendor could show some variations in performance.

System curves. A system curve describes the TDH required from a pump in a given process system as a function of the flow rate passing through the system. It is useful in determining the nominal pump size and the impeller diameter of a pump and in selecting the pump-motor size. Included in the evaluation of a system curve is an accounting of the maximum and minimum static-elevation differences between the level of the feed ahead of the pump and the termination point of the discharge flow as well as any static-pressure differences between the source and fluid-destination points. These values are independent of flow rate and are fixed by the process. The dependent variables are the head losses due to friction through the piping, fittings, and open valves in the process as well as through process equipment such as heat exchangers, filters, in-line mixers, etc., and instrumentation such as flow indicators, analyzers, restrictive orifices, etc. A further consideration is that of the additional loss through a control valve when it is throttling.

Pump calculation sheets similar to that presented as Table 11.2 are useful for evaluating system curves. Table 11.3 shows the calculation sheet in Table 7.12 carried further to determine the TDH for the situation.

The elements of a system curve are thus the fixed head requirements, the friction losses or pressure drops that are a function of flow rate, and head or pressure-drop allowances for control valves. Each of these elements outlined above is treated separately when constructing the curve. A typical system curve is illustrated in Fig. 11.22a.

The sum of the elevation and static-pressure differences in feet of fluid is entered as point A on the graph. Point B is the TDH, excluding the throttling drop of the control valve at the design flow rate for the system. The head difference between points A and B represents frictional losses in the system at the design flow rate. The frictional losses at the flow rates for points C and D, where flows are about 25 percent smaller and greater than the flow at B, can be obtained by applying the pressure-drop-versus-flow relationships. Care must be exercised when analyzing friction losses to segregate losses due to the laminar-flow regimes from those due to transient or turbulent flow. For the purpose of the system curves it may be assumed that friction losses of laminar flows vary as the first power of the flow rate while transient and turbulent drops vary as the square. Point E is then placed on the graph at the design flow rate B so that the head difference between the two points represents the desired head loss for the control valves if they are in the system.

When the system curve is superimposed on the performance curves for an available pump, as in Fig. 11.22b, it is possible to select the proper impeller and motor. In this case point E falls just below the 9-in-impeller curve and the 9-in impeller would be specified. The power requirements at the design flow rate on the 9-in-impeller curve

TABLE 11.3 Completed Pump Calculation Sheet*

FORM 5053

united engineers & constructors inc.

PUMP CALCULATION SHEET

PROJECT **MgO Regen.** DESIGN GPM **140** DESIGN TDH **82'** PUMP NO. **P-1515**
PUMP SERVICE **DILUTE H_2SO_4 TO COOLING TOWER** DATE **3-27-81**

PUMP NO. **P-1515**	SCH. & PIPE MATL. **10S, 316 SS**	VISCOSITY @ MIN P.T. **0.80** cpt (μ)
LBS/HR **61,200**	FLUID PUMPED **3% H_2SO_4**	SPEC.GRAV. @ MIN P.T. **1.02** (SG)
NORMAL GPM **120**	TEMP., MAX **140** MIN **70** °F	KINEMATIC VISCOSITY **0.78** cst (ν)
DESIGN GPM **140**	FROM **V-1515** TO **V-1522**	VAPOR PRESS. @ MAX P.T. **148** mmHg(VP)

REFERENCE DRAWINGS

SUCTION LINE

	LINE NO.	**1527**		
d	LINE SIZE, INCHES I.D.	3" (3.260)	4" (4.260)	
V	VELOCITY, FT/SEC	5.37	3.15	
F	FRICT. LOSS, FT.FLUID/100 FT	3.25	0.83	
Re	REYNOLDS NUMBER	173,000	132,000	

LINEAL FEET OF PIPE	@	1.6	3
GATE VALVES	@	2.1	4
GLOBE VALVES	@		
ELLS	@	7.7	31	10.2	41
TEES, THRU BRANCH	@	15.5	16	20.3	20
TEES, THRU RUN	@		
CHECK VALVE (0.5 psi min)	@		

TOTAL EQUIVALENT LENGTH EL **85** **100**

LINE FRICT., IF × EL/100)1.15	F_s	3.2	1.0
VESSEL PRES. psig × 2.31/SG	P_s^{**}	0	0
LIQUID ELEV. ABOVE PUMP, FT.	h_s^*	8.5	8.5
TANK CONTR. OR ENL. LOSS	F_s		
ORIFICE PL., psi × 2.31/SG	h_c +	0.2	0.2
CONTR. VALVE, psi × 2.31/SG			
EXCHANGER, psi × 2.31/SG			

ANALYSER

ALG. SUM = SUCT. OR DISCH. HEAD Hs **5.1** FT.I **7.2**
TOTAL DYNAMIC HEAD, Hd-Hs **78** FT.I **3"** SUCT., **3"** DISCH.

DISCHARGE LINE

	1529/1930	**1529/1930**	**1529/1930**	**1529/1930**	**1529/1930**
d	2" (2.157)	3" (3.260)	3" (3.260)	4" (4.260)	
V	12.2	5.37	5.37	3.15	
F	26.9	3.25	3.25	0.83	
Re	261,000	173,000	173,000	132,000	

d @	1.1	191	1.6	7	2.1	9
V @					
@	5.2	36	7.7	54	10.2	71
@	3.5	14	5.2	21	6.8	27
@	13.4	13	(3.5)	35	(13.9)	139

EL **254** **308** **437**

F_d	78.6	11.5	4.2
P_d^{**}	5.0	5.0	5.0
h_d^*	23.0	23.0	23.0
h_e	2.3	0.4	0.2
	8.0	8.0	8.0
	40.0	20.0	20.0
	11.0	11.0	11.0
	4.0	4.0	4.0

Hd **171.9** **82.9** **75.2**

2.26 FT. FLUID

FORMULAE:
- 1 psi = (2.31/SG)
- V = 408 GPM/d²
- $d = \sqrt{\dfrac{GPM}{2.45v}}$
- $z = \dfrac{\mu}{SG}$
- $h_c = V^2/128.8$
- $h_s = V^2/64.4$
- $Re = 3162 \times \dfrac{GPM \times SG}{d \times \mu}$

$F = \left(\dfrac{d_{SCH\,40}}{d_{SCH\,10}}\right)^5 F_{SCH\,40}$

NPSH AVAILABLE:
ABS. PRESSURE HEAD P_s^{**} + 34/SG **33.2** + **33.2**
STATIC HEAD h_s^* **8.0** + **8.0**
FRICTION HEAD $h_c + F_s$ **3.4** + **6.5**
VAP. PRES. HEAD .045 × VP/SG **6.5** +

ALG.SUM = NPSH,FT,FLUID **31.3** **33.5**

*MINUS FOR SUCTION LIFT. **MINUS FOR PRESSURE LESS THAN ATMOSPHERIC. FOR ELEVATIONS ABOVE SEA LEVEL, DECREASE 34 BY 1.1 FT FOR EACH 1000 FT OF ALTITUDE ABOVE SEA LEVEL.

*With permission of United Engineers & Constructors Inc., Philadelphia.

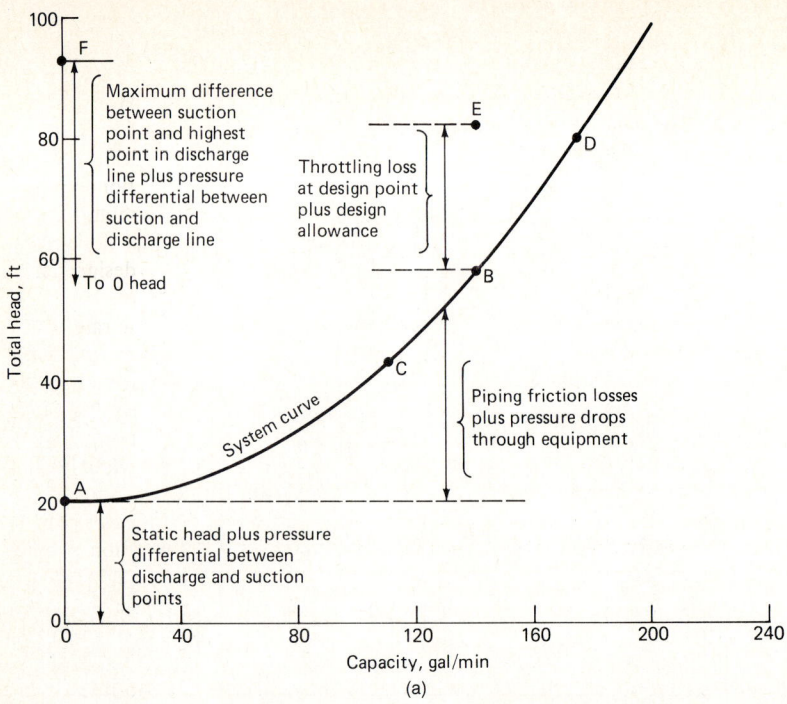

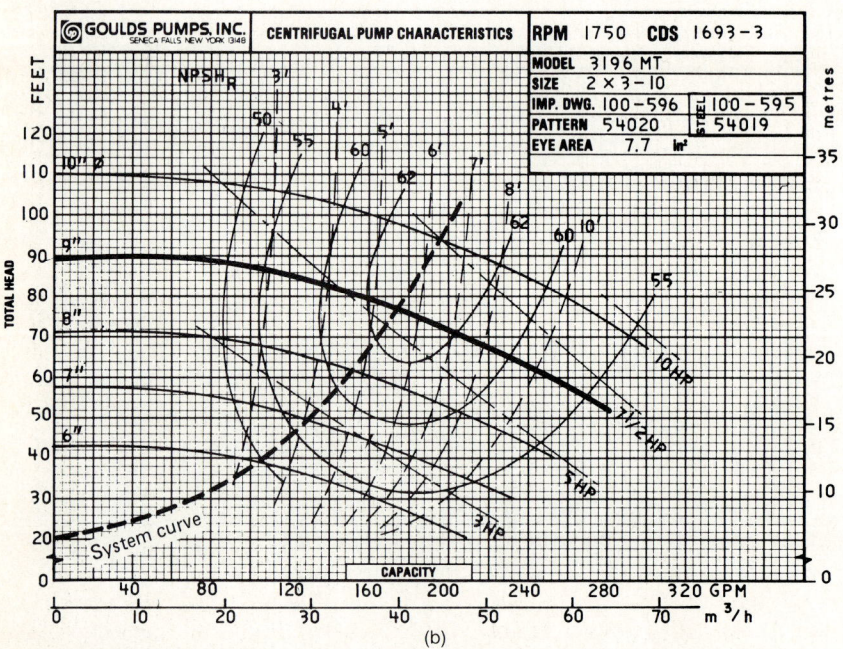

Figure 11.22 Typical system curves. (*a*) System curve. (*b*) System and pump performance curves, 1750 r/min. (*With permission from Goulds Pumps, Inc., Seneca Falls, N.Y.*)

is 5 bhp, so that a 5-hp motor could be used with water as the fluid. However, if the control valve were to open wide during a transient condition, the flow rate would increase to about 175 gal/min, the intersection of the system curve with the impeller curve, and the power requirement would be almost 6 hp. Normally a nonoverloading motor is specified for small- and medium-sized pumps so that the motor always is adequate. In this case a $7\frac{1}{2}$-hp motor would be required.

When large motors, i.e., of 75 hp and greater, are required for the design flow and head conditions, it is often not cost-effective to specify a nonoverloading motor if the system curve indicates that it is not possible to reach the runoff flow rate with the designated impeller. In that event, several considerations must be taken into account in selecting a motor that is not nonoverloading. It is important to review the system curve to account for safety factors included in friction factors, equivalent pipe lengths, and pressure drops through equipment as well as possible variations in static head and pressure differential. The system curve should then be revised to reflect minimum dynamic-head requirements. The intersection of this minimum curve with the performance curve for the designated impeller can then set the motor size.

When the designed impeller is not the maximum size, consideration should be given to selecting a motor to accommodate a larger-size impeller without changing the motor. If the motor is nonoverloading for the original impeller, the intersection of the system curve with the performance curve for a larger impeller will indicate whether or not it is suitable for the new conditions.

A separate but important item on a system curve is the point of no flow that represents the difference in static pressure between discharge and feed vessels plus the difference between the highest elevation in the discharge system and the lowest in the inlet vessel. If this value, point F in Fig. 11.22, is less than the design TDH point E at the design flow rate, its effect on pump selection may be ignored. However, if it is greater, a larger impeller size must be used in order to initiate flow in the system. In this case the end value of the pump must be equal to or greater than point F. A nonoverloading motor for this impeller size or the intersection of the "minimum" system curve with the performance curve may be used to size the motor.

Affinity relationships for performance curves. Many pump manufacturers list pump curves in their literature for several of the three common pump speeds of 1150, 1780, and 3550 r/min and for increments in impeller diameter from minimum to maximum size (see Fig. 11.21a and b). Others may show curves for the various diameters but only at one speed. In addition, certified curves for purchased pumps are usually given for the specified impeller diameter and at the purchased pump speed only. These limited curves are usually sufficient for the requirements for which the pump was purchased. However, at times a variable-speed drive is required, or it is necessary to change the diameter of an impeller or use a different pump speed to meet changed process requirements. If the necessary pump curves to make this transposition are not available, they can be approximated by applying the pump-affinity laws to any operating curve for the given pump.

The pump-affinity laws relate pump speed, impeller diameter, brake horsepower, and NPSHR. If all these variables are known for different flow rates and TDHs in a particular pump, the affinity laws can predict new flow rates, TDHs, and horsepower requirements for a change in impeller diameter, pump speed, or both. Similarly, if a different TDH is needed for a given flow rate, or vice versa, the pump-affinity laws can indicate the combination of impeller diameter and pump speed that could be used and the horsepower required for the particular pump. The NPSHR can also be approximated for the new conditions.

414 Equipment Selection, Sizing, and Related Subjects

The pump-affinity laws are simple ratios based on the dynamics of the impeller and are expressed as follows:

$$\frac{(\text{gal/min})_2}{(\text{gal/min})_1} = \frac{d_2}{d_1} = \frac{N_2}{N_1} \tag{11.12}$$

$$\frac{\text{TDH}_2}{\text{TDH}_1} = \left(\frac{d_2}{d_1}\right)^2 = \left(\frac{N_2}{N_1}\right)^2 \tag{11.13}$$

$$\frac{\text{bhp}_2}{\text{bhp}_1} = \left(\frac{d_2}{d_1}\right)^3 = \left(\frac{N_2}{N_1}\right)^3 \tag{11.14}$$

$$\frac{\text{NPSHR}_2}{\text{NPSHR}_1} = \left(\frac{d_2}{d_1}\right)^x = \left(\frac{N_2}{N_1}\right)^y \tag{11.15}$$

where gal/min = flow rate
TDH = total dynamic head, ft of fluid
bhp = brake horsepower
NPSHR = net positive suction head required, ft of fluid
d = impeller diameter, in
N = pump speed, r/min
subscript 1 = initial condition on known pump curve
subscript 2 = new condition
exponent x = applied to d for NPSHR, $-2.5 < x < 1.5$
exponent y = applied to N for NPSHR, $1.5 < y < 2.5$

The application of the affinity laws is straightforward with respect to pump capacity, head, and brake horsepower. The laws may be used for these parameters with any type of centrifugal pump to obtain fairly accurate results. However, changes in the NPSHR can only be estimated since the exponents x and y in Eq. (11.15) vary greatly with the type of pump and even with the initial impeller size in a given casing. However, the NPSHR can be established by using the highest exponents for x and y in solving Eq. (11.15) if the impeller or pump speed is to be increased or the lowest respective values if the diameter or speed is being reduced in magnitude. This method gives a worst-case condition for the NPSHR. If the NPSHR thus calculated is not suitable, it is necessary to contact the pump manufacturer to obtain the correct NPSHR value for the new conditions.

The use of the pump-affinity laws is best illustrated by an example which demonstrates how they can be used directly without resorting to a trial-and-error solution.

Example 11-2 An installed $2 \times 3 - 10$ pump with a $7\frac{1}{2}$-in impeller and a 5-hp, 1750-r/min motor has the following information from its performance curve:

gal/min	0	50	100	120	150	180	240
TDH, ft	65	64	61	59	55	48	35
bhp	2.0	2.5	2.9	3.1	3.5	3.6	4.2
Efficiency	...	0.32	0.53	0.57	0.60	0.60	0.51
NPSHR, ft	...	...	2.5	2.9	3.5	4.9	6.2	10

These are the only performance data available for the pump, and it is planned to reinstall the pump elsewhere in the process so that it will have to transfer 150 gal/min at 110-ft TDH intermittently. What changes should be made to the pump to meet these conditions? Prepare the original and final performance-curve data.

solution If the pump speed is kept at 1750 r/min, it is necessary to eliminate d_2 from Eqs. (11.12) and (11.13) by combining them to determine that

$$(\text{gal/min})_1/\sqrt{\text{TDH}_1} = (\text{gal/min})_2/\sqrt{\text{TDH}_2} = 150/\sqrt{110} = 14.3$$

The values of $(\text{gal/min})_1/\sqrt{\text{TDH}_1}$ for the $7\frac{1}{2}$-in performance curve are tabulated as follows:

$(\text{gal/min})_1$	0	50	100	120	150	180	240
TDH_1, ft	65	64	61	59	55	48	35
$(\text{gal/min})_1/\sqrt{\text{TDH}_1}$	0	6.2	12.8	15.6	20.2	26.0	40.6

From this table it can be seen that $(\text{gal/min})_1/\sqrt{\text{TDH}_1}$ equals 14.3 when $(\text{gal/min})_1$ is about 111 gal/min and TDH_1 is about 60 ft. By substituting these values into Eq. (11.12) or Eq. (11.13) it will be found that the new impeller diameter should be $10\frac{5}{8}$ in. However, the maximum impeller for the given pump model is only 10 in. Therefore, it is necessary to increase the speed of the pump. Since, the new service is intermittent, it is acceptable to use a 3550-r/min motor. The original performance curve for the $7\frac{1}{2}$-in impeller at 1750 r/min can be converted to a $7\frac{1}{2}$-in impeller at 3550 r/min by applying Eqs. (11.12), (11.13), and (11.14) as follows:

$(\text{gal/min})_{1,1750}$	0	50	100	120	150	180	240
$\text{TDH}_{1,1750}$, ft	65	64	61	59	55	48	35
$\text{bhp}_{1,1750}$	2.0	2.5	2.9	3.1	3.5	3.6	4.2
$(\text{gal/min})_{1,3550}$	0	101	203	243	304	365	487
$\text{TDH}_{1,3550}$, ft	267	263	251	242	226	196	144
$\text{bhp}_{1,3550}$	16.7	20.9	24.2	25.9	29.2	30.0	35.0
$[(\text{gal/min})_1/\sqrt{\text{TDH}_1}]_{3550}$	0	6.2	12.8	15.6	20.2	26.0	40.6

It should be noted that $(\text{gal/min})_1/\sqrt{\text{TDH}_1}$ is the same for the points on the $7\frac{1}{2}$-in-impeller curve that correspond to one another at the different motor speeds. The value $(\text{gal/min})_1/\sqrt{\text{TDH}_1}$ equals 14.3 when $(\text{gal/min})_1$ is 224 and TDH_1 is 246 ft at a pump speed of 3550 r/min. By substituting these values into either Eq. (11.12) or Eq. (11.13) it will be found that the new impeller diameter would be slightly over 5 in. However, the minimum impeller diameter for this casing is 6 in, and the motor would have to be sized for that unit. Calculations using all four affinity relationships show that the performance data expected for a 6-in impeller at 3550 r/min are:

$(\text{gal/min})_2$	0	81	162	194	243	292	390
TDH_2, ft	171	168	160	155	145	125	92
bhp_2	8.6	10.6	12.4	13.3	15.0	15.4	17.9
Efficiency	...	0.32	0.53	0.57	0.60	0.60	0.57
$\text{NPSHR}_{2,\text{max}}$	...	25.5	29.6	35.8	50.2	63.5	102

In addition to reducing the impeller size from $7\frac{1}{2}$ to 6 in, a new motor is required for the situation in Example 11.2. It should be 20 hp if it is to be nonoverloading.

However, it probably could be purchased as a 15-hp motor since the flow would have to be doubled before the safety factor, about 1.15, for a motor is exceeded. There is uncertainty about the NPSHR value. If the NPSHA is less than 29 ft, it will be necessary to consult the pump manufacturer to obtain the actual NPSHR.

The calculated data in Example 11.2 should be compared with the actual pump curves in Fig. 11.21. Such a comparison is shown in Fig. 11.23. The use of the revised

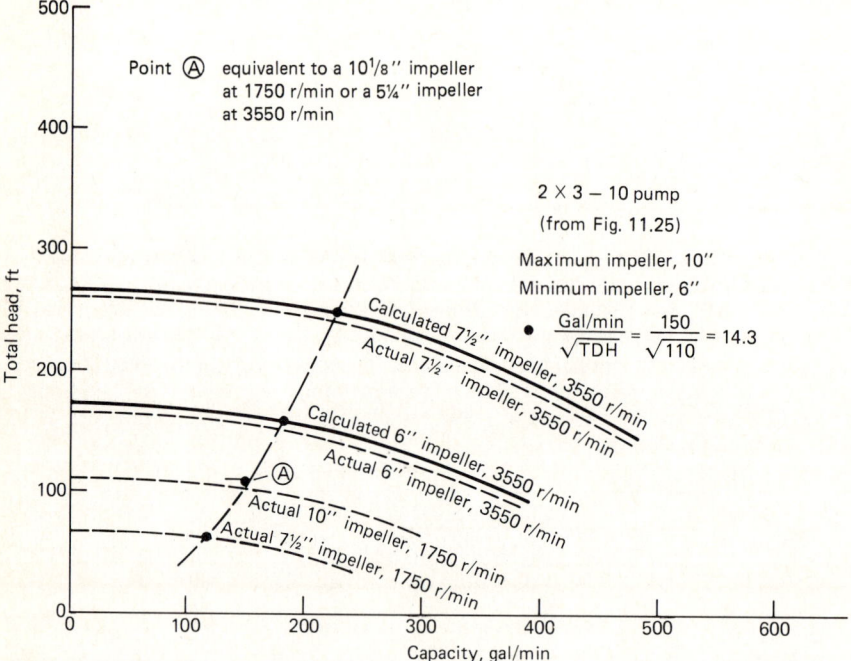

Figure 11.23 Application of affinity laws.

pump for intermittent operation is satisfactory. However, if the pump is to be in continuous operation, consideration should be given to the use of a more efficient pump at operating conditions.

Effect of viscosity on performance curves. The performance curves developed by most pump manufacturers are based on water at about 85°F, at which point its specific gravity is essentially 1.0 and its absolute viscosity is 0.80 cP. Its kinematic viscosity is, thus, 0.8 cSt, or 31 SSU. A variation in viscosity affects the operation of the pump since it changes the friction losses at the inlet and outlet nozzles as well as the casing and impeller skin friction losses as a direct function of the viscosity. Internal fluid bypassing, on the other hand, is an inverse function of viscosity. The net result is that the operating characteristics of flow rate, TDH, pump efficiency, and NPSHR can be changed when handling a fluid whose viscosity is other than 0.8 cSt. The change in the performance of a pump from that with 85°F water also depends on values of the desired flow rate and head as well as the fluid viscosity.

A fluid viscosity of less than the reference value of 0.8 cSt does not affect the capacity and TDH given by a 85°F-water performance curve. However, the efficiency of the pump is improved somewhat by the lowered viscosity in accordance with information from the Hydraulic Institute, as shown by the equation

$$\eta_v = C_\eta \times \eta_w \tag{11.16}$$

$$\text{and } C_\eta = \left(\frac{k_v}{0.8}\right)^a + \frac{1 - \left(\frac{k_v}{0.8}\right)^a}{\eta_w} \tag{11.17}$$

where C_η = correction factor for efficiency when viscosity is less than 0.8 cSt.
 η = pump efficiency
 k = kinematic viscosity, cSt
subscript w = water at 85°F
subscript v = viscous fluid
exponent a = value peculiar to pump design, usually between 0.05 and 0.1

Figure 11.24 presents the span of correction factors to 85°F-water pump efficiencies for fluid viscosities down to 0.1 cSt. It can be seen from Fig. 11.24 that the values of the efficiency correction factor could range up to 30 to 40 percent for very-low-

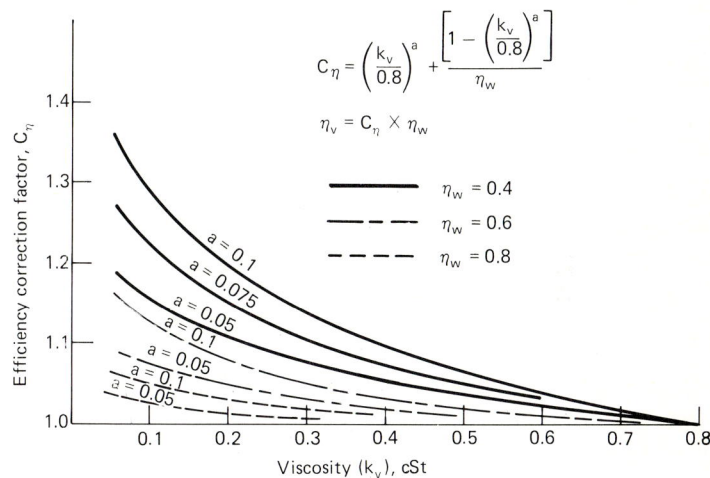

Figure 11.24 Viscosity correction factors for centrifugal-pump efficiency.

viscosity fluids depending on the initial water efficiency. For small-sized motors, this correction can be ignored for all practical purposes since a motor sized without it would be only slightly oversized. Consideration should be given to applying the correction when large-sized motors of several hundred horsepower are required since a somewhat smaller motor could be used to effect considerable savings.

As would be expected, the efficiency of a pump begins to decrease as the viscosity of the fluid exceeds 0.8 cSt. The flow-rate capacity of the pump starts to decline when the viscosity exceeds 3 to 10 cSt, and the effective dynamic head begins to fall as the fluid viscosity becomes greater than 50 to 100 cSt. There are no simple mathematical relationships to express the effect of the higher viscosities on pump capacity, dynamic head, and efficiency. Correlations have been made of correction factors for these parameters on the basis of pump test data, and these are presented as performance correction charts in Figs. 11.25 and 11.26.

Figures 11.25 and 11.26 may be used to predict required impeller size and motor horsepower for a viscous-flow situation when a set of conventional performance curves based on water is available. In this instance, the charts are used by entering them at the desired viscous-flow rate, following the capacity line vertically to the intersection with the inclined required-head line, moving horizontally right or left to the inclined line representing the fluid viscosity, and then proceeding vertically to obtain the correction factors as the respective curves are intercepted. It is obvious that the correct order of interception must be followed while proceeding through the figures to determine the correct factors. A review of the sequence is:

Capacity ⟶ head ⟶ viscosity ⟶ factors

```
              Factor
      TDH ⟋  ↑
          ⟋  ⎯⎯⟶ Viscosity
          ↑
          ↑
       Capacity
```

The use of performance correction charts to determine the water capacity and head equivalent to the viscous-fluid capacity and head and to calculate the brake horsepower required for the viscous flow is best illustrated by Example 11.3.

Example 11.3 Determine the approximate impeller size and brake horsepower required when pumping 150 gal/min of a fluid with 1.05 sp gr and a viscosity of 230 cP for a dynamic head of 80 ft. The $2 \times 3 - 10$ pump represented in Fig. 11.21b may be used if possible.

solution The kinematic viscosity of the fluid is the absolute viscosity divided by the specific gravity, or $230/1.05 = 219$ cSt. Enter Fig. 11.26 at 150 gal/min, go up to 80 = ft head, right to 219 cSt, and then vertically to read the corrections factors as $C_Q = 0.89$, $C_H = 0.88$ (for $Q_W = 1.0$), and 0.52 for C_η.

The required water capacity is 150/0.89, or 169 gal/min.
The required water head is 80/0.88, or 91 ft.

An impeller about $9\frac{5}{8}$ in in diameter for the $2 \times 3 - 10$ pump in Fig. 11.21b should be selected. Its water efficiency at 169 gal/min and 91-ft TDH is about 63 percent. Therefore, the efficiency for the viscous fluid is 0.52×0.63, or 0.328, and the brake horsepower required, taking into account the specific gravity of the fluid of 1.05, is $\dfrac{150 \times 80 \times 1.05}{3961 \times 0.328}$, or 9.7 bhp.

Figures 11.25 and 11.26 can also be used to transform a pump performance based on water to one for use with a viscous fluid. This is accomplished by determining the flow-rate capacity $Q_{\eta w}$ on the water performance curve corresponding to the highest efficiency point. Flow rates corresponding to 0.6 $Q_{\eta w}$, 0.8 $Q_{\eta w}$, 1.0 $Q_{\eta w}$, and 1.2 $Q_{\eta w}$ are then listed at the head of separate columns. Beneath each fractional-water-capacity heading are entered the water head and efficiency corresponding to the respective capacity. Figure 11.25 or 11.26 is entered for the water capacity equivalent to the maximum water efficiency, or 1.0 $Q_{\eta w}$. The correction factors to be used with the

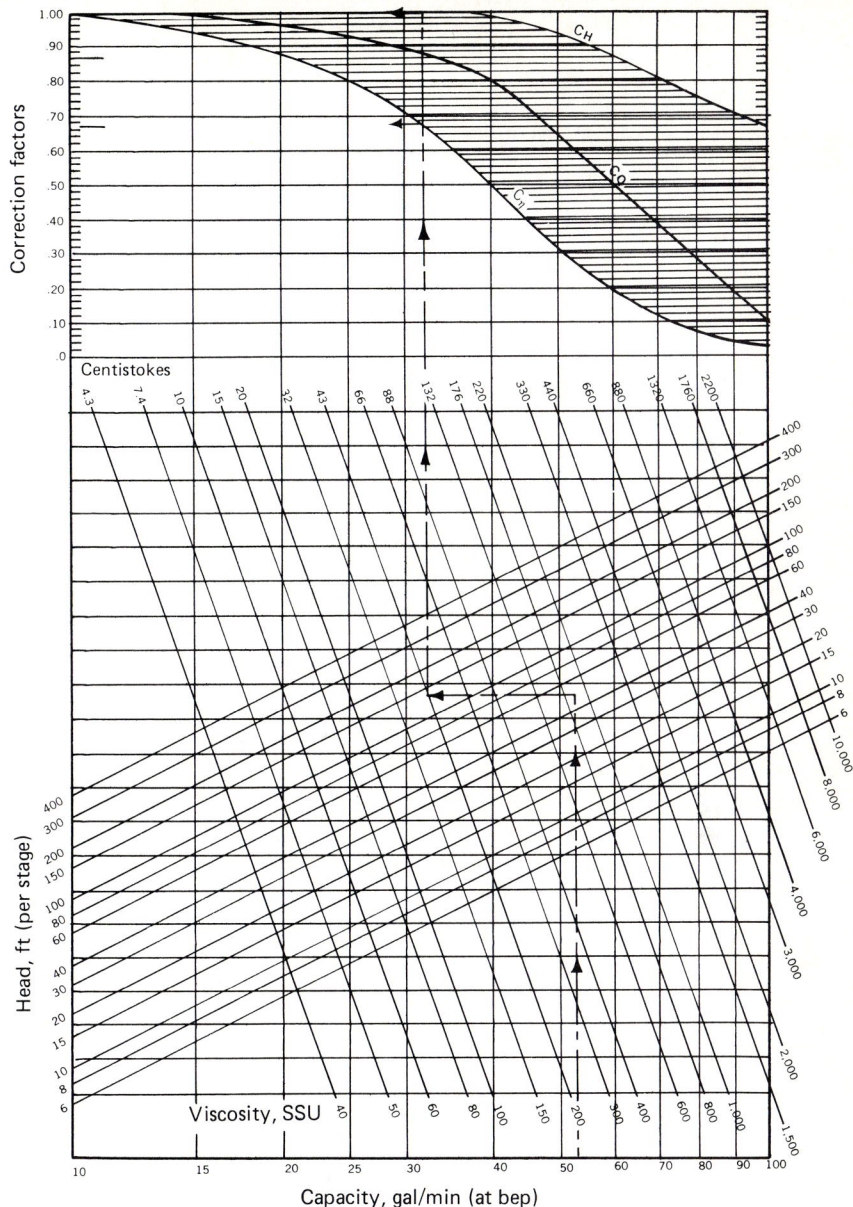

Figure 11.25 Performance correction chart for viscous liquids. *(The Hydraulic Institute, Cleveland, Ohio.)*

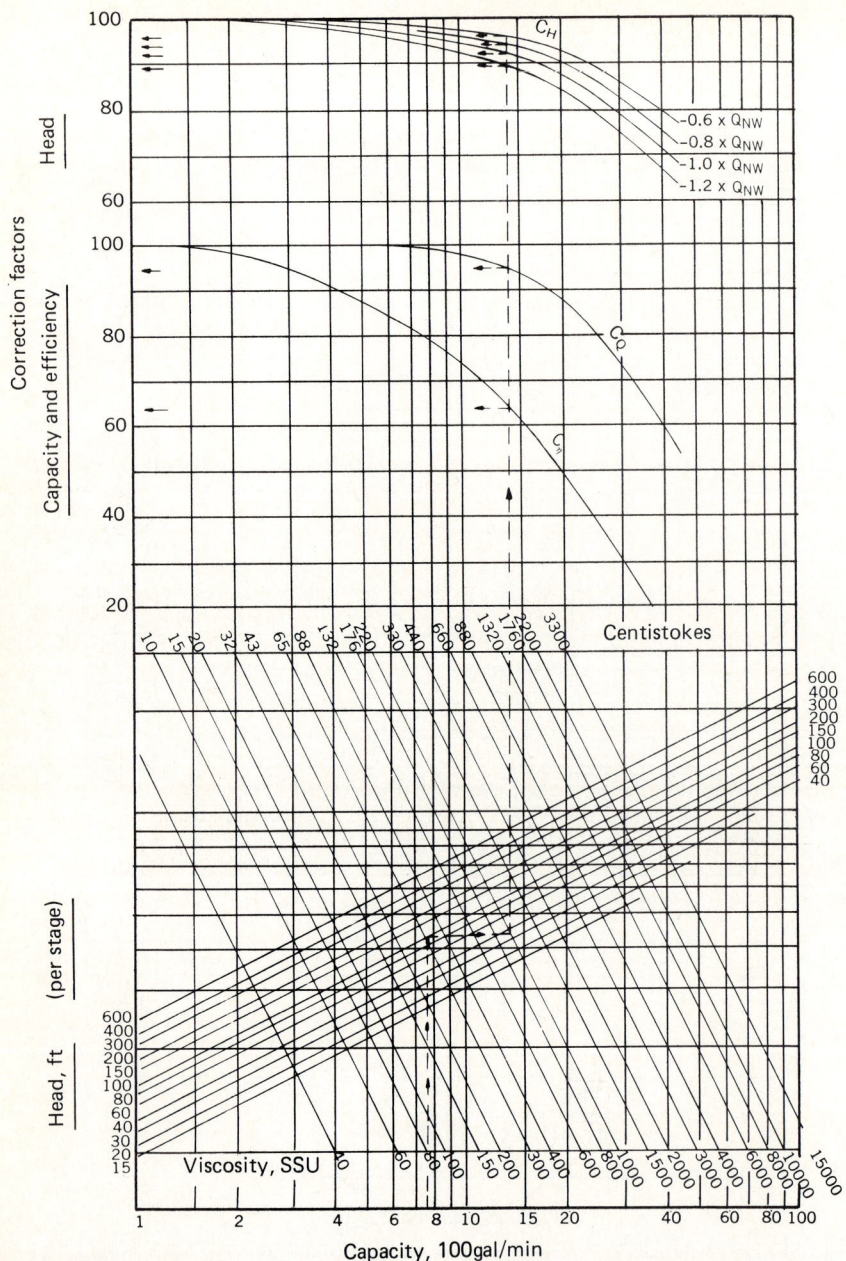

Figure 11.26 Performance correction chart for viscous liquids. *(The Hydraulic Institute, Cleveland, Ohio.)*

respective water capacities, heads, and efficiencies are obtained by proceeding to the water head corresponding to the maximum-water-efficiency capacity, then to the viscosity of the viscous fluid, and up to the correction factors. The correction factors thus obtained for capacity and efficiency are the same for all capacities and may be entered as such in each of the fractional-capacity columns. The head correction factor, on the other hand, must be taken according to the respective fractional-capacity head curves.

Once the correction factors have been determined, the viscous-fluid points on the transformed curve are obtained by multiplying the original water capacity, head, and efficiency values in each column by the respective correction factor. The viscous brake horsepower at each point is determined by using the viscous capacity and head, the viscous efficiency, and the specific gravity of the viscous fluid. The viscous NPSHR is taken as that for the water flow at the transposed water capacity and head, so that the correction factor for NPSHR is 1.0 Example 11.4 illustrates the performance-curve transformation process.

Example 11.4 In Example 11.3 it was found that a $9\frac{5}{8}$-in impeller for the pump in Fig. 11.20 should be suitable to pump 150 gal/min of the viscous fluid at a TDH of 80 ft. The fluid has a specific gravity of 1.05 and a viscosity of 219 cSt. Determine the performance curve for the viscous fluid.

solution The performance curves, efficiency, and horsepower curves for water are placed on a graph. The data for $0.6\ Q_{\eta W}$, $0.8\ Q_{\eta W}$, $1.0\ Q_{\eta W}$, and $1.2\ Q_{\eta W}$ as well as the respective correction factors and calculated viscous value are entered in tabular form as follows:

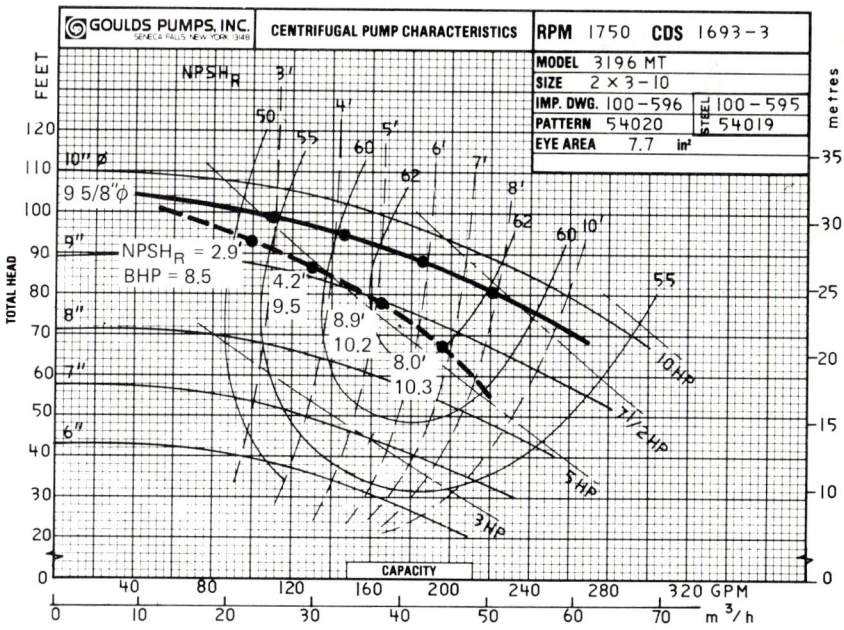

Calculations for Viscous-Performance Curve

	$0.6 \times Q_{\eta W}$	$0.8 \times Q_{\eta W}$	$1.0 \times Q_{\eta W}$	$1.2 \times Q_{\eta W}$
Q_{W1}, gal/min	111	148	185	222
H_{W1}, ft	99	95	89	80
η	0.54	0.60	0.63	0.615
NPSHR, ft	2.9	4.2	5.9	8.0
Viscosity, cSt	219	219	219	219
C_Q (Fig. 11.26)	0.88	0.88	0.88	0.88
C_H (Fig. 11.26)	0.93	0.90	0.87	0.84
C_E (Fig. 11.26)	0.52	0.52	0.52	0.52
C_{NPSHR}	1.0	1.0	1.0	1.0
$Q_{\text{vis}} = C_Q \times Q_{W1}$, gal/min	98	130	163	195
$H_{\text{vis}} = C_H \times H_{W1}$, ft	92	86	77	67
$\text{NPSHR}_{\text{vis}} = \text{NPSHR}_{W1}$, ft	2.9	4.2	5.9	8.0
Specific gravity	1.05	1.05	1.05	1.05
$\text{bhp}_{\text{vis}} = \dfrac{Q_{\text{vis}} \times H_{\text{vis}} \times \text{sp gr}}{3961 \times C_E \times \eta}$	8.5	9.5	10.2	10.8

The transformation points are plotted near the appropriate curves on the graph. Both the viscous head and efficiency curves are similar in shape to their water counterpart curves, while the brake-horsepower curves are essentially parallel. The transformed point for Example 11.3 should fall on the performance curve developed in this example. This occurs when the water capacity in Example 11.3 is at the maximum efficiency for the water performance curve at the impeller diameter determined in the example. However, if the transposed point did not correspond to the maximum water efficiency, the desired viscous head-capacity point will fall below or above the transformed viscous performance curve. If the difference is too large, another size of impeller should be chosen and a new viscous performance curve prepared.

Since Figs. 11.25 and 11.26 are based on empirical data, they should be applied only to centrifugal pumps based on conventional hydraulic design with open or closed impellers and operating in their normal range. The figures should not be applied with axial-flow or mixed-flow pumps or with nonuniform liquids.

Packing and Mechanical Seals

All pumps, as well as other rotating process equipment, are subject to leakage through the area between the rotating shaft and the housing containing the process fluid. Pumps must, therefore, be provided with sealing devices to reduce or eliminate leakage. Packing and mechanical seals are the two general types of sealing devices used for pumps.

Packing

Packing is a fibrous material woven into a rope, cut, and molded into rectangular sections which are wedged in a stuffing box located between the pump casing and the shaft. Figure 11.27 shows the basic arrangement for a stuffing box containing packing. Leakage is reduced as the packing gland is bolted against the packing. This packing method is limited to use with inexpensive pumps for noncritical and nontoxic

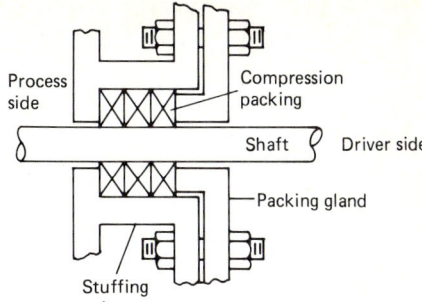

Figure 11.27 Simple compression-packing arrangement.

services, such as ambient-temperature water. A more complex packing is required for other applications.

There are numerous packing arrangements used on pumps and agitators. While a detailed description of each type of packing is beyond the scope of this book, a packing arrangement is usually composed of the following parts:

1. Throat-and-gland follower
2. Compression packing
3. Chevron or automatic packing
4. Loading spring
5. Lantern ring and lubrication port

A throat-and-gland follower is required for the packing used in reciprocating pumps such as piston pumps, plunger pumps, and some diaphragm pumps. The packing mounted on these pumps experiences a back-and-forth motion rather than a rotary motion. The throat-and-gland follower is placed on either side of the packing, as shown in Fig. 11.28, to prevent the packing from being extruded between the casing and shaft, thus losing its sealing effect.

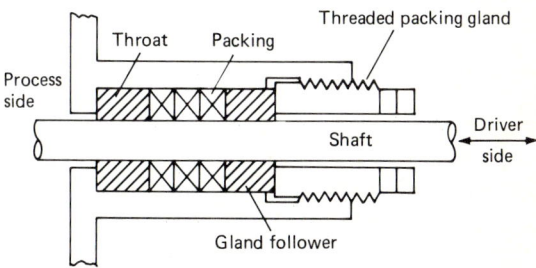

Figure 11.28 Simple packing arrangement for reciprocating pumps.

Compression packing as shown in Fig. 11.27 depends on the packing gland to force the packing to contact the shaft and reduce leakage. However, this increases frictional drag on the shaft, which in turn reduces the driver energy that can be transferred to the fluid. Lower frictional drag can be produced by using automatic or chevron packing as shown in Fig. 11.29. The automatic packing consists of chevrons or similarly

424 Equipment Selection, Sizing, and Related Subjects

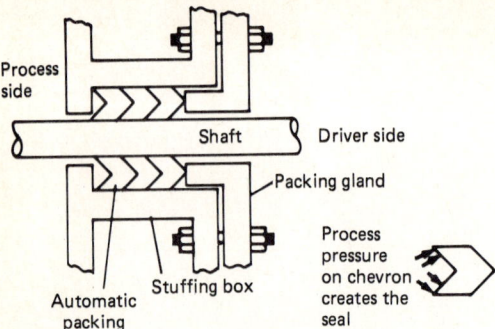

Figure 11.29 Simple automatic-packing arrangement.

shaped packing which depends on the process pressure to create a seal around the shaft. Automatic packing used on reciprocating pumps requires a throat-and-gland follower. Spring loading may also be required.

While compression packing and automatic packing may be spring-loaded, spring loading is more often used with automatic packing. The spring, as shown in Fig. 11.30, is always placed on the process side of the packing. The spring applies a force on the packing which is much lower than the process pressure but is additive to the

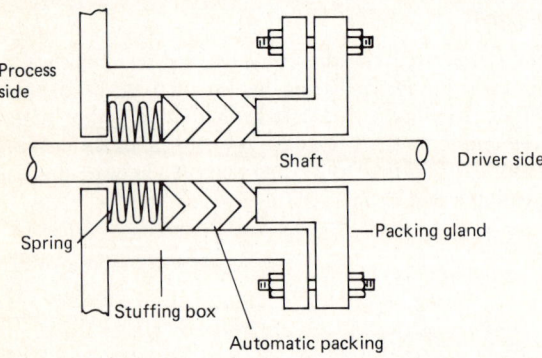

Figure 11.30 Spring-loaded packing arrangement.

process pressure and high enough to improve packing performance. Spring loading is used to reduce the need for adjusting the packing gland periodically to compensate for packing wear and compaction and to allow for thermal expansion of packing due to friction or process-temperature excursions.

Packing depends on the lubricating properties of the process fluid to reduce frictional drag on the shaft. Therefore, some leakage is desirable across the packing. If leakage is unacceptable owing to the hazardous properties of the process fluid, then either a lubricant is added to the packing or a mechanical seal must be used. Figure 11.31 shows two methods of adding lubrication to packing.

The packing lubricant is injected through a lubrication port in the stuffing box and into a lantern ring, or seal cage, which provides a free passage for the lubricant to be

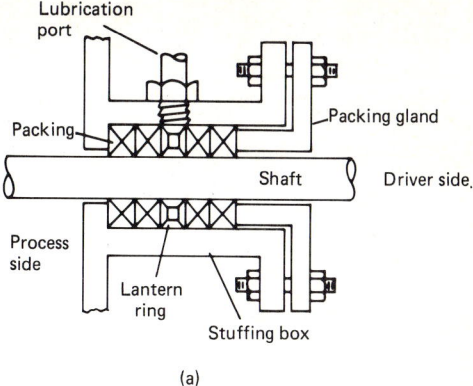

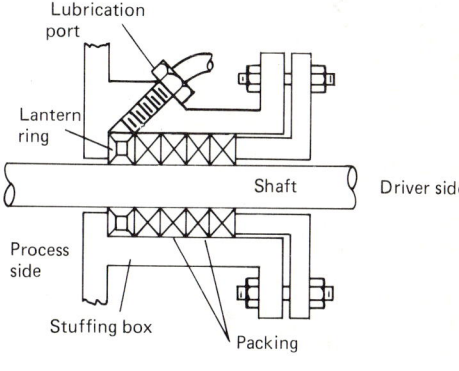

Figure 11.31 Lubricated-packing arrangement. (a) Normal service. (b) Abrasive service.

distributed around the shaft. The lantern ring is always located between two rings of packing.

Packing lubricants may be liquid or dry. Examples of liquid lubricants are petroleum oils, silicone oil for high-temperature service, grease, and tallow. Dry lubricants include graphite, molybdenum sulfide, tungsten disulfide, and tetrafluoroethylene (TFE). Selection of a lubricant depends on the operating temperature of the process and the compatibility of the lubricant with the process fluid.

Packing is not a perfect sealing device. Table 11.4 lists some of the advantages and disadvantages of packing. If packing is unsuitable for the process application, a mechanical seal should be considered.

Mechanical seals

Mechanical seals are the most frequently used sealing devices. Although they are more complex and expensive than packing, they are more effective in reducing or eliminating leakage from pumps. This is particularly important when the process

TABLE 11.4 Advantages and Disadvantages of Packing

Advantages	Disadvantages
1. Less expensive than mechanical seals. 2. Easy to replace. 3. Easy to maintain. Packing can be adjusted for wear by tightening the packing gland and can be replaced without disassembling the pump. 4. Packing fails gradually, not suddenly as would a mechanical seal. Unscheduled shutdown for packing failure is rare. 5. Useful in submersible service.	1. Non-spring-loaded packing requires frequent adjustment. 2. Higher leakage than a mechanical seal. A packing may leak at the rate of 60 drops per minute, while a properly installed mechanical seal may leak by as little as 15 drops per year. 3. Leakage is required as a lubricant for nonlubricated packing. 4. Packing failure causes shaft damage. 5. Packing has 6 times the frictional drag of a mechanical seal. 6. Packing life is shortened by exposure to abrasive materials.

fluid is costly, hazardous, or toxic. Table 11.5 lists some of the advantages and disadvantages of using mechanical seals.

There are more than 100 varieties of mechanical seals. However, mechanical seals can be characterized as to whether they are (1) inside or outside seals, (2) balanced or unbalanced seals, (3) rotating-head or rotating-seat seals, (4) single or multiple seals, (5) pusher or nonpusher types, and (6) single- or multiple-spring seals. Unlike packing, which is stationary, mechanical seals have a stationary and a rotating component. The primary sealing element, as shown in Fig. 11.32, is the face between the rotating head washer, which is between the stationary seat and the secondary sealing elements. The primary sealing elements, which prevent leakage around the shaft and around the space between the stationary seat and the housing, are the O rings. The spring located behind the head washer provides the force necessary to produce an effective primary seal.

TABLE 11.5 Advantages and Disadvantages of Mechanical Seals

Advantages	Disadvantages
1. Much lower leakage rate than packing. 2. No break in period required. 3. More practical for handling toxic or flammable fluids. 4. Some seals can be designed to include flushing and cooling features. 5. More stable operation in high-pressure and cyclical-pressure applications. 6. Lower frictional drag. 7. Lower shaft and sleeve wear.	1. More expensive than packing. 2. Higher maintenance costs. 3. Mechanical seals can fail suddenly, resulting in unscheduled shutdowns. 4. Mechanical seals are not self-lubricating unless a lubrication feature is added to the seal. 5. May not be effective in high-vibration service ($\sim$ 2 mils). 6. May require costly auxiliary equipment for cooling, flushing, or lubrication.

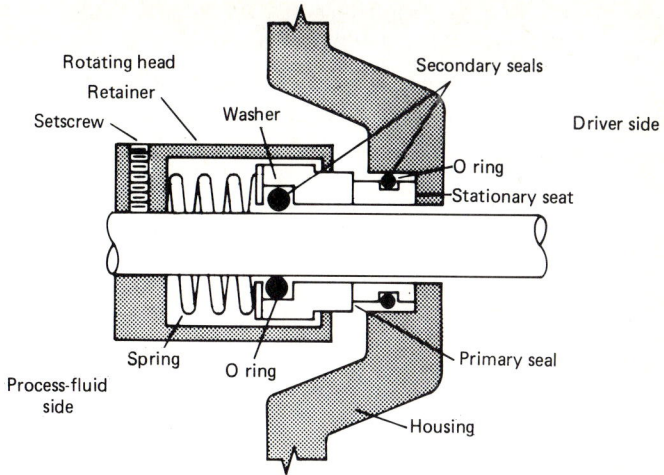

Figure 11.32 Basic inside-mechanical-seal arrangement.

The selection of the proper mechanical-seal arrangement is dictated by the process and maintenance requirements of the plant.

Inside versus outside seals. The seal shown in Fig. 11.32 is considered an inside seal because the mechanical seal is located inside the housing and is surrounded by the process fluid. Inside seals are more effective than outside seals because the hydraulic forces of the process fluid act in concert with the spring forces on the primary seal. However, inside seals are not recommended for mounting in corrosive fluids or abrasive slurries.

Outside seals are not as efficient as inside seals because the counteracting hydraulic forces of the process fluid work against the spring forces. The net force on the primary seal is the difference between these two forces as opposed to being the sum of the two forces if it were an inside seal. The greater the net force, the tighter the seal. However, outside seals are easier to maintain and are suited for corrosive and abrasive service. Flushing water can be added to an outside seal to reduce possible deterioration of the seal if it were in contact with the process fluid.

Balanced versus unbalanced seals. A balanced seal is a seal designed with a smaller annular area than an unbalanced seal. The reduced area results in less force being applied to the seal faces for a given pressure. Unbalanced seals are less costly and are easier to install. They are used for low-pressure applications; e.g., API standard 610 recommends the use of balanced seals only for pumps above 75 psig.

Rotating head versus rotating seat. While a mechanical seal can have either its head or its seat rotate with the shaft, a rotating-head arrangement is preferred for process service. The rotating-head seal has a simpler design and is easier to balance and maintain.

Single versus multiple seals. Single mechanical seals are normally simpler and more economical, but double mechanical seals are more practical when (1) seal faces

428 Equipment Selection, Sizing, and Related Subjects

must be lubricated with other than the process fluid, (2) the process fluid is abrasive and must be flushed from the seat, or (3) the process fluid is hazardous or toxic and must not be exposed to the atmosphere.

Secondary-seal arrangement. The simplest and most economical secondary seal is the O ring, but improved secondary sealing can be provided by using wedges, V rings, and bellows. O rings, V rings, and wedges are classified as pusher-type secondary seals because they push against the shaft to maintain the seal.

The wedge seal shown in Fig. 11.33 uses spring pressure to drive the wedge against the shaft, resulting in a better seal than an O ring, but it has less flexibility.

The V-ring seal shown in Fig. 11.34 is similar to chevron packing. V-rings can operate at higher pressures than O rings but occupy more space and have less flexibility.

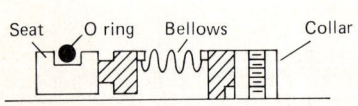

Figure 11.33 Mechanical seal with a wedge-type secondary seal.

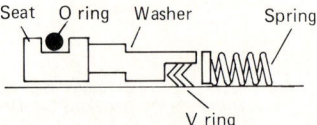

Figure 11.34 Mechanical seal with a V-ring secondary seal.

The bellows seal is a nonpusher seal which may be constructed of a metal or an elastomeric rubber. A typical bellows seal is shown in Fig. 11.35. The bellows seal does not depend on spring pressure or friction for sealing. Bellows seals are used in services experiencing high vibrations which might create fretting-type corrosion on the shaft if a pusher type of seal were used.

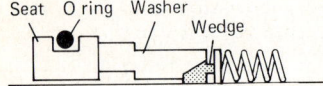

Figure 11.35 Mechanical seal with a bellows secondary seal.

Single versus multiple springs. Sealing pressure at the primary seal face may be provided by either a single large spring or a series of small springs around the retainer annulus. A single-spring seal is the most economical, but multiple springs provide a more even pressure distribution on the seal face. A single-spring seal is more practical for corrosive service because it presents a smaller surface area for corrosion and thus has a longer operating life.

Mechanical-seal materials. Mechanical seals can be constructed with a number of materials. The materials selected depend on their compatibility with the shaft, housing, and process fluid and on their wear resistance.

Materials in contact at the primary seal face experience the most wear. Primary seal faces are usually constructed of either tungsten carbide, ceramic, hard-faced cobalt alloy, soft carbon, or graphite. Ceramic on a carbon face offers the best wear resistance. However, ceramics are susceptible to thermal shock and are not suitable for hot or strong-caustic service.

O rings and V rings are usually constructed of TFE or a more resilient elastomer. TFE is limited to usage below 350°F. Bellows seals are usually constructed of either TFE or an elastomeric rubber, but metal bellows are used in high-temperature service.

Process engineers should provide complete process conditions to the seal manufacturer so that the most effective seal arrangement and materials of construction are specified.

Pumping-System Design

As the operating characteristics, TDH, and NPSHA for the pump are being determined and the pump is being specified and purchased, the physical layout of the piping system must be checked to see if any modifications to the system are needed to improve pump performance. The system must be checked not only for pump performance during normal operating conditions but also for such transient conditions as start-up or shutdown.

This subsection presents items that should be considered when laying out a piping system for a pump. It illustrates several physical arrangements and design parameters. In addition, the reader should refer to the subsection "Pumps" in the Chap. 3 checklist for discussions regarding other piping details for pumps.

Pump priming

A centrifugal pump cannot operate unless its casing is filled with liquid. If the pump casing is filled with gas or vapor, it circulates the fluid in the casing but draws no liquid. Therefore, a centrifugal-pump casing must be flooded with liquid, or "primed," before it can operate. Positive-displacement pumps and regenerative turbine pumps are self-priming. These pumps displace the gas or vapor space with liquid as they operate, provided the suction lift is less than about 28 to 32 ft, depending on the specific gravity and vapor pressure of the liquid and the NPSHR of the pump.

A pump located at a lower elevation than the liquid level of the vessel from which it draws its suction is primed by the static head of the liquid above it. However, a pump which has a suction lift, such as one drawing its suction from a sump or an underground tank, requires priming.

When priming is required, a specially designed self-priming centrifugal pump, as shown in Fig. 11.36, can be used, or auxiliary equipment is needed to prime the pump. The type of auxiliary priming system chosen depends on the specific operating characteristics of the system. The simplest priming system consists of a foot valve placed at the end of the suction line. A foot valve is a special check valve which prevents liquid from draining out of the suction line when the pump is shut off. A foot valve is normally fitted with a strainer to prevent debris from passing through the valve and damaging the pump.

A foot valve cannot maintain a column of liquid in the suction line for long periods of time. Therefore, pumps which have prolonged idle periods have provisions to flood the suction line and casing from another source. This can be accomplished with a line having a manual fill valve connected to the pump casing.

The fill valve and an air-vent valve on the pump casing are opened prior to starting the pump. As the pump and suction line are flooded, the air is displaced through the vent valve. An alternate location for the external supply line is the pump discharge line before the check and isolation valve. If an external supply is not available, a small

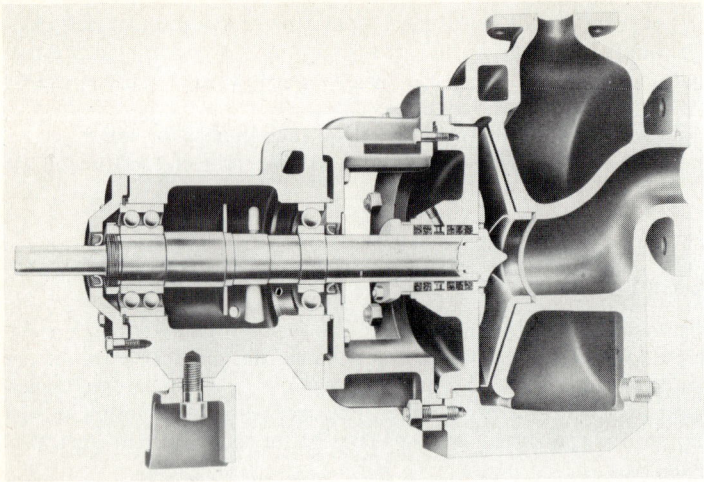

Figure 11.36 Self-priming centrifugal pump. *(The Duriron Company, Inc., Pump Division, Dayton, Ohio.)*

positive-displacement pump can be used to draw liquid from the same source to which the centrifugal pump is connected.

If the pump is connected to a vertical discharge line going up, the line will probably be full when the pump stops. To prime a pump with this arrangement, a small bypass line around the discharge check and gate valve can be used to fill the pump and the suction line with liquid from the vertical discharge line.

For other situations in which it is not practical to use a positive-displacement pump, a pump can be primed by drawing a vacuum on the pump casing and the suction line to lift liquid to the pump. The vacuum can be provided by an ejector or by a vacuum pump. A sight glass on the vacuum line is used to determine when liquid has filled the pump.

Another method for priming a centrifugal pump utilizes an automatic primer which consists of a partitioned tank like the one shown in Fig. 11.37. When the pump

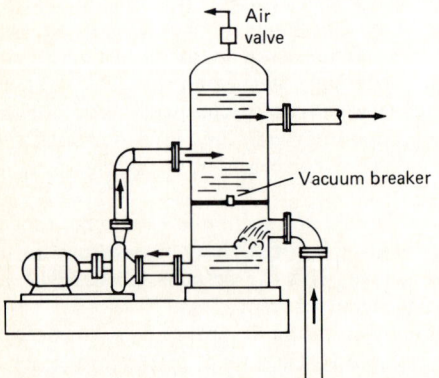

Figure 11.37 Automatic pump primer.

starts, it draws liquid from the lower partition in the tank. As the level drops, there is formed a vacuum which, although not sufficient to cause cavitation, is enough to draw liquid into the suction line, thus replenishing the liquid in the tank. Liquid is discharged into the upper partition in the tank, then out into the discharge piping downstream. When the pump stops, liquid flows back from the upper partition into the pump and the lower partition. The pump is primed with the liquid in the lower partition when it is restarted.

Except for self-priming pumps, the above systems are manually operated. However, automatic controls can be added to sense flow when a pump is filled, minimize the effect of fluctuating suction lift, and/or regulate vacuum levels in the system.

Minimum flow through pumps

Pumps are designed to do work on liquids. The work consists of increasing the pressure head of the liquid. The pump work measured in brake horsepower (kilowatts) is based on a given capacity and total dynamic head. If the flow through the pump is reduced, the efficiency of the pump decreases and the energy that would normally be used for work heats the liquid. If the liquid flow is reduced significantly, the heat generated in the pump could cause the liquid to flash, engendering cavitation, which can damage the pump. Therefore, a minimum flow through a pump must be maintained to prevent cavitation.

Some manufacturers provide minimum-flow requirements along with performance curves for their pumps. When minimum-flow information is not available, it must be estimated. Durco[1] has suggested a simple equation which equates heat generated to work done by the pump's driver for determining minimum flow through a pump:

$$Q_M = \frac{5 \, (\text{bhp}_0) C_{\text{vis}}}{C_p \Delta T}$$

where bhp_0 = end-point brake horsepower (nonviscous performance curve), hp
C_p = specific heat of the liquid, Btu/(lb·°F)
C_{vis} = horsepower correction factor (>1 for liquid viscosity > 1 cP)
Q_M = minimum flow through pump, gal/min
ΔT = maximum allowable temperature rise, °F

Once a minimum flow has been established, it is important to determine how it is to be achieved. There are many operating systems that require the pump to be in operation continuously regardless of the flow demands downsteram of the pump. For these systems, a recirculation line is usually added in the design. The recirculation line either recirculates the minimum flow back to a vessel from which the pump takes suction or directly into the pump's suction. If the flow is recycled into the pump's suction, the recycle must be cooled either by air-cooled fin-tube piping or with a regular heat exchanger. The type of heat exchanger depends on the amount of heat to be removed.

The minimum-flow line must be designed as simply and as economically as possible. The simplest design is based on continuous recirculation through a restriction orifice as shown in Fig. 11.38.

The restrictive-orifice method is useful for small systems with low operating costs. For large systems, the cost of operating a continous recirculation line may be excessive.[2] An automatic recirculation control system may be used for these systems. The

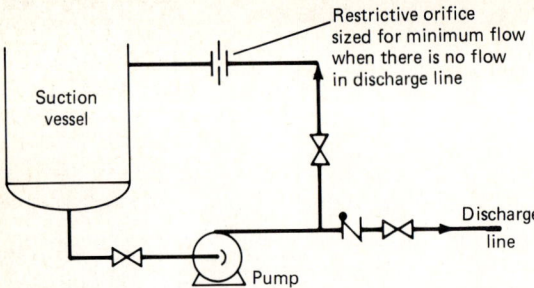

Figure 11.38 Recirculation piping system used to maintain minimum flow through a pump.

automatic recirculation may be a self-contained mechanical device as shown in Fig. 11.39 or a system consisting of a flow switch connected to a valve in the recirculation line as shown in Fig. 11.40.

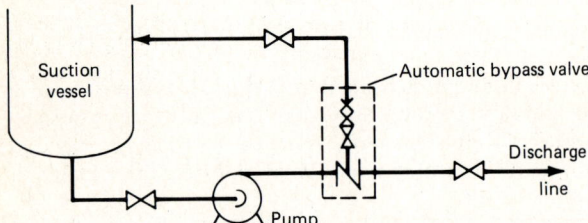

Figure 11.39 Automatic recirculation device used to maintain minimum flow.

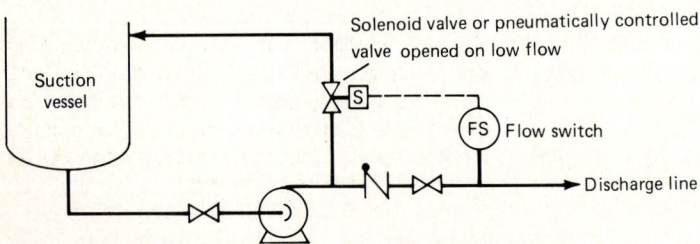

Figure 11.40 Minimum flow maintained by a low-flow switch.

Sump design

When horizontal or vertical centrifugal pumps are used in sumps, it is important that the sumps be designed to minimize vortex formation. This includes consideration of the minimum spacing between pumps, minimum pump submergence, and minimum velocity in the sump. The Hydraulic Institute provides a number of standards for the design of sumps. Figures 11.41 and 11.42 show the design standards recommended by the institute.

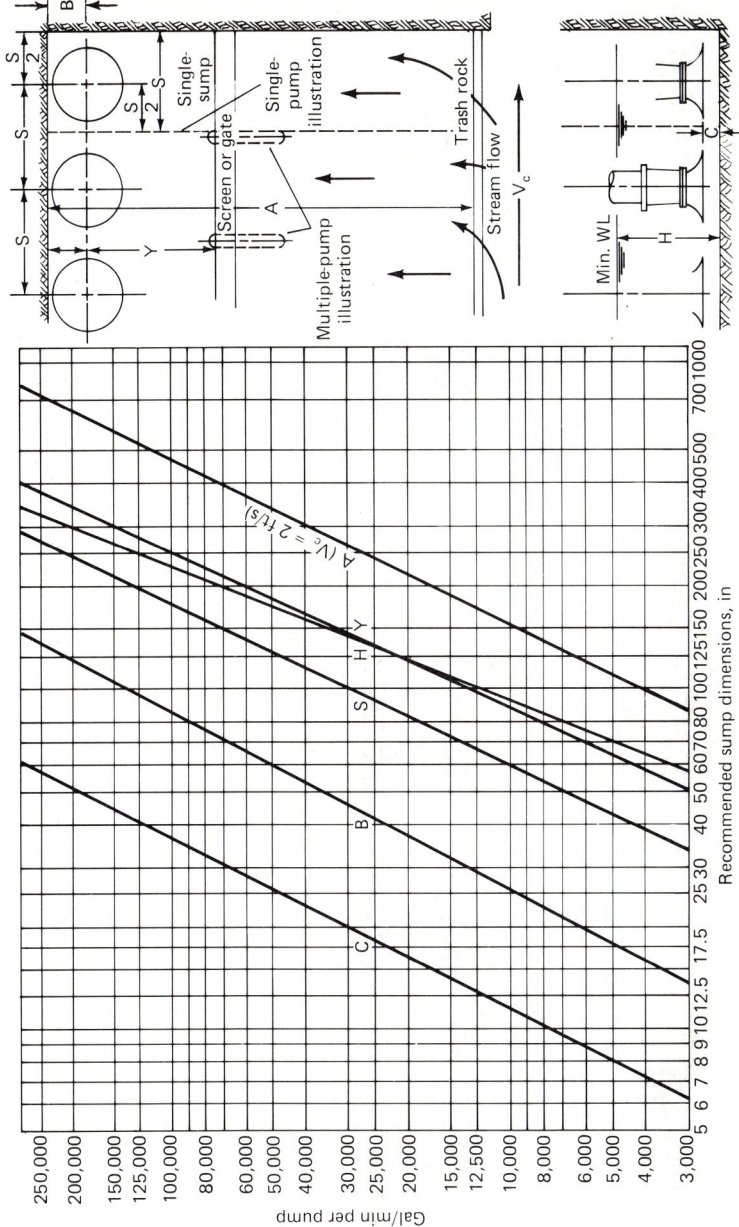

Figure 11.41 Hydraulic Institute chart for determining the size of a pump sump. Figures apply to sumps for clear liquid. For fluid-solids mixtures refer to pump manufacturer. *(The Hydraulic Institute, Cleveland, Ohio.)*

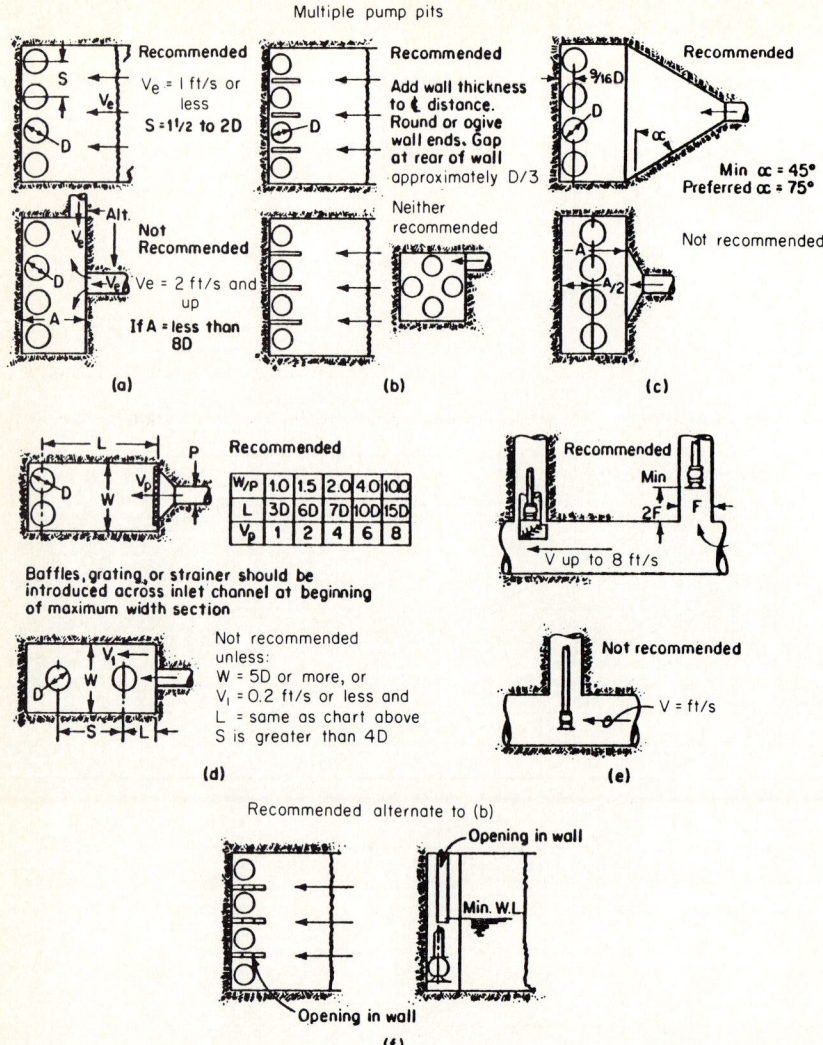

Figure 11.42 Hydraulic Institute standards for flow velocity and sump geometry. Figures apply to sumps for clear liquid. For fluid-solids mixtures refer to pump manufacturer. *(The Hydraulic Institute, Cleveland, Ohio.)*

REFERENCES

1. *Pump Engineering Manual,* ed. by R. E. Syska and J. R. Birk, The Durion Company, Inc., Pump Division, Dayton, Ohio, 1980, chap. 15.
2. *Fluid Movers: Pumps, Compressors, Fans and Blowers,* ed. by Jay Matley and staff of *Chemical Engineering,* McGraw-Hill Publications Company, New York, 1979, pp. 340–343.

BIBLIOGRAPHY

Centrifugal Pump Lexicon, Kleign, Schanzlin and Becker Aktiengesellschaft, Frankenthal, Pfalz, Federal Republic of Germany, 1975.
Equipment Design Handbook for Refineries and Chemical Plants, vol. 1, ed. by Frank L. Evans, Jr., Gulf Publishing Company, Houston, 1971, chap. 4.
Hicks, T. E., and T. W. Edwards: *Pump Applications Engineering,* McGraw-Hill Book Company, New York, 1971.
Pump Handbook, ed. by I. J. Karassik and W. C. Krutzch, McGraw-Hill Book Company, New York, 1976.

Chapter 12

Fans, Blowers, and Compressors

Fans, blowers, and compressors are used to pressurize and move compressible fluids through pipes and ducts. The difference in nomenclature lies in the range of differential pressure that each produces. While there is no precise delineation among these gas movers, fans are those gas movers that develop pressure rises of less than 60 in of water column, or about 2.0 psig, blowers operate in a range of about 1.0 to 30 psi, pressure rise, and compressors are used to produce higher pressure rises.

In addition to nomenclature, there are different design features, performance characteristics, and system operating characteristics for fans, blowers, and compressors. This chapter compares the differences and similarities of each gas mover and its performance characteristics as well as some system designs to be considered for the effective operation of the equipment.

Fans

Fans are used to move relatively large volumes of gas through a system and to operate at near atmospheric pressure. They are employed for such diverse applications as building ventilation, aiding combustion in furnaces, providing an air draft for cooling towers, and providing hot air for dryers.

The performance of a fan is represented differently from that of a blower or a compressor. The fan performance curve is graphically represented as a volumetric flow, cubic feet per minute versus static pressure, usually in inches of water, developed, while blowers and compressors use total pressure expressed as adiabatic head rather than static pressure.

The total pressure P_T of a flowing gas is the static pressure P_S and velocity pressure P_v; i.e.,

$$P_T = P_S + P_v, \text{ in of water} \tag{12.1}$$

where $P_v = \alpha\, u^2/2\, g_c$, in of water
u = linear gas velocity
α = unit conversion factor
g_c = dimensional constant, 32.17 (lbm/lbf)(ft/s^2)

Figure 12.1 shows that by pointing a pitot tube right into a stream of flowing gas, the total pressure and static pressure can be measured directly. If a fan is located upstream of the tube and flow is stopped by closing a set of dampers downstream of the tube, the total pressure and static pressure would be equal because velocity u would be zero and therefore velocity pressure P_v would be zero. As the dampers are

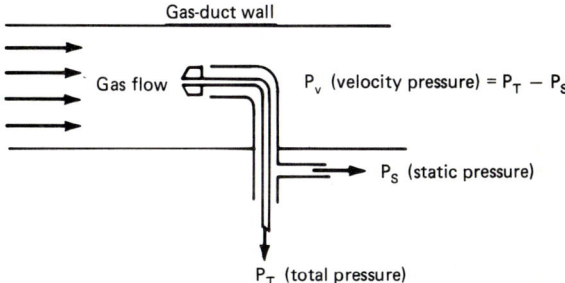

Figure 12.1 Pitot-tube measurement of flowing-gas pressures.

opened, the measured total pressure would become greater than the static pressure, the difference being the velocity pressure, which is proportional to the square of the gas velocity. Conversely, static pressure can be used to determine velocity pressure by difference if total pressure is also known.

Although total pressure is more useful, fan characteristic curves have traditionally been published in terms of static pressure. The use of static pressure rather than adiabatic head ignores the compressibility of the gas. This is not a problem, since the gas being moved through a fan, unlike that in a blower or a compressor, does not experience enough compression to effect a marked change in density. Therefore, the gas can be treated as an incompressible fluid, and the gas's total pressure and velocity pressure can be estimated, given the volumetric flow and static pressure.

There are three general classes of fans: the propeller fan, the centrifugal fan, and the axial-flow fan. The flow patterns within fan housings are complex, but they all produce a maximum output pressure in the neighborhood of peak efficiency. The peak efficiency that a fan can achieve depends on the class of fan and its blade configuration.

Propeller fans

Propeller fans are used to move large volumes of gas, usually air, but have very low output pressure. The static pressure of these fans is measured in fractions of an inch water column. Propeller fans are generally used as roof ventilators, as room exhausts, and on induced-draft cooling towers. The pressure developed by a propeller fan is a function of the angle at which the propeller blade impinges on the gas mass. These fans can also have speed controls or adjustable-pitch blades or adjustable-inlet vanes.

Centrifugal fans

Unlike propeller fans, centrifugal fans produce pressure by the centrifugal force of rotating air within the fan housing. The centrifugal force is created by a fan wheel within the housing. There are a variety of fan-wheel configurations, each having its own performance characteristics. A set of fan performance curves for different fan-wheel configurations is presented in Fig. 12.2.

Backward-inclined blades. Backward-inclined blades, shown in Fig. 12.2a, are inclined at an optimum angle backward to the wheel-rotation direction. The pressure curve produced by this fan shows a characteristic dip in pressure followed by a peak pressure, then by a steady drop. If the fan is operated in the region between the dip and the maximum-pressure point, the volumetric flow will oscillate between these points. The resulting volume pulsations can produce fan vibrations that could damage the fan. These fans can produce a maximum static efficiency of 77 to 80 percent and are typically used with process dryers, forced-draft boilers, and conveying systems.

Backward-curved blades. The backward-curved blades shown in Fig. 12.2b are used more often than backward-inclined blades because they promote smoother airflow, have a stronger construction, and do not have an output pressure dip which can cause unstable fan performance. These fans are used for the same applications as backward-inclined blades and have about the same peak efficiencies but are more costly to construct.

Radial blades. The radial blades shown in Fig. 12.2c are used in gas-flow applications containing dust or particulate matter such as pneumatic conveying systems. The radial-blade configuration does not produce a smooth airflow, as does a backward-inclined or -curved blade, but this configuration improves the blade's resistance to abrasion from entrained solids. The poorer airflow results in lower static efficiency (70 to 72 percent).

Open radial blades. The open radial blades shown in Fig. 12.2d are used for gases containing extremely abrasive solids. These blades have even lower static efficiency (65 to 70 percent).

Radial-tip blades. The radial-tip blades shown in Fig. 12.2e are also called backward-inclined, forward-curved blades. Radial-tip blades are used in lieu of radial blades because they have good abrasion resistance and higher static efficiency (78 to 83 percent). However, they are more costly and are not as easy to maintain as radial blades.

Forward-curved blades. Forward-curved blades, also called a squirrel-cage rotor, produce higher-velocity gas flow but at slower rotational speeds than other blades. These blades are limited to handling clean gases because particulate solids tend to accumulate on the blades. As shown in Fig. 12.2f, forward-curved blades also have an unstable region, as does a backward-inclined blade. The blades' maximum efficiency ranges between 72 and 76 percent.

Airfoil blades. Airfoil blades, as shown in Fig. 12.2g, produce extremely smooth airflow, resulting in very high static efficiencies (84 to 91 percent). However, this blade configuration is very costly to construct and is expensive to maintain or repair. The airfoil-blade design is used in clean-gas service and where low fan-noise levels are desired.

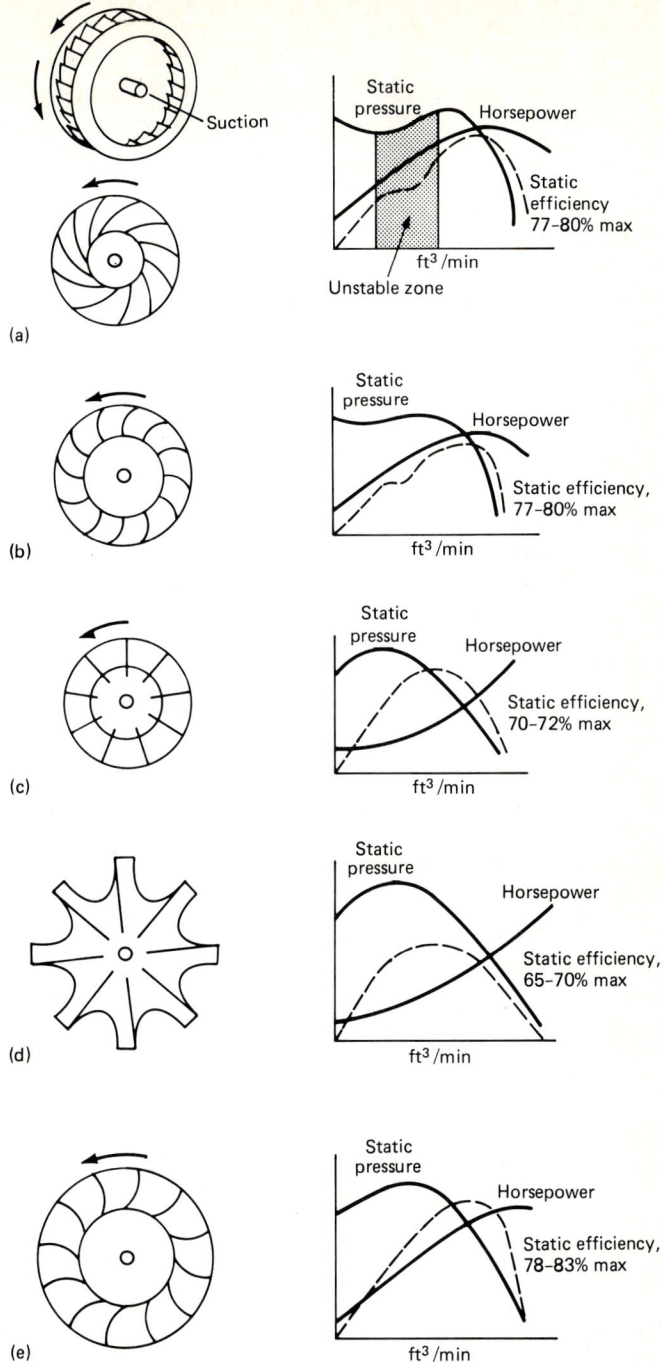

Figure 12.2 Centrifugal-fan blade types. (*a*) Backward-inclined blades. (*b*) Backward-curved blades. (*c*) Radial blades. (*d*) Open radial blades. (*e*) Radial-tip blades. (*f*) Forward-curved blades. (*g*) Airfoil blades.

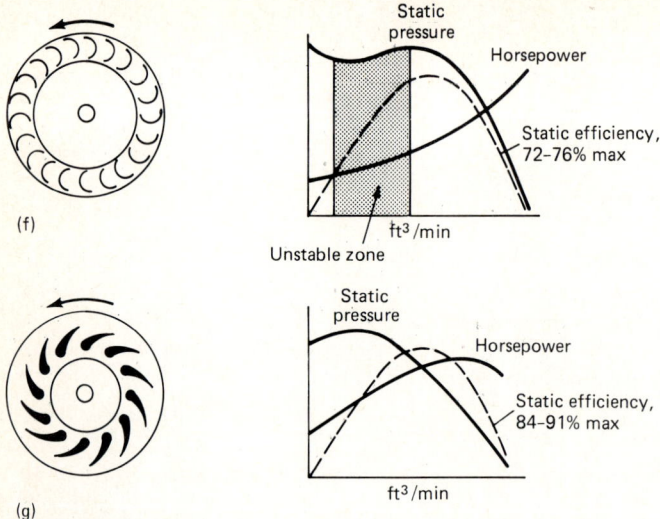

Figure 12.2 (*Continued*)

Axial fans

An axial fan is essentially a propeller fan housed in a tube or duct, but it can produce a much higher static pressure. There are basically two types of axial fans: tube-axial and vane-axial. A tube-axial fan (Fig. 12.3a) consists of a propeller fan housed in a tube or duct without any straightening vanes. The fan creates a swirling airstream. The energy consumed by the swirling action reduces the energy available to produce static pressure and linear-velocity pressure. Vane-axial fans (Fig. 12.3b) have air guide vanes which straighten the airflow, resulting in a higher static pressure than can be produced by tube-axial fans.

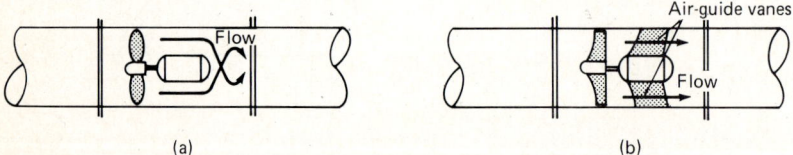

Figure 12.3 Axial fans. (*a*) Tube-axial. Static pressure = $\frac{1}{4}$ to $2\frac{1}{2}$ inH$_2$O. (*b*) Vane-axial. Static pressure = $\frac{1}{2}$ to 6 inH$_2$O, single-stage; <60 inH$_2$O, multistage (more than one wheel in series).

Centrifugal fans are preferred to axial fans for most process applications because they are easier to control, are less noisy, and have a more stable operating range and because their efficiencies do not drop off as rapidly as those of axial fans. Axial-fan instability and control difficulty arise from the typical dip in the pressure curve (Fig. 12.4), which is similar to that experienced by centrifugal fans with backward-inclined blades and forward-curved blades.

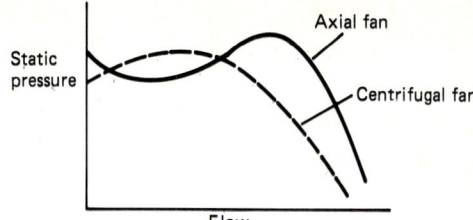

Figure 12.4 Comparison of performance curves of centrifugal and axial fans.

Some applications require the use of axial-flow fans in preference to centrifugal fans. Selection is based on fan efficiency and specific speed. Specific speed is an index used to characterize fluid movers and is defined as

$$\text{Specific speed} = \frac{N\sqrt{Q}}{h_a^{3/4}} \qquad (12.2)$$

where N = fan speed, r/min
Q = volumetric flow rate, ft³/s
h_a = adiabatic head, ft of fluid (for definition and calculations, see pages 450 and 451)

Equation (12.2) shows that the specific speed of a fan increases in proportion to the square root of its capacity, with the speed and head assumed to be held constant. Increased capacity requires a change in the design of the fan wheel. Fan-wheel design can be characterized by the diameter ratio D_s/D, which is the ratio of the shaft diameter D_s to the overall diameter of the wheel or fan blade. Figure 12.5 shows the relationship between the specific speed and the diameter ratio for centrifugal and axial-flow fans. As the specific speed of a centrifugal fan increases, the required diameter ratio also increases. This means that the eye of the fan approaches the diameter of the wheel itself. This makes the centrifugal fan impractical at higher capacities. However, axial-flow fans are more applicable at high specific speeds because of their more practical diameter ratios.

Fan controls

The controls on a fan are used not only to regulate the fan's flow and static-pressure characteristics in response to the process-system requirement but also to protect the fan from unstable operation. The normal operation of a centrifugal fan is typically controlled by using dampers, inlet vanes, a variable-speed drive, or a fluid drive. Axial-flow fans have variable-pitch blades to control performance.

The simple low-cost method for controlling fan operation uses dampers at the inlet or outlet of the fan. The damper position affects flow resistance. Figure 12.6 shows typical characteristic curves for a fan using dampers to control pressure and flow. An increase in damper-blade angle increases flow resistance, causing a decrease in static pressure and flow that the fan can supply to the system. Since the dampers create resistance, they cause the fan to operate less efficiently. It should be noted that in Fig. 12.6 the fan operates at 90 percent efficiency when the dampers are wide open (0°) but drops to 76 percent efficiency where the dampers are at a 30° angle.

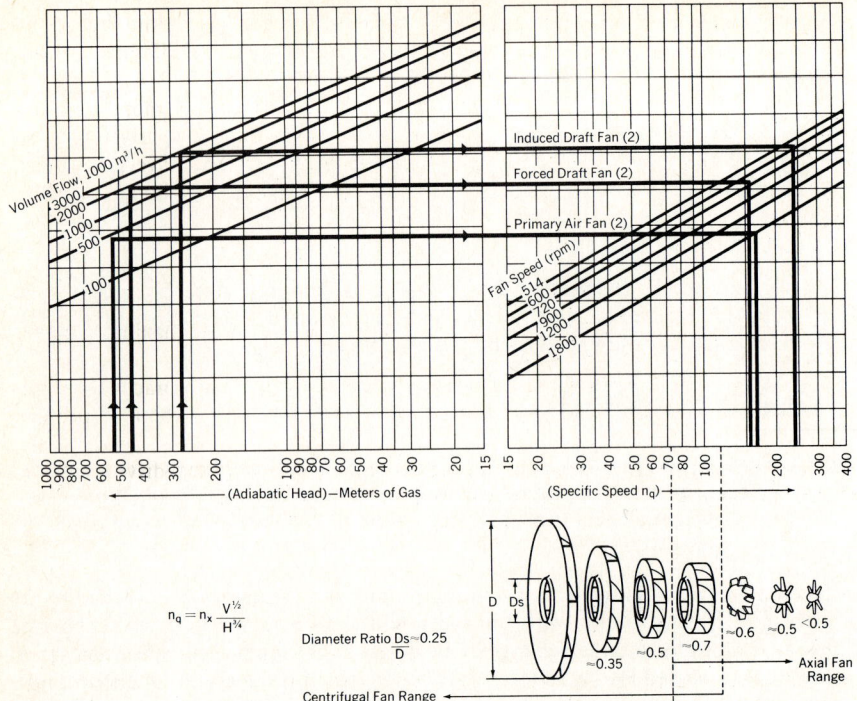

Figure 12.5 Fan-type-selection diagram. *(Babcock & Wilcox, Barberton, Ohio.)*

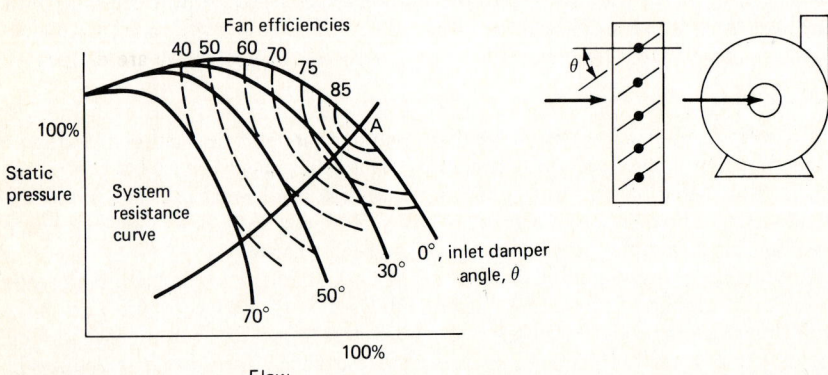

Figure 12.6 Fan characteristic curve using damper control; 100% = normal full-open operating conditions.

If adjustable-inlet vanes are fitted to the fan's inlet, they can be used to create a spinning motion to the air. They have the same effect as adjusting the fan speed. These vanes consume less power than do dampers, but the installation is more costly.

The use of variable-speed drivers to regulate fan operation is a more common method of controlling flow than using adjustable-inlet vanes. The driver may be either an electric motor or a steam turbine. With variable-speed drivers, the fan's mechanical construction can be much simpler and less expensive. Although variable-speed drivers are more costly, they are usually more efficient and easier to maintain than adjustable-inlet vanes.

A variable-speed electric driver may be either a dc motor, an ac wound-rotor motor, or an ac squirrel-cage motor with a silicon-controlled-rectifier (SCR) control system. A dc motor can control speeds down to about 35 percent of the fan's maximum speed, while ac motors are practical down to only about 50 percent.

If exhaust steam is available in process plants, a turbine drive is a practical option. Steam turbines can be effectively regulated for fan speeds down to about 35 percent of rated speed.

Other alternative speed controls for fans are a fluid- or a magnetic-drive coupling. These couplings provide a smooth transition over a wide range of speeds. Their costs are comparable with those of other variable-speed drive methods.

Fan-system arrangements

The proper installation of a fan in a gas-flow system can greatly improve a fan's performance. As a general rule the piping of ductwork connected to a fan should not promote any swirling action which can reduce the fan's static and velocity-pressure production. Therefore, right-angle bends or abrupt changes in airflow path should be avoided. Where these configurations cannot be avoided, flow-straightening vanes should be inserted in the ductwork. Figure 12.7 shows some examples of correct and improper ways of laying out a fan installation.

A convenient method for determining the number and design of straightening vanes has been proposed by Beissler.[1] Figure 12.8 shows a typical vane installation together with design details for the vanes. The letters in the figure are defined as follows:

$N = (6D/W) - 1 =$ total number of vanes, which is rounded off to next highest integer if N is a fractional number $\qquad$ (12.3)

where $D =$ duct width, in
$W =$ duct depth, in

and

$A = D/(N + 1) =$ axial width, in $\qquad$ (12.4)
$P = 1.41A =$ pitch, in $\qquad$ (12.5)
$R = 1.28A =$ radius, in $\qquad$ (12.6)
$L = 0.75A =$ straight length, in $\qquad$ (12.7)
$T \leq A/16 =$ flow-path thickness, in $\qquad$ (12.8)

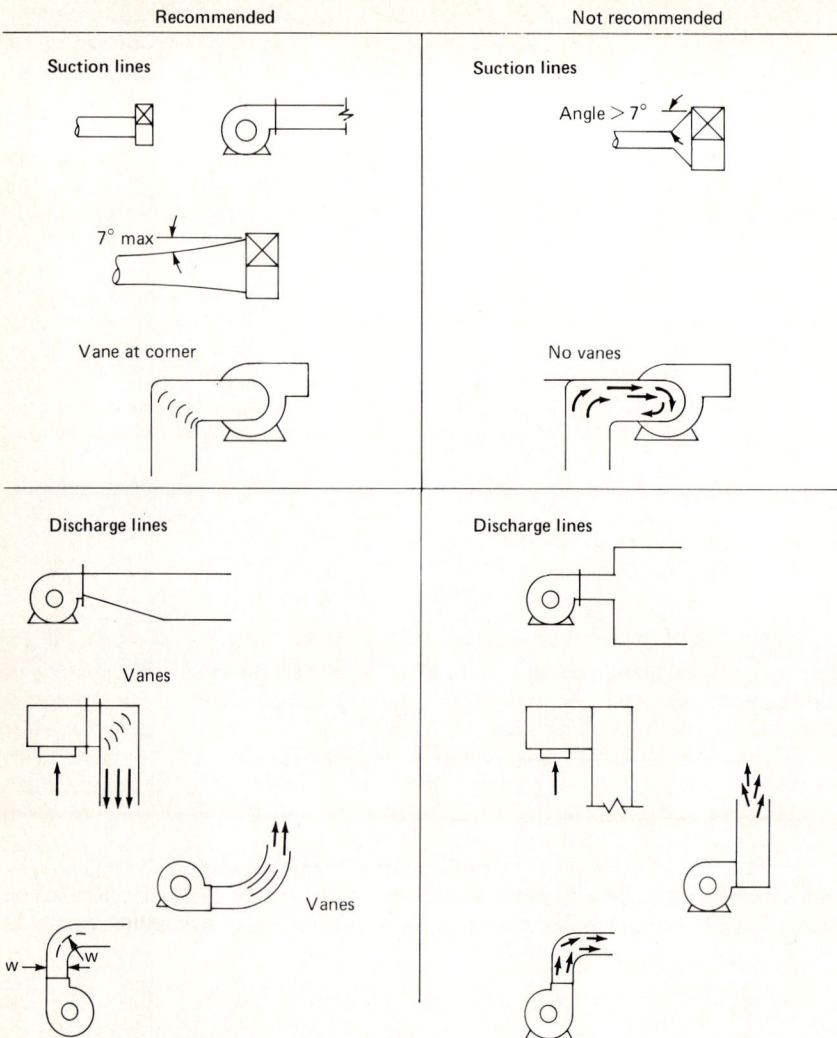

Figure 12.7 Right and wrong ways of laying out fans.

Blowers and Compressors

Blowers and compressors have more design similarities with each other than they do with fans, and it is for this reason that they were being discussed together. One factor which separates them from fans is the very narrow clearances between their housings and moving pressure-generating components. It is these close clearances which reduce blowback and produce their high pressures.

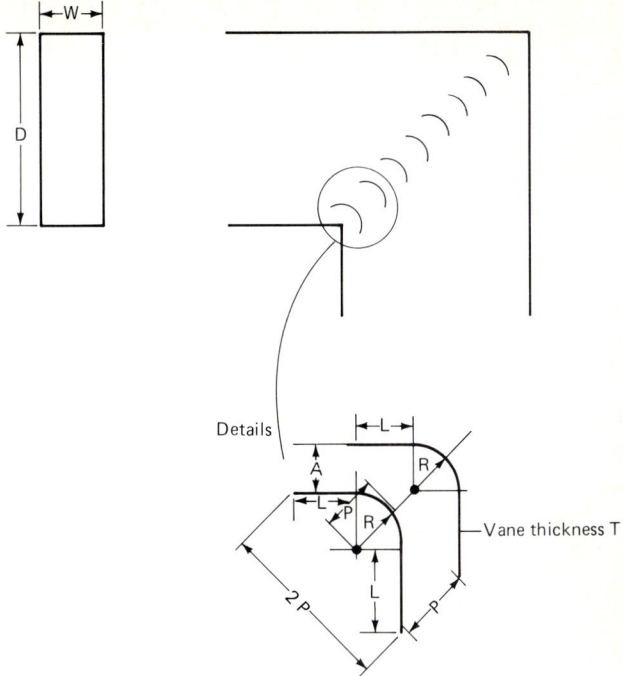

Figure 12.8 Airflow-vane installations.

Compressor mechanical design

Blowers and compressors fall into two general categories: centrifugal and positive-displacement units. The centrifugal units may be either radial-flow or axial-flow units, and the positive-displacement units may be either reciprocating or rotary units.

Characteristics of compressors. The performance curve produced by a compressor is dependent on the physical design of the machine. Reciprocating compressors are constant-speed machines. Therefore, these compressors can produce any discharge pressure for a given capacity fixed by the machine's speed. Curve 1 in Fig. 12.9 shows the performance of an ideal reciprocating compressor with no slip or clearance.

The mechanical design of a compressor requires some residual space between the head of a piston and the head of the compression chamber at the end of a compression cycle. This residual space is called "clearance." Clearance defines the maximum possible discharge pressure P_{max} for the compressor by

$$P_{max} = P_S \left(\frac{V_S}{V_C} \right)^n \qquad (12.9)$$

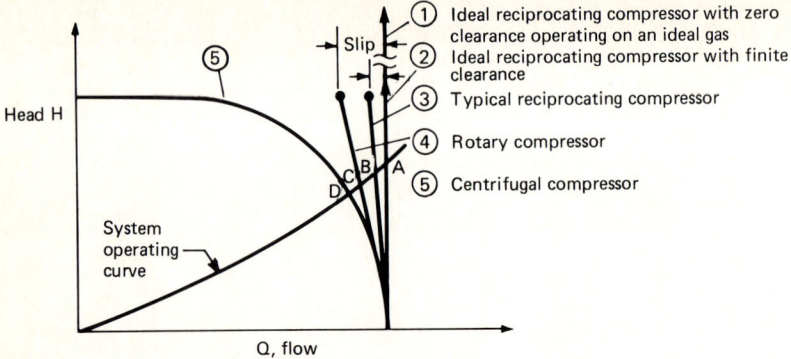

Figure 12.9 Generalized compressor operating curves.

where P_S = absolute pressure at suction
V_S = volume in compression chamber during the suction cycle
V_C = clearance volume
$n \leq k$, polytropic- or adiabatic-compression coefficient where $k = C_p/C_v$, the ratio of specific heats

The ideal compressor with a finite clearance is represented by curve 2 in Fig. 12.9.

Real reciprocating compressors also have finite clearances and thus leakage around valves and piston rings. The leakage causes a reduction in compressor capacity in proportion to the discharge pressure. The difference between ideal and actual compressor capacity is called "slip." Curve 3 in Fig. 12.9 shows the generalized performance curve for a real reciprocating compressor with slip.

Rotary compressors are not built to the close tolerances of reciprocating compressors. However, they behave similarly to reciprocating compressors but have greater slip. Their behavior is represented by curve 4 in Fig. 12.9.

Fans and centrifugal compressors use velocity to compress the gas and respond to the dynamics of the system in which they are installed. The performance of these machines roughly traces out a parabola as shown in curve 5 of Fig. 12.9. Since fans and centrifugal compressors are the most frequently encountered gas movers in a process plant, their operation will be discussed in greater detail.

Superimposed on Fig. 12.9 is a curve of a hypothetical system resistance. The system-compressor operating point is the intersection of the system with the compressor curve. Therefore, point A is the operating point for an ideal reciprocating compressor, point B for a real reciprocating compressor, point C for a rotary compressor, and point D for a centrifugal compressor. If system resistance increases, the system curve rises sharply, and if it decreases, the system curve becomes flatter.

There is a limited range of system resistances over which a centrifugal compressor can operate. As system resistance increases, the compressor-system operating point moves to the left in an orderly fashion. However, as the operating point moves up into a relatively flat portion of the compressor curve, operation becomes unstable. Flow and discharge pressure begins to pulsate rapidly. This phenomenon, called "surge," can cause machine vibration which may severely damage the compressor.

As the system resistance decreases, the compressor-system operating point moves to the right with an orderly increase in volumetric gas flow. However, gas flow through a compressor cannot increase without limit. The limit, called "stonewall," is caused by the choking of gas exiting from the compressor.

These two operating limits are represented on a generalized centrifugal-compressor curve in Fig. 12.10.

When choosing a fan or a centrifugal compressor, the process engineer should

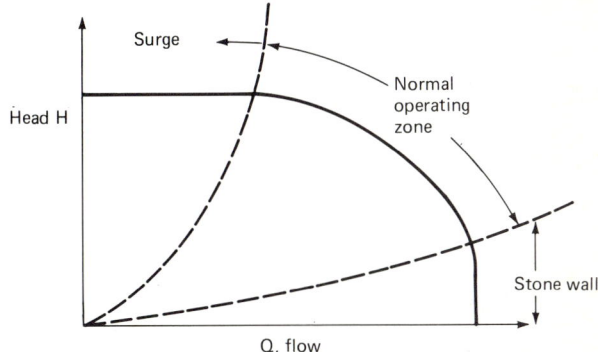

Figure 12.10 Characteristic operating zones for centrifugal compressors and fans.

ensure that the system-compressor curve-intersection point is well away from surge or stonewall for efficient operation.

Centrifugal compressors. Centrifugal compressors are the type most often chosen for process compression applications. They have a simple low-maintenance design which provides for continuous reliable service. Centrifugal compressors are available in two basic types, radial- and axial-flow units.

Radial-flow centrifugal compressors are more widely used for process applications because they can be employed in many harsh environments. The radial-flow centrifugal compressor has an open impeller which is less efficient than the closed impeller of an axial-flow compressor. The polytropic efficiency for a radial-flow centrifugal compressor ranges from about 77 percent at 200,000 ft^3/min down to 73 percent at 10,000 ft^3/min. Below 10,000 ft^3/min efficiency drops off markedly to about 65 percent at a capacity of 1000 ft^3/min. In this lower range, a rotary compressor may be considered as an alternative to a centrifugal compressor.

Axial-flow centrifugal compressors have closed or shrouded impellers. This allows a compressor to produce a high head at high efficiencies. A rough approximation of an axial-flow compressor's polytropic efficiency range is from 81 percent at 70,000 ft^3/min to about 83 percent at 600,000 ft^3/min. While axial-flow compressors can be operated below 70,000 ft^3/min, radial-flow compressors are usually preferred below this range. Axial-flow compressors are used in services handling clean noncorrosive gas because they are more susceptible to erosion, corrosion, and solids buildup than radial-flow centrifugals.

Rotary compressors. Rotary compressors are positive-displacement compressors which work on the principle of moving a fixed volume of gas by the rotating action of an internal rotor. They are classified according to the design of their rotating elements, being lobed, sliding-vane, rotary-screw, helical-screw, or liquid-ring units.

The general characteristics of the various rotary compressors can be found in Table 12.1. The discharge temperatures and horsepower requirements for these compressors, except for the liquid-ring type, can be calculated by assuming that adiabatic conditions exist. The liquid-ring unit is essentially isothermal at the liquid temperature.

Blowers and lobed compressors. The oldest and best-known units are the lobed compressors. These may have either two or three lobes and are commonly used as blowers. Lobed units have low capacities (2 to 20,000 ft^3/min) and generate discharge pressure in the range of 5 to 15 psig for air service. Some units have been designed to produce pressure up to 25 psig. These blowers and compressors are used for low-pressure services, such as pneumatic conveying, and can be employed as vacuum pumps.

Sliding-vane compressors. Sliding-vane rotary compressors have an eccentric rotor with slots containing vanes which slide in and out during rotation. The gas is trapped between the vanes and is compressed as the gas compartment rotates around the casing. The sliding vane allows the development of higher pressures than are generated by a lobed compressor. Sliding-vane compressors are low-capacity compressors with a maximum capacity of about 6000 ft^3/min. They can produce pressures up to about 50 psig in single-stage units and about 150 psig in two-stage units. Sliding-vane compressors are used in services similar to those of lobed compressors.

Helical-screw compressors. Helical-screw compressors consist of two screws, one male and one female, which compress the gas between the lobes of the screw and move it along their axes to an outlet port. This type of compressor can produce pressures up to about 60 psig in a single-stage unit and up to 400 psig in a multistage unit. These units may be either oil-flooded or dry. Dry compressors are used in the chemical-process industries where oil contamination is unacceptable. Oil-flooded units are used for plant air service and refrigeration.

Liquid-ring compressors. Liquid-ring compressors are used for vacuum service. Their operating characteristics are discussed in Chap. 13. Liquid-ring compressors have also been used to handle difficult gases such as chlorine, carbon disulfide, hydrogen sulfide, and sulfur dioxide. While the liquid ring is normally produced by water, other liquids can be used as a seal for special applications.

Reciprocating compressors. Reciprocating compressors are low-specific-speed compressors. Most reciprocating compressors used in process applications are of the piston type and operate at capacities of 3000 icfm (inlet cubic feet per minute) or less with

TABLE 12.1 Typical Operating Characteristics of Rotary Compressors

Type	Maximum capacity, icfm	P_D/P_S	Typical efficiency, %*
Lobed	30,000	2	68 (A)
Sliding-vane	6,000	2–4	72 (A)
Helical-screw			
Dry	20,000	1.5–4	75–85 (B)
Oil-cooled	3,500	3–15	55–80 (B)
Spiral axial	13,000	3	70 (A)
Liquid-ring	13,000	5	50–70 (B)

*A = overall efficiency; B = adiabatic efficiency.

discharge pressures up to 60,000 psig. Large-capacity compressors, however, are limited to pressures of about 3000 psig or less.

Diaphragm compressors are a type of reciprocating compressor which utilizes the reciprocating action of a flexible diaphragm in a compressor casing to pressurize a gas. Their capacities are limited to 200 icfm or less, and they can achieve discharge pressures up to 40,000 psia.

Reciprocating compressors, like other compressors, may be multistage units, depending on the required discharge pressure. The maximum compression ratio P_D/P_S per stage for a reciprocating compressor is limited to about 4. (See subsection "Performance Characteristics" below.)

The performance of a reciprocating compressor without interstage cooling is often assumed to approach adiabatic behavior. However, the frictional losses from compression-chamber valve openings increase the entropy of the existing gas. In the absence of test data, a compressor's adiabatic efficiency can be assumed to be about 80 percent.

Reciprocating compressors are constant-speed machines, in that they draw in the same volume of gas regardless of suction pressure, discharge pressure, or gas composition. This makes a reciprocating compressor practical for vacuum as well as pressure service.

Disadvantages of reciprocating compressors include high maintenance requirements compared with those of other compressors and pulsating discharge pressures due to the compressors' reciprocating action. Downtime due to maintenance can be reduced by having two or three half-capacity units in parallel. Pressure pulsation can be reduced by installing a surge tank on the compressor discharge. The volume of the surge tank should be at least 4 times the maximum displacement volume of the compression chamber of the compressor.

Compressor seals. Fans and compressors, like all rotating equipment, are susceptible to leakage along the shaft connecting them with their driver. Therefore, seals must be used to reduce or eliminate leakage. Four basic types of seals are used for compressors: labyrinth, restrictive-ring, liquid-film, and mechanical seals. Labyrinth seals (Fig. 12.11) and restrictive-ring seals (Fig. 12.12) are the least effective but are

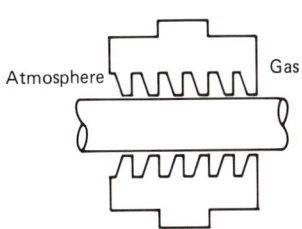

Figure 12.11 Labyrinth seal.

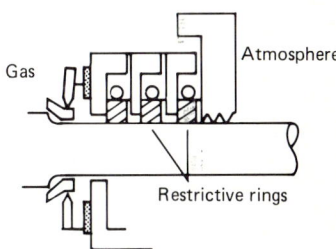

Figure 12.12 Restrictive-ring seal.

used where some leakage can be tolerated. Liquid-film seals (Fig. 12.13) use a liquid, usually an oil, to provide a liquid film around the seal to prevent leakage of hazardous gases or gases containing dust or other solids through the shaft's annular space. Compressor mechanical seals are essentially the same as those used on pumps and are explained in detail in Chap. 11. Mechanical seals are used for heavy gas service such as refrigerants and the higher-molecular-weight hydrocarbons.

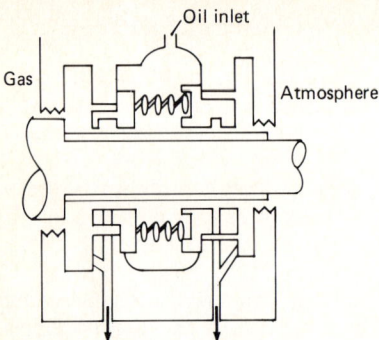

Figure 12.13 Liquid-film seal.

Performance characteristics

Blowers and compressors, as well as fans, have performance characteristics analogous to those of pumps. Their performance curves relate flow capacity to head. The primary difference is that gases are compressible and the head for blowers and compressors must be calculated differently from that of pumps.

The head developed by any fluid, liquid or gas, is derived from the general thermodynamic equation

$$h = \int_{P_S}^{P_D} \overline{V}\, dP \tag{12.10}$$

where h = head, ft of fluid or (ft·lbf)/lbm
P = pressure, lbf/ft^2
$\overline{V}$ = specific volume of the fluid, ft^3/lb
P_D = pump discharge pressure, lb/ft^2
P_S = pump suction pressure, lb/ft^2

Since liquids are essentially imcompressible, their specific volume may be considered as constant and the equation for liquid head h_L rearranges to

$$h_L = \overline{V} \int_{P_S}^{P_D} dP = \overline{V}(P_D - P_S) \tag{12.11}$$

The same equation can be used to estimate the differential head for gases in fans, since the change in specific volume of the gas is negligible. However, this is not the case for blowers and compressors because the pressures they develop cause the gas specific volume to change significantly. Therefore, the specific volume of the gas is dependent on pressure and must be integrated over the range from suction pressure to discharge pressure.

Adiabatic head. Ideally, a compressor can compress a gas either adiabatically or isothermally. In the adiabatic-compression process no heat is transferred into or out of the compressor. Therefore, an adiabatically compressed gas behaves according to the pressure-volume relationship:

$$PV^k = \text{constant} \tag{12.12a}$$

or by rearrangement

$$V = C_1/P^{1/k} \tag{12.12b}$$

where k = molal specific-heat ratio = $C_p/C_v = C_p/(C_v - R)$
$C_1 = VP^{1/k}$ = constant of proportionality
V = mole volume, ft³/(lb·mol)

V in Eq. (12.12b) can be substituted for $\overline{V}$ in Eq. (12.10) to give the adiabatic head h_a,

$$h_a = \int_{P_S}^{P_D} \left(\frac{C_1}{P^{1/k}}\right) dP \tag{12.13}$$

This equation on integration produces

$$h_a = C_1 \left(\frac{k}{k-1}\right) P_S^{(k-1)/k} \left[\left(\frac{P_D}{P_S}\right)^{(k-1)/k} - 1\right] \tag{12.14a}$$

From Eq. (12.12b) it is known that

$$V_S P_S^{1/k} = V_D P_D^{1/k} = C_1 \tag{12.14b}$$

Substituting this equation into Eq. (12.14a) to eliminate C gives

$$h_a = \left(\frac{k}{k-1}\right) P_S V_S \left[\left(\frac{P_D}{P_S}\right)^{(k-1)/k} - 1\right] \tag{12.15}$$

Using the gas-law relationship

$$P_S \overline{V}_S = Z_S RT_S/M \tag{12.16}$$

where Z_S = compressibility factor at suction
T_S = absolute temperature at suction, °R
M = molecular weight, lb/(lb·mol)
R = 1545 (ft·lbf)/(lb·mol·°R)

and substituting Eq. (12.16) into Eq. (12.15) give the equation used to calculate the adiabatic head of a compressor or blower:

$$h_a = \left(\frac{k}{k-1}\right) \frac{Z_S RT_S}{M} \left[\left(\frac{P_D}{P_S}\right)^{(k-1)/k} - 1\right] \quad \text{(ft·lbf)/lbm of gas} \tag{12.17}$$

Isothermal head. If all the heat of compression could be removed and the gas could be maintained at constant temperature, the pressure-volume relationship for isothermal compression would be based on

$$PV = C_2 \tag{12.18}$$

Substituting Eq. (12.18) into Eq. (12.10) and integrating the latter produce the isothermal head h_i equation:

$$h_i = \frac{Z_S RT_S}{M} \ln \frac{P_D}{P_S} \quad \text{(ft·lbf)/lbm of gas} \tag{12.19}$$

If the compressibility factor Z for the gas at discharge conditions is significantly different from that of the suction, then the average compressibility factor $\overline{Z}$ is used to calculate head; i.e.,

$$\overline{Z} = \frac{Z_S + Z_D}{2} \qquad (12.20)$$

as long as the reduced discharge pressure is less than 0.8.

There is always some heat exchange with the environment, so that a fan, blower, or compressor operates at less than adiabatic conditions. In addition, although a compressor may be completely water-jacketed or have a number of interstage coolers and an aftercooler, it never achieves true isothermal conditions. Therefore, both the adiabatic head and the isothermal head are ideal limits that are never reached in practice.

Because adiabatic and isothermal compession are ideal limits, real compression processes operate somewhere between these two limits. Actual compression processes are called polytropic processes because the gas being compressed is not at constant entropy as in the adiabatic process or at constant temperature as in the isothermal processes. The pressure-volume relationship for a polytropic process follows the equation

$$PV^n = C_3 \qquad (12.21)$$

This relationship when substituted into Eq. (12.10) produces the following polytropic-head equation, which is similar to Eq. (12.17):

$$h_p = \left(\frac{n}{n-1}\right) \frac{Z_S R T_S}{M} \left[\left(\frac{P_D}{P_S}\right)^{(n-1)/n} - 1\right] \quad \text{(ft·lbf)/lb of gas} \qquad (12.22)$$

The value of the exponent n lies somewhere between unity and k, the specific-heat ratio. Its value cannot be calculated but must be determined empirically. As a general rule most high-speed fans, blowers, and compressors without interstage coolers or aftercoolers have n values approaching k, while those gas movers with coolers have n values approaching unity. The actual n values for a gas mover must be obtained from the manufacturer of the unit.

Compressor horsepower. The calculation of fan, blower, or compressor horsepower is similar to the horsepower calculations for a pump. The formula used is

$$\mathcal{P} = Wh/33{,}000\, E \qquad (12.23)$$

where $\mathcal{P}$ = gas horsepower, hp
W = gas mass flow rate, lbm/min
h = gas head, (ft·lbf)/lb of gas
E = compressor efficiency

The horsepower calculated by this formula is based on the head for the type of compression process that the unit is assumed to follow, i.e., adiabatic head h_a, isothermal head h_i, or polytropic head h_p.

Mass flow rate. The data and performance curves from a manufacturer list a gas mover's capacity in volumetric units at inlet conditions, such as inlet cubic feet per minute, cubic feet per minute, or cubic feet per second. The gas volumetric flow, unless otherwise specified, is usually based on air at 14.7 psia and 60°F. Therefore, the actual mass flow is calculated by

$$W = Q \left[\frac{PM}{1545ZT} \right] \quad \text{lb/min} \tag{12.24}$$

where Q = volumetric flow rate at inlet conditions, icfm.

If the performance data are based on air at standard conditions but the compressor service is actually used for a different gas at different inlet conditions, then the actual volumetric flow shown in Eq. (12.24) must be calculated as follows:

$$Q = Q_{air} \left(\frac{P_{std}}{P_{actual}} \right) \left(\frac{M_{air}}{M_{actual}} \right) \left(\frac{Z_{actual}}{Z_{std}} \right) \left(\frac{T_{actual}}{T_{std}} \right) \tag{12.25}$$

Efficiency. Various efficiency relationships can be used to describe the performance of a fan, blower, or compressor, depending on the unit's geometry, its compression process, or the degree of energy conversion. Listed below are the various efficiency relationships used to describe compressor performance.

Volumetric efficiency. Volumetric efficiency is used for reciprocating compressors. The compression chamber of a reciprocating compressor has a maximum and a minimum volume during each compression cycle. Ideally the minimum volume is zero; however, an actual compressor has a finite minimum volume called clearance. This clearance reduces the net compressor throughput. The volumetric efficiency E_v is defined as

$$E_v = \frac{\text{inlet volume flow per cycle}}{\text{maximum volumetric piston displacement per cycle}} \tag{12.26}$$

Volumetric efficiency is also affected by "blowby," slippage, and leakage around valves, piston rings, and seals. These have a smaller effect than clearance, and their cumulative effect can only be determined by a performance test.

Mechanical efficiency. Mechanical efficiency measures the conversion of mechanical energy provided by the driver to the energy received by the gas. The difference between these two energies is lost in shaft friction and motor slip and is measured by efficiency E_m as defined by

$$E_m = \frac{\text{horsepower delivered to the gas}}{\text{brake horsepower of the driver}} \tag{12.27}$$

Typical mechanical efficiencies run between 90 and 95 percent.

Adiabatic efficiency. Adiabatic efficiency E_a is the ratio of theoretical adiabatic horsepower (100 percent efficiency) to actual horsepower delivered to the gas. The theoretical isentropic horsepower is calculated from Eq. (12.23) with $E = 1$. Adiabatic efficiency is expressed as

$$E_a = \frac{\text{theoretical adiabatic horsepower}}{\text{actual horsepower delivered to the gas}} \tag{12.28}$$

Isothermal efficiency. Isothermal efficiency E_I is the ratio of theoretical isothermal horsepower to actual delivered horsepower and is expressed as

$$E_I = \frac{\text{theoretical isothermal horsepower}}{\text{actual horsepower delivered to the gas}} \tag{12.29}$$

Polytropic efficiency. Polytropic efficiency is used to compare adiabatic with polytropic performance. Polytropic efficiency E_p is expressed as

$$E_p = \left(\frac{k-1}{k}\right) \bigg/ \left(\frac{n-1}{n}\right) \tag{12.30}$$

Overall efficiencies. When adiabatic, isothermal, and polytropic efficiencies are substituted into Eq. (12.23), the horsepower calculated is the actual horsepower delivered to the gas; i.e.,

$$\text{Adiabatic: } \mathcal{P}_a = \frac{Wh_a}{33,000\, E_a} \tag{12.31}$$

$$\text{Isothermal: } \mathcal{P}_i = \frac{Wh_I}{33,000\, E_I} \tag{12.32}$$

$$\text{Polytropic: } \mathcal{P}_p = \frac{Wh_p}{33,000\, E_p} \tag{12.33}$$

Note that adiabatic head must be used with adiabatic efficiency, isothermal head with isothermal efficiency, and polytropic head with polytropic efficiency to obtain the correct horsepower.

To get the brake horsepower for the driver, the overall efficiencies must be determined. These are the product of the compression process efficiency and the mechanical efficiency. The overall efficiencies for each process are calculated by

$$\text{Adiabatic: } E_{ao} = E_a E_m \tag{12.34}$$

$$\text{Isothermal: } E_{Io} = E_I E_m \tag{12.35}$$

$$\text{Polytropic: } E_{po} = E_p E_m \tag{12.36}$$

These efficiencies when substituted into Eq. (12.23) will calculate the brake horsepower of the driver.

Specific speed. All centrifugal pumps and compressors have an index number which characterizes their performances. This is termed the "specific speed N_s," and is defined as

$$N_s = \frac{N\sqrt{Q}}{h^{3/4}} \tag{12.37}$$

where N = actual speed, r/min
Q = capacity, ft^3/s
h = head (h_a, h_i, or h_p), (ft·lbf)/lb of fluid

In addition to specific speed, another dimensionless number is used to characterize pump and compressor performance. Called the "specific diameter D_s," it is defined as

$$D_s = \frac{Dh^{1/4}}{\sqrt{Q}} \tag{12.38}$$

where D = impeller diameter, ft.

These two parameters can be used to estimate the performance of a centrifugal gas mover for any change in speed, head, capacity, and diameter. Balje[2] has developed a chart of specific speed versus specific diameter which can be used to estimate performance and select single-stage compressors and pumps. This chart has been reproduced in Fig. 12.14.

The specific-speed equation appears to indicate that the head of a centrifugal compressor can be increased without limit by simply increasing its actual speed. This is

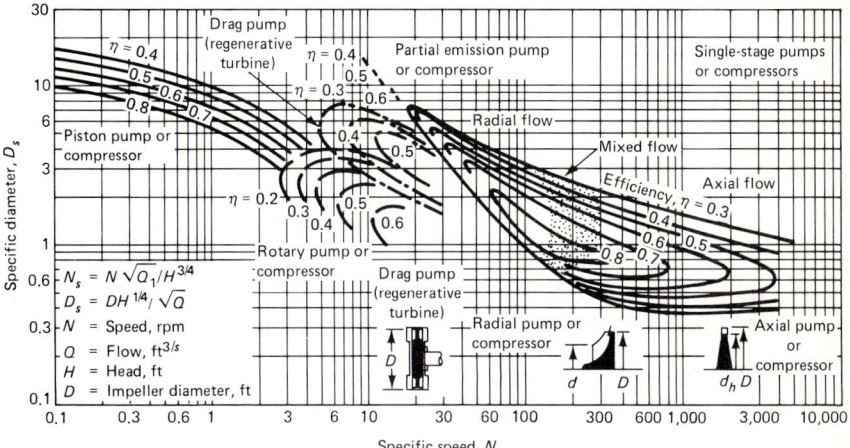

Figure 12.14 Specific-speed versus specific-diameter curve for single-stage pumps and compressors. (η = efficiency, E) (O. E. Balje, "A Study on Design Criteria and Matching of Turbomachines: Part B," Transactions of the American Society of Mechanical Engineers, Journal of Engineering for Power, *vol. 84, January 1962.*)

not true because the laws governing compressible-fluid flow put a finite limit on the capacity of a compressor. As the speed of a centrifugal compressor increases, the linear velocity of the gas exiting from the compressor also increases. When the gas velocity reaches the speed of sound in the gas, choking occurs and the gas head drops off. This results in stonewall, an effect previously discussed.

The speed of sound c for a gas can be calculated by

$$c = \sqrt{\frac{1545 g_c kZT}{M}} \quad \text{ft/s} \tag{12.39}$$

where k, Z, and T are determined at the discharge conditions of the compressor.

Multistage gas movers. A single-stage compressor may not be efficient enough or even practical to use in applications requiring high-pressure discharge. Therefore, a multistage unit must be selected. While the design of multistage compressors is beyond the scope of this book, a general rule for estimating the approximate pressure ratio in a multistage compressor is

$$(P_D/P_S) \text{ per stage} \cong \sqrt[N]{(P_D/P_S)_{\text{total}}} \tag{12.40}$$

where N = number of stages. It should be noted that there is a practical limit to the pressure ratio per stage depending on the compressor type. Typical ratios for different multistage compressors are:

Centrifugal compressors   1.2–1.4 per stage
Screw compressors   3–15 per stage

Interstage cooling and aftercooling. The temperature rise per compressor stage should normally be less than 250°F, with the maximum discharge temperature being less than 400°F. However, some axial-flow centrifugal compressors can tolerate discharge temperatures of up to 600°F. If higher temperatures are expected as a result of compression, interstage cooling is required.

Cooling removes the heat of compression. This allows the compressor to approach isothermal operation, which also reduces the brake horsepower required by the compressor. Cooling can be accomplished between stages in a multistage compressor with intercoolers and with an aftercooler at the compressor discharge. However, aftercooling and interstage cooling are not always desirable. If the gas must be heated prior to entering a reactor, downstream heaters can be designed for a lower duty if the heat of compression is not removed.

If a compressor with intercoolers is used to compress gases with condensable vapors, the combined effect of compression and cooling can condense out liquid. This liquid must be removed; otherwise, it could cause damage downstream of the compressor. Therefore, a knockout pot, as discussed in Chap. 3, must be added to collect and remove condensate after the intercooler.

Multicomponent-gas streams. When designing a gas mover for a gas mixture, the properties of the mixture must be estimated. The general procedure for estimating the properties of a gas mixture is to use the weighted molal average of the property. Some of the properties that must be estimated are:

1. Molecular weight $\quad M_{\text{mixture}} = \sum_{i=1}^{i=N} y_i M_i \quad$ (12.41)

2. Reduced pressure $\quad P_{r,\text{mixture}} = \sum_{i=1}^{i=N} y_i P_{r,i} \quad$ (12.42)

3. Reduced temperature $\quad T_{r,\text{mixture}} = \sum_{i=1}^{i=N} y_i T_{r,i} \quad$ (12.43)

4. Specific heat $\quad C_{p,\text{mixture}} = \sum_{i=1}^{i=N} y_i C_{p,i} \quad$ (12.44)

5. $k_{\text{mixture}} = C_{p,\text{mixture}}/(C_{p,\text{mixture}} - R)$ (12.45)

6. Compressibility factor $Z_{\text{mixture}} = f(T_r, P_r)$ for the mixture (12.46)

where y_i = mole fraction of component i.

Rerating gas-mover performance. Most gas movers, unless otherwise specified by the manufacturer, have their performance curves based on air at some standard conditions, usually 60°F, 14.7 psia. If the inlet gas to a fan, blower, or compressor is at other than standard conditions and/or is other than air, then the gas mover's performance curve must be rerated to those different conditions. The following transposition methods are valid for single-stage fans and centrifugal compressors. Analyses for other types of compressors are more complex, and the following calculation methods may give only an order-of-magnitude estimate of the machine's performance.

The transposition methods assume that the fan affinity laws, similar to the pump affinity laws given in Chap. 11, apply; i.e.,

1. The volumetric flow Q is directly proportional to the rotational speed N:

$$Q \propto N$$

2. The head h is proportional to the square of the rotational speed of the unit:

$$h \propto N^2$$

3. The power $\mathcal{P}$ required by the gas mover is proportional to the cube of the rotational speed:

$$\mathcal{P} \propto N^3$$

4. The specific-speed equation (12.36) is valid, so that conditions with equal specific speeds have the same efficiencies.
5. The adiabatic-head equation (12.17) is valid.

Variation in inlet temperature only. If the inlet gas temperature T_s is the only parameter to change from standard conditions $T_{s,\text{std}}$, then discharge pressure and power requirements change in accordance with the following equations:

$$P_D = P_{S,\text{std}} \left\{ \left(\frac{T_{S,\text{std}}}{T_S} \right) \left[\left(\frac{P_{D,\text{std}}}{P_{S,\text{std}}} \right)^{(k-1)/k} - 1 \right] \right\}^{k/(k-1)} \tag{12.47}$$

$$\mathcal{P} = \left(\frac{T_{S,\text{std}}}{T_S} \right) \mathcal{P}_{\text{std}} \tag{12.48}$$

Variation in inlet pressure only. If only the inlet pressure P_s of the gas changes and the discharge pressure remains constant, then mass flow rate, adiabatic head, and power change. The equation for adiabatic head is

$$h_a = \left(\frac{K}{K-1} \right) \frac{Z_S R T_S}{M} \left[\left(\frac{P_D}{P_{S,\text{std}}} \right)^{(k-1)/k} - 1 \right] \tag{12.49}$$

A greater inlet pressure increases the mass flow rate while maintaining the same volumetric flow rate through the compressor. The new mass flow is calculated by

$$W = \frac{Q P_s M}{Z R T_S} \tag{12.50}$$

and the horsepower would be calculated directly by Eq. (12.23).

If the pressure ratio remains constant and there is a change in suction pressure, then the adiabatic head would be constant but the mass flow rate would change as calculated by Eq. (12.50) and the power requirement would change in accordance with the following relationship:

$$\mathcal{P} = \mathcal{P}_{\text{std}} \left(\frac{P_S}{P_{S,\text{std}}} \right) \tag{12.51}$$

Change in molecular weight only. A molecular weight for a gas differing from that of air will affect the pressure ratio, mass flow rate, and power requirements of the gas mover. The gas ratio can be calculated by

$$\frac{P_D}{P_S} = \left\{ \left(\frac{M}{M_{\text{std}}} \right) \left[\left(\frac{P_{D,\text{std}}}{P_{S,\text{std}}} \right)^{(k-1)/k} - 1 \right] + 1 \right\}^{k/(k-1)} \tag{12.52}$$

The different mass flow rate can be calculated directly with Eq. (12.50), and the revised power consumption can be calculated by

$$\mathcal{P} = \left(\frac{M}{M_{std}}\right)\mathcal{P}_{std} \qquad (12.53)$$

Combined effects of nonstandard gas flow. Each of the above variations in gas conditions can occur in various combinations. In addition, there could be a difference in the compressibility factor Z or in the specific-heat ratio. For a compressor operating at a constant speed and thus having a given volumetric flow rate and head, the combined effect of all these terms on pressure ratio can be found by

$$\left(\frac{P_D}{P_S}\right) = \left\{\left(\frac{T_{S,std}}{T_S}\right)\left(\frac{Z_{std}}{Z}\right)\left(\frac{M}{M_{std}}\right)\left(\frac{k}{k-1}\right)_{std}\left(\frac{k-1}{k}\right)\left[\left(\frac{P_{D,std}}{P_{S,std}}\right)^{[(k-1)/k]_{std}} - 1\right] + 1\right\}^{k/(k-1)} \qquad (12.54)$$

The change in mass flow rate can be determined from Eq. (12.50), and power consumption can be calculated by

$$\mathcal{P} = \left(\frac{P_S}{P_{S,std}}\right)\left(\frac{T_{S,std}}{T_S}\right)\left(\frac{Z_{std}}{Z}\right)\left(\frac{M}{M_{std}}\right)\mathcal{P}_{std} \qquad (12.55)$$

Surge control

All centrifugal compressors and fans can go into surge if flow drops below the stable range (see Fig. 12.10). Surge can be remedied by installing a controlled-recycle line to recycle a portion of the discharge to the suction. The type of controls required depends on the variations in gas properties that can be expected in normal operation.

The simplest control system consists of a flow switch which opens a control valve in a recycle line on a low-flow signal. This scheme provides an on-off type of operation and is used only in small-capacity systems. A method which provides smoother control uses a flow transmitter to sense flow from the compressor. As the flow drops, the control valve in the recycle line opens in proportion to the flow reduction to the compressor system, thus providing sufficient recycle gas to prevent the unit from going into surge.

If the incoming gas varies significantly in temperature, a temperature-monitoring system must be added to the loop to correct the flow signal going to the controller.

Example 12.1 A centrifugal compressor must be purchased for a small refinery. It will be used to transfer the following vapor stream from gas processing to a fine-chemicals reactor:

Compound	Mole %
Hydrogen	30
Methane	45
Ethane	15
Propane	7
n-Butane	3
	100

The vapor at 200°F and 20 psia is to be used at the rate of 2000 icfm to the compressor. The required discharge pressure is estimated to be 80 psia. Assuming adiabatic compression, determine the mass flow rate, the adiabatic head, and the compressor horsepower by assuming an overall efficiency of 68 percent.

solution

a. The properties of the vapor mixture must be calculated by using Eqs. (12.41) through (12.46), and the physical properties for the pure compounds are:

Compound	Molecular weight	Critical properties		Specific heat, Btu/(lb·°F)
		Temperature, °F	Pressure, psia	
Hydrogen	2.01	−399.8	188.1	3.408
Methane	16.04	−115.8	673.1	0.5271
Ethane	30.07	90.3	709.8	0.4097
Propane	44.09	206.3	617.4	0.33885
n-Butane	58.12	305.6	550.7	0.3908

The mixture properties calculated are:

Average molecular weight = 17.16

Critical temperature = −134.9°F

Critical pressure = 525 psia

Mixture specific heat = 9.57 Btu/(lb·mol·°F)

Mixture specific-heat ratio = 9.57/(9.57 − 1.98) = 1.26

b. Calculate the mixture's compressibility factor from the mixture's reduced temperature and pressures:

Reduced pressure, suction = (20/525) = 0.308

Reduced pressure, discharge = (80/525) = 0.152

Reduced temperature, suction = (200 + 460)/(−134.9 + 460) = 2.03

From the compressibility-factor charts in Fig. 8.2, the estimated compressibility factor approaches unity; so the mixture can be treated as an ideal gas.

c. Calculate mass flow rate from Eq. (12.24). It is 96.9 lb/min.

d. Calculate adiabatic head from Eq. (12.17). It is 95,400 (ft·lbf)/lbm of gas.

e. Calculate brake horsepower by using Eq. (12.23). It is 412 hp. Motors in this range are available in increments of 100 hp in this size range; therefore, a 500-hp motor should be selected.

Example 12.2 In response to the compressor required in Example 12.1, a seller of used process equipment proposes to sell the refinery a centrifugal compressor which can provide 2000 icfm and a discharge pressure of 100 psig based on compressing air at 60°F from 1 atm. Should the refinery purchase this compressor?

solution

a. If constant speed is assumed, the compressor knows only that it must move 2000 icfm through itself. The resulting equivalent mass flow of air is 152.8 lb/min by using Eq. (12.24).

b. From Eq. (12.49) the adiabatic head for the compressor is calculated to be 77,431 (ft·lbf)/lb of air.

c. The adiabatic head at 2000 icfm for this compressor is lower than the specified head of 95,400 (ft·lbf)/lb of gas calculated in Example 12.1. If the specified conditions are treated as standard conditions in Eq. (12.54) and the air operating conditions for the proposed compressor are treated as nonstandard, then the required air-pressure ratio should be 10.0 and the discharge pressure should be 161.7 psia, or 147 psig.

The proposed compressor should have at least a 147-psig discharge pressure with air. Since it can provide only a 100-psig discharge pressure, it should not be considered for purchase.

REFERENCES

1. *Fluid Movers; Pumps, Compressors, Fans and Blowers,* ed. by Jay Matley and the staff of *Chemical Engineering,* McGraw-Hill Publications Company, New York, 1979.
2. O. E. Balje, "A Study on Design Criteria and Matching of Turbomachines: Part B," *Transactions of the American Society of Mechanical Engineers, Journal of Engineering for Power,* January 1962.

BIBLIOGRAPHY

Dimoplan, William: "What Process Engineers Need to Know about Compressors," *Hydrocarbon Processing,* May 1978, pp. 221–227.

Lapina, Ronald P.: *Process Compressor Technology,* vol. 1: *Estimating Centrifugal Compressor Performance,* Gulf Publishing Company, Houston, 1982.

Pollak, Robert: "Selecting Fans and Blowers," *Chemical Engineering,* Jan. 22, 1973, pp. 86–100.

Power's Handbook on Fans: Centrifugal, Propeller, Tubeaxial, Vaneaxial, ed. by Tyler Hicks, McGraw-Hill Book Company, 1951, pp. 88–102.

Scheel, Lyman F.: *Gas Machinery,* Gulf Publishing Company, Houston, 1972.

White, M. H.: "Surge Control for Centrifugal Compressors," *Chemical Engineering,* Dec. 25, 1972, pp. 54–62.

Chapter 13

Vacuum Equipment

The expression "vacuum" is generally understood to be pressures below 760 mmHg abs., but it is actually a relative term. It is expressed in pressure units and often compared to atmospheric pressure. It should be recalled that atmospheric pressure at any point on earth is the sum total of the weight exerted by all the gaseous molecules which are within the boundaries of a column that originates at the center of the earth, passes perpendicularly through a unit area on the surface of the earth's sphere, and then extends as far as the earth's gravitational field can capture and hold the gas molecules in outer space (see Fig. 13.1).

At mean sea level there are 14.67 lb of gas above a square inch of the earth's surface at mean sea level. If the area is taken as a square centimeter, the gas above it would be 1.0333 kg. Thus, 1 atm taken at sea level is 14.67 psia (pounds per square inch absolute) and also 1.0333 kg/cm^2 (kilograms per square centimeter absolute). Pressure, in addition to weight or force per unit area, is also expressed in terms of height of a column of fluid that has the same weight per unit area as does the atmosphere. If the fluid is water at 60°F, its density is 62.34 lb/ft^3, or 0.0361 lb/in^3. Thus, a column of water 407.2 in or 33.93 ft high is equivalent to 1 atm. Similarly, if the fluid is mercury, whose specific gravity at 0°C is 13.6, or 0.0136 kg/cm^2, the column would be 76.0 cm high at mean sea level. This latter value is given as 760 mmHg abs. or 29.92 inHg abs. for atmospheric pressure at mean sea level.

Many consider pressures below 760 mmHg to be vacuum. However, high-altitude sites such as Denver at 5280 ft and Mexico City at 7415 ft have atmospheric pressures of about 610 and 560 mmHg abs., respectively. Sherpas in the Himalayas live and work at altitudes of 16,000 ft where the pressure is less than 400 mmHg.

Differences in pressures are due to variations in the density of the atmosphere with altitude. The density changes in direct proportion to the hydrostatic weight of the gas column above it as shown in Fig. 13.1. The pressure or weight per unit area can be expressed as a function of height above sea level in accordance with Eq. (13.1)* for

---

*Based on data from ARDC Model Atmosphere 1959, AFCRC TR-59-267.

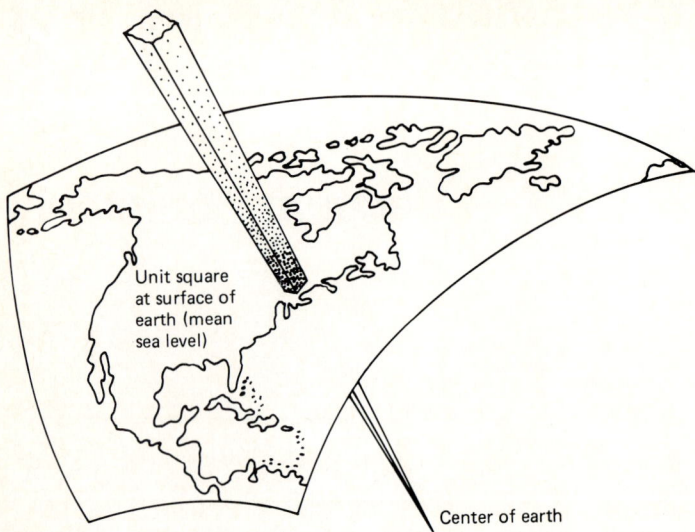

Figure 13.1 Atmospheric pressure.

altitudes up to about 80 mi above the earth.

$$\log P = \log 760 - 0.094H \qquad (13.1a)$$

or $\quad \log P = \log 760 - 1.77 \times 10^{-5}h \qquad (13.1b)$

where P = atmospheric pressure, mmHg abs
H = altitude, mi
h = altitude, ft

Atmospheric pressure gradually changes from 2.3×10^{-5} mmHg abs. at 80 mi above the earth's surface to about 9×10^{-7} mmHg at 100 mi. Above an altitude of 100 mi, the rate of change for pressure with height is less, so that between 100- and 430-mi altitude absolute pressure is expressed as

$$\log P = -(5.2 + 0.00848H) \qquad (13.2)\star$$

Beyond 430 mi, where the pressure is $10^{-8.85}$ mmHg, the pressure slowly declines to a value which approaches 10^{-11} mmHg for outer space.

Table 13.1 presents pressures as a function of miles and feet above mean sea level for several significant points relative to the earth's surface.

Although "vacuum" conditions, i.e., below 760 mmHg, exist at altitudes above mean sea level, the concept of vacuum in process engineering refers to subatmospheric pressure on earth which is achieved by some artificial means. Use of low pressure in space, although natural, for experimental purposes may be considered as vacuum since artificial means such as the space shuttles are required to put the experiment in orbit. For convenience of discussion, degrees of vacuum are classified according to pressure range. Table 13.2 summarizes the ranges as well as typical uses.

★Ibid.

TABLE 13.1 Pressure at Heights above Mean Sea Level

Height			Pressure		
mi	ft	Milestone	psia	mmHg	µHg
(−0.247)	(−1,302)	Lowest point on earth	15.5	801	
0	0	Mean sea level	14.7	760	
5.5	29,028	Highest point on earth	4.5	230	
6.9	....	Beginning of stratosphere	3.3	170	
23.4	123,800	Highest manned balloon ascent	0.1	5	
50	....	....	3×10^{-4}	0.016	16
100	....	....	....	9.0×10^{-7}	9.0×10^{-4}
200	....	....	....	1.3×10^{-7}	1.3×10^{-4}
210	....	Satellite orbits	....	1.0×10^{-7}	1×10^{-4}
500	....	....	....	3.6×10^{-10}	3.6×10^{-7}
...	....	Outerspace	....	1×10^{-11}	1×10^{-8}

TABLE 13.2 Vacuum Ranges*

Designation	Pressure range	Uses
Low vacuum	Atm to 50 torr	Pneumatic conveying Material handling by means of suction cups Filtration Shaping by deformation Chemical reactions to completion Distillations of thermal-sensitive materials
Medium vacuum	50 to 2.5×10^{-3} torr (2.5μ)	Chemical reactions to completion Distillation of thermal-sensitive materials Refrigeration Freeze drying Eliminating unwanted reactants Impregnation of porous substances Insulating layer
High vacuum	2.5×10^{-3} to 10^{-6} torr (2.5 to $10^{-3} \mu$)	Metal evaporation and coating Evacuation of electronic equipment and particle accelerators
Very high vacuum	10^{-6} to 10^{-9} torr (10^{-3} to $10^{-6} \mu$)	Special electronic tubes and accelerators Semiconductor-material-surface studies Environmental testing
Ultrahigh vacuum	10^{-9} torr and below ($< 10^{-6} \mu$)	Fusion reactors for power (outer space considered as $10^{-8} \mu$)

*Partial sources: R. R. LaPelle, *Practical Vacuum Systems*, McGraw-Hill Book Company, New York, 1972; H. A. Steinberg, *Handbook of High Vacuum Engineering*, Reinhold Publishing Company, New York, 1963.

A measure of vacuum often used is the torr,* which is equal to 1 mmHg abs. For high-vacuum work, the unit micron (μ), or 10^{-3} torr, is often used.

The degree of vacuum, as referred to in Table 13.2, is an inverse function of absolute pressure. Thus, the lower the absolute pressure, the "higher" the vacuum. Sometimes gauge pressures are used to describe vacuums as "2.5 psig vacuum," being actually 12.2 psia. Often reference to vacuum in terms of inches of mercury means that value below atmospheric such that 28.0 inHg vacuum is actually 1.92 inHg abs. The use of "gauge" values can be confusing at times, while the use of absolute values is more definite. The convention used in this chapter and book is to refer to all vacuums in terms of absolute pressure unless otherwise noted.†

Most process engineers for chemical plants, refineries, food-processing facilities, and boiler installations are primarily concerned with low- and medium-range vacuums. Low-range vacuum is also referred to as rough or crude vacuum, while medium-range vacuum is sometimes called fine or industrial vacuum. This chapter presents the methods used to create vacuums for such applications as pneumatic conveying, removing product gases from a reacting liquid, vacuum distillation, evacuating batch equipment before it is charged, freeze drying of food and other materials, and chilling water or other liquids. A review is also made of the methods used to produce vacuums in the high range, which are used in specialized metallurgy and semiconductor technology.

Equipment for Producing Vacuums

Vacuum equipment normally included in process engineering operations should be distinguished from that which is thought of as high-vacuum for specialized purposes or scientific investigation. High vacuums include environments at pressures less than 10^{-3} mmHg, i.e., $< 1\mu$, and are seldom utilized in normal plant operations. The vacuum conditions encountered in industrial situations are usually 1 mmHg abs. or greater.

High-vacuum equipment

Although high-vacuum conditions are normally not required for the operations usually encountered in the facilities with which most process engineers are associated, they are frequently used in analytical laboratories servicing plants. In addition, special processing conditions might require a process atmosphere less than 10^{-3} mmHg. Therefore, process engineers should be acquainted with the equipment needed to produce these vacuums.

Cryogenic panels. Cryogenic panels are used to produce ultrahigh vacuums. Arrays of panels are placed in the chamber whose pressure is to be lowered. The pressure in the chamber is reduced as low as possible (to about 10^{-3} torr) by using standard mechanical or steam-jet equipment, to be discussed later. The remainder of pressure reduction is accomplished by passing a cryogenic liquid through the panels. If the

---

*Named after Evangelista Torricelli, a friend of Galileo, who discovered the principle of the barometer in 1643 and correctly measured atmospheric pressure in terms of mercury.

†See discussion in the subsection "Line Tabulations" in Chap. 3 regarding reporting of slight negative gauge pressures in the pressure column of line tabulations.

cryogenic temperature is less than 50 K (90°R), most gases encountered in normal vacuum situations will solidify* when they strike the panel and a portion of them will stick to the panel. The pressure in the chamber, with no leakage from the outside environment, could become the vapor pressure of the solidified gas at its equilibrium temperature on the panels. Liquid nitrogen, hydrogen, or helium is used as the cryogenic fluid in the panels. With liquid helium as low as 4.3 to 5 K and a minimum of leakage, vacuums to 5×10^{-8} torr are attained. Under special conditions and precautions, vacuums less than 10^{-10} torr are possible.

Auxiliary to the cryogenic panels are the cryogenic apparatus to remove the heat absorbed by the circulating cryogenic liquid and the ancillary vacuum equipment needed for initial drawdown of pressure in the vacuum chamber. The capacity of an installation is limited only by the area of the cryogenic panels that can be installed in the chamber.

Molecular pumps. Molecular pumps are used to evacuate or maintain a vacuum in relatively small chambers and can attain vacuums down to 5×10^{-10} torr. The main components are high-speed disks, rotating at up to 16,000 r/min, which are positioned closely between two spirals. Molecules of gases from the vacuum chamber enter the center of the spiral. Momentum is imparted to them as they contact the rotating disk, and they are impelled around the spiral and compressed. The gases can leave the molecular pump at atmospheric pressure or go to an auxiliary vacuum pump to be boosted to atmospheric pressure. Compression ratios for air are as high as $5 \times 10^7:1$, and capacities of 300 ft^3/min at 10^{-2} torr and lower pressures are common. Cooling coils are used to remove the heat of compression.

Diffusion pumps. Diffusion pumps, like molecular pumps, employ momentum or impulse to remove gas from a confined space in order to create a high vacuum. In this case, the motivating force comes from the molecules of mercury or from other high-molecular-weight vapors being ejected from a series of annular venturi jets which accelerate the vapors to supersonic velocities. The top annulus is near the pump inlet, and the path of the emerging vapors is directed obliquely away from the inlet toward the next annulus or outlet. As gas molecules from the chamber randomly enter the pump, they are struck by an emerging vapor molecule and the transfer of momentum compresses the gas molecules closer together and impels them toward the next set of annular jets. The second set of jets is closer to the outer diameter of the pump, and the vapors from them impact on the gas molecules and further compress them. After one or more series of annular jets, a final jet of vapor expels the gas molecules from the pump.

The outer shell of the pump and the discharge nozzle are traced with water-cooling coils. These serve not only to remove the heat of compression but also to condense the mercury or oil vapors so that they can return to the reboiler section of the pump.

Low-vacuum equipment

The normal multistage rotary and centrifugal blowers that are described in Chap. 12 for moving large quantities of gases at moderate pressure rises may be used for producing the low vacuums required for pneumatic handling or filtration.

---

*Hydrogen solidifies at 11.7 K and helium at less than 0.8 K.

466 Equipment Selection, Sizing, and Related Subjects

Medium- or industrial-vacuum equipment

Process engineers are concerned mostly with the industrial-vacuum ranges for the application of vacuum to chemical reactions, diffusional operations, freeze drying, and removing gases from condensers. This range, which extends from 10^{-3} to 100 torr and spans the medium- and low-vacuum levels in Table 13.2, is attained by the use of rotating mechanical units for the direct compression of suction gas or by momentum transfer with steam-ejector apparatus.

A list of the major categories of vacuum equipment used for low- and medium-vacuum applications along with their practical operating vacuum ranges is given in

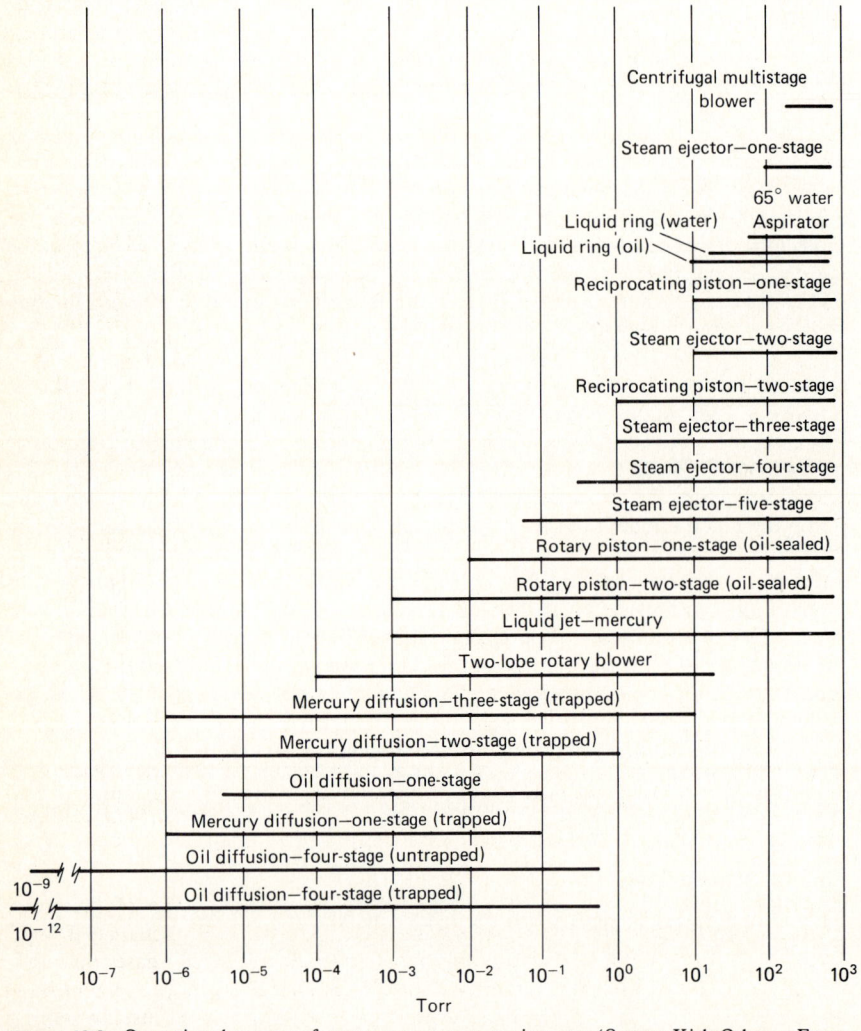

Figure 13.2 Operational ranges of common vacuum equipment. (*Source:* Kirk-Othmer Encyclopedia of Chemical Technology, *2d ed., vol. 21, Interscience Publishers, New York, 1970, p. 134.*)

Fig. 13.2. Combinations of different types of equipment can be used to produce higher vacuums more economically than one piece of equipment operating alone. For example, a rotary blower together with a liquid-ring pump can furnish and hold a vacuum of 1 torr, which neither could attain by itself.

A description of equipment used for vacuum in normal processing requirements follows.

Steam ejectors. Probably the most common method of producing an industrial vacuum, particularly if the required pressure is more than 1/10 atm, employs a single-stage steam ejector. The steam jet has the advantage of a very low first cost and reasonable operating costs for modest vacuums. The operation of a steam-jet ejector is similar in principle to that of the vapor-diffusion pump described earlier for high-vacuum equipment. In this case, however, there is no boiler, and there is one or more jets in the housing instead of an annulus (see Fig. 13-3). Motive steam at 60 to 150 psig enters the steam chest in the unit and passes through a convergent-divergent

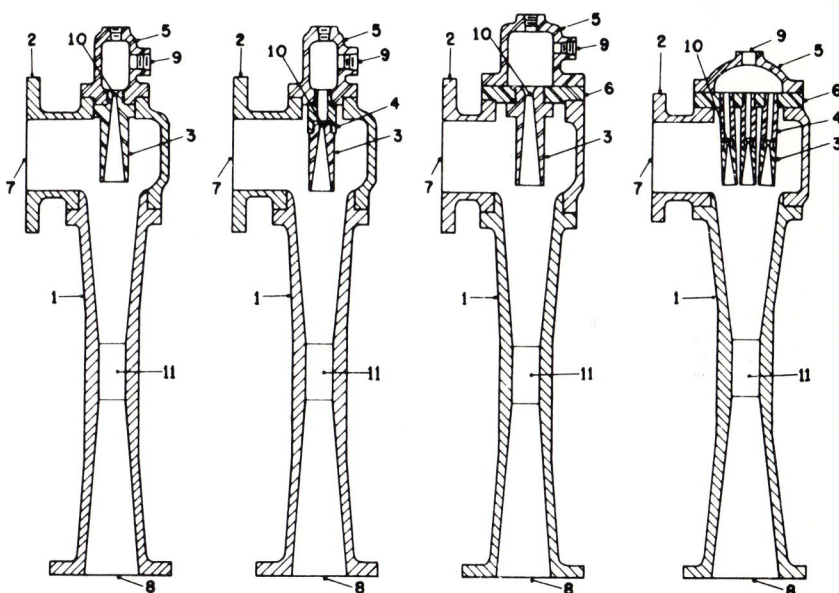

Figure 13.3 Typical steam-jet-ejector-stage assemblies. (1) Diffuser. (2) Suction chamber. (3) Steam nozzle. (4) Nozzle extensions, if used. (5) Steam chest. (6) Nozzle plate, if used. (7) Suction. (8) Discharge. (9) Steam inlet. (10) Nozzle throat. (11) Diffuser throat. (*With permission from the Heat Exchange Institute, Cleveland, Ohio.*)

diffuser nozzle projecting into the suction chamber. The pressure of the steam decreases while its velocity increases. As the high-velocity steam discharges from its diffuser nozzles, it collides with process vapors and noncondensable gases that enter the suction chamber through the intake port. The steam imparts its momentum to the vapors and gases from the process, and they all enter the main diffuser, where the velocity is reduced while the pressure of the mixture is increased to the pressure at the discharge nozzle.

Single-stage jets and aftercondensers. A single-stage steam-jet ejector normally operates with compression ratios* of 6.0 to 7.5:1, depending on the steam pressure and the configuration of the various components. Thus, a single-stage unit could be expected to hold a vacuum of 100 to 125 torr with discharge directly to the atmosphere or about 120 to 150 torr if there were a 3-psig back pressure at the discharge nozzle.

If only air is being handled by the jet ejector, the effluent steam-air mixture may be discharged directly to the atmosphere. However, it may be unsightly to have a plume emitting from the facility, or, more important, a discharge might contain condensable vapors which are worth recovering or which are toxic or hazardous. In these cases, an aftercondenser, similar to the intercoolers in Fig. 13.4, is included as a direct contact or a surface condenser in the system downstream of the ejector.

In direct-contact or barometric condensation, the steam and compressed vapors and gases pass through sprays or cascades of the cooling medium to be cooled and condensed. The vapor and gases usually approach within 3 to 5°F of the inlet cooling-medium temperature if sufficient cooling water is charged to a well-designed barometric condenser. Thus, the amounts of water vapor and process vapors leaving with noncondensables are reduced considerably, particularly if the process-vapor condensate is soluble in the cooling medium.

The cooling medium is heated somewhat by the sensible heat of the steam and vapor mixture, but most of the temperature rise as it passes through the barometric condenser is due to the latent heat of the condensing steam and vapor. The amount of water to a barometric aftercondenser should be such that the temperature of the total liquid effluent is less than 140°F or that its vapor pressure is less than 150 mmHg. Municipal water, well water, and even river water are used as cooling media in direct-contact condensers. Circulating cooling-tower water may also be used if it is acceptable to return the discharge to the cooling-tower system.

Surface condensers, even though more expensive than direct-contact condensers, are used when it is economical to recover and recycle condensed process vapors. If the process vapors and motive-steam condensate are immiscible, the effluent from either type of condenser is separated by decantation in a receiver. If the condensates are miscible, the use of surface condensation instead of direct-contact condensation reduces the water to be removed in a subsequent distillation step. A surface condenser is required if it is necessary to limit the amount of liquid from the condensing system in order to avoid overburdening a subsequent waste-treatment step.

Multistage jet systems and intercondensers. If the pressure required in the vacuum system is less than 100 to 125 torr, it is necessary to employ two or more steam-jet ejectors in series. If the system consists of two jets, a vacuum of 10 to 20 torr can be maintained. The intermediate pressure between adjacent ejector stages is designed so the compression ratio for both ejectors is nearly the same. Therefore, it can be approximated by the equation

$$P_i \cong P_s \times \left(\frac{P_d}{P_s}\right)^{i/n} \cong \sqrt{P_{i+1}/P_i} \tag{13.3}$$

where P_i = intermediate pressure after i^{th} ejector, mmHg abs
P_s = suction pressure, mmHg

---

*Ratio of discharge pressure from the jet ejector to the suction pressure to the ejector.

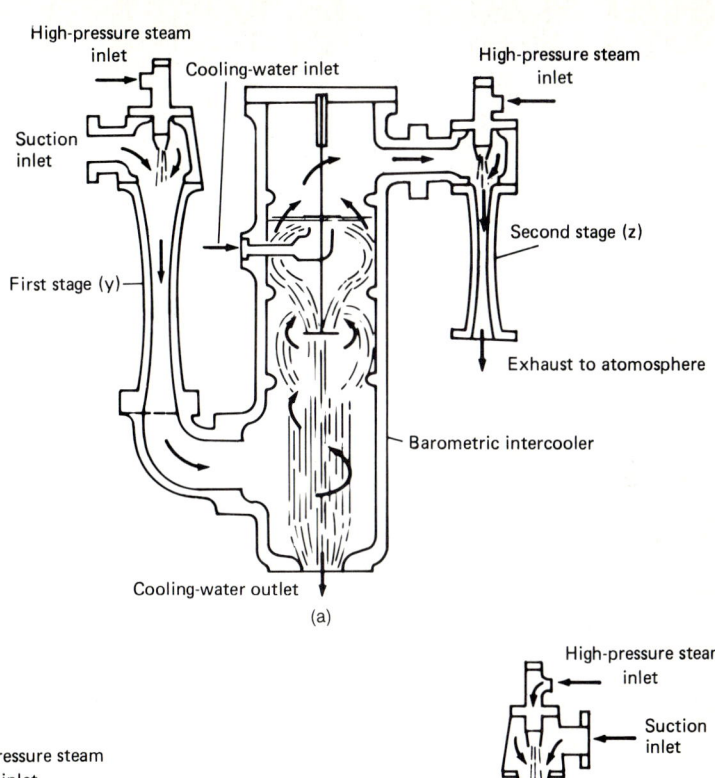
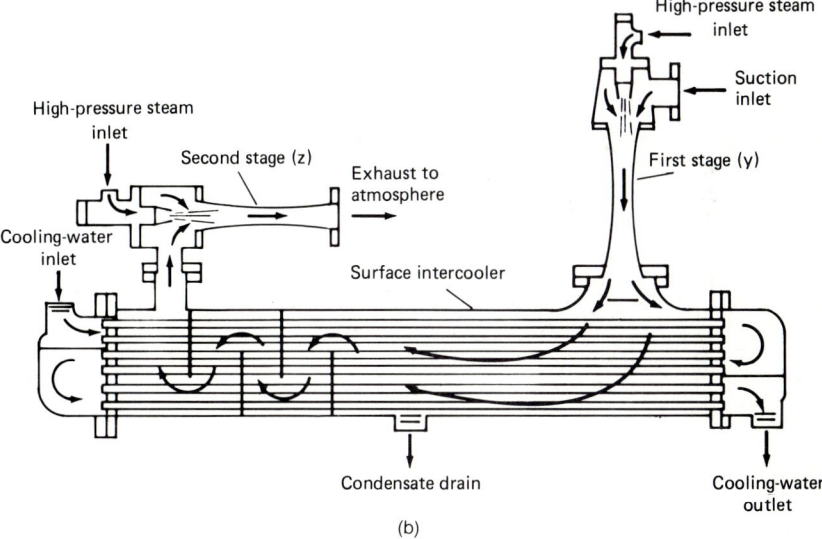

Figure 13.4 Condensers following an ejector. (*a*) Two-stage ejector with barometric direct-contact intercooler. (*b*) Two-stage steam-jet ejector with surface intercooler; shell-and-tube design with fixed tubesheet.

P_d = system discharge pressure, mmHg
n = total number of ejector stages in system

In other words, intermediate pressures are about equal to the geometric mean of the suction and discharge pressures. If an intermediate condenser is present, the ratio of discharge to inlet pressures for either stage is from 2 to 4 percent greater than the square root of the final discharge to first-stage inlet pressure to account for the pressure loss in the condenser. As a first estimate, the drop through a direct-contact condenser may be taken as 5 mmHg with that through a surface condenser as 20 mmHg.

The intermediate pressure in a two-stage system will be approximately 90 to 125 torr, which corresponds to steam dew points in the range of 120 to 135°F. Thus, it is easy to condense most of the motive steam and possibly some of the process vapors from the noncondensables when the discharge from the first-stage ejector passes through a direct or a surface condenser known as an intercondenser or intercooler (see Fig. 13.4). This considerably reduces the load that is placed on the second ejector. If a vacuum system is to be maintained at 70 torr, the motive steam to the second jet is reduced by a factor of from 1.2 to over 100, depending on the relative amount of noncondensables to condensables in the vapors from the process, by condensing and cooling the first-ejector discharge gases and vapors to 100°F or less.

Sufficient cooling water must be supplied to an intercondenser so that the vapor pressure of the liquid effluent from a barometric or surface condenser does not exceed 60 to 65 percent of the pressure of the intermediate stage. This rough rule of thumb is applicable since noncondensables usually constitute about 30 volume percent or less of the steam, vapor, and gas mixture in the effluent from the first stage to the intercondenser.

The following examples illustrate the type of analysis that can be done.

Example 13.1 Crystals containing a high-molecular-weight solvent are being dried under vacuum at 20 torr. The flow to the first jet at 150°F is 20-lb/h solvent (MW = 140) plus 30-lb/h air from the process including air leakage. Assume that the intercondenser is at 130 torr and that 200 lb/h of 125-psig saturated steam is the motive force. How much well water at 70°F is needed as the cooling medium for a barometric condenser?

solution With 70°F inlet contact water, it is a fair assumption that the discharge vapors and gas from the intercondenser will be at 75°F or less. For this example, it is also assumed that essentially all the solvent vapor will condense, is soluble in the contact water, and has a negligible partial pressure. At 75°F, water has a vapor pressure of 22.5 mmHg. Thus, the partial pressure of the air will be 107.5 mmHg (130 − 22.5), and about 0.22 mol/h (3.9 lb/h) of water vapor will pass with the 1.03-mol/h (30-lb/h) air to the second stage (1.03 × 22.5/107.5 = 0.22).*

The heat to be removed is that required for cooling the inlet vapors, gas, and motive steam from their respective inlet conditions to 75°F and for condensing the solvent and 196.1 (200 − 3.9) lb/h of steam. The temperature of the discharge from the condenser to the barometric leg should be such that the vapor pressure of the water in the effluent

---

*There are usually only minor amounts of process vapors leaving a barometric condenser since the concentration of condensed process vapor in the liquid effluent is very low. However, if the process condensate is not miscible with the cooling water, it too will exert its own vapor pressure. This, in turn, requires a colder cooling medium or greater loss of process and water vapor to the second stage.

is less than 0.60 of 130 mmHg, or 78 mmHg. This is equivalent to a water temperature of 116°F. If it is assumed that the effluent water leaves at 110°F, about 11 gal/min of 70°F inlet water is required for the cooling load of 216,000 Btu/h.

Example 13.2 Repeat Example 13.1, but determine how much water is required if a surface condenser with 70°F well water as the cooling medium is used.

solution A schematic diagram of the condenser follows.

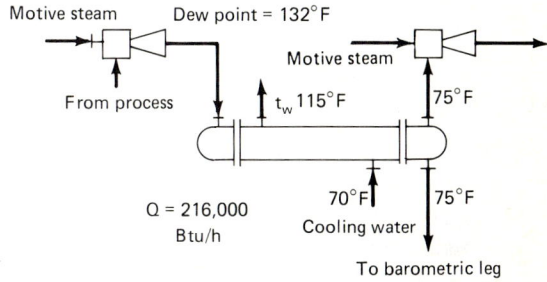

The solution is similar to the solution of Example 13.1, in that the outlet composition of the vapors from the surface condenser is essentially the same as that from the direct condenser and thus the cooling duties on both types are about the same (in this case, 216,000 Btu/h). It has been found that the use of a log-mean-temperature difference of about 10°F for the exchanger yields the most nearly economical combination of exchanger costs and water usage. In order to obtain this temperature difference for a process dew point of 132°F (calculated from the partial pressure of the water), the cooling water should exit from the exchanger at about 115°F. The water flow to the condenser would then be about 10 gal/min.

It should be pointed out that both the direct-contact condenser in Example 3.1 and the surface condenser in Example 3.2 use about the same amount of cooling water. However, a contact condenser is considerably less expensive than a surface condenser, and consideration is usually given first to the use of a contact condenser. Another important consideration is the concentration of condensed process vapor in the effluents from barometric and surface condensers. In Example 13.1, the concentration of the condensed process vapor is about 0.4 weight percent. With the surface condenser in Example 13.2, the concentration is about 9.1 weight percent. Since the example states that the process component has a high-molecular-weight component, it makes little difference to the vapor pressure of the condensate. However, if the solvent had instead been ethyl or a propyl alcohol, the vapor pressure of the condensate solution would have increased considerably and it would have been necessary for the condensate to leave the surface condenser at a lower temperature. This may not have been possible or reasonable with the available cooling water.

Hot wells and barometric legs. It should be remembered that the intercondenser between jet ejectors in a train operates at vacuum conditions. Thus, the liquid effluent from the condenser must be sealed from the atmosphere. The most common method is to discharge the condensate effluent line under a liquid seal in a vessel called a "hot well." The condensate either overflows from the hot well or is pumped from it under level control.

The pipe connecting the condenser to the hot well is known as the "barometric leg." There is a level of liquid in the leg above the liquid surface in the hot well which is a function of the degree of vacuum developed in the condenser plus the friction loss of the condensate flowing in the barometric leg itself. The friction loss for the liquid flowing in a properly designed barometric leg is negligble, so that for practical purposes the height of liquid in the leg is due essentially to the difference between atmospheric pressure and the pressure in the intercondenser. It also is a function of the density of the condensate in accordance with the equation*

$$z_l = \frac{\pi - P_c}{760} \times 14.7 \times \frac{2.31}{s} = \left(\frac{\pi - P_c}{760}\right)\left(\frac{33.96}{s}\right) \quad (13.4)$$

where z_l = height of liquid in barometric leg, ft
π = atmospheric pressure, mmHg abs.
P_c = pressure in intercondenser, mmHg abs.
s = specific gravity of condensate with respect to water at 60°F

For a 50-torr vacuum at sea level with water condensate at 60°F, the level of water in the barometric leg would stand at

$$z_l = \frac{760 - 50}{760} \times 14.7 \times \frac{2.31}{1.0}$$

or 31.72 ft above the surface of the hot well.

While the level in the leg in this illustration is calculated to be slightly under 32 ft, it is necessary to make the actual leg of pipe somewhat longer to ensure that the intercondenser does not become flooded in the event that the vacuum becomes lower or friction in the leg is not negligible. It should be noted that the intercondenser might be designed for use with cooling-tower water at its highest temperature, e.g., 85°F, in summer. However, the cooling water temperature might drop to 40°F, and the pressure in the intercondenser could fall as low as 25 torr. In this event, the leg height calculated by Eq. (13.4) increases to about 33 ft.

To maintain friction losses in the barometric leg at an acceptable level, the diameter of the leg must be oversized. Normal flow of liquids in pipes is such that friction loss is about 2 psi/100 ft of pipe. When the fluid is water, this is equivalent to 4.6 ft of water/100 ft of pipe. If the leg is straight, it will be 35 to 45 ft long and a normally sized pipe could add 20 to 25 in to the required height of the leg. Any additional height requirement for friction loss can be reduced considerably if the diameter of the barometric leg is made 40 to 50 percent greater than would be used in normal service.

The leg from the intercondenser to the hot well should be as straight as is practical for two main reasons. First, noncondensables will be entrained in the liquid, and they must be allowed to disengage easily. Second, bends or fittings in the barometric leg add considerably to the equivalent length of the leg and thus to friction loss. If it is necessary to offset the intercondenser from the hot well, the number of bends should be kept to a minimum and it is preferable that they be not less than 60° from the horizontal. If necessary, an offset made with two 45° bends often is practical. However, if the piping cannot be made with these particular bends, 30° ells may be used

---

*Equation (13.4) relates atmospheres to pounds per square inch per atmosphere to feet of water per pounds per square inch.

if they are placed in the lower section of the leg so as not to interfere with the rising of the disengaged gases.

An important consideration in determining the height of a barometric leg is the specific gravity of the condensate in the leg. In most conditions, the weight ratio of motive steam to process vapors in the stream to a surface condenser is great enough that the specific gravity of the condensate solution is close to that of water. In a direct condenser, there is even considerably more water in the effluent and its specific gravity approaches that of water. Care must be taken, however, if the process-vapor condensate is not miscible with water. If its gravity is less than 1.0, the lighter process condensate will be the continuous phase in the barometric leg and the water will fall through it as large drops. Because it is the continuous phase, the specific gravity of the lighter phase (or lightest phase if more than two liquid phases are present) should be used in Eq. (13.4). Thus, if heptane is the process vapor, the barometric leg would have to be almost 50 ft high since the specific gravity of heptane is 0.68. It is, therefore, important for process engineers to indicate the minimum height of the intercondenser above the hot well on the engineering flow diagram.

Pumped condensate. If it is not possible to accommodate the height of a barometric leg from a barometric condenser or a surface condenser, effluent can be removed from the vacuum system by allowing it to flow from the intercondenser to a receiver tank and then pumping it to disposal. Even though the line from the intercondenser to the receiver tank will be shorter than a barometric leg, similar precautions must be taken regarding its size and configuration. The line, in this case, must not only convey liquid from the condenser to the receiver but might also act as an equalizer line between the receiver and the condenser. Thus, the size of the line must be great enough to ensure that there is always a free passage of gas in it. Figure 7.12 and Table 7.14 in Chap. 7 are useful in determining the diameter of a vertical line that would just be completely closed at the liquid flow rate. As a rule of thumb, a vertically dropped line should be at least twice the diameter given by Fig. 7.12 or Table 7.14.

If the drop line must be offset, it should be treated as discussed previously. Figure 7.13 may be used to determine the minimum size of the inclined section of the pipe. The actual diameter should be at least twice that.

It should be noted that the net positive suction head (NPSH) available for the condensate pump will be low owing to the vacuum conditions in the receiver and will affect the specification of the pump.

Three and more ejector stages. The addition of a third steam-jet ejector to a system allows the maintenance of vacuums to about 2 torr. If the pressure at the suction chamber of the first ejector in the train is 6 torr or greater, it may be possible to utilize an intercondenser after this stage provided the cooling medium is 70°F or lower. If the cooling medium is warmer, excessive water vapor may accompany noncondensables to the second ejector or excessive amounts of cooling water may be required in the intercondenser. An analysis similar to those given for Examples 13.1 and 13.2, along with an economic evaluation of a system with and without the intercondenser between the first and second stages, should be made when the available cooling-water temperature is over 70°F. Without an intercondenser, the process vapors, noncondensables, and accompanying steam from the first ejector must all be compressed in the second ejector.

An ejector train with four stages is used when the vacuum to be maintained is from 0.2 to 0.7 torr and five stages can hold 32 to 100 μ, depending on the steam pressure, degree of superheat, and pressure drops between stages. The discharge pressures from the first stages in a four- or five-ejector train are such that it is not possible to use water as a coolant in a barometric condenser.* Thus, the first several ejectors in a four- or five-ejector train are coupled directly to one another until the discharge pressure is sufficiently great to be accommodated in an intercondenser.

System curves. Typical operating curves for a steam-jet ejector are shown in Fig. 13.5. The model in this case is representative of a single-stage ejector which would be suitable for a compression ratio of 10 or less. The capacity is given as inlet cubic feet per minute of 70°F air and is shown for various saturated-motive-steam pressures. As a rough rule of thumb it can be assumed that a 20 percent change in steam pressure has a 5 percent direct effect on capacity, while capacity is reduced by about 7 percent for every 200°F of superheat. Included in Fig. 13.5 is a straight line which represents the weight-volume ratio of 70°F air, i.e., pounds per hour of air per inlet

---

*The pressures to be expected after the first stage of a four-jet system are 2 to 4 mmHg abs. "Water" is actually ice at 12 to 29°F, respectively, when its vapor pressure is equal to these values.

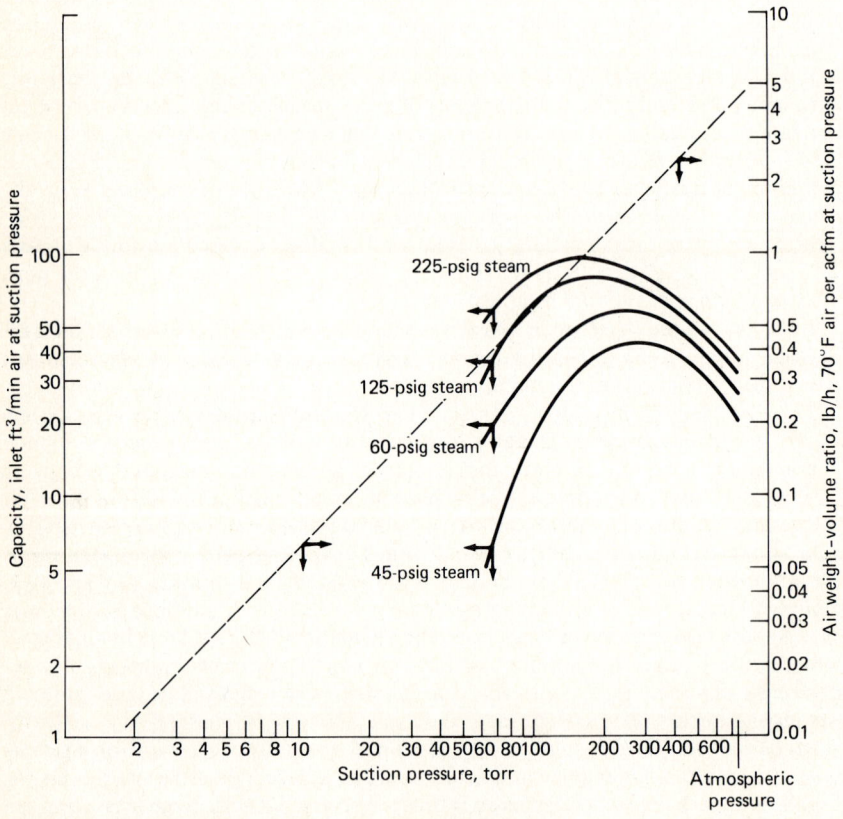

Figure 13.5 Typical performance curves of a single-jet ejector.

cubic foot per minute at suction pressure. Multiplying the system-curve capacity at any pressure by the weight-volume ratio of the gas defines the curve in terms of pounds-per-hour 70°F air capacity.

Figure 13.6 illustrates the relationship between the four individual curves in a four-train evacuation system. In this instance, it is assumed that the process is at 1 torr and

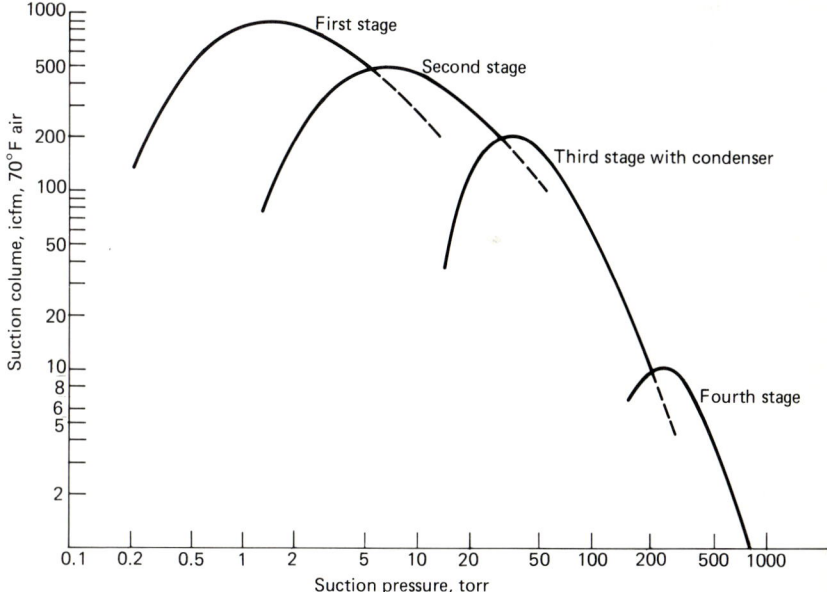

Figure 13.6 Performance curves of a four-jet-ejector system.

that discharge is at about 0.5 psig, so that the compression ratio per stage is about 5.45, taking into account a drop of 20 mmHg in an intercondenser. The first-stage jet is designed for capacity in terms of inlet cubic-feet-per-minute equivalent 70°F air at 1 torr such that the 1-torr pressure condition intersects the ejector performance curve to the left of its highest cubic-feet-per-minute point.

The second-stage jet in this case has an air capacity equal to the equivalent air rate in the first-stage suction plus the 70°F air equivalent of the motive steam to the first-stage ejector. The air-capacity point, when converted to inlet cubic feet per minute at 5.45 torr, should be to the left of the high point on the second-stage performance curve. Similarly, the third-stage jet has an air capacity equivalent to the equivalent air rate to the first-stage suction plus the 70°F air equivalent of the motive steam to the first- and second-stage ejectors. Again this air-capacity point, when converted to inlet cubic feet per minute at 29.7 torr, should lie to the left of the high point on the third-stage performance curve.

Since there undoubtedly will be a direct or surface intercondenser between the third- and fourth-stage ejectors, essentially all the process vapors to the first stage and most of the motive steam to the first-, second-, and third-stage ejectors are condensed and removed from the gaseous phase. Thus, the capacity of the last stage is the air equivalent to the noncondensables from the process and air leakage plus the equilibrium water and process vapors carried over from the intercondenser. The capacity is expressed as equivalent 70°F air, and this capacity point in cubic feet per minute at 160 torr should be to the left of the high point on the fourth-stage performance curve.

Since jet-ejector models come in discrete sizes and the various manufacturers produce their own sets of equipment with unique performance curves, it is best to consult equipment suppliers to obtain a balanced set of performance curves with intermediate pressures, steam usages, and condenser water requirements for the respective stages.

Steam and cooling-water usage. Process engineers should also consult the manufacturers of steam-jet-ejector equipment to obtain accurate values for steam consumption and cooling-water usage for the vacuum to be maintained at the operating and motive-steam parameters. Because various manufacturers make units in distinct sizes and throat nozzles which have differing efficiencies for varying conditions, it is rare that any two will offer the same-sized units or have the same steam and cooling-water usage for the given conditions of vacuum and flow. When actually recommending equipment to be purchased, the process engineer must weigh the initial offered costs against the respective costs of steam and cooling water and such intangible factors as the stability of the system with change in process flow rates or motive-steam conditions.

There are, however, guides to help process engineers approximate the utilities usage for a vacuum system. Figure 13.7 is a nomograph which relates the pounds of

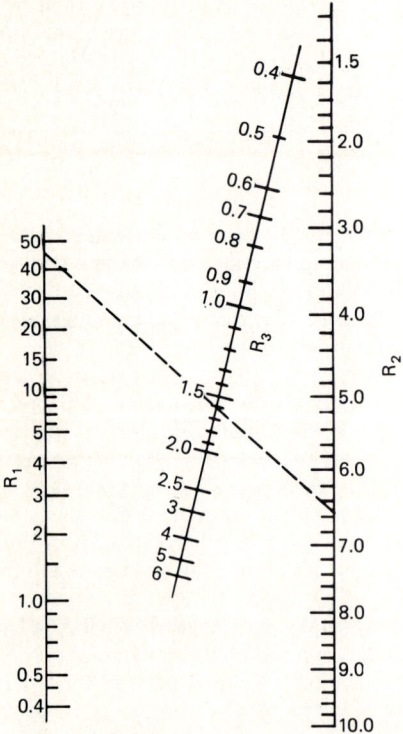

Figure 13.7 Steam consumption of ejectors. R_1 = steam pressure, psia, divided by suction pressure, torr; R_2 = discharge pressure, torr, divided by suction pressure, torr; R_3 = lb saturated motive steam per lb equivalent 70°F suction air. (Adapted from data by the Jet-Vac Corp., Waltham, Mass.)

dry saturated steam at various steam pressures required to compress 1 lb of dry air at 70°F through various compression ratios.

It is obvious that not all gases to the suction chamber are air at 70°F. Therefore, Fig. 13.8 is required to convert other gases or vapors to their 70°F air equivalency in order to use the steam-consumption nomograph. Figure 13.9 gives additional correction factors to be used with the air-equivalent rates for suction mixture temperatures

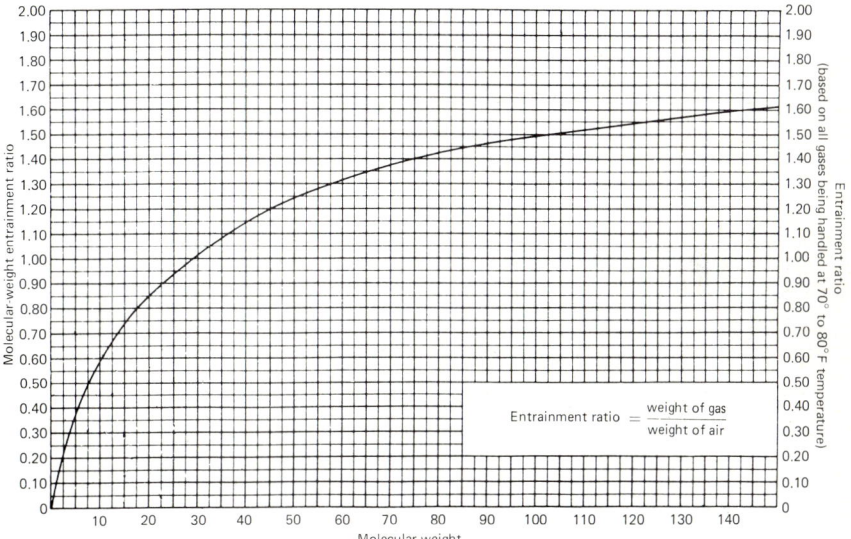

Figure 13.8 Air equivalencies. *(With permission from The Heat Exchange Institute, Inc., Cleveland, Ohio.)*

other than 70°F. As long as the temperature of a vapor mixture is below 400°F, it is good practice to apply the temperature correction factor as if the mixture were air. However, if the temperature is above 400°F and its average molecular weight is less than 26, the temperature correction factor for steam on Fig. 13.9 should be used. The following example illustrates the use of these graphs.

Example 13.3 What is the approximate usage of 125-psig saturated steam (353°F) for a three-jet system to compress 6-lb/h air, 22-lb/h CO_2, and 20-lb/h vapor (MW = 140) at 150°F from 3.0 torr to atmospheric pressure? Determine the water usage for well water at 70°F for the barometric intercondensers.

solution

a. Determine the number of ejectors and intermediate pressures so that the compression ratio doesn't exceed about 7. From Eq. (13.3), three stages are required. The compression ratio R_2 at each stage is about 6.45, allowing 2 percent excess for loss through an intercooler.
b. Calculate pressures after the first and second stages: 19.4 and 124.8 torr, respectively.
c. Determine the number of intercoolers.
   (1) Vapor pressure of 75°F discharge water is 22.2 mmHg, which is greater than the

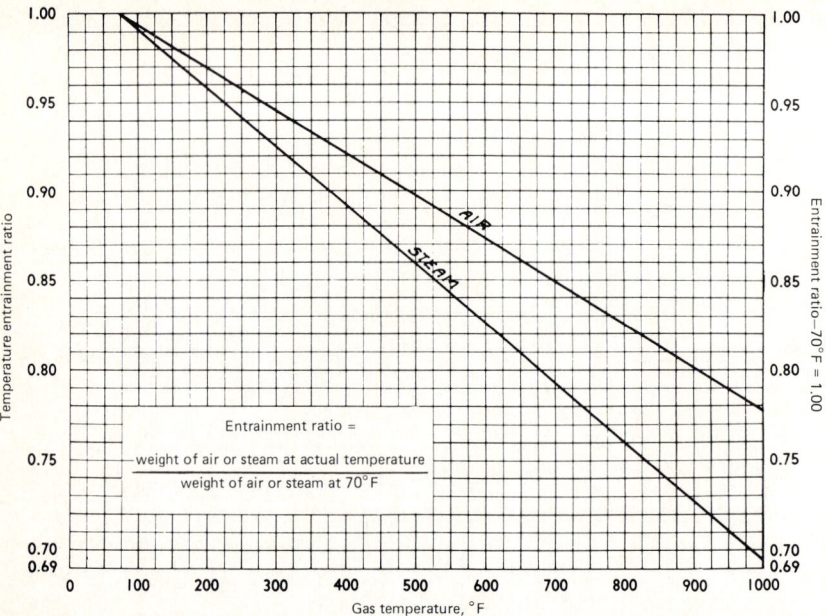

Figure 13.9 Temperature correction curves. (*With permission from The Heat Exchange Institute, Cleveland, Ohio.*)

pressure from the first-stage ejector, so that it is not possible to use an interstage condenser there.

(2) The 22.2-mmHg vapor pressure of discharge water is considerably below the discharge pressure from the second stage, so that an intercondenser is used between the second and third stages.

d. Determine steam to the first-stage ejector.
   (1) Calculate the average molecular weight of process vapors: average MW = 56.5.
   (2) Calculate the 70°F air equivalent to 150°F process vapors. From Figs. 13.8 and 13.9, by using either individual components or average molecular weight (MW correction is 1/1.29; temperature correction is 1/0.98), the equivalent 70°F air is 38.0 lb/h.
   (3) Calculate the motive-steam-to-suction-chamber pressure ratio: R_1 = 46.6 psi/torr.
   (4) For R_1 = 46.6 and R_2 = 6.45 R_3 (from Fig. 13.7) = 1.58 lb of steam/lb of 70°F air. Therefore, motive steam to the first stage is 60.0 lb/h.

e. Determine the steam to the second-stage ejector.
   (1) Do the heat and energy balance around the first-stage jet ejector to determine the temperature of the mixture from the first stage. This is most easily done by an iterative method by assuming a discharge temperature for the mixture and determining whether or not the energy given up by the motive steam is equal to that picked up by the feed vapors and gases to the suction chamber. The steam tables or a Mollier diagram for steam is useful in obtaining the temperature of the steam at its reduced enthalpy for the discharge pressure from the ejector. In this case, the temperature of the mixture can be deduced to be about 249°F.
   (2) Calculate the average molecular weight of discharge from the first stage. Discharge is the original process vapors plus the first-stage motive steam. Average MW = 25.8.

(3) Calculate the 70°F air equivalent to 249°F vapors. The molecular-weight correction factor from Fig. 13.8 is 1/0.95, and the temperature correction from Fig. 13.9 is 1/0.96. Therefore, the equivalent 70°F air is 118.4 lb/h.
(4) Calculate the motive-steam-to-suction-chamber pressure ratio: $R_1 = 7.2$ psi/torr.
(5) For $R_1 = 7.2$ and $R_2 = 6.45$, R_3 (from Fig. 13.7) = 2.35 lb of steam/lb of 70°F air. Therefore, motive steam to the second stage is 278 lb/h.

f. Calculate the composition of vapors from the barometric intercondenser.
  (1) Assume that there is a 5°F vapor approach to inlet cooling waters and that all process vapors are essentially condensed and dissolved so that they have a negligible partial pressure. Assume a 5-mmHg drop across the intercooler so that the discharge pressure is 120 torr.
  (2) From vapor pressure of 75°F water (22.4 mmHg) and total pressure (120 mmHg), determine the partial pressure of air and CO_2 together (97.6 mmHg). Calculate the mole-per-hour water vapor carried over with air and CO_2. Calculate the flow rate and molecular weight of the mixture. Total vapor flow = 30.9 lb/h, and MW = 35.6.

g. Calculate the energy balance for the intercondenser. Calculations similar to those for Example 13.1 show that the heat given up is about 340,000 Btu/h.

h. Determine the outlet temperature of liquid to the barometric leg and the amount of 70°F quench water.
  (1) The composition of the mixture from the second stage is such that it is about 96 volume percent steam and water vapor, so that the partial pressure of water in it is about 119 mmHg.
  (2) If the vapor pressure of the condenser effluent water is 60 to 65 percent of the partial pressure of water vapor in the second-stage-ejector discharge, it would be 69 to 75 mmHg. This is equivalent to a liquid effluent temperature of 111 to 114°F. Use 112°F.
  (3) For a duty of 340,000 Btu/h and a temperature rise of 42°F, 8100 lb/h, or 16 gal/min, of quench water is required at the intercondenser.

i. Calculate the steam required at the third or final stage.
  (1) In Step h it was determined that there was 30.9 lb/h of vapor, MW = 35.6 at 75°F, and 120 torr to the third-stage suction chamber.
  (2) The molecular-weight correction factor from Fig. 13.8 is 1/1.09. The temperature correction factor from Fig. 13.9 is 1/0.998. Therefore, the equivalent 70°F air is 28.4 lb/h.
  (3) Calculate the motive-steam-to-suction-chamber ratio: $R_1 = 1.16$ psi/torr.
  (4) Calculate the compression ratio for atmospheric discharge: $R_2 = 6.33$.
  (5) For $R_1 = 1.16$ and $R_2 = 6.33$, R_3 (from Fig. 13.7) = 6.04 lb of steam/h 70°F air. Therefore, motive steam to the third stage is 172 lb/h.

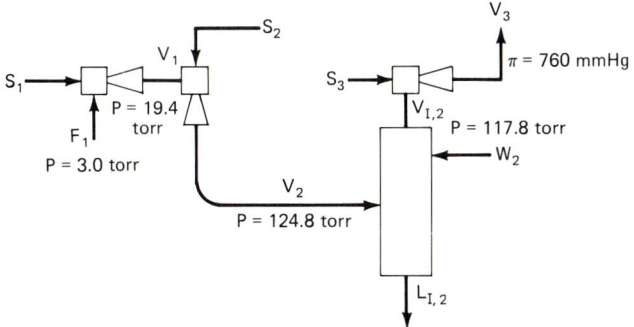

Summary of Material and Pressure Balance for Example 13.3

	F_1	S_1	V_1	S_2	V_2	W_2	L_2	$V_{1,2}$	S_3	V_3	Total steam
Air, lb/h	6.0	....	6.0	....	6.0	....	....	6.0	....	6	
CO_2, lb/h	22.0	....	22.0	....	22.0	....	....	22.0	....	22	
Process, lb/h	20.0	....	20.0	....	20.0	....	20	......	....	.....	
H_2O, lb/h	....	60.0	60.0	278	338.0	8100	8435	2.9	172	175	510
lb/h	48.0	60.0	108.0	278	386.0	8100	8455	30.9	172	203	510
MW	56.5	18	25.8	18	19.7	....	....	35.6	18	19.5	
Temperature	150°F	353°F	249°F	353°F	284°F	70°F	112°F	75°F	353°F	280°F	
Pressure	3	125	19	125	125	....	....	120	125	760	125
	torr	psig	torr	psig	torr	....	....	torr	psig	mmHg	psig

Preconditioners. It is obvious, from considering the steam requirements in Examples 13.1 and 13.3, that as much process vapor as possible must be eliminated from the gases going to the initial stage of a jet-ejector train. This is best accomplished by including a preconditioner between the process source and the first ejector. If the temperature of the process vapors is sufficiently high, i.e., 100 to 120°F or greater, normal cooling-water media may be used. If it is lower, it may be economical to use chilled water or refrigerated brines or glycols as the coolant in the preconditioner.

The vapor condensate may be refluxed back to the process (as in some reactions), collected and partially refluxed and partially removed (as in distillations), or collected and removed entirely.

Great care must be exercised, especially at pressures of 10 torr or less, that the pressure loss across the preconditioner does not increase the compression ratio across the ejector train to the point that it negates the effects of precondensing the process vapors.

Aspirators and eductors. The principle of jet ejectors may be employed for other purposes than compressing gases from vacuum to atmospheric conditions. In an apparatus known as an aspirator or siphon, the suction port of the ejector is connected to a solid pipe or hose which terminates below the liquid level in a tank, sump, or pit. The motive fluid initially creates a vacuum in the body of the aspirator, causing liquid to flow from the tank, sump, or pit to the suction of the ejector. Here the liquid is entrapped by the motive fluid and moved to the discharge, from which it is directed to its destination.

The motive fluid to an aspirator may be steam, high-pressure gas, or even a pressurized liquid such as water from a pumped system. The amount of motive fluid required to "pump" a given volume of liquid is a function of the nature of the fluid, its pressure, the specific gravity of the liquid, and suction lift of the fluid, i.e., the vertical height between the liquid level and the centerline of the aspirator. Units of this type operate with low efficiencies and are used only for special situations. Operating data for these units should be obtained directly from equipment suppliers or from their engineering-data literature.

The aspirator or siphon could be located under the liquid level in a tank or sump, but usually it is mounted above the liquid. However, the eductor (another type of jet ejector) is located under the liquid in a tank or sump and is used to circulate or mix the contents of the tank or sump. Such eductors mixed steam and deionizer blowdowns in an MgO plant. In this case the pressurized motive fluid, either a gas or a liquid, passes through the jet nozzles, and liquid from the tank is drawn into the eductor through one or more ports. If pumped liquid is the motive fluid, 3 to 5 gal/min of tank or sump fluid will be mixed with every gallon per minute of pumped liquid. Because of the relatively large amounts of tank fluids being mixed into the motive fluid, eductors are well suited not only for circulating tanks or sumps but for dissolving high-vapor-pressure liquid reactants into other liquids and for use as a noiseless means of combining steam with another fluid to heat the fluid.

Mechanical vacuum equipment. While steam-jet ejectors depend upon changing the pressure of a motive source to velocity head to compress suction gases and vapors, mechanical equipment produces a vacuum by physically entrapping the suction gas and compressing it by reducing its confining volume.

Reciprocating vacuum pumps. Machines that utilize a reciprocating-piston action and those which have rotary pistons, with and without sliding vanes, are described in the subsection "Compressor Mechanical Design" in Chap. 12. These units have close clearances so that they can achieve high compression ratios with a single-stage machine, effecting a vacuum down to 1 torr or less. Multistage units can be used to produce vacuums down to 1 μ. However, the capacities of reciprocating- and rotary-piston units are limited to about 1000 icfm at 1 torr.

Rotary vacuum pumps. The subsection "Compressor Mechanical Design" also reviews helical-screw compressors and lobed blowers. Because their clearances are more open, helical-screw machines can produce a vacuum of only 150 torr, while lobed blowers are limited to about 350 torr when discharging to atmospheric pressure. However, because of their high speeds these units have high capacities, up to 20,000 icfm for the lobed blower. The high capacity of lobed blowers at modest vacuums makes them well suited to serve as prime movers for negative pneumatic conveying systems which exhaust to the atmosphere or as fore pumps or booster pumps in tandem vacuum pumping systems.

Liquid-ring vacuum pumps. One of the more widely used pieces of equipment for vacuum service in the range of 50 to 550 torr is the liquid-ring vacuum pump, which is a special type of rotary unit. This unit consists of an eccentrically located impeller rotor having a number of blades that form compartments on it. The impeller spins inside a casing that is partially filled with a liquid sealant. The rotation of the impeller throws the liquid to the periphery of the casing, creating a void in the center of the pump. As a blade passes the inlet port, process gases are trapped in one of the compartments. Since the impeller moves eccentrically with respect to the turning liquid ring, the unwetted cross section of the blade compartment decreases and compresses the gas until it discharges at the exhaust port.

A continuous stream of liquid sealant enters the pump and leaves with the compressed gases to a separator tank following the pump. The liquid sealant also serves to absorb the heat of compression and any latent heats from condensing vapors. Evaporation of the sealant, if it takes place, also removes some of the heat of compression.

If the sealant is water and it remains uncontaminated by the compression of the process gases and vapors, the liquid effluent may be sent to the sewer. If contamination is such that it is easily treated, the liquid effluent may go to treatment before disposal. Sometimes contamination of the sealant is such that it is not feasible or economical to allow the sealant to go to treatment. In that event, a cooled recycle may be used. The sealant may be circulated by means of the differential pressure between the discharge and inlet ports, or a recycle pump may be employed, if required. Depending upon the nature of the process vapors to the unit, sealants other than water may be circulated.

The vapor pressure of the sealant limits the vacuum that can be effected by the liquid-ring pump and also reduces its capacity. For example, well water at 60°F has a vapor pressure of 13.2 mmHg abs. Thus, if 60°F well water were used as a once-through sealant, the highest vacuum that could be sustained, even with no flow from the process, would be 13.2 torr. For this reason, manufacturers of vacuum-ring equipment recommend that it not be operated below 30 torr. Even then, the capacity of the unit is reduced since a relatively large amount of water is vaporized into the incoming process stream at that pressure level.

If it is necessary to recycle the water sealant and it can be cooled only to 90°F with 85°F cooling water, the vapor pressure of the water sealant would be 35 mmHg. Thus, the unit would have to operate at a considerably lower degree of vacuum, about 105 to 120 torr, to prevent excessive vaporization of water into the incoming process stream.

The compression of process gases and vapors in intimate contact with the liquid sealant makes the liquid-ring vacuum pump well suited for applications such as degasification and deaeration and for filtration systems in which only modest vacuums are required and the process medium is essentially water so that water may be the sealant fluid. Figure 13.10 presents a typical system curve for a liquid-ring vacuum pump using 12 gal/min of 60°F water as the sealant.

If the dew point of the process gases and vapors is above that of the incoming sealant, the operating capacity of the pump can be increased with respect to the process gases and vapors by spraying all or part of the sealant fluid into the incoming process gases and vapors to precondense some of the process vapors, thereby reducing the suction volume of process vapors to the liquid-ring pump itself.

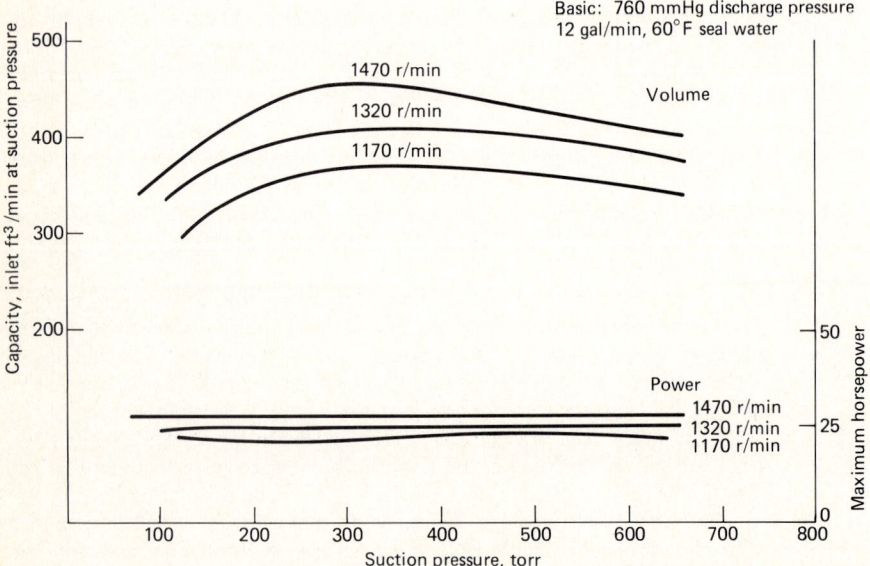

Figure 13.10 Typical liquid-ring-pump performance curves.

Vacuum-Equipment Sizing

The size or capacity of vacuum equipment depends upon several factors which must be designated when specifying vacuum equipment. They are:

1. The absolute pressure to be maintained at the inlet port to the initial stage of the vacuum equipment.
2. The total weight in pounds per hour as well as the molecular weights of the inlet vapors and noncondensables. Each component of the mixture should be identified and its molecular weight and hourly flow rate given.
3. The temperature of inlet gas mixture.
4. The pressure to which the outlet port of the last stage must discharge. Since this pressure is usually reported in psig or inches of mercury, the absolute barometric pressure should be given if the equipment is to be used at an altitude of 1000 ft or more above sea level.
5. System volume and initial composition of vapors and gases in the system if the system is to be pulled down from atmospheric pressure.
6. Maximum time for pulldown, if important.

The maximum temperature of available cooling water and its allowable temperature rise affect the sizes of vacuum units which have precondensers and intercondensers. The maximum and minimum available pressure of the water must be considered as well as its quality and limitations on quantity.

For those systems which utilize steam-jet ejectors, information regarding the maximum and minimum quantities of steam available and its maximum and minimum pressures at the steam inlet port, as well as the quality or degree of superheat, should be given.

Inlet pressure to vacuum equipment

The pressure in the process portion of the vacuum system is set by the operating conditions of the process. Only if the vacuum equipment is connected directly to the process vessel will the absolute pressure at the inlet to the vacuum-producing device be equivalent to the process requirements. Usually it is necessary to lead the vapors and gases from the process to the vacuum equipment through piping and fittings. In addition, heat-exchange equipment is sometimes interposed between the process and the vacuum device to recover or recycle process components or to reduce the load on the vacuum system. The pressure drop induced by the flow through piping and any heat-exchanger equipment must be deducted from the process pressure to determine the absolute inlet pressure to the vacuum equipment.

Great care must be taken in routing the piping to keep its length and number of bends to a minimum and in sizing the piping and heat-exchange equipment. Excessive resistance in the flow path from the process to the vacuum equipment could choke the flow and make it impossible to remove the specific amount of vapors and gases from the process. A high pressure drop in piping or an exchanger reduces the amount of condensation in the exchanger or requires excessively large vacuum equipment and considerably greater utility usage.

The first approximation for sizing the piping is that the friction losses for 100 equivalent ft of piping and fittings should not exceed 2 percent of the absolute process pressure. There are no rules of thumb for setting the pressure drop through the exchanger except that the combined losses through the piping and the precondenser should not exceed 30 to 40 percent of the process absolute pressure. If the pressure drop approaches this value, it may be necessary to prepare a cost analysis involving the annual fixed costs of the increased piping and exchanger sizes on the one hand versus the lower vacuum-equipment costs on the other hand and the reduced annual operating costs for steam or electricity as pressure losses are kept to a minimum.

While the normal flow of vapors and gases usually is in the transient- or turbulent-fluid-flow ranges, it is possible that flow in vacuum piping could approach or be in the viscous- or laminar-flow area. If so, the normal shortcut methods of calculating friction loss may not apply directly. It is recommended that Reynolds numbers be determined for these cases to establish the type of flow regimes so that the proper Moody-friction-factor correction can be used. It should be noted that at very high vacuums (10^{-3} torr and less—conditions which normally are not encountered in chemical, refinery, or power-plant facilities) the mean free path of gas or vapor molecules approaches or is greater than the diameter of the piping. In this case, friction losses are not described by laminar flow but by special analyses which are beyond the scope of this book. Reference should be made to a text on the kinetic theory of gases to evaluate such flows.

Flow rates of vapors and gases

Most vacuum equipment is utilized in continuous applications rather than in batch or intermittent operations. In these cases, it is important to know the steady-state flow of fluid to the vacuum equipment. It was noted earlier that the load to a steam-jet system is usually defined in terms of pounds of air per hour (lb/h) while inlet cubic feet per minute (icfm) is commonly used when referring to rotary or reciprocating vacuum machinery. In either case, it is necessary to define the composition of the fluid entering the vacuum equipment so that each component can be converted to its air equivalency by means of Fig. 13.8 or be expressed in icfm by means of the Charles and Boyle gas laws by knowing the inlet temperature and pressure. For all practical purposes, fluids entering a vacuum system are condensable vapors, noncondensable gases, or a combination of the two. A review of the effect of each on the sizing of vacuum equipment follows.

Vapor flow rates. The amount of vapors and noncondensables evolving from a process operation is ascertained from the material and energy balances about the operation. If no noncondensables enter the process with the reactants, are generated in the process, or are sparged into it and if the process equipment is completely airtight, then only process vapors pass to the vacuum equipment. Without a condenser between the process and vacuum equipment, the vapor flow rate and its temperature are set by the material and energy balances. If, on the other hand, a precondenser with sufficient heat-transfer area is included in the process and the temperature of the cooling medium is below the dew point of the vapors at the discharge end of the condenser, there is no theoretical need for vacuum equipment except to remove noncondensables from the process equipment and piping before the start of the operation.

However, it is virtually impossible to exclude noncondensables from a system under vacuum. Even if they do not enter as a reactant, are not dissolved in other reactants, or are not generated by a reaction or sparged into the process, it is inevitable that air will leak into the equipment, piping, and system connections. Extraordinary measures are needed to eliminate all air leakage. This is not normally required for vacuum systems for operations of the scope discussed here. Most of these operations can tolerate the amount of air leakage that enters the system after all reasonable care has been taken to make the system airtight. The rate of expected air leakage to a vacuum system is discussed in the next subsection.

Once the total rate of noncondensables inherent in the system has been established and the air leakage estimated, the flow rate of vapors directly to the vacuum equipment or to and from any precondenser is determined by Dalton's law of additive pressures. The partial pressures of the respective vapor components are those of the liquid or solid mixture in equilibrium with the vapors. The individual vapor flow rates are then given by the equation

$$M_i = \frac{pp_i}{P_v - \Sigma pp_i} \times M_{nc} \qquad (13.5)$$

where M_i = flow rate of vapor component i, mol/h
M_{nc} = flow rate of all noncondensables, mol/h
pp_i = partial pressure of component i, mmHg abs.
P_v = total pressure in vacuum system at equilibrium conditions, mmHg abs.

If there are no liquids or solids in the process system, the flow to the vacuum system is the process vapor plus noncondensables plus the air leakage.

Consideration must be given to the vapor pressure of gasketing or lubricant materials in high-vacuum systems. This could result in a small but steady flow of these materials, which increases the load on a vacuum system without a precondenser or even prevents the desired vacuum level from being attained. These problems normally are not encountered in the operations under discussion here unless the operations are at elevated temperatures. In that case, it is important to specify the proper gasketing and lubricant materials.

At a high vacuum of 10^{-2} to 10^{-4} torr or lower, the phenomenon of outgassing must be taken into account. This involves desorption of gases from the walls of the system. At very high vacuums and elevated temperatures, there is even the possibility of vaporizing small quantities of metals. Although outgassing and metal-vaporization problems occur only in specialized industrial operations, process engineers should at least be aware of the phenomena.

Air leakage. Although air leakage to a vacuum system is one of the more important elements in determining the size of vacuum equipment, it often is difficult to obtain an exact amount for that component. An educated estimate is often the best that can be done to obtain this critical value. However, there are empirical data which correlate leakage with the general volume of the system or with the number and type of connections as a function of system pressure.

An interesting observation can be made about the effect of the pressure differential, caused by the various vacuum levels, on air leakage. If the leakage were through a pinholelike orifice, the flow rate would increase until critical velocity is attained as the pressure decreases to about 400 mmHg abs. The weight rate of air leakage would then be constant even though the system pressure is lowered since critical velocity at the discharge of the orifice would be established at about 400 mmHg abs.

Leakage through a fissure in the vessel wall or across a connection, however, is not through a sharp orifice but by a tortuous path or a long, narrow channel. When the path length with respect to the hydraulic radius is relatively great, the friction-loss behavior is typically as a laminar flow. In this event, there is a nearly continuous pressure drop throughout the entire length of the path, and sonic velocity, if reached, takes place near the end of the path where the pressure is almost equal to the vacuum in the vessel. Thus, sonic velocity at a reduced pressure in a given path represents a diminished weight flow and a smaller air leakage.

Leakage as a function of system volume. Figure 13.11 may be used to estimate air leakage into a vacuum system on the basis of the system volume. The leakage rate thus determined is due essentially to air that enters through minuscule cracks and fissures which are inherent in equipment walls or the porosity of welds and which are not detectable during inspection or testing. Figure 13.11 is a general correlation which assumes that the system under consideration is geometrically similar in its surface-to-volume-area ratio to the average model used in the test work. Thus, it will be noted that the slope of the pressure-parameter curves is 0.66 on the log-log coordinates, which corresponds to the relationship

$$\frac{(\text{Surface area})_2}{(\text{Surface area})_1} = \left(\frac{\text{volume}_2}{\text{volume}_1}\right)^{2/3}$$

for geometrically similar objects.

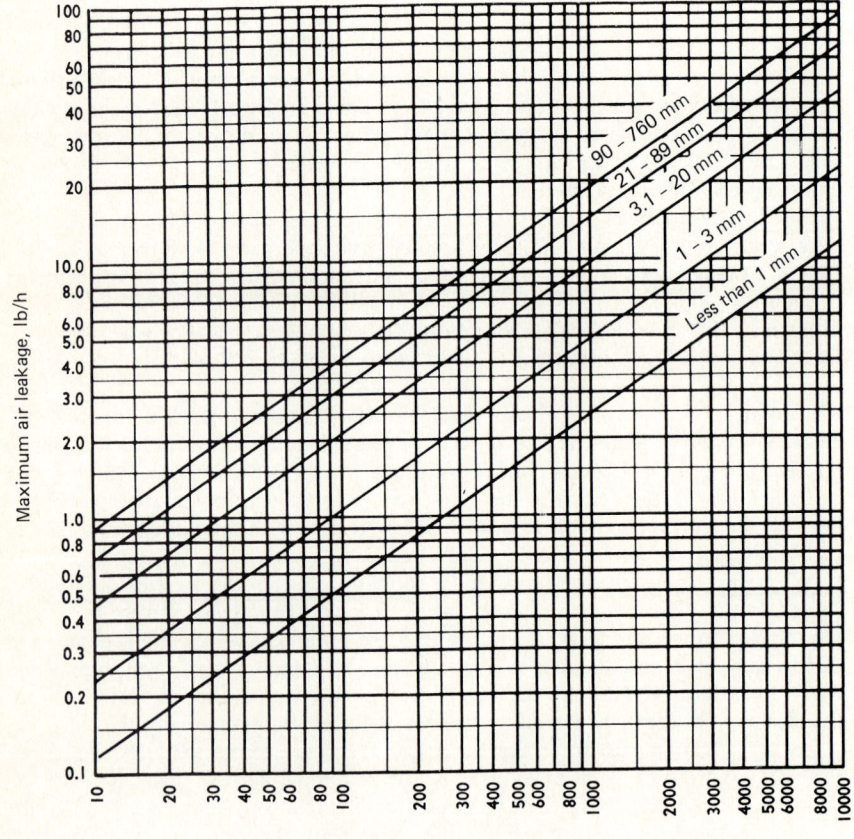

Figure 13.11 Air leakage due to system volume. Maximum air-leakage value for commercially tight systems. *(With permission from The Heat Exchange Institute, Cleveland, Ohio.)*

The correlations in Fig. 13.11 were developed by evacuating a commercially tight vessel with a minimum of connections to a predetermined vacuum, disconnecting the vacuum source, and then following the pressure rise in the unit with time. Simple gas-law equations then permit determination of the air-leakage rate for a known surface area in a given volume system.

Leakage at connections and shafts. Figure 13.11 accounts for only one portion of air leakage since it applies to vessels with a minimum number of connections. Most vessels and equipment in vacuum service contain many appurtenances other than the basic inlet and outlet connections. There are usually additional flanged or screwed connections to accommodate manholes and handholes, vacuum breakers, relief valves, level gauges, sight ports, indicators, and other instrumentation, as well as sampling points. In addition, the equipment itself may have been constructed in several sections which are bolted together. Rotary equipment, such as pumps and agitators, allows relatively large amounts of air leakage to pass through the shaft packing or mechanical seals, especially if the sealing mechanism is located in the vapor space of the system. Figure 13.12 presents typical leakage rates about various fittings that should be taken into account when sizing vacuum equipment. Suggested leakage rates are given as a function of the fitting or the type of shaft seal with system pressure as a parameter.

Leakage in submerged areas. The air-leakage values represented in Fig. 13.12 apply when the connection or shaft seal is located in the vapor space in the vacuum system. If an appreciable number of connections or seals is submerged in the liquid portion of the system, consideration should be given to the fact that the differential-pressure driving force for leakage from air or atmospheric pressure is reduced as a function of the liquid submergence. Thus, a correction factor should be applied for the connection or shaft leakage from Fig. 13.12 to account for submergence. This factor is given in Fig. 13.13 as a function of the degree of vacuum and the specific gravity of the liquid.

The correction values from Fig. 13.13 may also be applied to the air-leakage rate obtained from Fig. 13.11. In this case, Fig. 13.11 is used for that volume of the system which is entirely vapor or gas, while the correction factor that is applied to the volume of the submerged sections is modified as shown in Fig. 13.13 before it is applied to the value in Fig. 13.11.

Reduction of air leakage. The importance of minimizing air leakage can be appreciated by noting the effect of noncondensables on vapor carryover, vacuum-equipment sizing, and utility usage. Process engineers planning engineering flow diagrams should seek to minimize the number of potential leakage points. The design of equipment and piping should be such that welded, soldered, or brazed permanent connections are used wherever possible instead of welded or screwed disconnectable joints. A proper selection of gasketing, O rings, packing, and seal lubricants must be specified to reduce air leakage. Special flanges with surfaces which have concentric rather than spiral grooves may be used to reduce leakage.

In-house design of vessels and equipment must adhere to the correct techniques for preparing welded joints in vacuum service, and suitable welding rods must be specified. Tungsten–inert-gas (TIG) or metal–inert-gas (MIG) welding procedures with consumable bare electrodes are usually the specified procedures for joining metals. The subsection "Metals" in Chap. 5 discusses these methods as well as standards for electrodes, welding techniques, and solder qualifications for these methods. All equipment and prefabricated piping should be vacuum-tested for leakage prior to

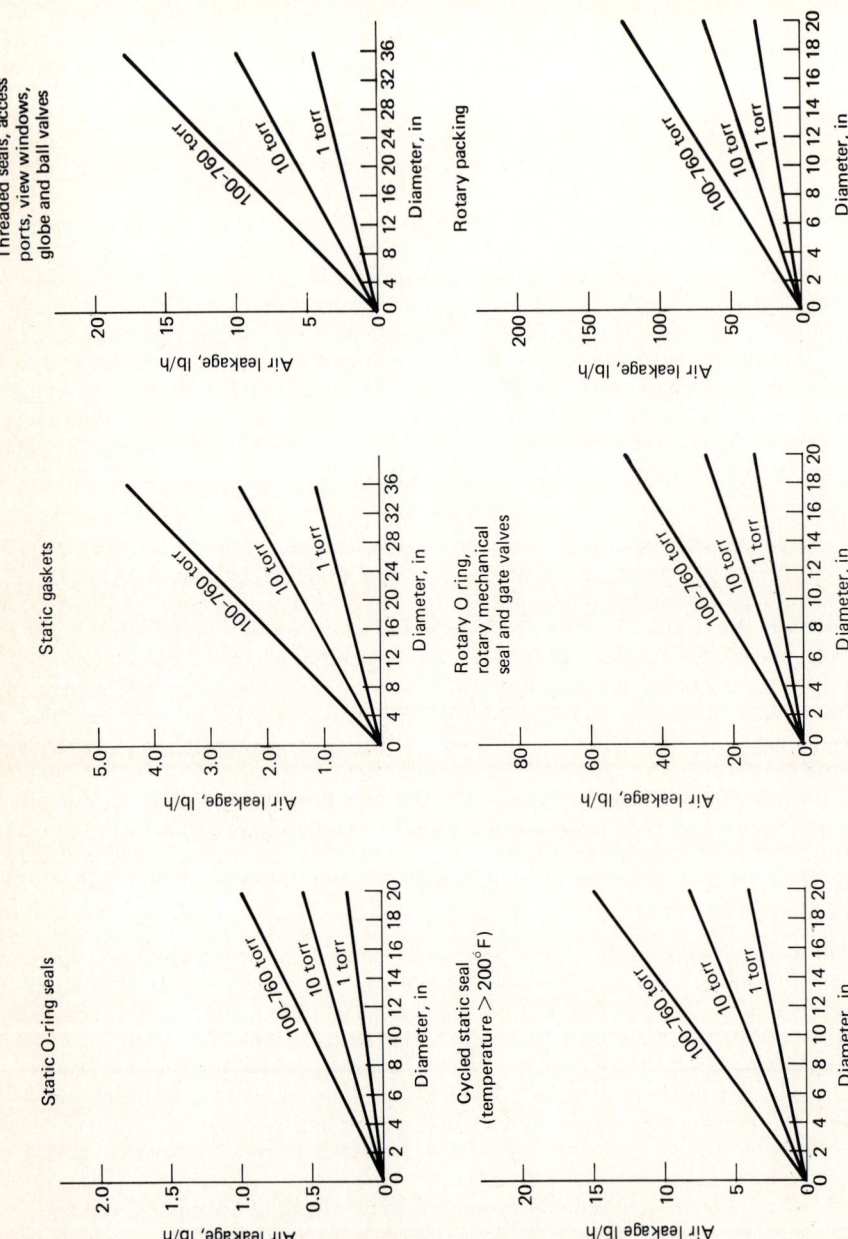

Figure 13.12 Air leakage at connections and shafts.

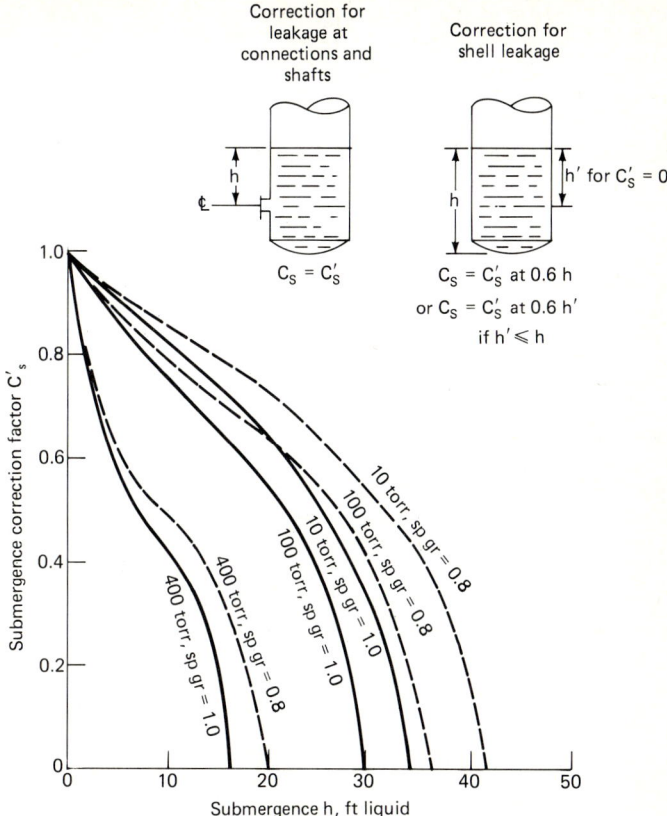

Figure 13.13 Leakage at submerged connections and shafts.

installation so that it can be corrected, if necessary, before it is put in place. All the piping is leak-tested after it is erected, corrections are made, and, if necessary, the entire system is then tested again.

A simple and effective leak test for piping and equipment is the soap-bubble test. The technique of applying a soap solution to all joints and visually inspecting for bubble formation with the vessel under slight pressure is an excellent way of detecting large leaks. Unfortunately, "small" leaks which would be essentially unnoticeable in pressure service could be unacceptable in a vacuum application. This factor can be appreciated by noting that gas leakage out at the rate of 0.1 scfm from a 1000-ft³ vessel, initially at 100 psig or more, changes the pressure in the vessel less than 1 percent in 24 h. On the other hand, a 0.1-scfm leakage of air into a vacuum vessel represents an additional burden of 80 to 300 icfm on a 1-torr-vacuum pumping system, depending on the amount of vapor that the air carries away with it. Thus, means other than soap-bubble testing are used to detect so-called minute leaks for correction.

Several methods for testing a vacuum system at vacuum conditions are available to detect locations where leakage, large or small, may occur. Liquid acetone or even water can be applied externally to local small areas of the surface. A small amount of liquid is drawn into the container through a fissure or a porous area and implodes under the vacuum conditions inside. The momentary conversion of a small volume of liquid to gas (1 oz of water becomes 1000 ft^3 of vapor at 1 torr) overwhelms the vacuum-producing equipment and results in a temporary jump in the pressure reading, denoting a leakage point in the area tested. Another test method directs a halogen gas onto local areas. Instruments sensitive to specific-heat changes can detect the presence of the halogen gas in the effluent from the vacuum equipment.

Once the area of the leak has been defined, it is necessary to grind out the surface metal or portion of the weld and weld over it under the proper conditions to repair the leak. If there is leakage at a flanged joint, it is best to replace the gasket and coat the surface with a sealant such as Glyptal®. Where possible, threaded connections should be backwelded or have a special sealant tape wrapped around the joint.

Attention must be paid to the auxiliary equipment. While it may be possible to test for leakage on the process side of a heat exchanger, it is difficult to locate the exact origin of a leak between the tubes and the tubesheet. It is therefore good practice for process engineers to request that tubes for vacuum service be backwelded to the tubesheets after they have been rolled into them.

Air to barometric condensers. A source of air entry into a vacuum system that is often overlooked is that of the air which is dissolved in the quench water fed to barometric condensers. Since the pressure in the last-stage barometric unit is about 1/5 atm or less, most of the air entering with the condensing water will come out of solution. If a barometric unit is included in the next-to-last stage, essentially all the dissolved air in the water to that unit will evolve. Table 13.3 gives the amount of dissolved air that can be expected in water saturated with air at various temperatures and an atmospheric pressure of 760 mmHg.

TABLE 13.3 Air Dissolved in Saturated Water at 760 mmHg

Water inlet temperature, °F	Air in 100-gal/min contact water, lb/h
50	1.49
60	1.32
70	1.19
80	1.07
90	0.98
100	0.89

Chapter 14

Heat Exchangers

Heat exchange plays a major role in almost every process involved in chemical plants, refineries, or boiler-house installations. This includes bringing reactants to the proper temperature for reaction, changing states in distillation, evaporation, or melting operations, production of steam for power generation, cooling of effluents to reduce thermal pollution, and similar operations. The subject of heat transfer is given much attention in the mechanical and chemical engineering curricula, and a large body of work in the reference and text literature deals with the subject. The intent of this chapter is to review some of the basic fundamentals of heat transfer and to present various factors to be considered in the practical selection and operation of heat-exchange equipment.

Heat-Exchange Fundamentals

Heat is transmitted from one body to another body by one or a combination of mechanisms: by conductance, by convection, or by radiation. The transfer is due to a temperature difference between the two bodies. In radiation, the heat transfer takes place owing to thermal-energy wave transport through space by which the respective bodies need not be in physical contact with each other; the intervening space may be vacuum or a real gas. Heat transfer within a body, regardless of its state, or between two or more bodies in contact with each other may be considered as conductive when there is essentially no tangible movement of particles within the body. Convection takes place in a fluid when there is actual movement of the liquid or gas molecules to transmit the heat from the source body to the body of the fluid. The movement of the fluid may be naturally induced by a density gradient or may be forced.

Conduction

Thermal conductivity k★ is an inherent property of a substance which is a measure of the heat flux or units of heat flow per unit time per unit area across a unit length

---

★Not to be confused with C_p/C_v, the ratio of specific heats for gases, which is also usually termed k.

of the material for one unit of temperature differential and may be expressed as

$$k = \frac{q}{A \, dt/dL} \tag{14.1}$$

where k = thermal conductivity $(\text{Btu} \cdot \text{ft})/(\text{h} \cdot \text{ft}^2 \cdot °\text{F})$
  q = heat transferred per unit time, Btu/h
  A = area at right angle to heat transfer, ft²
  t = temperature, °F
  L = length in direction of flow, ft

This is illustrated in Fig. 14.1, where the area of the surface perpendicular to the heat flow is taken as 1 ft². If L is 1 ft and the temperature difference is 1°F, the heat flow in 1 h represents thermal conductivity.

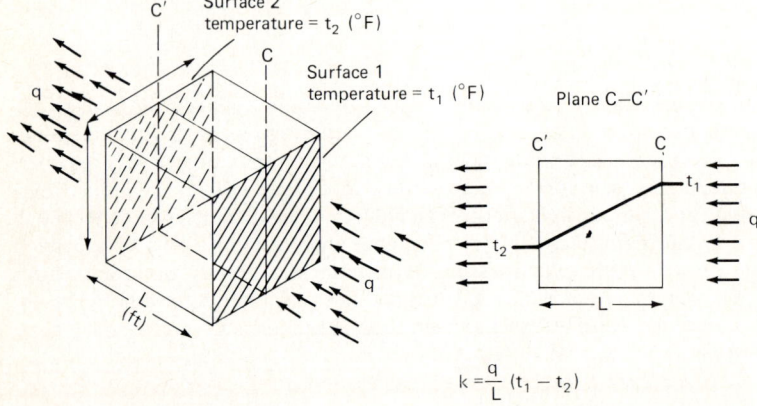

Figure 14.1 Thermal conductivity. All surfaces but 1 and 2 are completely insulated.

It should be noted that the thermal conductivity of a substance is not constant but varies with temperature. However, the variation in k over a small temperature difference is essentially negligible, so that, to all intents and purposes, the slope dt/dL is a constant also and a function of $1/k$. In the event of a large change in temperature, the value of k can be approximated by the natural-log-mean average of values at the two temperatures. However, the ratio of the larger k value to that of the lower one is usually small enough so that the arithmetic average may be used.★

If there is a change in heat-transfer area perpendicular to the line of heat flow or if the length L is not constant along the perimeter of the plane of heat transmission, Eq. (14.1) must be solved by higher-order differential equations. While treatment of these situations is not included in the scope of this work, the subsection "Usages for Thermal Insulation" in Chap. 15 investigates the common case in which the heat-transfer area across insulation on a pipe varies with the thickness of the insulation.

★At a ratio of 2, the arithmetic average is about 4 percent larger than the log-natural-mean average.

Values for thermal conductivity fall within a wide range, depending upon the physical state of the material, its crystalline or molecular structure, and its composition. In terms of British thermal units per hour per square foot per degree Fahrenheit for 1-ft thickness, the k of commonly encountered conductive metals at ambient temperatures is as low as 10 for stainless steel and as great as 240 for silver. Water has a thermal conductivity of 0.33 to 0.40 (ice is 1.3), with that of mercury being about 5 and that of liquid ammonia about 0.3. The k values for organic liquids are as low as 0.04 for some halogenated organics to 0.14 for methyl alcohol. Insulating solids have k values which range from 0.01 for rigid urethane to about 0.035 for calcium silicate. Gases are poor conductors since their k values span from 0.004 for chlorine to 0.015 for water vapor to 0.02 for methane. Hydrogen has a relatively high k value at 0.10 (Btu·ft)/(h·ft²·°F). Comprehensive listings of k's for individual substances at various temperatures are given in *Marks' Standard Handbook for Mechanical Engineers* and *Perry's Chemical Engineers' Handbook*.

At times, it is convenient to express thermal conductivity, especially for insulating materials, in terms of thickness in inches. In that event, the value of k is 12 times that when the unit is for a 1-ft thickness.

If two or more bodies are placed in intimate physical contact in series and heat transfer takes place between them through the same cross-sectional area owing to a thermal gradient across them, each body offers its own resistance to the heat transmission in proportion to its length perpendicular to the flow path and the inverse of its thermal-conductivity values. Thus, the resistance of the bodies shown in Fig. 14.2

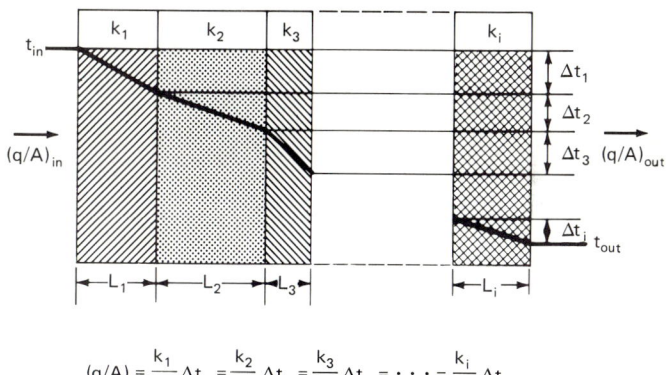

$$(q/A) = \frac{k_1}{L_1}\Delta t_1 = \frac{k_2}{L_2}\Delta t_2 = \frac{k_3}{L_3}\Delta t_3 = \cdots = \frac{k_i}{L_i}\Delta t_i$$

Figure 14.2 Resistance of bodies in series.

has the value of

$$R_i = \frac{L_i}{k_i} \qquad (14.2)$$

where R = resistance to heat flow, (h·ft²·°F)/Btu
L = thickness of body, ft
k = thermal conductivity, (Btu·ft)/(h·ft²·°F)
subscript i = body i

Eq. (14.1) may be rearranged and integrated for a constant slope of $(dt/dL)_i$ and Eq. (14.2) substituted into it to express the heat flux across any of the bodies by

$$(q/A)_i = -\frac{\Delta t_i}{R_i} \tag{14.3}$$

where Δt_i = temperature change across any body, °F.

The heat flux q/A is the same for each body at steady state; i.e., all the heat per unit time passing through the first body leaves the last body with no net accumulation or discharge in any of the bodies, and each has the same transfer area.

Thus

$$q/A = \frac{\Delta t_1}{R_1} = \frac{\Delta t_2}{R_2} = \cdots = \frac{\Delta t_i}{R_i} \tag{14.4}$$

so that $\Delta t_i = (q/A)R_i$. $\tag{14.5}$

However, the total temperature change Δt_T across all the bodies is equal to the sum of the individual temperature changes across the body, so that

$$\Delta t_T = \Sigma \, \Delta T_i = (q/A)(R_1 + R_2 + \cdots + R_i) \tag{14.6}$$

where Δt_T = overall temperature change across the bodies, °F.

The heat flux can be expressed as a function of an overall or total resistance and the overall temperature change as

$$q/A = \frac{\Delta t_T}{R_T} \tag{14.7}$$

where R_T = overall thermal resistance, $(h \cdot ft^2 \cdot °F)/Btu$.

By substituting Eq. (14.7) into Eq. (14.6) to eliminate q/A, it can be seen that

$$R_T = \Sigma R_i = R_1 + R_2 + \cdots + R_i \tag{14.8}$$

By replacing the individual resistance values with the expression from Eq. (14.2), Eq. (14.8) becomes

$$R_T = \frac{L_1}{k_1} + \frac{L_2}{k_2} + \cdots + \frac{L_i}{k_i} \tag{14.9}$$

Equations (14.8) and (14.9) are important since they allow calculation of an overall resistance as the direct sum of a series of resistances and permit a quick method of analyzing conductivity heat transfer. However, the reciprocal of resistance is conductance, and the overall thermal conductance can be expressed as

$$C = \frac{1}{R_T} = \frac{1}{\Sigma R_i} = \frac{1}{R_1 + R_2 + \cdots + R_i} \tag{14.10}$$

or $$C = \frac{1}{L_1/k_1 + L_2/k_2 + \cdots + L_i/k_i} \tag{14.11}$$

and $$C = \frac{1}{1/C_1 + 1/C_2 + \cdots + 1/C_i} \tag{14.12}$$

where C = conductance, $Btu/(h \cdot ft^2 \cdot °F)$.

Equation (14.12) is important since

$$C_i = k_i / L_i \qquad (14.13)$$

and thermal-conductivity problems may frequently be analyzed by using this form of the equation. However, equations similar to Eq. (14.12) are generally used to approximate heat transfer for convection problems and to treat problems involving combinations of conduction, convection, and radiation heat transfer. In these cases, resistance and conductance terms can be used in the same equation as long as they maintain their proper dimensional relationships.

By combining Eqs. (14.4) and (14.7), it can be seen that the fractional temperature change across a body is proportional to the fractional resistance of that body, or

$$\frac{\Delta t_i}{\Delta t_T} = \frac{R_i}{R_T} \qquad (14.14)$$

Equation (14.14) facilitates the apportionment of overall temperature change among the individual bodies. By grouping resistances in the proper order in a multibodies composition, the temperature of an intermediate boundary can be obtained from the initial temperature and use of the corollary equation

$$\frac{\sum_{i=1}^{i=n} \Delta t_i}{\Delta t_T} = \frac{\sum_{i=1}^{i=n} R_i}{R_T} \qquad (14.15)$$

The above relationships are based on steady-state conditions. However, they may be applied in their differential forms to unsteady-state and transient problems with the included mass and heat capacities of the various materials to determine the thermal inertia of the system.

Convection

Thermal heat transfer by convection is a special case of heat transfer by conductivity in that one or more of the bodies shown in Fig. 14.2 is a fluid, liquid, or gas in which there is considerable displacement of molecules within the body. The movement of the particles, therefore, facilitates the transfer of heat from one boundary of the body to the body itself or to another boundary. Thus, the conductance of heat across a body becomes considerably greater than could be accounted for by the thermal conductivity divided by the actual distance between boundary layers. The higher heat transfer in convection can be rationalized, however, if the heat transfer by thermal conductivity were considered to be across a thin layer of stationary matter at the boundary and then distributed to the bulk of the body by convective currents. The resistance to heat flow would then be that of the boundary layer plus a film of fluid that extends a short distance into the fluid. The boundary layer is composed of scale or foreign matter that is deposited from the fluid at the transfer surface, while the fluid resistance itself is due to a film of fluid in laminar flow at the boundary which, under certain conditions, gradually changes to turbulent flow.

Since the thicknesses of the boundary and film layers are influenced by a number of factors, they cannot be readily measured. However, the influence of these values on heat flux has been determined experimentally and correlated with well-known thermal dimensionless relationships. The results are known as scaling or fouling factors f, used to represent a stationary boundary layer as a resistance, and as film coefficients h, used to represent the film layers as a conductance. They are used in a combination of Eqs. (14.10) and (14.12), whereby the conductance term C is known as U, the overall heat-transfer coefficient, and is defined as

$$U = \frac{1}{1/h_i + f_i + L/k + f_o + 1/h_o} \tag{14.16}$$

where subscripts i and o = inside and outside, respectively.

The form of the equation for U in Eq. (14.16) is based on the assumption that the inside and outside heat-transfer areas are equal. If they are not, as is the case of a tube or pipe which has a relatively great thickness compared with its radius, then the overall coefficient U is referenced to one of the surfaces, usually the outer one, and the inner-film coefficient is corrected for the difference. The revised equation for U_o is then

$$U_o = \frac{1}{(1/h_i \times A_o/A_i) + (f_i \times A_o/A_i) + (L/k \times A_o/A_m) + f_o + 1/h_o} \tag{14.17}$$

where A_o = outer-surface area
A_i = inner-surface area
A_m = average surface areas of A_o and A_i

Evaluation of temperature differences. The classic expression for convective heat transfer in its differential form is

$$\frac{dq}{dA} = U \, \Delta T \tag{14.18}$$

and says that the heat transferred per unit time across a unit area dq/dA is a function of an overall heat-transfer coefficient U present at the differential area multiplied by the thermal driving force ΔT, which is the temperature difference between the fluid on one side of the area and the fluid on the other side.

There are only a few occasions in which the driving force ΔT stays constant throughout the entire area over which heat transfer takes place. A typical situation would be the loss of heat from an insulated line carrying saturated steam to its atmospheric surroundings. If the change in pressure of the steam along the pipe is negligible, the steam temperature remains essentially the same throughout. If the surrounding air is unconfined, the bulk-air temperature is little affected by heat from the line and is constant along the length of the line.

A second example of isothermal operations would be a situation in which a refrigerant is evaporated but not superheated from one side of an exchanger while a saturated second component condenses but is not subcooled on the other side and there is essentially no pressure differential for either fluid to affect the boiling point or the dew point of the respective sides. Another possibility of isothermal-operations difference is the counterflow of the two fluids with the coincidental situation that they both have the same product of the weight flow rate times specific heat at all points along the entire heat-transfer-area path.

In almost all other circumstances there is a variation in the thermal driving force ΔT along the path represented by the differential areas of heat transfer. In most situations, ΔT is a linear function of the heat transfer rate q itself. Likewise, in most instances, any variation in the overall heat-transfer coefficient U in turn can be approximated by a linear function of temperature difference. Thus, integration of Eq. (14.18) becomes

$$q = A \left[\frac{(U_1 \Delta T_2 - U_2 \Delta T_1)}{\ln\left(\dfrac{U_1 \Delta T_2}{U_2 \Delta T_1}\right)} \right] \tag{14.19}$$

where subscripts 1 = end of exchanger with greater ΔT
subscripts 2 = end of exchanger with smaller ΔT

In most cases there is only a small variation in the overall heat-transfer coefficient from one end of the exchanger to the other. As long as the U value at one end is not greater than twice that at the other, the overall heat-transfer coefficient may be treated as a constant equal to the arithmetic average of the two. Calculation results by using the arithmetic mean value are within acceptable engineering limitations the same as if the U's had been part of the natural-log mean $U\Delta T$. Equation (14.19) then can be expressed as

$$q = U_{avg} \, A \, \Delta T_m \tag{14.20}$$

where U_{avg} = arithmetic average of U_1 and $U_2 = \dfrac{U_1 + U_2}{2}$, Btu/(h · ft² · °F)

ΔT_m = natural-log-mean-temperature difference (LMTD), °F

If U_{avg} is considered as a constant and can be substituted for U_1 and U_2 in Eq. (14.19), then ΔT_m is the LMTD, which equals

$$\Delta T_m = \text{LMTD} = \frac{\Delta T_2 - \Delta T_1}{\ln\left(\dfrac{\Delta T_2}{\Delta T_1}\right)} \tag{14.21}$$

In many cases, there is a large variation between the discharge- and entrance-temperature differentials, and Eq. (14.21) must be used to calculate the LMTD. However, if the larger ΔT is less than twice the smaller one, the arithmetic mean of the two ΔT's may be used instead of the log-natural mean with results within acceptable engineering limits. If there is a significant variation in the overall heat-transfer coefficient as well as in the thermal differentials, use of the arithmetic averages for each of these values is acceptable as long as the ratio of the greater product $U_1 \Delta T_2$ or $U_2 \Delta T_1$ to the lesser is smaller than 1.5 and neither of the maximum ratios of the end values of overall heat-transfer coefficients or temperature differentials exceeds 2.

Parallel flow. Several flow patterns between the hot and cool fluids in an exchanger are illustrated in Fig. 14.3 for tubular-type exchangers. The simplest is parallel flow, or cocurrent flow, in Fig. 14.3a. In this case, both the hot and the cool fluids enter at the same end of the exchanger and proceed in the same direction to the other end. As one fluid transfers heat to the other, the temperature of the hotter fluid drops while that of the colder fluid increases. The largest temperature differential occurs at the entrance end of the exchanger. The temperature differential then decreases until it is at a minimum at the discharge end. If there were sufficient heat-transfer area in a parallel-flow exchanger, the temperatures of the two discharge fluids would approach one another. The value of that temperature would be a function of the

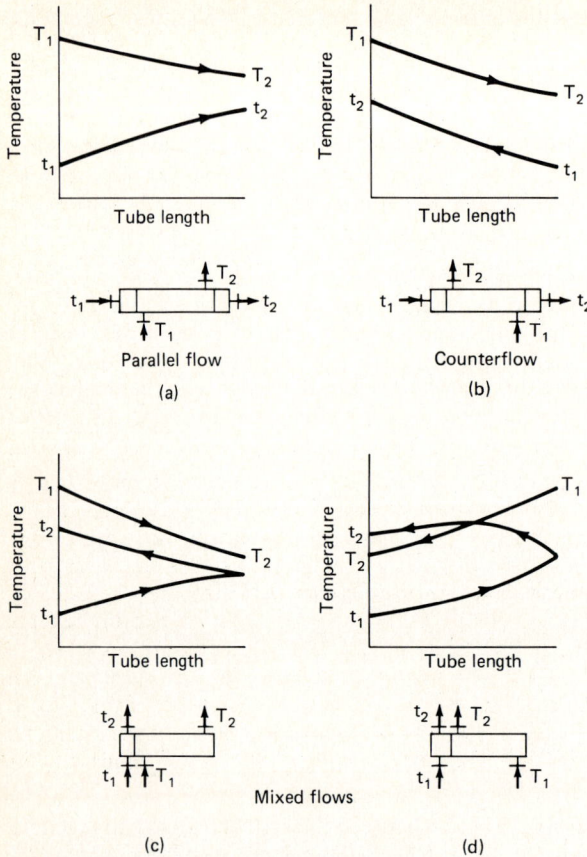

Figure 14.3 Fluid-flow patterns in heat transfer.

initial temperatures of the two streams and the product of their respective mass flow rates and specific-heat values with the same amount of heat being transferred to and from each other. In true parallel flow, there can never be a temperature cross of the two fluids; i.e., at no point in the exchanger will the temperature of the cooler fluid be greater than the lowest cooled hot-fluid temperature in the exchanger or the temperature of the hotter fluid be lower than that of the highest warmed cold-fluid temperature at any point in the exchanger.

Counterflow. Figure 14.3b illustrates counterflow, or countercurrent-flow, heat transfer. The hot fluid enters at one end of the exchanger, while the cold fluid enters at the opposite end, so that they flow countercurrently to one another through the entire exchanger. Depending on the available area in the exchanger, it is relatively easy to obtain a temperature cross in the exchanger. With sufficient transfer area, the temperature of one of the outgoing streams can approach the incoming temperature of the other stream. This is to be contrasted with parallel flow, for which the discharge temperature can only approach a temperature intermediate between the two inlet temperatures.

Mixed flow. Only certain types of heat exchangers such as the double-pipe type, spiral units, and single-pass shell-and-tube exchangers as well as some plate exchangers can achieve true parallel-flow or counterflow patterns. The majority of heat exchangers employ a combination of the two basic patterns in order to obtain the necessary fluid velocities to effect reasonable heat-transfer coefficients within limited dimensions. Figure 14.3c and d depict the temperature relationships for a mixed-flow situation in which there is a single pass for one fluid and a double pass for the other fluid. This results in a situation in which the flow of the fluids is initially parallel and then countercurrent, as in Fig. 14.3c, or is initially countercurrent and completes the second pass as a parallel flow, as in Fig. 14.3d. Despite the apparent difference in the shape of the curves in Fig. 14.3c and d, the discharge temperature of the respective fluids would be the same for the same entrance conditions and exchanger transfer area and design.

Because of the approach of the two fluid temperatures which occurs during the parallel-flow portion of the heat exchange, it is not possible to use the calculated LMTD from Eq. (14.21) alone with Eq. (14.20) to determine the area required for a given heat-transfer rate. A correction factor F_T must be used to account for deviations from true counterflow in the mixed-flow patterns. The LMTD, when multiplied by F_T, is known as the corrected natural-log-mean-temperature difference, or $\text{LMTD}_{\text{corr}}$.

The temperature correction factor is a function of the individual entering and discharge temperatures of the fluid stream and the flow patterns of the heat exchanger or series of exchangers required for the transfer. The factor may be calculated by means of a series of complex equations which are beyond the scope of this work (see Kern[1]). However, graphical solutions for F_T are available for a number of exchanger configurations, and several typical ones are presented in Fig. 14.4. F_T is found at the intersection of two parameters on the graph. One of them, termed R in Fig. 14.4, is the ratio of the mass heat contents of fluid a to fluid b. Since the heat transfers to or from fluids a and b are equal, the mass-heat-content ratio of the fluids may be

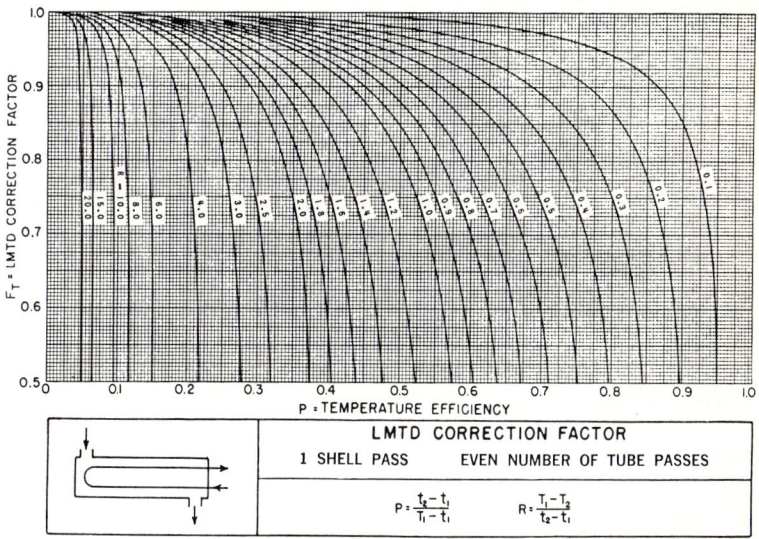

Figure 14.4 Temperature differential correction factors F_T. *(With permission from Tubular Exchanger Manufacturers Association, Tarrytown, N.Y., © 1978.)*

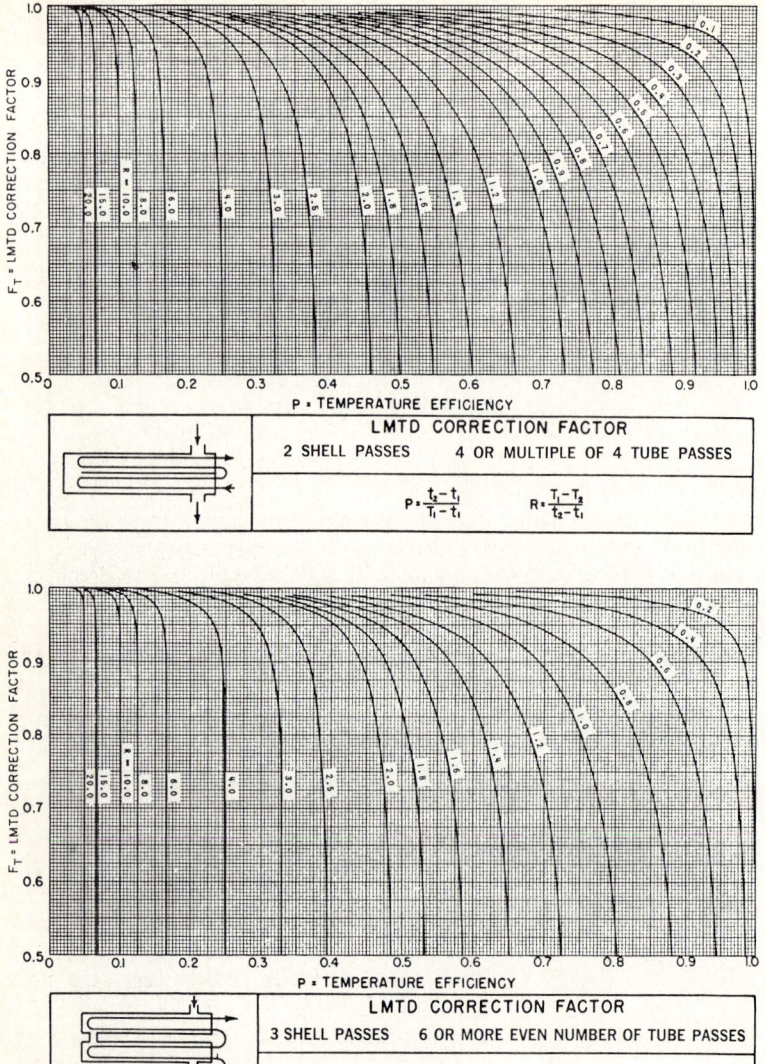

Figure 14.4 (*Continued*)

expressed as the inverse ratio of the temperature changes of the respective fluids as they pass through the exchanger, so that

$$R = \frac{(Wc_p)_a}{(Wc_p)_b} = \frac{(t_1 - t_2)_b}{(t_2 - t_1)_a} \tag{14.22}$$

where R = parameter in determining F_T
w = weight flow rate, lb/h

c_p = specific heat, Btu/(lb·°F)
t = temperature of fluid, °F
subscripts 1 and 2 = entering and leaving conditions respectively
subscripts a and b = fluid a and fluid b, respectively.

The companion factor P relates the temperature change in fluid a to the difference in entering temperature for the respective fluids, so that

$$P = \frac{(t_2 - t_1)_a}{(t_b - t_a)_1} \tag{14.23}$$

The correction factor F_T should be obtained from Fig. 14.4 and applied to the LMTD for all cases of mixed-flow patterns. It is of special importance in situations in which the discharge values of the two fluids approach one another or actually cross. In these cases, F_T may be 0.75 or lower, so that it is advisable to use a configuration involving an additional shell in order more nearly to approach counterflow conditions. The use of the additional shell passes of exchangers increases the F_T considerably, and the reduced total transfer area will offset the additional cost of a long baffle in the shell or that of a separate shell. The number of shell configurations in Fig. 14.4 should be chosen so that the F_T point is on or to the left of the shoulder of the R curve where the F_T is most stable. An F_T to the right of the shoulder or on the portion of the R curve that is nearly vertical could easily deteriorate under transient conditions so that the exchanger might be unstable in its response to these conditions.

Special conditions. There are sometimes situations in which the LMTD equation (14.21) is not valid when applied to the end values of fluids. This typically occurs in condensing situations, particularly for superheated multicomponent vapors containing a noncondensable. Typical temperature-duty curves for the vapors and the cooling medium are shown in Fig. 14.5. In this case, the area required for desuperheating is calculated separately from the condensing section not only for the expediency of obtaining the proper temperature-differential value but because the U for desuperheating is usually considerably lower than that for condensing. It should be noted that the transition between the desuperheating U and the condensing U takes place when the tube-wall temperature is below the dew point of the vapor.

While the LMTD equation may be used for the desuperheating zone, the remainder of the condensing curve should be divided into several convenient segments for which temperature change is nearly linear with duty and the area for each section is then evaluated separately. When noncondensables are present, as cited for the example in Fig. 14.5, there can be a considerable change in U as condensation progresses. Thus, the U values should be calculated for the beginning and end of each segment and used as the natural-log mean $(U\Delta T)$ shown in Eq. (14.19) for evaluating the transfer area for that segment.

Dimensionless groups. The values of heat-transfer coefficients for the various types of films that are encountered in convective-heat transfer are obtained from experimental data. The limited amount of data that has been obtained can be extended by means of several dimensionless groups which relate various thermal and physical properties of fluids as well as flow rates or velocities. The various dimensionless groups are combined in equations, and the experimental data are analyzed to obtain the exponents for each group and the coefficients of proportionality.

The name of a person who is being honored or who proposed the particular rela-

502 Equipment Selection, Sizing, and Related Subjects

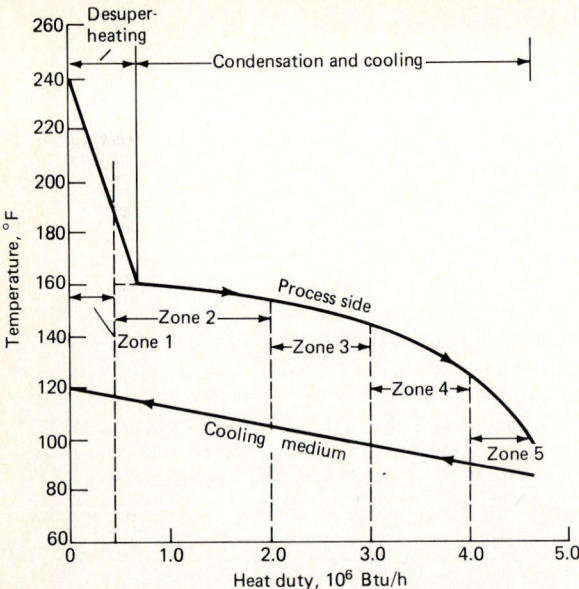

Figure 14.5 Typical desuperheating and condensing heat duty versus temperature curve with noncondensables.

tionship of physical properties is usually applied to the group. Table 14.1 lists several of the more common dimensionless groups used for analyzing heat transfer and for obtaining heat-transfer coefficients.

Natural convection. A distinction is usually made between natural and forced convection. Natural convection occurs when a solid material is in contact with an essentially stationary or still fluid, liquid or gas, and there is a temperature difference between the solid sink and the fluid. Because of the temperature differential, a movement is induced in the fluid adjacent to the solid. If the solid is cooler than the bulk-fluid temperature, the movement of the gas or liquid will be in the downward direction in the vicinity of the solid while the fluid will rise if the solid is warmer. The movement of the fluid reduces the resistance to heat transfer between the solid and the fluid. As the temperature differential increases, the fluid velocity near the solid face becomes greater so that the heat-transfer coefficient at the interface becomes larger.

The relationship among the various physical parameters of the fluid, the temperatures of the solid and the fluid, and the heat-transfer coefficient can be expressed in terms of three dimensionless groups: the Nusselt number, the Prandtl number, and the Grashof number. The correlation among the groups for a flat surface is

$$\frac{hL}{k} \propto \left[\left(\frac{c_p \mu}{k}\right)^m \left(\frac{L^3 \rho^2 \beta g_c \, \Delta t}{\mu^2}\right)^n \right] \qquad (14.24a)$$

where L = length of a flat surface.

TABLE 14.1 Dimensionless Groups for Heat Transfer

Name	Symbol	Group
Biot number	Bi	hr_m/k
Condensation number	Co	$(h/k)(\mu^2/\rho^2 g_c)^{1/2}$
For vapor-condensation equations	Cv	$L^3\rho^2 g\lambda/k\mu\Delta t$
Fourier number	Fo	$k\theta/\rho c_p r_m^2$
Graetz number	Gz	wc_p/kL
Grashof number	Gr	$D^3\rho^2 g_c\beta\Delta t/\mu^2$
Nusselt number	Nu	hD/k
Péclet number	Pe	$Du\rho c_p/k$
Prandtl number	ρ	$c_p\mu/k$
Reynolds number	Re	$Du\rho/\mu$
Schmidt number	Sc	$\mu/\rho\, D_v$
Stanton number	St	$h/c_p u\rho$

where the following definitions with English units satisfy the dimensionless characteristics:

c_p = specific heat, Btu/(lb·°F)
D = diameter, ft
D_v = volumetric diffusivity, ft²/h
g_c = dimensional constant, 4.17×10^8 (lbm/lbf)(ft/h²) ≈ 32.17 (lbm/lbf)(ft/s²)
G = mass velocity, lbm/(h·ft²)
h = coefficient of heat transfer, Btu/(h·ft²·°F)
k = thermal conductivity, (Btu·ft)/(h·ft²·°F)
L = length, ft
r = radius, ft
t = temperature, °F
u = velocity, ft/h
β = coefficient of expansion, 1/°R
Δ = prefix for finite difference
λ = enthalpy change, Btu/lbm
μ = absolute viscosity, lbm/(ft·h) = 2.42 cP
ρ = density, lbm/ft³
subscript m = mean

And for cylinders is

$$\frac{hD}{k} \propto \left[\left(\frac{c_p\mu}{k}\right)^m \left(\frac{D^3\rho^2\beta g_c\,\Delta t}{\mu^2}\right)^n\right] \qquad (14.24b)$$

where D = diameter of a cylinder.

The proportionality coefficients for use with Eqs. (14.24a) and (14.24b) have been established for various physical configurations by the analysis of experimental data by means of the dimensionless groups.

One of the more common applications for natural-convection relationships is the thermal-energy exchange between hot or refrigerated lines or vessels and the ambient surroundings. Of importance also is the prediction of heat transfer between coils within a vessel or from its jacketed walls to the contents within when the contents are not agitated. In the late 1920s, Wilhelm Nusselt conducted a wide range of experiments regarding natural convection for a wide variety of gases and liquids. T. H. Chilton and associates, in the early 1930s, interpreted Nusselt's data by using the quantity $k^3\rho^2 c_p\beta/\mu$ as a reference point and reported the variation of natural-convection heat-transfer coefficients for horizontal pipes at various ratios of temperature difference and pipe diameter for liquids and the same ratio multiplied by the square of the pressure for gases.

The correlation thus attained for liquids is

$$h_{L,n} = 85 \left(\frac{k^3 \rho^2 c_p \beta}{\mu} \right)^{0.25} \left(\frac{T_m}{d} \right)^{0.25} \tag{14.25a}$$

And for gases is

$$h_{g,n} = 3.6 \left(\frac{k^3 \rho^2 c_p \beta}{\mu} \right)^{0.5} \left(\frac{p^2 \Delta T_m}{d} \right)^{0.25} \tag{14.25b}$$

where h = heat-transfer coefficient, Btu/(h·ft²·°F)
 k = thermal conductivity, (Btu·ft)/(h·ft²·°F)
 c_p = heat capacity, Btu/(lb·°F)
 ΔT_m = mean temperature difference across film, °F
 d = diameter of pipe, in
 p = gas pressure, atm abs.
 β = coefficient of thermal expansion, 1/°F
 μ = viscosity, cP
 ρ = density at 1 atm, lb/ft³

subscripts L and n = liquid, natural convection
subscripts g and n = gas, natural convection

Since the temperature of the fluid affects all the parameters in the reference quantity, the heat-transfer coefficients derived from Chilton et al. for horizontal cylinders can be expressed by Eq. (14.26) for liquids and Eq. (14.27) for gases.

$$h_{L,n} = C(t_m)^a \left(\frac{\Delta T_m}{d} \right)^{0.25} \tag{14.26}$$

$$h_{g,n} = C(460 + t_m)^a \left(\frac{p^2 T_m}{d} \right)^{0.25} \tag{14.27}$$

where h = heat-transfer coefficient, Btu/(h·ft²·°F)
 C = proportionality coefficient
 t_m = mean temperature of fluid film, °F
 ΔT_m = mean temperature difference across film, °F
 d = diameter of pipe, in
 p = gas pressure, atm

subscripts L and n = liquid, natural convection
subscripts g and n = gas, natural convection

The proportionality coefficient C and the exponent a for many common fluids are given in Table 14.2, which also lists typical heat-transfer coefficients for some natural-convection situations for horizontal pipes.

It has been shown that natural-convection heat-transfer coefficients expected for vertical pipes and vertical plates are similar to those for horizontal pipes while those for horizontal plates with the interface to the fluid facing downward are only about 75 percent of the horizontal-pipe coefficients and those for horizontal plates facing upward are about 40 percent greater than that calculated for a horizontal pipe. When evaluating heat-transfer coefficients for flat surfaces, the length L is used in place of

TABLE 14.2 Natural-Convection Heat-Transfer Coefficients for Horizontal Pipes*

$$h_{g,n} = C(460 + t_m)^a \left(\frac{p \, \Delta T_m}{d}\right)^{0.25} \quad \text{gases}$$

$$h_{L,n} = C(t_m)^a \left(\frac{\Delta T_m}{d}\right)^{0.25} \quad \text{liquids}$$

[See Eqs. (14.26) and (14.27) for nomenclature.]

Gases	C	a	$h_{g,n}$ for $\left((t_m = 200), \frac{p^2 \Delta T_m}{d} = 100\right)$, Btu/(h·ft²·°F)
Air	2.5	−0.29	1.18
Ammonia	0.37	+0.02	1.34
Carbon dioxide	0.26	+0.02	0.97
Carbon monoxide	2.5	−0.29	1.23
Hydrogen	25.3	−0.50	3.20
Methane	71.0	−0.83	1.04
Nitric oxide	3.8	−0.37	1.09
Nitrogen	2.5	−0.29	1.20
Oxygen	3.2	−0.32	1.24
Sulfur dioxide	0.34	−0.04	0.83
Water vapor	0.21	+0.07	1.04

Liquids	C	a	$h_{L,n}$ for $(t_m = 200), \left(\frac{\Delta T_m}{d} = 100\right)$, Btu/(h·ft²·°F)
Acetic acid	11.3	0.128	70
Acetone	18.8	0.063	83
Aniline	3.8	0.30	59
Benzene	10.5	0.12	62
n-Butanol	3.9	0.29	58
Carbon disulfide	17.3	0.05	72
Carbon tetrachloride	13.9	0.01	47
Ethanol	5.8	0.27	77
Ethyl acetate	16.7	0.07	76
Ethyl ether	13.3	0.12	78
Hydrochloric acid (30%)	9.8	0.28	138
Methanol	12.2	0.16	89
n-Pentane	10.2	0.17	79
Sulfuric acid (60%)	8.3	0.27	109
Sulfuric acid (98%)	2.8	0.41	77
Toluene	11.7	0.10	62
Water	5.5	0.49	240

Convection multiplying factors for other configurations:
Vertical pipes   1.0
Vertical plates   1.0†
Horizontal plates (face down)   0.75†
Horizontal plates (face up)   1.40†

*Based on T. H. Chilton et al., *Transactions of the American Society of Mechanical Engineers, Petroleum Mechanical Engineering*, vol. 55, 1933, pp. 7–14 (still fluid).

†Use L in inches for d up to 12 in for vertical plates; use $d = 12$ in for longer vertical plates and all horizontal plates.

the diameter d. The units of L are inches; however, a maximum value of 12 should be used for L for all horizontal and vertical plates which exceed 12 in in length.

Heat transfer by natural convection from hot bodies to gases, especially atmospheric air, is usually accompanied by radiant heat transfer. In these instances, the radiant-heat-transfer coefficients described in the next subsection are added to the natural-convection coefficients to determine the overall heat-transfer coefficient h and the total heat transferred under the circumstances.

Forced convection. In natural convection a temperature difference and resulting specific-gravity differences within the fluid induce a movement of the fluid across the surface which lowers resistance to heat transfer. The greater the temperature difference, the greater the movement at the surface, and the heat-transfer rate is increased not only owing to the higher temperature differential but also because of a lower film resistance. Thus, increased heat-transfer film coefficients can be obtained by using means other than temperature difference to induce movement of a fluid past a surface across which heat transfer is to be effected. In other words, the movement of the fluid past the surface can be forced by mechanical methods such as pumps, blowers, compressors, or agitators or by nonmechanical means such as converting potential energy to kinetic energy. Flow of a liquid from a higher elevation to a lower one through a heat exchanger and expansion of a fluid through an ejector or eductor to induce flow past a heat-transfer surface are examples of nonmechanical forced flow. The transfer of heat by these "nonnatural" methods is known as forced convection.

The equations derived for natural convection are not suitable for indicating forced-convection heat-transfer film coefficients. The equations for use in heat transfer due to forced convection are reviewed in the following subsections.

Heat-transfer coefficients. The overall heat-transfer coefficient U used in Eq. (14.20) is a measure of the thermal conductance of a heat exchanger. This coefficient is the reciprocal of the total thermal resistance produced by the sum of the following individual resistances:

1. Film resistance for the hot phase, R_h
2. Fouling or scaling resistance on the hot-phase side, f_h
3. Thermal resistance of the solid boundary separating the two phases, x/k
4. Fouling or scaling resistance on the cold-phase side, f_c
5. Film resistance for the cold phase, R_c

The relationship of these resistances can be represented by a variation of Eq. (14.17):

$$\frac{1}{UA} = \frac{R_h}{A_h} + \frac{f_h}{A_h} + \frac{x}{kA_m} + \frac{f_c}{A_c} + \frac{R_c}{A_c} \tag{14.28}$$

where A_h = hot-side surface area, ft^2
A_c = cold-side surface area, ft^2
A_m = average area between A_h and A_c, ft^2
A = required heat-transfer area, usually based on the surface area that is the most convenient to measure, such as the outside area of tubing, ft^2
x = thickness of the solid boundary, ft
k = thermal conductivity of the material used as the solid boundary, (Btu·ft)/(ft^2·h·°F)

If all areas are assumed to be approximately the same, Eq. (14.28) reduces to

$$\frac{1}{U} = R_h + f_h + \frac{x}{k} + f_c + R_c \qquad (14.29)$$

Film resistance is normally reported in its reciprocal form, the film heat-transfer coefficient h. Therefore, Eqs. (14.28) and (14.29) are typically represented as

$$\frac{1}{UA} = \frac{1}{h_h A_h} + \frac{f_h}{A_h} + \frac{x}{kA_m} + \frac{f_c}{A_c} + \frac{1}{h_c A_c} \qquad (14.30)$$

and $\quad \dfrac{1}{U} = \dfrac{1}{h_h} + f_h + \dfrac{x}{k} + f_c + \dfrac{1}{h_c} \qquad (14.31)$

when $A = A_h = A_n = A_m = A_c$, respectively.

Table 14.3 lists typical ranges of overall heat-transfer coefficients for various types of heat-transfer equipment in different process applications. This table can be used to get a rough estimate of the overall heat-transfer coefficient for use in preliminary process designs or for cursory checks on offered areas for purchased equipment.

A closer approximation of the overall heat-transfer coefficient U can be obtained if individual film heat-transfer coefficients for different process fluids and operating conditions and fouling factors could be estimated. These could be inserted into Eq. (14.30) or Eq. (14.31) to calculate an overall heat-transfer coefficient. Tables 14.4 and 14.5 list the approximate film coefficients and fouling factors, respectively, for various process conditions. Thermal resistance for various engineering materials used for heat-exchange tubing is listed in Table 14.6.

The tables can be used to obtain a quick estimate of the overall heat-transfer coefficient for a heat exchanger. More accurate estimates of film coefficients require a rigorous analysis using Nusselt-type relationships of the form

$$\mathrm{Nu} = a(\mathrm{Re})^b (\mathrm{Pr})^c f(x) \qquad (14.32)$$

where $\mathrm{Nu} = hD/k$ = Nusselt number
$\quad\quad\quad \mathrm{Re} = D_t G/\mu$ = Reynolds number
$\quad\quad\quad \mathrm{Pr} = c_p \mu/k$ = Prandtl number
$\quad a, b, c$ = empirical constants which are a function of the heat-transfer equipment
$\quad\quad f(x)$ = empirical function of various key parameters which affect the film coefficient

The general form of Eq. (14.32) can vary considerably, depending on the heat-transfer equipment, equipment orientation, process fluid, and heat-transfer mechanism. For example, the Reynolds number is composed of different parameters depending on whether the film coefficient is calculated for a shell-and-tube heat exchanger or a jacketed reactor vessel. In heat-exchange devices for which natural convection is the dominant mechanism, the Reynolds number is replaced by the Grashof number, as noted earlier.

The empirical functions in Eq. (14.32) are not included in every relationship; for example, the Sieder and Tate equation includes the functions

$$f_1(x) = (D/L)^{0.33} \qquad (14.33)$$

and $f_2(x) = (\mu/\mu_w)^{0.14} \qquad (14.34)$

TABLE 14.3 Overall Heat-Transfer Coefficients for Typical Petrochemical Applications: U, Btu/(h·ft²·°F)*

			Velocities, ft/s		Overall	Temperature	Estimated fouling		
In tubes	Outside tubes	Type of equipment	Tube	Shell	coefficient	range, °F	Tube	Shell	Overall
Heating and cooling									
Butadiene mix, (superheating)	Steam	H	25–35		12	400–100			0.04
Solvent	Solvent	H		1.0–1.8	35–40	110–30			0.0065
Solvent	Propylene (vaporization)	K	1–2		30–40	40–0			0.006
C_4 unsaturates	Propylene (vaporization)	K	20–40		13–18	100–35			0.005
Solvent	Chilled water	H			35–75	115–40	0.003	0.001	
Oil	Oil	H			60–85	150–100	0.0015	0.0015	
Ethylene vapor	Condensate and vapor	K			90–125	600–200	0.002	0.001	
Ethylene vapor	Chilled water	H			50–80	270–100	0.001	0.001	
Condensate	Propylene (refrigerant)	K-U			60–135	60–30	0.001	0.001	
Chilled water	Transformer oil	H			40–75	75–50	0.001	0.001	
Calcium brine, 25%	Chlorinated C_1	H	1–2	0.5–1.0	40–60	−20–+10	0.002	0.005	
Ethylene liquid	Ethylene vapor	K-U			10–20	−170–(−100)			0.002
Propane vapor	Propane liquid	H			6–15	−25–100			0.002
Lights and chloro. HC	Steam	U			12–30	−30–260	0.001	0.001	
Unsat. light HC, CO, CO_2, H_2	Steam	H			10–2	400–100			0.3
Ethonolamine	Steam	H			15–25	400–40	0.001	0.001	
Steam	Air mixture	U (in tank)			10–20	−30–220	0.0005	0.0015	
Steam	Styrene and tars	H	4–7		50–60	190–230	0.001	0.002	
Chilled water	Freon-12	H	4–5		100–130	90–25	0.001	0.001	
Water	Lean copper solvent	H	3–5	1–2	100–120	180–90			0.004
Water	Treated water	H	2–3		100–125	90–110			0.005
Water	C_2-chloro HC, lights	H			6–10	360–100	0.002	0.001	
Water	Hydrogen chloride	H			7–15	230–90	0.002	0.001	
Water	Heavy C_2 chloro.	H			45–30	300–90	0.001	0.001	

Hot fluid	Cold fluid	Type		Temp 1	Temp 2			
Water	Perchlorethylene	H		55–35	150–90	0.001	0.001	
Water	Air and water vapor	H		20–35	370–90	0.0015	0.0015	
Water	Engine-jacket water	H		230–160	175–90	0.0015	0.001	
Water	Absorption oil	H		80–115	130–90	0.0015	0.001	
Water	Air-chlorine	U	4–7	8–18	250–90			0.005
Water	Treated water	H	5–7	170–225	200–90	0.001	0.001	
Condensing								
C_4 unsat.	Propylene refrigerant	K	v	58–68	60–35			0.005
HC unsat. lights	Propylene refrigerant	K	v	50–60	45–3			0.0055
Butadiene	Propylene refrigerant	K	v	65–80	20–35			0.004
Hydroden chloride	Propylene refrigerant	H		110–60	0–15	0.012	0.001	
Lights and chloroethane	Propylene refrigerant	KU		15–25	130–(−20)	0.002	0.001	
Ethylene	Propylene refrigerant	KU		60–90	120–(−10)	0.001	0.001	
Unsat. chloro. HC	Water	H	7–8	90–120	145–90	0.002	0.001	
Unsat. chloro. HC	Water	H	3–8	180–140	110–90	0.001	0.001	
Unsat. chloro. HC	Water	H	6	15–25	130–(−20)	0.002	0.001	
Chloro. HC	Water	KU		20–30	110–(−10)	0.001	0.001	
Solvent and non-cond.	Water	H		25–15	260–90	0.0015	0.004	
Water	Propylene vapor	H	2–3	130–150	200–90			0.003
Water	Propylene	H		60–100	130–90	0.0015	0.001	
Water	Steam	H		225–110	300–90	0.002	0.0001	
Water	Steam	H		190–235	230–130	0.0015	0.001	
Treated water	Steam (exhaust)	H		20–30	220–130	0.0001	0.0001	
Oil	Steam	H		70–110	375–130	0.003	0.001	
Water	Propylene cooling and cond.	H		25–50 / 110–150	30–45 (C) / 15–20 (Co)	0.0015	0.001	
Chilled water	Air-chlorine (part. cond.)	U		8–15 / 20–30	8–15 (C) / 10–15 (Co)	0.0015	0.005	
Water	Light HC, Cool and cond.	H		35–90	270–90	0.0015	0.003	
Water	Ammonia	H		140–165	120–90	0.001	0.001	
Water	Ammonia	U		280–300	110–90	0.001	0.001	
Air-water vapor	Freon	KU		10–50 / 10–20	60–10			0.01

TABLE 14.3 Overall Heat-Transfer Coefficients for Typical Petrochemical Applications: U, Btu/(h·ft²·°F)* *(Continued)*

In tubes	Outside tubes	Type of equipment	Velocities, ft/s Tube	Velocities, ft/s Shell	Overall coefficient	Temperature range, °F	Estimated fouling Tube	Estimated fouling Shell	Estimated fouling Overall
Reboiling									
Solvent, copper-NHs	Steam	H	7–8		130–150	180–160			0.005
C₄ unsat.	Steam	H			95–115	95–150			0.0065
Chloro. HC	Steam	VT			35–25	300–350	0.001	0.001	
Chloro. unsat. HC	Steam	VT			100–140	230–130	0.001	0.001	
Chloro. ethane	Steam	VT			90–135	300–350	0.001	0.001	
Chloro. ethane	Steam	U			50–70	30–190	0.002	0.001	
Solvent (heavy)	Steam	H			70–115	375–300	0.004	0.0005	
Monodiethanolamines	Steam	VT			210–155	450–350	0.002	0.001	
Organics, acid, water	Steam	VT			60–100	450–300	0.003	0.0005	
Amines and water	Steam	VT			120–140	360–250	0.002	0.0015	
Propylene	C₂, C₂⁻	KU			120–140	150–40	0.001	0.001	
Propylene butadiene	Butadiene, unsat.	H		25–35	15–18	400–100			0.02

*E. E. Ludwig, *Applied Process Design for Chemical and Petrochemical Plants*, vol. 3, 2d ed., Gulf Publishing Company, Houston, 1983, p. 70, Table 10-15.
†Unless specified, all water is untreated, brackish, bay, or sea.
NOTE: H = horizontal fixed or floating tubesheet; U = U-tube horizontal bundle; K = kettle type; V = vertical; R = reboiler; T = thermosiphon; v = variable; HC = hydrocarbon; (C) = cooling range Δt; (Co) = condensing range Δt.

TABLE 14.4 Design Coefficients for Shell-and-Tube Heat Exchangers: $(Btu \cdot h \cdot ft^2)/°F$*

Fluid	Liquid (no phase change)	Boiling liquid	Condensing vapor
Aromatic liquids			
Benzene, toluene, ethylbenzene, styrene	140	90	140
Dowtherm	140		
Inorganic solutions			
$CaCl_2$ brine (25%)	250		
Heavy acids	75		
NaCl brine (20%)	290		
Miscellaneous dilute solutions	290		
Light hydrocarbon liquids			
C_3, C_4, C_5	250	140	250
Chlorinated hydrocarbons	250	110	140
Miscellaneous organic liquids			
Acetone	140		
Amine solutions			
Saturated diethanolamine and monoethanolamine (CO_2 and H_2S)	140		
Lean amine solutions	200		
Oils			
Crude oil	65		
Diesel oil	90		
Fuel oil (bunker C)	55		
Gas oil			
Light	80		65
Heavy (typical of catalytic-cracker feed)	70		55
Gasoline (400° EP)	125	100	125
Heating oil (domestic 30° API)	100		
Hydroformate	165		
Kerosene	110		75
Lubricating oil stock	55		
Naphthas			
Absorption	125	100	165
Light virgin	140	100	140
Light catalytic	165	100	140
Heavy	125	90	120
Polymer (C_8's)	125	100	125
Reduced crude	55		
Slurry oil (fluid catalytic cracker)	65		
Steam			1000†
Water			
Boiler water	330		
Cooling tower (untreated)	140‡		
Condensate (flashed)	500		
River and well	140‡		
Seawater (clean and below 125°F)	250		
Gases in turbulent flow	*Coefficient*		
Air, CO, CO_2, and N_2	20		
Hydrocarbons (light through naphthas)	30		

*Data adapted from "Design Resistance for Shell-and-Tube Heat Exchangers," *Petroleum Refiner*, vol. 33, no. 7, 1954, p. 121.

†Condensing coefficients for steam may be much lower when noncondensables are present, as is often the case with low-pressure exhaust steam.

‡A value of 140 is suggested when doubt as to the quality of water exists.

TABLE 14.5 Typical Fouling Factors for Exchangers

	(ft²·h·°F)/Btu
Liquids	
Seawater	0.0005–0.001
River water, clear	0.001–0.003
River water, muddy	0.003–0.006
Well water, hard	0.003–0.005
Cooling-tower water, treated	0.001–0.002
Cooling-tower water, untreated	0.003–0.006
Boiler feedwater, softened	0.0005–0.001
Boiler feedwater, deionized	0.0005
Distilled water	0.0005
Brine	0.001–0.0015
Refrigerant liquids	0.001–0.0015
Organic liquids	0.001–0.002
Fuel oil	0.005–0.01
Gasoline	0.005–0.008
Tars, residual bottoms	0.01–0.02
Gases and vapors	
Steam, clean, oil-free	0.00–0.0003
Steam, exhaust	0.0005–0.001
Air	0.0015–0.003
Organic vapors	0.0005–0.001
Flue gas, clean	0.005–0.01
Flue gas, high-particulate-loading	0.01–0.1

for correlating the film heat-transfer coefficient for the inside of a pipe. The function $f_1(x)$ is required for laminar-flow conditions but not for turbulent flow. It is used to factor in the effect of pipe length, which is important when calculating the film coefficient during laminar flow. The function $f_2(x)$ is required in both the laminar and the turbulent correlations to correct for the difference in the bulk-fluid viscosity and the viscosity of the fluid at the pipe wall.

Table 14.7 lists various film heat-transfer-coefficient equations for shell-and-tube heat exchangers. Table 14.8 is a similar table for spiral and plate-type heat exchang-

TABLE 14.6 Thermal Resistances, x/k, of Metal Tubing of Different Wall Thicknesses

	Thermal resistance, (ft²·h·°F)/Btu				
BWG	12	14	16	18	20
in	0.109	0.083	0.065	0.049	0.035
Aluminum	0.000078	0.000059	0.000046	0.000032	0.000025
Brass	0.000142	0.000108	0.000084	0.000058	0.000045
Carbon steel	0.000363	0.000277	0.000217	0.000150	0.000117
Copper	0.000042	0.000032	0.000025	0.000017	0.000013
Hastelloy C	0.001817	0.001383	0.001083	0.000750	0.000583
Incoloy	0.001363	0.001038	0.000812	0.000563	0.000438
Inconel	0.001090	0.000830	0.000650	0.000450	0.000350
Monel	0.000727	0.000553	0.000433	0.000300	0.000233
Stainless steel	0.000948	0.000722	0.000565	0.000391	0.000304
Titanium	0.000991	0.000755	0.000591	0.000409	0.000318

TABLE 14.7 Film Heat-Transfer-Coefficient Equations for Shell-and-Tube Heat Exchangers*

Mechanism or restriction	Empirical equation
	Inside the tubes
No phase change (liquid), Re > 10,000	$\dfrac{h}{cG} = 0.023 \, (\text{Re})^{-0.2} \, (\text{Pr})^{-2/3}$
No phase change (gas), Re > 10,000	$h = 0.0144 \, G^{0.8} (D_i)^{-02} c_p$
No phase change (gas), 2100 < Re < 10,000	$h = 0.0059 \, [(\text{Re})^{2/3} - 125] \, [1 + (D/L)^{2/3}] \, (c_p/D_i) \, (\mu_f/\mu_b)^{-0.14}$
No phase change (liquid), 2100 < Re < 10,000	$\dfrac{h}{cG} = 0.116 \left[\dfrac{(\text{Re})^{2/3} - 125}{\text{Re}}\right] [1 + (D/L)^{2/3}] \, (N_{Pr})^{-2/3} \, (\mu_f/\mu_b)^{-0.14}$
No phase change (liquid), Re < 2100	$\dfrac{h}{cG} = 1.86 \, (\text{Re})^{-2/3} \, (\text{Pr})^{-2/3} \, (L/D_i)^{-2/3} (\mu_f/\mu_b)^{-0.14}$
Condensing vapor, vertical, Re < 2100	$h = 0.925 \, k \, (g\rho_i^2/\mu\Gamma)^{1/3}$
Condensing vapor, horizontal, Re < 2100	$h = 0.76 \, k \, (g\rho_i^2/\mu\Gamma)^{1/3}$
Condensate subcooling, vertical	$h = 1.225 \, (k/B) \, (cB\Gamma/kL_B)^{5/6}$
Nucleate boiling, vertical †	$\dfrac{h}{cG} = 4.02 \, (\text{Re})^{-0.3} \, (\text{Pr})^{-0.6} \, (\rho_L/P^2)^{-0.425} \, \Sigma$
	Outside the tubes
Nucleate boiling, horizontal †	$\dfrac{h}{cG} = 4.02 \, (\text{Re})^{-0.3} \, (\text{Pr})^{-0.6} \, (\rho_L/P^2)^{-0.425} \, \Sigma$
No phase change (liquid), cross-flow	$\dfrac{h}{cG} = 0.33 \, (\text{Re})^{-0.4} \, (\text{Pr})^{-2/3} \, (0.6)$
No phase change (gas), cross-flow	$h = 0.11 \, G^{0.6} \, D^{-04} c_p \, (0.6)$
No phase change (gas), parallel flow	$h = 0.0144 \, G^{0.5} \, D^{-02} c_p \, (1.3)$
No phase change (liquid), parallel flow	$\dfrac{h}{cG} = 0.023 \, (\text{Re})^{-0.2} \, (\text{Pr})^{-2/3} \, (1.3)$
Condensing vapor, vertical, Re < 2100	$h = 0.925 \, k \, (g\rho_L^2/\mu\Gamma)^{2/3}$
Condensing vapor, horizontal, Re < 2100	$h = 0.76 \, k \, (g\rho_L/\mu\Gamma)^{2/3}$
	Tube wall
Tube wall (sensible-heat transfer)	$h = (24 \, k_w)/(d_o - d_i)$
Tube wall (latent-heat transfer)	$h = (24 k_w)/(d_o - d_i)$
	Fouling
Fouling (sensible-heat transfer)	h = assumed
Fouling (latent-heat transfer)	h = assumed

*Excerpted by special permission from *Chemical Engineering* (Jan. 26, 1970). Copyright © 1970, by McGraw-Hill, Inc., New York, N.Y. 10020.

†$G = W\rho_L/A\rho_v$.

NOMENCLATURE:
- A heat-transfer area, ft²
- B film thickness, $(0.00187 Z\Gamma/g_c s^2)^{1/3}$, ft
- C core diameter, in
- c, c_p specific heat, Btu/(lb·°F)
- D_e equivalent diameter, ft

TABLE 14.7   Film Heat-Transfer-Coefficient Equation for Shell-and-Tube Heat Exchangers* *(Continued)*

D_H	helix or spiral diameter, ft
D_e	equivalent diameter, ft
D_i	inside tube diameter, ft
D_0	inside shell diameter, in
d	tube diameter, in
d_s	channel spacing, in
f	Fanning friction factor, dimensionless
G	mass velocity, lb/(h·ft² cross-sectional area)
g	gravitational constant, ft/[h²·(4.18 × 10⁸)]
g_c	dimensional constant, 4.18 × 10⁸ (lbm/lbf)/(ft·h²)
H	channel plate width, in
h	film coefficient of heat transfer, Btu/(h·ft²·°F)
k	thermal conductivity, Btu/[h·ft²·(°F/ft)]
L	total series length of tubes, $L_0 N_{PT}$ × number of shells, ft; or spiral plate length, ft
L_A	length of condensing zone, ft
L_B	length of subcooled zone, ft
L_0	length of shell, ft
M	molecular weight, lb/(lb·mol)
N_{PT}	number of tube passes per shell, dimensionless
n	number of tubes per pass (or in parallel), dimensionless
P	pressure, psia
P_B	baffle spacing, in
ΔP	pressure drop, lb/in²
p	plate thickness, in
Q	heat transferred, Btu
s	specific gravity (referred to water at 20°C), dimensionless
W	flow rate, (lb/h)/1000
Z	viscosity, cP
Γ	condensate loading, lb/(h·ft)
μ	viscosity, lb/(h·ft)
ρ_v	vapor density, lb/ft³
Σ, Σ'	surface-condition factor, dimensionless, 1.0 for copper and steel, 1.7 for stainless steel, and 2.5 for polished surfaces
σ	surface tension, dyn/cm
Subscripts	
b	bulk-fluid properties
c	cold stream
f	film-fluid properties
h	hot stream
i	conditions on shell side or outside tubes
L	liquid phase
o	conditions on tube side or inside tubes
s	scale or fouling material
v	vapor phase
w	wall or tube material
Dimensionless groups	
Nu	Nusselt number, hD/k
Pr	Prandtl number, $c\mu/k$
Re	Reynolds number, DG/μ
Re$_c$	critical Reynolds number = $20{,}000\,(D_e/D_H)^{0.32}$
St	Stanton number, h/cG

ers. Chapter 16 discusses methods to calculate film heat-transfer coefficients for agitated vessels.

Usually the equations for heat-transfer film coefficients may be applied as presented in Table 14.7. The coefficients so calculated are for pure components or for mixtures for which the physical parameters are a function of composition only. This is true for most types of heat-transfer applications. However, special care must be taken with heat transfer during condensation.

The condensation heat-transfer coefficients for pure components remain essen-

TABLE 14.8 Film Heat-Transfer-Coefficient Equations for Spiral and Plate-Type Heat Exchangers*

Mechanism or restriction	Empirical equation: heat transfer
Spiral flow	
No phase change (liquid), Re > Re$_c$	$h = (1 + 3.54\, D_e/D_H)\, 0.023\, cG\, (Re)^{-0.2}\, (Pr)^{-2/3}$
No phase change (gas), Re > Re$_c$	$h = (1 + 3.54\, D_e/D_H)\, 0.0144\, cG^{0.8}\, (D_e)^{-0.2}$
No phase change (liquid), Re < Re$_c$	$h = 1.86\, c\, G\, (Re)^{-2/3}\, (Pr)^{-2/3}\, (L/D_e)^{1/3}\, (\mu_f/\mu_b)^{-0.14}$
Spiral or axial flow	
Condensing vapor, vertical, Re < 2100	$h = 0.925\, k\, [g_c \rho_L^2/\mu \Gamma]^{1/3}$
Condensate subcooling, vertical, Re < 2100	$h = 1.225\, k/B\, [cB/kL_B]^{5/6}$
Axial flow	
No phase change (liquid), Re > 10,000	$h = 0.023\, c\, G\, (Re)^{-0.2}\, (Pr)^{-2/3}$
No phase change (gas), Re > 10,000	$h = 0.0144\, c\, G^{0.8}\, (D_c)^{-0.2}$
Condensing vapor, horizontal, Re < 2100	$h = 0.76\, k\, [g_c \rho_L^2/\mu \Gamma]^{1/3}$
Nucleate boiling, vertical†	$h = 4.02\, c\, G\, (Re)^{-0.3}\, (Pr)^{-0.6}\, (\rho_L \sigma/P^2)^{-0.425} \Sigma$
Plate	
Plate, sensible heat transfer	$h = 12\, k_w/P$
Plate, latent heat transfer	$h = 12\, k_w/P$
Fouling	
Fouling, sensible heat transfer	h = assumed (consult manufacturer)
Fouling, latent heat transfer	h = assumed (consult manufacturer)

*Excerpted by special permission from *Chemical Engineering* (May 4, 1970). Copyright © 1970, by McGraw-Hill, Inc., New York, N.Y. 10020. See also nomenclature for Table 14.7.
†$G = W_o\, \rho_L/A\rho_s$.

tially constant throughout the entire transfer area of a heat exchanger. However, if the feed is multicomponent, there could be a variation in U owing to a change in molecular weight and other physical properties in the condensation range. There is also a change in temperature during the progress of the multicomponent condensation. The variation in U and ΔT can usually be accommodated by means of Eq. (14.19). On the other hand, the presence of noncondensables in the vapors flowing to a condenser results in a considerable variation of U and ΔT values throughout the exchanger. Figure 14.5 presents a typical situation. The evaluation of the h's or U's in each of the separate zones set aside for ΔT calculation is required.

A rigorous method for evaluating the U's in each zone involves diffusivity considerations* of the various components. These are complex calculations and often best solved by the use of appropriate computer programs. However, an expedient which usually produces conservative results is to analyze the separate ΔT zones in Fig. 14.5 for the individual areas required by the three distinct heat-transfer functions that could take place in each zone, i.e., (1) reduction of sensible heat of the vapors and noncondensable gases, (2) latent heat of condensation, and (3) cooling of condensates.

In this case, an individual h is found for each of the transfer functions according to an applicable relationship in Table 14.7 for the gas sensible heat and for the latent

---

*See A. P. Colburn, and O. A. Harigen, *Industrial and Engineering Chemistry*, vol. 26, 1938, p. 1178.

heat transfer. The analysis of liquid subcooling film coefficients is difficult. However, it usually is acceptable engineering practice tacitly to assign a value of 50 Btu/(h·ft²·°F) for the subcooling film coefficients flowing within tubes and 40 Btu/(h·ft²·°F) for those on the shell side. The various process film coefficients can be combined with the appropriate fouling factors and coolant film coefficients to obtain the individual overall heat-transfer coefficient U for each of the various transfer functions, i.e., gas sensible heat, condensation, and liquid subcooling, at the beginning and end of each of the partial heat-interchange segments. The temperature differential between the process side and the coolant can be obtained by analyzing a diagram similar to Fig. 14.5. Equation (14.19) is then used to obtain the natural-log-mean $U \Delta T$ for each of the heat-transfer functions in the segment. Consequently, the pseudo area required for heat transfer for the individual heat changes in each segment can be calculated and summed across the entire exchanger to obtain the total area required for the process.

Temperature effects on heat-transfer coefficients. Because film and overall heat-transfer coefficients are based in part on the properties of the process fluids involved in the heat transfer, the calculated heat-transfer coefficients will vary as the fluid properties change with temperature. For most practical problems, the effect of temperature on a heat-transfer-coefficient change can be approximated by

$$h = a + bT \tag{14.35}$$

or

$$U = c + d\overline{T} \tag{14.36}$$

where T = bulk process-fluid temperature
$\overline{T}$ = average process-fluid temperature
a, b, c = regression coefficients to fit the data points

When there are significant changes in the overall heat-transfer coefficient with temperature, the overall heat-transfer coefficient and the temperature difference must be averaged together by using the natural-log-mean (ΔT) method according to Eq. (14.19) to obtain a more accurate estimate of the heat-transfer rate.

$$U_{\text{LM}} \text{LM}(\Delta T) = \frac{U_1 \Delta T_2 - U_2 \Delta T_1}{\ln\left(\dfrac{U_1 \Delta T_2}{U_2 \Delta T_1}\right)}$$

where U_{LM} = overall heat-transfer coefficient adjusted for temperature effects by the natural-log-mean (ΔT) method.

Fouling factors. Some thermal resistances which should be included as part of the overall heat-exchanger design are not readily calculable. These factors take into consideration such intangibles as dirt, scale, vapor barriers, corrosion, and general aging effects. These are known as scaling or fouling factors.

Two overall heat-transfer coefficients are usually listed when a heat exchanger is specified. A clean overall heat-transfer coefficient U_C accounts for all calculable heat-transfer coefficients and ignores the fouling factors, while the design (also called service or dirty) overall heat-transfer coefficient U_D consists of the clean overall heat-transfer coefficient adjusted for fouling; i.e.,

$$\frac{1}{U_D} = \frac{1}{U_C} + f_h + f_c \tag{14.37}$$

Table 14.5 lists some fouling factors that can be expected for different process conditions. The fouling factors are intended to ensure that the heat exchanger will deliver at least the duty required by process conditions over the life of the unit. They provide an excess design for a unit which increases the time span during which the unit can operate before maintenance is required.

Radiation

When a body is heated, it emits energy in the form of electromagnetic waves. This energy is emitted in all directions and can travel through a vacuum or a transmitting medium. When the energy comes in contact with another body, a portion of the energy is transmitted, a portion is reflected, and another portion is absorbed by the body receiving the energy (see Fig. 14.6). The fraction of energy absorbed heats the

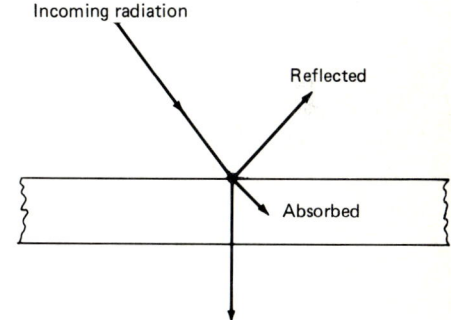

Figure 14.6 Distribution of radiant heat transferred to a body.

body. This phenomenon is called heat transfer by radiation. Radiant heat transfer is the dominant heat-transfer mechanism in high-temperature equipment such as furnaces and boilers. At lower temperatures, less than 1000°F, its proportional contribution to the total heat transferred decreases with temperature. Radiant heat transfer is the only mechanism for transferring heat in a vacuum. In a gaseous medium, radiant heat transfer acts in concert with the convective-heat-transfer mechanism.

Radiant-energy relationships. Radiant heat transfer can be defined in terms of radiant-energy relationships. Two important terms are used to describe the energy relationships: (1) total emissive power E, which is the total radiant energy emitted from the surface of a body into the entire volume surrounding the surface; and (2) total irradiation G, which is the total power incident or incoming to a surface. Both terms are in units of energy per unit area per unit time, such as British thermal units per square foot per hour.

Figure 14.6 describes the effect of incoming radiation on the surface. The energy balance at the surface can be defined in terms of G by

$$G = \alpha G + \rho G + \tau G \tag{14.38}$$
$$1 = \alpha + \rho + \tau \tag{14.39}$$

where α = absorptivity fraction, the fraction of energy absorbed by the surface
ρ = reflectivity fraction, the fraction of energy reflected by the surfaces
τ = transmissivity fraction, the fraction of energy transmitted through the surface

An opaque surface or body is one with zero transmissivity. Most engineering materials are opaque bodies but have varying degress of absorptivity and reflectivity. A convenient reference point for studying radiant heat transfer is the blackbody. The blackbody has an ideal surface which absorbs all radiant energy regardless of direction or wavelength. Therefore by definition, the properties of a blackbody are $\alpha = 1$, $\rho = 0$, and $\tau = 0$.

In addition to being an excellent absorber of heat, the blackbody is an excellent emitter of radiant energy. The total energy emitted by a blackbody is a function of the temperature of the body. However, the energy is emitted at various wavelengths. The amount of radiant energy emitted $E_{b\lambda}$ by a blackbody at various wavelengths and at various temperatures is shown in Fig. 14.7.

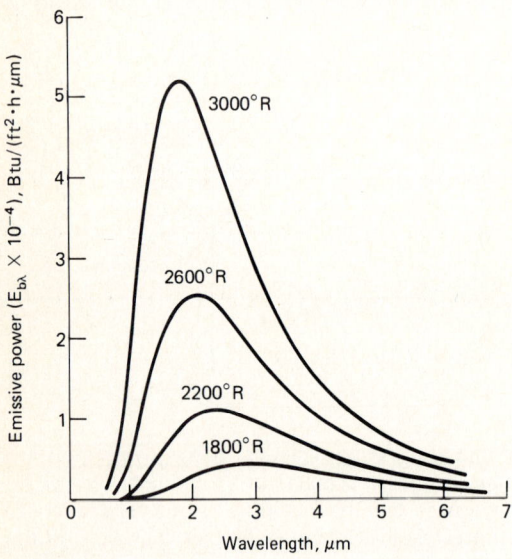

Figure 14.7 Emissive-power distribution from a blackbody as a function of temperature and wavelength.

These emissive-power curves have been fitted by Max Planck to the distribution-curve equation

$$E_{b\lambda} = \frac{c_1}{\lambda^5} \left[\frac{1}{e^{(c_2/\lambda T)} - 1} \right] \quad \frac{\text{Btu}}{\text{ft}^2 \cdot \text{h}} \left(\frac{1}{\text{ft}} \right) \qquad (14.40)$$

where $c_1 = 2\pi c^2 h$
$c_2 = ch/k$
c = speed of light = 3.535×10^{12} ft/h
h = Planck's constant = 1.746×10^{-40} Btu·h
k = Boltzmann constant = 7.278×10^{-27} Btu/°R
T = Absolute temperature, °R
λ = wavelength, ft

This equation gives the emissive-power flux from a blackbody for a given wavelength range between λ and $d\lambda$. Practical radiant-heat-transfer problems require the average emissive-power flux for a given temperature. Therefore, by integrating Eq. (14.40) over the range of wavelengths,

$$E_b = \int_0^\infty E_{b\lambda}\, d\lambda \qquad (14.41)$$

the average emissive power flux for a blackbody is calculated to be

$$E_b = \left(\frac{2\pi^5 k^4}{15 c^2 h^3}\right) T^4 \qquad (14.42)$$

Substituting

$$\sigma = \left(\frac{2}{15}\frac{\pi^5 k^4}{c^2 h^3}\right) = 1.322 \times 10^{-16} \quad \frac{\text{Btu}}{\text{ft}^2 \cdot \text{h} \cdot {}^\circ\text{R}^4} \qquad (14.43)$$

into Eq. (14.42) produces the widely used Stefan-Boltzmann equation; i.e.,

$$E_b = \sigma T^4 \qquad (14.44)$$

The Stefan-Boltzmann equation implies that all bodies radiate energy and that the amount of energy radiated is proportional to the fourth power of their absolute temperature. If two blackbodies are at two different temperatures, they will have two different emissive powers; i.e.,

$$E_{b1} = \sigma T_1^4 \qquad (14.45)$$
$$E_{b2} = \sigma T_2^4 \qquad (14.46)$$

The difference between these two body energies results in a net radiant-heat-transfer flux of

$$\frac{Q_r}{A} = E_{b1} - E_{b2} = \sigma T_1^4 - \sigma T_2^4 \qquad (14.47)$$

The heat transfer will continue until both bodies are in isothermal equilibrium. The actual radiant heat transfer is a complex process, in that as one body emits energy it can also absorb energy from another body. The relationship between emissive power and total irradiation can be explained by using Kirchhoff's law.

Kirchhoff's law. When a body is located in an evacuated isothermal enclosure such as the one shown in Fig. 14.8, it will emit and absorb energy at the same temperature as the other bodies in the enclosure. The energy balance for this system is

For body 1	$E_1 A_1 = \alpha_1 G_1 A_1$
For body 2	$E_2 A_2 = \alpha_2 G_2 A_2$
$\vdots$	$\vdots$
For body n	$E_n A_n = \alpha_n G_n A_n$
For a blackbody	$E_b A_b = \alpha_b G_b A_b$

At isothermal conditions, all bodies receive the same incident radiant energy; i.e.,

$$G_1 = G_2 = \cdots = G_n = G_b \qquad (14.48)$$

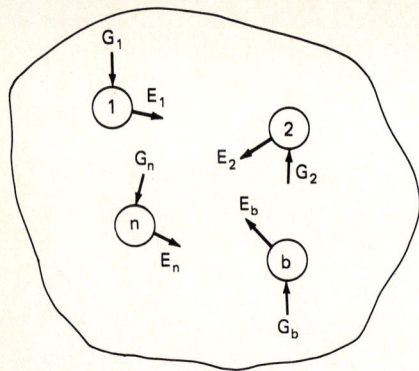

Figure 14.8 Bodies located in an evacuated isothermal enclosure simultaneously absorbing and emitting heat. b = blackbody; n = any body n.

Therefore,

$$\frac{E_1}{\alpha_1} = \frac{E_2}{\alpha_2} = \cdots = \frac{E_n}{\alpha_n} = \frac{E_b}{\alpha_b} \tag{14.49}$$

Since a blackbody absorbs all the energy it receives, its absorptivity is unity ($\alpha_b = 1$); therefore,

$$\frac{E_1}{\alpha_1} = \frac{E_2}{\alpha_2} = \cdots = \frac{E_n}{\alpha_n} = E_b \tag{14.50}$$

The above three equations define Kirchhoff's law, which states that at thermal equilibrium the ratio of total emissive power to absorptivity is a constant.

Emissivity. Since the absorptivity of a blackbody is unity, Kirchhoff's law states that it emits all the energy it absorbs. This makes the blackbody a useful reference point for measuring the emissive power for nonblackbody surfaces. By rearranging Eq. (14.50), a new term, emissivity ϵ, can be defined:

$$\epsilon_n = \frac{E_n}{E_b} \tag{14.51a}$$

Emissivity is defined as the ratio of the actual emissive power of a body to the blackbody emissive power at a given temperature. The value of emissivity is between zero and unity, with unity emissivity occurring with a blackbody. Table 14.9 lists the emissivity for some common engineering materials. The emissivity as defined by Eq. (14.51b, c, and d) is equal to the absorptivity; i.e.,

$$\frac{E_1}{E_b} = \alpha_1 = \epsilon_1 \tag{14.51b}$$

$$\frac{E_2}{E_b} = \alpha_2 = \epsilon_2 \tag{14.51c}$$

$$\frac{E_n}{E_b} = \alpha_n = \epsilon_n \tag{14.51d}$$

Since most engineering objects are opaque, the energy which is not emitted or absorbed must be reflected. This explains why highly polished surfaces and white surfaces have low or near-zero emissivities. As a general rule, a good absorber of

TABLE 14.9 Emissivities of Radiation for Selected Materials

Material	Temperature, °F	Emissivity
Metals		
Aluminum, bright rolled	250–950	0.039–0.050
Aluminum, polished, 98% pure	200	0.050
Aluminum, paint	212	0.20–0.40
Carbon, lampblack	206–520	0.952
Copper, polished	242	0.023
Copper, light oxidized	68	0.037
Copper, black oxidized	68	0.780
Iron and steel		
Polished iron	800–1880	0.144–0.377
Cast iron, polished	392	0.21
Cast iron, newly turned	72	0.435
Cast iron, oxidized	200	0.61
Steel, sheet, smooth	1650–1900	0.55–0.61
Steel, plate, rough	100–700	0.94–0.97
Steel, nickel-chrome	125–1894	0.64–0.76
Steel, oxidized at 1100°F	390–1110	0.79
Brass, polished	100–500	0.10
Brass, oxidized	100–500	0.46–0.75
Lead, pure polished	100–500	0.05–0.08
Lead, oxidized	68	0.28
Zinc, polished	100–2000	0.02–0.06
Zinc, oxidized	68	0.23–0.28
Nonmetals		
Asbestos	100–700	0.93–0.95
Brick, refractory, ordinary	2000	0.59
Brick, refractory, white	500–2000	0.89–0.29
Brick, red	70	0.93
Concrete	0–3000	0.91–0.60
Glass, Pyrex	500–1000	0.94–0.75
Graphite, polished	100–1000	0.42–0.097
Graphite, pressed	500–5000	0.44–0.73
Paints (except aluminum)	70–200	0.80–0.97
Paper	66	0.91–0.95
Rubber, hard	74	0.94
Rubber, soft	76	0.859
Water	32–212	0.95–0.963
Wood	70–150	0.82–0.94

radiant heat such as a black or a rough surface is also a good radiator (emitter) of heat.

Graybodies. Not only is the emittance of a body a function of temperature, it is also a function of wavelength (see Fig. 14.9). For most practical engineering problems, the emittance of the material is assumed to be constant with respect to wavelength within a given temperature range. Therefore, engineering materials are termed "graybodies." This greatly simplifies heat-transfer calculations. For example, if a material is assumed to be a graybody and radiates heat into surroundings that are effectively black, i.e., the surroundings do not reflect back radiation, the net heat transfer due to radiation is

$$q = \frac{Q}{A} = \epsilon\sigma(T_B^4 - T_S^4) \tag{14.52}$$

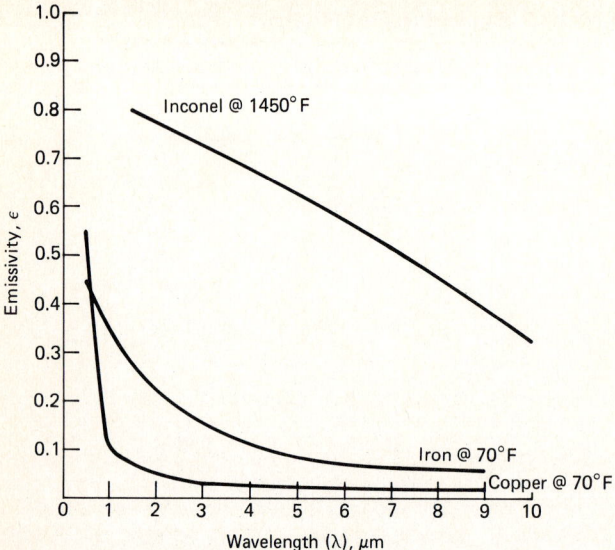

Figure 14.9 Emissivity of some engineering materials for different wavelengths.

where Q = total heat transferred, Btu/h
q = total heat flux, Btu/(ft$^2 \cdot$h)
A = area of radiating surface, ft^2
T_B = absolute temperature of radiating body
T_S = absolute temperature of surroundings

This equation is used typically for bodies radiating heat to the atmosphere or into a vacuum.

For situations in which the material and the surroundings have different emissivities at different temperatures, Eq. (14.52) becomes

$$q = \sigma(\epsilon_B T_B^4 - \epsilon_S T_S^4) \tag{14.53}$$

where ϵ_B = emissivity of body
ϵ_S = emissivity of surroundings

Radiant-heat-transfer coefficient. The radiant-heat-transfer Eqs. (14.52) and (14.53) are cumbersome to handle for performing heat-transfer calculations. This is particularly true when part of the heat is transferred by convection, as with boilers, or when heat is lost from a pipe to the air. For these situations, it is desirable to formulate the heat-transfer contribution from radiation in the form

$$Q_r = h_r A(T_1 - T_2) \tag{14.54}$$

where Q_r = radiant-heat flux, Btu/(ft$^2 \cdot$h)
A = heat-transfer area, ft^2
$T_1 - T_2$ = temperature driving force, °F
h_r = radiant-heat-transfer coefficient, Btu/(ft$^2 \cdot$h$\cdot$°F)

The radiant-heat-transfer coefficient can be conveniently combined with the convective-heat-transfer coefficient to get the total heat-transfer rate by

$$Q = h_T A(T_1 - T_2) = (h_c + h_r)A(T_1 - T_2) \tag{14.55}$$

where Q = total heat flux, Btu/(ft$^2 \cdot$h)
h_c = convective heat-transfer coefficient, Btu/(ft$^2 \cdot$h$\cdot$°F)
h_T = combined heat-transfer coefficient, Btu/(ft$^2 \cdot$h$\cdot$°F)

The radiant-heat-transfer coefficient can be derived from Eq. (14.52) as

$$h_r = \frac{q}{T_1 - T_2} = \frac{\sigma \epsilon}{T_1 - T_2}(T_1^4 - T_2^4) \tag{14.56}$$

This method can be used when T_1 and T_2 do not vary markedly during the operation.

Geometry factors. The radiant-heat-transfer arrangements discussed so far assume that all emitted radiation strikes the receiving surface and is completely absorbed or is emitted into a space which acts as a blackbody. This is not always true. In furnaces, heat radiates from a flame and is partially absorbed by the furnace walls and tubing, which in turn reradiate a portion of the energy back into the furnace volume or onto other furnace walls and tubing. The amount of heat energy absorbed by the various furnace components depends on their position relative to the flame and to other components. Therefore, a geometry factor F_{ij} must be included in the radiant-heat-transfer relationship.

The determination of geometry factors for simple geometries can be derived from theoretical relationships. However, complex geometries such as heat exchangers and furnaces usually derive their factors empirically from experiments or operating experience. The radiant-heat-transfer Eq. (14.52) can be rewritten to include the geometry factor in the form

$$Q_{ij} = \sigma A_i F_{ij} \epsilon_{ji}(T_i^4 - T_j^4) \tag{14.57a}$$

or $$Q_{ij} = \sigma A_j F_{ji} \epsilon_{ij}(T_i^4 - T_j^4) \tag{14.57b}$$

where Q_{ij} = net radiant heat transfer from surface i to surface j
F_{ij} = geometry factor for radiation from surface i to surface j
ϵ_{ji} = emissivity between surfaces
T_i = absolute temperature of surface i
T_j = absolute temperature of surface j

It should be noted that the emissivity of each surface can also be affected by geometry unless both or sometimes one of the surfaces can be assumed to be a blackbody. The emissivity for two parallel plates is calculated by

$$\epsilon_{ij} = \frac{1}{\dfrac{1}{\epsilon_i} + \dfrac{1}{\epsilon_j} - 1} \tag{14.58}$$

For concentric cylinders or concentric spheres the relationship is

$$\epsilon_{ij} = \frac{1}{\dfrac{1}{\epsilon_i} + \dfrac{A_i}{A_j}\left(\dfrac{1}{\epsilon_j} - 1\right)} \tag{14.59}$$

The geometry factor F_{ij} for the above cases is unity. However, more complex geometries such as for surfaces which are perpendicular to each other, for a cylinder parallel to an infinitely large plate, or for a tube bank have geometry factors less than unity. Wiebelt[2] and Rohsenow[3] have compiled a more comprehensive set of geometry factors for a variety of configurations. These geometry factors are for blackbody radiation. For applications to graybodies, the geometry factor must be adjusted by

$$A_i F_{ij} = \frac{1}{\left[\dfrac{1-\epsilon_i}{A_i \epsilon_i} + \dfrac{1}{A_i F_{ij}} + \dfrac{1-\epsilon_j}{A_j \epsilon_j}\right]} \tag{14.60}$$

Gas radiation. The radiant-heat-transfer contribution from nonluminous gases to a surface is significant at gas temperatures above 1000°F. Although simple monatomic and diatomic gases such as argon, helium, hydrogen, oxygen, and nitrogen have a negligible contribution, polyatomic gases like carbon dioxide, water vapor, carbon monoxide, sulfur dioxide, ammonia, and hydrocarbons absorb significant amounts of radiant energy at certain wavelengths. Because of the radiation effects of these gases, the actual heat transfer is greater than expected if only convective heat transfer between the gases and the surface is considered. The radiant-heat-transfer contribution from these nonluminous gases is important when analyzing the convective section of a furnace and in waste-heat boilers which recover waste heat in process gases.

H. C. Hottel[4,5,6] introduced a technique for calculating the radiant-heat-transfer contribution for a nonluminous gas. It is referred to as the gray-gas method. The radiant heat transfer between a gas at T_g and the T_S acting as a blackbody surface is

$$q = \sigma(\epsilon_g T_g^4 - \alpha_g T_S^4) \tag{14.61}$$

For a nonblackbody surface, the equation takes the form

$$q = \epsilon' \sigma(\epsilon_g T_g^4 - \alpha_g T_S^4) \tag{14.62}$$

where α_g = gas absorptivity
ϵ_g = gas emissivity
$\epsilon' = (\epsilon_S + 1)/2$
ϵ_S = surface emissivity

The absorptivity and emissivity of the gas are dependent on the type of gas, the gas temperature, the gas partial pressure, and the path length between the gas and the surface. The actual calculation is complex. As a general rule, radiant-heat contributions below 1000°F due to gases are negligible; however, as the gas temperature rises from 1000 to 2000°F, the radiant-heat contribution becomes the dominant heat-transfer mechanism.

Heat-Exchanger Design

There is no one configuration for heat-exchanger design since designs are predicated for specific process applications. There are, however, characteristics common to all heat-exchange devices: (1) they are composed of a hot phase and a cold phase separated by a solid boundary through which heat is exchanged, and (2) the rate of heat transfer, or duty, of a heat exchanger can be calculated by using the relationship given previously as Eq. (14.20). This equation determines the heat-exchanger physical design and must be solved simultaneously with the energy-balance equations for the hot and cold phases:

Hot-side energy balance $Q_H = W_H c_{pH}(T_1 - T_2)$ (14.63)

Cold-side energy balance $Q_C = w_c c_{pC}(t_2 - t_1)$ (14.64)

where Q_H = hot-side duty, Btu/h
  Q_C = cold-side duty, Btu/h
  T_1 = hot-side outlet temperature, °F
  T_2 = hot-side outlet temperature, °F
  W_H = hot-fluid mass flow rate, lb/h
  w_c = cold-fluid mass flow rate, lb/h
  c_{pH} = hot-fluid specific heat, Btu/(lb·°F)
  c_{pC} = cold-fluid specific heat, Btu/(lb·°F)
  t_1 = cold-fluid inlet temperature, °F
  t_2 = cold-fluid outlet temperature, °F

Mechanical design

The design of a heat-exchange device depends on its process application. The selection of a heat exchanger is based on the economy and efficiency of heat transfer.

Double-pipe exchangers. The double-pipe heat exchanger is the simplest and least costly design of all heat-transfer equipment. As its name implies, the double-pipe heat exchanger consists of two concentric pipes, as shown in Fig. 14.10. Heat is exchanged between the fluid flowing through the internal pipe and another fluid flowing through the annulus between the internal and the external pipes.

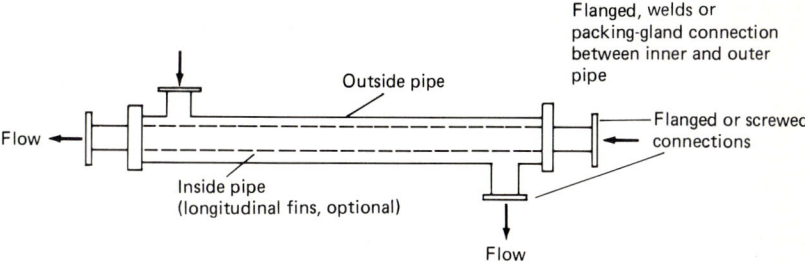

Figure 14.10 Double-pipe heat exchanger.

The double-pipe heat exchanger is used for applications requiring an effective heat-transfer area of less than 1000 ft^2. This type of exchanger comes in standard lengths of 12, 15, and 20 ft. A number of double-pipe heat exchangers can be connected together to increase the effective area as in Fig. 14.11. An advantage of the double-pipe exchanger is that it is easy to dismantle for cleaning. However, double-pipe exchangers have the disadvantage of requiring more space per effective heat-transfer area than other types of heat exchangers. Longitudinal fins on the outer surface of the internal pipe are sometimes used for double-pipe heat exchangers to increase their heat-transfer area.

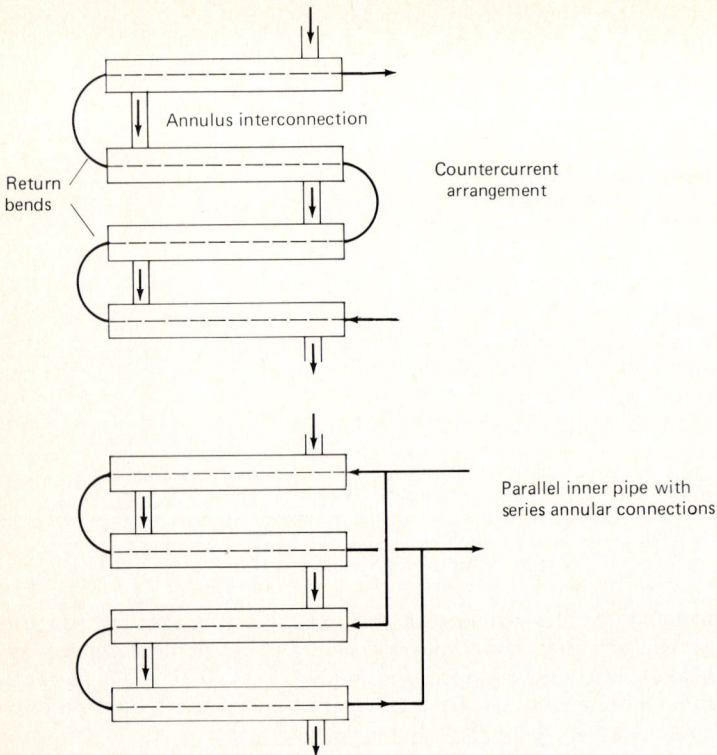

Figure 14.11 Multiple double-pipe-heat-exchanger arrangement.

Shell-and-tube exchangers. The most widely used heat-exchanger design is the shell-and-tube type. It has a much higher effective heat-transfer area per unit volume than does a double-pipe exchanger. Unlike the double-pipe exchanger, the shell-and-tube exchanger comes in a variety of design configurations and can be designed for cocurrent, countercurrent, cross-flow, and some other flow patterns. Shell-and-tube exchangers are available as off-the-shelf units with transfer areas from a few to several hundred square feet. Exchangers for use in the chemical and refinery industries commonly range up to 5000 to 6000 ft^2 each. Individual exchangers used as steam condensers in power-generating stations can be considerably larger.

The design criteria and configurations for most shell-and-tube exchangers used in the process industry in the United States have been standardized by the Tubular Exchanger Manufacturers Association.[7] This organization has established a simple classification system for designating the various types of shell-and-tube exchangers. Table 14.10 lists the three TEMA design classifications for shell-and-tube heat exchangers, and Fig. 14.12 shows the designation system used to describe a shell-and-

Figure 14.12 Designation system used to describe shell-and-tube heat-exchanger configurations. (*Taken from Fig. N-1.2 of the* Standards © *1978 by Tubular Exchanger Manufacturers Association, Tarrytown, N.Y.*)

TABLE 14.10 TEMA Classification for Shell-and-Tube Heat Exchangers

Class	Process application
B	Normal chemical-process service. The design provides maximum economy and overall compactness consistent with safety and service requirements. Most process heat exchangers are of this design class.
C	Moderate process service and commercial installations; used mostly for designing heat exchangers for building heating and cooling systems.
R	Used for severe services in petroleum and related process industries. The class is used to design safe and durable exchangers for generally "dirty" service.

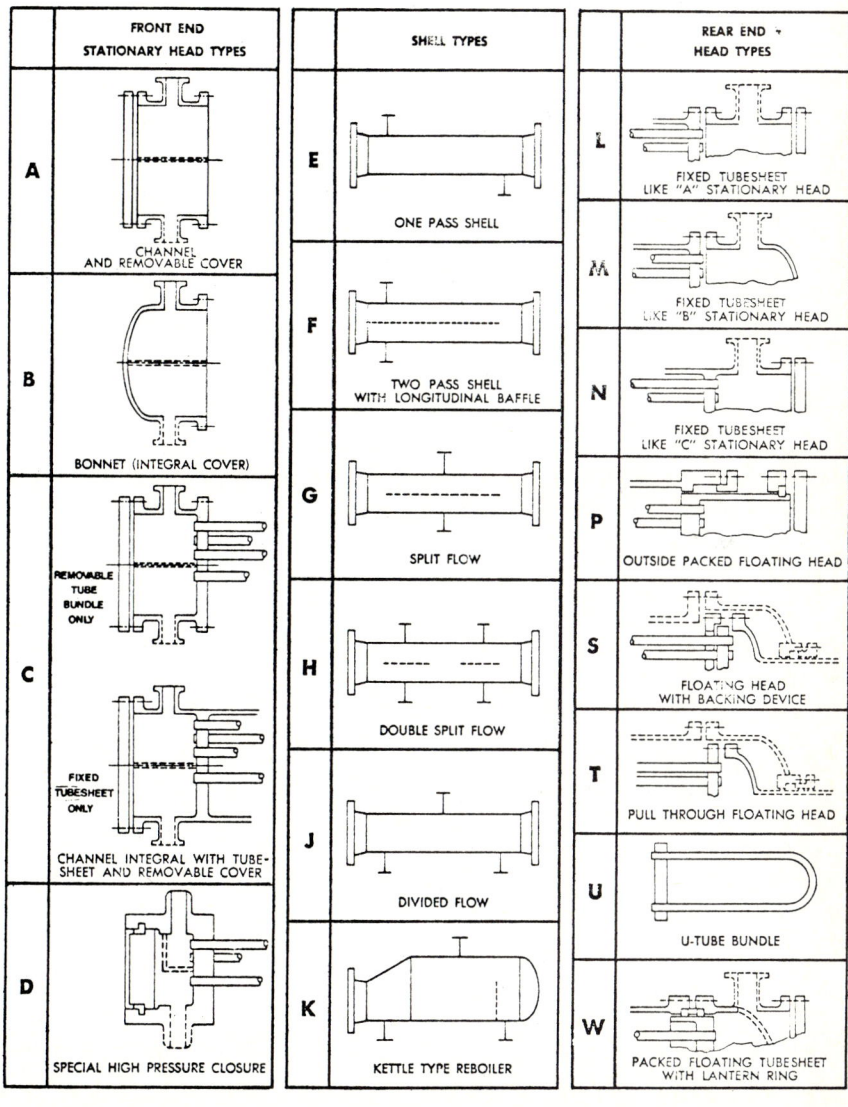

tube heat exchanger by its shell and head configurations. The TEMA standards have also provided an accepted nomenclature for the component parts of a shell-and-tube exchanger. Figure 14.13 shows typical exchanger configurations together with a cross-reference to their component parts.

The most economical exchanger configurations are AEU, AEM, BEU, BEM,* and CEU. These designs are used for most heat exchangers with moderate duties when thermal gradients are not large and the service is relatively clean.

If the fluid on the tube side enters the front head and exits from the back head in a single pass and the exchange has an E-type shell, the exchanger is termed a 1-1 exchanger. This means that the exchanger consists of one shell pass and one tube pass. This type of exchanger is used when low tube-side pressure drops are required and when a true countercurrent or cocurrent flow is needed. It is often used in processes in which condensation occurs on the tube side.

The 1-2 exchanger consists of one shell pass but two or more even numbers of tube passes. The AEU, BEU, and CEU exchangers are of this type of design. These are the most common designs and are the easiest to maintain. While these exchangers have baffles on the shell side perpendicular to the axis of the tube bank to increase fluid turbulence and improve shell-side heat transfer, they are still classified as 1-2 exchangers. However, when a baffle is placed longitudinally on the shell side, this produces a 2-4 exchanger, i.e., two shell passes and four or more tube passes. The 2-4 design is more costly but has improved performance over the 1-2 design. While three or more shell passes can also be designed into an exchanger, very often these are not cost-effective designs, and two or more exchangers can be connected in series to produce the desired miltiple-shell effect.

Tube pitch. The arrangement of tube arrays mounted on a tubesheet is called the "tube pitch." The arrangement, as shown in Fig. 14.14, may be an equilateral triangular, square, or staggered square pitch. Triangular pitch provides the highest tubesheet strength and the greatest shell-side turbulence, which in turn results in the highest heat-transfer rate and the largest pressure drops of the three arrangements. However, triangular pitch is not used if "dirty" service is expected on the shell side since it is the most difficult to clean. Square pitch is preferred for dirty service since the outsides of the tubes are easily accessible for cleaning with a removable bundle. Staggered square pitch is a compromise between the other two arrangements because while it produces a more turbulent flow than square pitch, it is still possible to clean the outside surface of the tubes.

Baffle arrangements. Baffles are inserted on the shell side of the exchanger to affect the shell-side flow pattern. The Type F shell with a single longitudinal baffle creates a two-shell-pass arrangement. The Type G shell, which has a single centrally located

---

*The BEM exchanger in Fig. 14.13 has an expansion joint, which is a costly nonstandard option for this exchanger configuration.

Figure 14.13 Nomenclature used to describe the component parts of some shell-and-tube heat exchangers. (*Extracted from Table N-2 and Fig. N-2 of the* Standards © *1978 by the Tubular Exchanger Manufacturers Association, Tarrytown, N.Y. Nomenclature is referenced to exchanger configurations AES, AKT, BEM, and CFO.*)

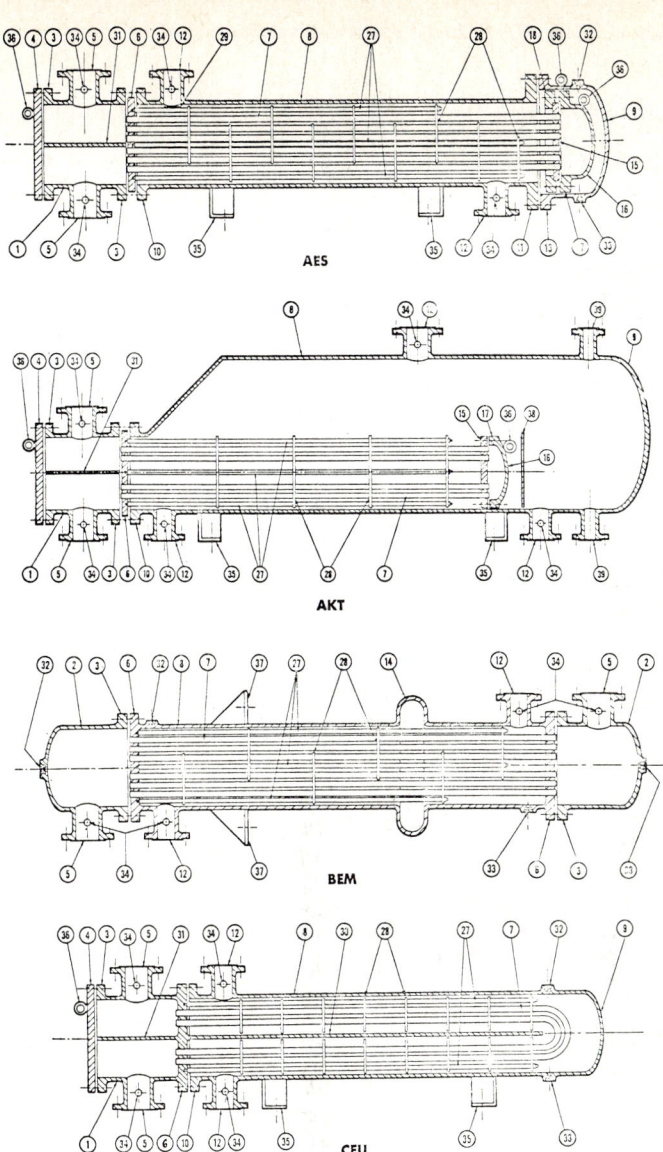

1. Stationary head—channel
2. Stationary head—bonnet
3. Stationary-head flange—channel or bonnet
4. Channel cover
5. Stationary-head nozzle
6. Stationary tubesheet
7. Tubes
8. Shell
9. Shell cover
10. Shell flange—stationary-head end
11. Shell flange—rear-head end
12. Shell nozzle
13. Shell-cover flange
14. Expansion joint
15. Floating tubesheet
16. Floating-head cover
17. Floating-head flange
18. Floating-head backing device
19. Split shear ring
20. Slip-on backing flange
21. Floating-head cover—external
22. Floating-tubesheet skirt
23. Packingbox flange
24. Packing
25. Packing follower ring
26. Lantern ring
27. Tie rods and spacers
28. Transverse baffles or support plates
29. Impingement baffle
30. Longitudinal baffle
31. Pass partition
32. Vent connection
33. Drain connection
34. Instrument connection
35. Support saddle
36. Lifting lug
37. Support bracket
38. Weir
39. Liquid-level connection

529

longitudinal baffle, causes the fluid to flow around either end of the baffle. This arrangement is used in feedwater heaters and thermosiphon reboilers in which improved fluid contact with low pressure drop is required.

Baffles are also inserted perpendicularly to the axis of the tubes, as illustrated in Fig. 14.15. These baffles are used to create a serpentine flow pattern as in Fig. 14.15b to increase the shell-side heat-transfer coefficient. However, it does result in increased pressure drop. Figure 14.15b illustrates baffles used to partition the shell side into compartments for a cross-flow arrangement. A dam baffle such as the one shown in Fig. 14.15a is provided to subcool condensed vapors. Another arrangement consists of a disk and doughnut baffles as shown in Fig. 14.15a.

Segmented baffles as shown in Fig. 14.15a are the most economical and the most often used baffle arrangement. The up-and-down flow pattern is used with liquids, gases, and noncondensing vapors. When rotated 90° from the vertical, the baffles create a side-to-side flow pattern used with mixtures of liquid and vapors.

Shell-side equivalent diameter. An equivalent diameter of the shell-side space must be determined when calculating the pressure drop and the heat-transfer coefficient for the shell side of an exchanger. By recalling Chap. 7, the equivalent diameter D_e is defined as

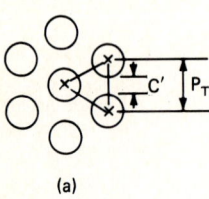

(a)

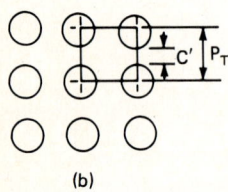

(b)

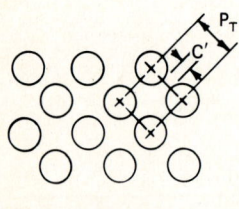

(c)

Figure 14.14 Tube-pitch arrangements. (a) Triangular pitch. (b) Square pitch. (c) Staggered (rotated) square pitch.

$$D_e = \frac{4 \text{ (wetted area)}}{\text{wetted perimeter}} \quad \text{ft}$$

The equivalent diameter d_e for a triangular-pitch and a square-pitch arrangement is calculated by

Triangular pitch $\quad d_e = \dfrac{3.44 P_T - 0.25 \pi d_o^2}{\pi d_o} \quad$ in $\hfill (14.65)$

Square pitch $\quad d_e = \dfrac{4 P_T^2 - \pi d_o^2}{\pi d_o} \quad$ in $\hfill (14.66)$

where P_T = tube pitch, in (shown in Fig. 14.14)
d_o = outside tube diameter, in

In addition to equivalent diameter, the mass flux or mass velocity must be calculated. The shell-side mass flux is determined by using

$G_s = W_s/A_s \quad$ lb/(ft²·h) $\hfill (14.67a)$

and $A_s = d_s\, C'\, B/(144\, P_T) \quad$ ft² $\hfill (14.67b)$

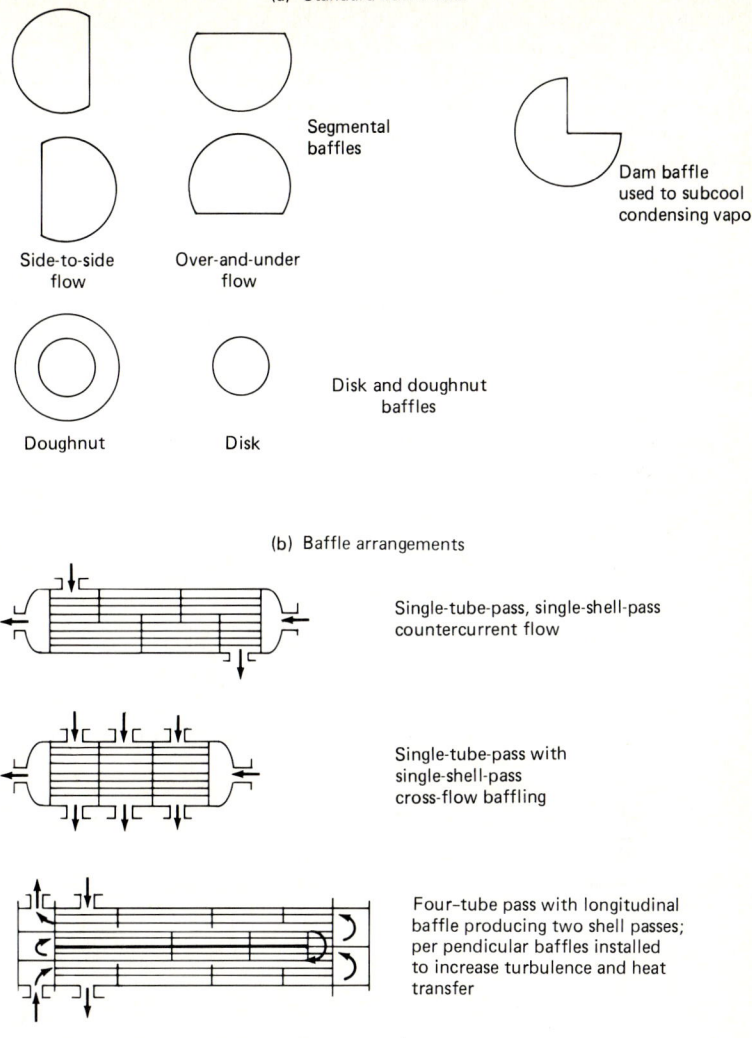

Figure 14.15 Exchanger shell-baffle geometrics.

where C' = tube clearance, in (shown in Fig. 14.14)
A_s = shell cross-flow area, ft^2
B = baffle spacing, in
W_s = total shell-side mass flow, lb/h
d_s = shell diameter, in

Tubesheets. Tubesheets form the boundary between the fluid in the shell and the fluid in the exchanger head which flows through the tubes. A tubesheet also acts as one of the supports for tubes. Tubesheets may be either fixed tubesheets or floating tubesheets. A floating tubesheet is one located on the rear head of an exchanger but not rigidly connected to it such as head Types P, S, T, and W shown in Fig. 14.12. Fixed tubesheets are used when thermal stresses are low and the shell-side fluid is clean so that the tube bundle need not be removed to clean the outside of the tubes. If fixed tubesheets are used at both ends of an exchanger, an expansion joint may be required for the exchanger shell to reduce thermal stresses.

Tubes are joined to the tubesheet by rolling or by welding. Rolling consists of inserting the open end of the tube into the tubesheet and then rotating and expanding the tube until its diameter has expanded sufficiently to create residual compressive stress which locks it tightly against the tubesheet. The rolled joint provides sufficient leaktightness for moderate operating temperatures and pressures. For high-temperature and -pressure service or when handling hazardous or toxic materials, the tubes should be welded to the tubesheet. For exceptionally corrosive designs or to avoid cross-contamination, a double tubesheet is used; leakage from either side is then to the atmosphere.

Reboilers. A reboiler is a heat exchanger designed to supply heat to a distillation column. Earlier distillation columns contained tube bundles inserted into the base of the column to supply the process heat. However, this created a maintenance problem so that the reboiler type of heat exchanger was developed. Figure 14.16 shows several reboiler designs together with their installation on columns.

The kettle-type reboiler is similar in design to a small evaporator. The process fluid enters the shell side of the exchanger, and the heating medium passes through the tube side. The reboiler contains a weir which provides the necessary liquid holdup time in the exchanger and to cover the tubes. The reboiler in effect becomes the equivalent of a theoretical tray on the column with the vapor leaving through the top nozzle in equilibrium with the liquid spilling over the weir into a discharge nozzle at the base of the shell. For batch-type distillations, the column is usually attached directly to the reboiler. The kettle type is one of the more expensive reboiler designs.

The thermosiphon reboiler is a more common design. This type of reboiler uses the density difference produced by heating the fluid to circulate the heat fluid back into the column, where it flashes into a vapor-and-liquid phase. A vertical thermosiphon reboiler such as shown in Fig. 14.16d is used where space limitations exist. A horizontal heat exchanger with a G type of shell, as in Fig. 14.16e, is used when low pressure drops are required.

The once-through natural-circulation configuration, as shown in Fig. 14.16d, takes liquid directly from the downcomer, heats it, and then flashes it into the bottom of the column. This design differs from the thermosiphon arrangement, which continuously recirculates the column bottoms through the reboiler.

The forced-circulation design is used when fluid viscosity is a problem or in installations in which more sensitive temperature control is required. The pump-through design reduces the possibility of boiling in the tubes, which could create a vapor barrier that increases thermal resistance and reduces the reboiler's efficiency while giving a high skin temperature on the process side. Vaporization in the tubes can be reduced by higher recirculation rates or by pressurizing the fluid by the use of a restriction orifice in the downstream piping.

Spiral heat exchangers. The spiral exchanger, also called a spiral-plate heat exchanger, was first developed in the 1930s in Sweden for use in that country's large

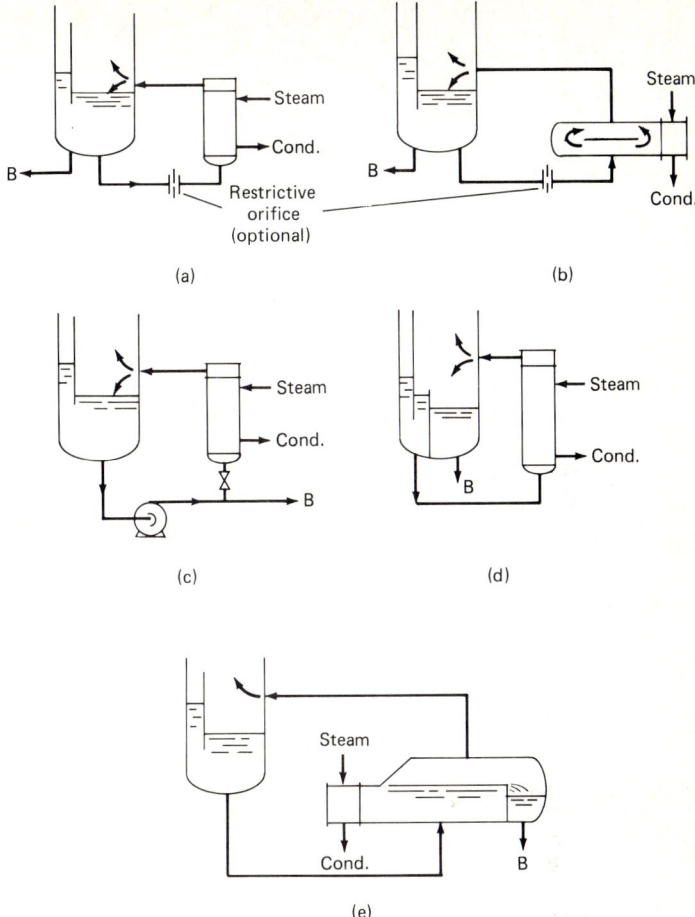

Figure 14.16 Reboiler designs. Cond. = condensate; B = bottoms product. (a) Vertical thermosiphon. (b) horizontal thermosiphon. (c) Forced circulation. (d) Natural circulation. (e) Kettle-type reboiler.

pulp-and-paper industry. The unit is designed primarily to handle slurries or other highly fouling fluids containing large amounts of solids, crystallizing materials, polymerizing materials, or fibers. The spiral unit, shown in Fig. 14.17, consists of two flat strips of steel which are butt-welded and rolled around a split mandrel to generate two concentric spiral passages, one for the hot fluid and one for the cold fluid. Each fluid flows through its own single, curving, rectangular cross-sectioned passage in a truly countercurrent or cocurrent manner.

Instead of the multitude of passages found on the tube side of shell-and-tube exchangers, the spiral exchanger has only a single passage through which the fluid travels. This makes the unit self-scrubbing. Thus, if solids begin to deposit in a channel, the flow cross-sectional area becomes constricted. The resulting constriction

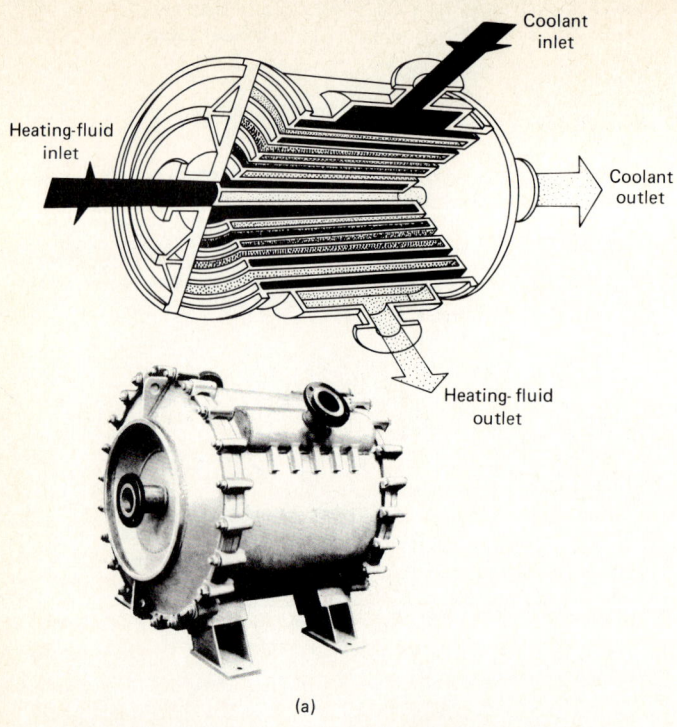

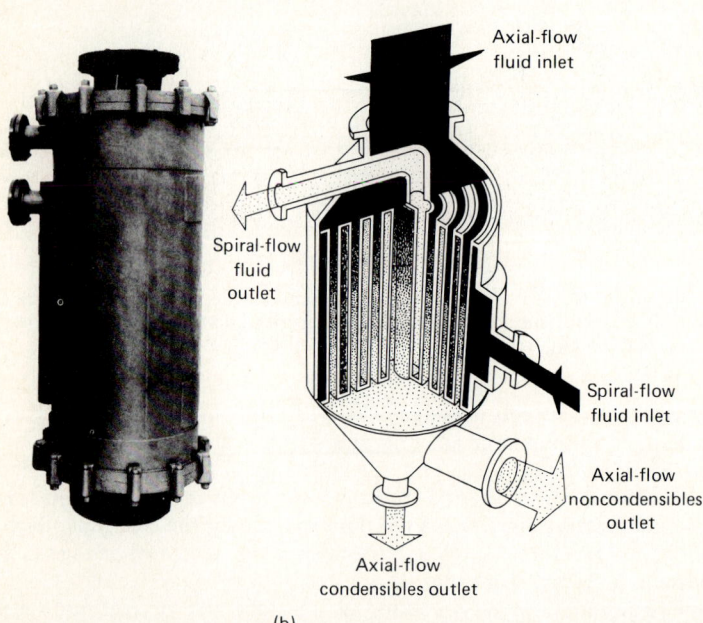

Figure 14.17 Spiral heat exchangers and their flow patterns. (*a*) Spiral flow, both channels; standard design for liquid-to-liquid heat exchange and gas cooling. (*b*) One channel spiral and the other axial; used for vertical condensers and reboilers; gas heating and cooling and liquid-to-

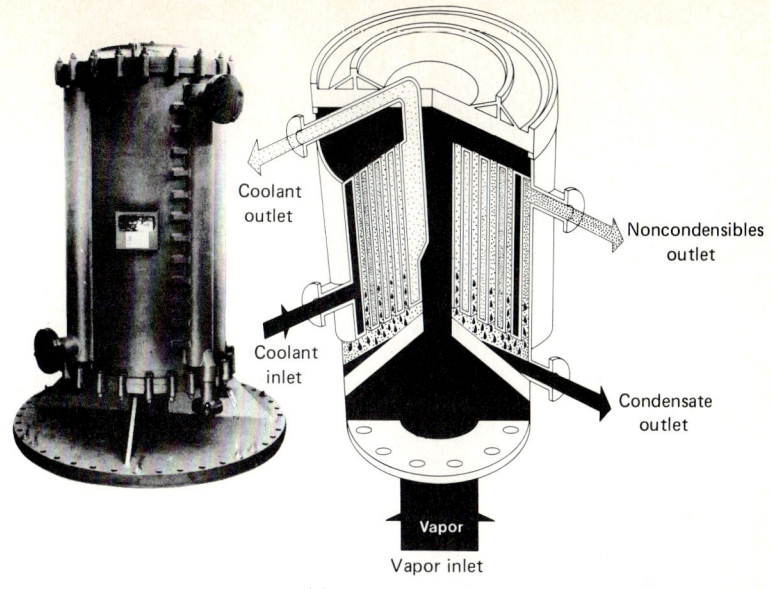

(c)

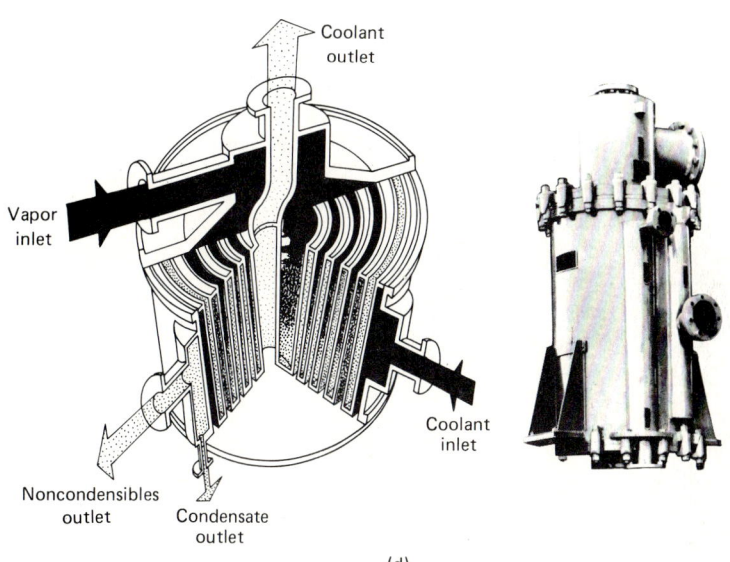

(d)

liquid heat transfer. (*c*) Overhead condenser. (*d*) Overhead condenser. Vapor flows through axially with condensate subcooling. *(Courtesy of Alfa-Laval, Inc., Fort Lee, N.J.)*

increases the local velocity, causing the built-up solids to be resuspended or scrubbed away. Although the self-scrubbing is not complete, the channel remains open and heat transfer still takes place.

Only one end of each channel is welded shut, so that there is complete access to either the hot channel or the cold channel, thus leaving the other end sealed between a heavy-duty cover by a sheet gasket of any appropriate material. For cases in which one fluid is highly fouling and the other is relatively clean, e.g., city water, chilled water, or steam, only the clean-fluid channel is welded shut, so that complete access is obtained from either cover to the dirty-fluid channel. When frequent cleanout is anticipated, the exchanger is constructed with hinged covers for easy maintenance.

Figure 14.17 shows the different flow configurations within a spiral exchanger. The flow pattern in Fig. 14.17 used for liquid-to-liquid exchange is designed with a horizontal axial flow through the center of the spiral. Exchangers with a vertical spiral axial flow are used for the low-volume condensing of steam or vent vapors from storage tanks. This vertical configuration permits easy, gravity-assisted flow of condensate. Other variations of the spiral are available for vacuum condensing service or direct column-mounted service in which the vapors are in cross-flow while the heat-transfer medium in the other channel travels through a spiral pattern. The vertical configuration develops extremely low pressure drops (1 to 5 mmHg). Because of their low pressure drops, these exchangers have been used as condensers mounted directly above the top tray of a distillation column. Condensate produced in these exchangers can be gravity-drained directly back to the top tray or removed peripherally to a reflux splitter or surge drum. Spiral exchangers can also be designed to permit condensing and subcooling within the same unit.

Typical heat-transfer coefficients for spiral exchangers are about the same as those found in shell-and-tube exchangers. An exception consists of those exchangers used for condensing or thermosiphon-condensing performance, in which case the coefficients are greater than those for smaller shell-and-tube exchangers.[8] Heat-transfer-coefficient correlations for spiral heat exchangers are listed in Table 14.8. These film coefficients are based on the standard Nusselt correlations, using the equivalent mean diameter for the spiral based on a rectangular passage. Fouling factors have been reported by independent observers to be one-sixth to one-tenth of those of conventional shell-and-tube units when both fluids are in spiral flow. Therefore, the lower fouling factors result in better overall heat-transfer coefficients.

A disadvantage of spiral-plate exchangers is that the plates are heavier-gauge than the corresponding tubes in a shell-and-tube exchanger. As a result, spiral exchangers are more costly but can be used in abrasive services. However, the added cost must be weighed against reduced fouling potential or improved condensing performance. In addition, spiral-plate exchangers' compactness and ease of maintenance for certain applications should be considered.

Plate heat exchangers. The plate heat exchanger, also called a plate-and-frame exchanger because of its similarity to that type of filter, is one of the most efficient liquid-to-liquid heat exchangers available today. The exchanger, as shown in Fig. 14.18, consists of a series of flat, corrugated, gasketed plates compressed between two thick carbon steel covers which are not wetted by the heat-transfer fluids. Gaskets on each plate are used to control fluid-flow direction, making the unit a partially or totally countercurrent or cocurrent heat exchanger. In addition, the gasketing can be used to create a variety of complex flow distributions. Figure 14.19 shows the various

Figure 14.18 Plate heat exchanger. *(APV Equipment, Inc., Tonawanda, N.Y.)*

flow patterns developed for series- and parallel-flow arrangements. As a general rule, the parallel-flow arrangement is used when pressure drops must be limited. Exchangers usually have a maximum of 350°F design temperature and with pressure ranging from 150 to 300 psig. The pressure rating is limited by the size of the plate exchanger and material selected for the plates, while the temperature limitation is generally due to the fixed gasket materials. For temperatures above 300°F, certain high-temperature elastomers or compressed asbestos can be used for the gaskets.

Plate heat exchangers have highly turbulent flow, and the resulting scouring action produces low fouling. Within these design conditions, they have an overall heat-transfer coefficient which is 2 to 3 times greater than that of a corresponding shell-and-tube exchanger, with U values of 700 to 1000 Btu/(ft$^2 \cdot$ h $\cdot$ °F) common.

Two plates and the gasket between them form intermediate contact points across the surface of the plate, permitting rather high design pressures. These contact points also permit the plates to be constructed out of rather thin material (0.6- to 0.8-mm metal) while still generating a plate pack of high design pressure and extreme rigidity when held between the frame covers. High heat-transfer coefficients combined with the use of thinner construction material result in a much lower-cost heat exchanger per British thermal unit of heat transferred than a comparable shell-and-tube exchanger. Consequently, plate heat exchangers are often preferred in heat-recovery applications.

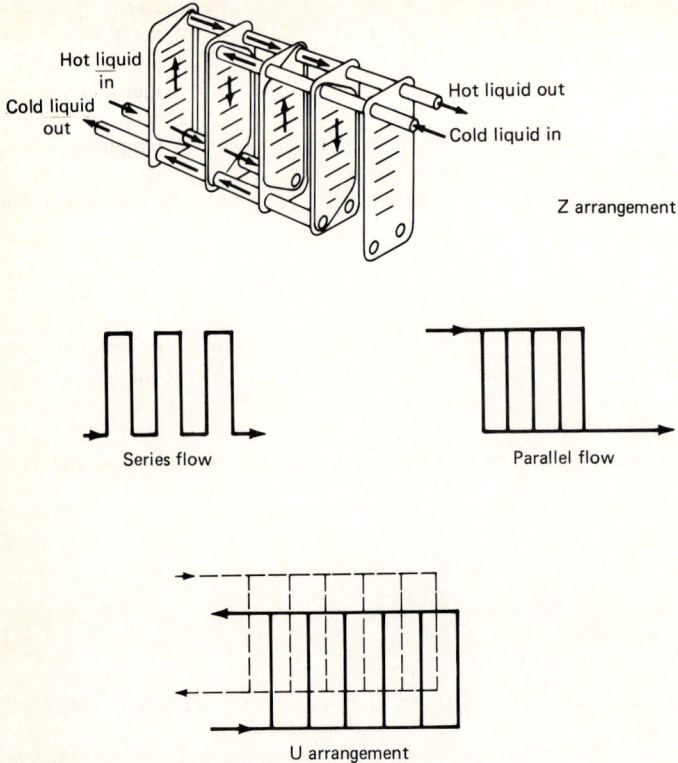

Figure 14.19 Plate-heat-exchanger flow patterns.

Plates for these exchangers are available in a wide range of materials such as stainless steel, Incoloy, titanium, palladium-stabilized titanium, and Hastelloy. Gasket materials are chosen for their compatibility with the fluid being handled. Gaskets are commonly available in nitrile rubber, resin-cured butyl rubber, ethylene-propylene diene monomer (EPDM), Viton, silicone-filled Teflon, and compressed asbestos.

Plate heat exchangers are generally used for liquid-to-liquid applications because they do not have sufficient internal volume to permit handling gases and, thus, are not recommended in certain steam-heating applications. They should not be used with fluids having a solids content greater than 3 to 5 percent by volume because of the possibility of erosion corrosion. Similarly, fluids containing any quantity of fibers should not be used in plate exchangers because the greater number of contact points across the surface of each plate will cause these fibers to "hang up" and eventually block the plate channels.

There are no widely accepted correlations for sizing plate heat exchangers; this is one of the drawbacks to their gaining wider acceptance. However, published reports indicate much lower fouling factors observed in plate exchangers than in tubes for sedimentation fouling. Independent investigators[9] have also reported on the extremely high film coefficients available in plate exchangers and the ability to enter a turbulent zone at much lower Reynolds numbers, 20 to 200, than are found in flow-through tubes, 2000 to 4000. Generally, plate heat exchangers use Nusselt-type equations with corrections for the "waviness factor" caused by the corrugation pattern embossed in each plate.* Pressure drops are about the same as those used for shell-and-tube designs. Although the heat-transfer coefficient of an exchanger increases in proportion to the pressure drop, plate heat exchangers generally do better at pressure drops greater than 2 to 3 psi than do comparable shell-and-tube exchangers.

Pressure-drop calculations. The pressure drop of the fluid flowing through either side of a heat exchanger coincides in theory with the Darcy formula

$$\Delta P_f = \frac{fG^2L}{2(144)\,g_c\rho D_e} \quad \text{psi} \tag{14.68}$$

However, the complex geometry of a heat exchanger together with the changing fluid properties with changing temperature requires empirical modifications to the pressure-drop relationships and the individual variables used in them.

When there is no phase change and when entrance losses and losses due to bends can be ignored, the mass velocity used in the Darcy equation for tube-side pressure drops is calculated by

$$G = \frac{4(144)W}{N\pi d_t^2} \tag{14.69}$$

where W = total mass flow, lb/h
d_t = tube inside diameter, in
N = number of tubes per tube pass

The length in the Darcy equation is equal to the product of the tube length and the number of tube passes.

The Moody friction factor can be approximated by using the relationship found in Chopey and Hicks:[10]

$$\text{Re} < 2100 \quad f = 64/(\text{Re}) \tag{14.70}$$

$$\text{Re} \geq 2100 \quad f = 0.216/(\text{Re})^{0.2} \tag{14.71}$$

Chopey and Hicks include a simple relationship for calculating pressure drops due to entrance and exit losses for the nozzles on the heads and the head-to-tube connection; i.e.,

Head entrance or exit losses $\quad \Delta P_N = K\rho u_n^2/9266 \tag{14.72}$

Tube entrance or exit losses $\quad \Delta P_e = KN_p\rho u_t^2/9266 \tag{14.73}$

where u_t = linear velocity at the tube nozzle, ft/s
u_n = linear velocity at the shell nozzle, ft/s
N_p = number of tube passes
ρ = fluid density, lb/ft^3
K = 0, head-nozzle fluid entrance loss; 1.25, head-nozzle fluid exit loss; 1.8, tube entrances and exits

---

*Corrugation patterns vary from one manufacturer to another.

The three pressure drops can be combined to get the total tube-side pressure drop by

$$\Delta P = \Delta P_f + \Delta P_N + \Delta P_e \tag{14.74}$$

Other pressure-drop calculations are more complex. Table 14.11 lists the pressure-drop formulas for the more complex geometries of shell-and-tube heat exchangers and for plate and spiral-plate exchangers.

The acceptable range of pressure drops in a heat exchanger depends on the system operating pressure. Typical pressure-drop ranges are as follows:

System pressure	Pressure-drop range
Vacuum	5–10% of absolute system pressure
1–10 psig	0.5–5 psi
> 10 psig	5 psi up to 50% of system gauge pressure

For high-pressure-drop designs, fluid velocity may be the more limiting criterion. For example, liquid velocity in excess of 15 ft/s can cause erosion corrosion in exchangers.

Specialty heat-exchange equipment

In addition to the heat-exchange equipment previously discussed there are a number of other types of equipment. These are considered specialty equipment because they are used for specific process applications or they employ unique mechanisms for heat transfer such as combustion-reaction kinetics used by boilers and fired heaters or mass transfer used by cooling towers. The following subsections provide brief descriptions of these devices and how they are used in process plants. Agitated vessels requiring heating or cooling are covered in Chap. 16.

Heat-transfer panels. An auxiliary method of providing heating or cooling to a piece of equipment employs heat-transfer panels. They are fabricated from thin metallic plates to which are welded half pipes or other embossed plates to form channels for the flow of the heating or cooling medium. The dimensions of standard individual plates or panels vary from 12 to 43 in in width and from 23 to 143 in in length. They can be fabricated to conform to almost any internal or external configuration of a vessel, tank, or specialized piece of equipment.

The overall heat-transfer coefficients for plates and panels are a function of the type of transfer within the process unit, i.e., free convection or agitation of the fluids, boiling or condensation. The U's are a particular function of the intimacy of contact of the panel with the fluid if the panel is internal to the vessel or with the surface of the vessel if the panel is external to the vessel. For those situations in which the panel is internal to the vessel or the panel is integral with the vessel's external surface, the overall heat-transfer coefficients U are similar to those discussed in the earlier subsections on natural and forced convection. However, if the panel is merely clamped to the exterior of the vessel, there is a considerable reduction in the U value. A heat-transfer mastic placed between the panel and the exterior surface of the metal improves the U considerably. The manufacturer of the panel equipment should be consulted to select overall heat-transfer U values for the particular application and the type of panel configuration and construction to be used.

TABLE 14.11 Pressure-Drop Formulas for Various Heat Exchangers*

Mechanism or restriction	Empirical equation
	Shell-and-Tube Exchangers
	Inside the tubes
No phase change Re > 10,000	$\Delta P = \dfrac{(Z_i)^{0.2}}{S_i}\left(\dfrac{W_i}{n}\right)^{1.8}\dfrac{N_{PT}[(L_o/d_i)+25]}{(5.4d_i)^{3.8}}$ (See Note 1)
No phase change, 2100 < Re < 10,000	$\Delta P = \left(\dfrac{Z_i}{S_i}\right)\left(\dfrac{W_i}{n}\right)\dfrac{N_{PT}[(L_o/d_i)+25][(\text{Re})^{2/3}-125]}{(50.2d_i)^3}$ (See Note 1)
No phase change, Re < 2100	$\Delta P = \dfrac{(Z_b)^{0.526}(Z_f)^{0.14}}{S_i}\left(\dfrac{W_i}{n}\right)^{4/3}\dfrac{N_{PT}(L_o)^{2/3}}{(5.62d_i)^4}$
Condensing	$\Delta P = \dfrac{(Z_i)^{0.2}}{S_i}\left(\dfrac{W_i}{n}\right)^{1.8}\dfrac{N_{PT}[(L_o/d_i)+25]}{(5.4d_i)^{3.8}} \times 0.5$ (See Note 1)
	Shell side
No phase change, cross-flow	$\Delta P = \dfrac{0.326}{S_o}(W_o)^2\dfrac{L_o}{P^3 D_o}$
No phase change, parallel flow	$\Delta P = \dfrac{(Z_o)^{0.2}}{S_o}\left(\dfrac{W_o}{n}\right)^{1.8}\left[\dfrac{n^{0.366}L_o}{(N_{PT})^{1.434}(4.912d_o)^{4.8}}\right.$ $\left. + \dfrac{0.31n^{0.0414}(W_o)^{0.2}L_o}{d_o(N_{PT})^{1.76}(4.912d_o)^4 Z^{0.2}B_o^2}\right]$ (See Notes 2 and 3)
Condensing	$\Delta P = \left(\dfrac{0.081}{S_o}\right)(W_o)^2\left(\dfrac{L_o}{P^3 D_o}\right)$
	Spiral and Plate-Type Exchangers
	Spiral flow
No phase change, Re > Re$_c$	$\Delta P = 0.001\dfrac{L}{s}\left[\dfrac{W}{d_s H}\right]^2\left[\dfrac{1.3 Z^{1/3}}{(d_s+0.125)}\left(\dfrac{H}{W}\right)^{1/3}+1.5+\dfrac{16}{L}\right]$ (See Note 4)
No phase change, 100 < Re < Re$_c$	$\Delta P = 0.001\dfrac{L}{S}\left[\dfrac{W}{d_s H}\right]^2\left[\dfrac{1.035 Z^{1/2}}{(d_s+0.125)}\left(\dfrac{Z_1}{Z_b}\right)^{0.17}\left(\dfrac{H}{W}\right)^{1/2}+1.5+\dfrac{16}{L}\right]$ (See Note 4)
No phase change, Re < 100	$\Delta P = \dfrac{LsZ}{3{,}385(d_s)^{2.75}}\left(\dfrac{Z_1}{Z_b}\right)^{0.17}\left(\dfrac{W}{H}\right)$
Condensing	$\Delta P = 0.0005\dfrac{L}{s}\left[\dfrac{W}{d_s H}\right]^2\left[\dfrac{1.3 Z^{1/3}}{(d_s+0.125)}\left(\dfrac{H}{W}\right)^{1/3}+1.5+\dfrac{16}{L}\right]$
	Axial flow
No phase change, Re > 10,000	$\Delta P = \dfrac{4\times 10^{-5}}{sd_s^2}\left(\dfrac{W}{L}\right)^{1.8}\left[0.0115 Z^{0.2}\dfrac{H}{d_s}+1+0.03H\right]$
Condensing	$\Delta P = \dfrac{2\times 10^{-5}}{sd_s^2}\left(\dfrac{W}{L}\right)^{1.8}\left[0.0115 Z^{0.2}\dfrac{H}{d_s}+1+0.03H\right]$

See Table 14.7 nomenclature for customary English units.
1. For U bands, use $[(L_e/d_i)+16]$ instead of $[(L_o/d_i)+25]$.
2. B_o is equal to fraction of flow area through baffle.
3. Number of baffles $(N_s) = 0.48 (L_e/d_o)$.
4. $\text{Re}_c = 20{,}000 (D_e/D_w)^{0.32}$.

*Excerpted by special permission from *Chemical Engineering* (Jan. 26 and May 4, 1970.) Copyright © 1970, by McGraw-Hill, Inc., New York, N.Y. 10020.

Air-cooled heat exchangers. Air coolers, or air-cooled heat exchangers, are often described as dry cooling towers. This common misconception occurs because air coolers resemble cooling towers and often are used for similar applications; however, an air cooler's heat-transfer mechanism is more in line with that of a shell-and-tube heat exchanger.

The air cooler is used primarily for cooling or condensing process fluids. The cooling medium is air, which passes over the outside surface of a tube bundle containing the process fluid. There are two types of air coolers: mechanical-draft and natural-draft. The mechanical-draft air cooler shown in Fig. 14.20 uses a fan to drive air

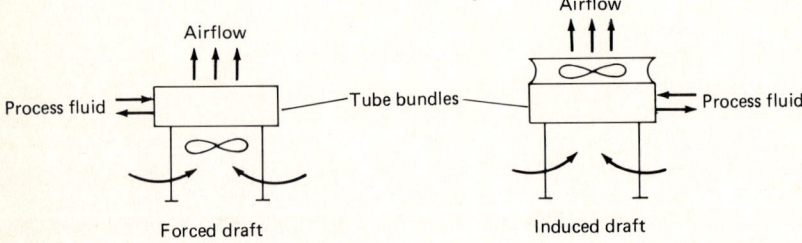

Figure 14.20 Mechanical-draft air cooler.

across the tube bundles. The natural-draft air cooler shown in Fig. 14.21 uses the convective action of natural drafts through the cooler to carry air across the tube bundle.

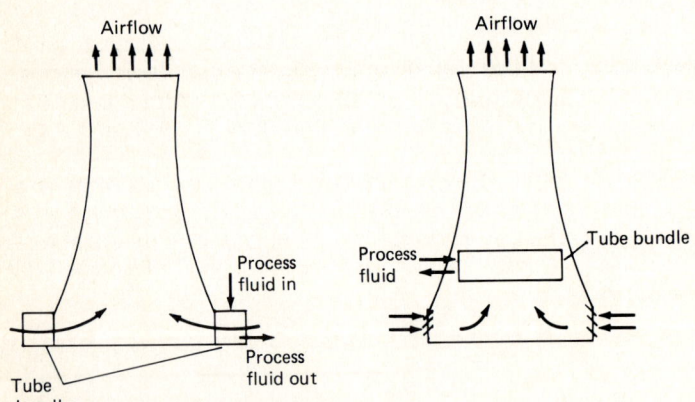

Figure 14.21 Natural-draft air cooler.

Theoretically, the process fluid can be cooled to a temperature equal to the ambient-air temperature, but this would require an infinitely large surface area. The difference between the process fluid's outlet temperature and the ambient-air temperature is called the approach of the cooler. The smaller the approach, the larger and more expensive the air cooler. Typical approaches for air coolers range from 5 to 20°F.

The overall heat-transfer coefficients for air coolers are low. The typical range of heat-transfer coefficients is:

Application	Overall heat-transfer coefficient, $Btu/(h \cdot ft^2 \cdot °F)$
Liquid cooling, cP	
< 5	60–100
5–10	10–60
10–100	5–10
Gas cooling, psig	
< 50	30–50
> 50	50–90
Partial condensing	60–100
Total condensing	70–130

Because of the relatively low heat-transfer coefficients for air-cooled heat exchangers, finned tubes are normally used to increase the heat-transfer area per unit volume of the exchanger. Coefficients above are for finned pipe based on base-pipe area.

Air-cooled heat exchangers are used mainly in areas where there is a scarcity of water or where an approach of more than 50°F to the summer ambient dry-bulb temperature is acceptable. Air coolers can also produce colder process outlet temperatures during the cold winter months than would be obtained from a wet cooling tower. The colder process temperatures may be an advantage or a disadvantage depending on the application. The heat-transfer rate on an air cooler can be reduced during colder weather by reducing the speed or shutting off the fans in a mechanical-draft unit or by changing the position of the air louvers.

Cooling towers. Cooling towers are used for waste-heat removal. They are usually a component part of a process plant's cooling-water loop. Cooling towers use evaporative cooling to cool the water circulated through them. Therefore, they use a combination of heat and mass-transfer mechanisms between water and air for their operation.

The operation of a cooling tower can be described by using a psychometric chart such as the one shown in Fig. 14.22. If the chart is reduced to air enthalpy versus cooling-water temperature and the operating line of a cooling tower is placed in the chart (Fig. 14.23), the operation of the tower can be determined. The theoretical relationship for tower operation is

$$\frac{KaV}{L} = \int_{T_1}^{T_2} \frac{dT}{H^\star - H} \qquad (14.75)$$

where a = water-surface area per unit tower volume, ft^2/ft^3
V = active tower volume per cross-sectional area with respect to water flow, ft^3/ft^2
L = water flow rate, $lb/(h \cdot ft^2$ of cross-sectional area)
T = water temperature, °F (1 = outlet, 2 = inlet)
$H^\star$ = saturated air enthalpy, Btu/lb of dry air
H = enthalpy of actual air mass, Btu/lb of dry air
K = empirically derived mass-transfer rate of evaporated water, $lb/(h \cdot ft^2$ of water-surface area)

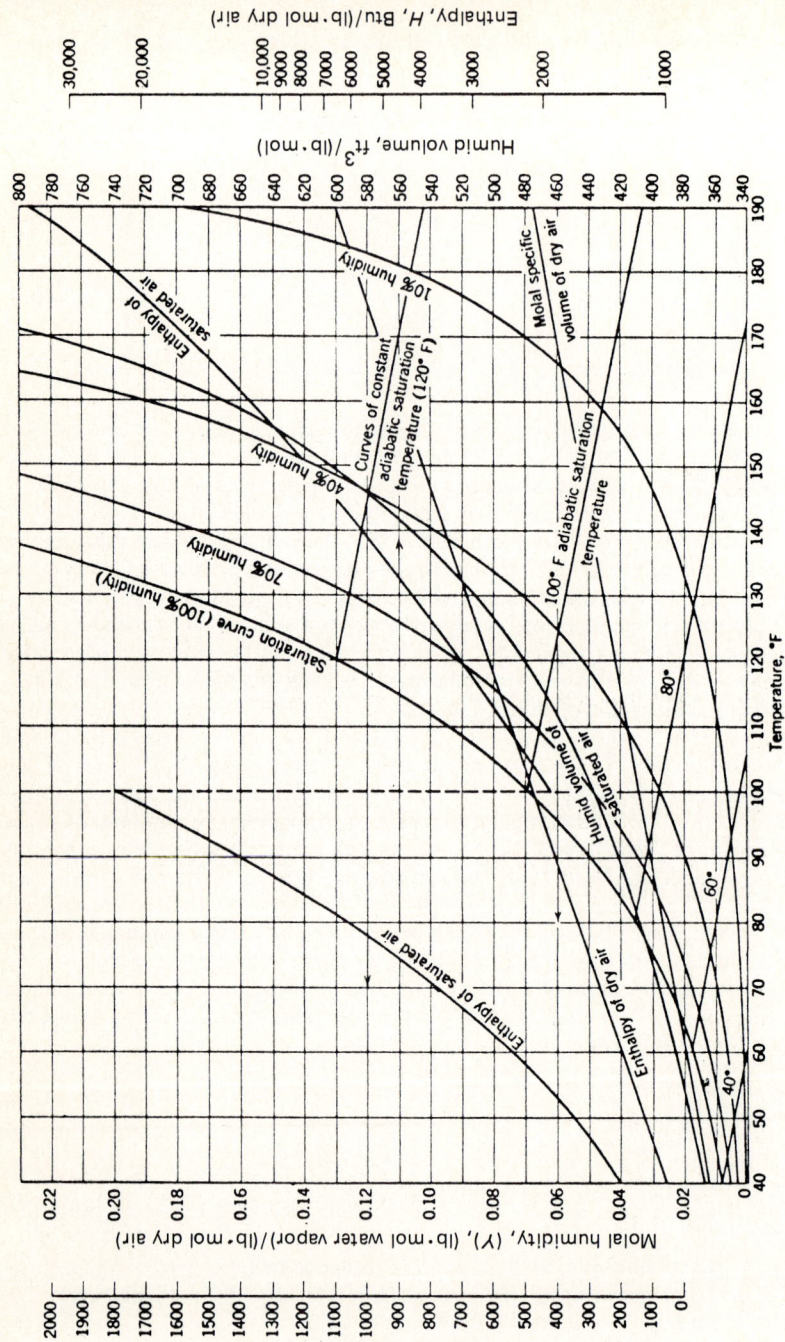

Figure 14.22 Psychometric chart for air-water system; 1-atm total pressure.

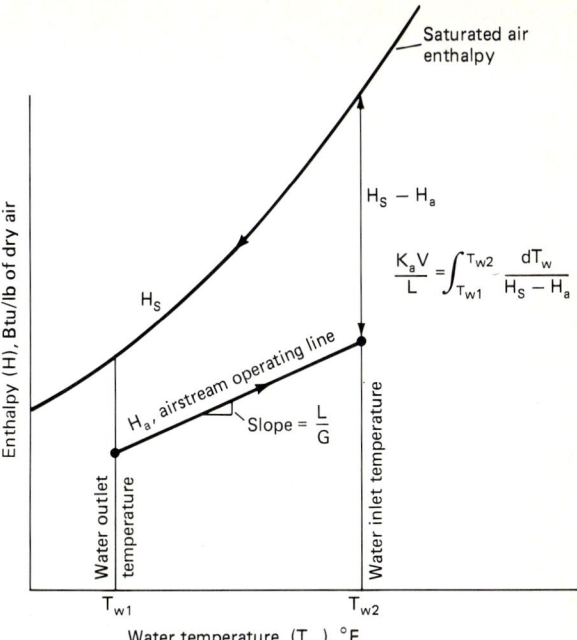

Figure 14.23 Graphical representation of cooling-tower operations. L = water mass flow rate; G = air mass flow rate; K_a = mass-transfer (evaporation) flow rate; V = active volume, ft³/horizontal cross-sectional area.

The term KaV/L is called the tower characteristic. If the saturation line is approximated by a straight line as in Fig. 14.24, the average enthalpy difference through the tower can be approximated by a log-mean difference

$$\overline{\Delta H} = \overline{(H^* - H)} = \frac{\Delta H_2 - \Delta H_1}{\ln\left(\dfrac{\Delta H_2}{\Delta H_1}\right)} \tag{14.76}$$

and the tower characteristic can be approximated by

$$\frac{KaV}{L} = \frac{T_2 - T_1}{\overline{\Delta H}} \tag{14.77}$$

Fraas and Ozisik[11] suggest subtracting a correction factor (δh) from each enthalpy difference for a more accurate estimate of the tower characteristic. This correction factor is defined as

$$\delta h = \frac{H_1^* + H_2^* - 2H_M^*}{4} \tag{14.78}$$

where H_M^* is the enthalpy at the temperature equal to $(T_1 + T_2)/2$.

By comparing Eq. (14.77) with Figs. 14.23 and 14.24, it can be seen that the tower characteristic KaV/L increases as the operating line approaches the saturation line

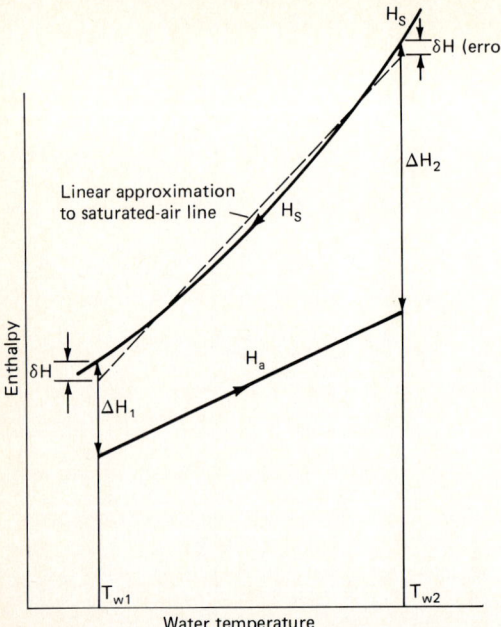

Figuer 14.24 Approximation method for calculating cooling-tower operation.

for the same temperature difference. This means that a larger tower is needed for locations with high-humidity environments than in drier climates.

The slope of the operating line is the liquid-to-gas (air) ratio. If the gas flow is changed while the water flow is maintained constant, the operating characteristics of the tower are changed. Figure 14.25 shows the effect of this change for three airflow rates A, B, and C. Airflow is low for condition A, producing a higher air enthalpy because a more humid air is produced. The airflow rate at condition C is the greatest, and the outlet air is the least humid.

The heat-transfer theory just presented provides a general description of how a cooling tower operates but has limited application to its practical design. Tower design requires the determination of a value of K and a. It is difficult, if not impossible, for some tower designs to determine separate values for K and a. Therefore, the combined value of Ka is obtained by empirical correlations developed by various tower manufacturers from operating data and experimentation. Cooling-tower design is thus the responsibility of the manufacturer, and the process engineer is required to provide the necessary process data for the design. The basic information which should be submitted to the cooling-tower manufacturer is:

1. The estimated cooling-tower duty, i.e, the amount of heat that is to be dissipated by the tower in British thermal units per hour.

2. The inlet and outlet water temperatures. The expected outlet water temperature must be greater than the maximum summer wet-bulb temperature. Most practical tower designs have a 5 to 10°F approach, or the difference between water temperature and the air wet-bulb temperature. Manufacturers can provide towers with a 1°F approach but at a premium price.

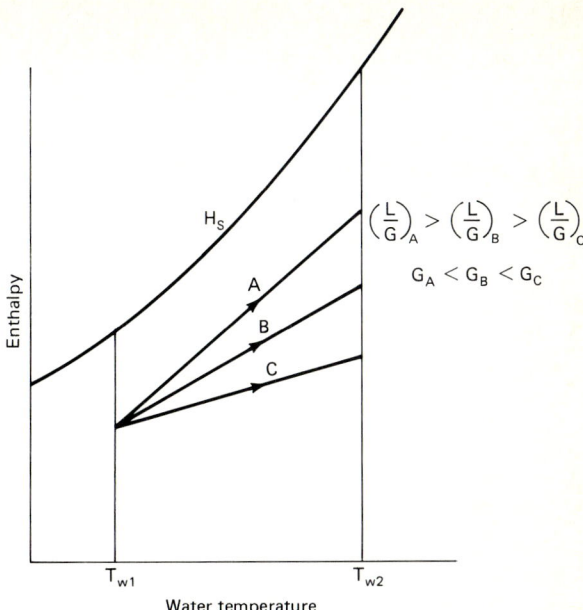

Figure 14.25 Effect of airflow through a cooling tower in operation.

3. The wet-bulb and dry-bulb temperature variation for the tower site. The design wet-bulb temperature is based on the maximum expected summer wet-bulb temperature which is not exceeded more than 5 percent of the time.

4. The prevailing wind direction and wind speed for the site and the tower orientation relative to these winds.

5. The expected seasonal heat-load variation. If the heat load is not related to seasonal changes, then the process engineer should provide the manufacturer with the appropriate heat cycle.

Cooling-tower types. While cooling towers may be provided in a variety of designs, they can generally be classified on the basis of the air-movement mechanism and the flow arrangement. A cooling tower uses either a mechanical-draft or a natural-draft design to produce air movement through the tower. Mechanical-draft towers utilize fans to provide air movement. If the fan is used to draw ambient air into the tower, the design is termed a forced-draft design. If the fan is placed at the air exhaust, the tower has an induced-draft design. In addition to the draft mechanism, the flow arrangement between the water stream and the airstream may be either crosscurrent or countercurrent flow. Figure 14.26 shows the different types of cooling-tower design.

The most prevalent design used in process plants is the mechanical-induced-draft crosscurrent-flow cooling tower. For very large heat loads, with large variations in loading, a series of these units or cells is placed side by side. With this arrangement, a cell can be taken in and out of service on the basis of plant demand. For moderate variations in heat load, a single unit may be used with a variable-speed drive on the fan. This allows the unit to change the water-to-air ratio with changes in heat load.

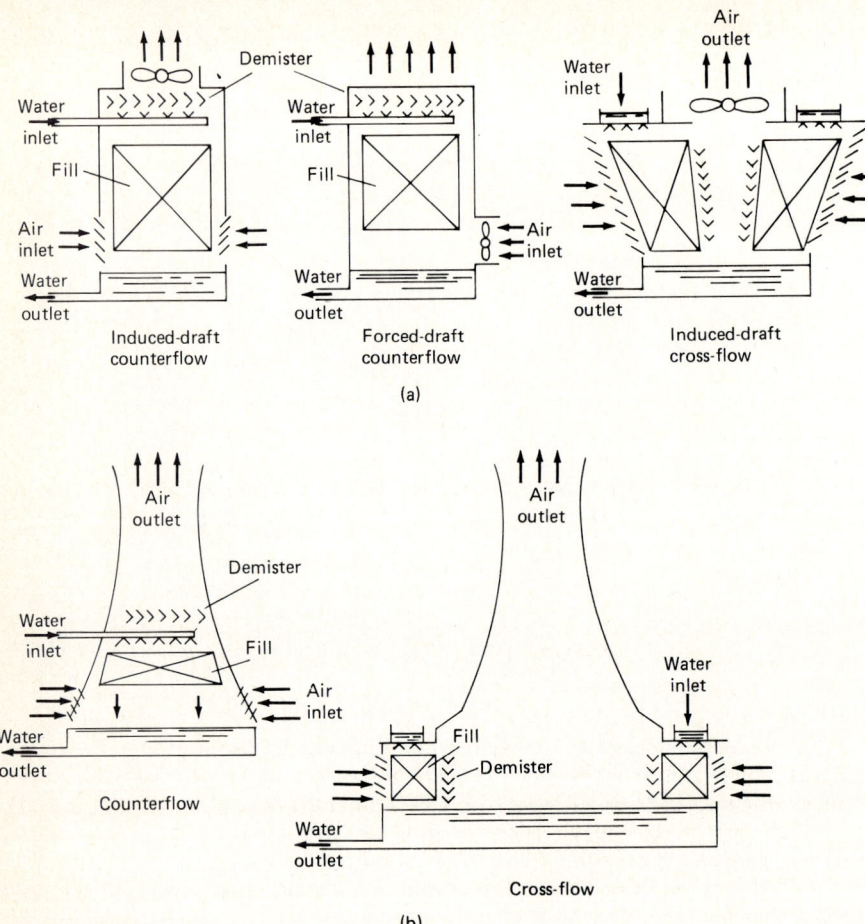

Figure 14.26 Typical cooling-tower designs. (*a*) Mechanical draft. (*b*) Natural draft (hyperbolic towers).

Mechanical countercurrent-flow cooling towers are more thermally efficient than cross-flow towers, but they have a higher airflow pressure drop. This results in greater horsepower requirements for the fan. These towers are usually taller and take up less space than equivalent cross-flow towers. Cross-flow towers are normally preferred over countercurrent units because of their lower operating costs, but countercurrent towers are used where space limitations are a key factor.

Natural-draft towers, with their characteristic hyperbolic chimneys, are used when very large quantities of heat must be dissipated, as in power plants of roughly 300 or more MW. These units have higher installed costs and take up more space than mechanical-draft towers. However, they have lower operating costs because they use natural air currents rather than fans to drive the air through the tower.

Tower materials of construction. Traditionally, cooling towers were constructed completely of wood, but modern towers may be constructed of a variety of plastics, elastomers, ceramics, metals, and concrete. Wood is still used in some tower designs, but it is specially treated with a fire retardant and other chemicals to prevent biodegradation.

The body of a standard-size tower, which includes the louvers and fan stacks in mechanical-draft units, is constructed of metal or fiberglass-reinforced plastic (FRP). Larger units are usually constructed of reinforced concrete with FRP or metal louvers. The tower fill or packing may be either plastic, wood, metal, or ceramic. The fill material is selected for its corrosion-resistance, fire-resistance, and biodegradability properties.

Tower orientation. Because cooling-tower performance depends on airflow, the orientation of the tower is important. The tower should be located as far as possible from any structures which can obstruct airflow. If the process is located in a densely populated area, the tower must be located atop the roof of the plant. However, such installations may require aesthetically pleasing tower designs or shields used to conceal the tower and to direct airflow.

Barring any area or aesthetic restrictions, the tower should be oriented in the general direction of the prevailing summer wind. This orientation not only reduces the work required of the tower fans but also reduces the chance of recirculation. Recirculation is an effect whereby moist air exiting from the tower returns to the intake of the same tower or another tower, thus reducing the respective tower's performance (see Figs. 14.27 and 14.28).

Another phenomenon to consider in a multiple-tower orientation is interference. Interference is the recirculation effect produced by locating a group of multiple-cooling-tower units in close proximity to another cooling-tower group, thus affecting the performance of the latter group. When a new cooling tower is installed, it should not be located in such a way that it could interfere with an existing cooling tower's performance. Figure 14.29 shows some general arrangements of multiple-cooling-tower installations with reduced interference. As a rule, recirculation and interference are not a problem with natural-draft units because the moist air is exhausted at a great height.

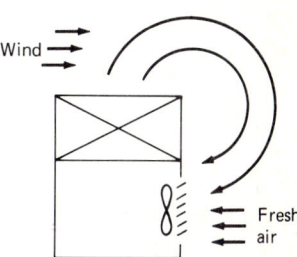

Figure 14.27 Recirculation in a mechanical-forced-draft tower.

Fogging. From time to time a fog or mist is produced by the exiting air. This is an undesirable condition which can create hazards such as obscured visibility if the tower

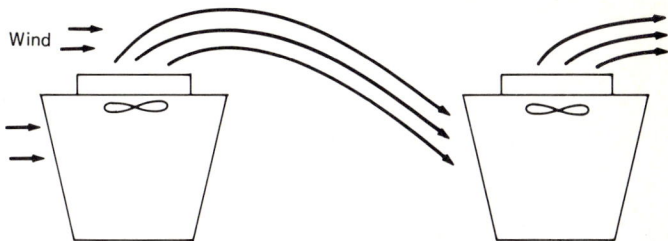

Figure 14.28 Recirculation caused by locating one cooling tower in front of another with respect to the prevailing wind.

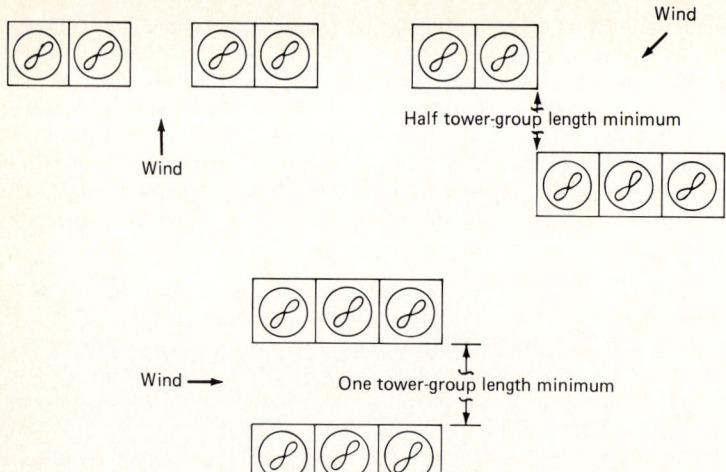

Figure 14.29 Preferred tower-group orientation with respect to prevailing winds.

is located near an airport or a highway and icy roads and equipment when the fog condenses during subfreezing temperatures.

Fogging occurs when the moisture content in the air is greater than the air can absorb. Figure 14.30 shows how fogging can be produced in a cooling tower. The operating curve A represents a condition in which the air passing through a tower is

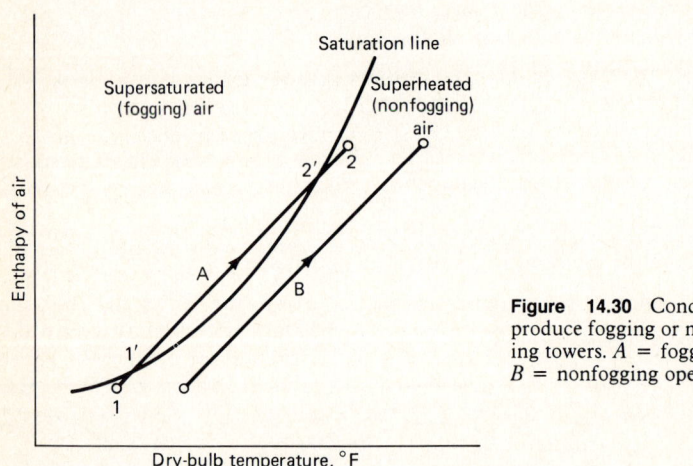

Figure 14.30 Conditions which produce fogging or misting in cooling towers. A = fogging operation; B = nonfogging operation.

heated and takes up more moisture than it can absorb. This occurs between points 1' and 2', where the operating line enters the supersaturated region of the air. In this region, the unabsorbed moisture condenses into droplets, forming a fog.

Fogging normally occurs during cool, humid weather or at nights when the air temperature drops faster then the moisture content of the air. If these conditions occur frequently at the plant site, a standard cooling-tower design is not practical. However, a combined wet-dry mechanical-draft cooling tower such as shown in Fig. 14.31 is generally used for these situations. The wet-dry tower uses an air-cooled exchanger in combination with the cooling tower to reduce the chance of fogging. The air from the cooling tower either passes through an air cooler or is mixed with fresh air which has passed through an air cooler. This allows the exit air to be heated, driving it into the superheated region and eliminating the fog.

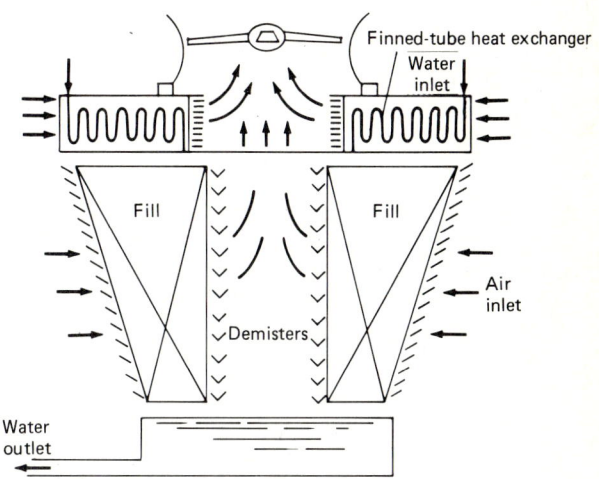

Figure 14.31 Wet-dry cooling tower.

Water treatment. The operation of a cooling tower is also affected by the quality of the water passing through it. Because the cooling tower works on the basis of evaporative cooling, any impurities in the water become concentrated in the tower as water is evaporated. Each time that the water is recycled, the total volume of water in the cooling-water loop decreases but the impurity concentration increases. Therefore, the cooling tower must be periodically or continuously blown down and fresh water added to reduce solid buildup. If a cooling tower is not blown down, the increased impurity concentration will cause fouling and reduced performance in plant heat exchangers using this cooling water.

In addition to adjusting impurity concentrations in the cooling water, chemicals are added to the cooling water in the tower to inhibit corrosion and prevent biological growth in the tower. There is a balance between chemical addition and blowdown that should be maintained for effective tower performance. A water-treatment expert should be consulted to determine the proper water treatment for a cooling tower.

Boilers. Steam is one of the major plant utilities in a process plant. If the process plant is located in a large urban area, the steam may sometimes be purchased from a local utility, but steam is often generated by a boiler located at the plant site. A boiler is a furnace which uses the heat of combustion of a fuel to evaporate water under pressure to form steam.

Most methods used to calculate the heat-transfer rate of a boiler are variations on a method developed by Lobo and Evans.[12] Some of these methods are proprietary and are used exclusively by boiler manufacturers for their specific boiler designs. Usually the process engineer need only specify the performance requirements of the boiler, while the manufacturer is responsible for the actual boiler design.

Boilers generally are of two types, fire-tube and water-tube. Fire-tube boilers (see Fig. 14.32) have a simpler boiler design and are used for installations requiring a steam capacity of 25,000 lb/h or less with an operating pressure not exceeding about 250 psig. Fire-tube-boiler designs consist of a large cylindrical chamber into which fuel, normally oil or gas, is injected and ignited. The tube is surrounded by a chamber

Figure 14.32 Fire-tube boiler. (*Courtesy Eclipse Inc., Rockford, Ill.*)

containing the water being boiled. Most modern fire-tube boilers recirculate the flue gas through tubes (see Fig. 14.33) which pass through the boiling water. Recirculating the flue gas reduces the amount of heat lost out the stack.

Fire-tube boilers are rated in terms of boiler horsepower. One boiler horsepower is defined as the ability to evaporate 34.5 lb of water from the boiler at 212°F. It is equivalent to a boiler heat output of 33,475 Btu/h. For example, a 20,000-lb/h steam-generating capacity is roughly equal to about 600 boiler horsepower.

Fire-tube boilers have low installed costs and, because of their simple design, are easy to operate. These boilers are generally used in process plants having fluctuating loads.

Water-tube or water-walled boilers are used for high-steam-capacity applications. Figure 14.34 shows the basic arrangement of a water-tube power boiler which consists of a series of tubes surrounding a chamber into which fuel is injected and ignited. Heat is transferred to the tubes from the flame primarily by radiation. Water entering the tubes is heated and becomes less dense. High-temperature, low-density water travels up the tubes together with bubbles of steam by natural convection into a drum at the top of the boiler. The top drum, called the steam drum, is used to separate steam

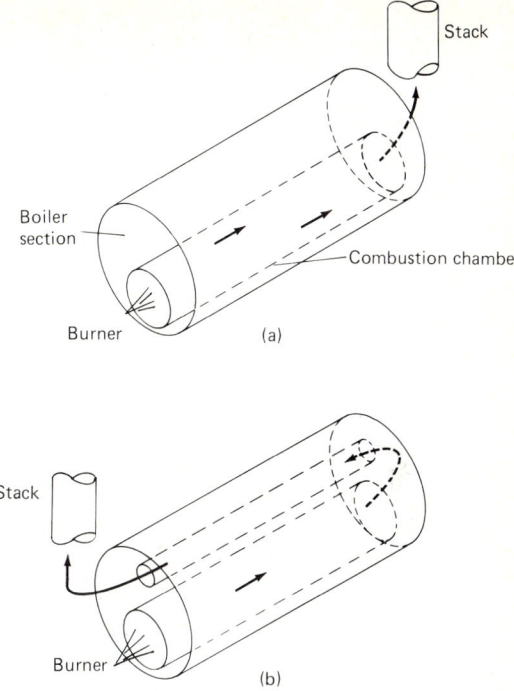

Figure 14.33 Fire-tube-boiler arrangements with flue-gas recirculation. (*a*) Once-through. (*b*) Two-pass.

from water. The steam then exits from the boiler, and the water returns to a different set of tubes which circulates water to another drum at the base of the boiler. This lower drum, called the mud drum, collects impurities concentrated in the recirculated water when steam is produced. The mud drum is drained or blown down to reduce solids buildup in the boiler which can foul the boiler and lower its performance. Boiler blowdown may be either continuous or intermittent. The boiler-tube arrangement shown in Fig. 14.35 is a Type D boiler. The boiler-tube arrangement in Fig. 14.36 has two mud drums and is a Type A boiler.

Water-tube boilers are used for pressures as high as 1500 psig and can be designed to produce superheated steam. Boilers with capacities in excess of 350,000 lb/h are likely to be field-erected units rather than package units. Units of this size are also used to generate electricity as well as steam for a processing plant (see Fig. 14.34). The larger-capacity units generally require additional equipment to increase the boilers' efficiency. This equipment is covered in the subsequent subsection on boiler auxiliaries.

Waste-heat boiler. There are a number of chemical processes which either generate great heats of reaction or require high temperatures to initiate reactions. Once the reaction has been completed, this heat is wasted if it is not used. Therefore, waste-heat boilers have been designed which recover this by-product heat by employing it to generate steam that can be used in the plant.

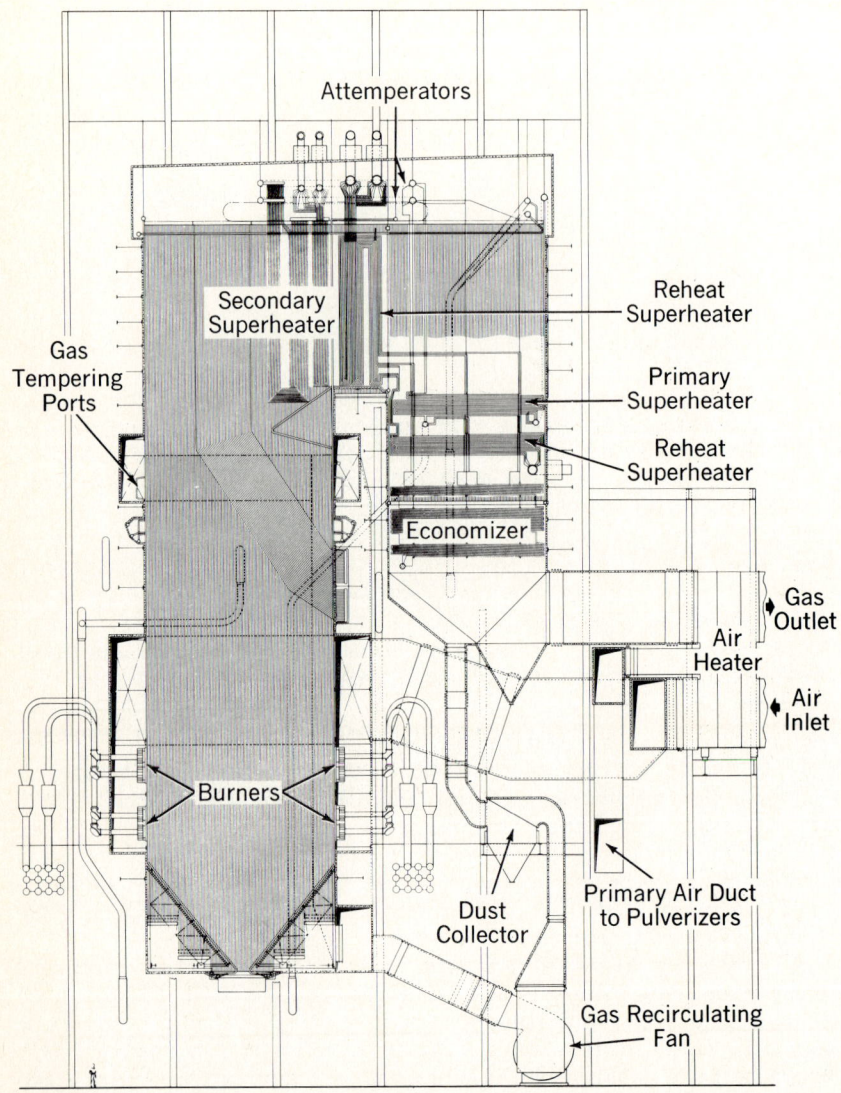

Figure 14.34 1300-MW universal-pressure boiler capable of generating 9,775,000 lb steam/h. This is a specially designed water-tube boiler used for power generation. *(Courtesy of Babcock & Wilcox, Barberton, Ohio.)*

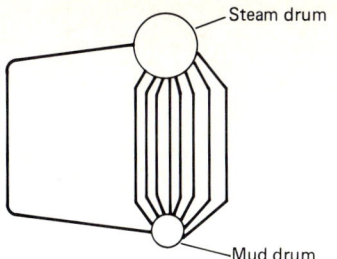

Figure 14.35 Type D water-tube boiler.

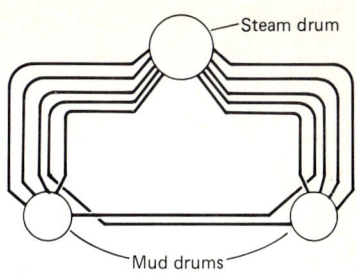

Figure 14.36 Type A water-tube boiler.

Most waste-heat boilers are designed differently from fire-tube and water-tube boilers. Fire-tube and water-tube boilers include fuel-handling systems, and their designs are based on radiant heat transfer being the dominant mechanism. Waste-heat boilers use hot exhaust gases as the source of heat energy and are designed with convective heat transfer being the dominant mechanism. Thus, waste-heat boilers often use extended-surface tubes in nonfouling applications to increase the heat-transfer area. Bare tubes in a square-pitch arrangement for easy cleaning are used in services with high fouling potential.

Boiler auxiliaries. All boilers require auxiliary equipment or accessories which improve their performance or operation. The following are some of the more important auxiliary equipment used with boilers.

Economizers. An economizer is used to preheat the feedwater entering the boiler with heat taken from the flue gas leaving the boiler. The water is heated in bare or finned tubes, depending on the type of fuel used in the boiler.

Feedwater heaters. Feedwater heaters use exhaust or low-pressure steam to preheat the feedwater. Low-heating-value steam is used to preheat the water to prevent thermal shocking of boiler parts while recovering as much heat energy as possible. The feedwater heater is located downstream of an economizer if the latter is used.

Air heaters. Large boilers use air heaters to recover still more heat from the flue gas leaving an economizer. The air heater has either a recuperative or a regenerative design using thin plates or tubes to transfer heat from the flue gas to the air entering the furnace portion of the boiler.

Superheaters. Process plants containing turbine-driven equipment require dry superheated steam for their operation and to mitigate erosion corrosion. The design of the superheater depends on its proximity to the furnace section of the boiler. Superheaters close to the furnace section have radiant-type heat-transfer designs, while those further away are of the convective type. Large boilers used for power generation have multiple superheaters.

Desuperheaters. Process equipment such as an evaporator requires saturated rather than superheated steam for efficient operation. Therefore, a desuperheater must be provided for a boiler producing superheated steam. The desuperheater may be either a shell-and-tube heat exchanger or a water spray injected into the steam line.

Soot blowers. Many fuels, especially solid fuels, produce noncombustible dust or soot which collects on the surface of boiler tubes. These deposits build up and act as an insulator, reducing the boiler's efficiency. Therefore, they should be removed periodically. Soot blowers are mechanical devices which are used to remove soot or dust from heat-transfer surfaces while the boiler is in operation. They employ pulses of air or steam or a combination of these to clean the heat-transfer surface. There are also sonic soot blowers, which use sound waves to shake soot off boiler surfaces.

Fans. Most small package boiler designs depend on natural circulation to draw air through the unit. However, large power-generating boilers require fans to provide adequate circulation of air and flue gas through the boiler. The fans used to push (force) air into the furnace are called forced-draft fans, and the fans used to draw flue gas out of the boiler and out the stack are called induced-draft fans.

Feedwater pumps. Feedwater pumps are used to provide a positive flow of feedwater into the boiler. They may be centrifugal pumps, but most are positive-displacement pumps. Positive-displacement pumps are used in high-pressure boiler operations.

Deaerators. Deaerators are needed to remove air dissolved in returned condensate and makeup feedwater that is to be fed into the boiler. The presence of air in a boiler inhibits heat transfer and increases the possibility of corrosion.

Water-treatment equipment. In addition to deaeration, water must be pretreated before it enters the boiler. The water treatment is used to scavenge oxygen and to reduce scale-forming impurities in the water. Most low-pressure boilers (100 psig or less) usually require softened water; however, high-pressure units require demineralization to reduce total solids, hardness, and silica. The higher the boiler's operating pressure, the greater the efficiency of the demineralization required. Adjustment of pH may also be required, since boilers are less susceptible to corrosion at moderately basic pH's.

Fuel handling. Boilers can be designed to use a variety of fuels. While natural gas, oil, and coal are the most frequently used fuels, some boilers have been designed to use wood chips, bagasse, trash, and sludge as fuels. Each fuel requires storage and handling that are specific to the properties of the fuel.

Gas-fired boilers require the simplest handling system. This usually consists of pressure flow controls and a storage tank. However, natural gas, liquefied natural gas (LNG), and substitute natural gas (SNG) are often the most costly while being the cleanest fuels to use.

Oil-fired units, particularly those using No. 2 and No. 4 oils, require a storage tank, pumps, and flow controls. However, the high-viscosity and less expensive No. 6 or bunker C oils require heated lines and storage facilities to lower the viscosity of the oil, allowing it to flow and atomize more easily. Figure 14.37 shows a schematic diagram of a typical fuel-handling system for bunker C oil.

Coal- and pulverized-coal-handling systems are the most costly and complex handling systems. Yet coal is usually a less expensive fuel source. Figure 14.38 shows a schematic diagram for a typical coal-handling system.

As coal is burned, it produces noncombustible ash, which must be removed and treated. Pulverized coal produces a fine-particulate ash called fly ash and must be treated differently.

REFERENCES

1. Donald Q. Kern, *Process Heat Transfer,* McGraw-Hill Book Company, New York, 1950, chaps. 7 and 8.
2. J. A. Wiebelt, *Engineering Radiation Heat Transfer,* Holt, Rinehart and Winston, Inc., New York, 1966, app. 4.

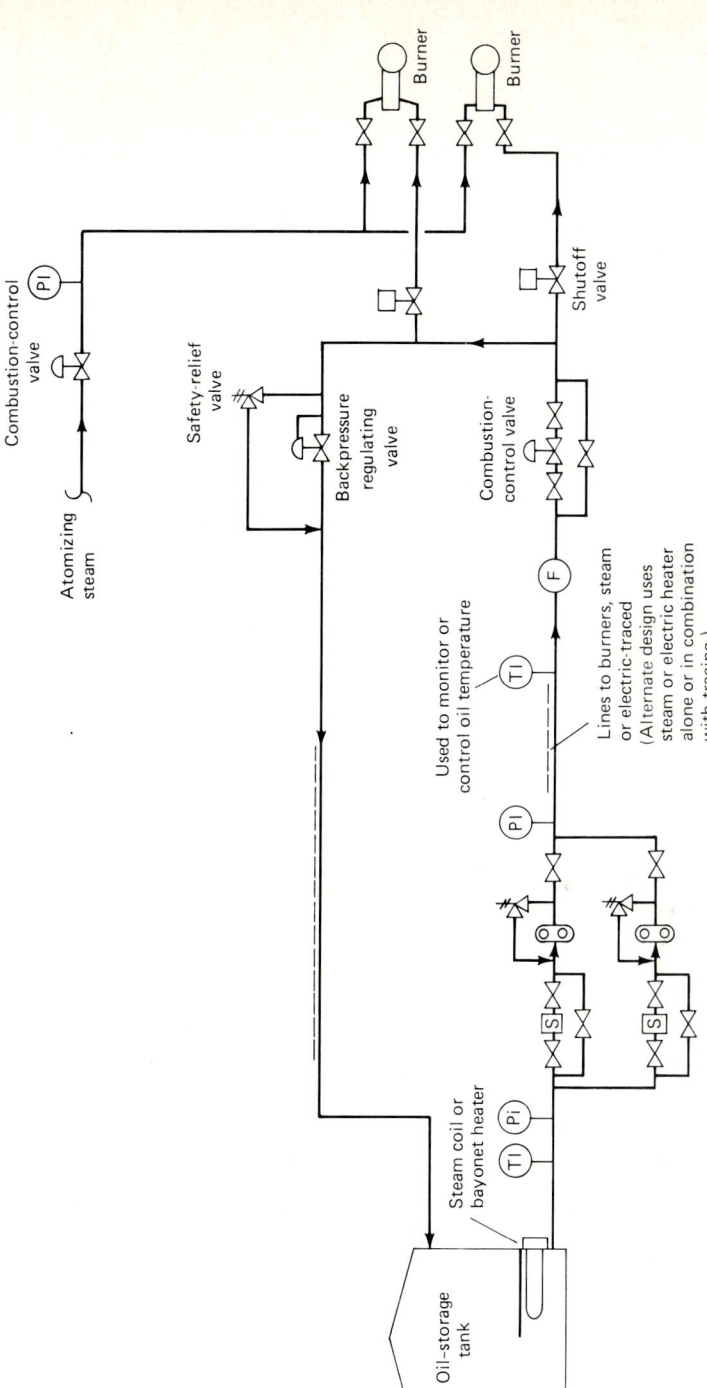

Figure 14.37 Boiler fuel-oil-handling system. TI = temperature indicator; PI = pressure indicator; F = flow indicator, or totalizer; S = strainer.

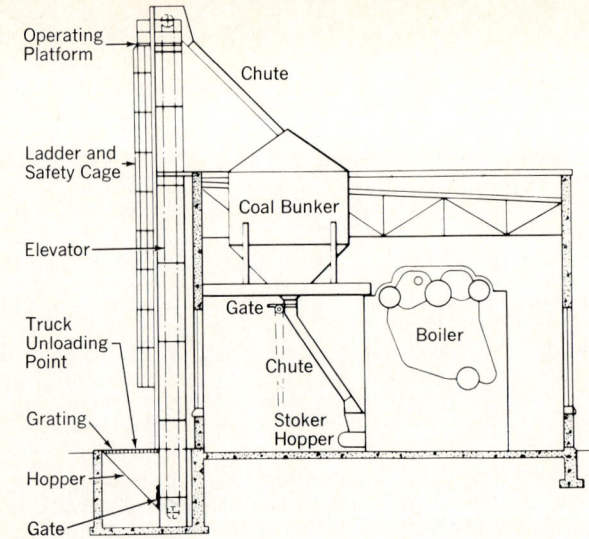
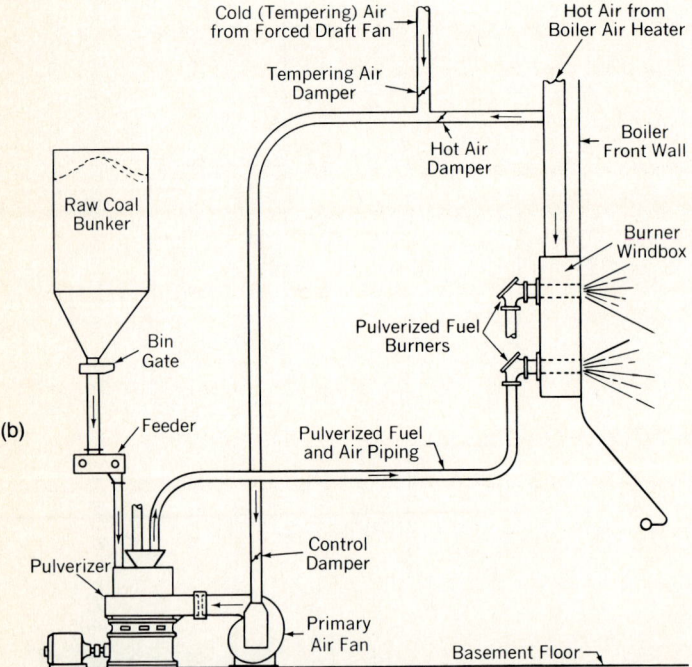

Figure 14.38 Coal-handling system. (*a*) Coal-handling equipment for truck delivery. (*b*) Direct-firing system for pulverized coal. (*Courtesy of Babcock & Wilcox, Barberton, Ohio.*)

3. *Handbook of Heat Transfer,* ed. by W. M. Rohsenow and J. P. Hartnett, McGraw-Hill Book Company, New York, 1973, chap. 15, part B, pp. 15-29–15-32.
4. Wiebelt, op. cit., pp. 168–193.
5. W. H. McAdams, *Heat Transmission,* McGraw-Hill Book Company, New York, 1954, chap. 4.
6. Kern, op. cit., pp. 689–696.
7. *Standards of Tubular Exchanger Manufacturers Association,* 6th ed., Tubular Exchanger Manufacturers Association, Inc., 25 North Broadway, Tarrytown, N.Y. 10591.
8. S. Yilmaz, A. Moliterno, and B. Samuelson, "Vertical Thermosiphon Boiling in Spiral Plate Heat Exchangers," Twenty-First National Heat Transfer Conference, Seattle, July 1983.
9. J. Marriott, "Where and How to Use Plate Heat Exchangers," *Chemical Engineering,* Apr. 5, 1971.
10. *Handbook of Chemical Engineering Calculations,* ed. by N. P. Chopey and T. G. Hicks, McGraw-Hill Book Company, New York, 1984, pp. 7–86.
11. A. P. Fraas and M. N. Ozisik, *Heat Exchanger Design,* John Wiley & Sons, Inc., New York, 1965, p. 247.

BIBLIOGRAPHY

Buffinton, M. A.: "How to Select Package Boilers," *Chemical Engineering,* Oct. 27, 1975.
Campbell, John C.: "How to Prevent Cooling Tower Fog," *Hydrocarbon Processing,* December 1976.
Combustion: Fossil Power Systems, ed. by Joseph G. Singer, Combustion Engineering Inc., 1000 Prospect Hill Road, Windsor, Conn. 06095, 1981.
Cooling Tower Fundamentals and Application Principles, Marley Company, Kansas City, Mo., 1969.
Coulson, J. M., and J. F. Richardson: *Chemical Engineering,* vol. I, 3d ed., Pergamon Press, Elmsford, N.Y., 1977.
DeMonbrun, J. R.: "Factors to Consider in Selecting a Cooling Tower," *Chemical Engineering,* Sept. 9, 1968.
Fair, J. R.: "What You Need to Design Thermosiphon Reboilers," *Petroleum Refiner,* February 1960.
Gutterman, C.: "Specify the Right Heat Exchanger," *Hydrocarbon Processing,* April 1980.
Klein, A.: "Cooling Towers," *Specifying Engineer,* May 1979.
Kolflat, T. D.: "Cooling Tower Practices," *Power Engineer,* January 1974.
Kothandaraman, C. P., and S. Subramanyan: *Heat and Mass Transfer Data Book,* 2d ed., Halsted Press, John Wiley & Sons, Inc., New Delhi, 1975.
Lerner, J. E.: "Simplified Air Cooler Estimating," *Hydrocarbon Processing,* February 1972.
Lydersen, A. L.: *Fluid Flow and Heat Transfer,* John Wiley & Sons, Inc., New York, 1979.
North American Combustion Handbook, 2d ed., North American Manufacturing Co., Cleveland, Ohio 44105, 1978.
Raju, K. S., and J. Chand: "Consider the Plate Exchanger," *Chemical Engineering,* Aug. 11, 1980.
Steam: Its Generation and Uses, 39th ed., Babcock & Wilcox Company, 161 East Forty-Second Street, New York, N.Y. 10017, 1978.
Wimpress, Norman: "Generalized Method Predicts Fire-Heater Performance," *Chemical Engineering,* May 22, 1978.

Chapter 15

Thermal Insulation and Tracing

Chapter 14 concerned itself with the transfer of heat from one medium to another, the principles affecting such transfer, and the equipment used to effectuate the process. As important as it is to accomplish heat transfer, there are situations in which it is equally important to reduce the amount of heat transferred between process equipment or piping and its surroundings or to compensate for heat losses as they occur. Insulating materials are interposed between process components and their surroundings to reduce heat losses or gains. Since it is not feasible to eliminate these heat transfers completely, it is sometimes necessary to introduce an outside source of heating or cooling in intimate contact with the process equipment or piping to maintain a requisite process temperature. The application of these external sources of energy is known as tracing, jacketing, and bundling.

Thermal Insulation

In the selection of heat-transfer equipment, every attempt should be made to obtain as high an overall heat-transfer coefficient U as possible. In order to restrict heat transfer, the overall heat-transfer coefficient should be reduced to a minimum. Figure 15.1 illustrates a typical situation for evaluating heat loss without and with insulation.

The heat loss per unit area of pipe or equipment is given by the classic equation

$$Q = UA(t_p - t_{amb}) \tag{15.1}$$

For uninsulated pipe or equipment, U is essentially equal to the heat-transfer coefficient h for natural convection and radiation from the surface of the pipe or equipment. For all practical purposes, it may be assumed that the surface of a bare pipe is at a temperature t_p equal to that of the fluid for heat flow to or from the surroundings, at ambient temperature t_{amb}. Table 15.1 lists radiation and natural-convection heat-transfer coefficients for bare pipe in still air at 70°F. The radiation and natural-convection coefficients are added together to obtain the overall coefficient U. For lines

Thermal Insulation and Tracing   561

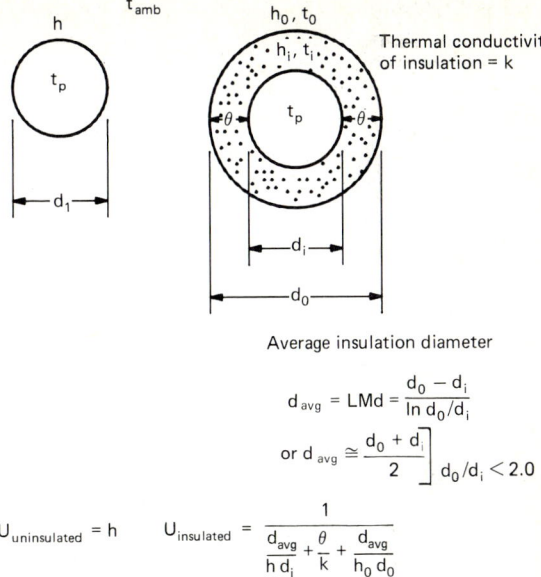

Figure 15.1 Overall heat-transfer coefficients for uninsulated and insulated pipes.

located outdoors, the natural-convection coefficient should be corrected by the factors given in Table 15.1 to account for wind effects.

The overall coefficient U_{ins} for insulated pipe or equipment, on the other hand, consists essentially of three elements: h_i the convection heat-transfer coefficient at the temperature conditions at the inner-surface layer between the pipe or equipment and the insulation; k/θ, the conductance across the insulation itself, which is a function of its thermal conductivity divided by the thickness of the insulation; and h_o, the sum of the convection and radiation heat-transfer coefficients at the temperature conditions at the outer surface of the insulation. The terms in the expression for U_{ins} have to be corrected since they can all refer to different diameter values. The reference diameter is the natural-log-mean diameter for the insulation.* The expression for the overall coefficient for insulation pipe or equipment is

$$U_{ins} = \frac{1}{\dfrac{d_{avg}}{h_i d_i} + \dfrac{\theta}{k} + \dfrac{d_{avg}}{h_o d_o}} \tag{15.2}$$

where U_{ins} = overall heat-transfer coefficient for insulated pipe or equipment, Btu/(h·ft²·°F)

h_i = heat-transfer coefficient at inner surface of insulation, Btu/(h·ft²·°F)

---

*For practical purposes, the natural-log-mean diameter is equal to the arithmetic mean diameter as long as the ratio of the outer and inner diameters is less than 2. At that ratio the arithmetic average diameter is about 4 percent larger than the natural-log-mean value.

h_o = heat-transfer coefficient at outer surface of insulation, Btu/(h·ft²·°F)
d_i = inner diameter of insulation, in
d_o = outer diameter of insulation, in
d_{avg} = natural-log-mean diameter, in
k = thermal conductivity of insulation, (Btu·in)/(h·ft²·°F)
θ = thickness of insulation, in

It should be noted that in most instances the resistance to heat flow due to the inner and outer film coefficients in Eq. (15.2) is an order of magnitude less than that of the heat-flow resistance of the insulation. If only a close approximation of heat loss is required, the inner and outer film coefficients may be ignored. However, they must

TABLE 15.1 Film Coefficients for Radiation and Natural Convection

Radiation coefficient, h_r	Surface temperature; emissivity $\epsilon = 1.0$					
	100°F	200°F	300°F	400°F	500°F	600°F
	1.11	1.47	1.90	2.44	3.08	3.84

Basis: $h_r = \dfrac{0.172\,\epsilon\left[\left(\dfrac{T}{100}\right)^4 - \left(\dfrac{530}{100}\right)^4\right]}{T - 530}$, Btu/(h·ft²·°F); T = surface temperature, °R; 70°F ambient

Natural convection, h_c		Surface temperature, still air					
Nominal size, in	O.D., in	100°F	200°F	300°F	400°F	500°F	600°F
1	1.315	0.88	1.24	1.39	1.49	1.56	1.61
2	2.375	0.76	1.07	1.20	1.29	1.35	1.39
3	3.50	0.69	0.97	1.09	1.17	1.22	1.26
4	4.50	0.65	0.91	1.02	1.10	1.15	1.19
6	6.625	0.59	0.83	0.93	1.00	1.04	1.08
8	8.625	0.55	0.77	0.87	0.93	0.98	1.01
10	10.75	0.52	0.73	0.82	0.88	0.92	0.95
12	12.75	0.50	0.70	0.79	0.85	0.89	0.92
16	16.0	0.47	0.66	0.75	0.80	0.84	0.86
20	20.0	0.45	0.63	0.71	0.76	0.79	0.82
24	24.0	0.43	0.60	0.67	0.72	0.76	0.78
32	32.0	0.40	0.56	0.63	0.67	0.70	0.73
Flat surface (vertical)		0.63	0.91	1.05	1.15	1.23	1.32

Basis: $h_c = 2.55\left(495 + \dfrac{t}{2}\right)^{-0.293}\left[\dfrac{(t-70)}{d}\right]^{0.25}$, Btu/(h·ft²·°F); d = outside diameter, in; surface temperature, °F; 70°F ambient

Correction factor to still air h_c for wind rates:

	still air	2.5 mi/h	5.0 mi/h	10 mi/h	20 mi/h
$\dfrac{h_{c,\text{corr}}}{h_{c,\text{still air}}} =$	1.0	2.0	2.7	3.8	5.9

Basis: Data from L. Clarke and R. L. Davidson, *Manual for Process Engineering Calculations*, 2d ed., McGraw-Hill Book Company, New York, 1962, p. 279.

be taken into account when it is necessary to determine the temperature of the inner or outer surfaces of the insulation.

The temperatures of the inner and outer surfaces of the insulation may be calculated from the resistance of the individual portions of the overall heat-transfer-coefficient equation and the total temperature drop in accordance with the expressions

$$\frac{t_p - t_i}{t_p - t_{amb}} = \frac{\dfrac{d_{avg}}{h_i d_i}}{\dfrac{d_{avg}}{h_i d_i} + \dfrac{\theta}{k} + \dfrac{d_{avg}}{h_o d_o}} \tag{15.3}$$

and

$$\frac{t_o - t_{amb}}{t_p - t_{amb}} = \frac{\dfrac{d_{avg}}{h_o d_o}}{\dfrac{d_{avg}}{h_i d_i} + \dfrac{\theta}{k} + \dfrac{d_{avg}}{h_o d_o}} \tag{15.4}$$

where t_p = temperature of bare pipe surface, °F
 t_i = temperature of inner insulation surface, °F
 t_o = temperature of outer insulation surface, °F
 t_{amb} = temperature of surroundings, °F

The overall coefficient calculated from Eq. (15.2) may be used to calculate heat loss by means of Eq. (15.1). However, it should be noted that the transfer area A is based on the average diameter of the pipe or equipment with insulation.

In order to appreciate the effects of insulation it would be well to compare the heat losses from a pipe without and with insulation in the following examples.

Example 15.1 What is the heat loss per linear foot of a 4-in-diameter piping system containing saturated steam at 400°F with ambient temperature at 70°F? Assume that emissivity ϵ of the bare pipe surface is 0.95.

solution For a bare pipe $U_{ins} = h_o$ or (from Table 15.1) (0.95 × 2.44) + 1.10 for 3.42 Btu/(h·°F·ft²). The area of a linear foot is $\pi\, d/12$ or $\pi \times 4.5/12 = 1.18$ ft²/ft. Therefore, heat loss Q is $UA\,(t_p - t_{amb})$ or 3.42 × 1.18 × (400 − 70), which equals 1330 Btu/(h·ft).

Example 15.2 What would be the loss per linear foot for the pipe in Example 15.1 if 3 in of calcium silicate insulation is placed around the pipe? Calculate the temperature at the inner and outer surfaces of the insulation. The average thermal conductivity of calcium silicate is about 0.43 (Btu·in)/(h·ft²·°F). Assume that insulation cover has an emissivity of 0.25.

solution The diameter of the bare pipe is 4.50 in, and the diameter of the pipe with 3 in of insulation is 10.50 in. Therefore, the natural-log average diameter equals $\dfrac{(10.5 - 4.5)}{\ln(10.5/4.5)}$, or 7.08 in. The radiation plus natural-convection heat-transfer coefficient h_o at the outer insulation surface is about 0.93 Btu/(h·ft²·°F). The inner coefficient is about 1.44. The overall heat-transfer coefficient from Eq. (15.2) is $\dfrac{1}{\dfrac{7.08}{1.44 \times 4.5} + \dfrac{3}{0.43} + \dfrac{7.08}{0.93 \times 10.50}}$, or $\dfrac{1}{8.80}$, which equals 0.117 Btu/(h·ft²·°F). The average surface area per linear foot is $\pi \times 7.08/12$, or 1.85 ft²/ft, and therefore the heat loss per linear foot equals $0.117 \times 1.85 \times (400 - 70)$, or 70 Btu/(h·ft). The temperature at the inner surface, from Eq. (15.3), is $400 - \left(\dfrac{7.08}{\dfrac{1.44 \times 4.5}{8.80}}\right)(400 - 70)$, or 359°F, and that at the outer insulation surface, from Eq. (15.4), is $70 + \left(\dfrac{7.08}{\dfrac{0.93 \times 10.5}{8.80}}\right) \times (400 - 70)$, or 97°F.

While Example 15.2 shows that there still is a heat loss of 70 Btu/(h · linear ft) from a 4-in line provided with 3 in of calcium silicate insulation, that value is considerably less than the loss of 1330 Btu/(h · linear ft) if the pipe had been left bare, as in Example 15.1. In this case, the 3 in of insulation effects a savings of 1260 Btu/h per linear foot of piping. The effect of the insulation is to reduce condensation of the 400°F saturated steam (235 psig, with a latent heat equivalent to 825 Btu/lb) within the pipe from 1.61 to 0.084 lb/h per linear foot. If the 4-in steam line is part of a steam-distribution system, it could likely be several hundred feet long and, if uninsulated, would have to be fed with about 500 extra lb/h of steam. A 4-in steam section normally carries about 9000 lb/h of 235-psig steam, so that the insulation reduces the load on the boiler and the cost of usable steam by about 6 percent.

Usages for thermal insulation

There are many situations in which it is advantageous to apply thermal insulation to piping or equipment. The major usages are reviewed in the following several subsections.

Heat and energy conservation. The word "insulation" is most commonly associated with heat or energy conservation, as was seen in Example 15.2. Most applications of insulation are made in that context. Traced piping is normally insulated to reduce heat loss to the surroundings so that the heat output of the tracing may be kept within a reasonable value to reduce first costs and operating expenses. Lines carrying superheated vapors or mixtures are insulated to ensure that they are kept in that state. In the case of steam, high rates of heat loss from lines or headers would result in excessive condensation, which requires not only a larger boiler but more line traps and higher operating costs. Similarly, piping or equipment with refrigerants, chilled water, or other cold fluids is insulated to prevent its absorbing heat from its surroundings and requiring additional energy at the chilling or refrigeration equipment. Insulation for heat or energy conservation is also required for process lines transferring heated or chilled fluids from one piece of equipment to the next where a processing step or storage is to take place and it is not practical or desirable to include additional heating or chilling at that point.

Personnel protection. Thermal insulation is required for several purposes other than heat or energy conservation. One of the prime uses is for personnel protection, i.e., to prevent skin burns if a hot surface is touched. In this case, it is common practice to provide portions of equipment or piping which contains fluids at 140°F or greater with insulation or guards for those sections with which personnel could come into contact during normal operating duties. This is considered to be within 8 ft of operating or passage area levels. It is recommended that sufficient insulation be applied to the piping or equipment so that the surface temperature of the insulation does not exceed 140°F when a canvas jacket is used over the insulation for indoor applications or does not exceed 113°F with metallic jackets used in outdoor service.

As can be seen from the results of Example 15.2, the thickness of insulation for personnel protection could be less than that needed for heat conservation. However, the amount of protective insulation required depends on ambient temperature. For Example 15.2, the surface of the 3 in of insulation was 97°F with an ambient temperature of 70°F and the fluid in the 4-in pipe at 400°F. Similar calculations show that the insulation need only be $1\frac{1}{2}$ in thick to drop the outside surface to 113°F, whereas the surface temperature would be 127°F with insulation of 1-in thickness. However, if ambient temperature is taken as 100°F instead of 70°F, 2 in of insulation is required to reduce the outer surface to less than 140°F while the full 3 in is needed to reduce a metal-jacket surface to 113°F. As a result, the thickness of insulation for personnel protection is usually taken as that used for heat or energy conservation.

If nonmetallic pipe of such materials as resins, plastics, or glass is used, the temperature drop across the wall of the pipe may be sufficient to bring the pipe's outer skin temperature below 140°F. A heat balance across the pipe wall similar to Eq. (15.4) would be used. However, the denominator would have to include a term $(\theta/k)_w$ for the thickness and conductivity of the pipe wall. Despite the greater insulating properties of these nonmetallic materials, it is usually necessary to include insulation for personnel protection if the temperature of the fluid within them is 175°F or greater.

Protection of fluids from hostile environments. Fluids in pipes may, at times, need protection from a hostile environment within the plant itself. An example is that of a pipe passing close to a bare piece of equipment operated at an elevated temperature. Radiation or conduction to the pipe could possibly raise its temperature above the maximum allowable fluid or pipe temperature. In this event, a portion of the piping near the equipment should be insulated. If the process engineer can foretell this eventuality while preparing the engineering flow diagrams, a note can be included with the line to such effect.

Antisweat protection. Pipe insulation is also used to prevent condensate from dripping on personnel, equipment, instrumentation, or other items in an environment where the dew point of the atmosphere surrounding the pipe or equipment is greater than the temperature of the fluid contained within the pipe or equipment. The presence of condensate within the insulation also increases its thermal conductivity and voids its effectiveness. Insulation to prevent condensation is known as antisweat protection and should be considered for all lines or equipment in outdoor or indoor service where the normal operating temperature is less than the highest ambient dew point. The highest ambient dew point normally occurs in the summer season but can take place at other times if water vapors are present from process or other sources.

Most lines handling refrigerants, chilled water, or brines will already be insulated for energy conservation. Such insulation also serves as antisweat protection. However, there may be lines or equipment containing chilled process fluids which do not require energy protection but still need antisweat insulation. Normally cooling-tower water lines need not be provided with antisweat insulation since the cooling-water temperature only approaches the ambient adiabatic saturation temperature. However, if the lines pass through a closed building with an elevated humidity, they may have to have the antisweat insulation. Well water is usually at a lower temperature than summer ambient temperature, and lines carrying it indoors are usually provided with antisweat protection.

Equation (15.4) may be used to calculate the degree of insulation needed to ensure that the temperature of the outer insulation surface exceeds the maximum ambient dew point.

Reduction of ventilation or air-conditioning requirements. The presence of bare pipes or equipment containing fluids at high temperatures can add considerably to the burden placed on ventilating closed areas. For example, if the 4-in 400°F pipe in Example 15.1 were to run bare the length of a closed area that is 30 ft wide and 20 ft high, a ventilating rate of about 12 changes of air an hour would be required to maintain the air temperature rise at 10°F or less. This rate is to be compared with the 3 to 5 changes for normal ventilation purposes. Similarly, if the pipe were to run through an air-conditioned area, about 12 additional tons of refrigeration would be required for every 100 ft of bare pipe. As noted in Example 15.2, the addition of 3 in of insulation reduces the loads by a factor of 16 or more.

Types of insulation

A large variety of insulation is on the market, some materials being for general purposes and others for specific applications. An excellent guide to more than 20 thermal-insulation materials and their characteristics is given by H. F. Rase and M. H. Barrow.[1] The *Thermal Insulation Handbook* by W. C. Turner and J. F. Malloy[2] also describes the more important insulation materials. It should be noted that asbestos-fiber insulations, long favorites, must now be used with extreme caution owing to their health hazards.

Most operating companies have standardized on insulation types for the services most frequently encountered in their operations. Table 15.2 lists several standard insulation materials and their applications.

Fiberglass (FG) insulation is popular for moderate-temperature (70 to 350°F) installations. However, it should not be used on the heads of vertical vessels (see "Cellular glass" below).

Mineral wool (MW) is used for most high-temperature (350 to 1100°F) process and hot-oil applications. It should not be used on heads for vertical vessels (see "Calcium silicate" below).

TABLE 15.2 Standard Insulation Materials

Material	Typical symbol	Thermal conductivity, k, (Btu·in)/(h·ft²·°F)	Temperature range for use
Fiberglass	FG	0.22–0.26	Moderal temperature (70 to 350°F)
Mineral wool	MW	0.30–0.45	High temperature (350 to 1100°F)
Calcium silicates	CS	0.40–0.50	High temperature (350 to 1200°F)
Rigid urethanes	RU	0.12–0.19	Low and intermediate temperatures (−50 to 100°F)
Cellular glass	CG	0.38–0.48	Low and moderate temperatures (−50 to 400°F)

Calcium silicate (CS) insulations, with and without mixtures of magnesium oxides, are employed with high-temperature (350 to 1250°F) piping and equipment, especially when steam is being contained. CS is used as the insulating material for the heads on vertical vessels which are otherwise insulated with mineral wool.

Rigid urethane (RU) is employed for outdoor installations of brine and chilled-water services and for cryogenic services to $-50°F$.

Cellular glass (CG) serves as insulation for cold applications (down to $-50°F$) and also where there may be alternately moderate temperatures (up to 400°F) for process fluids or oils in the same line. Cellular glass also serves as the insulating material on the heads of vertical vessels which are otherwise insulated with fiberglass.

It should be noted that the temperature limitation for the insulating materials discussed in this subsection is about 1250°F. Special materials such as refractory brick are required for temperatures beyond that limit. The refractory is usually installed internally to the piping or equipment.

Recommended thickness of insulation

The thickness of insulation to be applied for a given internal temperature for various diameters of pipe and equipment should ideally be set by an economic balance. Such an economic balance reflects the annual fixed costs for the installed costs of various thicknesses of insulation as a function of the equivalent annual heat or energy savings for the individual thicknesses. Figure 15.2 shows the form of such an analysis.

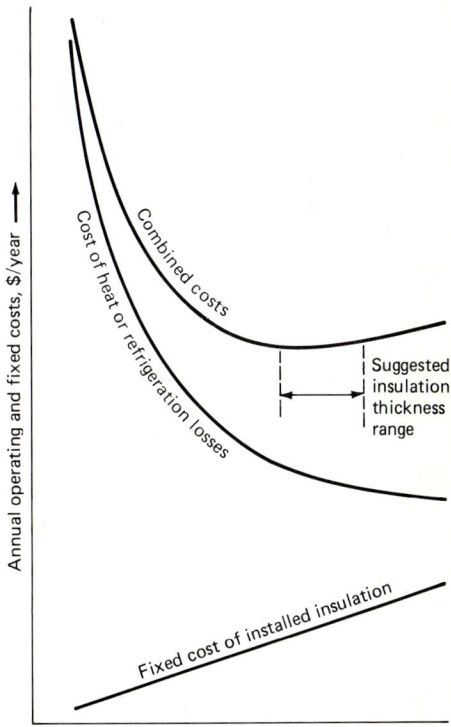

Figure 15.2 Optimum insulation thickness.

It will be observed that the total-cost curve used to determine optimum insulation thickness is usually relatively flat near the optimum point. Thus, the optimum thickness of insulation is not very sensitive to changes in energy costs and installed-insulation charges.

Standard tables are available for insulating materials which suggest thicknesses for various pipe diameters at several fluid-temperature ranges for various insulating materials. These tables are usually based on average economic conditions so that they can be used to specify optimum insulation thicknesses for most normal process situations. However, if there are unusual circumstances of extraordinarily high or low energy charges, abbreviated annual hours of operation, or extremely long or short payback times, it may be necessary to recalculate the thicknesses for the particular case.

The Thermal Insulation Manufacturers Association, Inc. (TIMA)* has available computer programs for maximizing energy conservation while attaining the lowest annualized cost of insulation. It also publishes brochures which give standardized tables for use in typical design situations for average economic conditions. The TIMA data and other information have been rearranged and presented as Tables 15.3 through 15.7. The tables may be used to obtain recommended thicknesses of the commonly used types of insulation.

It should be noted that Tables 15.3 through 15.7 were predicated on a 70°F ambient temperature. Thus, it may be necessary to increase the thickness of insulation to limit heat loss from warm fluids for winter design temperatures of less than 70°F or to

TABLE 15.3 Recommended Thicknesses for Fiberglass (FG)

Nominal pipe size, in	70–100°F	101–200°F	201–300°F	301–400°F
1/2	1″	1″	1½″	1½″
3/4	↓	↓	↓	2″
1	↓	↓	↓	↓
1½	↓	↓	↓	↓
2	↓	↓	2″	2½″
3	↓	1½″	↓	↓
4	↓	↓	2½″	↓
6	↓	↓	↓	↓
8	↓	2″	↓	3″
10	↓	↓	↓	↓
12	↓	↓	↓	↓
14	↓	↓	↓	↓
16	↓	↓	↓	↓
18	1½″	↓	↓	↓
20	↓	↓	↓	↓
24	↓	↓	3″	3½″
30	↓	↓	↓	↓
36 and greater	↓	↓	↓	↓
Flat surface	1½″	2″	3″	4″

Basis: 70°F ambient still air.

---

*Mount Kisco, N.Y. 10549.

TABLE 15.4 Recommended Thicknesses for Mineral Wool (MW)

Nominal pipe size, in	70–200°F	201–300°F	301–400°F	401–500°F	501–600°F	601–700°F	701–800°F	801–900°F	901–1000°F	1001–1100°F
½	1″ →	1½″ →	2″ →	2½″ →	2½″ →	3″ →	3″ →	3½″ →	4″ →	5″ →
¾					3″ →	3½″ →	3½″ →	4″ →	4½″ →	5½″ →
1	1½″ →		2½″ →	3″ →	3½″ →	4″ →	4″ →	4½″ →	5″ →	6″ →
1½		2″ →		3½″ →	3½″ →	4½″ →	4½″ →	5″ →	5½″ →	6″ →
2			3″ →		4″ →		5″ →	5½″ →	6″ →	7″ →
3					4″ →			6″ →	7″ →	7½″ →
4		2½″ →			4½″ →	4½″ →	5½″ →	6½″ →	7″ →	7½″ →
6			3½″ →				6″ →	7″ →	7½″ →	8″ →
8	2″ →						6½″ →	7½″ →	8″ →	8½″ →
10							7″ →	8″ →		9″ →
12										
14										
16						5½″ →		8″ →		9½″ →
18				4″ →						
20	2″ →	3″ →	4″ →	5″ →	6″ →	7″ →	8″ →	9″ →	9″ →	10″
24										
30										
36 and greater										
Flat surface	2″	3″	4″	5″	6″	7″	8″	9″		

Basis: 70°F ambient still air.

TABLE 15.5 Recommended Thicknesses for Calcium Silicate (GS)

Nominal pipe size, in	70–200°F	201–300°F	301–400°F	401–500°F	501–600°F	601–700°F	701–800°F	801–900°F	901–1000°F	1001–1100°F
½	1" →	1½" →	2" →	2½" →	2½" →	3" →	3½" →	3½" →	4" →	4" →
¾	1½" →	2" →	2½" →	3" →	3" →	3½" →	4" →	4" →	4½" →	4½" →
1	2" →	2" →	2½" →	3" →	3½" →	4" →	4½" →	4½" →	5" →	5" →
1½	2" →	2½" →	3" →	3½" →	4" →	4" →	4½" →	5" →	5½" →	5½" →
2	2" →	2½" →	3" →	3½" →	4" →	4½" →	5" →	5" →	5½" →	6" →
3	2" →	3" →	3" →	4" →	4½" →	4½" →	5" →	5½" →	5½" →	6" →
4	2½" →	3" →	3½" →	4" →	4½" →	5" →	5" →	5½" →	6" →	7" →
6	2½" →	3" →	3½" →	4" →	4½" →	5" →	5½" →	6" →	6½" →	7½" →
8	2½" →	3" →	3½" →	4" →	4½" →	5" →	6" →	6½" →	7" →	8" →
10	2½" →	3" →	3½" →	4" →	4½" →	5½" →	6½" →	7" →	7½" →	8½" →
12	2½" →	3" →	3½" →	4" →	4½" →	5½" →	6½" →	7½" →	7½" →	8½" →
14	2½" →	3" →	3½" →	4" →	4½" →	5½" →	7" →	7½" →	8" →	9" →
16	2½" →	3" →	3½" →	4" →	4½" →	5½" →	7" →	7½" →	8" →	9" →
18	2½" →	3" →	3½" →	4" →	4½" →	5½" →	7" →	8" →	8" →	9" →
20	2½" →	3" →	3½" →	4" →	5" →	5½" →	7" →	8" →	9" →	10" →
24	2½" →	3" →	3½" →	4" →	5" →	6" →	7" →	8" →	9" →	10" →
30	2½" →	3" →	3½" →	4" →	5" →	6" →	7" →	8" →	9" →	10" →
36 and greater	2½" →	3" →	3½" →	4" →	5" →	6" →	7" →	8" →	9" →	10" →
Flat surface	2½" →	3" →	3½" →	4" →	5" →	6" →	7" →	8" →	9" →	10" →

Basis: 70°F ambient still air.

TABLE 15.6 Recommended Thicknesses for Rigid Urethane (RU)

Normal pipe size, in	−50 to −25°F	−24 to 0°F	1–25°F	26–100°F
½	1½″	1½″	1″	1″
¾	↓	↓	↓	↓
1	↓	↓	1½″	↓
1½	2″	↓	↓	↓
2	↓	↓	↓	↓
3	↓	2″	↓	↓
4	↓	↓	↓	↓
6	↓	↓	↓	↓
8	↓	↓	↓	↓
10	↓	↓	↓	↓
12	2½″	↓	↓	↓
14	↓	↓	↓	↓
16	↓	↓	↓	↓
18	↓	2½″	↓	↓
20	↓	↓	↓	↓
24	↓	↓	2″	↓
30	↓	↓	↓	↓
36 and greater	↓	↓	↓	↓
Flat surface	2½″	2½″	2″	1½″

Basis: 70°F ambient still air. (See page 572 for cellular glass.)

reduce heat picked up by cold or chilled fluids when the summer design temperature is above 70°F. Normally, it is recommended that insulation-thickness requirements be recalculated for heat transfer at the revised ambient conditions. However, Tables 15.3 through 15.7 can readily be altered to reflect the change in the design air temperature. This compensation is easily accomplished by the expedient of adding the difference between 70°F and the new design air temperature to the fluid temperatures represented at the top of each table. Thus, for a design air temperature of −10°F, the fluid temperature should be mentally shifted upward by 80°F [70°−(−10°)] when conserving heat with hot fluids. Similarly, for a design air temperature of 110°F, the fluid temperature would be shifted down by 30°F when conserving energy with cold fluids.

Example 15.3 What thickness of insulation is required for Example 15.2 if the insulation is to be designed for −10°F ambient temperature instead of 70°F?

solution The difference between 70 and −10°F is 80°F. Therefore, consider the fluid temperature to be 480°F instead of 400°F. From Table 15.6 it can be seen that the recommended thickness of calcium silicate insulation is still 3 in. Had the initial fluid temperature been 450°F, the revised temperature would have been 530°F. In this case, the insulation thickness would have increased to 3½ in.

Equation (15.4) indicates that the insulation surface temperature would decrease as the design ambient air temperature is reduced and that the insulation surface temperature would increase as the design ambient-air temperature is increased. Thus, it could be argued that insulation requirements might be relaxed and still provide adequate personnel protection when the design air temperature is lower than 70°F with

TABLE 15.7 Recommended Thicknesses for Cellular Glass (CG)

Nominal pipe size, in	−50–−25°F	−24–0°F	1–25°F	26–100°F	101–200°F	201–300°F	301–400°F	401–500°F	501–600°F	601–700°F	701–800°F
½	3" →	2" →	2" →	1" →	1½" →	2" →	2" →	2½" →	3" →	3½" →	4" →
¾		2½" →					2½" →				
1	3½" →	3" →	2½" →	1½" →		2½" →		3" →	3½" →	4" →	4½" →
1½							3" →	3½" →	4" →		5"
2	4" →		3" →	2" →	2" →	3" →			4½" →	4½" →	5½"
3		3½" →									6"
4	4½" →					3½" →	3½" →	4" →			6½"
6		4" →									7" →
8	5" →							4½" →	5" →	5" →	
10											
12											
14											
16											
18											
20											
24	5" →	4" →	3" →	2" →	2" →	3½" →	4" →	5" →		5½" →	
30											
36 and greater											
Flat surface	5"	4"	3"	2"	2"	3½"	4"	5"	5"	6"	7"

Basis: 70°F ambient still air.

hot fluids or still ensure antisweat protection when the design air temperature is above 70°F with cold fluids. However, any insulation system must still function to give personnel or antisweat protection for those portions of the year when the ambient temperature is 70°F. Thus, those portions of Tables 15.3 through 15.7 dealing with thicknesses above the minimum in each fluid-temperature range should be retained when dealing with personnel and antisweat protection only.

Tracing

Tracing, i.e., the application of an outside source of heat to piping or equipment, is required to maintain the temperature of the fluid in the line or equipment at or above a certain minimum temperature for various process or physical reasons. One of the most obvious is freeze protection of water, aqueous solutions, and certain organic fluids in stagnant or potentially stagnant lines located in an environment which may be below freezing or at a temperature which could cause precipitation or separation in the fluid if allowed to remain in that condition for considerable intervals of time. Other liquids such as heavy oils do not "freeze" per se, but, as shown in the subsection "Viscosity" in Chap. 6, their viscosities increase considerably. An increase in viscosity could cause an excessive pressure drop in a line or result in a reduced flow rate. Excessive viscosities could also lead to malfunctions in pieces of equipment which handle the fluid or use it. Tracing of lines in these instances is required to maintain the fluids within the proper temperature range for the desired viscosity requirements.

In some instances, insulation may be sufficient to prevent the contents of a line from dropping below a given temperature when the ambient temperature passes somewhat beneath that temperature for a short time. However, a time-temperature transient analysis of the situation is required before tracing can be omitted.

If a line contains an expendable fluid such as cooling water and a minimum flow can be maintained through its entire length, those portions containing the running flow need not be traced. A heat-balance calculation is required to determine the minimum continuous flow. As a rough rule of thumb, a flow of 2 gal/min plus an additional 1 gal/min for each nominal inch of pipe diameter for every 100 ft of piping is deemed sufficient to keep a water line from freezing when exposed to $-20°F$ ambient temperature with a 25-mi/h wind if the initial water temperature is 40°F or greater. If 2 in of insualtion is on the pipe, the flow of water can be reduced by a factor of 10.

Frequently gas flows contain water or organic vapors which could condense within the line or equipment should the ambient temperature fall below the dew point of the vapor mixture. The condensate could then constitute a problem if it is corrosive, e.g., water vapor condensing from hydrogen chloride or sulfur dioxide, or if it were not possible to remove the condensate from the piping or equipment without incurring excessive expenses. Freezing of the condensate could cause large pressure drops or operational difficulties. Tracing is recommended in these instances to maintain the affected piping network or equipment at a temperature sufficiently above the dew point to prevent condensation.

The most common types of tracing systems employed in chemical and refinery industries use steam or electricity. Often a client expresses a predilection for one type over the other on the basis of past experience. In general, both types are suitable for most applications, but there are some cases in which one type of tracing or the other may be preferred. For example, it is often easier to get better temperature control with electric tracing than with steam tracing. On the other hand, there are certain temperature limitations on electrical-tracing wire that don't apply to steam tracing.

Process engineers should be familiar with the advantages and drawbacks of the various forms of steam and electric tracing so that they can provide informed guidance to their clients on the matter. Factors to be taken into account are attainable temperature levels, safety, installed and operating costs, and convenience.

Steam tracing

In most instances, one or more heat-tracing lines, usually $\frac{3}{8}$- or $\frac{1}{2}$-in-O.D. copper tubing, are wrapped about the equipment and process or utility pipe or are attached parallel to it. The various methods are as shown in Fig. 15.3. Special mastics are often applied when the tracing tubing is laid parallel to the pipe to increase the heat transfer.

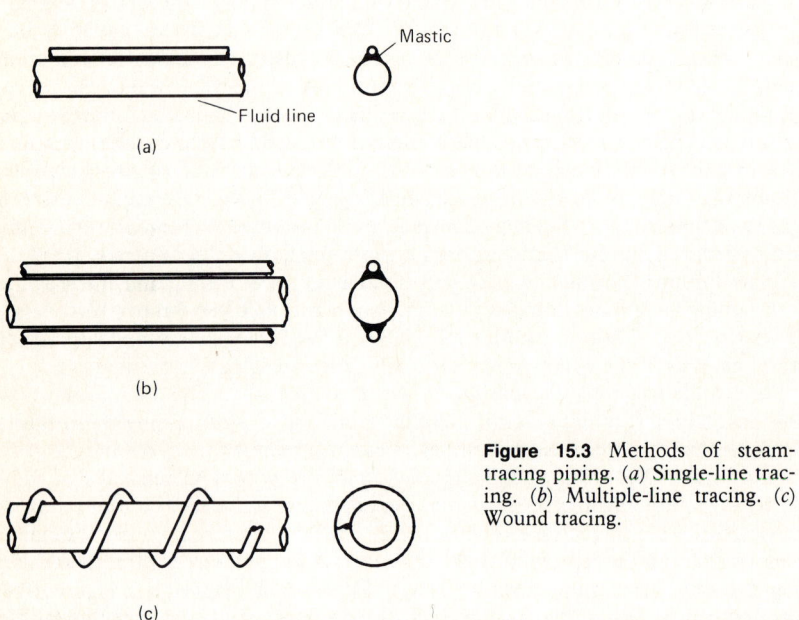

Figure 15.3 Methods of steam-tracing piping. (*a*) Single-line tracing. (*b*) Multiple-line tracing. (*c*) Wound tracing.

Steam at the temperature required by the process is fed to the tracing at several points along the length of the traced piping. Tracing-steam pressures will vary from 15 to 100 psig or higher. Similarly, condensate is withdrawn from the tracing tubing at several locations. The mechanical specifications for the project should contain a section devoted to tracing. The exact tracing tubing, methods of applying it, lengths of tubing between steam feed points, and condensate-withdrawal locations, as well as controls, are delineated in the mechanical specifications and accompanying drawings.

Steam should be applied to the tracing tubing whenever the ambient temperature approaches a point which would be deleterious to the fluid in the pipe. In some instances, steam is applied manually as the calendar season approaches when there is danger of reaching the lower temperature limit and is then shut off during those months when ambient temperature is expected to be above the lower temperature limit.

Alternatively, steam could be applied by an automatic controller sensing the ambient temperature. This effects a savings in steam for those portions of the days in the heating season when the ambient temperature is above the set point. Further savings may be effected if the piping is divided into sections and a temperature sensor placed on the outside skin of each piping section to control the steam to the tracing tubing in that section. Care must be taken in this case to ensure that all stagnant sections containing liquids have thermostats separate from those on a line that is normally flowing. The annual fixed costs of an installed automatic control system for sensing ambient temperature or for separate piping sections must be compared with the resultant annual savings in steam.

Some fluids may be sensitive to the temperature of the saturated steam at the pressure of the steam in the tracing tubing. In that event, it is necessary to use pipe-skin-temperature sensors which control tracing steam to segments of the piping. The problem of high temperature may be avoided by isolating the tracing tubing from the piping with fiberglass rope. Since the tracing tubing is then no longer in contact with the metal piping, the temperature of segments of pipe containing stagnant fluid will no longer approach the temperature of saturated steam in the tubing. Instead, the temperature of the fluid in the pipe will be approximately equivalent to the temperature of the air layer between the pipe and the insulation surrounding the piping and tracing. A temperature calculation using a variation of Eq. (15.4) must be made in order to assure that the temperature in the dead-air space does not exceed the maximum temperature to which the fluid may be subjected. The warmest ambient temperature during which the steam may be turned onto the tracing and the highest steam temperature should be used in the calculations.

Electric tracing

Two basic categories of wire or tracers can be used if the tracing of pipe or equipment is to be accomplished by electric current. One type uses constant-resistance wire, while the resistance of the wire in the other type varies with its temperature.

Tracing with constant-resistance wire. This type of wire has a nearly constant-wattage output of heat per unit length of wire for a given voltage regardless of the temperature which the wire might attain in its position between the pipe or equipment and the insulation. It may be used at temperatures from -50 to $1100°F$, depending on the type of sheathing. Tracing wires of this sort are available for a wide variety of wattage-per-foot ratings, ranging from 2.5 to 100 W/ft. This is equivalent to heat outputs of 8.5 to 340 Btu/(h·ft) when using a single tracer parallel to the pipe. Additional wattage per foot of pipe can be obtained, if required, by using multiple tracer lengths in parallel or by wrapping the tracer about the pipe in a spiral fashion.*

Concern should be given to the possibility of overheating a process fluid if the fluid in a section of piping is stagnant. Its temperature eventually equals that of the outer diameter of the pipe which is in contact with the tracer. The determination of this temperature is a simple heat-balance calculation across the insulation. However, the temperature of the pipe wall of a stagnant line electrically traced with constant-wattage wire varies with the ambient temperature. Since the wattage load is based on the

---

*Parallel tracers are normally used since it is difficult to spiral long lengths of tracer wire.

minimum ambient temperature, it is, therefore, important to note the expected pipe-wall temperature at the maximum ambient temperature. This, of course, must not exceed the maximum allowable temperature for the fluid in the pipe or for the tracing wire itself.

If it is not possible to provide one wattage to satisfy the ambient-temperature extremes, sufficient wattage must be used to provide protection to prevent the fluid in the pipe from going below its minimum allowed temperature. Thermostatic control devices are then used to regulate the power or voltage to the tracing wire to avoid excessive fluid temperatures. Several types of controllers are available. Some simply switch the current to the tracing wire on or off, while others regulate the voltage to the wire to vary the wattage output. Selection of the type should be a joint decision of the electrical, instrument, and process engineers.

Most tracer assemblies consist of two conducting wires covered with and separated by resinous or plastic materials. These substances limit the service temperature of the tracing-wire assembly. Typical limitations are 150 to 225°F maximum for long exposure and 360 to 400°F for short-term usage, depending upon the tracer materials. These values must also be taken into account not only when deciding whether or not to use electric tracing but in selecting a wattage per linear foot of pipe.

In order to guard against overheating the tracer wires there must be excellent contact between the tracer and the piping or equipment. Any air gaps could result in isolated hot spots on the wire which could cause failure of the tracing system.

When contact between the tracer and a metallic pipe is intimate, conductance of heat through the cross section of the pipe wall is so great that there is only a negligible temperature difference between the tracer and the pipe and the temperature around the perimeter of the pipe is essentially the same. However, this is not the case when the tracing is on nonmetallic pipe or equipment such as plastic, fiberglass-reinforced plastic (FRP), or resinous materials which have relatively low thermal conductivities. Typical values of thermal conductivity, long-term-service temperature, and maximum-exposure temperature for several nonmetallic piping and equipment materials are given in Table 15.8. Thermal conductivities of sample metals are given as a contrast.

TABLE 15.8 Thermal Conductivities of Various Piping and Equipment Materials

	Thermal conductivity, $(Btu \cdot in)/(h \cdot ft^2 \cdot °F)$	Typical service rating, °F	Maximum temperature, °F
Plastics			
PE (polyethylene)	2.3	140	180
PVC (polyvinyl chloride)	0.84–1.2	140	221
CPVC (chlorinated polyvinyl chloride)	0.96	140	221
ABS (acrylonitrile butadiene styrene)	1.44–2.16	180	200
PP (polyproylene)	1.2–1.32	160	200
FRP (fiberglass-reinforced plastic)	1.32–1.68	200	250
Metals			
Copper	2350–2710		
Aluminum	804–1620		
Cast iron	336–360		
Carbon steels	300–324		
Stainless steels	98–174		

In cases with low thermal conductivities there can be large temperature differences around the perimeter of the pipe which result in a hot spot in the area where the tracing wire is in contact with the pipe. The temperature in that area could well exceed the maximum temperature allowed for the fluid, the piece of equipment, or the wire. The problem may be relieved by the use of tracing tape which is attached to metallic foil to facilitate adherence to the pipe wall or equipment. The foil also assists heat distribution. However, it is of limited width and thickness. Thus, calculations involving temperature isotherms within the pipe wall and insulation are required to assure that temperature limitations are not exceeded. The vertical surfaces of nonmetallic vessels are sometimes wrapped with wider bands of metallic foil before electric tracing is applied to provide improved heat distribution.

It is usually not difficult to remain below the upper sensibility temperature of the fluid when tracing metallic piping which has a flow through it. However, segments of a piping system such as that used for spared equipment or piping manifolds and which cannot readily be drained when they are not in use can become stagnant in normal operation. If they need tracing, they could require individual thermostatic controls. An alternative to the use of multiple thermostatic regulators for these stagnant sections is the use of a dummy line to simulate fluid conditions in a large number of segments. A dummy line is a pipe of the same specifications as the stagnant lines to be simulated. It is made in a reasonable length, about 6 to 10 ft, and is filled with the same fluid as the stagnant lines. It is then traced and insulated in the same manner prescribed for the pipe diameter. The dummy line is located in an area typical of most of the piping. A sensor on the dummy length can then be tied into a regulator to control the tracing on all pipes of similar size. It can be shown that if the only concern is maintaining the stagnant lines above a lower temperature, only one dummy line with its diameter equivalent to the largest-diameter pipe to be simulated need be constructed. Similarly, if the only concern is the upper fluid sensitivity temperature, the dummy length is made equivalent to the smallest diameter.

Self-regulating tracing wire. A popular type of tracing wire is that for which the resistance of the tracing assembly varies with its temperature rather than remaining effectively constant. Self-regulating wire is useful for many applications when it is necessary to maintain only a relatively low temperature level for the fluid. Typical uses for self-regulating tracing are in freeze protection of water or aqueous solutions or to ensure that a maximum fluid viscosity is not exceeded. Typical curves for wattage versus pipe temperature are shown in Fig. 15.4.

For the tracers shown in Fig. 15.4, power outputs are at their nominal values at 50°F. However, the electrical resistance of the material separating the conductors increases with temperature. Thus, the wattage output per foot of tracing wire drops as its temperature rises. This can be seen in the curves in Fig. 15.4 in which the wattage per foot is almost halved in going from 50 to 95°F for the 10-W tracer and halved again in raising the temperature further to 125°F. The materials in this wire assembly usually limit its application to systems which do not exceed 150 to 160°F. Other types of self-regulating tracers have a constant nominal power output from −50 to +50°F and then decrease in power output until it is about one-third of its nominal value when the system is at about 300°F. In this case, the system applications may be as great as 375°F. Manufacturers' engineering data should be consulted and limitations information applied for individual situations.

The design temperature for the tracing system is usually taken as a minimum of 10 to 30°F above the lower temperature at which the fluid may change state, form a

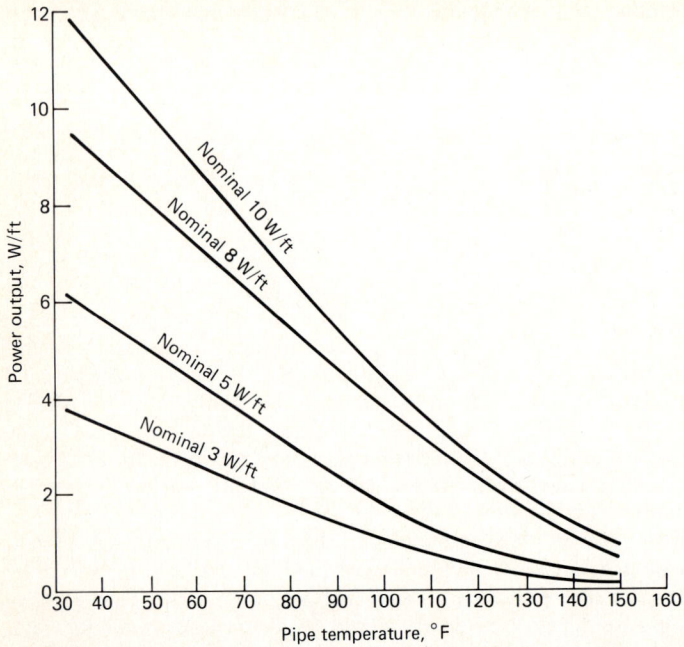

Figure 15.4 Self-regulating tracing wire.

condensate or precipitate, or become too viscous. Thus, a self-regulating system is generally designed for 50°F with water or dilute aqueous solutions which freeze at or near 32° and for 85°F, for example, with a 50 percent caustic solution which begins to crystallize at 55°F. The design temperature selected is used with the simple heat-balance calculation for a given diameter of pipe for a given insulating thickness to determine the wattage output per linear foot of pipe required from the tracer. The wattage available from any self-regulating wire can be obtained at the system design temperature from curves similar to those shown in Fig. 15.4. The wattage required by the system can then be supplied by one wire or by several lower-wattage parallel wires.

In the event that there is an upper sensibility temperature for the fluid in the pipe, a temperature profile for the pipe and insulation should be made. The wattage output of the wire at the sensibility temperature and the maximum outer-skin temperature of the insulation or its covering should be used in determining whether or not the maximum allowable temperature of the fluid would be exceeded at the pipe wall.

Normally, a system using self-regulating electrical-tracing wire does not require thermostatic devices, except possibly a device to sense ambient temperature in order to turn the system on or off. A system of thermostatic regulators, however, may have to be installed if calculations show that protection is required to prevent the upper sensibility limit of the fluid from being exceeded. In that case, the sensing points for regulation may be taken from individual sections of the piping or from a dummy line, as discussed for constant-wattage tracing wire.

Jacketing and bundling

There are situations in which steam or electric tracing cannot provide sufficient heating. At other times, cooling of piping or equipment is required. In these cases, the piping or equipment is jacketed. Basically, pipe jacketing is accomplished by having the process line centered inside a larger-diameter pipe with a heating or cooling medium flowing in the annulus. Equipment must be specified with a special external jacket. Steam, special heating oils, tempered water, brines, and glycols are used as heat-transfer fluids.

When long runs of fluids such as oils parallel a steam line, they may be bundled together with the steam line inside the same insulation covering. The temperature of the hot fluid must be less than the maximum temperature allowable for the fluid if there is a possibility that the fluid may be stagnant at times. However, if it is certain that there will always be a minimum flow of fluid through the bundled line, a heat balance is required to determine whether or not the heat absorbed in the pipe run will raise the temperature of the minimum flow beyond its maximum desirable temperature. If bundling can be used, it eliminates the cost of tracing the process or utility line and reduces the total cost of insulation.

REFERENCES

1. H. F. Rase and M. H. Barrow, *Project Engineering of Process Plants,* John Wiley & Sons, Inc., New York, 1957, pp. 476–479.
2. W. C. Turner and J. F. Malloy, *Thermal Insulation Handbook,* Robert E. Krieger Publishing Co., Malabar, Fla., and McGraw-Hill Book Company, New York, 1981, pp. 212–263.

Chapter 16

Mixers and Agitators

The mixing process is one of the oldest type of operations in the process industry. Unlike other unit operations, the mixing process may be required to perform several tasks simultaneously. These tasks include pumping, heat transfer, mass transfer, and effecting reaction kinetics.

Although the mixing process is one of the oldest unit operations, it is probably one of the least understood. However, there are some general rules that can be used to analyze and evaluate a mixing system. These rules are presented in this chapter together with guidelines for specifying mixing equipment. When selecting and purchasing mixing equipment, the process engineer should consult a reputable mixing-equipment manufacturer to determine the most effective method to be used and the proper installation of the equipment.

There are many types of mixing equipment, but all this equipment can be divided into two categories. One category consists of impeller-type mixers, also called agitators, which are installed in mixing vessels; the other type is the motionless mixers which are installed directly in pipes.

Agitators: Impeller-Type Mixers

Impeller-type mixers are the most frequently used mixing devices in process plants. These mixers operate in accordance with the basic principles of fluid mechanics. The rotating action of the impeller imparts kinetic energy to the fluid in which it is immersed. This creates a velocity gradient throughout the fluid. The velocity gradient may be characterized by either a smooth, swirling action creating a vortex along the agitator's shaft, or multiple eddy-current swirls throughout the vessel, or some combination between these two extremes. The single smooth swirls are desirable for applications in which the mixer is required to pump the fluid around a vessel or through a draft tube. However, this flow pattern has poor mixing characteristics and requires a high blending time. The flow pattern characterized by multiple eddy-current swirls is more desirable for mixing applications such as emulsification, absorption, and solubilization requiring short blending times and efficient mixing. However, this pattern consumes most of the agitator energy for mixing, leaving little if any energy left for pumping or circulating the fluid.

Fluid mechanics of agitators

The operation of an agitator in a vessel can be described in theory by using the Bernoulli equation; i.e.,

$$\Delta\left(\frac{u^2}{2\alpha g_c}\right) + \Delta z \frac{g}{g_c} + \Delta(PV) + \Sigma F = -W_f \qquad (16.1)$$

where u = fluid velocity
 g = acceleration due to gravity
 g_c = dimensional constant
 P = system pressure
 V = system volume
 W_f = work done on the fluid
 z = static head
 F = friction
 α = unit conversion factor

The driver provides work through the rotation of the attached agitator, which, in turn, causes fluid movement in the region around the agitator blade and a shearing action resulting in friction. Therefore, the work done by the agitator contributes to the kinetic-energy term and the friction term in the Bernoulli equation. Since agitators are used on liquids which are assumed to be incompressible, the pressure-volume-term contribution is negligible except where a gas is liberated when the mixing is done in a closed vessel. This leaves the static-head term to be accounted for.

If the agitator is placed in a smooth-walled cylindrical vessel, its rotation imparts a velocity to the fluid, causing it to move away from the centerline of the agitator shaft. As a result, the liquid at the center is depleted while fluid accumulates against the vessel walls, producing a higher liquid level at the walls than when the liquid is at rest. Therefore, a static head is built up in the vessel, as shown in Fig. 16.1a, as well as the creation of a vortex about the centerline of the agitator shaft.

When baffles are placed in a cylindrical tank, they produce the equivalent of surface roughness in a pipe, or drag. The resulting drag creates turbulence and localized eddy currents which consume the kinetic energy that would otherwise induce a vortex and a static head. Since the energy is used to overcome the frictional drag caused by the liquid, the surface becomes choppy rather than a smooth vortex. Figure 16.1b shows an idealized representation of the effect of adding baffles in the vessel.

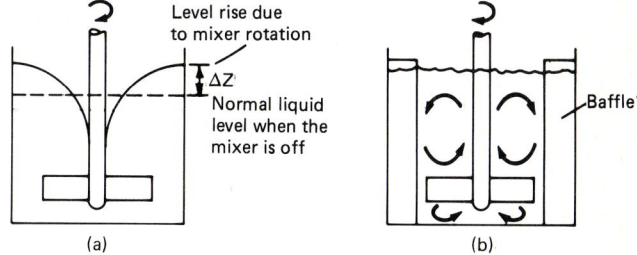

Figure 16.1 Fluid mechanics of mechanical agitation. (a) Vortex formation in a cylindrical tank without baffles. (b) The effect of baffles on mechanical agitation.

Agitator mixing correlations. The Bernoulli equation can only be used to explain how the energy input in the mixing process is distributed between velocity, static-head, and frictional terms in a qualitative manner. More useful relationships for analyzing the mixing process utilize dimensional analysis.

As in other fluid systems, the liquid being mixed can be in a laminar- or a turbulent-flow regime. Therefore, the fluid characteristics can be expressed as a Reynolds number. The velocity of a fluid in a mixing vessel is a function of the agitator-impeller diameter and of the speed expressed in terms of revolutions per unit time. Therefore, the Reynolds number for the mixing process takes the form

$$\text{Re} = 0.1723 \frac{ND^2\rho}{\mu} \tag{16.2}$$

where N = impeller speed, r/min
D = impeller diameter, in
ρ = liquid density, lb/ft^3
μ = liquid viscosity, cP

The generation of a velocity and the associated Reynolds number requires the consumption of power. Therefore, a dimensionless quantity termed the Power number (Po) is another factor used to analyze the mixing process; it is defined as

$$\text{Po}_0 = 9.5 \times 10^{14} \frac{\mathcal{P}}{N^3 D^5 \rho} \tag{16.3}$$

where D = impeller diameter, in
N = impeller speed, r/min
$\mathcal{P}$ = agitator horsepower
ρ = liquid density, lb/ft^3

Power consumption is a function of the Reynolds number; i.e.,

$$\text{Po} = f(\text{Re}) \tag{16.4}$$

This equation shows that power consumption and Reynolds number are related but gives no indication as to how much power is needed for different process applications. Table 16.1 presents the range of power consumption for different process applications.

Once the expected power consumption for the application has been approximated, the process engineer can determine the specific parameters to be considered when designing a mixing system. Table 16.2 lists some of these parameters.

Agitator-system design

The way in which the factors listed in Table 16.2 affect the design of an agitator system must be understood. Each of these terms is covered in this subsection.

Impeller types. There are numerous designs for agitator impellers, but they all can be classified into the following general types: propellers, turbines, paddles, anchors, and helical ribbons. These impellers, shown in Fig. 16.2, may be installed singly or in multiples on a single agitator shaft, depending on the mixing-vessel geometry. In

TABLE 16.1 Power-Consumption Ranges for Mechanically Agitated Systems

Process condition	Typical power consumption, hp/1000 gal
Very high agitation	
Emulsification	15–25
Dissolving solids	10–12
Dissolving low-solubility gases	3–10
High agitation	
Rapid heat transfer	1.5–2.5
Contacting	1.5–2.0
Medium agitation	
Dissolving moderately soluble gases	1.0–2.0
Solid suspension	1.0–1.6
Washing	1.0–1.5
Normal heat transfer	0.9–1.3
Low agitation	
Liquid extraction	0.7–1.0
Crystallization	0.8–1.2
Stirring	0.5–0.9
Blending	0.5–0.8
Dissolving highly soluble gases	0.5–0.8

addition, the geometry of turbine impellers and paddles varies with the specific needs of a process application. Figure 16.3 shows some of the various geometries.

The selection of an impeller type is usually predicated on the viscosity of the process fluid being mixed. Propeller- and turbine-type impellers are generally used for fluids with viscosities of less than 2000 cP. For fluids with viscosities between 2000 and 50,000 cP, a turbine is usually preferred, and anchors, helical ribbons, and paddles are selected when viscosities are between 10,000 and 1 million cP, with paddles being preferred for the lower portion of the viscosity range. Above 1 million cP, special mixing devices such as Banbury mixers, kneaders, extruders, sigma mixers, and similar devices are more practical.

TABLE 16.2 Parameters to Be Considered When Specifying Agitators

Process parameters
  Fluid viscosity
  Solute solubility
  Thermal conductivity of fluid and solute if heat transfer is required
  Fluid density
  Solid-particle size
Mechanical parameters
  Impeller diameter
  Impeller revolutions per minute
  Impeller geometry
  Vessel volume
  Vessel geometry
  Agitator location with reference to the vessel
  Vessel internals: baffles, sparger, draft tubes, cooling coils, etc.

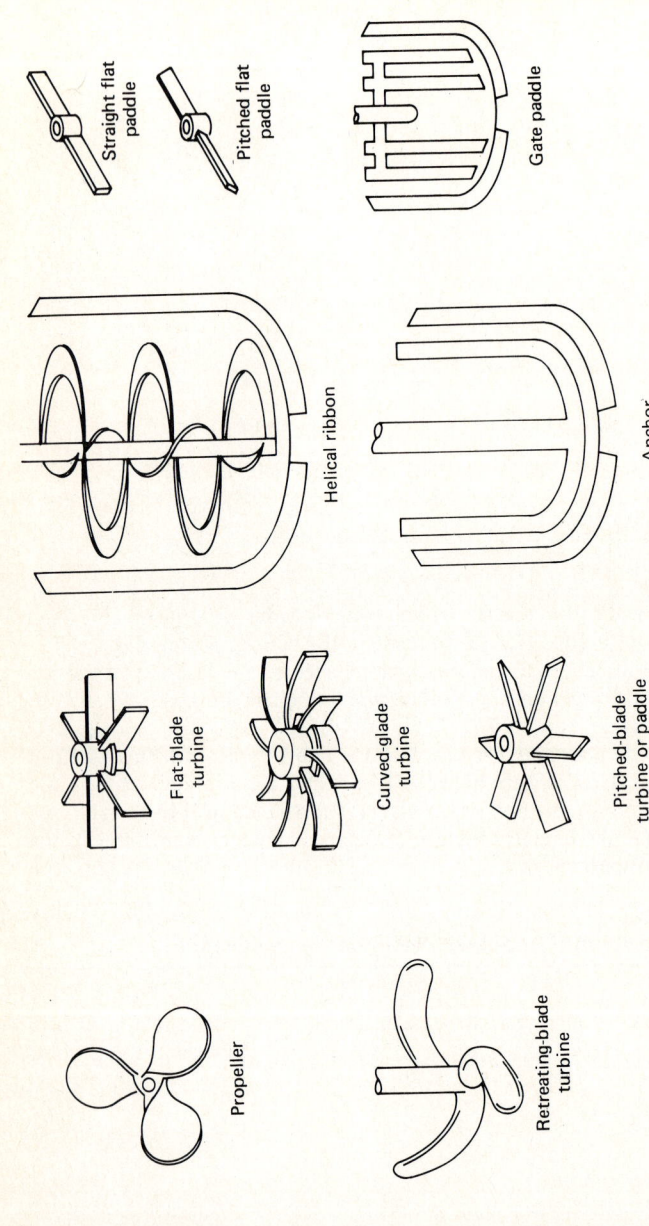

Figure 16.2 Typical impeller designs for mechanical agitation.

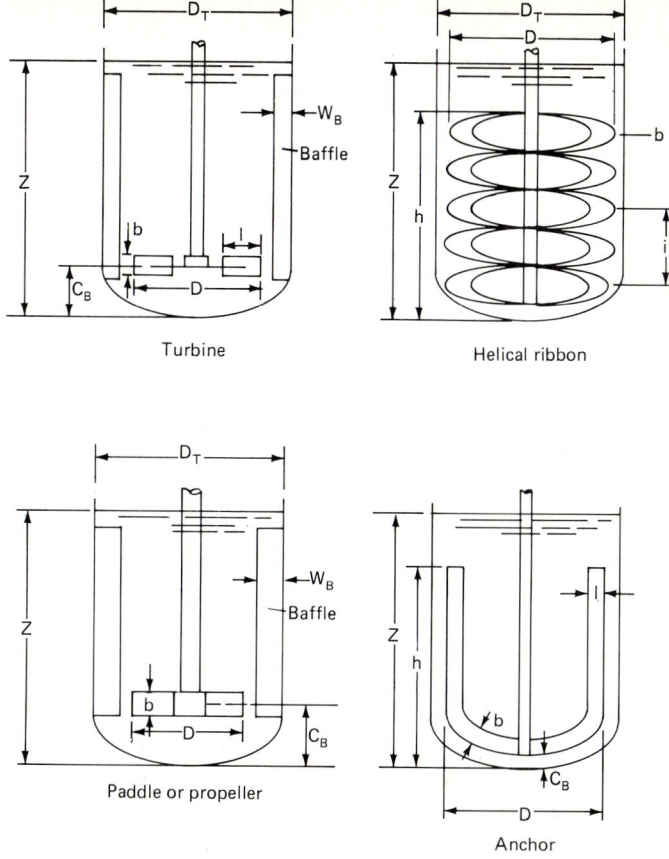

Figure 16.3 Turbine and paddle impeller geometries.

In addition to viscosity considerations, the selection of propeller, paddle, or turbine impeller is predicated on the process application. Table 16.3 lists some typical process applications for these impellers.

Relationship of impeller and vessel geometry. Not only does the impeller geometry have an effect on mixing but also the impeller's size relative to that of the vessel, its orientation in the vessel, and the vessel's geometry and internals have an effect on mixing. Figures 16.4, 16.5, and 16.6 show the Power number versus the Reynolds number for a variety of impellers of different diameters and pitches and different vessel diameters with and without baffling.

Baffles. Baffles are installed in mixing vessels to increase turbulence (see Fig. 16.1). Although baffling raises power consumption, it has the advantage of improving heat transfer and reducing mixing time. Figure 16.5 shows the relative effect on power consumption as a function of vessel diameter, baffle width, and number of baffles.

TABLE 16.3 Impeller Selection Based on Application

	Impeller type*		
Application	Propeller	Turbine	Paddle
Blending	1	2	3
Dispersion	2	1	3
Solid suspension	2	1	3
Reactions	2	1	3
Gas dispersion	3	1	3
Heat exchange	2	1	2
Crystallization	2	1	1

Typical impeller speeds
  With direct drive: 1150 or 1750 rpm
  With reduced-gear drive: 300 or 1750 rpm

*1 = most often used; 2 = occasionally used; 3 = rarely used.

Square vessels without baffles have more turbulent mixing than do similar cylindrical tanks. This increased turbulence is the result of eddy currents being developed at the corners of the vessel. The turbulence in square vessels can be increased by placing baffles at the centerline of each wall. If baffles are required in rectangular vessels, then the most practical location is on the longest sides of the vessel with the baffles being aligned with the centerline of the agitator shaft.

Vessel geometry. The most practical vessel geometry for mixing is the vertical cylindrical vessel. Vessels with dished heads, flat bottoms, and conical heads with

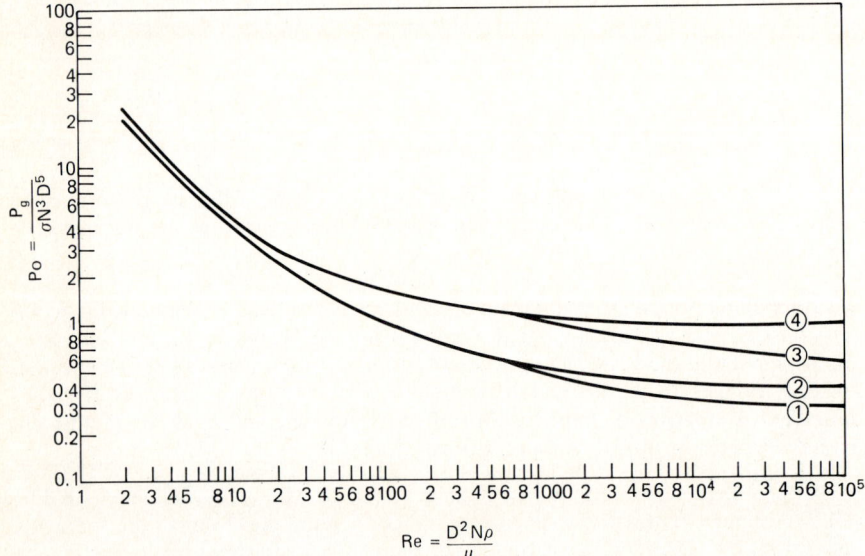

Figure 16.4 Power number versus Reynolds number for propeller impellers. (1) Pitch = D, no baffles. (2) Pitch = D, four baffles, $W_B = 0.1 D_T$. (3) Pitch = $2D$, no baffles. (4) Pitch = $2D$, four baffles, $W_B = 0.1 D_T$.

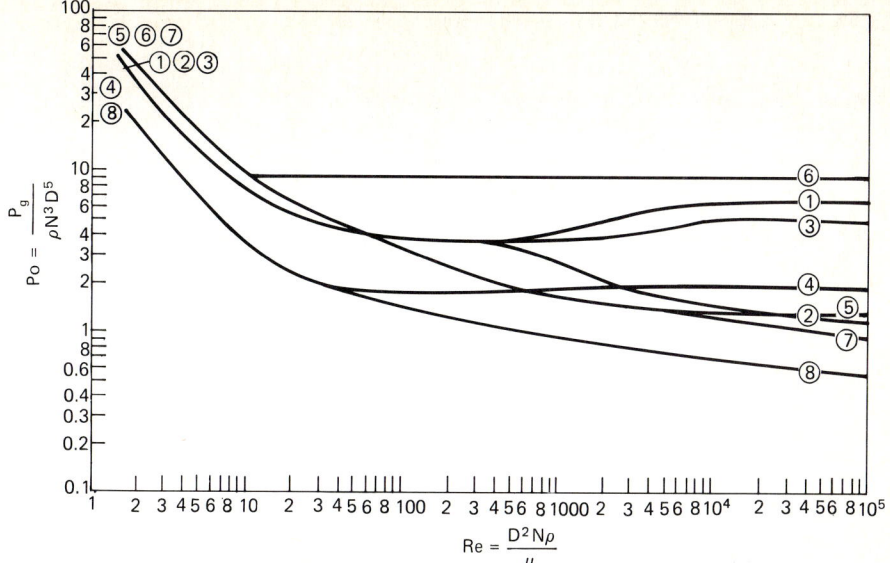

Figure 16.5 Power number versus Reynolds number for turbine and paddle impellers. (1) Flat six-blade turbine, four baffles, $W_B = 0.1D_T$. (2) Flat six-blade turbine, no baffles. (3) Curved six-blade turbine, four baffles, $W_B = 0.1D_T$. (4) Flat two-blade paddle, four baffles, $W_B = 0.1D_T$. (5) Shrouded six-blade turbine, four baffles, $W_B = 0.1D_T$. (6) Flat paddle, no baffles, $b = 0.2D_T$, $D = 0.3D_T$. (7) Flat paddle, four baffles, $W_B = 0.15D_T$, $b = 0.2D_T$, $D = 0.3D_T$. (8) Flat paddle, baffles unknown, $b = 0.05D_T$, $D = 0.3D_T$. (Nos. 6, 7, and 8: S. Nagata, Mixing: Principles and Applications, *John Wiley & Sons, Inc., New York, 1975.*)

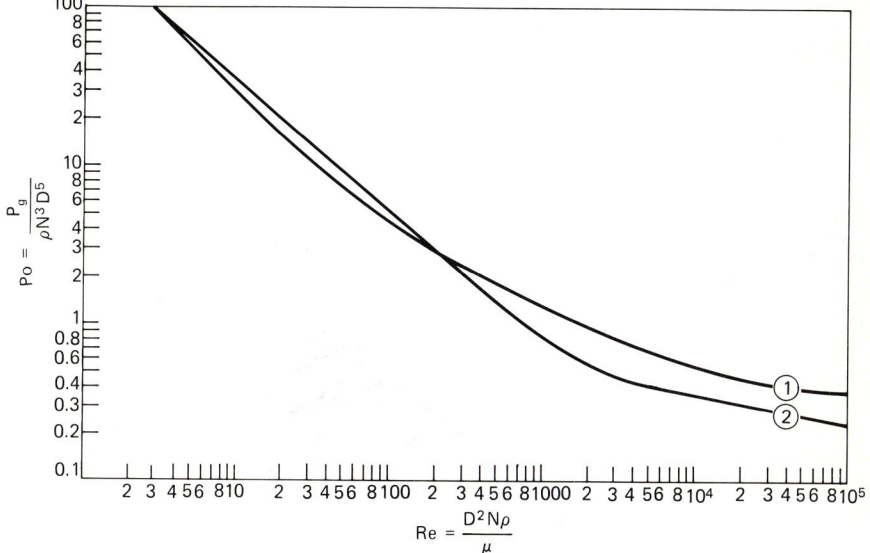

Figure 16.6 Power number versus Reynolds number for miscellaneous geometries. (1) Anchor impeller, $(D/12) \leq 1 \leq (D/8)$, $D_T - D = 0.55$. (2) Double helical spiral, pitch $= 0.5D$, $b = D/6$, helix length $= 2i$, $i = D/2$, $D_T - D = 0.075D$. (James Y. Oldshue, "How to Specify Mixers," *Hydrocarbon Processing*, October 1969, pp. 73-80.)

588 Equipment Selection, Sizing, and Related Subjects

angles less than 15° behave similarly. Although other arrangements are used, their mixing performances are not as effective. The most common arrangement alternates are spherical vessels, horizontal cylindrical vessels, and rectangular vessels. These vessels may require special baffling to improve mixer performance. The location and design of the baffling cannot be determined precisely by analysis but may require experimentation or reliance on historical information provided by an agitator manufacturer on similar installations.

Side-entering agitators. When an agitator must be installed in the side of a vessel rather than through the top, the agitator shaft should not be in line with the center of the vessel. Figure 16.7a shows the fluid motion for this installation, which consists of a single large swirl. The work of Oldshue[1,7] has indicated that improved mixing performance can be achieved by placing the agitator at an angle of 7 to 10° off the centerline of the vessel as in Fig. 16.7b.

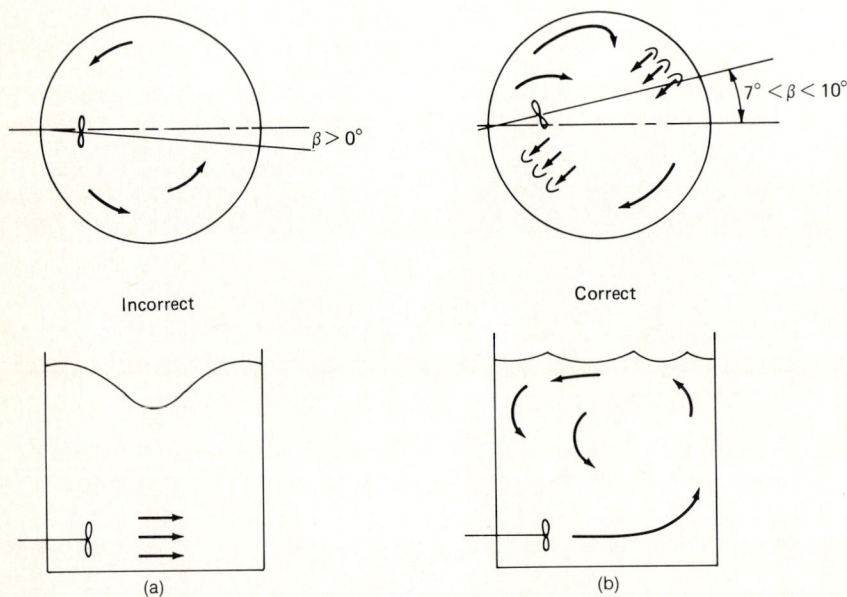

Figure 16.7 The effect of orientation of a side-entering agitator on the mixing characteristics in a vessel. (*S. Nagata*, Mixing: Principles and Applications, *John Wiley & Sons, Inc., New York, 1975; James Y. Oldshue*, "How to Specify Mixers," Hydrocarbon Processing, *October 1969, pp. 73–80.*)

Heat transfer in agitated vessels. Many mixing processes require either heating or cooling. Figure 16.8 shows the various vessel arrangements for heat-transfer applications.

The rate of heat transfer in agitated vessels can be calculated by using a variation of the classic Sieder and Tate equation discussed in Chap. 14, i.e.,

$$\text{Nu} = a(\text{Re})^b(\text{Pr})^c \tag{16.5}$$

where Nu = Nusselt number = hD_T/k
Pr = Prandtl number = $c_p\mu/k$

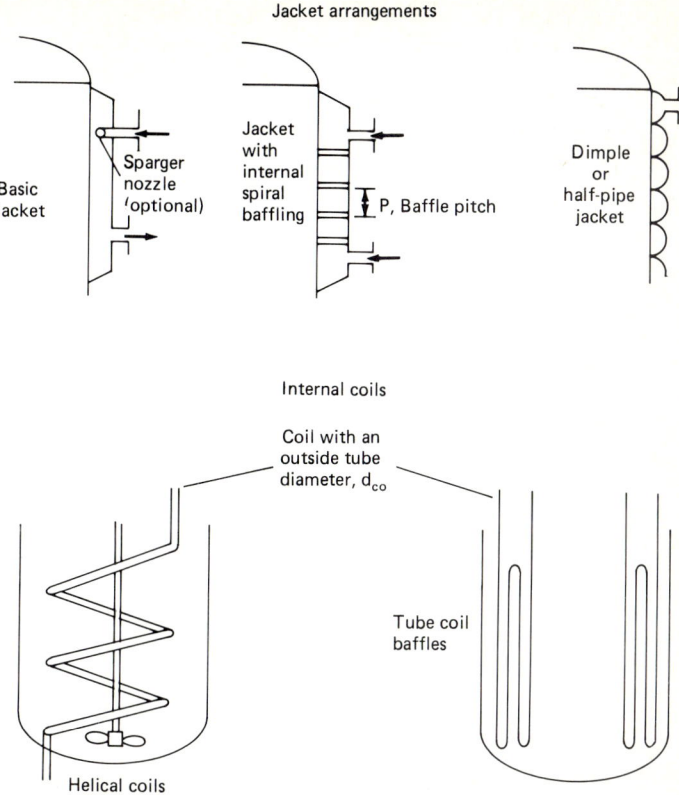

Figure 16.8 Methods of heating and cooling agitated vessels.

Re = Reynolds number = $ND^2\rho/\mu$
a, b, and c = empirical constants

The empirical constants used in Eq. (16.5) depend on the type of impeller, vessel configuration, and baffling and the method of heat transfer. The Nusselt number calculated from the above relationship produces a heat-transfer film coefficient h which is used to calculate an overall heat-transfer coefficient U by Eq. (14.16).

An agitated vessel, like any other heat-transfer equipment, has an inside and an outside film coefficient. Table 16.4 lists the correlations for these coefficients, depending on whether jackets, coils, or coil baffles are used for heat transfer. Methods for estimating fouling factors and typical wall conductivities can be found in the subsection "Convection" in Chap. 14.

High-viscosity materials, solids and slurries, requiring heating or cooling need more specialized equipment for heat transfer. These include such devices as troughs with jackets, hollow screws or agitators through which a heat-transfer fluid is carried, or sigma agitators with jacketed shells.

Scaleup of mixing systems. The problem of scaling up or scaling down a mixing process is more complicated than for other process equipment because an agitator

TABLE 16.4 Heat-Transfer Correlations for Agitated Vessels

(h_i) Inside film coefficient
1. Jacketed vessels
    a. Propeller $\qquad$ $Nu = 0.54(Re)^{0.67}(Pr)^{0.25}(\mu/\mu_w)^{0.14}$
    b. Turbine, flat
    $\quad Re > 400 \qquad Nu = 0.74(Re)^{0.67}(Pr)^{0.33}(\mu/\mu_w)^{0.14}$
    $\quad Re < 400 \qquad Nu = 0.54(Re)^{0.67}(Pr)^{0.33}(\mu/\mu_w)^{0.14}$
    If $Z/D_T = 1$ and $D_T/D = 3$; otherwise, use
    $\quad Re > 400 \qquad Nu = 0.85(Re)^{0.66}(Pr)^{0.33}\left(\dfrac{D_T}{z}\right)^{0.56}\left(\dfrac{D}{D_T}\right)^{0.13}\left(\dfrac{\mu}{\mu_w}\right)^{0.14}$
    c. Paddle
    $\quad 20 < Re < 4000 \qquad Nu = 0.415(Re)^{0.67}(Pr)^{0.33}(\mu/\mu_w)^{0.14}$
    $\quad Re > 4000 \qquad Nu = 0.36(Re)^{0.67}(Pr)^{0.33}(\mu/\mu_w)^{0.14}$
    d. Turbine, retreating-blade
    $\quad$ Three-blade, glass-lined $\qquad Nu = 0.33(Re)^{0.67}(Pr)^{0.33}(\mu/\mu_w)^{0.14}$
    $\quad$ Three-blade, metal-alloy $\qquad Nu = 0.37(Re)^{0.67}(Pr)^{0.33}(\mu/\mu_w)^{0.14}$
    $\quad$ Six-blade $\qquad Nu = 0.68(Re)^{0.67}(Pr)^{0.33}(\mu/\mu_w)^{0.14}$
    e. Anchor
    $\quad 30 < Re < 300 \qquad Nu = 1.0(Re)^{0.67}(Pr)^{0.33}(\mu/\mu_w)^{0.18}$
    $\quad 300 < Re < 4000 \qquad Nu = 0.38(Re)^{0.67}(Pr)^{0.33}(\mu/\mu_w)^{0.18}$
    f. Helical ribbon
    $\quad Re < 130 \qquad Nu = 0.248(Re)^{0.5}(Pr)^{0.33}\left(\dfrac{D}{e}\right)^{0.22}\left(\dfrac{D}{i}\right)^{0.28}\left(\dfrac{\mu}{\mu_w}\right)^{0.14}$
    $\quad Re > 130 \qquad Nu = 0.238(Re)^{0.67}(Pr)^{0.33}\left(\dfrac{D}{i}\right)\left(\dfrac{\mu}{\mu_w}\right)^{0.14}$
    where e = clearance = $(D_T - D)/2$
    $\quad\quad i$ = agitator-ribbon pitch
2. Internal coils
    a. Propeller $\qquad h_i d_{co}/k = 0.078(Re)^{0.62}(Pr)^{0.33}(\mu/\mu_w)^{0.14}$
    b. Turbine, flat
    Helical coils, six blades and
    $0.018 \leq d_{co}/DT \leq 0.036$
    $400 < Re < 1.5$ million $\qquad h_i d_{co}/k = 0.017(Re)^{0.67}(Pr)^{0.37}\left(\dfrac{D}{D_T}\right)^{0.1}\left(\dfrac{d_{co}}{D_T}\right)^{0.5}\left(\dfrac{\mu}{\mu_w}\right)^{\eta}$
    where $\eta = 0.71(\mu)^{-0.21}$
    Vertical tubes used as baffles, four blades
    $1300 < Re < 2$ million $\qquad h_i d_{co}/k = 0.09(Re)^{0.65}(Pr)^{0.3}\left(\dfrac{D}{D_T}\right)^{0.33}\left(\dfrac{2}{B}\right)^{0.2}\left(\dfrac{\mu}{\mu_w}\right)^{0.14}$
    c. Paddle $\qquad Nu = 0.87(Re)^{0.62}(Pr)^{0.33}\left(\dfrac{\mu}{\mu_w}\right)^{0.14}$
    d. Turbine, retreating-blade (six-blade only)
    $\qquad Nu = 1.40(Re)^{0.62}(Pr)^{0.33}\left(\dfrac{\mu}{\mu_w}\right)^{0.14}$

(h_o) Outside coefficients: jacketed tanks
NOTE: $Re = D_e u \rho/\mu$
1. Spiral baffling in jacket
    $Re < 2100 \qquad h_o D_e/k = 1.86\left[(Re)(Pr)\left(\dfrac{D_e}{L}\right)\right]^{0.33}\left(\dfrac{\mu}{\mu_w}\right)^{0.14}$
    $Re > 10,000 \qquad h_o D_e/k = 0.027(Re)^{0.8}(Pr)^{0.33}\left(\dfrac{\mu}{\mu_w}\right)^{0.14}\left[1 + 3.5\left(\dfrac{D_e}{D_c}\right)\right]$
    $2100 < Re < 10,000 \qquad$ In transition region, use j_t factor from Colburn analogy in Chap. 14.
    where D_e = equivalent diameter based on jacket width and pitch of spiral.
2. No baffles in jacket $\qquad \dfrac{h_o D_e}{k} = 1.02(Re)^{0.45}(Pr)^{0.33}\left(\dfrac{D_e}{L}\right)^{0.4}\left(\dfrac{D_{jo}}{D_{ji}}\right)^{0.8}\left(\dfrac{\mu}{\mu_w}\right)^{0.14}(Gr)^{0.05}$

may be required to perform more than one unit operation. However, different unit operations are affected by different mixing parameters. Oldshue,[2] Nagata,[3] and Penney[4] discuss this problem, and each recognizes that it is not possible to formulate a general method for scaling up all mixing processes. However, scaleup procedures for specific processes can be obtained by testing and experience.

There are similarity parameters that can be used as a guideline to scaleup. These are geometric, kinematic, and dynamic similarities. Geometric similarity requires that all dimensions be proportionally the same, kinematic similarity requires that all velocities be proportionally the same, and dynamic similarity requires that all force ratios be proportionally the same. In most mixing systems, the force ratios or dynamic similarity appears to be the most important.

Dynamic similarity utilizes four different force relationships in three dimensionless numbers:

$$\text{Re} = \text{Reynolds number} = \frac{ND^2\rho}{\mu} = \frac{\text{inertia force}}{\text{viscous force}} \qquad (16.6)$$

$$\text{Fr} = \text{Froude number} = \frac{N^2D}{g} = \frac{\text{inertia force}}{\text{gravitational force}} \qquad (16.7)$$

$$\text{We} = \text{Weber number} = \frac{N^2D^3\rho}{\sigma} = \frac{\text{inertia force}}{\text{surface-tension force}} \qquad (16.8)$$

The Reynolds number and the Froude number are the most important for all mixing systems, while the Weber number is useful in mixing systems concerned with gas or liquid miscibility.

When estimating power consumption in a mixing system, Nagata[5] found that the Froude number should be considered when analyzing agitated vessels with a gas space above the liquid but that a relationship between Power number and Reynolds number such as suggested by Eq. (16.4) is adequate for all other systems. For vessels without baffles but still experiencing turbulent flow, the general relationship is

$$\phi = (\text{Po})(\text{Fr})^g \qquad (16.9)$$

where $g = (a - \log \text{Re})/b$ (16.10)

      ϕ = unspecified function of Reynolds number (Figs. 16.4, 16.5, and 16.6 are the result of such a function)
      D = vessel diameter
      Z = pitch
a and b = empirical constants, which for a three-bladed propeller in an unbaffled tank are:

Z/D	a	b
1	2.1	18
1.05	2.3	18
2.0	1.7	18

Solid suspension. When the agitator system is used to fluidize solids in a liquid, the critical (minimum) speed N_c must be calculated. This is the speed above which all the solids will be suspended. It is determined indirectly from the Reynolds number, which is a function of critical speed, and the Froude number. By using dimensional analysis, the general relationship takes the form

$$\text{Re} = k(\text{Fr})^a \left(\frac{\rho_S - \rho_L}{\rho_L}\right)^b \left(\frac{D_p}{D}\right)^c \tag{16.11}$$

where a, b, c, and k = empirical constants
ρ_S = solids density
ρ_L = liquid density
D_p = particle diameter
D = vessel diameter

Oldshue[2] has tabulated a variety of critical-speed correlations which are diverse in form. The correlations are somewhat consistent, in that the density-ratio exponent b ranges between 0.5 and 0.7, while the exponent for the particle-to-vessel-diameter ratio c, ranging between 0.2 and 0.9, is considerably more variant. However, the constants a and k do not appear to follow any apparent correlation and must be determined from experimental data for a given operating condition.

Dissolving solids and solution formation. When a mixing system is used for dissolving solids, the process can be correlated by using the following general mass-transfer correlation:

$$\text{Sh} = c(\text{Re}')^a(\text{Sc})^b \tag{16.12}$$

where Re' = Reynolds number = $\left(\dfrac{D_T^2 N \rho_L}{\mu}\right)$

Sc = Schmidt number = $\left(\dfrac{\mu}{\rho_L \mathcal{D}}\right)$

Sh = Sherwood number = $\left(\dfrac{k D_T}{\mathcal{D}}\right)$

k = mass-transfer coefficient
$\mathcal{D}$ = diffusivity
D_T = vessel diameter

Nagata[3] used this relationship to correlate the solution of a number of organic and inorganic chemicals and produced the following equations.

For a propeller,

$$3300 < \text{Re}' < 330{,}000 \quad \text{Sh} = 0.0043(\text{Re}')(\text{Sc})^{0.5} \tag{16.13}$$

For a four-pitched-blade paddle,

$$\text{Re}' \leq 67{,}000 \quad \text{Sh} = 2.7 \times 10^{-5}(\text{Re}')^{1.4}(\text{Sc})^{0.5} \tag{16.14}$$

$$\text{Re}' \geq 67{,}000 \quad \text{Sh} = 0.16(\text{Re}')^{0.62}(\text{Sc})^{0.5} \tag{16.15}$$

Gas absorption. For applications such as a fermenter or a waste-water-treatment aerator requiring solubilization of a gas in a liquid, a form of Eq. (16.12) is used to determine the mass-transfer coefficient of the gas to the liquid. The total mass of gas being transferred is calculated by

$$N = kA(c^{\star} - \bar{c}) \tag{16.16}$$

where N = mass of gas transferred, (lb·mol)/s
k = mass-transfer coefficient calculated by Eq. (16.12), (lb·mol)/(ft²·s) [(lb·mol/ft³)]
c^* = equilibrium concentration of gas in the liquid, (lb·mol)/ft³
$\bar{c}$ = average concentration of gas in the liquid, (lb·mol)/ft³
A = total interface area between the gas and the liquid phase, ft²

The equilibrium concentration can be found using the Henry's-law constant. The interface area, however, depends on the average gas-bubble diameter D_p formed in the mixing vessel. The ratio of average bubble diameter to impeller diameter has been found to be a function of the Weber number (We):

$$D_p/D = f(\text{We}) \tag{16.17}$$

Nagata[6] has proposed the following general relationship:

$$D_p/D \propto (\text{We})^{-3/5}(\text{Po})^{-2/5}(D^3/V)^{-2/5} \tag{16.18}$$

where V = mixing-vessel volume.

Mechanical design of agitators

It is important to understand the basic mechanical design of an agitator system. The agitator can be divided into five basic components: motor, reducing gears, drive shaft and bearings, mechanical shaft seals, and impeller blades.

Motor size is determined from estimated power consumption. The motor enclosure depends on the location of the mixing system within the plant. Motor sizing and enclosure selection are discussed in Chap. 17.

The use of reducing gears for mixing is optional for low-viscosity (<100-cP) liquid. It is more economical to use a direct-drive system so that the impeller revolutions per minute equals the motor revolutions per minute. This is typically 1150 or 1750 r/min with lower speeds being used in larger vessels. However, it is more practical to use reducing gears for high-viscosity liquids, with 420 r/min being the most frequently selected low speed.

The use of mechanical seals is optional; seals are employed for those services which require the prevention of fluid leakage. They are used when mixing fluids (gases or liquids) are hazardous, toxic, or costly. Mechanical seals used on agitators are essentially identical to those used on pumps. Chapter 11 explains the various mechanical seals that are available and the advantages and the disadvantages of each.

The design of shafts and bearings is the responsibility of the agitator manufacturer. They must be designed so that the bearings reduce friction to a minimum and the shaft is compatible with the process material and has a minimum of vibration during operation.

In addition to complying with process conditions, the agitator-impeller blades must be constructed of a material compatible with the process fluid. The material must also be easy to maintain or replace. Many impellers can be unbolted from their shafts.

Mounting of agitators must also be considered. Figure 16.9 shows some of the conventional mounting procedures for agitators.

Motionless Mixers

One of the more recent developments in mixing technology is the motionless mixer. It has no moving parts and is constructed of a series of elements placed in a pipe or

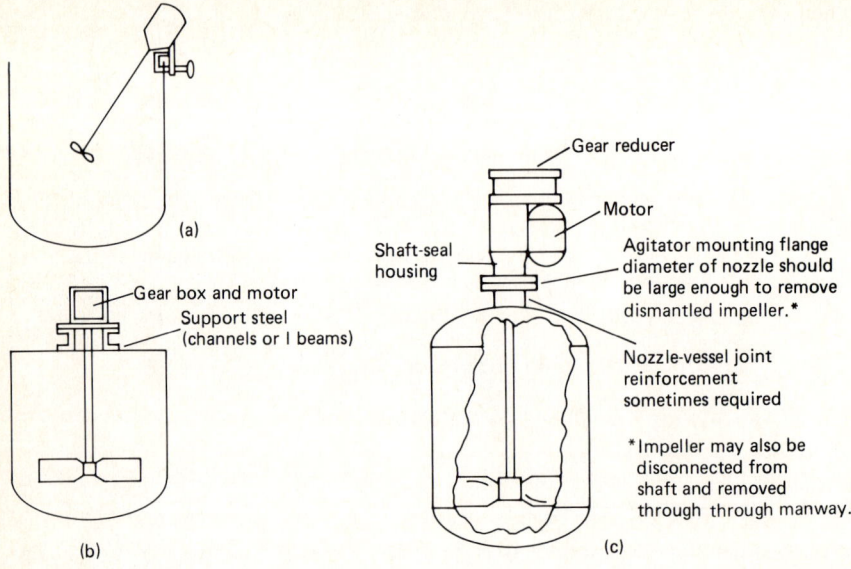
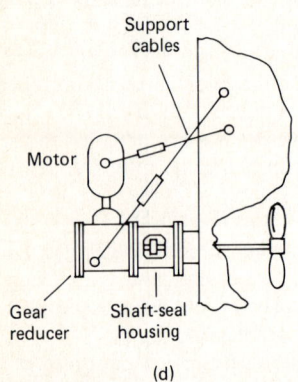

Figure 16.9 Agitator mountings. (*a*) Clamped portable agitator. (*b*) Open-topped vessel or sump. (*c*) Pressure-vessel mounting. (*d*) Side-mounted agitator.

conduit. Figure 16.10 shows some of the element configurations used in motionless mixers.

Mixing in a motionless mixer is produced by the turbulence caused by the elements in the mixer as the fluid passes over it, the division and redivision of the fluid into finer and finer layers, or a combination of both. The motionless mixer in Fig 16.10*a* uses turbulence as the primary mechanism for mixing, while the mixer in Fig. 16.10*b* utilizes the division into layers, or the strata method.

The performance characteristics of a motionless mixer are significantly different from those of an agitator-type mixer. When a fluid enters a mechanically agitated

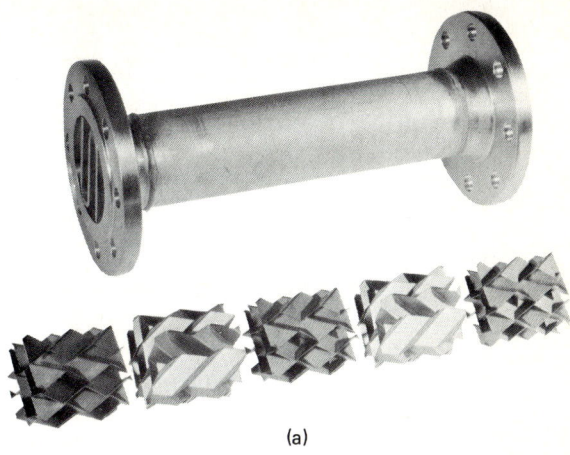

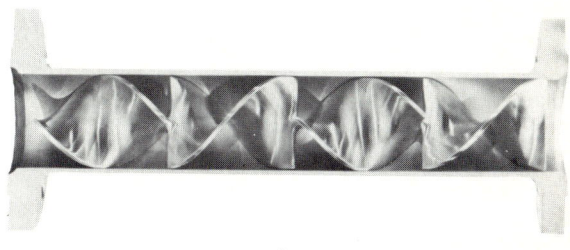

Figure 16.10 Two typical element configurations for motionless mixers. (*a*) A static mixing unit with corrugated elements. *(Koch Engineering Company Inc., Wichita, Kans.)* (*b*) A Kenics static mixing unit with radial mixing elements. *(Chemineer, Inc., Dayton, Ohio.)*

tank, the fluid particles are dispersed throughout the tank, which results in complete back mixing. The concentration ratio between input concentration and output concentration is a function of residence-time ratio $t/\bar{t}$ in the vessel, where t is the actual time that the fluid spends in the vessel and $\bar{t}$ is the time constant for the vessel, defined as

$$\bar{t} = \frac{\text{vessel volume}}{\text{flow rate of fluid through the vessel}} \qquad (16.19)$$

The characteristic curve for an instantaneous step change in fluid concentration entering a completely back-mixed, mechanically agitated vessel is shown in Fig. 16.11. Figure 16.11 also shows the characteristic curves for an empty pipe, a motionless mixer, and an ideal plug-flow reactor, given the same step change in concentration. A plug-flow reactor is an ideal model of a reactor in which the concentration at the entrance to the reactor remains the same at the exit of the reactor when it leaves

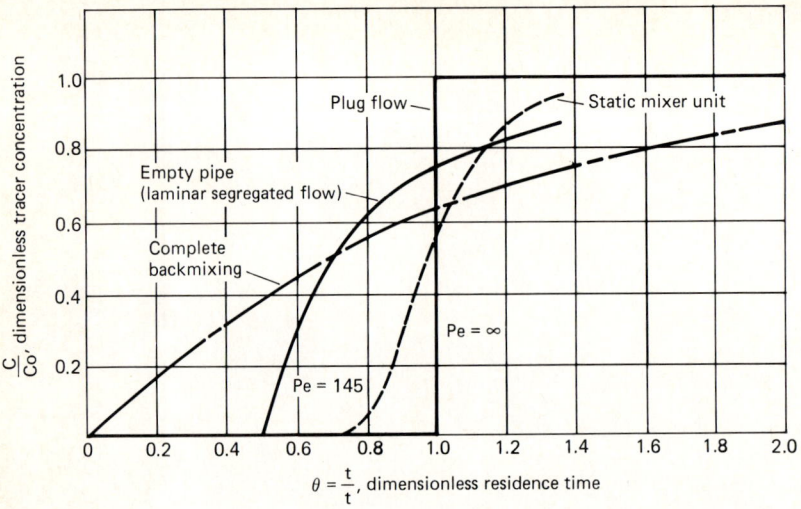

Figure 16.11 Concentration distribution versus fluid residence time in various reactor and mixing-vessel designs owing to a sudden step increase in the inlet concentration. Pe = Péclet number; t = residence time. (*Chemineer, Inc., Dayton, Ohio.*)

one residence-time increment later. Real flow systems have turbulence which disperses fluid backward and forward in a vessel or pipe. A motionless mixer comes closest to an ideal plug-flow reactor.

Figure 16.11 indicates that motionless mixers have a more predictable performance than a mechanical agitator. For example, the unit shown in Fig. 16.10a, when handling low-Reynolds-number flow, divides the stream into ever-decreasing strata at each element. The number of strata can be determined by the equation

$$L = 2^N \tag{16.20}$$

where L = number of strata
N = number of elements

Although the motionless mixer has no moving parts, it does consume power in the form of friction. The pressure drop for a motionless mixer is expressed in terms of some multiple of the pressure drop through an empty pipe of the same diameter and length as the mixer; i.e.,

$$\Delta P_{mixer} = K \, \Delta P_{pipe} \tag{16.21}$$

If the motionless mixer is jacketed or heat-traced, the heat-transfer performance of the mixer can be calculated by the dimensionless-number equation:

$$Nu = 0.078(Re)^{0.8}(Pr)^{0.33} \tag{16.22}$$

A motionless mixer will have roughly 3 times the internal heat-transfer rate of an empty pipe of the same diameter and length. Manufacturer's engineering data should be consulted for information on pressure drop and heat-transfer rate.

Motionless mixers have been used for many of the same applications as mechanical agitators. However, they are most often used for handling high-viscosity polymer streams and for mixing solid streams when uniform blending and temperature control are critical.

REFERENCES

1. James Y. Oldshue, *Fluid Mixing Technology,* McGraw-Hill Book Company, New York, 1983, pp. 78–81.
2. Ibid., chap. 9.
3. S. Nagata, *Mixing: Principles and Applications,* John Wiley & Sons, Inc., New York, 1975, sec. 10.7, pp. 444–447.
4. Roy W. Penney, "Recent Trends in Mixing Equipment," *Chemical Engineering,* Mar. 22, 1971, pp. 86–88.
5. Nagata, op. cit., p. 20.
6. Ibid., p. 364.
7. James Y. Oldshue, "How to Specify Mixers," *Hydrocarbon Processing,* October 1969, pp. 73–80.

BIBLIOGRAPHY

Bondy, F., and S. Lippa: "Heat Transfer in Agitated Vessels," *Chemical Engineering,* Apr. 4, 1983, pp. 62–71.
Chen, S. J., and A. R. MacDonald: "Motionless Mixers for Viscous Polymers," *Chemical Engineering,* Mar. 19, 1973, pp. 105–111.
Ho, F. C., and A. Kwong: "A Guide to Designing Special Agitators," *Chemical Engineering,* July 23, 1973, pp. 94–103.
Lydersen, A. L.: *Fluid Flow and Heat Transfer,* John Wiley & Sons, Inc., New York, 1979.
Mixing: A Chemical Engineering Report, published by *Chemical Engineering,* June 8, 1964.
Root, W. L., and R. A. Nichols, "Heat Transfer in Mechanically Agitated Units," *Chemical Engineering,* Mar. 19, 1973, pp. 98–104.

Chapter 17

Electrical Power and Motors

The energy source for most process-equipment prime movers and controls is electricity. Therefore, familiarity with electrical equipment and its interaction with mechanical equipment is necessary to make logical decisions on equipment design, control, and operation. Although the electrical engineer is responsible for the design of the electrical systems in a process plant, the process engineer should understand the data requirements of the electrical engineer and provide necessary information wherever possible. This chapter outlines some important points to be considered in laying out electrical systems and selecting electrical equipment for a process plant.

Power Distribution

Electricity generated by a public utility at a central power plant is normally at a voltage from 18 to 24 kV. Depending on the distance between the user and the power plant, the voltage may be stepped up to a level between 115 and 700 kV; 1000 V/mi of transmission has been used as a rule of thumb.[1] The voltage is stepped down to a standard distribution voltage at local substations for distribution to local users. Depending on a user's needs, the voltage may require further reduction by transformers located near or within a user's facilities. The selection of primary voltage supplying a process plant is determined by an economic evaluation based on plant size, power availability, utility rate structure, and anticipated plant growth.

Figure 17.1 shows a simplified schematic of a power-distribution network. The same network is shown in the lower portion of Fig. 17.1 as a one-line diagram. A one-line diagram is analogous to the process engineer's flow diagram, but it is used by electrical engineers to represent in a simplified manner the flow of electric power from a source to the users. Figure 17.2 shows some of the symbols used on a one-line or single-line diagram. The reader should consult Refs. 2, 3, and 4 for a more detailed list and explanation of the symbols.

Although Fig. 17.1 shows a range of volts on a given line, an actual system has only

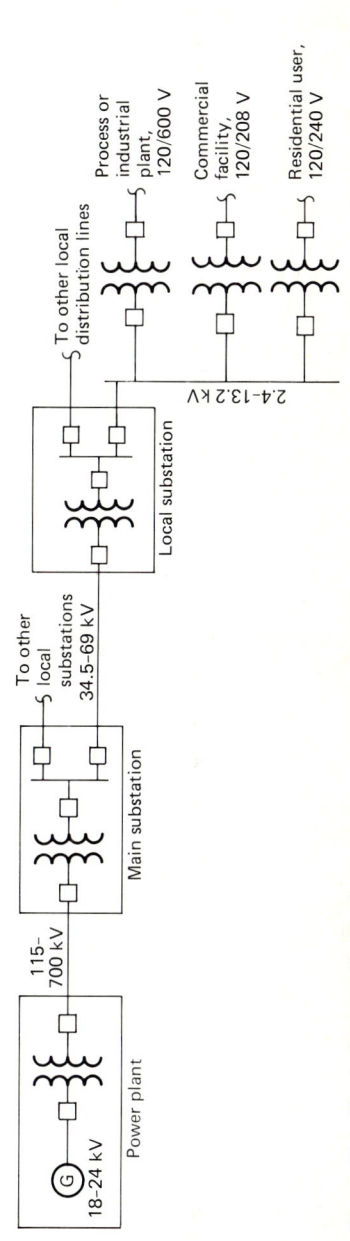

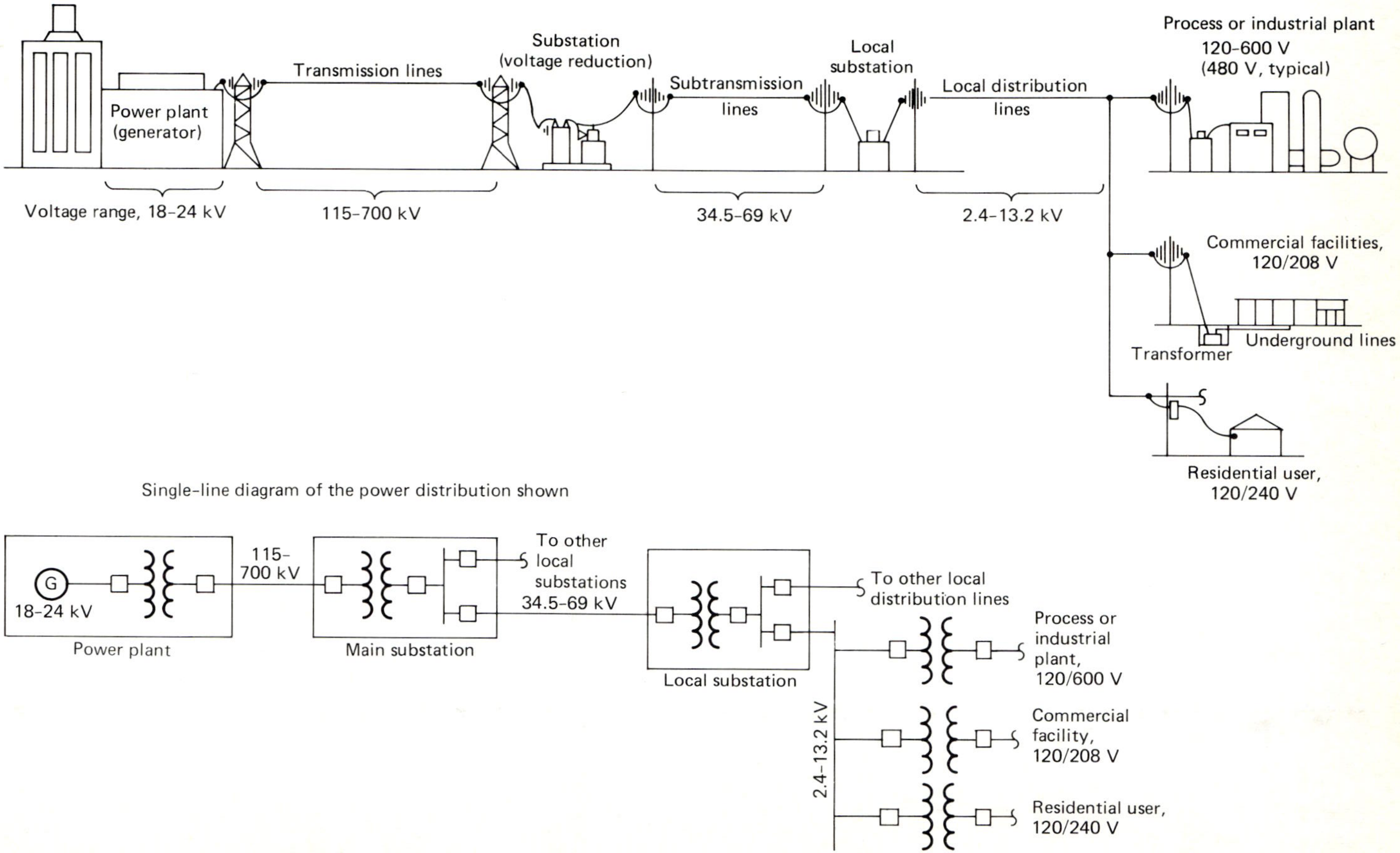

Figure 17.1 Typical power-distribution system.

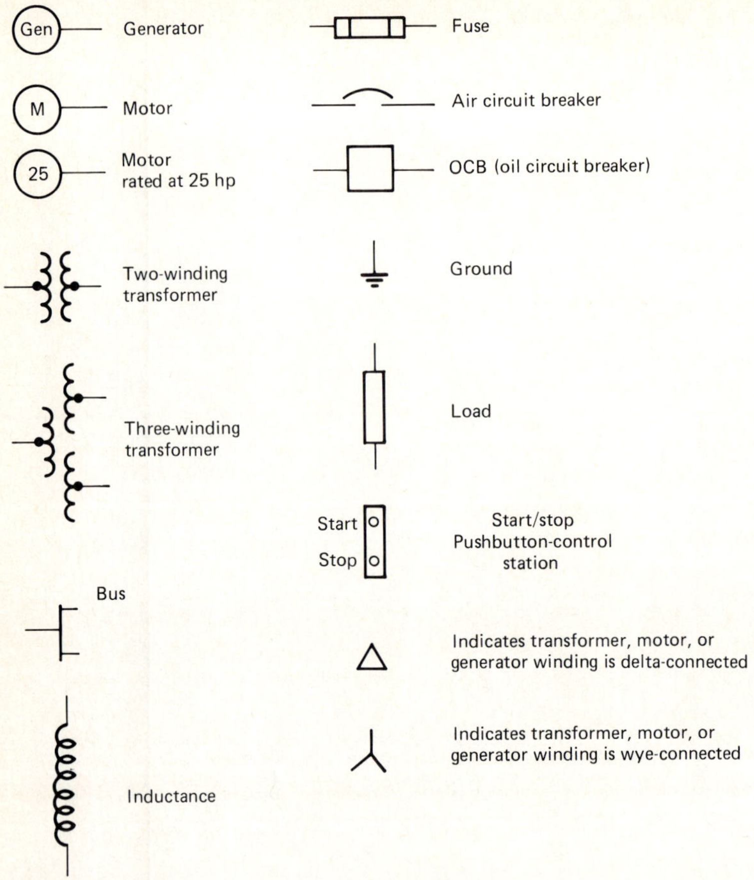

Figure 17.2 Some common one-line-diagram symbols.

only a single voltage on a line. The voltages that are typically found in a power-distribution system are listed in Table 17.1. The voltages found in process plants are usually 13.8 kV or less with 120-, 480-, and 4160-V levels used for most motors in the plant. Table 17.2 lists typical applications for the 120-, 240-, 480-, and 4160-V levels. The 208- and 277-V levels are usually used in commercial buildings for high-power lighting systems.

TABLE 17.1 Typical Voltage Levels for Different Points in a Power Distribution System

Generated voltage	18, 22, and 24 kV
Transmission-line voltages	115, 230, 345, 500, and 700 kV
Subtransmission-line voltage	34.5 and 69 kV
Distribution lines	2.4, 4.16, 8.32, and 13.2 kV
User voltages	120, 208, 240, 480, and 600 V

TABLE 17.2 Voltage-Level Applications

Voltage level	Typical application
120 V, single-phase	Typical residential voltage level, low-power office and laboratory equipment, ac motors less than 1 hp
240 V, single-phase	Residential and commercial building appliances and equipment with high-power requirements such as electric heating and air-conditioning units
480 V, three-phase	Industrial motors used in the range from 1 to about 500 hp
4160 V, three-phase	Motors with horsepowers in excess of 200 hp

Single-phase and multiphase alternating-current systems

The power-supply systems mentioned thus far are alternating-current (ac) systems. The standard frequency used in the United States and Canada is 60 Hz (cycles per second). In addition, a power supply may be either single-phase or three-phase (Fig. 17.3). Three-phase power systems are more efficient and more economical than single-phase systems for large motors. As a general rule, equipment having a power con-

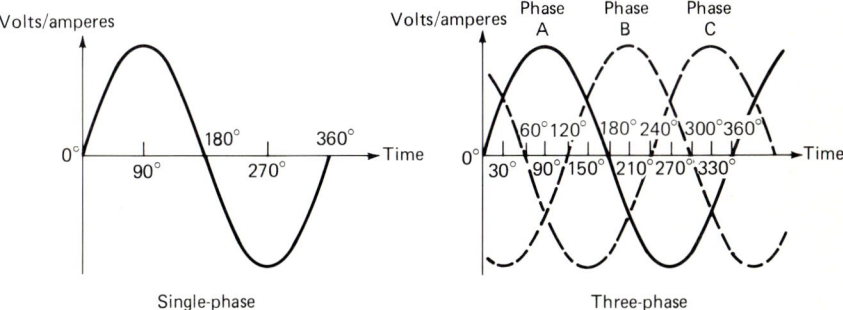

Figure 17.3 Single- and three-phase waveforms.

sumption in excess of 1 kW requires three-phase service. In addition, a three-phase power-transmission line can carry 1.73 times more energy per pound of electrical wiring than a single-phase transmission line. Polyphase systems in excess of three phases are available, but the three-phase and single-phase systems are the dominant systems in process plants. The three-phase system is quite versatile and can be connected to transformers to provide a variety of voltages within the plant.

Delta and wye three-phase connections. All three-phase ac systems can be interconnected between each of the individual three-phase power lines. These connections take the form of a delta or wye configuration as shown in Fig. 17.4. The reason for these configurations is economics and can be explained in a series of simplified diagrams for a single-phase electrical generator system and some three-phase generator systems.

There are two basic methods of generating a single-phase alternating current. One is the rotating-armature method, and the other is the rotating-field method. In the

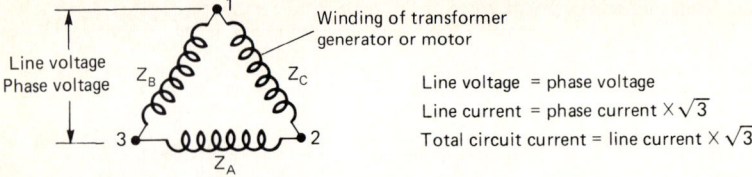
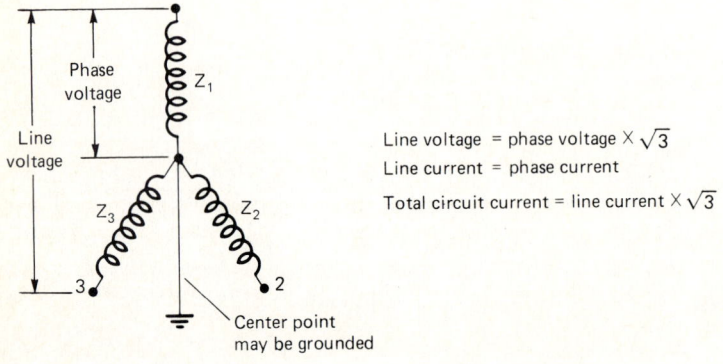

Figure 17.4 Delta and wye connections.

rotating-armature method, a magnetic field is created by direct-current (dc) voltage applied to a conducting wire wound around the stationary pole pieces called the stator in the generator. As the conductor windings on the rotor shown in Fig. 17.5 revolve through the magnetic field, a single-phase alternating electric current which takes the form shown in Fig. 17.3 is generated. The current travels through the rotor windings to slip rings. It leaves the generator through the rings and travels down the transmission lines until it reaches an electrical load. The configuration for the rotating-armature method is opposite that of the rotating-field method. The rotating-field method applies a magnetic field to the rotor conductors and generates an electric current in the stator windings. Most industrial motors are constructed similarly to generators and are designed to operate by the rotating-field method.

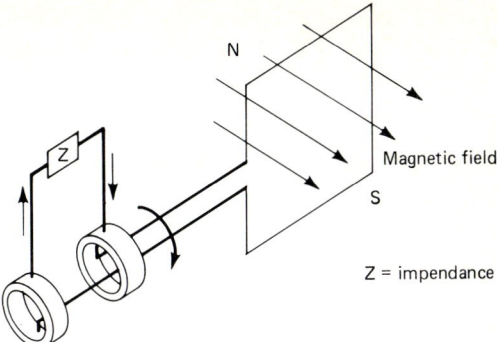

Figure 17.5 Generating a single-phase alternating current with a stationary magnetic field.

A three-phase generator can be constructed of essentially three single-phase windings placed 120° apart on the same rotor just as shown in Fig. 17.6. This configuration requires six slip rings (three pairs) and six leads leaving the generator. A more economical design for a generator can be produced by connecting different pairs of leads as shown in Figs. 17.7 and 17.8. Each pair of leads acts as an independent single-phase circuit. As a result, only three slip rings and three leads are needed to create a three-phase voltage.

Figure 17.7 shows a generator with a wye configuration, and Fig. 17.8 shows a delta configuration. These same delta and wye connections can also be found in other electrical equipment such as motors and transformers.

Both delta and wye configurations can be connected to each other to create different voltage and current characteristics across electrical loads. When dealing with three-phase systems, one must understand the difference between line voltage or current and phase voltage or current. Line voltage is the voltage measured between a phase line and the ground or neutral line. Phase voltage is the voltage measured between two phase lines. If the voltage measured between one pair of phase lines is

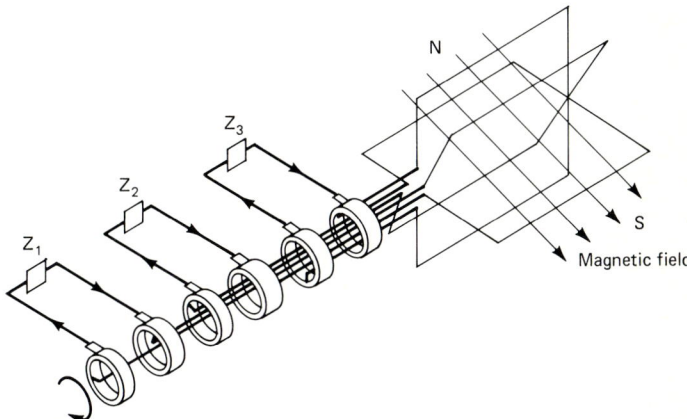

Figure 17.6 A three-phase generator with a six-slip-ring design.

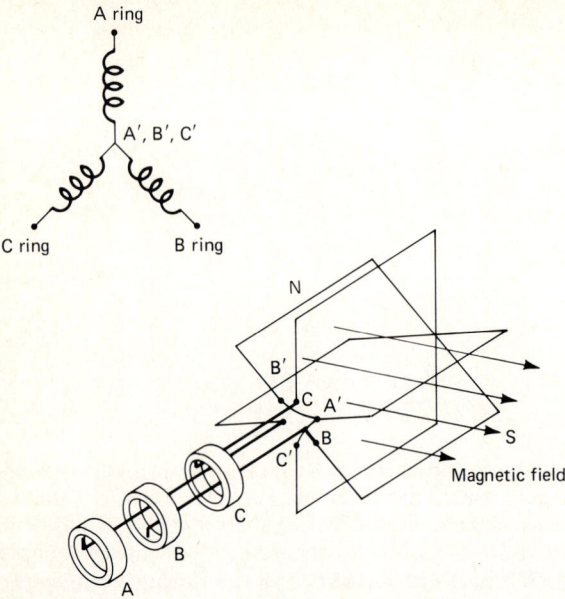

Figure 17.7 A three-phase generator connected in a wye configuration.

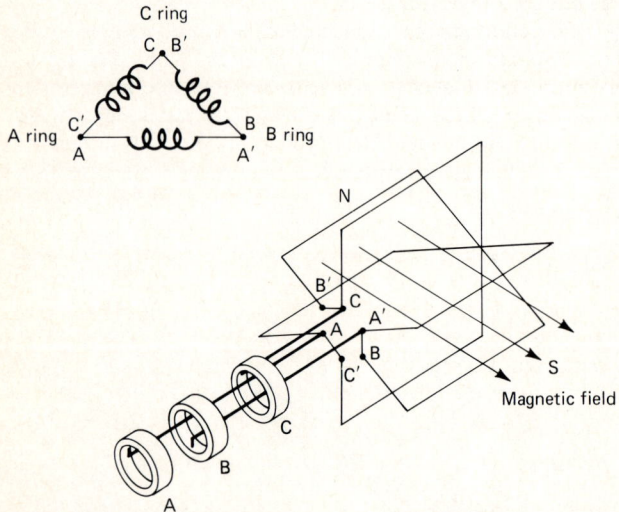

Figure 17.8 A three-phase generator connected in a delta configuration.

different from the voltage between another pair of phase lines, the system is unbalanced. Therefore, a fourth line to ground is sometimes added to a wye configuration to balance each phase. The voltage current, and impedance relationships between delta and wye connections are shown in Fig. 17.4.

Transformers in single-phase and multiphase systems. Transformers are the most important electrical devices in power-distribution systems. Although they are used to regulate voltage levels up and down in power systems, they also have many applications such as power monitoring and circuit protection.

A transformer consists of primary and secondary windings connected magnetically through either an iron core or an air core, the iron core being more efficient. Depending on the ratio of the number of turns of wire in the primary and secondary windings, the transformer may be either a step-up or a step-down transformer. The step-up transformer increases the voltage in the secondary winding to that in the primary winding. The step-down transformer accordingly decreases voltage.

Although a step-up transformer increases voltage, it simultaneously decreases the current. This is used to advantage in long-distance power-distribution systems. The high voltages produced for transmission lines have relatively low currents and therefore less line losses (I^2R) for the same resistance than do lower-voltage transmission lines.

Since three-phase power-distribution systems supply power to process plants, special three-phase transformers are used to supply the various voltage levels required in the plants. The winding of these transformers can be connected as delta-delta, delta-wye, wye-delta, or wye-wye. In addition, by tapping only two of the three phases of the distribution system, a single-phase power supply can be produced. Figure 17.9 shows the different transformer configurations and how they are represented schematically. Figure 17.10 shows one of the distribution systems frequently used in process plants and commercial facilities. The three-phase four-wire wye connection is the more practical distribution system because it can provide the variety of voltages required for industrial motors, industrial lighting, and offices and laboratory equipment.

Direct-current power systems

Direct-current electricity is sometimes used in a process plant. However, dc power is not normally a part of the power-distribution system but is produced by the rectification of an ac source directly at the equipment. Rectification of three-phase power supplies produces smaller ripples in the dc power and thus requires less filtering than single-phase systems. Typical users of dc power are traction motors for elevators, conveyors, and cranes, stepping motors, and electrical instrumentation.

Electrical Codes and Standards

A process engineer should have an understanding of electrical codes and standards to assist the electrical engineer in selecting the proper electrical equipment and its enclosure. The process engineer must indicate to the electrical engineer the type and the location of electrical-hazard conditions throughout the plant.

In the United States the two most often used sources for electrical-equipment installations and enclosures are the National Electrical Code (NEC)[5] and the National Electrical Manufacturers Association (NEMA) Standards. NEC Articles 500, 501, and

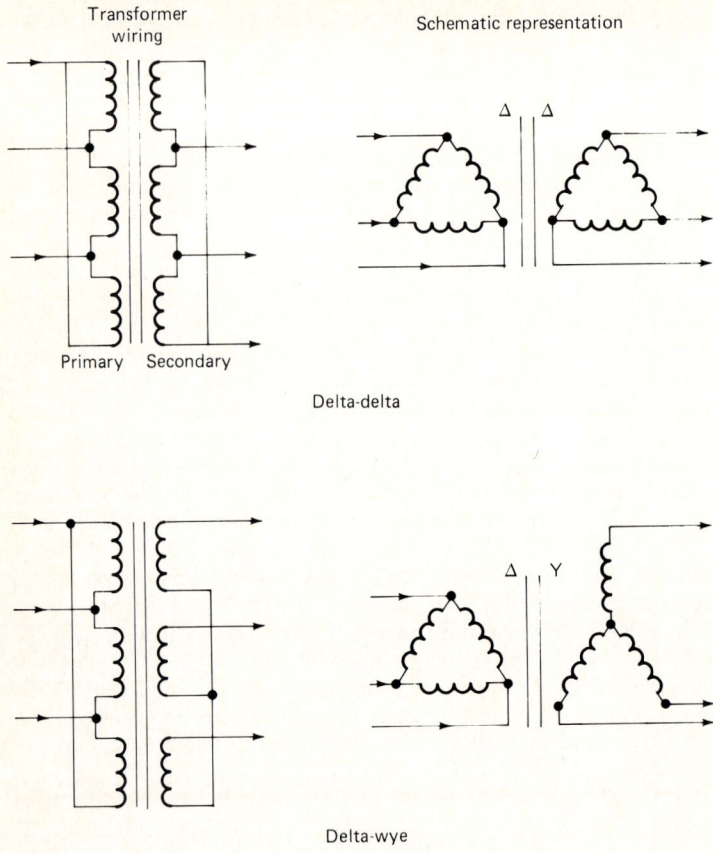

Figure 17.9 Some typical three-phase transformer connections. Transformers can also be connected in wye-delta and wye-wye, as required.

502 are of particular interest because they define what are considered electrically hazardous locations and provide a method of classification for these areas. NEMA provides a classification system for induction-motor types and for electrical-equipment enclosures corresponding to the environment in which they will be located. Table 17.3 outlines the NEC hazardous-grouping and hazardous-area classifications, while Table 17.4 lists the NEMA enclosures with typical applications and enclosure construction. Motor classifications will be discussed later.

Determination of class and hazardous group is made jointly by the plant operator and the insurance carriers. The responsibility of the electrical engineer and the process engineer consists of reviewing the type and location of hazardous materials to be processed, stored, or handled within the plant and making recommendations to the plant operator and insurance carriers as to where the boundaries of electrically hazardous areas should be established. Segregating equipment by electrical hazard, where possible, can produce a more economical electrical system for the plant; e.g., if water-softening equipment which is generally nonhazardous and may use either a driptight or a watertight NEMA enclosure for its electrical equipment were placed in a room

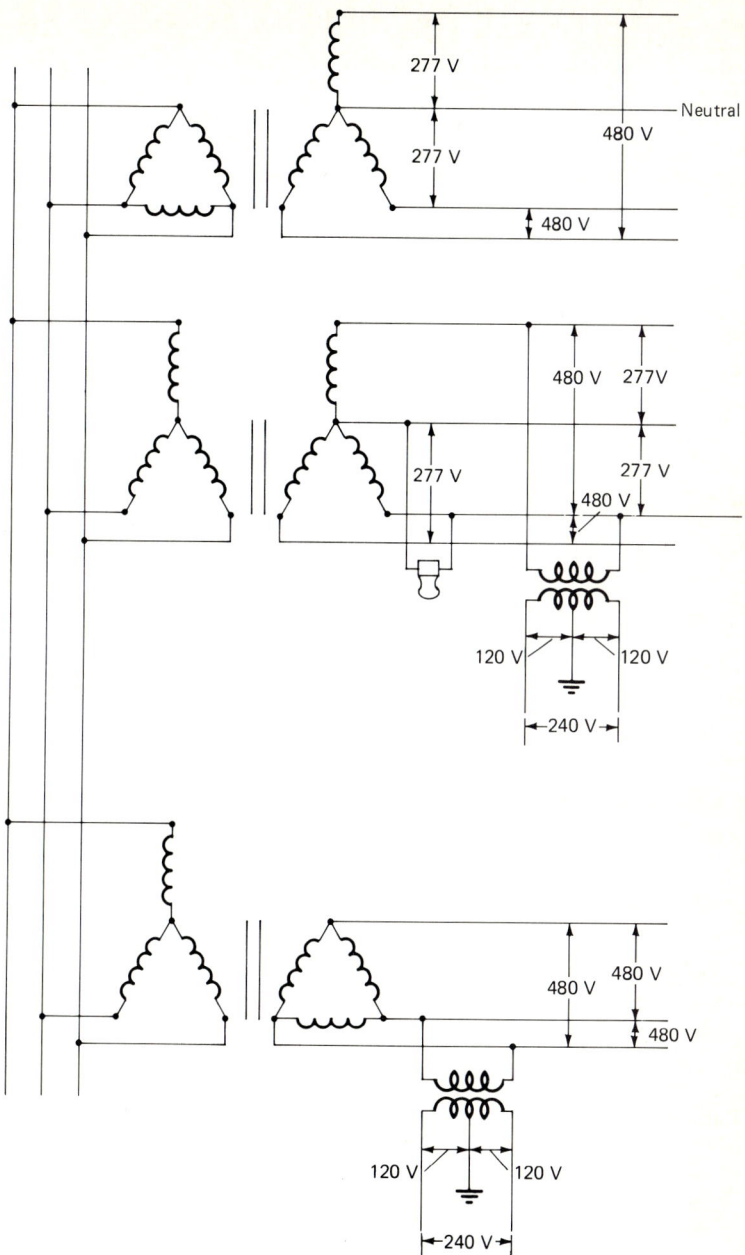

Figure 17.10 Some of the more frequently used transformer connections to provide the various voltages in a process plant.

TABLE 17.3 Outline of Articles 500, 501, 502, and 503 of the 1984 National Electrical Code (NEC) Standards*

	Hazardous Atmosphere Grouping		
Class	Atmosphere type	Hazard group	Typical materials in group
I	Gases and vapors	A†	Acetylene
		B†	Hydrogen, arsine, ethylene oxide, propylene oxide, etc.
		C‡	Carbon monoxide, ethylene, cyclopropane, ethyl ether, hydrogen cyanide, etc.
		D†	Acetic acid, ammonia, benzene, ethane, alcohol, gasoline, methane, etc.
II	Combustible dust	E	Combustible dusts having a resistivity of less than $10^5 \, \Omega \cdot cm$
		G	Combustible dusts having a resistivity of more than $10^5 \, \Omega \cdot cm$
III	Flammable fibers and flyings	...	Carbonized or excessively dry organic materials, ignitable fibers

	Definition of Classes and Divisions	
Class	Division 1	Division 2
I	Hazardous under normal operations and normal maintenance	Hazardous only during abnormal or unusual conditions, e.g., an accident; applies to gases and vapors in closed or well-ventilated systems
II	Same as Class I or where dust may cause abnormal operation such as a mechanical failure or a short circuit	Hazardous only when dust is in sufficient quantities to accumulate, causing either a heat buildup by its insulating qualities or arcing of contacts or causing a fire around the equipment
III	Same as Class I with the qualification that the hazard is caused by combustible fibers or flyings produced during normal handling and manufacturing	Combustible fibers and flyings not expected to be suspended in hazardous quantities in the air such as in storage and handling (not manufacturing) facilities

National Electrical Code, 1984 ed., NFPA 70-1984, published by the National Fire Protection Association, Batterymarch Park, Quincy, Mass. 02269.
†Assumed lowest ignition temperature of materials is 280°C (536°F).
‡Assumed lowest ignition temperature of materials is 180°C (356°F).

containing process equipment handling ethylene oxide, classified in Hazard Group B, the water-softening electrical equipment could require upgrading to explosion-proof enclosures as required for Class I, Division 1 locations.

Electric Motors

Electric motors are very often the preferred choice as drivers for equipment in a process plant. Plants having large steam-generation facilities may use steam turbines in

TABLE 17.4 NEMA Classification of Electrical Enclosures

Type number	Description	Application
1	General-purpose	Suitable for normal indoor atmospheres; used to prevent accidental contact with the enclosed equipment.
2	Driptight	Same protection as above except that the enclosure must also protect equipment from dripping water and condensate.
3	Weatherproof (weather-resistant)	Suitable for outdoor use; protects equipment from specified weather hazards.
3R	Raintight	Protection from heavy rain (not sleet) exposure.
4	Watertight	Protection for equipment subject to direct streams of water used during cleaning operations.
5	Dusttight	Protection from nonexplosive or combustible dust.
6	Submersible	Permits successful operation of equipment while completely submerged in water for specified conditions of time and pressure.
7*	Hazard Groups A, B, C, or D, Class I (air break)	Suitable for use in NEC Class I hazardous locations. Contact operations and circuit interrupts occur in air environments.
8*	Hazard Groups A, B, C, or D, Class I (oil-immersed)	Same as NEMA 7 enclosure above except that the equipment within the enclosure is immersed in oil.
9*	Hazard Group E or G, Class II	Suitable for NEC Class II dusty atmospheres.
10	Bureau of Mines	Explosion-proof enclosure suitable for use in gassy coal mines.
11	Acid- and fume-resistant	Equipment within this enclosure is immersed in oil. Enclosure provides protection from acid and other corrosive fumes.
12	Industrial use	Enclosure provides protection from dust, fibers, flyings, oil seepage, or coolant seepage experienced in manufacturing plants.
13	Dustproof	Enclosure is designed to prevent accidental contact or interference of the enclosed equipment from entering dust. The dust and enclosed equipment must be specified when specifying this enclosure.

*NEC Standards should be consulted for the proper application of these enclosures.

lieu of motors. However turbines are limited to high-horsepower equipment and emergency standby equipment, such as fire-water pumps.

While both ac and dc motors are used in process plants, ac squirrel-cage induction motors are the most commonly used motor type.

DC motors

DC motors have limited applications in process plants because they are usually more expensive and require higher maintenance than ac motors. They are used primarily on specialty equipment requiring variable speeds within narrow limits of speed and torque.

Figure 17.11 shows a cutaway of a typical dc motor. On the basis of their starter-control schemes, there are three types of dc motor configurations. These are the

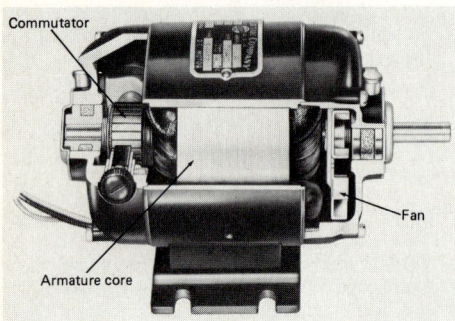

Figure 17.11 Cutaway view of direct-current motor. *(Bodine Electric Company, Chicago.)*

series, the shunt, and the compound motor. The schematic arrangement for each configuration together with its respective torque and speed curves is shown in Fig. 17.12. References 1 and 6 provide more detailed designs and applications of dc motors, starters, and control circuits.

AC motors

AC motors are used extensively in process plants. These motors are either single-phase or three-phase. Single-phase motors (Fig. 17.13) are usually fractional-horsepower motors (less than 1 hp). The standard sizes for fractional-horsepower motors are $\frac{1}{20}, \frac{1}{12}, \frac{1}{8}, \frac{1}{6}, \frac{1}{4}, \frac{1}{3}, \frac{1}{2}, \frac{3}{4}$, and 1 hp. Single-phase motors are also built in standard horsepower sizes of 1, $1\frac{1}{2}$, 2, 3, 5, $7\frac{1}{2}$, and 10 hp for 120- and 240-V service, but these larger horsepowers are normally provided by three-phase motors in a process plant.

Three-phase motors. Three types of three-phase ac induction motors are found in a process plant: the squirrel-cage motor, the wound-rotor motor, and the synchronous motor.

Squirrel-cage motors. Squirrel-cage motors are the most extensively used motors in a process plant. The name comes from the rotor, which resembles a rotating squirrel's cage (see Fig. 17.14). The rotor has windings consisting of embedded copper bars connected by metal rings at each end. When a three-phase current is applied to the stationary field winding or the stator of the motor, a rotating magnetic field is created. The revolving magnetic field cuts across the squirrel-cage rotor, causing rotary motion. Because rotation is caused by the revolving magnetic field, the mechanical rotation will be lower than the field's angular speed since there is not a 100 percent energy conversion. This difference in speed is referred to as slip.

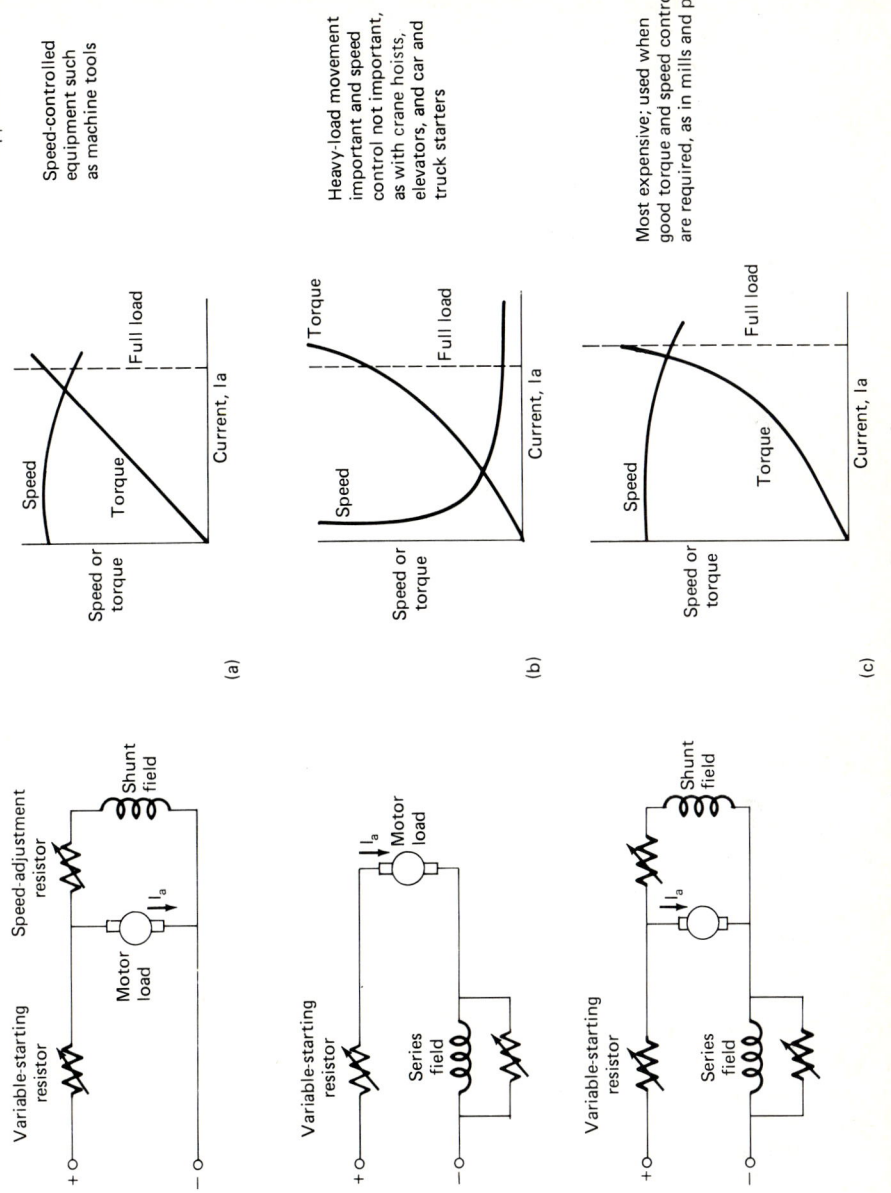

Figure 17.12 Direct-current motor controls and characteristics. (*a*) Shunt motor. (*b*) Series motor. (*c*) Compound motor.

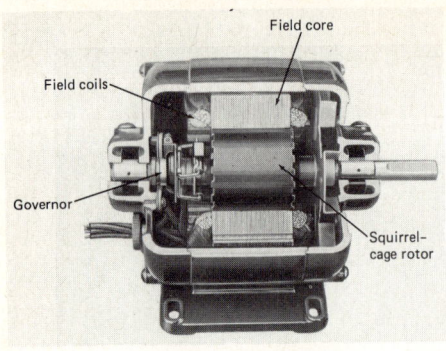

Figure 17.13 Cutaway view of a single-phase alternating-current motor. This is a typical fractional-horsepower induction motor. *(Bodine Electric Company, Chicago.)*

The speed of a revolving field is calculated by the equation

$$\text{rpm} = \frac{120f}{P} \tag{17.1}$$

where f = frequency, Hz
P = number of poles (number of stator windings)

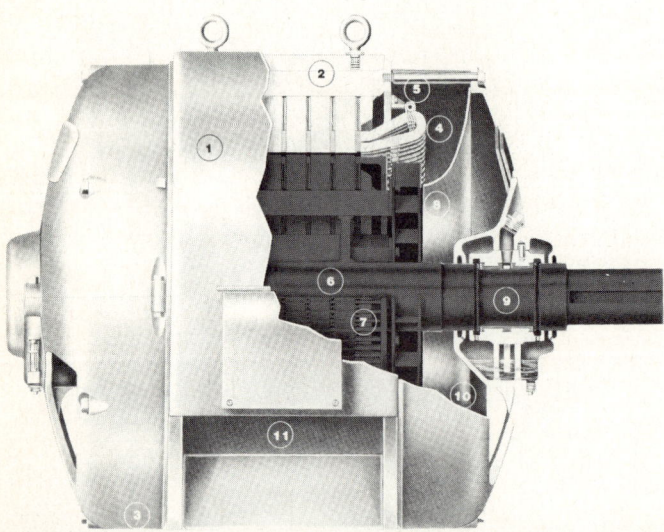

Figure 17.14 Cutaway view of a three-phase induction motor, (1) Motor enclosure. (2) Enclosure cross section. (3) Support feet. (4) Stator windings. (5) Stator-windings support. (6) Rotor. (7) Squirrel-cage windings. (8) Cooling-air intake. (9) Bearings. (10) Cooling-air intake. (11) Cooling-air discharge. *(Electric Machinery Manufacturing Company, Minneapolis.)*

This is the synchronous speed of the motor. Actual speed can be calculated from the synchronous speed and percent slip by the equation

$$\text{Actual speed} = (\text{synchronous speed}) \times \frac{100\% - \% \text{ slip}}{100\%} \qquad (17.2)$$

There is considerable variation in the speed and torque produced in a squirrel-cage motor. Figure 17.15 shows a typical performance characteristic curve for a squirrel-cage motor when connected to a piece of rotating mechanical equipment such as a

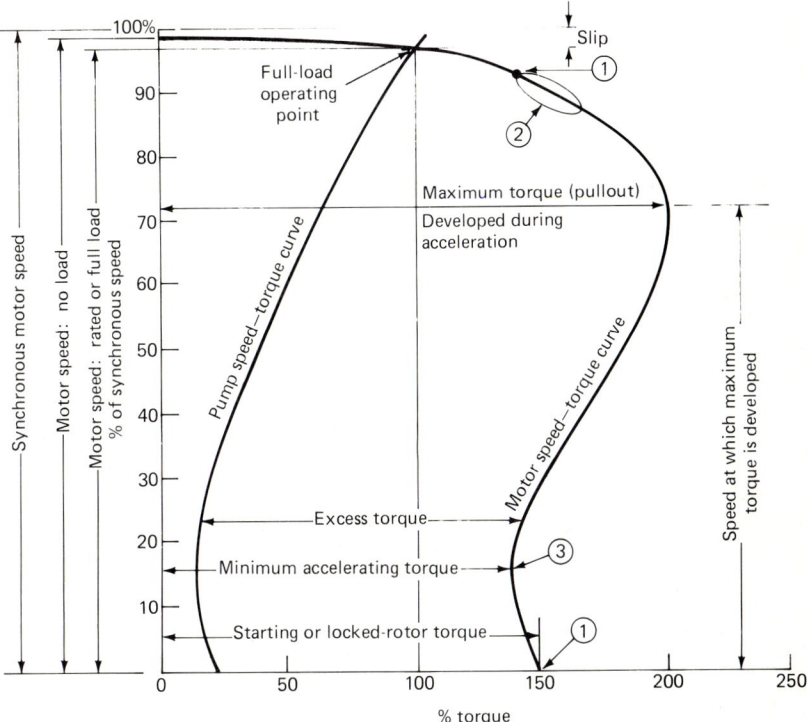

Figure 17.15 Performance characteristics of a typical squirrel-cage induction motor when connected to a piece of rotary mechanical equipment such as a pump. (1) Speed-torque range limits during initial acceleration. (2) Maximum torque range recommended for overloads of short duration. (3) Reverse-curvature point in motor speed-torque curve.

pump. The pump has a speed-versus-torque characteristic curve different from that of the motor. The point of intersection between the pump torque curve and the motor curve is the operating point at full-load conditions. The lower speed and higher motor torque conditions are the transient conditions of the motor that occur during pump start-up and shutdown. Figure 17.15 indicates that the operating speed of the motor is lower than the motor speed at no load and lower still than the synchronous speed. This explains why centrifugal-pump curves are defined for speeds like 1750 rpm while the motor is nominally at 1800 rpm.

NEMA has developed a code-letter system for the standard design of motors.[7,8] The general speed-torque characteristic curves for each class of motor can be found in Fig. 17.16. The characteristic curve in Fig. 17.15 is for a NEMA Class B motor. A detailed explanation of each class is beyond the scope of this book, but Table 17.5 lists some typical applications for each class.

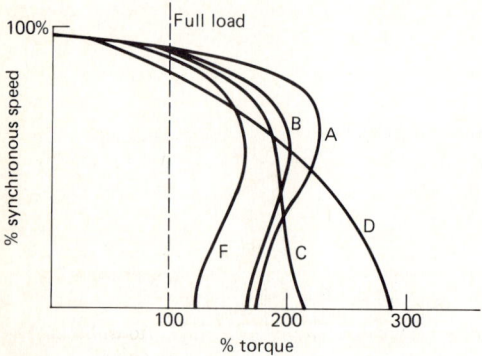

Figure 17.16 Typical speed-torque characteristics for different NEMA class squirrel-cage motors. Class A = normal; Class B = general-purpose; Class C = double-cage, high-torque; Class D = high-resistance-rotor; Class F = double-cage, low-torque.

In addition to classifying motors by torque characteristics, NEMA defines the different types of motor enclosures (motor housings). Table 17.6 lists some of the more important enclosures used in most industrial facilities.

Three-phase induction motors come in the following standard horsepower ratings: $\frac{3}{4}$, 1, $1\frac{1}{2}$, 2, 3, 5, $7\frac{1}{2}$, 10, 15, 20, 25, 30, 40, 50, 60, 75, 100, 125, 150, 200, 250, 300, 400, 500, 600, etc. Motors of 200 hp or less generally use 480-V power supplies, while 4160-V power supplies should be considered for motors of 250 hp or greater.

TABLE 17.5 Typical Applications for Different NEMA Class Motors*

NEMA class	Type	Typical application
A	Normal	Shredder, rotary vacuum pump
B	General purpose	Pumps, blowers, centrifugal and rotary compressors, low-viscosity liquid mixers, fans
C	Double-cage, high-torque	Ball mill, loaded reciprocating compressor, large conveyors, hammer mill, high-viscosity liquid mixer, kneader, rotary kiln
D	High-resistance-rotor	Centrifuge, crusher with flywheel, drum dryer, oil-field pumps with flywheel, rotary kiln
F	Double-cage, low-torque	Uneconomical for most industrial loads because of their high starting-torque-to-breakdown-torque ratio

*Extracted from Charles C. Libby, *Motor Selection and Application*, McGraw-Hill Book Company, New York, 1960, Table 5-5, pp. 208–217.

Electrical Power and Motors 615

TABLE 17.6 NEMA Motor-Enclosure Types

I. Open enclosures	Ventilated by openings for air cooling of winding.
A. Dripproof	Ventilation openings designed to prevent most liquids or solid particles falling on enclosure at an angle $<15°$ from the vertical from entering enclosure.
B. Splashproof	Ventilation openings designed to prevent most liquids or solid particles falling on enclosure at an angle $<100°$ from the vertical from entering enclosure.
II. Totally enclosed motors	Totally enclosed to prevent exchange with the atmosphere; however, enclosure is not airtight.
A. Totally enclosed nonventilated (TENV)	No external means of cooling are included with the enclosure.
B. Totally enclosed fan-cooled (TEFC)	The motor is cooled by a fan integrally mounted to the enclosure blowing cooling air over the enclosure.
C. Explosion-proof	The motor enclosure is designed to withstand any internal explosion caused by ignition of flammable gases or vapors from the surrounding atmosphere.
D. Dust-explosion-proof	Motor enclosures suitable for Class II locations, which withstand explosions or ignition of dust on or around the enclosure.
E. Waterproof	The enclosure prevents the entrance of water in the form of a directed stream, such as a hose, with the exception of leakage through the shaft. It may also include automatic draining facilities.

Wound-rotor motors. A wound-rotor motor is considered a squirrel-cage motor whose rotor resistance can be varied for starting and limited speed control. The rotor has a three-phase insulated winding on its rotor (see Fig. 17.17). The windings are joined to slip rings called collector rings which can be connected to external variable resistors through brushes. Figure 17.18 shows a typical wiring arrangement and the resulting speed-torque characteristics for these motors. Wound-rotor motors are more complex and more expensive than squirrel-cage motors. However, they require lower

Figure 17.17 Rotor for a wound-rotor induction motor. *(Westinghouse Electric Corporation.)*

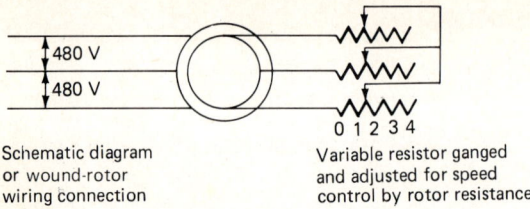

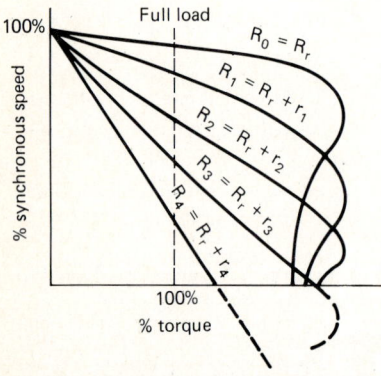

Figure 17.18 Wiring and characteristic behavior of a wound-rotor induction motor. R_o = rotor resistance with zero external resistance. $R_o < R_1 < R_2 < R_3 < R_4$. v_i = external variable resistor at position i. R_i = total rotor resistance when external resistor is at position i.

current for starting and are used in place of squirrel-cage motors when high starting currents are undesirable. For example, a rule of thumb estimating induction motor-current draw in amperes during normal loads is given by Roe[9] as

$I(A) \approx hp \times 1$	for 500 V, three-phase
$I(A) \approx hp \times 1.25$	for 440 V, three-phase
$I(A) \approx hp \times 2.5$	for 230 V, three-phase
$I(A) \approx hp \times 5$	for 220 V, single-phase
$I(A) \approx hp \times 10$	for 110 V, single-phase

These equations show that high-horsepower motors which use high-voltage multiphase power supplies are more efficient and draw less current than single-phase low-voltage motors. Starting current may be 125 to 275 percent of the above normal load current, depending on the NEMA motor class. If a wound-rotor motor is started with a high rotor resistance and if the resistance is reduced as the motor approaches normal load and operating speed, the current draw can be maintained at or below normal load current. The resulting lower current draw requires a smaller electrical-cable size than a squirrel-cage motor having a higher starting current.

Synchronous motors. Synchronous motors are the most expensive type of induction motor. A synchronous motor is different from other induction motors in that it runs without slip, i.e., at synchronous speeds. This is achieved by locking the rotor magnetically to the rotating magnetic field of the stator. The rotor (Fig. 17.19) of a synchronous motor is excited by an external direct current which produces alternate

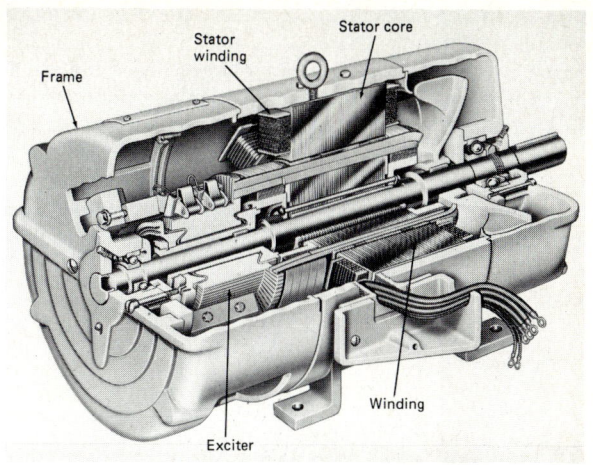

Figure 17.19 Cutaway view of a synchronous motor. (*Louis Allis, Milwaukee.*)

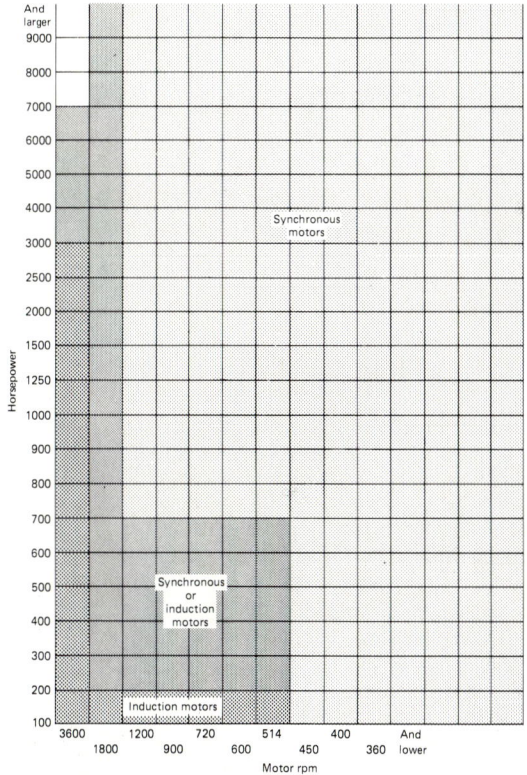

Figure 17.20 Operating range for selecting synchronous motors and induction motors. (*"Special Motors Report,"* Power, *June 1969.*)

north and south magnetic poles that are attracted and locked to the rotating magnetic field in the stator.

Synchronous motors are used for large, slow-speed machinery with constant speeds and steady loads such as large-capacity fans, pumps, compressors, mills, and grinders. Figure 17.20 shows the typical range of horsepower and revolutions per minute where synchronous motors would be selected over squirrel-cage induction motors.

Motor controls. The switching of motors on and off, depending on differing process operations, and the associated wiring and motor-control logic require a different representation from those used for process-control loops. Figure 17.21 shows some of the

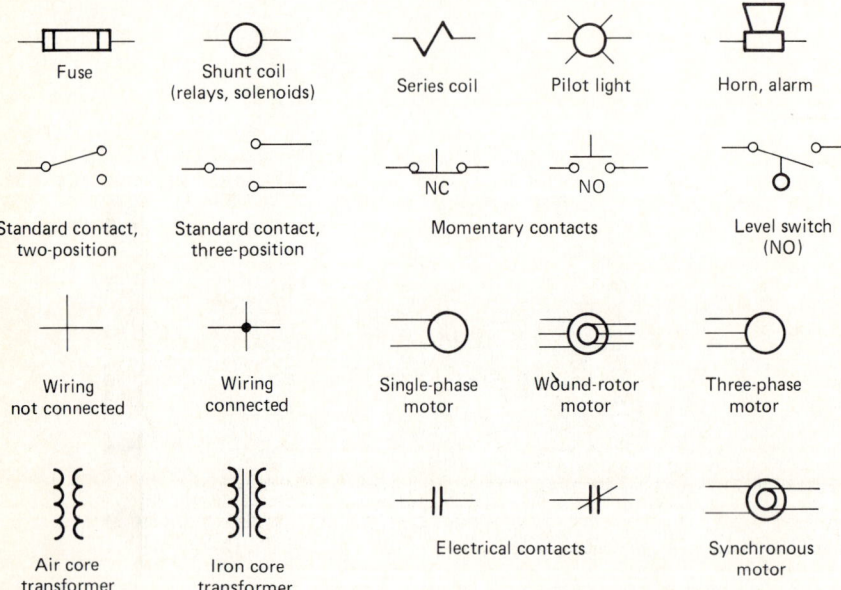

Figure 17.21 Some standard electric-motor-control symbols. NC = normally closed; NO = normally open.

symbols used in motor-control diagrams. Figure 17.22 illustrates actual motor-control equipment together with the symbols representing the equipment on motor-control diagrams.

Motor-control diagrams. The motor-control diagram, also referred to as a ladder diagram or an elementary wiring diagram, presents the electrical devices in a motor-control circuit on horizontal lines which give the appearance of rungs on a ladder. Figure 17.23 shows two simple motor-control circuits for a sump pump which can be

Figure 17.22 Some standard electric-motor-control apparatus and its symbols. NC = normally closed; NO = normally open. (a) Cutaway view of a bimetallic overload relay; OL = overload. (b) Control relay with eight normally closed contacts. (c) Pressure switch. (d) Fluid dashpot timing relay. (e) Limit switch. (f) Temperature-control switch with vertical immersion bulb. (*Square D Company, Palatine, Ill.*)

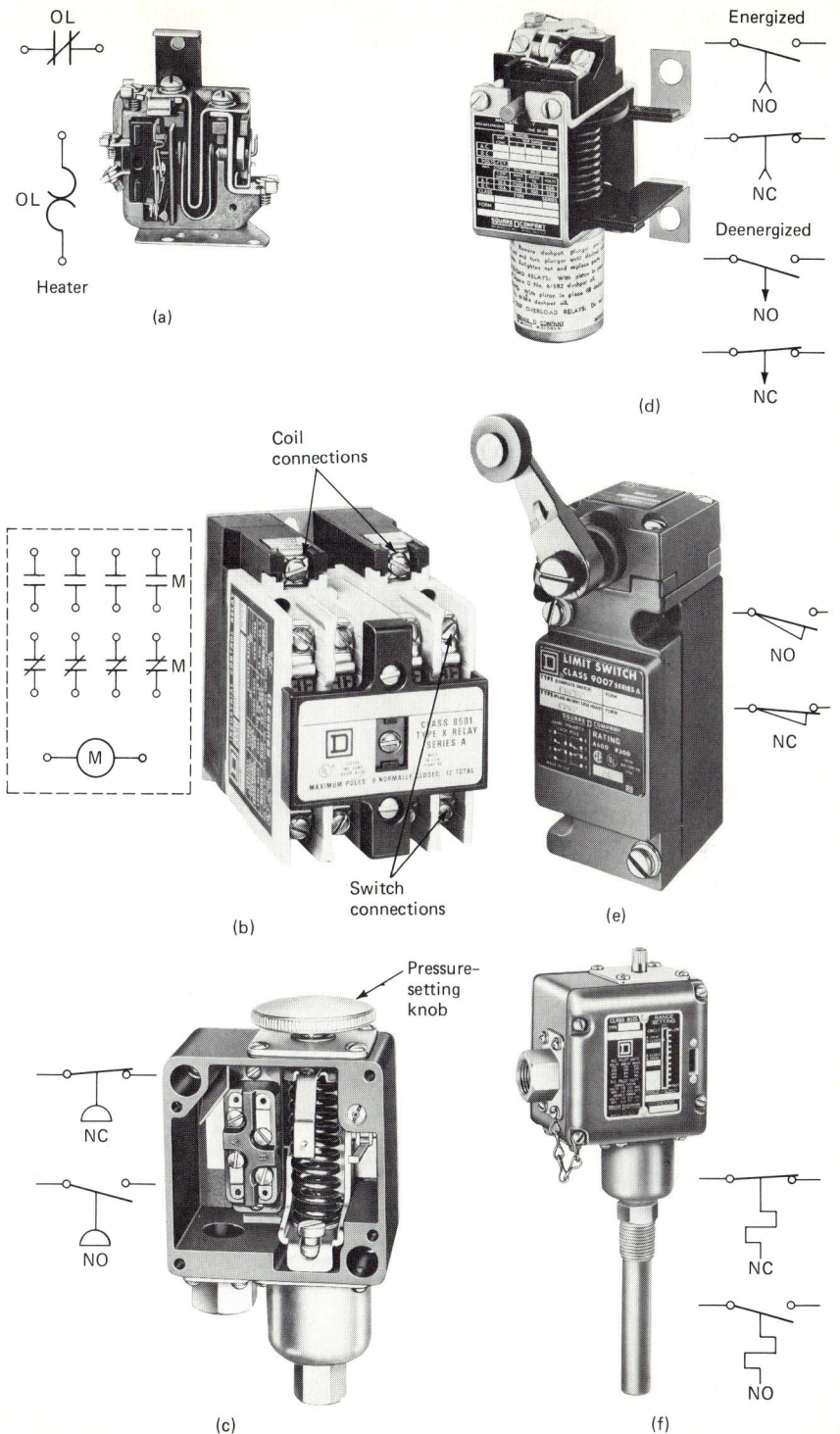

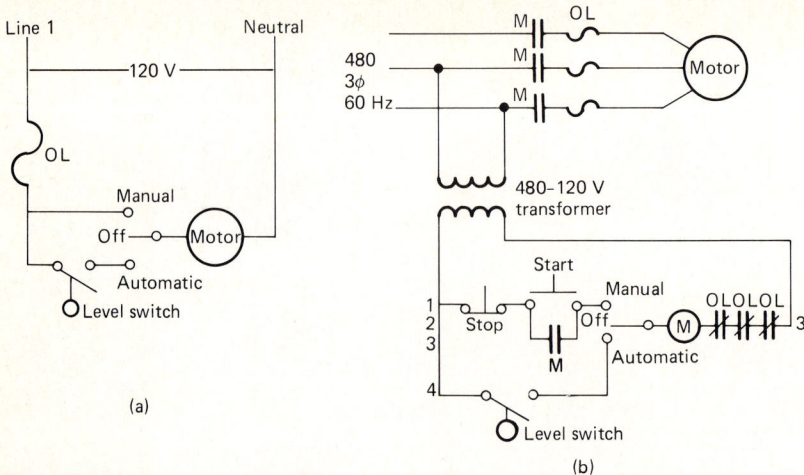

Figure 17.23 Electrical ladder logic diagram for automatic sump-pump operation. (*a*) Single-phase motor control. (*b*) Three-phase motor control.

operated manually or automatically on level. One scheme presents the controls for a pump with a single-phase motor, and the other scheme is for a pump with a three-phase motor.

Single-phase motor control consists of a three-position switch, a level switch, and an overload (OL) relay to protect the pump motor from overheating. In this circuit, the pump runs when the pump switch is either in the manual or in the automatic position and the level switch is closed on high level. The pump shuts off when either the three-position switch is in the off position or when the switch is in the automatic position with the level switch opened on low level. The pump will also shut off if an overload is sensed by the OL relay.

The three-phase circuit is more complicated. Since motor controls operate at 120 V, the 120 control voltage must be derived from the power circuit of the motor which is connected to the primary windings of a transformer. The transformer's secondary windings produce the required 120 V for the control circuit.

Just as in the single-phase pump motor, the selector switch may be in the manual, automatic, or off position. In the manual position, the pump is started by pushing the start button, which closes the circuit energizing the motor contactor *M* of a relay. This, in turn, closes the normally open contacts *M* which are placed across the start button and on each three-phase line going to the motor. The *M* contact across the start switch keeps the circuit closed and allows current to continue to flow through the control circuit.

The pump can be shut off by pushing the stop button to open the circuit by deenergizing the motor contactor and returning all *M* contacts to their normally open position. The pump also shuts off if an overload condition exists in any of the feeder lines to the pump motors.

When the three-position switch is in the automatic position, the manual start-stop circuit is bypassed and the pump operation is controlled by the level switch alone. However, the pump can still be shut off if an overload is sensed in the feeder lines.

More complex motor-control circuits than those shown in Fig. 17.23 are beyond the scope of this book. However, the reader is referred to Refs. 1, 4, 6, and 10 for a more thorough discussion of motor-control circuit design.

REFERENCES

1. Walter M. Alerich, *Electric Motor Control,* Van Nostrand Reinhold Company, New York, pp. 141–167.
2. *Handbook of Electric Power Calculations,* ed. by Arthur H. Seidman, Haroun Mahrous, and Tyler G. Hicks, McGraw-Hill Book Company, New York, 1984, sec. 10.
3. Lionel B. Roe, *Practices and Procedures of Industrial Electrical Design,* McGraw-Hill Book Company, New York, 1972, pp. 15–17, 249–250.
4. Charles W. Snow, *Electrical Drafting and Design,* Prentice-Hall, Inc., Englewood Cliffs, N.J., 1976, chap. 3.
5. *National Electrical Code,* 1984 ed., NFPA 70-1984, published by the National Fire Protection Association, Batterymarch Park, Quincy, Mass. 02269.
6. Irving L. Kosow, *Control of Electric Machines,* Prentice-Hall, Inc., Englewood Cliffs, N.J., 1973, pp. 40–84.
7. Irving L. Kosow, *Electric Machinery and Transformers,* Prentice-Hall, Inc., Englewood Cliffs, N.J., 1972, pp. 342–347.
8. Charles C. Libby, *Motor Selection and Application,* McGraw-Hill Book Company, New York, 1960, chap. 2.
9. Roe, op. cit., p. 254.
10. William S. Holmes, "An Introduction to Electrical Control Circuits," *Chemical Engineering,* Nov. 15, 1971, pp. 176–182.

BIBLIOGRAPHY

Adams, James E.: *Electrical Principles and Practices,* 2d ed., McGraw-Hill Book Company, New York, 1973.

Fardo, Stephen W., and Dale R. Patrick: *Electrical Power Systems Technology,* Howard W. Sams & Co., Inc., Indianapolis, 1979.

1981 Code Digest, Bull. 2939, published by Crouse-Hinds, Syracuse, N.Y., 1984.

Index

Abbreviations on lead sheets, 36
Absolute pressure, 464
Accounting and item-number assignment, 50
Acoustic velocity, gas, 282–284
Acronyms on lead sheets, 36
Acrylics, 157
Adiabatic-fluid-flow process, 204, 280–300
   isothermal conversion for, correlation
      factors to, (*illus.*) 295
Aerosols, 200, 201
Agencies, regulatory, for pressure vessels,
   (*table*) 380
Agitators, 580–593
  baffles for, 581, 585–586
  checklists on engineering flow diagrams,
     77–78
  design of, 582–593
  for dissolving solids, 592
  fluid mechanics of, (*illus.*) 581
    calculations for, 581–582
  for gas absorption, 592
  heat transfer and, 588–593, (*illus.*) 589
    calculations for, (*table*) 590
  horsepower requirements for, (*table*) 583
  impellers for, 582–585, (*illus.*) 584, 586,
     587

Agitation (*Cont.*):
  mechanical components of, 593
  mountings for, (*illus.*) 594
  parameters for selection, (*table*) 583
  scaleup of, 589–591
  seals for, 78
  side-entering, (*illus.*) 588
  for solid dissolution, 592
  for solid suspensions, 591–592
  types of, 77, (*illus.*) 585
  and vessel geometry, 586–588
Air:
  dissolved in water, (*table*) 490
  pressure drop of, 246, (*tables*) 248–250
Air gap in piping systems, 271–273
Air leakage:
  in fan and compressor systems, 449
  in vacuum systems, 485–490
    at connections and shafts, 487, (*illus.*)
      488
    estimation of, (*illus.*) 486
    halogen test, 490
    liquid-acetone test, 489
    prevention of, 488–489
    soap-bubble test, 489
    in submerged sections, 487, (*illus.*) 489

Air leakage, in vacuum systems (*Cont.*):
    testing of, 489–490
Alloy 20, 138
Alloys in piping systems, 190, 195
Aluminum, alloys of, 136–139, (*tables*) 137
    treatment methods for, (*table*) 138
    weldability of, 136
American codes and standards, 144
    (*See also* Codes and standards)
American Society of Mechanical Engineers
        (ASME), 376
    Boiler and Pressure Vessel Code, 373
    Section VIII: standards for vessels, 378
    Unfired Pressure Vessel Code, 378
American Society for Testing and Materials
        (ASTM):
    alphanumeric and subject index of, 147–148
    standards of, (*table*) 146
    viscosity indices of, 204
American Standards Association (ASA), 187
Annual fixed cost of piping systems, 193
Annual production:
    capacity exponent, (*table*) 6
    in proposed installation, 5
    unit costs of chemicals, (*table*) 6
Archives of work in progress, 119–120
ASME (*see* American Society of Mechanical
        Engineers)
Aspirators, 480–481
ASTM (*see* American Society for Testing
        and Materials)
Atmospheric pressure, 67, 461–463, (*table*) 463
Available net positive suction head (*see* Head,
        net positive suction available)

Baffles, 72, 581, 585–586
Baker-plot analysis of two-phase flows, 337–338, (*illus.*) 338, 346, 347
Barometric legs, 471–473
    and fluid pressure loss, 198
Bench-scale experiments, 29
Bernoulli, Daniel, 223
    relationships for fluids, 223–224
Bernoulli equation, 275, 396
Blackbodies, 518–519, (*illus.*) 518
Block diagrams, 23–28
Blowers, 85–87, 440–460
    (*See also* Compressors; Fans)
Boilers, 551–556, (*illus.*) 557, 558
    auxiliary equipment, 555–556

Boilers (*Cont.*):
    types of, 551–555
        fire-tube, 552, (*illus.*) 552, 553
        waste-heat, 14–15, 553–555
        water-tube, 552–553, (*illus.*) 554
Boundary definitions, 99
Brake horsepower:
    of compressors, 452
    in process flow sheets, 29
    of pumps, 401–402
Brittleness, definition of, 123
Bronowski, J., 1
Bundling, pipe, 579

Calcium silicate, 567
    as insulation, 570
Calculation sheets, filing of, 119
Cameron Hydraulic Data Book for calculation
        of friction loss, 234, 246, 255–258
Casings for centrifugal pumps, 391
Cast iron, ductile, 129–131
    gray, 129
    silicon, 131
    white, 129
    wrought, 131
Cathodic protection, 167
Catsup as example of thixotropic fluid, 210
Cavitation, 399
Cellular glass as insulation, 567, 572
Centipoise, 202, 215
Centistokes, 202–203, 211
Centrifuges, checklists of, for engineering
        flow diagrams, 89–90
Ceramic materials, 158–160
    ceramics and cermets, 159
Channels, open, for liquid transport, 192
Checklists for engineering flow diagrams,
        68–108, (*table*) 69
Chemical engineers, process-flow-sheet
        preparation by, 40–41
Choking, gas-flow rate and, 282–284, 298–299, 314–316
Chromium in iron alloys, 133–135
Clearances:
    for reciprocating compressors, 445–446
    for reciprocating pumps, 402, (*illus.*) 403
Codes and standards:
    agencies, regulating, (*table*) 380
    electrical, 606–609
    instrument, 37
    materials selection, 144–149
    non-U.S., 149

Codes and standards (*Cont.*):
  for piping, 187
  sump pump, 434
  U.S., 144
  valves, 100
  vessels, 373, 376, 378, 380
Complex fluids (*see* Two-phase flows)
Compressibility factors for gases, 278, 287
Compressible-fluid flow:
  calculations for: rigorous approach, 299–333
    shortcut approach, 285–298
  critical-state, 301–302
  enthalpy, 299–301, 314–316
  frictionless, 274–333
  maximum equivalent length and flow rate, 297–299, 314–316
  property ratios of, 302
  shock-wave effect, 330–333, (*table*) 333
  stagnation-state, 299–303
  zero linear velocity, 299–301
Compressible fluids:
  definition of, 199–200, 274–277
  properties of, 277–284
  use of liquid tables for friction loss, 239–246
Compression, heat of, in vacuum pumps, 481
Compressors, 85–87, 445–460
  adiabatic head, 450–451
  centrifugal, 447
    axial-flow, 447
    radial-flow, 447
  clearance, 445–446
  efficiency of, 453–454
  isothermal head, 451–452
  knockout pots, 87
  mass flow rate of, 452–453
  multistage, 455–456
  performance characteristics of, 450
  polytropic pressure-volume relationship, 452
  positive-displacement units, 445, 447–449
  reciprocating, 446, 448–449
    diaphragm-type, 449
  rotary, 446–448
    blowers and, 448
    helical-screw, 448
    liquid-ring, 448
    lobed, 448
    operating characteristics of, (*table*) 448
    sliding-vane, 448
  screens, filters, and silencers for, 87

Compressors (*Cont.*):
  seals for, 449
  specific speed N of, 454–456
  surge control of, 458–459
  system curves of, (*illus.*) 446
  (*See also* Blowers; Fans)
Computer software, xiv, 4
  for drafting, 37
  for graphics: CADD, 42
    CGS, 42
  for insulation estimates, 568
Concrete, reinforced, 128–129
Condensation, barometric, 468–471, 490
Condensers:
  cooling water for, 468, 470–471, 477–479
  types of: direct-contact, 468–471
    intercondensers, 468–479
    precondensers, 480, 484–485
    surface, 468–471
Conduction in heat transfer, 491–495
  definition of, 491
Conductivities in heat transfer, piping and equipment materials for, (*table*) 576
Conductivity, thermal:
  definition of, 492–495
  values, reference sources for, 493
Conduit, noncircular, 224
Conference notes, filing of, 119–120
Controllers, instrumentation for, 42
Controls (*see* Instrumentation)
Convection in heat transfer, 495–517
  coefficients of: film, (*table*) 562
    for horizontal pipes, (*table*) 505
    natural, (*table*) 502
  definition of, 491, 495
  forced, 506–517
  natural, 502–506
Conversion tables for dimensional units, 17–20
  reverse conversion, 20
Cooling towers, 468, 543–551, (*illus.*) 546, 547, 551
  air-water system curves, (*illus.*) 544
  recirculation, (*illus.*) 549
  types of, 547–548, (*illus.*) 548
  water for, 472
Cooling water for condensers, 468–480
Copper alloys, 139–143
  brasses, 139
  bronzes, 140
  copper-beryllium, 139
  nickel, 140, (*table*) 141
  pure, 139

Index

Corrosion, 133, 162–185, (*tables*) 179–185
  environmental parameters, factors affecting, 163
  types of, 163–174
    cavitation, 169
    crevice, 170
    dealloying, 171
    erosion, 344
    exfoliation, 170
    fatigue, 173
    fretting, 173
    galvanic, 165
    hydrogen embrittlement, 173
    intergranular, 171
    microbial, 174
    pitting, 169
    stray-current, 167
    stress cracking, 172
    thermogalvanic, 167
    uniform, 165
Corrosion allowances:
  and alloy piping, 190
  and piping-schedule selection, 188
  stainless-steel piping, 190
Cost estimates:
  annual fixed costs: for equipment or process, 11–12
    and pipe-size selection, 192
  annual operating costs, 13
    and pipe-size selection, 192
  final detailed, 8–9
  installed cost: plus annual cost, (*table*) 15
    plus operating requirements, (*table*) 14
  for major-equipment installation, 6–8, (*table*) 8
  precision of, 5
  preliminary, 4–8, (*table*) 4
  product-based, 5–6
  types of, (*table*) 4
Cost indexing systems, 10, (*illus.*) 11
Costs:
  of equipment, 6–8
  of materials, 16
  of power, 192
  reference sources for, 7, 16
  of utilities, 16
Crane Company tables, friction loss for air in pipes, 246, (*tables*) 248–250
Creep, definition of, 123
Curricula for engineers, 1–2

Dampers, for fan controls, 441–443
Darcy relationship for pressure drop, 254
Deformation, plastic, 366
  of vessels, 366
Density of atmosphere, 461
Denver, atmospheric pressure of, 461
Dimensional units:
  conversion tables for, 18–20
    English, metric, and SI, 19–20
    English units, 16–17
  nomenclature for: English, 16–18
    International System of Units (SI), 16, 19–20
  reverse conversion, 20
Dimensionless groups, 501–502, (*table*) 503
Dip leg and head loss, 270
Distribution and wiring drawings, 43
Drag coefficient of fluid-solid systems, (*illus.*) 350, (*table*) 351
Drawings:
  checking of, 3
  for cost estimates, 9
  distribution and wiring, 43
  filing of, 120
  piping and instrumentation (*see* Piping and instrumentation diagrams)
  (*See also* Block diagrams; Engineering flow diagrams)
Drivers, pump, 79–80
Drums, reflux, 382
Ductility, definition of, 123

Earthquakes, vessel loading and, 373–376
Economic analysis (*see* Cost estimates)
Economic diagram for pipe sizing, 192
  velocity constraints, 197
Eddy currents, 213–218, 229, 254
Eductors, 480–481
Elastomers, 158
Electric current, use of, in pipe tracing, 575–579
Electric-generating stations, 64
Electric motors (*see* Motors)
Electric power distribution systems, 598–605, (*illus.*) 599
  connection for, (*illus.*) 602
  symbols for, 600
  voltage levels for, (*tables*) 600–601
Electrical codes and standards, 605–608
  National Electrical Code (NEC), 605, (*table*) 608
  National Electrical Manufacturers Association (NEMA) standards, 605, (*table*) 609

Electrical diagrams, 599, 620
Electrical enclosures, classification of, 609
Electrical engineer:
  P&ID preparation by, 43
  responsibilities of, 606
Electrical generators, 64, 601–605, (*illus.*) 603, 604
Electrical substations, 43
Electrical transformers, 605, (*illus.*) 606, 607
Elevation, effect of, on fluid pressure, 224
Emissivity, 520–521
  of various materials, (*table*) 521
Emulsions, stability of, 201
Energy balance, in process flow sheets, 27
Engineering flow diagrams, 35–120
  checklists for, 68–108
  correction of, 115–119
    (*See also* updating of, *below*)
  dating of: formal 117–118
    provisional, 117–119
  definition of, 35–36
  filing system for, 116–120
  issues of: construction, 109–120
    interim, 115–116, 118–119
    preliminary, 110–112
    revision, 114–119
    zero issue, 112–114
  master sets of, 109, 116–118
  preparation of, 43–120
    equipment identification, 50–52
    example of, (*illus.*) 47
    formatting of, 44–49
    information basis for, 44
    piping identification, 52–62
    processing sections, 37, 39
    scale of, 46, 49
    special legends and tables, 63
    specialty flow diagrams, 40
    utilities sections, 39–40, 45
  relationship of: to engineering disciplines, 40–43
    to project objectives, 109
  as road maps, 35, 52
  scope of, 110
  title-block record, 110, 113–115
  tracings, 116–118
  types of, 36–40
  updating of, 3, 42, 45, 109, 115–119
    (*See also* Piping and instrumentation diagrams)
Enthalpy of compressible-fluid systems, 299–301, 314–316
Epoxies, 158

Equipment:
  depicted on engineering flow diagrams, 41, 46–52
  depicted on process flow sheets, 30
  description of, 52
  listing of, 51
  names of, 51–52
  purchase of, 41
  scale of, on diagrams, 46, 49
  sizing of, 6–7, 41, 44
  symbols for, 48–49
  vendor drawings of, 119
Equipment costs:
  estimation of, 6–8
  information sources for, 7
Equivalent lengths for friction loss, 257–260
Erosion of pipes:
  and flow rate, 196, 197
  and two-phase flows, 196
Erosion corrosion, 344

Failure alarms, 106
Fan systems, 443–444
  arrangements of, (*illus.*) 444
Fanning friction factors, 228–230, 254
Fans, 437–444
  blade arrangements of, 438–439
  checklists for, on engineering flow diagrams, 85–87
  controls for, 441–443
  diameter of, vs. speed, (*illus.*) 442
  filters for, 87
  knockout pots for, 87
  screens for, 87
  silencers for, 87
  speed of, 441
  types of, 437
    axial-flow, 440–441
    centrifugal, 437–439, (*illus.*) 439
    propeller, 437
  (*See also* Blowers; Compressors)
Fatigue, definition of, 125
Ferrous alloys, 129
Fiberglass as insulation, 566, 568
Files:
  for calculation sheets, 119–120
  for work in progress, 118–120
Film coefficients:
  heat transfer, (*table*) 513–515
  for natural convection, (*table*) 562
  radiation, (*table*) 562

Index

Filter checklists for engineering flow diagrams, 89–90
Fire protection, 107
Fittings and components, pipe, 256–266
Flow diagrams (*see* Engineering flow diagrams)
Flow lines on process flow sheets, 30–32
Flow patterns, 211–214
  critical flow, 213
  eddy currents, 213–214
  heat exchangers, 497–501, (*illus.*) 498
  linear velocity, 212–218, (*illus.*) 212
    profile, 212–214, (*illus.*) 212
    wall effect, 212
  Reynolds number: vs. flow type, (*illus.*) 216
    vs. wall roughness, (*graph*) 216, (*table*) 216
  types of, 211–218
    laminar, 213, 218, 226–228, 230
    transition, 214, 218, 228–230
    turbulent, 214, 218, 228–230
    viscous, 213
    water, 216–217, (*illus.*) 217
  (*See also* Two-phase flows)
Flow rate:
  in compressors, 452–453
  erosion effects in pipes, 196, 197
  gases, 196, 282–284, 437
  line-tabulation values, 65–66
  pipe sizing: carbon steel pipe, (*table*) 194
    and parameters for, 194
  piping systems: annual fixed cost, 193–195
    annual operating cost, 194–195
  partially filled pipes, (*illus.*) 272
  on process flow sheets, 33–34
  in pumps, 396
  velocity constraints, 195–197
  and viscosity, 201–222
Flow regimes (*see* Flow patterns)
Flue-gas-desulfurization plant, as example for: block diagram, 24
Fluid-solid systems, 348–362
  design criteria, 361
  pressure gradients, 356–361
  terminal velocity in, 348–351
    drag coefficients, 350–351
    sphericity values, (*tables*) 351
  types of, (*illus.*) 353
    asymmetrical, 352–355
    moving-bed, 352–355
    stationary, 352–355
    symmetrical, 352–355

Fluids:
  compressible vs. incompressible, 199–200, (*illus.*) 276
  dilatant, 208
  ideal, 211–212
  newtonian, 207–208
  nonnewtonian, 208–210
  nonperfect, 224
  plastic, Bingham, 208
  pseudoplastic, 208
  rheopectic, 210
  shear stress on, 208–210, 212
  single-phase flows, 198–200
  thixotropic, 208–210
  two-phase flow (*see* Two-phase flows)
  viscosity of, (*illus.*) 209
  (*See also* Compressible-fluid flow; Incompressible-fluid flow)
Fluorocarbons, 156
Foams, 200, 201
Fouling factors, 495, 516
Friction, definition of, 224
Friction factors, 223–230
  Churchill equation for, 229
  Colebrook equation for, 229
  laminar flow, 225–228
  Moody equation for, 229
  transition flow, 228–230
  turbulent flow, 228–230
  (*See also* Darcy relationship for pressure drop; Fanning friction factors; Hazen and Williams roughness friction factors; Moody friction factors)
Friction loss:
  barometric leg of condenser, 198, 472
  for compressible fluids, 274–334
  for fittings and pipe components, 256–266, (*table*) 259
    orifices, 260–261, (*table*) 263
    valves, 261–266
  and flow type: laminar, 226–228, 230
    transition, 228–230
    turbulent flow, 228–230
  and fluid pressure, 224
  fluid-solid system, 357–361
  head loss, shortcut calculations for, 230–254
  for incompressible fluids, 223–273
  kinetic-energy change, 286–290
  mnemonic devices for calculation of, 251–254
    Rule of Fours, 253

628   Index

Friction loss (*Cont.*):
  in other than clean steel pipe, 254–257
  pipe flow diagrams for, 251, (*illus.*) 253
  and pipe sizing, 198–199
  in pipe systems, 67, 193, 223–334, 483–484
  slide rule for, 234, (*illus.*) 235
  tables and charts for calculation of, 228–258
    Cameron, 234–238
    Crane, for air, 248–250
    Genereaux, 232
    Hydraulic Institute Engineering Data Book, 230–234
    Moody charts, 228–233
  in vacuum service, 198–199
  viscous fluids in new steel pipe, (*tables*) 242–245
  water in new steel pipe, (*tables*) 236–237
  (*See also* Head loss; Pressure drop)
Fuel oils, 40

Gas:
  absorption of, in liquid, 592–593
  adiabatic flow: of ideal gases, 287–289
    of nonideal gases, 289
    property ratios vs. Mach numbers, (*table*) 308–309
  analysis of piping for gas flow, 200
  compressibility factors for, 278, 287
  as compressible fluid, 199–200
  critical pressure of, 278
  critical temperature of, 278
  heat ratio, specific, 282
  ideal behavior of, 286
  isentropic flow of, property ratios vs. Mach numbers, (*table*) 310–311
  isothermal flow: of ideal gases, 286–287
    of nonideal gases, 287–294
    property ratios vs. Mach numbers, (*table*) 306–307
  linear velocity of: effect of elevation on, 276
    effect of friction on, 274–297, 314–318
  nonideal behavior of, 278
  polytropic flow of, 289
  pressure on, effect of, 274–334
  pressure-drop relationships, 274–297
  radiant heat transfer, 524
  shock-wave effect of, 330–333, (*table*) 333
  temperature of, effect of, 274–334
  viscosity of, (*illus.*) 206

Gas-liquid flow systems, 335–348
  design of, 343–348
  linear velocities of, (*table*) 337
  pressure drop, 338–347
    Martinelli moduli, (*table*) 341
  types of, (*illus.*) 336
    annular, 337, 339
    bubble, 337, 339
    dispersed, 337, 340, 345–346
    plug, 337, 339
    slug, 337, 339, 345
    stratified, 337, 339
Gelatins as example of rheopectic fluids, 210
Generators, electrical, 64, 601–605, (*illus.*) 603, 604
Glasses, 160–162
Grade line, 46
Graphite, 159–160
Gravity flow and fluid pressure loss, 199
Graybodies, 521–522

Hardness, definition of, 125
Hazen and Williams roughness friction factors, 225–257, (*table*) 256
Head:
  adiabatic, 450–452
  discharge in pumped systems, 396–399
  hydrostatic, in fluid-solid systems, 356–357, (*illus.*) 360
  isothermal, 451–452
  net positive suction available (NPSHA):
    for centrifugal pump, 407
    for condensate pump, 473
    in pumped systems, 396–401
  net positive suction required (NPSHR), 407–408, 416
  polytropic, 452
  total dynamic (TDH), for pumped systems, 396–398, 411
  velocity, 218–222
Head loss:
  calculation of, 230–254
    alignment charts, 232–234
    friction-loss tables, 234–247
    mnemonic devices, 251–254
      flow-diameter correlations, 251–254
      Rule of Fours, 254
    Moody charts, 232
    nomographs, 232–234
    slide rule, 234, (*illus.*) 235
  cumulative, 266–273
  (*See also* Friction loss; Pressure drop)

Index

Heads for vessels, (*illus.*) 372
  conical dished, 371
  ellipsoidal dished, 371
  flanged and dished, 371
  head-wall thickness, (*table*) 371
  hemispherical, 371
  torispherical, 371
Heat capacity ratio (*see* Specific heat ratio)
Heat duty, (*illus.*) 502
Heat exchange (*see* Heat transfer)
Heat exchangers, 491–559
  checklist for engineering flow diagrams, 82–85
  cost estimate plus operating requirement, (*table*) 14
  design of, 425–559
  flow patterns, 497–501, 538, (*illus.*) 498
  fouling factors, (*table*) 512
  heat-transfer coefficients, (*tables*) 513–515
  insulation, 85
  mechanical design, 525–540
  pressure-drop calculations, 539–541
  tracing, 85
  transfer panels, 548
  tubing in, use of, 190
  types of: air-cooled, 541–543, (*illus.*) 542
    cocurrent, 536–539
    double-pipe, 525, (*illus.*) 526
    plate heat, 536–539, (*illus.*) 537
    shell-and-tube, 526–532, (*illus.*) 529
    spiral-plate, 532–536, (*illus.*) 534–535
  (*See also* Cooling towers; Heat transfer)
Heat transfer:
  with agitation, 589, (*table*) 590
  calculation of, 493–524, 561, (*table*) 590
  coefficients of, (*tables*) 502, 508–515
    calculation of, 522–524, 561
  dimensionless groups, 502, (*table*) 503
  fundamentals of, 491–524
  insulated and uninsulated pipe, 561
  in motionless mixers, 596
  panels on vessels, 75
  (*See also* Conduction in heat transfer; Convection in heat transfer; Heat exchangers; Radiation in heat transfer)
Himalayas, the, atmospheric pressure of, 461
"HOLD" designation on engineering flow diagrams, 110, 112, (*illus.*) 111
Horsepower:
  for agitators, (*table*) 583
  for compressors, 452–459

Horsepower (*Cont.*):
  for pumps, 401–402
Hot wells, 471–472
Hydraulic Institute Engineering Data Book, tables for friction loss, 225, 230–234
Hydraulic radius, 224
Hydrodynamica (Bernoulli), 223

Impact strength, definition of, 125
Impellers:
  for agitators, 582–585, (*illus.*) 584, 586, 587
  for centrifugal pumps, 391, (*illus.*) 394, (*table*) 395, 404–422
    axial-flow, 391
    mixed-flow, 391
    radial-flow, 391
  for liquid-ring vacuum pumps, 481
  in lobe pumps, 388
Incompressible-fluid flow, friction loss for, 223–273
  (*See also* Friction loss)
Inspection of piping and vessels, (*table*) 377
Instrument engineers, 41–42, 112, 116–117
Instrument Society of America:
  standard nomenclature of, 37
  standard symbols of, 37
Instrumentation:
  on flow diagrams, 32, 37, 41–42, 112
  on lead sheets, 37
  on P&ID, 41–42
Instrumentation and safety checklists for engineering flow diagrams, 99–107
Instrumentation application (flow) diagram, 35, 42
  process flow sheet as basis for, 112
  scope of, 110
Insulation, thermal, 560–573
  coefficients for, 561
  on engineering flow diagrams, 60–62
  flow-rate analysis, 280
  on line tabulations, 60–62, 67–68
  materials for, 566–572, (*tables*) 566, 568–572
  recommended thickness of, 567–573, (*tables*) 568–572
  types of, 566–567
  usages for: air-conditioning burden reduction, 566
    antisweat protection, 565–566
    heat conservation, 564

630 Index

Insulation, thermal, usages for (*Cont.*):
    heat exchangers, 85
    personnel protection, 564–565
    pipe fluid protection, 565
    pumps, 81
International System of Units (SI), 16
Interstage cooling and aftercooling, 456, 470
IPS (iron pipe sizes), 187
Iron (*see* Cast iron)
Iron-carbon equilibrium, 130
Iron pipe sizes, 187
ISA (*see* Instrument Society of America)
Isothermal fluid flow, 280–304
    definition of, 280
Item numbers, 50–51
    standardization of, 50
    symbols for, (*table*) 50

Jacketing:
    pipe, 98–99, 579
    pumps, 81, (*illus.*) 82
    vessels, 74
Joints:
    expansion, 104
    welded, 376

Kinetic energy, effect of, on friction loss of gas, 286–290
Kirchhoff's law, 519–520

Laminar flow:
    flow patterns, 213, 218, 226–228, 230
    friction loss, 225–228, 230
    L/D values, 259–260, (*table*) 259
Layering, use of, for updating flow diagrams, 42
Lead sheet:
    definition of, 36–37
    example of, 38
    and piping specifications, 58
    prototypes of, 37
    tracings for, 37
Leakage, fluid, 270–271
    in pumping systems, 422–429
    (*See also* Air leakage)
Legends on flow diagrams, 63
Legs for vessels, 71, 381
Length-to-diameter (L/D) ratios for fittings, 257–260
Level indicators, 102–103

Line numbers, 54–57, 63
    changing of, (*illus.*) 59
    continuation of, (*illus.*) 55
Line-size reductions, 53–54, (*illus.*) 54, 64
Line tabulations, 45, 56, 63–68, (*illus.*) 64
    placement of, 45
Linear velocity, 211–214
    effect of wall roughness on, 212, 215
    profile, flow patterns of, 212–214, (*illus.*) 212
Liquid seals, 270–271
Liquids:
    effect of temperature on density, 274
    pressure-drop relationships, 275–277
    for self-lubricating pump, 387
    thermal-conductivity values, 493
    viscosity of, 203–206
    (*See also* Flow patterns; Flow rate; Fluid-solid systems; Gas-liquid flow systems)
Loading:
    seismic, on vessels, 373–376
    wind, on vessels, 373
Loads, flow:
    to rotary and reciprocating vacuum systems, 484
    to steam-jet-ejector vacuum systems, 484
Log-mean-temperature difference (LMTD), 497, 501
Lubricants, packing, 424–425
Lugs, 381

Mach number, (*tables*) 306–313
    definition of, 302
    examples of use of, 303
    and shock waves, 330–333
Malleability, definition of, 125
Material balances in process flow sheets, 29, 32–34
Materials, construction, 129–162
    equivalent of, in various countries, (*tables*) 150–155
    mechanical properties of, 123–129
Mechanical engineers:
    engineering-flow-diagram preparation by, 41
    responsibilities of, 40, 41, 58
Mechanical-piping drawings, 41
Metals, 129–154
    vaporization of, 485
    (*See also* Aluminum; Cast iron; Chromium in iron alloys; Copper alloys;

Metals (*Cont.*):
    Nickel alloys; Steel; Tantalum;
        Titanium; Zirconium)
Mexico City, atmospheric pressure of, 461
MgO regeneration plant, as example for:
    block diagram, (*illus.*) 25, 26
    economic decisions, 14, 15
    equipment requirements, 15, 43
    flow diagram, 40
    gas flow in pipes, 200
    isometric diagram, 28
    lead sheet, (*illus.*) 38
    line numbers, 56–57
    piping lines, 53, 186
    piping specifications, (*table*) 58, (*illus.*) 59
    piping-system descriptors, 57
    process comparisons, 14
    process engineer staffing, 43
    process lines, 53
    process water, 39
    schematic diagram, (*illus.*) 27
    two-phase flow systems, 200
    utilities for, 39
    utility flow diagram, 40
Mineral wool as insulation, 566, 569
Mixers:
    impeller-type (*see* Agitators)
        checklists for, 77–78
    motionless, 593–596, (*illus.*) 595
        heat transfer in, 596
        performance curves, (*illus.*) 596
    (*See also* Agitators)
Mnemonic devices for calculating pressure
        drop, 251–254
Moody friction factors, 225, 226, 232, (*illus.*)
        229, (*table*) 230, (*illus.*) 231, 484
Motor-control centers, 43
Motors, 608–621
    applications for, (*table*) 614
    controls of, 618–620, (*illus.*) 619
        symbols for, (*illus.*) 618
    enclosures for, (*table*) 615
    sizes of, single-phase, 610
        three-phase, 614
    types of: alternating-current (ac), 610–620
        direct-current (dc), 610, (*illus.*) 611
        induction, 616–618
        single-phase, 610, 620, (*illus.*) 612
        squirrel-cage, 610–614
        synchronous, 616–618
        three-phase, 610, 620, (*illus.*) 612
        wound-rotor, 615–616
Muntz metal, 139

National Electrical Code (NEC), 605–608
National Electrical Manufacturers
        Association (NEMA), 605–606,
        609, 614–615
Newtonian fluids, 207–208
Nickel alloys, 141–143
    Hastelloys, 142, (*table*) 143
    Incoloys, 142
    Inconels, 142
    Monels, 142
    pure nickels, 142
Nomenclature:
    dimensional units, 16–20
    of Instrument Society of America, 37
Nonnewtonian fluids, 208–210
Notes:
    on engineering flow diagrams, 62–63, 107–
        108
    on process flow sheets, 34
Notes of conference, 119–120
Nozzles:
    compressible-fluid flow, 318–330, (*illus.*)
        318
    friction loss, 260–261, (*table*) 262–263
    pressure drop, 270–271
NPSHA (*see* Head, net positive suction
        available)
NPSHR (*see* Head, net positive suction
        required)
Nusselt number applied to agitators, 589

Oil paints, as example of Bingham plastic
        fluids, 208
Opaque bodies, 518
Orifices:
    compressible-fluid flow, 318–330
    friction loss, 260–261, (*table*) 262–263,
        342–343
Outgassing in vacuum systems, 485

P&ID (*see* Piping and instrumentation
        diagrams)
Packing, 422–429
    advantages and disadvantages of, (*table*)
        426
    lubricant for, 424–425
    for pumps, 422–425
    types of, 423–424
        chevron, (*illus.*) 424
        compression, (*illus.*) 423, 424
        spring-loaded, (*illus.*) 424

632 Index

Packing, types of (*Cont.*):
    throat-and-gland, (*illus.*) 423
Payback time, 12, 15–16
Petroleum products, 205
Phenolics, 157
Pilot-plant experiments, 29
Pipe sizing:
    and annual fixed costs, 192–193
    economic analysis for: Alloy 20 system, (*illus.*) 192
      alloy pipes, 195
      batch operation, 195
      carbon steel pipes, 192–195, (*illus.*) 192
      effect of flow rates, 192–195
      effect of friction loss, 198–199, 257–266
      effect of insulation, 195
      lined or glass pipes, 195
      stainless-steel pipe, 195
    economics diagram, 192–195, (*illus.*) 197
    internal diameter, 187
    nominal sizes, 188
    outside diameter, 187
    schedule dimensions, (*table*) 189
    specifications on flow diagrams, 58–60, 63–68, (*table*) 58
    standards of, 187
    tubing dimensions, (*table*) 191
    wall thickness, 188
    and water-flow pattern, 217, (*table*) 217
Piping:
    for compressible-fluid flow, uniform pipe, 297
    control-valve bypasses, 97
    corrosion of, 188–190
    on engineering flow diagrams: checklists for, 90–99
      designation, 57
      formatting, 53
      identification, 52–62
      insulation documentation, 60–62
      specifications, 58–60, 63–68, (*table*) 58
    fittings and components, 257–266
    inspection of, 377
    jacketing, 98–99
    limitations on, 187
    metal: alloy, 190
      carbon steel, 188
      copper and brass, 191
      lined steel, 190
      stainless steel, 190
    noncircular, 224
    nonmetallic, 190
    outer dimensions vs. inner dimensions, (*table*) 189

Piping (*Cont.*):
    and pressure rating, 188
    and schedule dimension, (*table*) 189
    scheduling of, 187
    specialty, (*table*) 191
    specifications for, (*table*), 58
      change markers, (*illus.*) 59
    standard sizes of, 187
    supports for, 188
    valves for, 90–97
    wall roughness of, 212, 215–217, 224–225, (*illus.*) 227, 254–257
    wall thickness of, 188, (*table*) 189
    (*See also* Line numbers; Line-size reductions; Piping and instrumentation drawings)
Piping and instrumentation diagrams (P&ID), 35–36, 41–42, (*illus.*) 47, 109, 120
    definition of, 35–36
    logistics of, 109–120
    relationship to project objectives, 109
    revision of, 114–119
    (*See also* Engineering flow diagrams)
Piping and instruments symbols diagram (*see* Lead sheet)
Piping systems:
    air gap in, 271–273
    costs of, 192–193
    design of, 186–222
      limitations on, 186–222
      theoretical concepts, 186–222
    erosion corrosion, 344
    friction loss (*see* Friction loss)
    head loss, cumulative, 266–273
    installation of, 193–195
    velocity constraints, 195–198
Plant capacity in product-based cost estimates, 5–6
Polymeric materials, 149–158
    polyamines, 157
    polycarbonates, 157
    polyesters, 158
    polyethers, 157
    polyolefins, 155
    polystyrene, 156
    polyvinyl chloride (PVC), 156
    polyvinylidene chloride (PVDC), 156
Polytropic fluid flow, 280–304, (*table*) 280
Power distribution, 598–601, 605
Pressure:
    atmospheric, 67, 461–463, (*table*) 463
    and design of vessels, 383
    effect of, on liquid viscosity, 204–206

Pressure (*Cont.*):
  effect of high temperature on, 198
  gas-law thermodynamic relationships, (*table*), 280
  hydrostatic, 67
  in pipelines, 58–59, 66–67
    (*See also* Pressure drop; Pressure gradients for fluid-solid systems
  pump discharge, 67
  in reciprocating pumping systems, 402–404
Pressure drop:
  air in pipes, 246
  compressible vs. incompressible fluids, (*illus.*) 276
  control valves, 265–266
  gas-liquid flow, 338–343
    annular, 339
    bubble, 339,
    dispersed, 340
    slug, 339
    stratified, 339
  in heat exchanger, 539–540, (*table*) 541
  hydrostatic head, effect of, 341–342
  kinetic-energy change, 289–290, 293–294, (*illus.*) 291
  liquid vs. gas, 297
  mnemonic devices, 251–254
  nozzle, 270–271
  orifice, 342–343
  in piping systems, 193, 199, 483–484
    compressible fluids, 274–334
    incompressible fluids, 223–273
  Rule of Fours for, 254
  two-phase flows, 347–348, (*table*) 347
  valves, 263–266
  wave-flow, 340–341
  (*See also* Friction loss; Head loss)
Pressure gradients for fluid-solid systems, 356–361, (*illus.*) 353, (*illus.*) 357, 359, 360
Pressure ratings and piping-schedule selection, 188
Pressure sensors (*see* Sensors)
Priming, pumping systems, 430
Problem solving, general approaches to, 4
Process data books, 27–29
  based on literature data, 29
Process design, data for, 27–30
Process documentation, 23–120
Process engineers:
  as consultants, 29
  as problem solvers, 3–4
  responsibilities of, 2, 35–43, 51, 59, 109,

Process engineers (*Cont.*):
  116–117
  role of, 2–3, 111, 112
Process flow sheets, 23, 27–34, 44, 109–120
  basis for cost estimates, 6, 8
  data required, 27–30
  example of, 31
  as information basis for engineering flow diagrams, 44, 110
  presentation of, 29–34
Professional associations, (*table*) 145
  AISI, 145
  ANSI, 145
  ASME, 145
  ASTM, 145
  CDA, 145
  SAE, 145
Project status, 109
Property ratios for sonic velocities:
  critical-state, 302–303
  stagnation-state, 302–303
Pumping system: design of, 429–459
  minimum-flow, 431–432, (*illus.*) 432
  priming of, 429–431
Pumps, 384–459
  centrifugal, 390–395, (*illus.*) 392–394
    advantages and disadvantages of, 391
    affinity laws, 413–416
    axial-type impeller, 391
    casing design for, 391
    definition of, 390–391
    efficiency of, 401–402, (*illus.*) 405, 407, (*illus.*) 408–409, 412–422, (*illus.*) 412
    impeller classification, (*table*) 395
    impeller design for, 391
    mixed-flow-type, 391
    performance curves, 404–422, (*illus.*) 405, 408, 409
    priming of, 429, (*illus.*) 430
    radial-type impeller, 391
    specific speed of, 454–456
    sumps for, 432–435
    system curves, 410–413, (*illus.*) 412
    viscosity effects on, 416–422
  checklists for engineering flow diagrams, 78–82
  classification of, 384
  for condensates, 473
  definition of, 384
  drivers, types of, 79–80
  dynamic head, calculation sheet for, 398, 411
  flow rate of, 396

Pumps (*Cont.*):
  flushing and seal fluid system, 81
  jacketed, 81
  liquid-ring vacuum, 481–482
    applications for, 482
    impellers for, 481
    performance curve, 482
  performance of, 396–422
    general parameters, 396–402
  positive-displacement, 429
  quench systems, 80
  reciprocating, 384–387
    action curves, 385
    calculation of NPSHA, 399–400
    capacity vs. head curves, 403
    chambers of, 384–386
    definition of, 384
    diaphragm, 386–387
    direct-acting, 386
    double-acting, 384
    efficiency of, 401–402
    performance characteristics of, 402, 404
    piston, 386
    plunger, 386
    power, 386
    single-acting, 384
  recycle, 81, 482
  regenerative, 395
    priming of, 429
  rotary, 387–390
    advantages and disadvantages of, 387
    cam-and-piston-type, 387
    definition of, 387
    efficiency of, 401–402
    gear-type, 388–389, (*illus.*) 388
    lobe-type, 387–388, (*illus.*) 387
    screw-type, 389, (*illus.*) 389
    vane-type, 389, (*illus.*) 390
  tabulation sheet for, 267–268
  tracing and insulation for, 81
  turbine, 395
  types of, 78, 384, (*illus.*) 385
  vacuum, 481–482
  valved vents and drains, 80
  vertical turbine, 395
Purchase contracts, equipment specifications in, 3

Radiant energy, 517–524
Radiation in heat transfer:
  definition of, 491, 524
  distribution of, (*illus.*) 517

Radiation in heat transfer (*Cont.*):
  film coefficients of, (*table*) 562
Reboiler, 532, (*illus.*) 533
Recycling:
  of process vapors, 468
  of water sealant, 482
  (*See also* MgO regeneration plant)
Relief valves in pump piping, 402, 404
Required net positive suction head (*see* Head, net positive suction required)
Resistance:
  coefficient K for friction loss, 257
  heat transfer, 493, 495
  metal tubing, (*table*) 512
Reynolds, Osborne, 214
Reynolds number (Re), 214–218
  applied to fluids in agitators, 582, 586, 587, 590–592, 596
  and flow patterns, 215
  and friction factors, 225, (*illus.*) 226
  and friction loss: in pipes, 234
    in vacuum equipment, 484
  and laminar-flow L/D values, 259
  and prediction of eddy currents, 215–218, 254
  quick calculation of, 239
  transition and turbulent flow, 228–230
Rose and Duckworth functions, (*illus.*) 359–361
Roughness:
  of various materials, (*table*) 232
  (*See also* Wall roughness for pipes)
Roughness factor, 215–218
  effect on friction factors, 229
  effect on flow regimes, (*illus.*) 216, 217
Roughness friction factor, 225–257, (*table*) 256
Rule of Fours, 253–254
Rupture disks, codes for, (*table*) 100, 101

Safety:
  checklists, 99–107
  showers and eyewashes, 106
Sand, footprints in, as example of dilatant fluids, 208
Scaling factors, 495
Schedule dimensions, piping, (*table*) 189
Sea level, atmospheric pressure of, 461 (*table*) 463
Sealing flow rates in horizontal lines, 270–273

Seals:
  liquid: in vacuum pumps, 481
  mechanical, 425–429, (*illus.*) 427
    advantages and disadvantages of, 426
    for agitators, 593
    for compressor systems, 449
    fluid system for, 81, 107
    materials for, 428–429
    secondary, 428
      bellows, 428
      O-ring, 428
      V-ring, 428
      wedge, 428
    spring, 428
Seismic stress, 373–376
Sensors:
  instrumentation for, 42
  pressure and temperature, 101–102, 575
Service stations, 107
Shock waves:
  calculation of, 331–333
  formation of, 330–333
  property ratios vs. Mach numbers, (*table*) 312–313
Shortcut analysis methods:
  for friction loss, 230–254
  for pressure drop, 292–294, 296–297
SI (International System of Units), 16
Single-phase flow, 199–200
Siphon effects, 226–273
Sizing (*see* Equipment, sizing of: Pipe sizing; Sump pumps, sizing of; Vacuum-producing equipment, sizing of)
Skirts, vessel, 381–382
Slip:
  in compressor systems, 446
  motor, 610
  in pumping systems, 402
Slurries:
  definition of, 200
  flow rates of, 196
  pump selection for, 386–387
Software (*see* Computer software)
Sonic velocity, gas, 282–284, 294, 314–319
  and critical states, 301–302
  and Mach numbers, 302
  vacuum systems, 485–486
Soot blowers, boiler, 556
Space shuttles, 462
Specific gravity:
  effect of: on friction loss, 247
    on pressure: gas, (*illus.*) 221
      liquids, (*illus.*) 219

Specific gravity (*Cont.*):
  viscosity, liquids, (*table*) 203
Specific-heat ratio (k):
  definition of, 280
  effect on thermodynamic relationships, 280, (*table*) 281, 282
  gases, 282
Specific speed, 454–455, (*illus.*) 455
Specification sheets, 3
Specifications:
  for pipe-line change, 59, (*illus.*) 59
  for purchase of equipment, 3, 41
Standards (*see* Codes and standards)
Start-up documents, 35
Steam, use of, in pipe tracings, 574–575
Steam-jet-ejector systems, 467–482
  multijet system curves, 475
  performance curves, 476
  single-jet system curves, 475
Steel:
  carbon, 131
    alloying elements, (*table*) 132
    in piping systems, 188
  ductile, stress-strain curve, (*illus.*) 366
  low-alloy, 131
  stainless, 132–136
    austenitic, 135
    ferritic, (*illus.*) 135
    martensitic, (*illus.*) 134, 135
    in piping systems, 190
Stokes, metric unit for viscosity measurement, 202, 203, 211
"Stonewall" (limit of gas flow in compressor systems), 447
Stream numbers on process flow sheets, 32
Strength, 125–129
Stress(es):
  allowable, 366
  bending, on vessel, (*illus.*) 369
  combined, (*table*) 376
  compressive, 367
  deadweight, 368, 381
  external-pressure, 373
  internal-pressure, 368–371
  rupture, 366
  seismic, 373–376
  shear, on fluids, 202, 208–210, 212–214
  tensile, 367
  ultimate, 366
  on vessels, 365–376
  yield, 366
Stress-strain curves: concrete, 128
  ductile steel, (*illus.*) 366

Stress-strain curves (*Cont.*):
  gray cast iron, 128
  polycrystalline copper, 127
Structural engineers:
  and engineering flow diagrams, 43
  responsibilities of, 40, 43
Stuffing box, 422–423
Sump pumps:
  design of, 432–435
  electrical diagram for, 619
  sizing of, 433
  standards for, 434
Sumps, design of, 432–433
Supersonic velocities in nozzles, orifices, and pipelines, 330–333
Surge in compressor systems, 446
Symbols:
  in computer-assisted drafting, 48, 49
  of Instrument Society of America, 37
  instrumentation, 37
  insulation, 61–62, (*illus.*) 62
  for item numbers, (*table*) 50
  on lead sheets, 36–37
  for line-size reducers, 53–54, (*illus.*) 54
  for motor controls, 618
  for physical parameters, 32
  on piping and instrumentation diagrams, 47
  for process equipment, 26–27, 48, 49
  for process flow sheets, 32
  semigraphical, 26
  for valves, 91

Tanks:
  day, 382
  expansion, 106
  for intermediate storage, 382
  surge, 386, 449
  (*See also* Vessels)
Tantalum, 143
TDH (*see* Head, total dynamic)
Temperature:
  ambient, 66
  effect of: on compressible and noncompressible fluids, 274
    on metal pipes, 198
  and heat-transfer coefficients, 516
  of pipelines, 58–59, (*table*) 58, 66
  and piping-schedule selection, 188, 190
  steam, 66
  and thermal conductivity, 492–493
  and vessel design, 382–383

Temperature (*Cont.*):
  and viscosity: of gas, 206
    of liquid, 204–206
  (*See also* Adiabatic fluid-flow process; Isothermal fluid flow; Polytropic fluid flow)
Temperature differentials in heat exchangers, 496–517
  correction factors for, 499, 500
Temperature sensors (*see* Sensors)
Terminal velocity, 348–351
Thermal conductivity (*see* Conductivity, thermal)
Thermal insulation (*see* Insulation)
Thermal resistance of metal tubing, (*table*) 512
Thermodynamic laws applied to gas pressure, 278–284, (*table*) 281
Thermostats for pipe tracing, 578
Titanium, 143
Title-block record, (*illus.*) 110
  location on engineering flow diagram, 110, 113–115
  on revision issues, 114, 116
  on zero issues, 113, (*illus.*) 113
Toughness, defintion of, 125
Tracing (electric, heat, and steam), 38, 573–579
  depicted on engineering flow diagrams, 60, 61
  electric-current, 575–579
  steam, 574–575
  thermostat for, 578
  wire, 575–578
    constant-resistance, 575–577
    self-regulating, 577–579
Transition flow:
  friction loss, 228–230
  L/D values, 258–261, (*table*) 259
Transmitters, instrumentation for, 42
Trenches, 192
Tubesheets, 532–533
Tubing, 190–192
  dimensions of, 190, (*table*) 191
  materials for, 190
  for steam tracing, 574–575
  wall thickness of, 190
Tubular Exchanger Manufacturers Association (TEMA), standards for tubing of, 190
Turbulent flow:
  friction loss, 228–230
  L/D values, 258–261, (*table*) 259

Index 637

Two-phase flows, 335–362
  Baker plot, 337–338, (*illus.*) 338, 346, 347
  definition of, 199
  and pipe erosion, 196
  and species type, 200–201
  (*See also* Flow patterns; Fluid-solid systems; Gas-liquid flow systems)

Units of measurement (*see* Dimensional units)
Urethane, 567, 571
Utilities:
  costs of, 16
  engineering flow diagrams, 39–40, 45
    formatting on, 45
  on process flow sheets, 30, 32
  water, 39
Utility-distribution-header systems, 65
Utility line, piping-system designation for, 57, (*illus.*) 57
Utility service stations, 107
Utility-use diagram, 32
Utility water, 39

Vacuum:
  crude, 464
  definition of, 461–464
  fine, 464
  at high altitudes, 461
  industrial, 464
  rough, 464
Vacuum conditions, 67
Vacuum-producing equipment, 87–89, 464–490
  aspirators, 480–481
  checklists on engineering flow diagrams, 87–89
  cost analysis of, 484
  eductors, 480–481
  friction loss in, 198–199
  high-vacuum equipment, 464–465
    cryogenic panels, 464–465
    diffusion pumps, 465
    molecular pumps, 465
    problems of, 485
  industrial, 466–483
  inlet pressure, 483–484
  knockout pots, 89
  low-vacuum equipment, 465
  mechanical-type, 481–482
    liquid-ring vacuum pumps, 481–482

Vacuum-producing equipment, mechanical type (*Cont.*):
    reciprocating vacuum pumps, 481
    rotary vacuum pumps, 481
  medium-vacuum equipment, 466–483
    analyzing water needs for, 470–471
    steam-jet-ejector systems, (*illus.*) 467
      multistage, 468–483
      single-stage, 468
      two-stage, (*illus.*) 469
    operational ranges of, (*table*) 463
    vs. use of, (*table*) 463
  sizing of, 483–490
  steam requirements, 477–479, (*nomograph*) 476
Vacuum systems:
  air leakage to, 485–490
  barometric leg, 471–473
  fluid pressure loss, 198
Valves:
  capacity of, 261–266
  checklists for, on engineering flow diagrams, 90–107
  codes for, (*table*) 100
  coefficient C_v for pressure drop, 263–266
  common types of, (*table*) 93
  materials for construction for, (table) 92
  pressure drop, 263–266, 397
  sizes of, (*illus.*) 92
  symbols for, 91
  types of, (*illus.*) 91
    ball, 94, 95
    butterfly, 95
    check, 97
    diaphragm, 96
    fire, 104
    flush-bottom, 96
    foot, 429
    gate, 94
    globe, 94
    needle, 96
    pinch, 96
    in piping, 90–97
    plug, 95
    relief, 99–100, 402, 404
    slide, 96
    spiral sock, 97
    spring-closure, 104
    squeeze, 96
Vanes, fan-control, 441–443
Vapor: bubble formation, 399
  condenser effluent, 471, 473, 475–476, 480–482

638 Index

Vapor: bubble formation (*Cont.*):
    flow rate of, in vacuum equipment, 484–490
    in two-phase flows, 335–348
Velocities: acoustic, 282–284
    fluid, quick calculation of, 193, 239
    linear (*see* Linear velocity)
    solid, 354–356
    sonic (*see* Sonic velocity)
    supersonic, 330–333
Velocity constraints on fluids in pipes, 195–197
Velocity head, 218–222
Vendor drawings, 119
Venturis, compressible-fluid flow, 315
Vessels, 365–383
    agitators for, 586–588
    checklists for engineering flow diagrams, 68–77
    day tanks, 382
    design of, 365, 382–383
    design codes for, 378–380
    drains for, 71
    elevation of, 382
    horizontal tank, 368–369
    inspection of, 376–377
    live bottoms for, 74
    pressure on: external, 373
        internal, 383
    standardization of, (*illus.*) 379
    storage, types of, (*table*) 380
    stresses on, 365–376
        seismic, 373–376
        wind-loading, 373
    supports for, 381–382
    types of, 379–380
    volume of, 382
    vortex breakers for, 71
    wall thickness, (*illus.*) 374
    (*See also* Tanks)
Viscosimeters, 210–211
    conversion factors, (*table*) 211
Viscosity, 201–222, 238–246

Viscosity (*Cont.*):
    absolute, 203, 204, 215
    classification of fluids, 207–210
    (*See also* Fluids)
    correction curves for centrifugal pumps, 417–422, (*illus.*) 419, 420
    effect on friction loss, 224
    friction loss, (*table*) 244
    gases, (*illus.*) 206
    indices, 204, (*illus.*) 205
    kinematic, 202–203, 206
        typical gases, (*illus.*) 206
    liquids, (*table*) 203, (*illus.*) 209
    measurement of, 210–211
        bubble-method, 210
    nonnewtonian fluids, 208–210
    temperature effects: on petroleum products, (*illus.*) 205
        and pressure effects on gas, (*illus.*) 206

Wall roughness for pipes, 212, 215–217, 224–225, (*illus.*) 227, 254–257
Waste-heat boiler, cost estimate plus operating requirements for, 14–15, (*table*) 15
Water:
    for condensers, 468–480
    for cooling tower, 551
    flow pattern of (*illus.*) 217
    flow rate of, in metallic pipes, 195–196
    friction loss of, (*tables*) 236–237, 240–241
    for vacuum equipment, 470–471
    velocity of, 216–217
Water hammer, 195
Water utilities, 39
Welded joints, 376
Wind, effect of, on cooling tower, (*illus.*) 550
Wind loading of vessels, 373
Wind-loading stress on vessels, 373

Zero linear velocity, 299–301
Zirconium, 144

ABOUT THE AUTHORS

HENRY J. SANDLER earned his B.S. in chemical engineering and his M.S. in chemical engineering practice from the Massachusetts Institute of Technology. He also received an advanced degree from the City and Guilds College of the Imperial College of Science and Technology in London for research in chemical engineering.

A registered professional engineer, Mr. Sandler has more than 35 years' experience in applying process engineering principles working with such companies as Atlas Powder Company, Catalytic Construction Company, and TRW Products Company. Now a senior process engineer, he has been with the Chemical Division of United Engineers & Constructors Inc., a subsidiary of Raytheon Company, since 1960, with the exception of 3 years' service as a consulting engineer with the Institut Français du Pétrole near Paris.

Mr. Sandler's duties at United Engineers include supervising process engineering for inorganic and organic chemical projects, which encompass a large variety of unit operations including waste treatment. He has also assisted on power plant design and on cogeneration studies. During the past decade, Mr. Sandler has been responsible for training process engineers entering United Engineers.

EDWARD T. LUCKIEWICZ, also a registered professional engineer, has diverse experience in the fields of mechanical engineering, nuclear engineering, instrumentation and control engineering, and computer software design. He holds a B.S. degree in chemical engineering and an M.B.A. in financial management and marketing management from Drexel University, where he is adjunct professor of chemical engineering, teaching courses in plant design, engineering economics, and phase-separation processes.

Mr. Luckiewicz has more than 20 years' experience in the design and construction—including computer simulation—of process and power plants with such companies as Day & Zimmermann Inc., Singer Link Simulation Systems Division, Stearns-Catalytic Inc., and United Engineers & Constructors Inc. He is presently with Singer Link Simulation Systems Division in Cherry Hill, New Jersey, and specializes in power plant simulators.

DATE DUE